5th International Conference on Advances in Solidification Processes and 5th International Symposium on Cutting Edge of Computer Simulation of Solidification, Casting and Refining (ICASP5-CSSCR5)

IOP Conference Series: Materials Science and Engineering
Volume 529

Salzburg, Austria
17 – 21 June 2019

ISBN: 978-1-5108-8917-0
ISSN: 1757-8981

Printed from e-media with permission by:

Curran Associates, Inc.
57 Morehouse Lane
Red Hook, NY 12571

Some format issues inherent in the e-media version may also appear in this print version.

No changes have been made to the content of these proceedings. There may be changes to pagination and minor adjustments for aesthetics.

Printed by Curran Associates, Inc. (2019)

Additional copies of this publication are available from:

Curran Associates, Inc.
57 Morehouse Lane
Red Hook, NY 12571 USA
Phone: 845-758-0400
Fax: 845-758-2633
Email: curran@proceedings.com
Web: www.proceedings.com

Table of contents

Volume 529

Joint 5th International Conference on Advances in Solidification Processes (ICASP-5) & 5th International Symposium on Cutting Edge of Computer Simulation of Solidification, Casting and Refining (CSSCR-5)

17–21 June 2019, Salzburg, Austria

Accepted papers received: 11 April 2019
Published online: 17 June 2019

Preface

Joint 5th International Conference on Advances in Solidification Processes (ICASP-5) & 5th International Symposium on Cutting Edge of Computer Simulation of Solidification, Casting and Refining (CSSCR-5)

Peer review statement

Papers

Additive manufacturing: experiments and modeling

Particle scale modelling of porosity formation during selective laser melting process using a coupled DEM – CFD approach

Akash Aggarwal and Arvind Kumar.....1

Level-set modelling of Laser Beam Melting process applied onto ceramic materials – Comparison with experimental results

Q. Chen, L. Moniz Da Silva Sancho, G. Guillemot, Ch.-A. Gandin, M. Bellet, J.-D. Bartout, M.-H. Berger and C. Colin.....8

Phase field study of spacing evolution during wire and laser additive manufacturing under transient conditions

Zheng Wang, Tao Jing and Hongbiao Dong.....14

Solidification microstructure during selective laser melting of Ni based superalloy: experiment and mesoscopic modelling

Yuze Li, Antonio Olmedilla, Miha Založnik, Julien Zollinger, Lucas Dembinski and Alexandre Mathieu.....21

Thermo-mechanical simulation of track development in the Laser Beam Melting process - Effect of laser-metal interaction

Alexis Queva, Yaasin Mayi, Michel Bellet, Gildas Guillemot, Patrice Peyre, Morgan Dal, Clara Moriconi and Charlotte Metton.....27

Solidification of niobium-silicide-based alloys during laser additive manufacturing process

A. Allen, A. C. Douglas, L. M. Feitosa, H. Qian, J. Li, H. B. Dong, X. Lin and Y. Li.....35

Eutectic/peritectic/intermetallics

Investigation of the microstructure adjustment by velocity variations during the directional solidification of Al-Ag-Cu with the phase-field method

J. Hötzer, M. Kellner, W. Kunz and B. Nestler.....41

Solidification path and phase transformation in super-austenitic stainless steel UNS S31254

Raphaël Marin, Hervé Combeau, Julien Zollinger, Moukrane Dehmas, Bernard Rouat, Aude Lamontagne, David Cardinaux and Lucile Lhenry-Robert.....47

Prediction of Carbide Precipitation Using A Combined Partial Equilibrium - Para-Equilibrium - Lever Rule Approximation in Automobile Gear Materials

Hongwei Zhang, Keiji Nakajima, Mengmeng Su, Shaoshao Gao, Hong Lei, Qiang Wang and Jicheng He.....53

Effects of interphase boundary anisotropy on the three-phase growth dynamics in the β(In) – In_2Bi – γ(Sn) ternary-eutectic system

S Mohagheghi, U Hecht, S Bottin-Rousseau, S Akamatsu, G Faivre and M Şerefoğlu.....60

The formation of Al_6 (Fe, Mn) phase in die-cast Al-Mg alloys

Xiangzhen Zhu and Shouxun Ji.....67

Optimization strategies for identifying the controlling mechanism for solid-state transformation in FeCrNi during rapid solidification

D M Matson, X Xiao and S Jeon.....72

Investigation using 4D-CT of massive-like transformation from the δ to γ phase during and after δ-solidification in carbon steels

Hideyuki Yasuda, Takahiro Hashimoto, Naoki Sei, Kohei Morishita and Masato Yoshiya.....79

Numerical modelling of infiltration of Al into additive manufactured Fe preforms in order to obtain Fe-Al intermetallics

R. Berger, S. Jana, N. Focks, H. Michels, W. Kochanek and C. Charitidis.....86

Recent advances in the understanding of the role of vanadium carbonitride precipitation to improve surface edge cracking on continuous casting of blooms

E Erdem Hornauer, U Grafe, U Plociennik, W Klos, M Reifferscheid and M Luettenberg.....93

Experimental investigation/characterization/analysis

Control of Solidification Pattern of Cast Irons by Simultaneous Thermal and Contraction / Expansion Analysis

I Riposan, S Stan, M Chisamera, L Neacsu, A M Cojocaru, E Stefan and I Stan.....100

Chunky graphite in spheroidal graphite iron: review of recent results and definition of an predicting index

Jon Sertucha, Jacques Lacaze and Rodolfo González-Martínez.....106

Cellular-to-Dendritic and Dendritic-to-Cellular Morphological Transitions in a Ternary Al-Mg-Si Alloy

C. Brito, H. Nguyen-Thi, N. Mangelinck-Noël, N. Cheung, J.E. Spinelli and A Garcia.....112

Impact of gravity-related phenomena on the grain structure formation: comparative study between horizontal and vertical solidification of a refined Al-20wt.%Cu alloy

H Soltani, G Reinhart, M C Benoudia, M Zahzouh and H Nguyen-Thi.....118

Direct observation of dendrite fragmentation in the solidification of undercooled melts

Andrew M Mullis and Nafisul Haque.....124

Rapid solidification of Al-Cu droplets of near eutectic composition

A-A. Bogno, J. Valloton, D.D. Jimenez, M. Rappaz and H. Henein.....130

Primary Dendrite Trunk Diameter in Al-7wt% Si Alloy Directionally Solidified Aboard the International Space Station

SR Upadhyay, SN Tewari, M Ghodes, RN Grugel, DR Poirier and M Lauer.....136

Characterization of dendritic growth in Fe–C system using time-resolved X-ray tomography and physics-based filtering

H Yasuda, T Kawarasaki, Y Tomiyori, Y Kato and K Morishita.....142

Microstructure in A356/AA6xxx after bimetallic casting with flux coating

Aina Opsal Bakke, Lars Arnberg and Yanjun Li.....151

Formation of micro-plumes at a planar solid/liquid interface in a temperature gradient

J P Mogeritsch, T Peifer and A Ludwig.....157

Solidification cracking during welding of steel: In situ X-ray observation

L. Aucott, J. Li, A. Rack and H. B. Dong.....163

Microstructure modeling at different length scales

Capillary-mediated solid-liquid energy fields: their detection with phase-field method

M Glicksman and K Ankit.....169

Efficient model for the prediction of dendritic grain growth using the lattice Boltzmann method coupled with a cellular automaton algorithm

Stephan Jäger and Andreas Ludwig.....177

Finite diffusion microsegregation model applied to multicomponent alloys

V. Maguin, G. Guillemot, V. Jaquet, N. T. Niane, L. Rougier, D. Daloz, J. Zollinger and Ch.-A. Gandin.....183

Three-phase numerical modeling for equiaxed solidification of Sn–10 wt.%Pb alloy under forced convection driven by electromagnetic force

T Wang, E Wang, Y Delannoy, Y Fautrelle and O Budenkova.....191

A 3D discrete-element model for simulating liquid feeding during dendritic solidification of steel

Yi Feng, Miha Založnik, Brian Thomas and André Phillion.....197

A novel route to the coupling of molecular dynamics and phase-field simulations of crystal growth

P C Bollada, H Men, C Fang, P K Jimack, Z Fan and A M Mullis.....203

Effect of the Coriolis force on the macrosegregation of aluminum in the centrifugal casting of Ti-Al alloys

M. Cisternas Fernández, M. Založnik, H. Combeau, C. Huang, J. Zollinger and U. Hecht.....209

A Study of the Mass Transfer Kinetics during the Dissolution of Ti-N Particles in Liquid Titanium

L Yao, S L Cockcroft and D M Maijer.....215

Kinetics of rapid crystal growth: phase field theory *versus* atomistic simulations

P K Galenko, A Salhoumi and V Ankudinov.....221

Prediction of solidification structures in a 9.8 t steel ingot

Benjamin Gerin, Hervé Combeau, Miha Založnik, Isabelle Poitrault and Maya Cherif.....227

Semi-grand canonical Monte Carlo simulation for derivation of thermodynamic properties of binary alloy

Kensho Ueno and Yasushi Shibuta.....234

Numerical simulations of solidification structures and macrosegregation by a cellular automaton model coupled with flow calculations

Y Natsume.....240

Analysis of columnar-to-equiaxed transition experiment in lab scale steel casting by a multiphase model

S. Sachi, M. Založnik, H. Combeau, Ch.-A. Gandin, M. Gennesson, J. Demurger, M. Stoltz and I. Poitrault.....246

Three-dimensional mesoscopic modeling of equiaxed dendritic solidification in a thin sample: effect of convection flow

A. Olmedilla, M. Založnik, M. Cisternas Fernández, A. Viardin and H. Combeau.....252

On the Modelling of Macrosegregation during Twin-Roll Casting

C M G Rodrigues, A Ludwig, A Kharicha, M Wu and A Vakhrushev.....258

Atomistic simulation of crack propagation along γ-TiAl lamellar interface

W Li, H Nan, X W Qian, Y J Yin, Q Xu, X Shen, W Yu, X Feng, L P Zhu and J X Zhou.....265

Nucleation and grain refinement

Numerical simulation of wave-like nucleation events

Lang Yuan, Arvind Prasad, Peter D Lee and David StJohn.....270

Increasing the Liquidus Temperature by Employing the Controlled Diffusion Solidification (CDS) Process: A Potential Route to Improved Castings

Mohammad Pourgharibshahi, Mehdi Divandari, Hassan Saghafian, Alberto Fabrizi and Giulio Timelli.....276

Influence of minor additions on Icosahedral Short-Range Order (ISRO) and its consequences on nucleation and solidification growth kinetics in fcc alloys

J. Zollinger and M. Rappaz.....282

Mechanisms of TiAl alloys isomorphic inoculation from cryo-milled Ti-Al-Nb powders

J. Zollinger, J.R. Kennedy, E. Brodu, B. Rouat and E. Bouzy.....288

Molecular dynamics simulation of heterogeneous nucleation via grain refiner inoculated in aluminium melt

Takuya Fujinaga and Yasushi Shibuta.....293

Segregation of Ca at the Mg/MgO interface and its effect on grain refinement of Mg alloys

S H Wang, F Wang, Y Wang, Q M Ramasse and Z Fan.....299

Grain refinements of magnesium alloys inoculated by additions of external SiC particles

Yuanding Huang, Jian Gu, Sihang You, Karl Ulrich Kainer and Norbert Hort.....306

Solidification processing

Optimization of ultrasonic cavitation processing in the liquid melt flow

T Subroto, D G Eskin, I Tzanakis, G S B. Lebon, A Miranda and K. Pericleous.....312

Online Modelling of Heat Transfer, Solidification and Microstructure in Continuous Casting of Steel

S Louhenkilpi, J Miettinen, J Laine, R Vesanen, I Rentola, J Moilanen, V-V Visuri, E-P Heikkinen and A Jokilaakso.....319

Characterization of heat transfer and its effect on solidification in water cooled LPDC of wheels

AU Saadah, C Reilly, DM Maijer and SL Cockcroft.....325

Effects of strain rate on hot tear formation in Al-Si-Cu alloys

S Bhagavath, B Cai, R Atwood, PD Lee and S Karagadde.....331

Interfaces at internal chills in solidifying steel sections

Charles Monroe, Robin Foley, John Griffin, Sean Gibbons and Alec Saville.....337

On Analytical Concepts of Novel Multi-Resolution Casting Simulations

Amir M Horr.....343

Quantification of microstructure to reveal the solidification path of an alloy

A-A. Bogno, W. Hearn, J.E. Spinelli, J. Valloton and H. Henein.....349

Effect of cooling rate on the porosity defect in the thick aluminum casting by 3D computed tomography analysis

I S Cho, S Y Kwak, Y H Kim, H K Lee and S M Yoo.....355

Numerical simulation of fluid flow, solidification and defects in high pressure die casting (HPDC) process

K Dou, E Lordan, Y J Zhang, A Jacot and Z Y Fan.....362

Solidification processing of scrap Al-alloys containing high levels of Fe

J Lazaro-Nebreda, J B Patel, I T H Chang, I C Stone and Z Fan.....368

Microstructure evolution in an Al-Si piston alloy under ultrasonic melt processing

Suwaree Chankitmunkong, Dmitry G. Eskin and Chaowalit Limmaneevichitr.....374

Simulation of casting filling process using the lattice Boltzmann method

Y J Zhang, X W Qian, J X Zhou, Y J Yin, X Shen, X Y Ji and J L Huang.....380

Using morphology-equivalent method to simulate the evolution of shrinkage in Ti6Al4V alloy castings during HIP

Q Xu, J X Zhou, W Li, X Feng, H Nan and Y J Yin.....386

Steelmaking and continuous casting

Advances in Modeling of Steel Solidification with IDS

J Miettinen, S Louhenkilpi, V-V Visuri and T Fabritius.....391

Investigation on Band Segregate Formation during the Electroslag Remelting of H13 Die Steel

Q Wang, X J Wang, G Q Li and Y Liu.....397

Dendrite growth direction measurements: understanding the solute advancement in continuous casting of steel

A. SenGupta, B. Santillana, S. Sridhar and M. Auinger.....403

Utilizing inclusion data in characterization of oxide-sulfide stringers in hot-rolled plates

T Alatarvas, T Antola and T Fabritius.....409

A Novel Powder Metallurgy Technique for Introducing Synthetic Inclusions into Liquid Steel

A Smith, T Tran, F Ji, W Tiekink, G Abbel and S Verdier.....415

Numerical Simulations of Electromagnetic Counteractions to Mold Fluid Flow Asymmetry Deviations

Martin Sedén and Bengt Rydholm.....421

Continuous Casting of High Carbon Steel: How Does Hard Cooling Influence Solidification, Micro - and Macro Segregation?

A. Nicholas Grundy, Steve Münch, Stephan Feldhaus and Johan Bratberg.....427

Numerical simulations of liquid steel alloying in the three strand continuous casting bloom tundish

A Cwudziński.....434

Evaluation of AHSS concepts with a focus on the product properties and appropriate casting characteristics of *Arvedi ESP thin slab casters*

C. Bernhard, B. Linzer, P. Presoly, I. Watzinger and J. Watzinger.....441

Effect of different process parameters on non-metallic inclusions during electro-slag remelting of a tempering steel

R Schneider, S Schroft, M Molnar, G Klösch and C Schüller.....451

Solidifying shell waviness during continuous casting of AHSS slabs

Hongbin Yin.....457

Thermomechanics and properties

Study of elasto-plastic deformation in a cast AlCu7 alloy

M Schöbel, R Fernández and HP Degischer.....464

Effect of Pores on Tensile Fracture of Die-cast AlSiMgMn alloys with 3D X-ray μ-CT and FE Simulation

H D Zhao, F Liu, R S Yang and F Z Sun.....470

On the use of heterogeneous thermomechanical and thermophysical material properties in finite element analyses of cast components

Johan Jansson, Jakob Olofsson and Kent Salomonsson.....476

Determination of pattern allowances for a steel casting using an inverse elastoplastic deformation analysis

D Galles, J Lu and C Beckermann.....483

Three-dimensional study of nodule clustering and heterogeneous strain localization for tailored material properties in ductile iron

Jakob Olofsson, Kent Salomonsson, Arne K. Dahle and Ragnvald H. Mathiesen.....490

Shape of the in-situ (Al, Zn)-Ti reinforcing particles and their influence on structure and structural stability of selected Zn-Al and Al-Zn cast alloys

P K Krajewski, A L Greer and W K Krajewski.....496

An investigation of the chill-casting interface dynamics in production of sand-cast A319 engine blocks

Farzaneh Farhang Mehr, Steve Cockcroft and Daan Maijer.....502

On modelling viscoplastic behavior of the solidifying shell in the funnel-type continuous casting mold

A Vakhrushev, A Kharicha, M Wu, A Ludwig, G Nitzl, Y Tang, G Hackl, J Watziger and C M G Rodrigues.....508

A partitioned solution algorithm for fluid flow and stress-strain computations applied to continuous casting

Shaojie Zhang, Gildas Guillemot, Charles-André Gandin and Michel Bellet.....515

Thermo-elasto-viscoplastic constitutive laws for metallic alloys during their solidification

G M Laschet and H Behnken.....523

Solid-liquid interfacial energy of Al-Zn solid-solutions in equilibrium with Al-Zn liquid

E Yılmaz, H Erol, E Acer Erol and M Gündüz.....530

IOP Conf. Series: Materials Science and Engineering **529** (2019) 011001 doi:10.1088/1757-899X/529/1/011001

5th International Conference on Advances in Solidification Processes (ICASP5)

5th International Symposium on Cutting Edge of Computer Simulation of Solidification, Casting and Refining (CSSCR5)

June 17-21, 2019

The Pitter Event Center, Salzburg, Austria

The fifth International Conference on Advances in Solidification Processes (ICASP5) and the fifth International Symposium on Cutting Edge of Computer Simulation of Solidification, Casting and Refining (CSSCR5) are held as a joint event in Salzburg, Austria, on June 17-21, 2019. The conference location, Salzburg, is known as the birth town of Wolfgang Amadeus Mozart, the UNESCO World Heritage city, at the heart of Europe. Solidification science and technology deserve the destination as we, experts in the field of solidification, place it as high as a Wolfgang Amadeus Mozart's partition.

History in brief. The conference series ICASP started in Stockholm (Sweden) in 2005, continued in Graz (Austria) in 2008, Aachen (Germany) in 2011, London (UK) in 2014. The conference series CSSCR started in Osaka (Japan) in 1999, continued in Sapporo (Japan) in 2010, Stockholm (Sweden) in 2013, Xi'an (China) in 2016.

Statistics. The joint event ICASP5-CSSCR5 attracted over 200 participants from 29 countries worldwide. From 255 abstracts received, the final program included 158 oral presentations (including 11 plenaries, 9 keynotes and 18 highlights) and over 60 poster presentations.

Topics and highlights. The joint event ICASP5-CSSCR5 covered a broad spectrum of solidification related topics. There were 27 parallel sessions, dealing with nucleation and grain refinement, in-situ observation and experimental characterization of microstructure (primary dendritic, eutectic and peritectic, intermetallic and composites); numerical modelling at different length scales (atomistic, microscopic, mesoscopic and macroscopic), casting processes (continuous casting, ingot, shape casting, electroslag remelting, …) and cast alloys (steel, cast iron, aluminum, etc.), thermo-mechanics and properties. For the first time of this conference series, a new topic, additive manufacturing, with two parallel sessions was added to the program. Freeze casting was also for the first time introduced. The plenary lectures were carefully defined while accounting for the recommendations of the International Scientific Committee. They covered an historical review on the development route leading to modern solidification science, progress in the microstructure modelling which focuses on bridging different length scales, in-situ observation of metal alloy solidification, discussions on the facing challenges such as the missing physical properties and coupled flow-structure mechanics in solidification, and interdisciplinary research bridging solidification science beyond metallurgy.

Proceedings. After the peer-review procedure, 84 articles were included in the proceedings. Each article was reviewed, heavily relying on the members of the International Scientific Committee or senior authors. The proceedings were published before the conference.

Acknowledgments. The editors and the conference chairmen gratefully acknowledge the International Scientific Committee for promoting the conference, proposing topics of plenary lectures, and reviewing the papers. Special thanks go to ASMET, The Austrian Society for Metallurgy and Materials, for taking the responsibility of the conference organization.

Charles-André GANDIN
MINES ParisTech, France
charles-andre.gandin@mines-paristech.fr

Menghuai WU
University of Leoben, Austria
menghuai.wu@unileoben.ac.at

IOP Conf. Series: Materials Science and Engineering **529** (2019) 011002 doi:10.1088/1757-899X/529/1/011002

Peer review statement

All papers published in this volume of *IOP Conference Series: Materials Science and Engineering* have been peer reviewed through processes administered by the proceedings Editors. Reviews were conducted by expert referees to the professional and scientific standards expected of a proceedings journal published by IOP Publishing.

Particle scale modelling of porosity formation during selective laser melting process using a coupled DEM – CFD approach

Akash Aggarwal and Arvind Kumar*

Department of Mechanical Engineering, Indian Institute of Technology Kanpur, Kanpur 208016, India.

*Corresponding author: arvindkr@iitk.ac.in

Abstract. Selective Laser Melting (SLM), the most popular metal additive manufacturing (AM) process, is well suited for making complicated parts which are difficult to manufacture by conventional manufacturing techniques. Currently, the main bottlenecks inhibiting the usage of the Selective Laser Melting (SLM) parts include the problems, such as porosity, low resolution, low surface finish quality and low build rate. In order to overcome the aforesaid problems, latest SLM machines are now being equipped with laser having small spot radius for enhanced resolution and surface finish, and high power to increase the build rate. The combination of high power and small spot radius leads to high energy density, exceeding the threshold value, resulting in transition of melting mode in the SLM process from conduction mode to keyhole mode and a formation of porosity due to collapsing of keyhole. In this study, high fidelity particle scale model is developed using open-source codes LIGGGHTS and OpenFOAM to understand the formation of porosity and to describe the physical phenomena (convection, melting, evaporation and solidification), melt flow dynamics and melting mode transition occurring in the SLM process.

1. Introduction

In SLM, similar to laser welding, if the melting mode is keyhole, due to extremely high energy density, metal vaporizes and plasma is formed. The formation of vapour cavity substantially increases the beam absorption enabling the laser to penetrate deeper resulting in melt pools that are deeper and narrow [1,2]. Also, if the energy density is very high then recovery of the collapsed deep keyhole will not be achieved leading to formation of spherical porosity. The collapse of the cavity can leave voids in the propagation path of the laser beam.Although there are some experimental work reported about the porosity formation with increasing energy density during the SLM process [3], but its formation mechanism is not understood mathematically.

In this paper, a particle scale model is developed using open-source codes LIGGGHTS and OpenFOAM to study the porosity formation due to keyhole collapsing in the SLM process. Discrete element method is used to determine the spatial arrangement of the particles in the powder layer. Thereafter, the volume of fluid approach using the finite volume method is used to identify and track the interface of the powder particles undergoing phase transition. Subsequently, geometrical characteristics of the porosity formation obtained by computational modelling are qualitatively validated with the in-house single track experiments.

IOP Conf. Series: Materials Science and Engineering **529** (2019) 012001 doi:10.1088/1757-899X/529/1/012001

2.Numerical modelling and methodology

To understand the interaction of high energy beam with the powder layer and the resulting porosity formation, single track scanning over a powder layer is simulated with the help of a thermo-fluidic powder scale model developed using an integrated discrete element method (DEM) – computational fluid dynamics (CFD) approach. The developed thermo-fluidic powder scale model uses a realistic powder bed for CFD modelling and incorporates the physics of laser irradiation on the powder bed, melt flow due to thermo-capillary force, effect of the recoil pressure and the phase transition (melting, solidification, evaporation and condensation).

2.1. Single layer powder bed Generation

The present work uses a DEM based modelling approach to simulate the powder spreading process. DEM, a Lagrangian based approach, calculates forces acting on granular material from the initial conditions, governing physical laws and contact models. The computational domain considers each particle as an individual entity with its own properties interacting with other particles and boundaries in its vicinity. The motion of each particles is described using Newton's law of motion for conservation of momentum. In this work, an open source DEM code LIGGGHTS (LAMMPS Improved for General Granular and Granular Heat Transfer Simulations) is being used for powder spreading. Firstly, particle size distribution (PSD) was experimentally measured by performing image analysis on the SEM (scanning electron microscopy) micrographs. Then, a cloud of randomly generated particles with the experimentally measured PSD was generated inside the DEM computational domain and allowed to fall under gravity, after all the powder particles were settled (Fig. 1). Thereafter, using a MATLAB script the particles information of 40 micron thickness was exported to an open source CFD C++ code OpenFOAM (Open field operation and manipulation). The CFD model of thermo-fluidic phenomena, presented in the following, was developed in OpenFOAM.

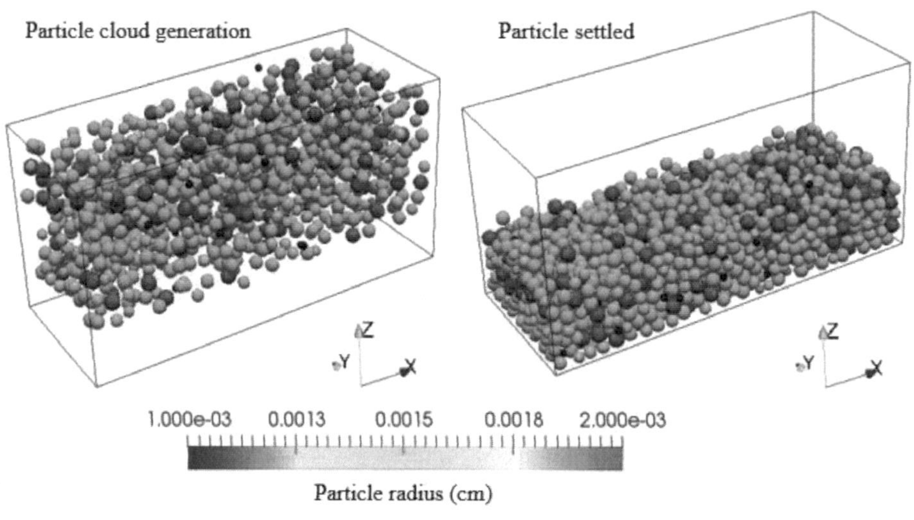

Fig. 1.Discrete element method simulation results using open source code LIGGGHTS

2.2. Free Surface Thermo-Fluidic Modelling

Figure 2 shows the computational domain considered for the single track thermo-fluidic simulation. The computational domain consists of an Inconel 718 substrate, an argon inert gas region and a 40 μm thick Inconel 718 powder layer. In the model, transient heat transfer and fluid flow dynamics in the melt pool are considered using the volume of fluid surface tracking method (VOF). The VOF transport equation for interface tracking between two immiscible phases (Inconel 718 and argon gas) is given by [4]

$$\frac{\partial \gamma}{\partial t} + \nabla \cdot (\gamma \vec{U}) + \nabla \cdot ((1-\gamma)\gamma \vec{U_r}) = 0 \quad (1)$$

where γ is the phase fraction, $\vec{U}$ is the velocity vector and $\vec{U_r}$ is the compression velocity.

In the simulation, the thermo-physical properties of the two immiscible phases are calculated using the continuum formulation based on the classical mixture theory. Assuming the flow to be incompressible, laminar and Newtonian, the governing conservation equations for mass, momentum and energy conservation are formulated.

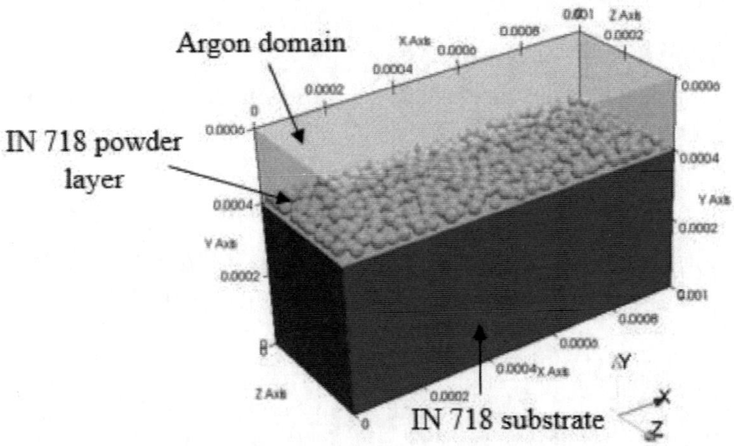

Fig. 2. Computational domain used for CFD simulation (labels in X, Y, Z axis are distance in m)

The governing energy conservation equation is given by

$$\frac{\partial(\rho C_p T)}{\partial t} + \nabla \cdot (\rho \vec{U} C_p T) = \nabla \cdot (k \nabla T) + S_{Latent} + Q_{Losses} + Q_{Laser} \quad (2)$$

The source term S_{Latent} in Eq. 2 accounts for the evolution of the latent heat during phase change. The source term Q_{Losses} in Eq. 2 represents evaporation heat loss, radiation heat loss and convective heat loss due to the flowing argon gas. It is defined as

$$Q_{Losses} = (q''_{evaporation} + q''_{radiation} + q''_{conv})|\nabla \gamma| \frac{2\rho C_{peff}}{\rho_{metal} C_{pmetal} + \rho_{gas} C_{pgas}} \quad (3)$$

The evaporative heat loss is the product of the latent heat of vaporization L_v and the vaporized mass flow rate $\dot{m}_v$.

$$q''_{evaporation} = -L_v \dot{m}_v \quad (4)$$

The vaporized mass flow rate of the escaping vapour $\dot{m}_v$ [5] is given by:

$$\dot{m}_v = (1-\beta) \frac{M P_v}{\sqrt{2\pi M R T}} \quad (5)$$

where M is the molar mass, R is the ideal gas constant and P_v is the recoil vapor pressure which is given by

$$P_v = 0.54 P_0 exp\left[\frac{L_v M (T - T_v)}{R T T_v}\right] \quad (6)$$

where P_0 is the atmospheric pressure, L_v is the latent heat of vaporization and T_v is the vaporization temperature. β in Eq. 5 is the retro diffusion coefficient and it represents the extent of condensation of escaping vapour. In this study, $\beta = 0.18$ was assumed.

The source term Q_{Laser} in Eq. 2 accounts for heating by the moving laser beam. The energy input from the laser beam is approximated by a Gaussian distribution and is defined as

$$Q_{Laser} = \frac{2\eta P}{\pi R_{spot}{}^{2}} exp\left[-2\frac{(x-x_i-vt)^2+(z-z_i)^2}{R_{spot}{}^{2}}\right] f_{top}|\nabla\gamma|\frac{2\rho C_{peff}}{\rho_{metal}C_{pmetal}+\rho_{gas}C_{pgas}} \tag{7}$$

where $\eta$is the absorption coefficient, v is the laser traversal velocity, (x_i, z_i) denotes the beginning point of the laser melting and $f_{top}$is the unit function representing the top portion of the powder particle where heat flux is applied.

The momentum conservation equation is given by

$$\frac{\partial(\rho\vec{U})}{\partial t} + \vec{U}\cdot\nabla(\rho\vec{U}) = -\nabla p + \nabla\cdot(\mu\left(\nabla\vec{U} + (\nabla\vec{U})^T\right) + \overrightarrow{F_N} + \overrightarrow{F_D} + \overrightarrow{F_S} \tag{8}$$

The buoyant force due to the natural convection is implemented in the current model with the help of a source term $\overrightarrow{F_N}$ in Eq. (8) which is given as

$$\overrightarrow{F_N} = \rho_l\vec{g}\beta_T(T-T_{ref})(9)$$

where ρ_l,β_T and T_{ref} are the liquid metal density, thermal expansion coefficient and reference temperature, respectively.The source term $\overrightarrow{F_D}$ appearing in Eq. (8) is defined in Eq. (10). It aids to smoothly bring down the velocity of the fluid at the liquid-solid interface and makes the fluid velocity in the un-melted solid zone as zero.

$$\overrightarrow{F_D} = \frac{\left(1-f_{liquid,metal}\right)^2}{f_{liquid,metal}{}^{3}+b} C\vec{U} \tag{10}$$

The constant C in Eq. (10) represents mushy zone constant and a value of 160,000 kg m^{-3} s^{-1} is considered in the current model [6].The term b is another constant having a small value ($\sim 10^{-6}$) and is used to prevent division by zero when the liquid fraction ($f_{liquid,metal}$) becomes zero.The source term $\overrightarrow{F_S}$ in Eq. 8 term represents the forces which are acting at the interface and is given by

$$\overrightarrow{F_S} = \left\{\sigma\kappa\hat{n} + \frac{d\sigma}{dT}[\nabla T - \hat{n}(\hat{n}\cdot\nabla T)] + \hat{n}(P_v)\right\}|\nabla\gamma|\frac{2\rho}{\rho_m+\rho_g} \tag{11}$$

where σ is the surface tension coefficient,$\kappa = -(\nabla\cdot\hat{n})$is the mean curvature of free surface, $\hat{n} = \nabla\gamma/|\nabla\gamma|$ is the interface normal unit vector and $d\sigma/dT$ is the temperature coefficient of surface tension. The first term is the surface tension force acting normally to the interface, the second term is the force due to Marangoni convection acting tangentially at the interface and the third term is the recoil pressure exerted by the metal vapour on the top surface of the molten pool.

For thermo-fluidic CFD simulation, an element size of 4 μm provides mesh independent and computationally.To make sure the solution is stable, self-adaptive time step based on the Courant–Friedrichs–Lewy (CFL) condition was used.For each time step, volume fraction advection, continuity, momentum, and energy transport equations were solved. The material properties of IN718 used in this simulation is taken from Ref. 7.

3.Results and discussion

Single track scanning over a powder layer has been simulated for two scanning velocity(v = 0.5 m/s and v = 0.7 m/s)keeping power (P = 100 W) to be the same. Fig. 3 shows the top view of the deposited single tracks at 1175 μs for v = 0.5 m/s.In the figure the substrate, the unmelted powder layer, and a penetrated melt pool caused by recoil pressure can be clearly seen. Due to heating by the laser beam the powder layer melts and forms a bulk melt. Once the laser beam traverses forward the bulk melt re-solidified in to bulk solid.

Figure 4 shows maps of temperature field along a cross section of the computational domain for the scanning velocity of v= 0.7 m/s and v= 0.5 m/s illustrating the elongated melt pool (melting and solidification). It can be clearly seen for the both cases that due to recoil pressure, a topologically depressed cavity starts forms. As the temperature exceeds the boiling point, recoil pressure starts to act and due to melt displacement from the centre of the melt pool, a bulge is formed. As the laser beam moves forward, the melt pool size increases and finally it reaches a quasi-steady state. It can be seen that substantial melting of the substrate has taken place. The melting in the substrate is crucial in determining surface morphology and metallurgical bonding of the solidified powder layer with the substrate. As laser beam moves away, molten metal starts filling the depressed region.

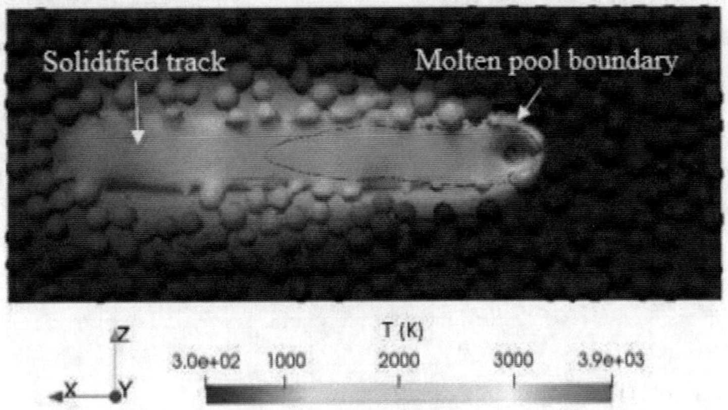

Fig .3.Top view of the deposited single tracks at 1175 μs (P = 100 W, v = 0.5 ms^{-1}).

The filling of molten metal causes recovery of the topologically depressed region. If the depressed region is substantially deep then it may collapse with gas entrapped within it. As shown in Fig. 4, porosity-free single track deposition is achieved for the case of v= 0.7 m/s. But if the energy density is further increased (v= 0.5 m/s) then recovery of the collapsed deep keyhole will not be achieved leading to formation of spherical porosity. The collapse of the cavity leaves voids in the propagation path of the laser beam. Therefore, high energy density is not preferable as they will lead to porosity.

In order to qualitatively validate the simulation results and to obtain understanding of the porosity formation, single tracks were deposited by varying the scanning velocity. The tracks were deposited on a 40 μm thick Inconel 718 powder placed on a 1 mm thick Inconel 718.

Figure 5 shows the top view and longitudinal cross-section of the deposited single tracks at v= 0.5 m/s.Clearly, as observed in simulation (Fig. 4), for the case of v= 0.5 m/s spherical porosity forms along the laser traversal direction.

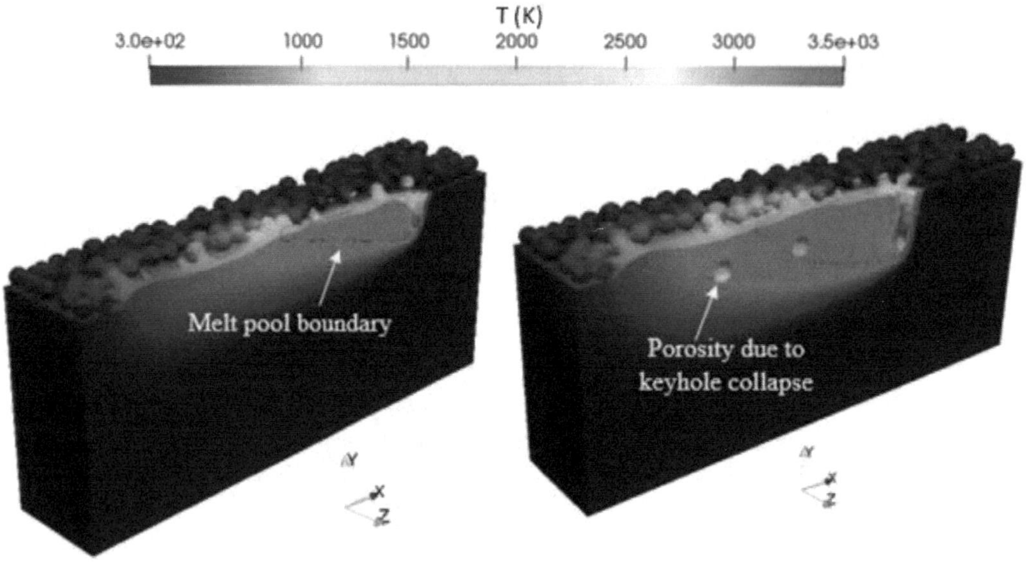

Fig. 4. Simulation results showing the temperature field (longitudinal cross-sectional view), (a) $v = 0.7$ ms^{-1} and (b) $v = 0.5$ ms^{-1}

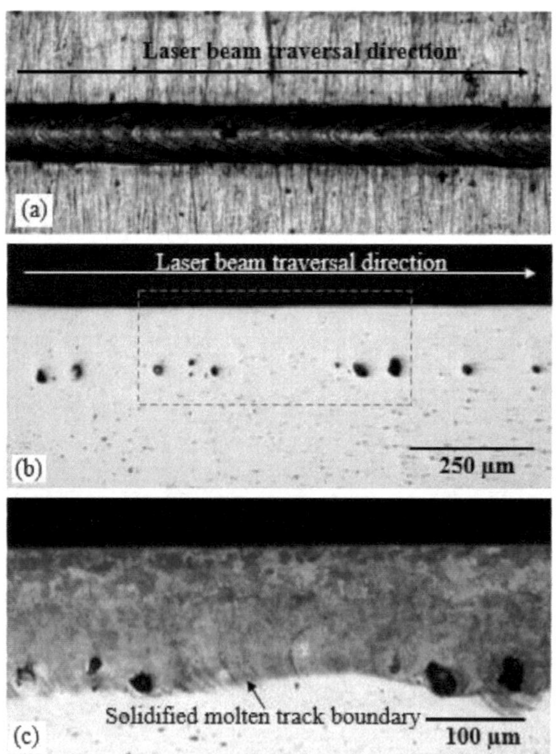

Fig. 5. Longitudinal cross section of anInconel 718 single track with porosity formed at the bottom of solidified molten track (P = 100 W, $v = 0.5$ ms^{-1}).

4.Conclusion

A three-dimensional open-source DEM – CFD simulation incorporating the physics of melting mode transition, melt flow dynamics and the physical phenomena (convection, melting, evaporation and solidification) has been developed. For focused laser beam, keyhole mode of melting occurs where both thermo-capillary force and recoil pressure play a dominant role resulting in a narrow and deeper melt pool. In keyhole mode of melting, if the depressed region is substantially deep then it collapse leading to spherical porosity formation. The collapse of the cavity can leave voids in the propagation path of the laser beam. Therefore, deep keyhole is not preferable as they will lead to porosity, but if stable keyhole is formed then better mechanical and microstructural properties is achieved.

References

1. Fabbro, R. (2010). Melt pool and keyhole behaviour analysis for deep penetration laser welding. *Journal of Physics D: Applied Physics*, *43*(44), 445501.
2. Courtois, M., Carin, M., Le Masson, P., Gaied, S., & Balabane, M. (2013). A new approach to compute multi-reflections of laser beam in a keyhole for heat transfer and fluid flow modelling in laser welding. *Journal of Physics D: Applied Physics*, *46*(50), 505305.
3. Dilip, J. J. S., Zhang, S., Teng, C., Zeng, K., Robinson, C., Pal, D., & Stucker, B. (2017). Influence of processing parameters on the evolution of melt pool, porosity, and microstructures in Ti-6Al-4V alloy parts fabricated by selective laser melting. *Progress in Additive Manufacturing*, *2*(3), 157-167.
4. Samkhaniani, N., & Ansari, M. R. (2017). Numerical simulation of superheated vapor bubble rising in stagnant liquid. *Heat and Mass Transfer*, *53*(9), 2885-2899.
5. Khairallah, S. A., Anderson, A. T., Rubenchik, A., & King, W. E. (2016). Laser powder-bed fusion additive manufacturing: Physics of complex melt flow and formation mechanisms of pores, spatter, and denudation zones. *Acta Materialia*, *108*, 36-45.
6. Pathak, N., Kumar, A., Yadav, A., & Dutta, P. (2009). Effects of mould filling on evolution of the solid–liquid interface during solidification. *Applied Thermal Engineering*, *29*(17-18), 3669-3678.
7. Lee, Y. S., & Zhang, W. (2016). Modeling of heat transfer, fluid flow and solidification microstructure of nickel-base superalloy fabricated by laser powder bed fusion. *Additive Manufacturing*, *12*, 178-188.

IOP Conf. Series: Materials Science and Engineering **529** (2019) 012002 doi:10.1088/1757-899X/529/1/012002

Level-set modelling of Laser Beam Melting process applied onto ceramic materials – Comparison with experimental results

Q. Chen[1], L. Moniz Da Silva Sancho[2], G. Guillemot[1], Ch.-A. Gandin[1], M. Bellet[1], J.-D. Bartout[2], M.-H. Berger[2], C. Colin[2]

1. CEMEF – Centre de Mise en Forme des Matériaux, CNRS UMR 7635 – PSL Research University, MINES ParisTech, 06904 Sophia Antipolis, France

2. CdM – Centre des Matériaux, CNRS UMR 7633 – PSL Research University, MINES ParisTech, 91003 Evry Cedex, France

E-mail: gildas.guillemot@mines-paristech.fr

Abstract. Laser Beam Melting (LBM) processes benefit from significant progress in recent years. Currently, manufacturing of ceramic parts for applications at high temperature in aeronautical industries can be planned. However, understanding of defect formation is required in order to optimize manufacturing strategy. In this work, level-set modelling is proposed to simulate tracks development during LBM processes. Thermo-mechanical solution is performed in both powder and dense domains. Fluid flow is computed considering the surface tension and Marangoni forces. In addition mechanical resolution is achieved to investigate stress evolution in the rear part of the track. Applications are developed on alumina material. The influence of laser power, scanning velocity and physical properties are investigated and discussed. Validations of the heat source model are proposed by comparisons of melt pool dimensions and shapes with experimental measurements. A coherent evolution of the track morphology is shown when varying process parameters or material properties.

1. Introduction

In the last decade, Additive Manufacturing (AM) processes have emerged in industries. These processes reduce the time gap between design and manufacturing to deliver parts for direct end-use applications. In addition, they give the possibility to produce complex shapes as limited constraints are imposed on the parts geometry. Different AM processes have been developed by industries with various manufacturing strategies and applications covering polymer, metallic or ceramic materials [1]. Among these processes, LBM appears as attractive for high value-added products as encountered in aerospace, energy or medical industries. In such applications, high level of mechanical properties is required and conventional forming technologies are inefficient to develop expected pieces.

In order to develop part layer by layer, powder is melted in LBM. This is achieved by a laser focused on a narrow domain and moving at large velocity. A track then develops during cooling and solidification stages. Fully dense parts can thus be achieved and near net shape is possible. However, defects like cracks, delamination, porosities or poor surface quality are still encountered and induced by the use of non-optimized process parameters. These defects have detrimental consequences on final properties for direct end-use applications [2] and limit the dissemination of LBM processes. The required tracks should be regular and free from porosities. Efforts are consequently made to master LBM and achieve expected final properties. Clear improvements in the knowledge and influence of physical mechanisms occurring at the track scale are expected in future years. For that purpose, the development of modelling tools is an answer to the need to provide a better understanding and control of the phenomena impacting track shape and mechanical properties. In addition, such tools would permit to propose optimized process parameters and scanning strategy.

However, less development is made on the application of LBM to ceramics compare to metals. Indeed, ceramics like alumina or zirconia have outstanding mechanical strength and excellent thermal and wear resistance [3]. Nevertheless, some difficulties are known in the application of LBM technologies to ceramic materials explaining this situation. Ceramics have a low energy absorption to Nd:YAG laser used in LBM causing energy loss and reducing production rate. Moniz [4] proposed to overcome low absorption of powders by adding a specific amount of IR-absorbers as carbon or silicon carbides particles. In addition, ceramics have high melting point, high-temperature strength and low thermal conductivity [5]. Cracks formation in solidified tracks are also regularly reported due to the low fracture toughness of ceramics [6,7]. Solutions are proposed in order to limit cracks developments mainly based on powder preheating before melting as developed by Hagedorn [3] on alumina-zirconia (Al_2O_3-ZrO_2) or Liu [5] on zirconia-yttria (ZrO_2–Y_2O_3) ceramics using an auxiliary laser.

In order to provide tools to master and control LBM at narrow scale, a level-set (LS) modelling has been developed and applied on pure alumina powders to follow tracks development [8,9]. Thermo-mechanical evolution of material is investigated in both liquid and dense domains. Fluid flow evolutions in the melt pool are considered induced by the effects of surface tension and Marangoni forces. The same resolution gives access to the velocity field at the liquid/gas interface and track shape evolution. In addition the mechanical resolution is achieved considering a temperature dependent constitutive law in the solidified domain. This model provides the tools to investigate the influence of process parameters and physical properties of materials on track evolution and defects development.

2. Model

2.1. Level Set Method

The finite-element (FE) model presented in this paper has been extensively described by Chen *et al.* [8,9,10]. The approach is briefly underlined hereafter in order to describe the points of interest and focus on the originally of the model. The system is made of two domains (material and gas) (Fig. 1a) where powder ceramic is assimilated to a continuum. Material domain is made of pure alumina and corresponds to the powder bed in its initial state before melting. Previous consolidated layers are considered as the substrate domain on which powder layer is deposited. The powder is melted by laser during scanning step and rapidly cooled down by thermal diffusion to form a compact solid domain. During this one-way transformation, densification is modelled through the local apparent density variation from powder to liquid state. The surface evolution then depends from forces acting in the liquid bath. Its evolution is tracked with a LS method depending from the signed distance function, ψ, attached to the material / gas interface. This function is considered as positive (resp. negative) in the gas (resp. material) domain (Fig. 1 a). The Heaviside function, $\mathcal{H}$, is then deduced from the distance function in the whole system and evolves continuously in the transition domain $[-\varepsilon, +\varepsilon]$:

$$\mathcal{H}(\psi) = \begin{cases} 0 \text{ if } \psi < -\varepsilon \\ \frac{1}{2}\left[1 + \frac{\psi}{\varepsilon} + \frac{1}{\pi}\sin\left(\frac{\pi\,\psi}{\varepsilon}\right)\right] \text{ if } |\psi| \leq \varepsilon \\ 0 \text{ if } \psi > +\varepsilon \end{cases} \quad (1)$$

The local properties $\langle\chi\rangle^{D_i}$ are computed in each domain $D_i (i \in \{1,2\})$ from the intrinsic properties of phase φ (liquid, domain, powder or gas) and of their volume fraction, $g^{\varphi}_{D_i}$. The system properties $\{\chi\}$

are then deduced in any position by using the Heaviside function value, $\mathcal{H}$, and the properties associated to each domain. These relations aims at defining the thermo-physical properties and the set of equations averaged over the system corresponding to energy, momentum and mass conservation.

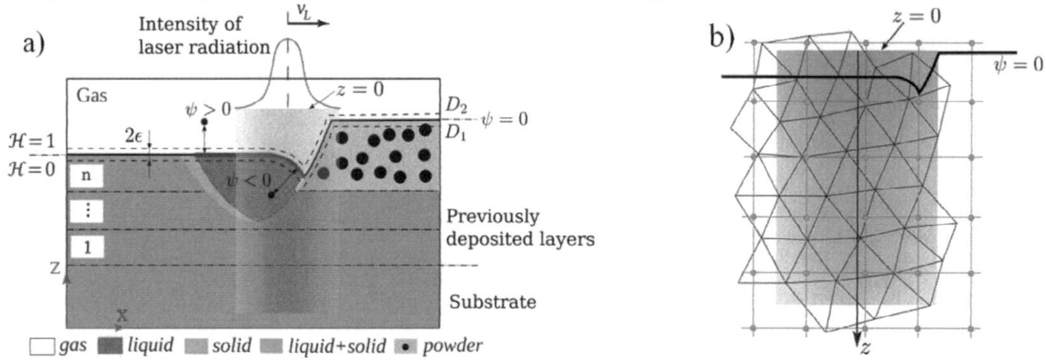

Figure 1 : a) LBM modelling by LS approach used to track the distance function ψ associated to the gas/ceramic interface b) Implementation of the heat source / FE mesh (black) and regular grid (blue).

2.2. Energy conservation

The temperature distribution during LBM is computed by the resolution of the non-steady energy conservation equation (Eq. 2) where T is the temperature, ρ the density, h the specific enthalpy and λ the conductivity. The velocity field, $\boldsymbol{u}$, is the solution of the momentum conservation equation.

$$\frac{\partial\{\rho h\}}{\partial t} + \nabla \cdot \{\rho h \boldsymbol{u}\} - \boldsymbol{\nabla} \cdot \{\lambda \boldsymbol{\nabla} T\} = \{\dot{q}_L\} - \{\dot{q}_r\} \quad (2)$$

The Nd:YAG Gaussian laser source used in LBM provides low interaction with ceramics and pure alumina is considered as transparent with deep propagation contrary to metals. The ceramic powder thickness and several layers are consequently heated. This phenomenon should be limited to prevent large melting/solidification history. Consequently carbon particles (dopants) are added in the powder bed [4] to increase interaction. A continuous modelling is proposed where the local effective laser absorption, α, is considered and the laser heat source, $\{\dot{q}_L\}$, is computed from a Beer-Lambert law [11]. For a given depth z, this source is derived from the spatial integration of the local absorption coefficient considering material properties variations. A regular grid covering the impacted domain is used to develop computation (Fig. 1 b) before interpolation onto the initial FE mesh. The radiation loss, $\{\dot{q}_r\}$, is estimated assuming a pure radiative heat exchange from the surface.

2.3. Momentum conservation

During melting, forces acting onto the liquid domains influence the liquid movement, model the interface and change the melt pool geometry which has significant effect on the final bead shape. Combined effect of surface tension, Marangoni forces and buoyancy forces are consequently considered in the momentum conservation equation in order to compute velocity field, $\boldsymbol{u}$:

$$\{\rho\}\left(\frac{\partial\{\boldsymbol{u}\}}{\partial t} + (\{\boldsymbol{u}\} \cdot \boldsymbol{\nabla})\{\boldsymbol{u}\}\right) - \boldsymbol{\nabla} \cdot \{\underline{\underline{\boldsymbol{\sigma}}}\} = \boldsymbol{f}_s + \boldsymbol{f}_m + \{\rho\}\boldsymbol{g} \quad (3)$$

$\underline{\underline{\boldsymbol{\sigma}}}$ is the stress tensor and $\boldsymbol{g}$ is the gravity vector. $\boldsymbol{f}_s$ and $\boldsymbol{f}_m$ are the forces associated to surface tension and Marangoni forces applied onto the liquid/gas and liquid/powder interfaces. These forces are respectively developed in the normal and tangential direction of the interface and expressed in (Eq. 4).

$$\boldsymbol{f}_s = \delta_{ST}\, \gamma\, \kappa\, \boldsymbol{n}\,, \boldsymbol{f}_m = \delta_{ST} \frac{\partial \gamma}{\partial T} \boldsymbol{\nabla}_s T \quad (4)$$

δ_{ST} is the Dirac function computed onto the interfaces where forces act. γ is the temperature-dependent surface tension coefficient and κ is the curvature radius. $\boldsymbol{n}$ is the normal vector to interface and $\nabla_s T$ is the surface temperature gradient. In addition, shrinkage is considered as apparent ceramic density sharply evolves during melting and consolidation of powder. Thus a compressible Newtonian constitutive law is used in the whole system including material and gas domains:

$$\underline{\underline{\sigma}} = \underline{\underline{s}} - p\underline{\underline{I}} \qquad \underline{\underline{s}} = 2\mu\left(\underline{\underline{\dot{\epsilon}}} - \frac{1}{3}tr\left(\underline{\underline{\dot{\epsilon}}}\right)\underline{\underline{I}}\right) \qquad tr\left(\underline{\underline{\dot{\epsilon}}}\right) = \nabla\cdot\{\boldsymbol{u}\} \tag{5}$$

where $\underline{\underline{s}}$ is the deviatoric stress tensor, p is the pressure and μ is the dynamic viscosity. The velocity field divergence is estimated from the temporal derivative of the effective density in the expression of mass conservation equation. After resolution, the velocity field is used in the transportation equation to update the position of the distance function and LS interface associated to track shape.

2.4. *Solid mechanics*

In LBM process, the stresses at the track scale are generated by the large temperature gradients and associated density variation during heating and cooling stages. The modelling of solid mechanics consists in the resolution of equilibrium equations associated to momentum conservation as:

$$\nabla\cdot\underline{\underline{\sigma}} + \{\rho\}\boldsymbol{g} = \boldsymbol{0} \tag{6}$$

with given boundary conditions corresponding to imposed velocity or normal forces. The modeling of solid mechanics requires consequently a suitable constitutive law which can be used in the full temperature range in order to cover the whole temperature evolution of ceramics. As viscous phenomena become important at large temperature, a general model is required. Three regimes are consequently distinguished on ceramic materials: an elasto-viscoplastic (EVP) behavior at temperature lower than a critical temperature T_c close to the solidus temperature, T_S, where viscous effects are negligible; a viscoplastic (VP) behavior in the mushy zone between liquidus (T_L) and solidus (T_S) temperature and a Newtonian behavior for temperature higher than the liquidus (T_L).

3. Applications and discussion

First applications are proposed in order to follow track development of a 50 μm powder layer deposited on a 1 mm-thick and 3 mm-long substrate. Materials properties of ceramic materials are detailed by Chen *et al.* [8,9,10]. The first investigation is proposed for a laser power, P_L, of 84 W and a scanning velocity of 200 mm·s^{-1}. Laser interaction radius r_{int} is fixed to 50 μm and same absorption coefficient for solid and liquid phase ($\alpha_s = \alpha_l = 5\ \text{mm}\cdot\text{s}^{-1}$) are considered. Fig. 2 a) illustrates the evolution of liquid velocity field and melt pool size for various Marangoni coefficients, $\partial\gamma/\partial T$. Centrifugal convection is observed in first (#1) and second (#2) simulations induced by the negative coefficient corresponding to forces acting in the opposite direction of temperature gradient (Eq. 4). The melt pool is elongated with length close to 400 μm. In the opposite case, centripetal velocities are obtained with a shorter melt pool. In both case, the melt pool is stable and feed by droplets formed in the irradiated domain. In addition, it is observed that Marangoni effect is restricted to the upper part of the liquid bath with few convection in the bottom part due to the high viscosity of liquid ceramic.

The mechanical behaviour has been computed with various strategies in order to investigate additional heating effect on stress distribution (Fig. 2 b). A principal laser is used with $P_L^{(1)} = 84$ W and $r_{int}^{(1)} = 50\ \mu m$. Scanning velocity is fixed to 300 mm·s^{-1} and stress distribution are shown in first case (Single laser). When the material is heated, the density decreases inducing thermal expansion and local compressive stress. Local tensile stresses are observed in upper part which may lead to cracks formation. An additional coaxial laser is used later with lower density energy ($P_L^{(2)} = 60\ \text{W}, r_{int}^{(2)} = 200\ \mu m$) to enlarge the heating zone and control cooling regime. Due to temperature effect, the thermal gradient is reduced and lower tensile stresses are observed. This may limit defect development and provide tracks of better quality. The last case corresponds to the shift of the same auxiliary laser toward the tail at a distance of 250 μm. This configuration affects the tail of the melt pool rather than the front. Consequently, the melt pool is enlarged with a similar depth. As the additional heating is imposed directly on the melt pool tail, the thermal gradient and cooling rate are reduced. This successfully decreases the stress near the track surface and in the cooling domain. These simulations provide valuable information in the heating strategy to limit defect development in LBM processes.

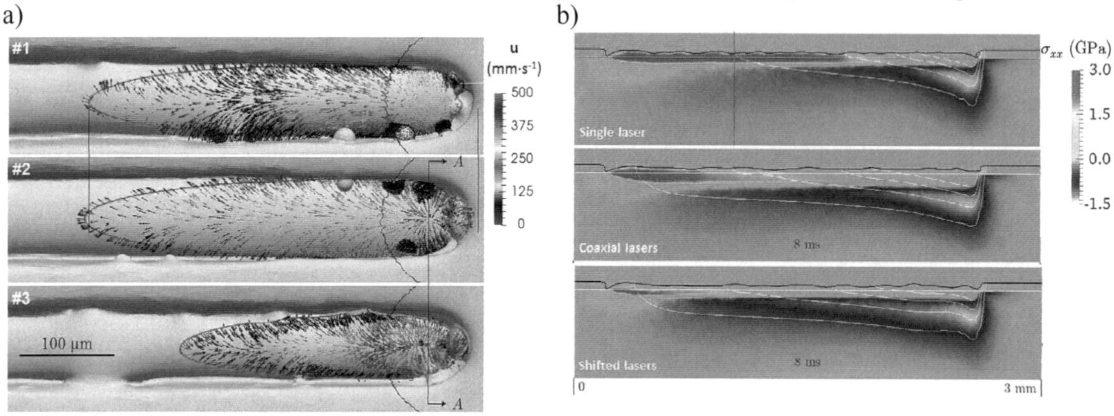

Figure 2 : a) Melt pool (red contour) and liquid velocity field when the laser arrives to position $X = 1.416\ mm$ (top view) for various Marangoni coefficient reported in literature equal to (#1) – 8.2 × 10^{-5} [12] and (#2) – 48 × 10^{-5} N·m^{-1}·K^{-1} [13]. An added case (#3) is proposed with + 48 × 10^{-5} N·m^{-1}·K^{-1}. Irradiated domain is highlighted by the black circle with radius r ($r = 1.5 \times r_{int} = 75$ µm). b) Stress distribution σ_{xx} (*building direction*) at $t = 8$ ms for tree heating strategy: single laser, coaxial laser and shifted laser. White lines correspond to iso-temperature lines in ceramic material.

In addition, comparisons of simulated melt pool shape with observations have been developed based on experimental results produced by LBM on Al_2O_3 ceramic [4,14]. Dedicated experiments were carried out under different laser powers and scanning speeds. Melt pool shape were measured on cross sections for a selected set of process parameters (Fig. 3). A careful calibration procedure has been afterward developed in order to estimate laser interaction radius and absorption coefficients from observations also including an analytical model. Comparisons show good agreements with melt pool depths, H_{ZR}, similar to observations (Fig. 3 a) despite the simplified approach proposed. Numerical shapes are also similar to experiments in particular for medium linear energy (Fig. 3 b).

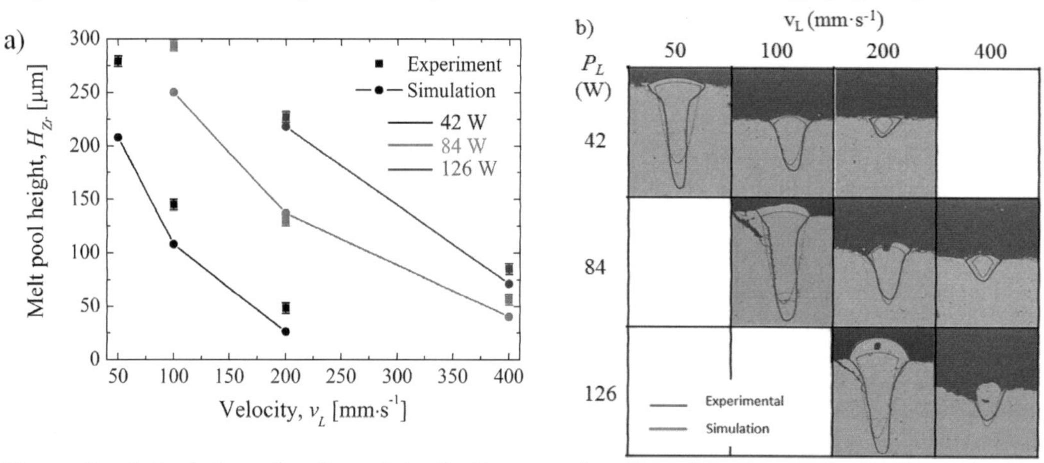

Figure 3 : a) Evolution of melt pool depth, H_{ZR}, as a function of the laser velocity for various power. Comparisons are developed between experiments and numerical simulations for calibrated values ($r_{int} = 64.6$ µm, $\alpha_s = 8.98\ mm^{-1}$, $\alpha_l = 3.88$ mm^{-1}) b) Transverse cross section of tracks showing melt pool shape (blue line) and process parameters influence [10] with simulation results (red line).

4. Conclusion

A levet-set modelling of LBM process has been developed to investigate track development. A volume heat source model is applied considering local absorption coefficient of ceramic powder and dense phases. The fluid flow computation considers the surface tension, Marangoni and gravity force effect to estimate velocity field in melt pool. Marangoni effect is highlighted on the flow evolution at the surface of the liquid bath. The mechanical resolution provides also valuable estimation of stress field in the track. The interest of an auxiliary laser to decrease stresses and limit cracks formation is demonstrated. This point may provide optimized scanning strategy for future applications. An experimental comparison is proposed for a single track development and a set of selected process parameters. A calibration procedure is developed in order to estimate unknown parameters associated to laser interaction radius and absorption coefficients. The thermo-mechanical model provides track morphology similar to experimental observations. Melt pool depth show also good agreement with simulations although some differences are visible for low velocity or high power. Future development should consider several superimposed tracks or complex scanning strategy, also including more physical phenomena. Coupling with current developments at macro- scale will also be considered.

5. Acknowledgements

This work has been conducted within the framework of the CEFALE project, part of the ACLAME program funded by the Institute CARNOT M.I.N.E.S (Paris, France).

References

[1] C. Y. Yap, C. K. Chua, Z. L. Dong, Z. H. Liu, D. Q. Zhang, L. E. Loh and S. L. Sing, Appl. Phys. Rev. 2, 041101 (2015).
[2] E. O. Olakanmi, R. F. Cochrane and K. W. Dalgarno, Prog Mater. Sci. 74 (2015), 401.
[3] Y.-C. Hagedorn, J. Wilkes, W. Meiners, K. Wissenbach and R. Poprawe, Physcs Proc. 5 (2010), 587.
[4] L. Moniz, C. Colin, J.-D. Bartout, K. Terki, M.-H. Berger, JOM. 70 (2018), 1.
[5] Q. Liu, Y. Danlos, B. Song, B. Zhang, S. Yin, H. Liao, J. Mater. Process. Tech. 222 (2015) 61.
[6] S.L. Sing, W.Y. Yeong, F.E. Wiria, B.Y. Tay, Z. Zhao *et al.*, Rapid Prototyping J., 23 (2017), 611.
[7] M.X. Gan, C.H. Wong, J. Eur. Ceram. Soc., 37 (2017), 4147.
[8] Q. Chen, G. Guillemot, C.A. Gandin, M. Bellet, Addit. Manuf. 16 (2017) 124.
[9] Q. Chen, G. Guillemot, C.A. Gandin, M. Bellet, Addit. Manuf. 21 (2018) 713.
[10] Q. Chen, Thermomechanical numerical modelling of additive manufacturing by selective laser melting of powder bed : Application to ceramic materials. PhD Thesis, PSL University, 2018.
[11] J. F. Li, L. Li and F. H. Stott, Int. J. Heat Mass Transf. 47 (2004) 1159
[12] P.-F. Paradis and T. Ishikawa, JSAP (2005), 5082.
[13] L. Battezzati and A. L. Greer, Acta Metall. 37(7) (1989), 1791.
[14] L. Moniz, Q. Chen, G. Guillemot, M. Bellet, Ch.-A. Gandin, C. Colin, J.-D. Bartout, M.-H. Berger, J. Mater. Process. Tech. 270 (2019) 106.

Phase field study of spacing evolution during wire and laser additive manufacturing under transient conditions

Zheng Wang,[1] Tao Jing,[1,*] Hongbiao Dong[2,*]

[1]Key Laboratory for Advanced Materials Processing Technology, School of Materials Science and Engineering, Tsinghua University, Beijing 100084, China
[2] Department of Engineering, University of Leicester, Leicester LE1 7RH, UK

E-mail: jingtao@mail.tsinghua.edu.cn, hd38@leicester.ac.uk

Abstract. Understanding the dynamic evolution of primary dendritic spacing in the laser melt pool is significant from a technological viewpoint because primary spacing is one of the foremost parameters that control the final mechanical properties of additive manufactured products. In this work, a multi-scale computational framework that couples FEM and a developed quantitative phase field method is employed to simulate the evolution of microstructure and primary spacing of a nickel-based superalloy during wire and laser additive manufacturing (WLAM) solidification. Transient conditions in the laser melt pool are considered in which both temperature gradient G and solidification speed V_P are made time-dependent. Through the use of this model, the dendritic morphology, tip velocity and spacing evolution during the solidification are investigated to provide the relationship between the laser processing parameters and the final spacing. Moreover, we attempted to clarify the intrinsic mechanism of spacing adjustment under different laser processing parameters from a novel perspective. This work provides meaningful understanding of spacing evolution in nickel-based superalloy and demonstrates the potential of controlling the complex microstructure morphologies and final primary spacing during wire and laser additive manufacturing process.

1. Introduction

Metal-based additive manufacturing (AM), also known as three-dimensional (3D) printing, is a novel manufacturing technique across multiple industry fields, including the automotive, biomedical and aerospace industries [1, 2]. Wire and laser additive manufacturing (WLAM) is a new type of AM process in which large metallic materials can be fabricated layer by layer [3-5]. In this AM process, wire is fed at a constant rate into a laser beam and is melted onto a substrate or the previously deposited layer. After completion of one layer, the wire feeding–scanning–melting–solidification process is repeated until the whole component is built. Solidification in the laser melt pool controls the dendrite sizes, dendritic orientation, dendritic morphology, segregation of secondary elements and ultimately the properties of the product. Therefore, it is critical to investigate the solidification behaviour in the laser melt pool.

In the solidification environment generated by the laser additive manufacturing process, the microsegregation patterns and primary arm spacing are two primary components of columnar dendrites growth. These features have critical effects on both the mechanical properties of the as-built products and the subsequent heat treatment processes [6]. However, the primary arm spacing across the dendrites during laser additive manufacturing have not received enough attention yet. To improve the properties of the final product, a fundamental knowledge of the growth of dendritic morphologies during laser melt pool solidification is required.

To date, few studies have been done to simulate the dendritic morphologies developed in laser additive manufacturing processes using phase-field methods. In 2008, A Farzadi et al. [7] investigated the solidification conditions and dendritic morphology during the directional solidification in gas tungsten arc welding of Al–3wt%Cu alloy using the phase-field model. However, the simulated primary dendrite spacing was found not to be in a reasonable agreement with the predictions of the theoretical model proposed by Hunt et al [15]. In the study conducted by V. Fallah et al. [8], a thin-interface phase-field model was coupled to a new heat transfer finite element model of the laser powder deposition process (during this process, a fiber laser was used to deposit the injected powder). The accuracy of the coupled model was validated by studying spacing evolution of Ti–Nb alloys. This work demonstrated the potential of simulating topologically complex microstructures present in laser additive manufacturing process. Using three modeling techniques - Finite element method, phase-field, and CALPHAD-based solidification models, Trevor Keller et al. [9] systematically investigated the microsegregation between dendrite arms, the primary arm spacings and the composition profiles during laser powder bed fusion (LPBF) process of IN625 alloys. However, the so-called steady-state solidification conditions throughout the entire process with constant temperature gradient and solidification velocity were used in all of the investigations above. This indicates that none of these studies has considered the variation in solidification conditions along the S/L interface within the solidifying melt. Although steady-state solidification is an important academic issue, it is far from representing the complex conditions really occur in the laser melt pool during WLAM process conducted in the current work. Therefore, understanding of the melt pool solidification behaviour under transient conditions in which both the pulling speed and thermal gradient are time-dependent variables is essential.

The objective of this paper is to perform numerical simulations, in two dimensions, of the microstructure evolution, and spacing evolution during wire and laser additive manufacturing of nickel-based superalloy GH3039 using the finite element method and phase-field model.

2. Model description

A double ellipsoidal volumetric heat source model is used to describe the laser melting process. To numerically solve the equations of three-dimensional heat transfer, a general FEM (finite element method) computer software, ANSYS 18.0, is utilized. In the present work, numerical results are presented for the case of a single-track laser deposition, however these can be translated to multilayer geometry by iterative solution over different layers. The thin interface 2D Karma phase-field model of alloy solidification [10] is used to investigate microstructure evolution in the laser melt pool. This model has been developed by Nikolas Provatas and co-workers [11]. Alloy solidification during WLAM process can be regarded as directional solidification which is assumed to be two-dimensional and with surface-tension anisotropy γ, moves in the z direction at a time-dependent solidification velocity V_P in an externally imposed time-dependent temperature gradient G. A phase-field parameter φ is employed here, which takes the value $\varphi = 1$ (-1) in the solid (liquid). The complete set of the phase-field equations including crystalline anisotropy can be found in our previous work [12]. For the microscopic phase-field simulations, the diffuse interface parameters are chosen as $W = 0.27$ μm and $dx = 0.8\ W$, yielding $\lambda = 20.913$and $W\ /\ d_0 = 23.66$, which have been tested for convergence study. The dimensionless discretization time $dt = 0.004$. The GH3039 alloy (Nickel-based superalloy) is employed in this paper. Chemical composition (wt.%) of the selected GH3039 alloy is shown in Table 1. It should be noted that we have approximated the alloy GH3039 to be a binary alloy Ni-20.835 wt.% Cr. The thermophysical parameters of Ni-Cr binary alloy are as follows [13, 14]: the melting point of pure Ni $T_m = 1455$ K and the liquidus temperature $T_L = 1693$ K, equilibrium partition coefficient $k = 0.714$, slope of the liquid line $m = -2.1$ K/wt.%, liquid diffusion coefficient, $D_L = 2.0 \times 10^{-9}$ m²/s and Solid diffusion coefficient, $D_S = 3.7 \times 10^{-13}$ m²/s, Gibbs-Thomson coefficient $\Gamma = 2.0 \times 10^{-7}$ K/m, anisotropy $\varepsilon = 0.03$.

In this work, the FEM heat transfer model is used to simulate temperature distribution of the laser molten pool. The time-dependent temperature gradient and pulling velocity can be obtained from the FEM simulation results. Microstructure evolution during transient solidification will be simulated using the developed quantitative phase-field model. Following this, the multi-scale computational framework will be used to predict the spacing evolution of GH3039 alloy during WLAM process.

Table 1. Nominal chemical composition of the GH3039 alloy (wt.%).

Cr	Mo	Ti	Nb	Al	Fe	C	Ni
20.835	2.14	0.62	1.10	0.61	1.09	0.0474	Bal.

Table 2. Conditions for the simulations of the laser melt pool.

Parameter	P (KW)	V_S (mm/s)	b (mm)	h (mm)	T_h–T_L (K)
Case 1	2	10	2.8	1.5	1010
Case 2	2	8	3.2	1.8	1100
Case 3	2	6	3.6	2.1	1250
Case 4	2	4	4.9	3.2	1360
Case 5	1.5	10	2.3	1.2	950
Case 6	2.5	10	3.9	2.1	1080
Case 7	3	10	5.2	2.8	1200

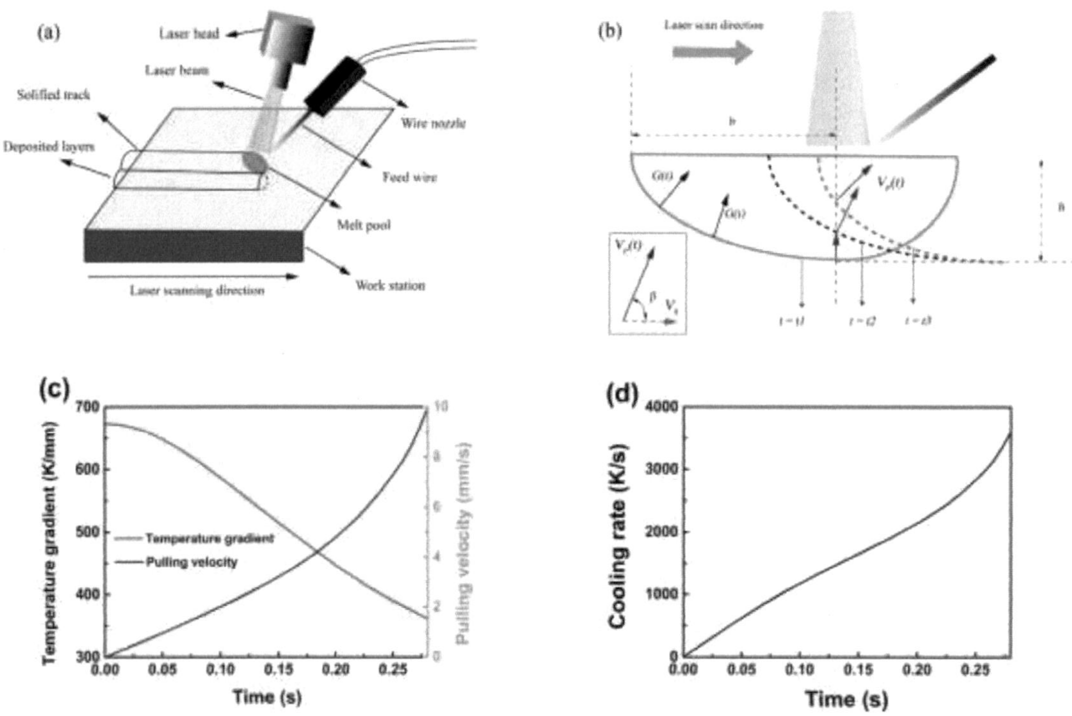

Figure 1. (a) Schematic illustration of the experimental setup for WLAM process. (b) Solidification in the laser melt pool. (c) FEM simulation results of $G(t)$ and $V_P(t)$. (d) Time evolution of the cooling rates. In (c) and (d), a typical case of P = 2 KW, V_S = 10 mm/s was used.

3. Results and discussion

3.1. FEM simulations of the laser melt pool

Figure 1(a) shows the schematic illustration of the experimental setup and processing procedure during WLAM process. In Figure 1(b), the illustration of the solidification in the laser melt pool is presented, macro-scale morphology of the laser molten pool can be regarded as a quarter of ellipse, V_S is the laser scan speed, h and b are depth and width of the laser molten pool respectively. The conditions for the FEM simulations of the laser melt pool are listed in Table 2, T_h is the highest temperature of the melt pool. As shown in Figure 1(b), $V_P(t)$ increases rapidly from zero at the bottom of the laser melt pool to a larger value at the rear of the melt pool interface. We can calculate the time-dependent $G(t)$ and $V_P(t)$

under transient conditions in the laser melt pool [12]: $V_P(t) = hV^2t/[V^2t^2(h^2 - b^2) + b^4]^{1/2}$ and $G(t) = (T_h - T_L)/[V^2t^2 + h^2(1 - (V^2t^2/b^2))]^{1/2}$. Figure 1(c) and (d) shows the time evolution of simulated $G(t)$, $V_P(t)$ and the cooling rate under transient conditions. The cooling rate R is a significant variable in determining the microstructures in the laser melt pool which can be defined as: $R = G(t) \times V_P(t)$. It is reasonable to conclude that compared to traditional steady-state conditions, this transient solidification conditions where the solidification speed and temperature gradient are both time-dependent values are closer to the real solidification condition in the laser melt pool.

3.2. Phase-field simulations of microstructure evolution

Understanding the evolution of microstructure morphology and interface velocity is critical for analysis of spacing evolution. Figure 2(a) shows simulated evolution of microstructure during the solidification of GH3039 alloy under Case 4 listed in Table 2. The interface morphologies transform into a relatively uniform cellular array (a3) through the stages of planar interface, rudiment of cells (a1) and dendrite submerging stage (a2). The interface dynamics and microstructure selection are determined by the instantaneous interface velocity. Figure 2(b) shows the dynamic evolution of interface velocity, where stage 1, 2, 3 and 4 are used to denote the planar growth stage, the initial competitive stage, submerging stage and short-term stable stage respectively. After the initial planar instability, the tip velocity increases sharply, and becomes even larger than the growth velocity. In stage 3, the tip velocity is still larger than the pulling speed though the tip velocity decreases with time. As shown in Figure 2(a2), if the lagged dendrite with small spacing cannot release its solute around the dendritic tip due to the interdendritic interactions, this lagged dendrite will be eliminated by its neighbours. This process is usually called cell/dendrite submerging [11]. In stage 4, the survived cellular tip velocity is almost equal to the pulling velocity, and so as its growth acceleration. The surviving cells keep the primary spacing uniform for certain time by lateral adjustment as shown in Figure 2(a3). Under transient growth conditions with time-dependent solidification velocity, dimensionless tip undercooling Δ varies with time even after the initial transient stage as shown in Figure 2(c).

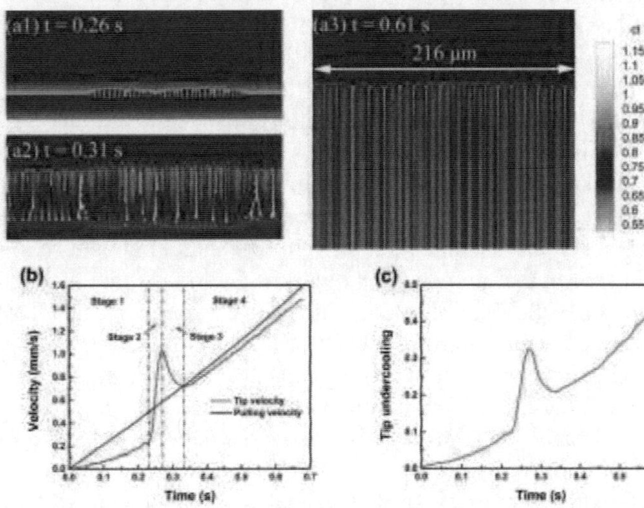

Figure 2. A typical simulation case of Case 4 in Table 2 (P = 2 KW, V_S = 4 mm/s). (a) Phase field simulation results of microstructure evolution. The color contrast shows the normalized concentration field $ct = ck/c_\infty$, where c is the actual concentration field and c_∞ is the initial alloy concentration. (b)Time evolution of the tip velocity. (c) Time evolution of dimensionless tip undercooling under transient conditions.

3.3. Primary spacing evolution of the cellular dendrites

Boundary conditions will strongly affect numerical results from phase-field simulations, especially the primary spacing of the columnar dendrites [11]. Figure 3(a) shows the dependence of the average primary spacing on the system width under Case 4 and 5. Phase-field simulations of the spacing evolution for Case 1-4 were carried out using the system width of 550 μm. For Case 5-7, the system width is chosen as 400 μm. In this section, effects of the laser processing parameters on the final primary spacing are investigated. As depicted in Figure 3(b), when the laser scan speed ranges from 4 mm/s to 10 mm/s with the same laser power given by 2 KW, the primary spacing decreases with increasing the laser scan speed. In addition, when the same laser scan speed is given by 10 mm/s and the laser power ranges from 1.5 KW to 3 KW, the primary spacing increases with increasing the laser power as shown in Figure 4(a). In the following, we will try to investigate the intrinsic reason of spacing adjustment under different laser processing conditions.

First, effect of the initial spacing should be considered. Figure 3(c1) and (c2) show the variations of the crossover time from the planar morphology to cellular patterns and the initial spacing at the crossover point with different laser scan speeds. When the laser scan speed is increased, the crossover time decreases, while the average initial spacing remains stable. This indicates that the increasing scan speed could promote the initial instability while having limited effect on the initial spacing. The same conclusions could be obtained in Figure 4(b1) and (b2) when the laser scan speed is fixed and the laser power varies. In both cases, it is difficult to study the selection mechanism of final spacing through the initial spacing, the evolution process after planar instability need to be considered.

In section 3.2, Figure 2(b) shows the dynamic evolution of interface velocity under Case 4. It is noteworthy that although all cases presented in Table 2 exhibit similar qualitative behaviour for the time evolution of the interface velocity, the duration of stage 1, 2, 3 and 4 is quite different. As shown in Figure 2(b), the duration of the stage 3 is defined as *t*3, the maximum value of the interface velocity at stage 2 and the minimum value of the interface velocity at stage 3 are defined as V_{imax} and V_{imin}, respectively. As seen from Figure 3(d1), the values of *t*3 and *t*3 / *t*2 increase when the laser scan speed is decreased from 10 mm/s to 4 mm/s. This indicates that the role of the submerging stage (stage 3) is becoming more essential in the whole evolution process. And in Figure 4(c1), both *t*3 and *t*3/*t*2 increase with the increase of the laser power, showing that the submerging stage occupies a more significant status. The microstructures evolve much faster in stage 2 than in other stages and the interface velocity reaches its maximum value at the end of stage 2. In the submerging stage, the interface velocity is decreased and a minimum value occurs at the end of this stage. Figure 3(d2) shows $(V_{imax} - V_{imin})/V_{imax}$ as a function of the laser scan speed. This value increases with decreasing the laser scan speed, which indicates that the gap between the maximum and minimum is getting larger. While in Figure 4(c2), the value of $(V_{imax} - V_{imin})/V_{imax}$ keeps nearly zero with increasing the laser power, indicating that the interface velocity is almost uniform in this stage. Let us turn to the primary spacing adjustment in stage 3. In the submerging stage, though the interface velocity decreased, the interface velocity is still larger than the solidification speed. The primary spacing is smaller than the lower limit of the allowable range in stage 4 (short-term stable stage), accordingly, the spacing will be enlarged by the submergence of the cells. In stage 4, the surviving cells keep the spacing uniform for certain time by lateral adjustment. Although the growth conditions vary with time, the average spacing keeps nearly constant in the late stage of stage 4. Therefore, it is reasonable to believe that the final primary spacing will get larger with increasing the duration of stage 3 which can be measured by *t*3 and *t*3 / *t*2. From the analysis above, when a fixed laser power is used, the smaller the laser scan speed (when the laser scan speed is fixed, the larger the laser power), the longer the duration of the submerging stage. Thus, resulting in larger final primary spacing within the scope of the present study. This conclusion is consistent with the results shown in Figure 3(b) and Figure 4(a), indicating that this criterion is effective in predicting the final primary spacing for different laser processing parameters.

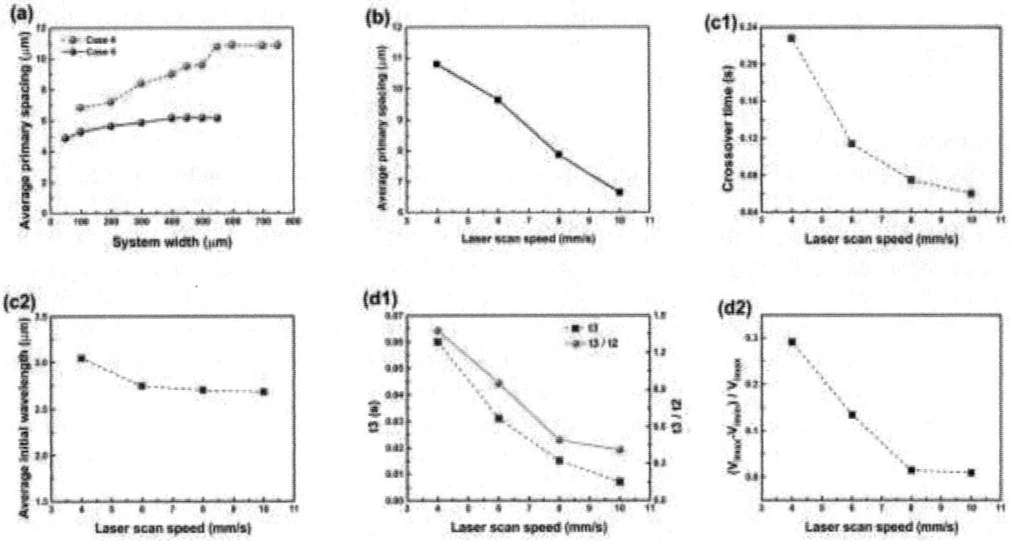

Figure 3. (a) Simulated primary spacing vs. the system width. (b) Average final primary spacing vs. laser scan speed. (c1) The crossover time and (c2) the initial spacing at the crossover point from four different laser scan speeds. (d1) Variations of $t3$ and $t3$ / $t2$ with the laser scan speed. (d2) $(V_{imax} - V_{imin})/V_{imax}$ as a function of the laser scan speed. It should be noted that in all cases the same laser power is given by 2 KW.

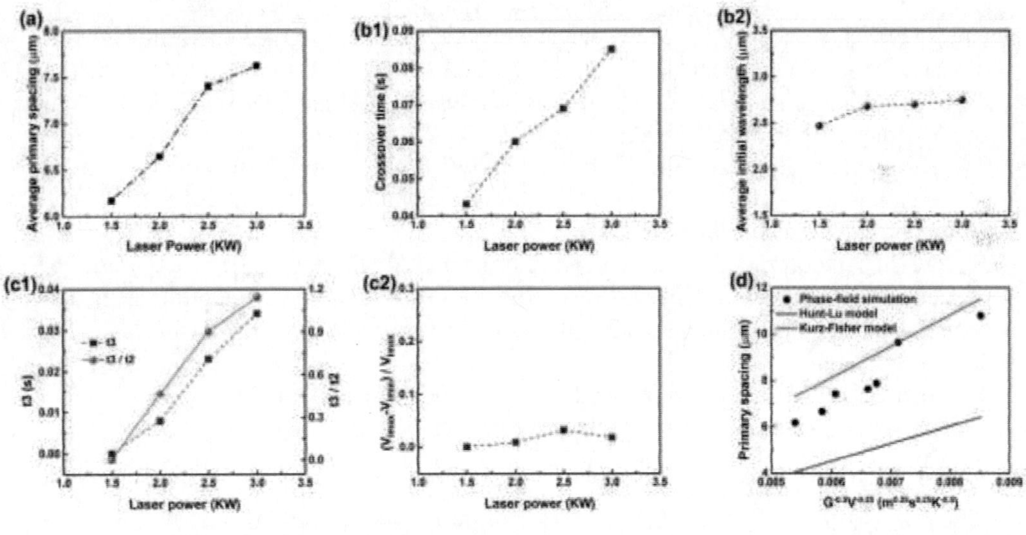

Figure 4. (a) Final average primary spacing vs. laser power. (b1) The crossover time and (b2) the initial average spacing at the crossover point for different laser power. (c1) Variations of $t3$ and $t3$ / $t2$ with the laser power. (c2) Variations of $(V_{imax} - V_{imin})/V_{imax}$ with the laser power. It should be noted that in (a) – (c) the same laser scan speed is given by 10 mm/s. (d) Comparation of simulated primary spacing with the analytical models of Hunt and Lu [12] and Kurz and Fisher [13].

The comparison between the simulated and analytical mean primary spacing is illustrated in Figure 4(d), the simulated values are larger than the Hunt-Lu model and are slightly smaller than the Kurz and Fisher model [15, 16]. Although a completed quantitative agreement between the simulated primary spacing and the two analytical models is difficult to obtain, the appropriate tendency and the similar values demonstrate that the phase-field model can be a powerful tool for quantitative simulation of spacing evolution under transient conditions.

4. Conclusion

In this work, primary spacing evolution during wire and laser additive manufacturing process of GH3039 alloy were studied using a developed multi-scale model. The transient condition which is distinctively different from the traditional steady-state condition was derived by making the solidification speed and temperature gradient both time-dependent variables. Effects of the laser processing parameters on the final primary spacing were investigated. Primary dendritic spacing predicted by this multi-scale model agrees well with the analytical model proposed by Hunt and Kurz. We attempted to clarify the intrinsic mechanism of spacing adjustment under different laser processing parameters from a novel perspective. Results revealed that the final primary spacing will get larger with increasing the duration of the submerging stage which can be measured by $t3$ and $t3$ / $t2$. This criterion demonstrates the potential of predicting the final primary spacing under different solidification conditions present in wire and laser additive manufacturing and other relevant casting conditions.

Acknowledgement

This work is financially supported by The National Key Research and Development Program of China No. 2017YFB1103700.

References

[1] J.H. Martin, B.D. Yahata, J.M. Hundley, J.A. Mayer, T.A. Schaedler and T.M. Pollock 2017 *Nature* **549** 365.
[2] D. Ma, A.D. Stoica, Z. Wang and A.M. Beese 2017 *Mater. Sci. Eng. A.* **684** 47.
[3] T.E. Abioye, J. Folkes and A.T. Clare 2013 *J. Mater. Process. Technol.* **213** 2145.
[4] B. Wu, Z. Pan, D. Ding, D. Cuiuri, H. Li, J. Xu and J. Norrish 2018 *J. Manuf. Process.* **35** 127.
[5] Z. Nie, G. Wang, J.D. McGuffin-Cawley, B. Narayanan, S. Zhang, D. Schwam, M. Kottman and Y. Rong 2016 *J. Mater. Process. Technol.* **235** 171.
[6] E.C. Kurum, H.B. Dong and J.D. Hunt 2005 *Metall. Mater. Trans. A* **36** 3103.
[7] A. Farzadi, M. Do-Quang, S. Serajzadeh, A.H. Kokabi and G. Amberg 2008 *Modelling Simul. Mater. Sci. Eng.* **16** 065005.
[8] V. Fallah, M. Amoorezaei, N. Provatas, S.F. Corbin and A. Khajepour 2012 *Acta Mater.* **60** 1633.
[9] T. Keller, G. Lindwall, S. Ghosh, L. Ma, B.M. Lane, F. Zhang, U.R. Kattner, E.A. Lass, J.C. Heigel, Y. Idell, M.E. Williams, A.J. Allen, J.E. Guyer and L.E. Levine 2017 *Acta Mater.* **139** 244.
[10] B. Echebarria, R. Folch, A. Karma and M. Plapp 2004 *Phys. Rev.* E **70** 061604.
[11] M. Amoorezaei, S. Gurevich and N. Provata 2010 *Acta Mater.* **58** 6115.
[12] Z. Wang, S. Ma, W. Sun, M. Zhang, T. Jing and H. Dong 2018 *Comput. Mater. Sci.* **155** 364.
[13] P. Nie, O.A. Ojo and Z. Li 2014 *Acta Mater.* **77** 85.
[14] X.H. Zhan, Z.B. Dong, Y.H. Wei and R. Ma 2009 *J. Cryst. Growth* **311** 4778.
[15] J.D. Hunt and S.Z. Lu 1996 *Metall. Mater. Trans. A* **27** 611.
[16] W. Kurz and D.J. Fisher 1981 *Acta Metall.* **29** 11.

IOP Conf. Series: Materials Science and Engineering **529** (2019) 012004 doi:10.1088/1757-899X/529/1/012004

Solidification microstructure during selective laser melting of Ni based superalloy: experiment and mesoscopic modelling

Yuze Li[1,2], Antonio Olmedilla[1], Miha Založnik[1], Julien Zollinger[1], Lucas Dembinski[3], Alexandre Mathieu[4]

[1] Institut Jean Lamour, CNRS - Université de Lorraine, F-54011 Nancy, France
[2] IRT M2P, F-57078 Metz, France
[3] LERMPS, ICB UMR 6303, CNRS, Univ. Bourgogne Franche-Comté, UTBM, F-90010 Belfort, France
[4] Laboratoire Interdisciplinaire Carnot de Bourgogne (ICB) UMR 6303 CNRS-Université Bourgogne-Franche Comté, 12 rue de la Fonderie, F-71200 Le Creusot, France

E-mail: yuze.li@univ-lorraine.fr

Abstract. A set of single track laser melting experiments was performed in a selective laser melting (SLM). The tracks were done on an Inconel 718 plate with various laser scan velocities at a constant laser power of 150 W. The geometries of the molten pool (MP), as well as the solidified dendrite structures, i.e., primary and secondary dendrite arm spacing (PDAS and SDAS), in the cross sections of the molten path were characterized to evaluate the effect of the laser scan velocity during SLM. Moreover, the local solidification thermal conditions (cooling rate R*, tip growth velocity V* and temperature gradient G*) at the MP bottom were deduced from the SDAS and the geometries of the molten pool. Finally, the mesoscopic envelope model was used to simulate the PDAS selection of the columnar dendrite growth in the molten pool. The simulated results were compared with the experimental data, and a good agreement was achieved under different laser scan velocities.

1. Introduction

Inconel 718, a nickel based superalloy, is widely used in many industrial applications, such as turbine blades in aerospace and automobile industries, combustion chambers and nuclear reactors, due to its excellent mechanical properties, oxidation resistance and great corrosion resistance at high temperatures [1–3]. However, it is very difficult to manufacture products of Inconel 718 by conventional machining techniques due to its high hardness and low thermal conductivity [4,5]. Thus, the novel additive manufacturing (AM) technologies will have a significant potential in the applications of Inconel 718, especially for the components with complex structures and high dimensional specifications.

Selective laser melting (SLM) is a commonly used additive manufacturing technique, which generally consists of the melting of small volumes of powders in a localized region and the subsequent rapid solidification of the molten pool (MP) [6,7]. Many aspects of SLM and other AM processes are being widely investigated using both experimental and modelling methods. Until now, most of these research focused on the heat transfer and liquid flow in the MP [8], on defects [9], on the properties of the final product [1–3] and also on the microstructure evolution during AM process [6], which are on a scale the grains. However, there is a lack of studies on the formation of the solidified dendrite structure (substructure of the grain) during SLM, which is crucial for controlling the grain texture, the segregation, and ultimately the material properties [10]. The solidification during SLM is a highly complex process and is significantly affected by various conditions, such as the spot size, the power and the velocity of the laser [11]. Therefore, it is necessary to elucidate the effects of these conditions on the solidification microstructure.

The traditional approach for researching the SLM process is using the final layer-by-layer buildup, which is highly time consuming and costly. As a substitute, the single track laser melting experiment was used in some investigations such as the correlation of the MP geometries and the SLM parameters, which greatly reduced the time and cost [12]. In the aspect of solidification characterization, due to the epitaxial dendrite growth between the layers [7], the dendrite structure of the final buildup is mostly controlled by the first layer. Thus, it is reasonable to utilize the single track experiment to investigate the solidification behavior in AM process, and also to estimate the solidification conditions which could be very difficult to be obtained from experiment but would be necessary to be used in the simulation.

In this study, a set of single track laser melting experiments was performed on Inconel 718 plate under various laser scan velocity at a constant laser power of 150 W to evaluate the effect of the laser scan velocity on the MP geometries and on the characterization of the solidified dendritic structures. Moreover, the mesoscopic envelope model was used to simulate the primary arm spacing (PDAS) selection of the columnar dendrite growth in the MP, using the solidification thermal conditions deduced from the MP geometries and the solidified structures. The simulation results were compared with the experimental data.

2. Single track laser melting experiments

A set of single track laser melting experiments was performed on the same Inconel 718 plate under six various laser scan velocities from 50 to 300 mm/s using a commercial laser powder bed fusion machine ReaLizer SLM 250. A continuous laser with the wave length of 1060-1100 nm was used. The spot size of the laser was around 35 μm and the laser power was constant at 150 W. The dimensions of the plate were approximately 150 × 150 × 5 mm. The original microstructure of this plate was characterized as equiaxed grains structure without specific texture. The length for each laser scan velocity is around 140 mm and the distance between the neighbor single tracks is 2 mm to avoid the interference from each other. The samples were named by the scan velocity, i.e., V50 means the sample with the scan velocity of 50 mm/s. After the SLM process, the samples for metallographic analysis were cross sectioned perpendicular to the laser scan direction, at the position of more than 15 mm from the start of the laser, at which the MP reached a steady regime. The samples were mechanically polished with diamond suspension as the final step to identify the dendritic microstructure and then examined by scanning electron microscopy (SEM).

3. Mesoscopic envelope model

The mesoscopic envelope model, developed by Steinbach et al. [13], is a powerful simulation tool for dendritic solidification which bridges the gap between microscopic and macroscopic approaches. The core idea of this model is the description of a dendritic grain by its envelope - a virtual smooth surface that links the tips of the actively growing dendrite branches. The growth velocity of the envelope growth can thus be calculated from the velocities of the dendrite tips. The growth of the dendrite tips is determined from the local supersaturation of the liquid in the proximity of the envelope by a stagnant-film formulation of a tip model [14]. The branched dendritic structure inside the envelope is not resolved; the interior of the envelope is rather described in a volume-averaged sense by a phase-fraction field and other volume-averaged quantities. The phase change that determines the evolution of the structure inside the growing envelope is controlled by the solute exchange with the surroundings of the grain. We used the in-house mesoscopic model code CrystalFOAM, developed on the OpenFOAM finite-volume platform. More detailed description about this model could be found in our previous studies [14].

4. Results and discussion

4.1. Effect of the laser scan velocity on the MP morphologies

Figure 1(a) shows the SEM image of the laser track cross section of V50. Only V50 was chosen as a representative one because the MP shape is almost the same for all the parameters used in this study. A typical semi-ellipse-shaped MP could be observed as shown by the red dash line in figure 1(a). Thus, the width (W) and depth (D) of the semi-ellipse were used to describe the MP geometry. Figure 1(b) shows the profiles of the MP geometry as a function of the laser scan velocity (V_L). It's clear that the MP size is highly sensitive to V_L. With the increase of V_L, both W and D gradually decrease: from W=320 μm, D=94 μm at V_L=50 mm/s, to W=180 μm, D=60 μm at V_L=300 mm/s. This can be easily explained by the decrease of heat input with the increase of V_L.

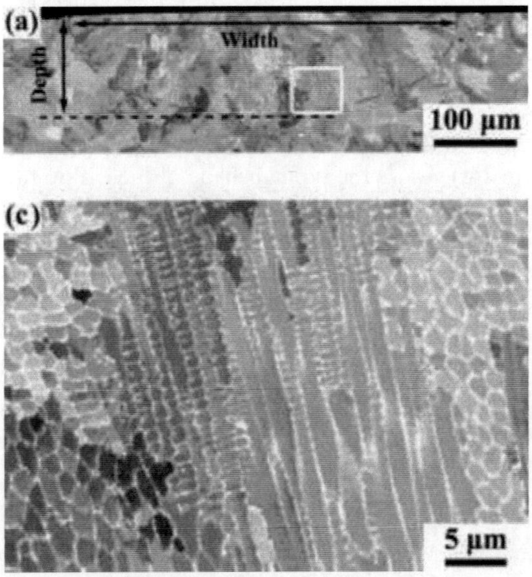

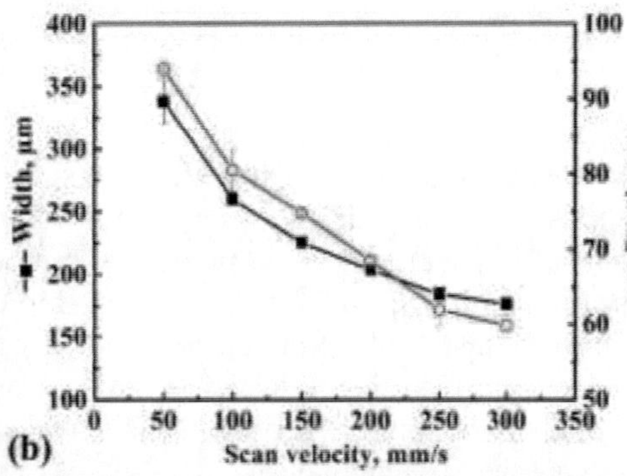

Figure 1. (a) SEM image of the laser track cross section of V50. The boundary of the MP was depicted by the red dash line. (b) The profiles of the width and depth of the MP as a function of the laser scan velocity (V_L). (c) Solidified dendrite structure zoomed in from the yellow box in (a).

4.2. Dendrite structure in the MP

The solidified grain structure in the MP of V50 can also be observed in figure 1(a). Figure 1(c) shows the corresponding microstructure zoomed in from the bottom of the MP as indicated by the yellow box in figure 1(a). It clearly shows a dendritic solidification structure under the present laser remelting conditions. Accordingly, the mean primary and secondary dendrite arm spacing (PDAS and SDAS) could be measured from the images. The measured results of the PDAS and SDAS are shown in table 1. Only V300 and V50 are included because only the samples with the highest and lowest V_L were representatively chosen for the characterization and the simulation in the following sections.

Table 1. The measured PDAS and SDAS and the calculated solidification conditions.

Sample	V_L (m/s)	PDAS (μm)	SDAS (μm)	R* (K/s)	D/L	V* (m/s)	G* (K/m)
V50	0.05	1.20	0.78	5.0e5	0.40	0.02	2.5e7
V300	0.30	0.56	0.32	6.7e6	0.30	0.09	7.5e7

From the measured SDAS, the local cooling rate R* could be estimated by [10]:

$$\mathrm{SDAS} = 5.5\left(-\frac{\Gamma_{sl} D_l \ln\left(\frac{C_{eut}}{C_0}\right)}{m_l(1-k_0)(C_{eut}-C_0)}\right)^{\frac{1}{3}}\left(\frac{\Delta T_0'}{R^*}\right)^{\frac{1}{3}} \quad (1)$$

where Γ_{sl} is the Gibbs-Thomson coefficient, D_l the diffusion coefficient in liquid, C_0 the concentration in the liquid, C_{eut} the concentration in the liquid at the eutectic point, m_l the liquidus slope, k_0 the binary equilibrium partition coefficient and $\Delta T_0'$ the non-equilibrium solidification range. By simplifying the Inconel 718 as a Ni-5.2Nb binary system, these parameters are shown in table 2, which were calculated by fitting the solidification path of Inconel 718 using Thermocalc and from reference [15]. The estimated R* by equations (1) are shown in table 1. The results indicate that both the PDAS and the SDAS decrease, while R* increases, with increasing V_L, which could be expected.

Table 2. The thermophysical properties of Inconel 718.

Γ_{sl} [15] (Km)	D_l [15] (m^2/s)	C_0 (at.%)	C_{eut} (at.%)	m_l (K/at%)	k_0	$\Delta T_0'$ (K)	K [15] (W/m·K)	α [15] (m^2/s)
3.65e-7	3e-9	5.2	18.56	-16.00	0.33	145	30.1	5.5e-6

*4.3. Estimation of the local tip growth velocity V**

In order to simulate the columnar solidification structure with the mesoscopic model, it is important to get one more solidification conditions except R*. Figure 2(a) shows the typical shape of the MP in the longitudinal section during the SLM process. It can be clarified that time the solidification front needed from the bottom of the MP to the surface (from B to C with the distance of the MP depth, D) should be the same with the time needed from the left boundary to the same position (from A to C with the distance of the MP length, L), as shown in figure 2(a). By assuming a constant tip growth velocity, the local velocity at the MP bottom, V*, could be estimated from V_L, L and D by $V^*=DV_L/L$. Thus, it will be necessary to get the MP length during SLM process.

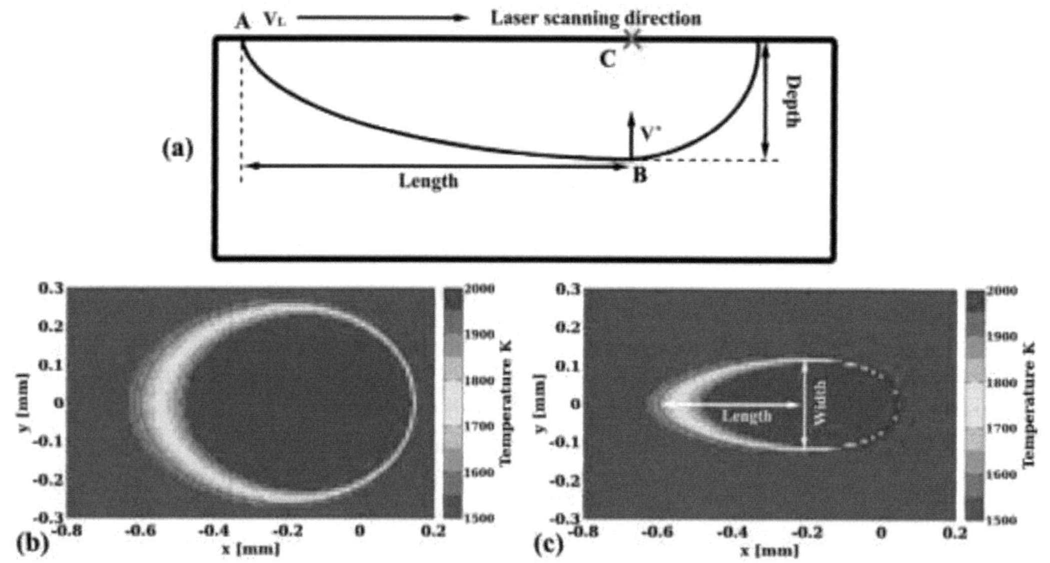

Figure 2. (a) Schematic diagram of the MP in the longitudinal section during SLM. (b, c) The Rosenthal's solution of the MP on the plate surface for V50 and V300 during SLM, respectively. The red lines in (b) and (c) are the liquidus temperature of Inconel 718.

Rosenthal proposed an analytic solution for the temperature field around a moving point heat source, which travels over a semi-infinite medium along the positive x direction at velocity v [16]:

$$T = T_0 + \frac{P}{2\pi Kr}\exp\left(-\frac{v(x+r)}{2\alpha}\right) \tag{2}$$

where T_0 is the room temperature 298 K, P the laser power, K the thermal conductivity of the medium, r the radial distance from the moving point, α the thermal diffusivity. Accordingly, the width,

W, and length, L, of the MP can be approximated. Figure 2(b) and (c) show Rosenthal's solutions for V50 and V300 using the thermophysical properties of Inconel 718 in table 2, from which the ratio L/W of the MP was roughly estimated. With the experimentally measured W in section 4.1, the MP length L could be deduced. Correspondingly, the V* could be estimated by $V^*=DV_L/L$, followed by the local temperature gradient, G*, by the equation $G^*=R^*/V^*$, as shown in table 1. As expected, for both samples, V* is lower than V_L, and decreases with decreasing V_L.

4.4. Simulation of the dendrite structure by mesoscopic envelope (MS) model

Two simulations were performed on the simplified Ni-Nb binary system, according to the corresponding local solidification thermal conditions at the MP bottom of V50 and V300 in table 1 and the thermophysical properties of Inconel 718 in table 2. The widths of the domain (Ws) were chosen as 5 μm and 2.4 μm, with the grid sizes of 0.025 μm and 0.006 μm, for V*=0.02 m/s and V*=0.09 m/s, respectively. The moving frame computational domain is used, in which minimum distance between the most advanced tip and the top boundary is maintained. A symmetry boundary condition was used at the vertical sides and a fixed concentration C_0 was imposed at the top boundary.

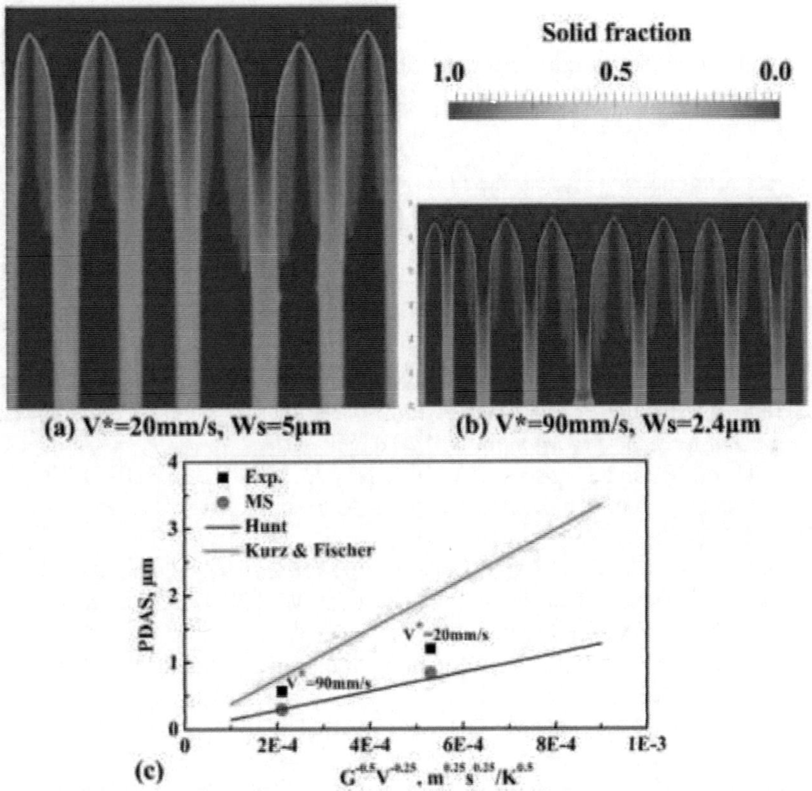

Figure 3. (a) and (b)The simulated dendritic structures by MS model at the steady state for V*=0.02 m/s and V*= 0.09 m/s, respectively. (c) The comparison of the predicted PDASs by MS model with the two geometrical models and the experiments (Exp.).

The dendrites growth initiated from a planar envelope placed at the bottom of the domain at an initial supersaturation corresponding to the required V*. During an initial transient, the initial planar envelope destabilizes, protuberances with a small spacing appear and then the competition of the branches takes place until achieving a stable PDAS. The growth lengths to reach the steady state in the simulations are 20.5 μm and 4.3 μm for the cases V*=0.02 m/s and V*=0.09 m/s, respectively. The simulated PDASs in the steady state are shown in figures 3(a) and (b). The predicted PDASs are 0.84 μm and 0.28 μm for the cases V*=0.02 m/s and V*=0.09 m/s, respectively. The results by MS model can be compared to classic geometrical models, such as two models developed by Hunt [17] and Kurz and Fisher [18], which give an average PDAS for steady growth as a function of G* and V*, i.e., proportional to $G^{*-0.5}V^{*-.025}$. The comparison of the predicted PDASs by MS model with the two geometrical models is shown in figure 3(c), as well as the experimental ones. The results indicate that the MS model can predict the PDAS values and the PDAS variation with the V* in the selected regions and parameters.

5. Conclusion and perspectives

The effect of the laser scan velocity on the MP geometries and the solidified dendrite structure in the MP during the SLM of Inconel 718 was investigated by the single track laser melting experiments. Moreover, a simplified estimation of the local solidification conditions, such as cooling rate R*, tip growth velocity V* and temperature gradient G*, was proposed. In this estimation, the R*, V* and G* were deduced from the SDAS and the MP geometries. Finally, using the estimated R*, V* and G* as the input, the mesoscopic envelope model was used to simulate the PDAS selection of the columnar dendrite growth in the MP under two laser scan velocities. A good agreement between the experiment and the mesoscopic envelope model was achieved for both cases.

Acknowledgments
We gratefully acknowledge the support to this research by the Institut de Recherche Technologique - Matériaux Métallurgie Procédés (IRT M2P) and by the French National research Agency (ANR) via the project RESEM 2018.

References

[1] Popovich A A, Sufiiarov V S, Polozov I A and Borisov E V 2015 *Key Engineering Materials* **651–653** 665–70
[2] Jia Q and Gu D 2014 *Journal of Alloys and Compounds* **585** 713–21
[3] Amato K N, Gaytan S M, Murr L E, Martinez E, Shindo P W, Hernandez J, Collins S and Medina F 2012 *Acta Materialia* **60** 2229–39
[4] Sharman A R C, Hughes J I and Ridgway K 2004 *Machining Science and Technology* **8** 399-414
[5] Wang X, Gong X and Chou K 2017 *Proceedings of the Institution of Mechanical Engineers, Part B: Journal of Engineering Manufacture* **231** 1890–903
[6] Acharya R, Sharon J A and Staroselsky A 2017 *Acta Materialia* **124** 360–71
[7] Saedi S, Shayesteh Moghaddam N, Amerinatanzi A, Elahinia M and Karaca H E 2018 *Acta Materialia* **144** 552–60
[8] Chen Q, Guillemot G, Gandin C-A and Bellet M 2018 *Additive Manufacturing* **21** 713–23
[9] Khairallah S A, Anderson A T, Rubenchik A and King W E 2016 *Acta Materialia* **108** 36–45
[10] Dantzig J A and Rappaz M 2009 (EPFL Press)
[11] Perevoshchikova N, Rigaud J, Sha Y, Heilmaier M, Finnin B, Labelle E and Wu X 2017 *Rapid Prototyping Journal* **23** 881–92
[12] Ghosh S, Ma L, Levine L E, Ricker R E, Stoudt M R, Heigel J C and Guyer J E 2018 *JOM* **70** 1011–6
[13] Steinbach I, Beckermann C, Kauerauf B, Li Q and Guo J 1999 *Acta Materialia* **47** 971–82
[14] Viardin A, Založnik M, Souhar Y, Apel M and Combeau H 2017 *Acta Materialia* **122** 386–99
[15] Nastac L and Stefanescu D M 1996 *MMTA* **27** 4061–74
[16] Darmadi D, Norrish J and Tieu A 2011 *Faculty of Engineering - Papers (Archive)* 154–62
[17] Hunt J 1979 Proc. Conf., Sheffield, England, July 1977, pp 3–9
[18] Kurz W and Fisher D J 1981 *Acta Metallurgica* **29** 11–20

IOP Conf. Series: Materials Science and Engineering **529** (2019) 012005 doi:10.1088/1757-899X/529/1/012005

Thermo-mechanical simulation of track development in the Laser Beam Melting process - Effect of laser-metal interaction

Alexis Queva[1,3], Yaasin Mayi[2,3], Michel Bellet[1], Gildas Guillemot[1], Patrice Peyre[2], Morgan Dal[2], Clara Moriconi[3], Charlotte Metton[3]

[1] *MINES ParisTech, PSL Research University, CEMEF – Centre de mise en forme des matériaux, CNRS UMR 7635, CS 10207 rue Clause Daunesse, 06904 Sophia Antipolis Cedex, France*
[2] *PIMM Laboratory, UMR 8006 CNRS – Arts et Métiers ParisTech, 151 Bd de l'Hôpital, 75013 Paris, France*
[3] *Safran, site de Paris-Saclay, Rue des Jeunes Bois - Chateaufort, CS 80112 78772 Magny-les Hameaux*

Production Editor, *Journal of Physics: Conference Series*, IOP Publishing, Dirac House, Temple Back, Bristol BS1 6BE, UK

E-mail: `alexis.queva@mines-paristech.fr`

Résumé. Interest has recently emerged for the manufacture of aeronautical parts by Laser Beam Melting (LBM) additive process. This energy efficient process can for instance be used to build complex geometries, which cannot be made with traditional processes. However, complex phenomena occur during powder melting and track development : vaporisation phenomena influence laser-matter interaction by creating metal vapours that are responsible for the reduction of absorbed energy. The recoil pressure generated by the vaporisation counteracts the surface tension between the melt pool and the inert gas, also inducing liquid instabilities. The study of laser-matter interaction and induced phenomena can help understand the origin of defects such as porosities or cracks. In this approach, a level-set modelling of the LBM process at a mesoscopic scale is proposed to follow melt pool evolution and track development during build. A volume heat source model is used for laser/powder interaction considering the material absorption coefficient. A surface heat source is used to take into account the high laser energy absorption by dense metal alloys. An energy solver is coupled with thermodynamic database and pre-determined solidification path. Shrinkage during consolidation from powder to liquid and compact medium is modelled by a compressible Newtonian constitutive law. An automatic remeshing adaptation is also used to save time and avoid high computational cost. In the future, the computation of multiple beads or the build of a wall in a context of lattice structures will have to be considered.

IOP Conf. Series: Materials Science and Engineering **529** (2019) 012005 doi:10.1088/1757-899X/529/1/012005

1. Introduction

The LBM process is attractive for aeronautical applications. More specifically, manufacturing of complex shapes or one-pieces with significant added values in metallic alloys is the main interest in this field. The LBM process involves a focused laser beam which selectively melts a thin bed of spread powder material. This step is repeated track-by-track and layer-by-layer until the final product is obtained. During the laser scan, fast melting and solidification phenomena occur. The cooling rate has a major impact on the microstructure and mechanical properties of the final part [1, 2]. As a consequence, it is required to have an accurate estimation of these terms in order to determine the final state of the pieces. Numerical simulation represents a useful tool to predict the occurence of defects, such as porosities or cracks, regarding the cost to determine optimal operating parameters.

A numerical Finite Element (FE) 3D model for LBM process adapted for metals is presented [3]. This model is focused on the thermomechanical evolution of material at the scale of the bead. The laser-matter interaction is modeled by two different terms. A volume Beer-Lambert source for the laser-powder interaction and a surface source for the laser-dense material interaction. During the development of the melt pool, capillary and boiling effects occur, having influence on the melt pool shape [4]. In addition, the model considers an automatic remeshing procedure which avoids huge CPU cost due to mesh constraints [3].

2. Modeling

This FE model is based on the work of Chen at CEMEF [3]. The system is made of a material and a gas domain. The main assumptions of the model are as follows :

- The powder bed is considered as a continuum domain with a given porosity ;
- The densification of the powder bed is a one-way transformation.

During the heating stage, the powder bed melts continually to form a compact deposition. This transformation occurs by the evolution of the apparent density of the material powder, depending of the temperature, also considering a transition domain.

2.1. Level-set Method

The total system is made of a material D_1 and a gas domain D_2 (Figure 1). A signed distance function ψ is used respecting the eikonal condition $||\nabla\psi|| = 1$. The evolution of the position of the interface is tracked by the resolution of the transport equation of this distance function [5] :

$$\frac{\partial\psi}{\partial t} + \mathbf{u}\cdot\nabla\psi = 0 \tag{1}$$

where $\mathbf{u}$ is the velocity field induced by the powder shrinkage and the melt pool evolution defined on the whole domain. A geometric reinitialisation method is used in order to respect the eikonal condition after the transport step [6]. The local average properties are computed with the Heaviside function corresponding to the transition domain [5].

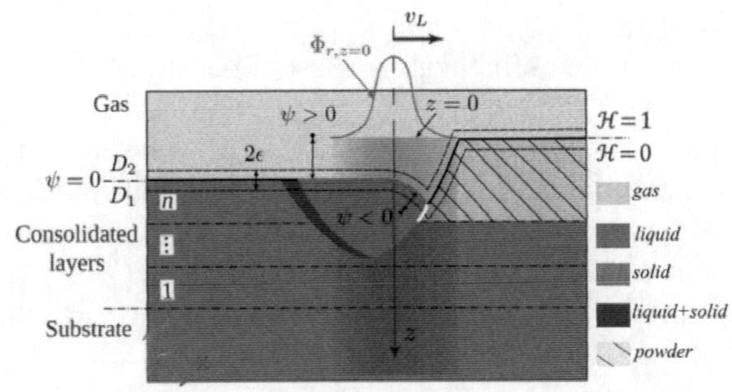

Figure 1. Schematic of the LBM process with the level-set (LS) method to track the interface associated to the isovalue $\psi = 0$ of the signed distance function [3].

2.2. Governing equations

2.2.1. Energy conservation and heat source modeling

The computation of the temperature field is required to propose accurate prediction of the different phenomena that occur during the LBM process. We may consider the influence of this field on phase transformations, the development of a recoil pressure or Marangoni forces. These different phenomena affect the shape of the bead. Considering the fast thermal evolutions observed in AM processes, it is necessary to take into account the temperature variation of physical properties. The modeling of the heat transfer in LBM is obtained as the solution of the non-steady equation of energy conservation :

$$\frac{\partial \{\rho h\}}{\partial t} + \nabla \cdot (\{\rho h\}\, \mathbf{u}) - \nabla \cdot (\{\lambda\}\, \nabla T) = \dot{q}_L \tag{2}$$

$$\{\rho h\} = f(T) \tag{3}$$

where ρ is the density, h the specific enthalpy, λ the thermal conductivity and T the temperature. $\dot{q}_L$ represents the heat source associated to the laser.

The laser interacts mainly with the powder bed and the melt pool. The laser follows a radial Gaussian distribution also considering a reflection coefficient R. Consequently, the heat source can be shared as two terms depending on the state of the surface :

$$\dot{q}_L = \begin{cases} (1-R)\frac{2P_L}{\pi r_{int}^2} e^{-\frac{2r^2}{r_{int}^2}} \left(\alpha e^{-\int_0^z \alpha dl}\right) & \text{[Laser - Powder bed interaction] [3]} \\ (1-R)\frac{2P_L}{\pi r_L^2} e^{\frac{-2r^2}{r_L^2}} \cdot \delta_2(\psi) & \text{[Laser - Dense material interaction] [7]} \end{cases} \tag{4}$$

Where P_L is the laser power, r_{int} the interaction radius, r_L the laser radius and α the local value of absorption coefficient. A volumetric Beer-Lambert law has been chosen to take into account the multiple reflections taking place in the powder bed and the progressive absorption of laser radiation in the material. It is assumed that the laser energy is totally absorbed after four reflections ($R^{n_r} \cdot (1-R) < 0.07$, $\forall n_r > 4$ where n_r is the number of reflections) [8]. For the laser - dense material interaction (liquid and solid phase), the surface heat source model is more representative regarding the high absorption of laser energy by dense metallic material.

2.2.2. Momentum conservation

The convective flow has a major influence on the temperature field and the melt pool shape. Consequently, the fluid flow evolutions in the liquid domain are limited to the dimension of the melt pool. Fluid dynamics equations are established by the mass and momentum conservation equations (Navier-Stokes). To take into account the change of apparent density in the powder bed during the melt stage, compressible flow has to be considered in the moment conservation equation :

$$\nabla \cdot \mathbf{u} = \dot{\theta} = -\frac{1}{\{\rho\}} \cdot \frac{\partial \{\rho\}}{\partial t} \tag{5}$$

$$\{\rho\} \left(\frac{\partial \mathbf{u}}{\partial t} + (\mathbf{u} \cdot \nabla)\mathbf{u} \right) - \nabla \cdot \bar{\bar{\boldsymbol{\sigma}}} = \mathbf{f_v} \text{ where } \bar{\bar{\boldsymbol{\sigma}}} = 2\mu \left(\bar{\bar{\dot{\boldsymbol{\epsilon}}}} - \frac{1}{3} tr(\bar{\bar{\dot{\boldsymbol{\epsilon}}}}) \bar{\bar{\mathbf{I}}} \right) - p\bar{\bar{\mathbf{I}}} \tag{6}$$

where $\dot{\theta}$ is the variation rate due to the shrinkage of the powder to dense matter, $\bar{\bar{\boldsymbol{\sigma}}}$ is the stress tensor, μ is the dynamic viscosity and $\mathbf{f_v}$ are the volumetric forces. These forces contribute to the dynamic of the melt pool and have several origins. Firstly, capillary effects represented by the surface tension and the Marangoni forces have to be considered. The recoil pressure during the vaporisation phase is also considered and induces the deflection of the interface by the high velocity vapour metal that is ejected. In addition, buyoancy forces are introduced. The whole forces are then written as :

$$\mathbf{f_v} = \mathbf{f_v^{STn}} + \mathbf{f_v^{STt}} + \mathbf{f_v^{recoil}} + \mathbf{f_v^{g}} \tag{7}$$

where :

- $\mathbf{f_v^{STn}} = \gamma\kappa\delta\mathbf{n}$ represents the surface tension force [3].
- $\mathbf{f_v^{STt}} = \delta\frac{\partial\gamma}{\partial T}\nabla_s T$ is the Marangoni force induces by the temperature dependence of the surface tension force. This dependency induces a shear stress at the liquid/gas interface in an parallel direction to the temperature gradient [3].
- $\mathbf{f_v^{recoil}}$ is the force generated by the intense ejection of metallic vapour [9].
- $\mathbf{f_v^{g}} = \{\rho\}\,\mathbf{g}$ are the buoyancy forces due to local variations of density with temperature.

3. Numerical tests

Different configurations are presented using material data for Inconel 718 alloy. The first cases correspond to static laser shooting and the second case to track development.

3.1. Static laser shooting on solid substrate

Static shooting on a substrate and secondly on a powder bed (LBM) are considered. The different parameters (machine and properties of Inconel 718) are detailed in Figure A1. The expression of the recoil pressure has been also simplified for this study with a surface force independent of temperature and following a radial Gaussian distribution :

$$\mathbf{f_v^{recoil}} = -\left(\frac{1+\beta_r}{2} \right) \cdot 1.63 \cdot p_0 \cdot e^{\frac{-2r^2}{r_L^2}} \delta(\psi)\mathbf{n} \tag{8}$$

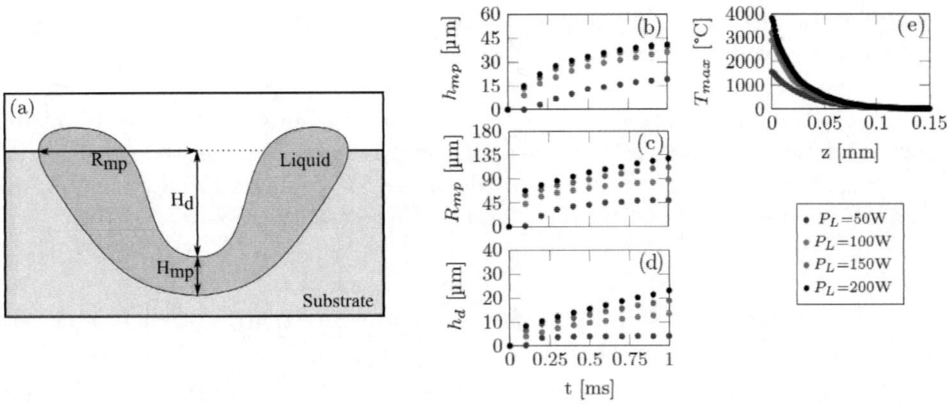

Figure 2. (a) Definition of liquid dimensions. (b), (c) and (d) are respectively the evolution of the melt pool height h_{mp}, the melt pool radius at the surface R_{mp} and the deflection of the interface h_d due to recoil pressure force. (e) Evolution of the maximal of temperature field observed in the material with depth.

where $\beta_r = 0.17$ is the retrodiffusion coefficient. The different simulations are compared in Figure 2 when considering various laser power. The vaporisation still does not occur in the material for a laser power of 50 W after 1 ms of laser shooting. Slight variations in the evolutions of the melt pool height h_{mp} and the radius of the melt pool R_{mp} are observed after 1 ms of laser shooting. A balance is observed between the heat introduced by the laser and the one extracted through the substrate by thermal conduction. As the convection in the melt pool is dominant regarding the other phenomena, the maximum of temperature reached after 1 ms of laser shooting increases more and more slowly when the laser power increases.

3.2. Static laser shooting on a powder bed

The second case investigated is a static shoot on a powder bed with a thickness h_p and a porosity of 50 %. The laser power is equal to 250 W with a radius of 50 µm. the different properties of the material are detailed in the Appendix (Figure A1). The different stages

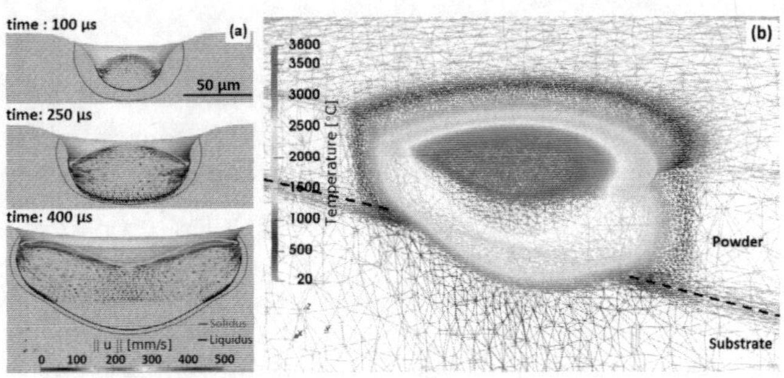

Figure 3. (a) Stages obtained during the SLM process in the case of a laser shooting. (b) Evolution of the temperature field and the anisotropic mesh at t=400 µs.

during the SLM process are predicted by the model as shown on Figure 3.(a) : the beginning of the compaction of the powder bed into a melt pool, the spheroidization of the melt pool occurs due to the surface tension and finally, the deflection of the interface when the vaporisation stage is reached due to the recoil pressure. On the Figure 3.(b), the maximum of temperature observed is approximatly 3800°C. This temperature is much greater than the vaporisation temperature chosen in the model. The maximum in velocity magnitude observed in this simulation is close to 500 mm/s.

3.3. Single pass on a powder bed

The configuration investigated is similar to the first case except that the domain is larger (1500 mm × 1000 mm × 1100 mm) with a continuous displacement of the laser. The different properties in the Appendix are used for this numerical simulation. The different results are shown in Figure 4. The development of a single bead is computed. For this

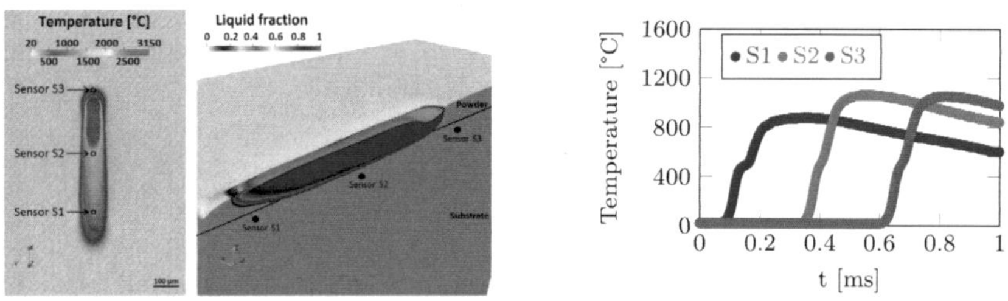

Figure 4. Temperature field and liquid fraction g_l at t=0.65 ms (left) and evolution of the temperature with time measured on three virtual sensors S1, S2 and S3 (right).

configuration, the powder is completly melted whereas the substrate is barely melted. The different virtual sensors shown in Figure 4 are located 10 µm below the surface of the substrate to observe if melting occurs. The temperature reached at this depth is 200°C below the liquidus. The bead obtained is not acceptable regarding SLM processes since it is crucial to merge the bead with the substrate. The input linear energy is not high enough to melt the substrate more deeply. Nevertheless, it should be relevant to get experimental results since it is possible that the reflection coefficient is overestimated or the material properties are not fully realistic.

4. Conclusion

A level-set model of SLM process applied to metals is presented to follow track evolution. The laser-matter interaction is described by a volume heat source between the laser and the powder bed and a surface source between the laser and the dense matter. Surface tension, Marangoni forces and the recoil pressure at the vaporisation stage are considered in order to provide an accurate modelling of the melt pool development.

Two configurations are investigated in this context. The first case corresponds to a laser shooting on a substrate and on a powder bed and the second case to a single track simulation on a powder bed. These configurations will be compared at a laser stage with the results provided by the commercial software Comsol Multiphysics® and experiments in a context of a benchmark with Arts et Métiers ParisTech.

The next step is to compute several unidirectional single passes to understand the influence of previous solidified tracks on the surface morphology. Finally, an interesting perspective is to simulate the build of a wall to observe the influence of the previous powder bed layer on the results obtained for the following one and compare these results with experimental data observations on surface stage of pieces and interaction between passes.

References

[1] Z. Wang, K. Guan, M. Gao, X. Li, X. Chen and X. Zeng. *Journal of Alloys and Compounds*, 513 :518–523, 2012.
[2] T. Antonsson and H. Fredriksson. *Metallurgical and Materials Transactions B*, 36(1) :85–96, 2005.
[3] Q. Chen, G. Guillemot, CA. Gandin and M. Bellet. *Additive Manufacturing*, 16 :124–137, 2017.
[4] F. Verhaeghe, T. Craeghs, J. Heulens and L. Pandelaers. *Acta Materialia*, 57(20) :6006–6012, 2009.
[5] M. Sussman, P. Smereka and S. Osher. *Journal of Computational physics*, 114(1) :146–159, 1994.
[6] M. Shakoor, B. Scholtes, P-O. Bouchard and M. Bernacki. *Applied Mathematical Modelling*, 39(23) :7291–7302, 2015.
[7] V. Bruyere, C. Touvrey, P. Namy and N. Authier. *Journal of Laser Applications*, 29(2) :022403, 2017.
[8] X. Jin, L. Li and Y. Zhang. *Journal of Physics D : Applied Physics*, 35(18) :2304, 2002.
[9] S. Anisimov and V. Khokhlov. CRC press, 1995.
[10] Web site. Dry air properties. `https://www.engineeringtoolbox.com/dry-air-properties-d_973.html`.
[11] C. Mills, C. Kenneth. Woodhead Publishing, 2002.
[12] P. Zehner and E.U. Schlünder. *Chemie Ingenieur Technik*, 42(14) :933–941, 1970.

Appendix :

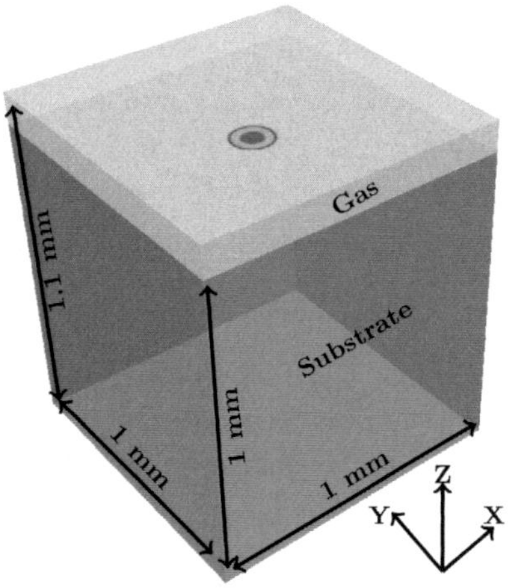

Welding	Property	Symbol	Value	Unity
Laser	Power	P_L	[50−200]	W
	Radius	r_L	75	µm

LBM	Property	Symbol	Value	Unity
Laser	Power	P_L	250	W
	Radius	r_{int}	50	µm
	Speed	v_L	1.1	$\mathrm{m\,s^{-1}}$
Powder Bed	Thickness	h_p	50	µm
	Porosity	ϕ_p	0.5	−
	Absorption	h_p	20	$\mathrm{mm^{-1}}$
	Thermal conductivity	λ_p^{ZS}	Fig 5(b)	$\mathrm{W\,m^{-1}\,K^{-1}}$
	Viscosity Dynamic	μ_p	Fig 5(c)	Pa s

Material	Property	Symbol	Value	Unity	Ref.
Argon	Density	ρ_g	1.78	$\mathrm{kg\,m^{-3}}$	
	Thermal conductivity	λ_g	0.024	$\mathrm{W\,m^{-1}\,K^{-1}}$	[10]
	Enthalpy per unit mass	$C_{p,g}$	Fig 5	$\mathrm{J\,kg^{-1}\,K^{-1}}$	
	Absorption coefficient	α_g	0	$\mathrm{m^{-1}}$	[−]
IN718	Solidus	T_s	1443	K	[11, 12]
	Liquidus	T_l	1609	K	
	Boiling	T_v	3184	K	
	Absorption (liquid)	α_l	120	$\mathrm{mm^{-1}}$	
	Density solid, liquid	ρ_s, ρ_l	8190, 7300	$\mathrm{kg\,m^{-3}}$	
	Thermal conductivity	λ	26.9	$\mathrm{W\,m^{-1}\,K^{-1}}$	
	Enthalpy per unit mass	h_m	Fig 5(a)	$\mathrm{J\,kg^{-1}\,K^{-1}}$	
	Latent heat of fusion	L_f	270	$\mathrm{kJ\,kg^{-1}}$	
	Latent heat of vaporisation	L_v	6690	$\mathrm{kJ\,kg^{-1}}$	
	Surface tension	γ	1.89	$\mathrm{N\,m^{-1}}$	
	Marangoni coefficient	$\frac{\partial\gamma}{\partial T}$	$-1.1\cdot 10^{-4}$	$\mathrm{N\,m^{-1}\,K^{-1}}$	
	Dynamic viscosity (liquid)	μ_d (μ_l)	Fig 5(c) ($7.8.10^{-3}$)	Pa.s	

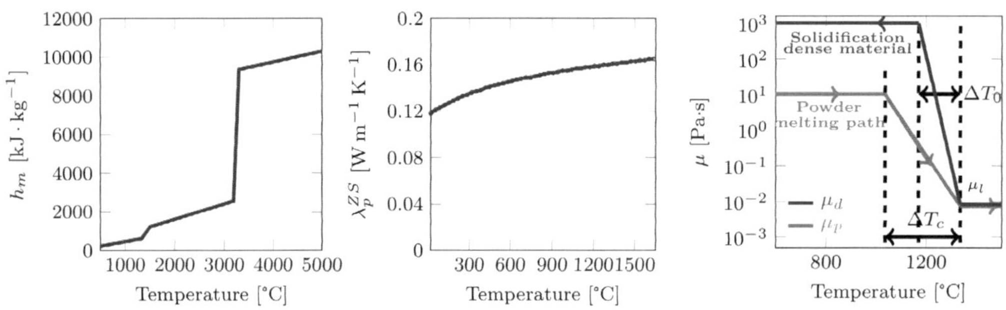

(a) (b) (c)

Figure A1. Configuration studied (domain and machine) for a welding or SLM simulation, properties chosen for Inconel 718 alloy and evolution of the specific enthalpy h_m, thermal conductivity in the powder λ_p^{ZS} and the dynamic viscosity μ with temperature. ΔT_0 and ΔT_C are respectively the temperature range of the melting stage and compaction (larger than the melting stage one to avoid numerical convergence issues) of the powder bed in a melt pool [3].

IOP Conf. Series: Materials Science and Engineering **529** (2019) 012006 doi:10.1088/1757-899X/529/1/012006

Solidification of niobium-silicide-based alloys during laser additive manufacturing process

A. Allen[1], A. C. Douglas[1], L. M. Feitosa[1], H. Qian[1], J. Li[1], H. B. Dong[1]*, X. Lin[2], Y. Li[2]

[1] Department of Engineering, University of Leicester, Leicester, LE17RH, UK

[2] School of Materials Science and Engineering, Northwestern Polytechnic University, Xi'an, P.R. China

Corresponding author: hd38@le.ac.uk

Abstract. Niobium silicide-based composites, in the application of gas turbine blades, promise significant efficiency improvements compared to current Ni-based alloys. The higher temperature capability would allow the engine to run at a higher temperature than that of current alloys, increasing engine efficiency. Nb-Si based composites possess a lower density, due to the presence of ceramic phases such as Nb_5Si_3 and/or Nb_3Si. This would reduce the weight of the rotating blades. However, improvements in certain properties, such as room temperature toughness and oxidation resistance are needed.

This study focuses on the manufacturability aspect of the powder feeding laser additive manufacturing (LAM) process to engineering Nb-Si based alloy samples. LAM has the advantage of forming near-net shapes without the use of expensive cores and moulds for the reactive Nb-Si melt. Fine microstructure and even chemical composition distribution with reduced macro-segregation are obtained. With the use of power feeding system, new Nb-Si based alloys are LAMed with varying atomic composition. Microstructures of the LAMed alloys will be presented, and the relationship between the microstructure and the alloy chemistry will be reported.

Keywords: microstructure evolution; laser additive manufacturing; niobium-silicides; rapid freezing; diffusion.

1. Introduction

Niobium silicide-based composites, in the application of gas turbine blades, promise significant efficiency improvements compared to current Ni-based alloys [1]. Higher temperature capability would allow engines to run at higher temperatures, increasing their efficiency [2]. These composites also possess lower density further increasing engine efficiency. [3]. The high promise of Nb-silicide based alloys combined with the manufacturing processes based on AM methods have been proven to be highly innovative, as it opens a new window of opportunities for efficiency, the reduction of energy consumption and the waste of raw materials [4].

Additions of alloying elements such as Al, Ti, Cr, Hf, V and Mo into the composition of Nb-silicide alloys seeks a balance of properties to produce a composite useful in applications at high temperatures [5]. Al and Cr are used to improve the oxidation behaviour by reducing oxygen diffusivity and forming a protective Al-oxide layer and Laves phase [6]. Ti can increase the kinetics of transformation from Nb_3Si to Nb_5Si_3, and it increases the fracture toughness and the ductility of the niobium solid solution [7]. The silicide phase can form as a tetragonal high temperature β-Nb_5Si_3, the tetragonal low temperature α-Nb_5Si_3 [8], or as the metastable hexagonal γ-Nb_5Si_3 silicide. The γ-Nb_5Si_3 is not desirable owing to its creep properties [9]. The Nb_{ss} can present high concentration of Ti with very low amounts of Si [10]. In relatively large amounts, it pushes the Nb-silicide composites to a ternary system, therefore a fair understanding of the phase equilibria for this is needed [11]. Additions of V and Mo control the

IOP Conf. Series: Materials Science and Engineering **529** (2019) 012006 doi:10.1088/1757-899X/529/1/012006

formation of α-Nb_5Si_3 phase, which increases the yield strength at room temperature. Additions of Hf improve the fracture toughness, room temperature strength and the oxidation behaviour [12].

2. Experimental

Nb-18Si-24Ti-5Cr-5Al-2Mo (2Mo) and Nb-18Si-24Ti-5Cr-5Al-2Mo-5Zr-0.4Y (5Zr) bars were prepared using a LAM method. The laser process requires a substrate, in this case titanium. A focused laser beam is aimed onto the substrate to create a molten pool, with a constant flow rate of metallic powder being directed at the beam through a nozzle. A computer numeric control (CNC) motion system assists the movement of the substrate, creating the final piece shape with a computer assisted design (CAD) model. As the substrate moves, the molten pool is dislocated, leaving a rapidly solidified layer behind. Figure 1 shows the schematic for LAM.

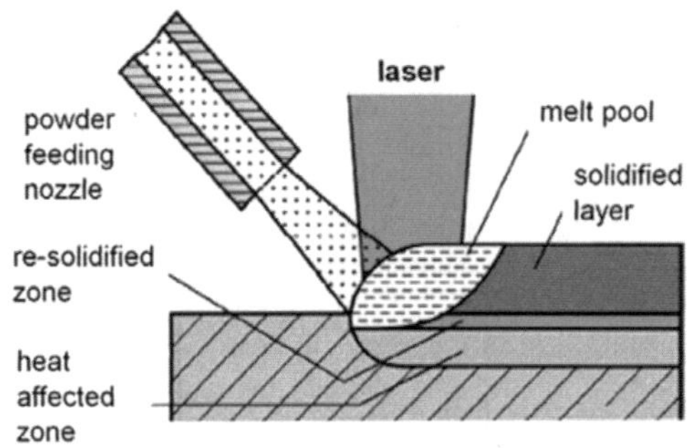

Figure 1 - Schematic of the LAM processing method.

The second step of this process involves the deposition of a following layer of metal on top of the initial layer, creating a wall structure. The repetition of this step is designated to form complex shapes by the movement of the laser beam, which controls the direction of the molten pool, and therefore the design of the shaped part without any mould or die.

The parameters used during the preparation and forming of the materials are detailed in Table 1. The nozzle was fed with a pre-mixed powder of the elements needed to produce the compositions aimed. With a laser melting the powder onto a Ti substrate, a protective argon gas was used at all times to prevent oxidation of the molten pool.

Table 1 - Parameters used for the LAM equipment during processing.

Parameter	Value
Laser power	3.0 kW
Laser spot diameter	5.0 mm
Powder gas flow rate	6.0 L/min
Scanning speed	900 mm/min

The compositions and microstructures were characterised using an FEI FEG scanning electron microscope (SEM), equipped with energy-disperse X-ray spectroscopy analysis (EDS), with which phases were identified. A Bruker D8 Advance with DaVinci X-ray diffractometer was used to analyse crystal structure.

3. Results and Discussion

3.1. 2Mo

The as-formed microstructures are shown in Figure 2. 2Mo appeared to exhibit a three-phase microstructure with a Nb_5Si_3 primary phase, a Nb_{ss} phase and a further Ti-rich Nb_5Si_3 phase. The elemental composition of the phases and volume fraction is shown in Table 2.

Table 2 - Phase elemental composition (at%) and volume fraction for 2Mo

Phase	Nb	Si	Ti	Al	Cr	Mo	Volume Fraction (%)
Nbss	46.39	6.65	28.36	6.98	8.82	2.81	33.73
γ-Nb_5Si_3	45.42	32.35	17.31	3.03	1.00	0.88	35.38
Ti rich γ-Nb_5Si_3	34.85	28.30	28.84	4.45	2.76	0.79	30.89

XRD analysis using MAUD [13], shown in Figure 3, suggest that the only silicide present was the γ-Nb5Si3 crystal structure, therefore indicating that the Ti-rich Nb5Si3 phase was also of the γ crystal structure. A similar alloy was produced by Geng et al [16] using vacuum arc remelting but produced a microstructure with β-Nb_5Si_3 and another silicide Nb_3Si. It is thought that the relatively rapid cooling rate of additive manufacture compared to vacuum arc remelting alters the solidification path, bypassing the formation of Nb_3Si. It is also known that high concentration of Ti can include the formation of Ti_5Si_3 [14], which possesses the same crystal structure as γ-Nb_5Si_3. It is known that high Hf content can promote the formation of the γ-silicide [10], and Ti and Hf are in the same column of the periodic table, indicating similar properties. It is likely therefore that the high Ti content combined with rapid cooling resulted in the formation of the γ-Nb_5Si_3 rather than the β-Nb_5Si_3 observed by Geng. The morphology shows a uniform microstructure with smoothed blocks of the primary silicide in a matrix of Nb_{ss}. The Ti-rich silicide was dispersed amongst the Nb_{ss}. The proposed solidification sequence is as follows:

L → γ-Nb_5Si_3

L → Nb_{ss} + γ-Nb_5Si_3 (Ti saturated)

The γ-Nb5Si3 forms first as the liquid cools. This reduces the amount of Si in the liquid and increases the amount of Ti as the primary silicide only contains 17 at.% Ti. As the liquid cools further, the Si depleted liquid begins to form Nb_{ss} and further amounts of γ-Nb5Si3. However, the forming Nb_{ss} is ejecting Ti as it becomes saturated, resulting in the formation of a Ti-rich Nb_5Si_3 phase.

3.2. 5Zr

Initially, 5Zr appeared to be exhibit a two-phase microstructure of a primary Nb_5Si_3 phase inside a continuous matrix of Nb_{ss}. Closer inspection shows there was still small amounts of Ti-rich Nb_5Si_3 present but was integrated into the primary γ-Nb_5Si_3. The elemental composition and volume fraction are shown in Table 3, clearly showing that Zr partitions primarily to the γ-Nb_5Si_3 phase.

IOP Conf. Series: Materials Science and Engineering **529** (2019) 012006 doi:10.1088/1757-899X/529/1/012006

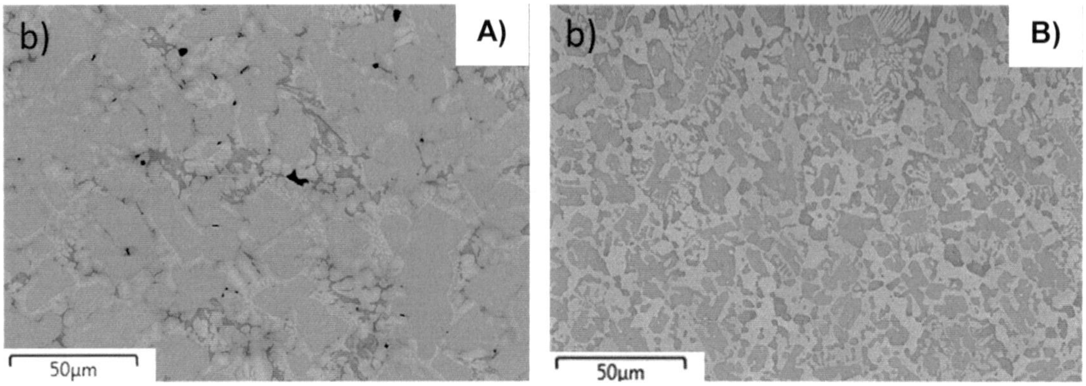

Figure 2 – Backscattered electron images of a) 2Mo and b) 5Zr alloys. Images were captured at the centre of the samples.

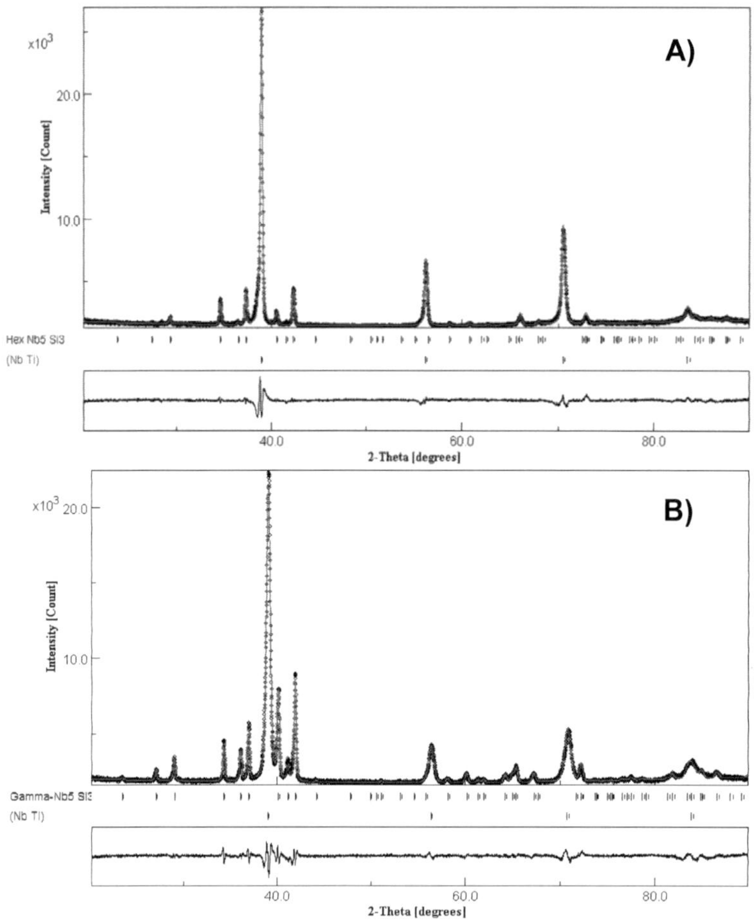

Figure 3 – Backscattered electron images of a) 2Mo and b) 5Zr alloys. Images were captured at the centre of the samples.

This also shows a reduced amount of Ti-rich silicide present. MAUD analysis [13], again in Figure 3, shows that the only the γ-Nb_5Si_3 is present. This is logical as the newly introduced Zr is also in the same column of the periodic table as Ti and Hf, which both appear to induce the formation of γ-Nb_5Si_3. It is likely that Zr also induced the formation of γ-Nb_5Si_3.

Table 3 - Phase elemental composition (at%) and volume fraction for 5Zr

Phase	**Nb**	**Si**	**Ti**	**Al**	**Cr**	**Mo**	**Zr**	**Y**	**Volume Fraction (%)**
Nb_{ss}	58.98	1.86	20.95	7.85	6.52	2.64	0.81	0.42	51.06
γ-Nb_5Si_3	36.61	32.79	17.31	3.69	1.00	0.08	9.49	0.14	48.94

The morphology of the microstructure was markedly different. The overall microstructure appeared refined with smaller blocks of silicide phase. In other alloy systems Zr is known to aid grain refinement by creating nucleation sites [15]. The silicide blocks were more angular and exhibited a wide variety of shapes including 'U' and 'H' shapes. Angular shapes are often the result of more brittle phases, indicating the addition of Zr has reduced the ductility of the silicide. Areas of eutectic Nb_{ss}/Nb_5Si_3 were present near the larger of the silicide blocks. The Ti-rich silicide only appears to be present on the edges of the primary silicide and not through-out the Nb_{ss} matrix as with 2Mo alloy. These occurrences are explained in proposed solidification sequence:

L → γ-Nb_5Si_3

L → Nb_{ss} + γ-Nb_5Si_3

L → Nb_{ss} + γ-Nb_5Si_3 eutectic

As with the 2Mo alloy, the first phase to solidify is the primary γ-Nb_5Si_3. The remaining liquid is less saturated in Ti as the Zr partitions preferably to the γ-Nb_5Si_3 at the expense of Nb, resulting in more Nb in the Nb_{ss}. A higher density of grains are formed thanks to the Zr additions forming more nucleation sites. As the liquid cools further Nb_{ss} and γ-Nb_5Si_3 start to form. With a reduction in the amount of Nb present you would expect the equilibrium to shift away from the eutectic. However, it appears the addition of Zr shifts the system towards the equilibrium and once the liquid is cooled sufficiently eutectic Nb_{ss}/γ-Nb_5Si_3 forms.

4. Conclusions

(1) Proposed reaction sequences for the 2Mo and 5Zr alloys both show that the γ-Nb_5Si_3 phase forms initially. This is followed by Nb_{ss}/Ti-rich γ-Nb_5Si_3 for 2Mo and by Nb_{ss}/γ-Nb_5Si_3 for 5Zr with a small amount of eutectic forming.

(2) γ-Nb_5Si_3 was formed rather than other crystal structures for both 2Mo and 5Zr as Ti and/or Zr content and rapid cooling drive the formation of the hexagonal crystal form.

(3) Adding 5 at% Zr at the expense of Nb appears to have refined the resulting microstructure. This is due to Zr forming more nucleation sites. Zr has also moved the system closer to the Nb_{ss}/Nb_5Si_3 eutectic, with eutectic formations observed near the larger γ-Nb_5Si_3 blocks. Finally, Zr appears to have made the γ-Nb_5Si_3 blocks more angular in shape. This is likely due to Zr decreasing the ductility of the silicide.

Acknowledgements

A. Allen and A.C. Douglas wish to acknowledge Rolls-Royce plc. and EPSRC CDT in Innovative Metal Processing for their financial contribution to their studies. L.M. Feitosa would like to thank CNPq/Brazil for funding his PhD project. All authors wish to thank X. Lin and W.D. Huang from the Northwestern Polytechnic University, in China, for their support on preparing and producing the laser additive manufactured samples.

References

1. B. P. Bewlay, M.R.J., J. C. Zhao, P. R. Subramanian, M. G. Mendiratta, and J. J. Lewandowski, Ultrahigh-Temperature Nb-Silicide-Based Composites. Mrs Bulletin, 2003: p. 646-653.
2. N. D'Souza, B.K., G.D. West, L.M. Feitosa, H.B Dong, Key aspects of carbide precipitation during solidification in the Ni-base superalloy, MAR M002. Journal of Alloys and Compounds, 2017. 702: p. 6-12.
3. B. P. Bewlay, M.R.J., J. C. Zhao, P. R. Subramanian, A Review of Very-High-Temperature Nb-Silicide–Based Composites. Metallurgical and Materials Transactions A, 2003. 34: p. 2043-2052.
4. Attaran, M., The rise of 3-D printing: The advantages of additive manufacturing over traditional manufacturing. Business Horizons, 2017. 60: p. 677-688.
5. I. Grammenos, P.T., Study of the role of Mo and Ta additions in the microstructure of Nb-18Si-5Hf silicide-based alloy. Intermetallics, 2010. 18(1524-1530).
6. S. Kashyap, C.S.T., K. Chattopadhyay, Effect of Gallium on microstructure and mechanical properties of Nb-Si eutectic alloy. Intermetallics, 2011. 19: p. 1943-1952.
7. Q. Huang, X.G., Y. Kang, J. Song, S. Qu, Y. Han, Microstructures and mechanical properties of directionally solidified multi-element Nb-Si alloy. Progress in Natural Science: Materials International, 2011. 21: p. 146-152.
8. Okamoto, H., Phase diagrams for binary alloys. Materials Park, OH: ASM International, 2000.
9. B.P. Bewlay, M.R.J., and M.F.X. Gigliotti, Intermetallic Compounds—Principles and Practice, . Vol. 3. 2001: John Wiley & Sons,.
10. Jie Geng, P.T., Guosheng Shao, A study of the effects of Hf and Sn additions on the microstructure of Nbss/Nb5Si3 based in situ composites. Intermetallics, 2007(15): p. 69-76.
11. Y. Li, S.M., K. Ohsasa, C. Ma, H. Zhang, Ultrahigh-temperature Nbss/Nb5Si3 fully-lamellar microstructure developed by directional solidification in OFZ furnace. Intermetallics, 2011. 19: p. 460-469.
12. Jie Geng, P.T., Guosheng Shao, Oxidaion of Nb-Si-Cr-Al in situ composites with Mo, Ti and Hf additions. Materials Science and Engineering, 2006. 441: p. 26-38.
13. Lutterotti, L., Bortolotti, M., Ischia, G., Lonardelli, I. & Wenk, H. R.,, Rietveld Texture analysis from Diffraction Images. Zeitschrift fur Kristallographie, 2007. 1: p. 125-130.
14. Li, Y., et al., As-cast microstructures and solidification paths of the Nb–Si–Ti ternary alloys in Nb5Si3–Ti5Si3 region. Rare Metals, 2013. 32(5): p. 502-511.
15. Wang, F., et al., The grain refinement mechanism of cast aluminium by zirconium. Acta Materialia, 2013. 61(15): p. 5636-5645.

IOP Conf. Series: Materials Science and Engineering **529** (2019) 012007 doi:10.1088/1757-899X/529/1/012007

Investigation of the microstructure adjustment by velocity variations during the directional solidification of Al-Ag-Cu with the phase-field method

J. Hötzer[1,2], M. Kellner[1,2], W. Kunz[1] and B. Nestler[1,2]

[1]Institute of Applied Materials (IAM), Karlsruhe Institute of Technology (KIT), Straße am Forum 7, 76131 Karlsruhe, Germany Tel.: +49 721 608-45315

[2]Institute for Digital Materials (IDM), Hochschule Karlsruhe Technik und Wirtschaft (HSKA), Moltkestr. 30, Karlsruhe, Germany

E-mail: johannes.hoetzer@kit.edu

Abstract. Directional solidification is a favored process to manufacture homogeneous microstructures which lead to macroscopically unique properties for a material. The dependence of the spacing and type of the arising microstructure from the solidification velocity for constant velocities is extensively investigated. However the effect of changes in the solidification velocity on the resulting microstructure adjustment processes is still unclear. Therefore large-scale (3D+t) simulations of the ternary eutectic system Ag-Al-Cu with changing solidification velocities are conducted with a phase-field model based on the grand potential approach. To study the spatially complex rearrangement process during velocity changes in statistical representative volume elements, simulations with different velocity profiles are calculated in large-scale domains. The results show, that the evolving microstructure continuously rearranges by splitting and merging of the rods despite constant growth conditions. By increasing the velocity, the microstructure refines by splitting of the Al_2Cu phase. Whereas by decreasing the velocity, the microstructure coarsens by overgrowing events of both intermetallic phases.

1. Introduction

The process of directional solidification allows to produce different tailored materials with long ranges of homogeneous microstructures and consistent properties. The microstructure characteristics can be controlled by the applied process conditions. Especially the spacings and phase arrangements in the microstructure can be influenced by the externally applied solidification velocity. Despite the well understood correlation between the spacings of the evolving microstructure and the solidification velocity [1], the adjustment processes in the microstructure during a variation of the velocity is in focus of current research. An increase of the solidification velocity leads to a refinement whereas a decrease leads to a coarsening of the microstructure [2, 3, 4, 5]. In ternary eutectic systems a velocity change during directional solidification leads to split, merge, overgrowth and nucleation events as well as to a possible change of the overall evolving microstructural arrangement of the rods and lamellea. To investigate these complex three-dimensional adjustment processes, 3D samples of the microstructure are needed. Experimentally, a high technical effort like synchrotron tomography [6, 3] is required, to observe these rearrangement processes. Beside experiments and theory, simulations have become a powerful method for the scientific investigation of physical phenomena. For the three-dimensional investigation of the diffusion driven phase transformations in the microstructure during solidification processes, the phase-field method has been established [7].

To resolve representative volume elements with complex spatial microstructure arrangements comparable to experimental micrographs [8, 9, 10], large simulation domains are needed. Further, the calculation on current high performance computers is crucial to solve the compute-intensive models in the necessary large-scale domains in reasonable time. Therefore massive-parallel and highly optimized solvers, like PACE3D [11, 12] and WALBERLA [13, 14], are required. By using optimized and scalable solvers on high performance systems, the directional solidification of binary [8, 5] and ternary eutectic reactions [15, 2, 10, 16, 17, 18, 19] is investigated in large-scale phase-field simulations. A detailed description of the used model can be found in [20, 21, 10]. The implementation of the model is presented in [10, 14, 22, 12].

Two- and three-dimensional simulations of the microstructure adjustment processes due to velocity variations during directional solidification for the two-phase eutectic system NiAl-34Cr are performed in [4] and [5], respectively. In these works, separate acceleration and deceleration steps of the temperature gradient velocity are used to realize variations in the solidification velocity. A comparison between three-dimensional phase-field simulations and two experiments of the ternary system Ag-Al-Cu with velocity variations is given in [2]. In the experiments a single velocity increase and decrease is executed, respectively, whereas the conducted simulations include multiple subsequent velocity variations. A pure experimental observation via synchrotron tomography of a decelerated solidification of Ag-Al-Cu is published in [3]. For this study the same analysis methods using a graph based approach are used as for the investigation of the three-dimensional phase-field simulations in [2] and in this work.

Due to the complexity of the microstructure adjustments in simulations with multiple velocity variations, large-scale 3D simulations of the system Ag-Al-Cu with single controlled acceleration and deceleration steps are conducted in this work to identify the underlying mechanism controlling the rearrangement processes in ternary eutectic systems.

2. Results and discussion

To investigate the microstructure rearrangement mechanisms during a velocity change, the ternary eutectic system Ag-Al-Cu is used. The simulation setup is presented together with a summary of the material and process parameters in [10, 2]. During the solidification the matrix phase Al and the two intermetallic phases Ag_2Al and Al_2Cu evolve simultaneously. The velocity changes are realized by variations of the temperature gradient velocity v_G. The solidification velocity adjusts according to the changes in the analytic temperature field. The velocity variations are performed similar to [5] with deceleration and acceleration steps instead of a jump function like in [2]. Due to the unnaturally enlarged temperature gradient, which is used in the conducted system, to reduce the computational effort [2], the acceleration and deceleration step can be performed in less than 20 000 time steps. One simulation is decelerated from $0.325\,\mu m/s$ to $0.25\,\mu m/s$, which corresponds to a velocity decrease of 30 %. The other simulation is accelerated from $0.175\,\mu m/s$ to the same end velocity of $0.25\,\mu m/s$, resulting in a velocity increase of 30 %. It is assumed that the microstructure refines during an acceleration of the solidification by splitting and nucleation of the rods and coarsens during a deceleration by merging and overgrowing of rods. To investigate these rearrangement events, the graph-based analysis approach described in [3] is used. Despite rarely observed nucleation events found in [3] it is assumed that for low solidification velocities the nucleation barrier is to high [2, 23]. Hence no nucleation mechanism is considered in the model. The simulations with a computational domain size of $800 \times 800 \times 275$ voxel cells are calculated on 11 264 cores for approximately 40 h on the Hazel Hen supercomputer

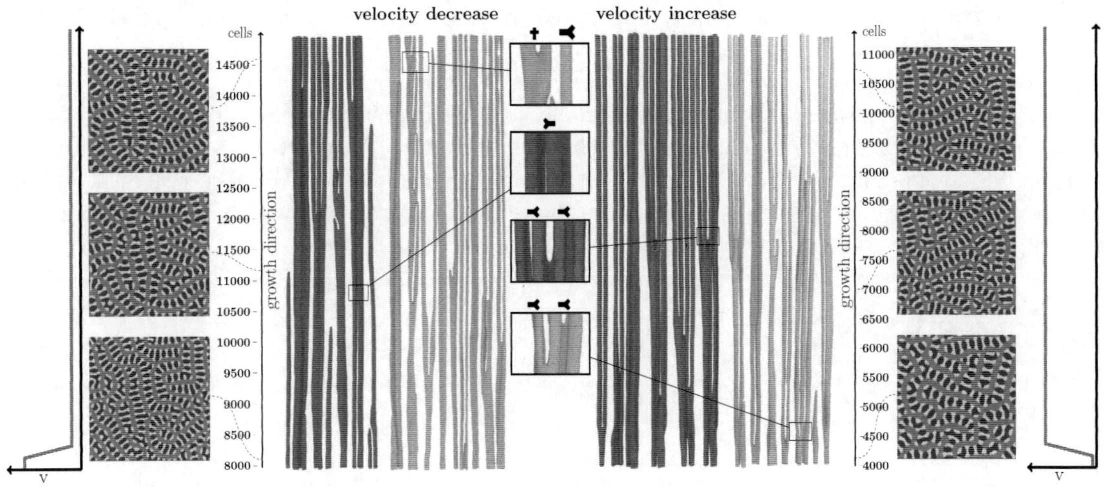

Figure 1. Computed micrographs perpendicular to the solidification direction (left and right images) and selected rods (central images) showing the effect of velocity variations.

[24]. By applying a moving window approach, this results in a growth height of 14 672 voxel cells for the deceleration and 10 950 voxel cells for the acceleration simulation. The difference in the growth heights is caused by the different initial velocities.

An overview of the two simulation results is composed in fig. 1. The two intermetallic phases, Ag_2Al and Al_2Cu are shown in green and blue, respectively, and the matrix phase Al is indicated in red. The rod diameters adjust following the velocity-undercooling-spacing relationship described by Jackson and Hunt [1]. For both simulations, computed micrographs parallel to the solidification front are depicted before and after the velocity change and at the end of the simulation. The overall evolving pattern remains the same for both simulations. Selected rods of the intermetallic phases over the growth height are highlighted in the center of fig. 1. For the deceleration simulation (left), an increase of the rod size is observed, whereas the rod size decreases for the acceleration (right). In both simulations the spacing in the microstructure adjusts by multiple split, merge and overgrowth events. Also multiple sequences of splitting and afterwards overgrowing of one splitted arm are observed in both intermetallic phases. For a better visualization of the microstructure adjustment events selected examples are magnified. The selected rods indicate that the adjustment processes are not only driven by the corresponding events for refinement and coarsening. Splitting events, leading to a refinement of the microstructure, are also multiple times observed in the deceleration simulation as well as overgrowing events, leading to a coarsening of the microstructure, in the acceleration simulation.

To analyze the microstructure events quantitatively, the area of the two intermetallic rods perpendicular to the solidification direction at the events is plotted over the growth height in figs. 2 and 4. The type of event is indicated by the symbols and the phases are marked by the colors. Due to the continuous structure of the Al matrix, split, merge and overgrowing events lead not to separate parts of this phase and hence they are not detected. In fig. 2 more events can be observed before the velocity decrease. Despite the similar phase fractions of the intermetallic phases of $\sim 1/3$ and the similar number of rods, the rod area at which they split and merge differs. The area of the Al_2Cu rods is approximately twice as large as of the Ag_2Al rods. For a better comparison of the event occurrence rate, the average number of events per cell over the growth height is shown in fig. 3. Whereas in the beginning the microstructure adjustment is dominated by splitting and merging events, after the velocity decrease, mostly splitting and overgrowing events occur. For a growth height less than 1 000 μm a higher number of events is observed due to the evolution of the artificially used Voronoi tessellation as initial condition to a stable chain-like pattern [25]. After self similar growth is reached for a constant velocity, the number of events is stable. After the velocity decrease, the number of overgrowing events increases, leading to a coarsening of the microstructure. Due to a rod area of zero for overgrowing events, the symbols in fig. 2 are strongly overlapping. Therefore the peak of the overgrowing events in fig. 3 is not visible identifiable in fig. 2. A similar behavior of the events as in fig. 3 is also observed experimentally in [3, 2].

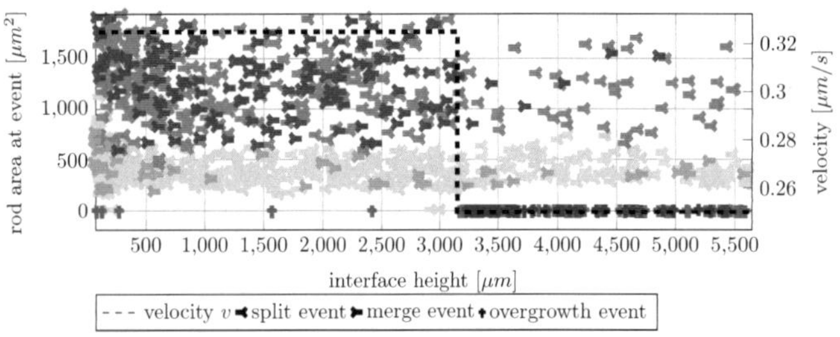

Figure 2. Area of the rods perpendicular to the solidification direction at which the microstructural adjustment events take place over the growth height for a growth velocity decrease.

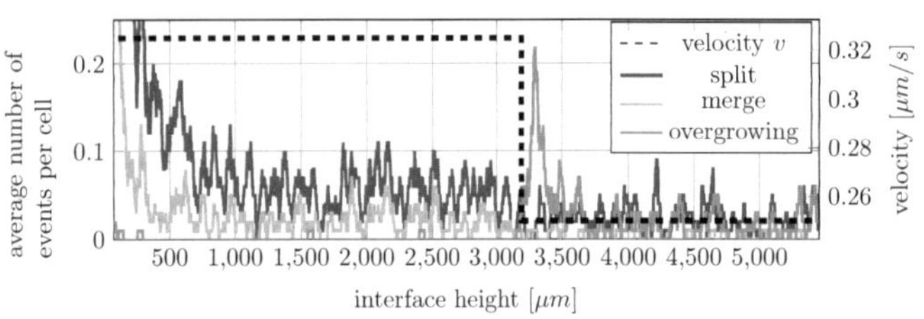

Figure 3. Average number of the microstructural adjustment events per cell over the growth height using a moving average approach for a growth velocity decrease.

In fig. 4 the rod size at the occurring events over the growth height is plotted for the velocity acceleration. Before the velocity increase, the number of events is reduced in contrast to the higher velocity afterwards. Similar to the deceleration simulation, different rod areas at which the events occur can be observed depending on the phase. Also more events occur initially due to the Voronoi tessellation until stable growth is established. Despite the overall refinement of the microstructure, multiple overgrowing events can be observed after the velocity increase. This is also indicated in fig. 5 by the average number of events per cell over the growth height. To refine the microstructure, various rods split as shown by the peak directly after the velocity change. Thereby mostly the blue Al_2Cu rods split as displayed in fig. 4.

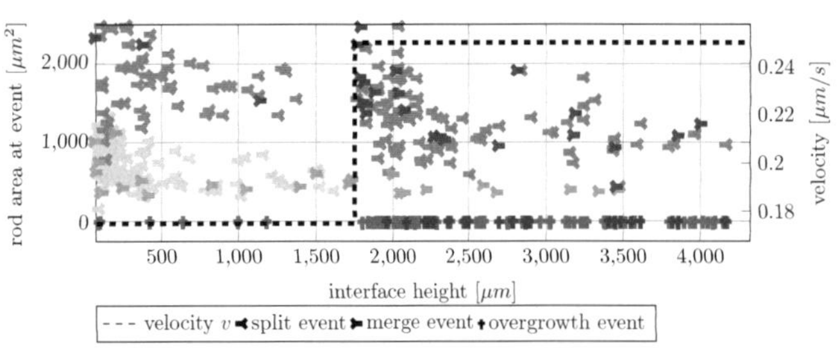

Figure 4. Area of the rods perpendicular to the solidification direction at which the microstructural adjustment events take place over the growth height for a growth velocity increase.

For both simulations, the events to adjust the spacing occur with a delay after the new solidification velocity has been established. This behavior is in accordance with the observations of [5]. The number of the microstructural events is found to be dependent on the growth rate

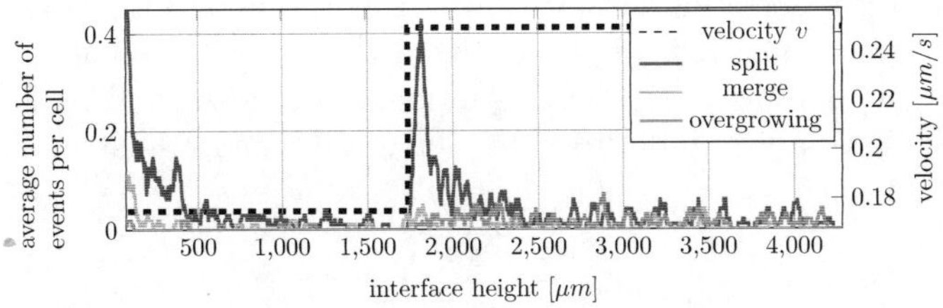

Figure 5. Average number of the microstructural adjustment events per cell over the growth height using a moving average approach for a growth velocity increase.

and increases with higher solidification velocity. Despite constant growth velocities, an ongoing rearrangement process of the two intermetallic rods by splitting and merging events is observed. Thereby a split event of one rod often leads to a merge of rods of the other phase and vice versa. This is also reported in experimental data in [3] and in phase-field simulations [2]. Independent of the initial velocities, both simulations have a similar number of ongoing events for the same imprinted final velocity. For all investigated velocities the size of the rods at which the event occurs is larger for the Al_2Cu rods than for the Ag_2Al rods. Despite the ongoing rearrangement at constant velocities, the coarsening of the microstructure is dominated by overgrowing events and the refinement by splitting events. This is in accordance to phase-field simulations with velocities variations of the two-phase eutectic system NiAl-34Cr as presented in [5] and of the ternary system Ag-Al-Cu shown in [2]. Similar findings are found as well in experimental data of Ag-Al-Cu containing a velocity decrease [3]. In contrast to the investigations of the pseudo-binary two-phase eutectic system NiAl-34Cr, the number of events is higher and the rearrangement process is more complex in ternary systems. In order to understand the basic mechanisms, more studies of defined and idealized setups are needed.

3. Summary and conclusions

In this work the influence of single independent velocity variations on the microstructure rearrangement process of the ternary eutectic system Ag-Al-Cu during directional solidification is investigated. This investigation leads to the following points: (i) An ongoing rearrangement process of the two intermetallic rods by splitting and merging events is observed for constant growth velocities. (ii) Depending on the observed phase the same microstructure events take place at different rod sizes, despite similar phase fractions and number of rods. (iii) During a velocity decrease the coarsening of the microstructure is dominated by overgrowing events, whereas (iv) during a velocity increase the refinement is dominated by splitting events. (v) The events occur delayed to the velocity adjustment, caused by the spatial rearrangement of the material. (vi) However splitting events are also observed during the deceleration simulation and overgrowing events during the acceleration simulation. This indicates that the ongoing rearrangement processes are still continuing through the microstructure coarsening and refinement processes. (vii) For the same velocity at the end of the simulations, both microstructures show a similar behavior of the events. (viii) The results show that the used phase-field model is able to reproduce experimental phenomena of the microstructure rearrangement.

These investigations lead to a better understanding of the rearrangement and adjustment processes in the microstructures required for the development of tailored materials with locally defined microstructures.

Acknowledgments

We are grateful for the provided computational resources at the High Performance Computing Center Stuttgart and thank Martin Bauer from the Friedrich-Alexander-Universität Erlangen-Nürnberg for helping to implement the solver in the waLBerla framework. The authors thank

Henrik Hierl, Christopher Seer and Marco Seiz for their support in the generation and analyses of the data. Further the authors thank for the financial support within the project SKAMPY (Ultra-scalable multiphysics simulations for solidification processes in metals) funded by BMBF and the cooperative graduate school Gefügeanalyse und Prozessbewertung by the ministry of Baden-Wuerttemberg.

References

[1] Jackson K A and Hunt J D 1966 *Transactions of the Metallurgical Society of AIME* **236** 1129–1142
[2] Hötzer J, Steinmetz P, Dennstedt A, Genau A, Kellner M, Sargin I and Nestler B 2017 *Acta Materialia* **136** 335–346
[3] Steinmetz P, Hötzer J, Dennstedt A, Serr C, Nestler B and Genau A 2018 *Journal of Crystal Growth* **498** 230–243
[4] Kellner M, Hötzer J, Steinmetz P, Dargahi Noubary K, Kunz W and Nestler B 2017 *Proc. of the 6th Decennial Int. Conf. on Solidification Processing (Old Windsor)* 372–375
[5] Kellner M, Kunz W, Steinmetz P, Hötzer J and Nestler B 2018 *Computational Materials Science* **145** 291 – 305
[6] Dennstedt A, Helfen L, Steinmetz P, Nestler B and Ratke L 2015 *Metallurgical and Materials Transactions A* 1–4
[7] Hötzer J, Kellner M, Steinmetz P and B N 2016 *Journal of the Indian Institute of Science* **96** 235–256
[8] Kellner M, Sprenger I, Steinmetz P, Hötzer J, Nestler B and Heilmaier M 2017 *Computational Materials Science* **128** 379 – 387
[9] Steinmetz P, Yabansu Y C, Hötzer J, Jainta M, Nestler B and Kalidindi S R 2016 *Acta Materialia* **103** 192–203
[10] Hötzer J, Jainta M, Steinmetz P, Nestler B, Dennstedt A, Genau A, Bauer M, Köstler H and Rüde U 2015 *Acta Materialia* **93** 194 – 204
[11] Vondrous A, Selzer M, Hötzer J and Nestler B 2014 *International Journal of High Performance Computing Applications* **28** 61–72
[12] Hötzer J, Reiter A, Hierl H, Steinmetz P, Selzer M and Nestler B 2018 *Journal of Computational Science* **26** 1–12
[13] Godenschwager C, Schornbaum F, Bauer M, Köstler H and Rüde U 2013 *Proc. on Int. Conf. for High Performance Computing, Networking, Storage and Analysis (Denver, Colorado)* (ACM) p 35
[14] Bauer M, Hötzer J, Jainta M, Steinmetz P, Berghoff M, Schornbaum F, Godenschwager C, Köstler H, Nestler B and Rüde U 2015 *Proc. on Int. Conf. for High Performance Computing, Networking, Storage and Analysis (Austin, Texas)*(ACM) p 8
[15] Steinmetz P, Kellner M, Hötzer J, Dennstedt A and Nestler B 2016 *Computational Materials Science* **121** 6–13
[16] Hötzer J, Steinmetz P, Jainta M, Schulz S, Kellner M, Nestler B, Genau A, Dennstedt A, Bauer M, Köstler H and Rüde U 2016 *Acta Materialia* **106** 249 – 259
[17] Noubary K D, Kellner M, Steinmetz P, Hötzer J and Nestler B 2017 *Computational Materials Science* **138** 403 – 411
[18] Steinmetz P, Hötzer J, Kellner M, Dennstedt A and Nestler B 2016 *Computational Materials Science* **117** 205–214
[19] Steinmetz P, Kellner M, Hötzer J and Nestler B 2017 *Metallurgical and Materials Transactions B*
[20] Plapp M 2011 *Physical Review E* **84** 031601
[21] Choudhury A and Nestler B 2012 *Physical Review E* **85** 021602
[22] Hötzer J, Tschukin O, Said M B, Berghoff M, Jainta M, Barthelemy G, Smorchkov N, Schneider D, Selzer M and Nestler B 2015 *Journal of Materials Science* **51** 1788–1797
[23] Zhao P 2012 *Acta Metallurgica Sinica* **48** 33–40
[24] HLRS 2016 Cray XC40 (Hazel Hen) https://www.hlrs.de/systems/cray-xc40-hazel-hen [Online; accessed March 2016]
[25] Dennstedt A and Ratke L 2012 *Transactions of the Indian Institute of Metals* **65** 777–782

IOP Conf. Series: Materials Science and Engineering **529** (2019) 012008 doi:10.1088/1757-899X/529/1/012008

Solidification path and phase transformation in super-austenitic stainless steel UNS S31254

Raphaël Marin [1,2,3] Hervé Combeau [1], Julien Zollinger [1], Moukrane Dehmas [2], Bernard Rouat [1], Aude Lamontagne [3], David Cardinaux [3], Lucile Lhenry-Robert [3]

[1] Université de Lorraine, CNRS, IJL, F-54000 Nancy, France;

[2] CIRIMAT, Université de Toulouse, CNRS, Toulouse, France;

[3] Industeel, Centre de Recherche des Matériaux du Creusot (CRMC), 71200 Le Creusot, France;

Raphaël Marin, raphael.marin@univ-lorraine.fr

Abstract. The solidification path and the σ-phase precipitation mechanism of UNS S31254 alloy were studied on the basis of directional solidified experiments accompanied by scanning electron microscopy observations and energy dispersive X-ray a nalysis. The resulting temperatures of solidification paths and phase transformation were compared with Gulliver-Scheil and equilibrium calculations predicted using ThermoCalc© software. It was confirmed that the experimental solidification path was in agreement with the thermodynamic calculations. The complementarity of the results have made it possible to propose a solidification path and a σ-phase precipitation mechanism for the UNS31254 steel.

1. Introduction

Stainless steels are an important class of materials. They are used everywhere, from daily life applications as cooking utensils to space vehicles. These steels are rich in chromium in order to get an oxide layer on the surface, which acts as protection against rusting and pitting. However, most of the stainless steels that are used in applications as seawater desalination and paper manufacturing plants require a better corrosion resistance [1–3]. Consequently, molybdenum is added to enhance corrosion resistance. During the processing route, the high amount of alloying elements contained in these steels can lead to the formation of chromium and molybdenum rich intermetallic precipitates and therefore to a significant lower corrosion resistance by removing these elements from the matrix. Especially, as-cast ingots exhibit σ-phase (space group 136, tetragonal [4]) in interdendritic region, that has to be eliminated by heat treatments after solidification. Numerous studies in the literature were focused on the σ-phase formation during heat treatment of different classes of stainless steels (ferritic, austenitic, duplex). There are two well established σ-phase precipitation mechanisms: (i) σ-phase nucleation in austenite [5–9], (ii) σ-phase nucleation due to the eutectoid decomposition of the δ-ferrite [5,7,10–13]. However, to our knowledge, few studies report on the σ-phase precipitation mechanism during the solidification stage, especially in the S31254 (UNS norm) super-austenitic stainless steel ("super" = steel containing ≈ 50 wt.% iron – 50wt.% alloying elements [5]). Two cases are formulated in the literature for the σ-phase nucleation in cast stainless steel ingots; (i) it can be formed during the last stage of solidification due to the enrichment of the liquid in σ-forming elements [14], (ii) the interdendritic regions are composed of δ-ferrite that decomposes through a solid-state phase transformation [15]. Both of these two hypotheses are related to the segregation of the molybdenum in the liquid during the solidification process [16].

This consideration prompted the present study that is focused on the study of the solidification path and the phases transformations of the UNS S31254 steel. A quenching directional solidification (QDS) set-up was used in parallel with ThermoCalc© simulations to establish the solidification path and the σ-phase transformation mechanism of the steel.

2. Material and experimental procedures

2.1. Material

The used steel was hot rolled grade, namely S31254 (UNS designation) and provided by INDUSTEEL company. The nominal chemical composition (wt.%) is given in Table 1.

Table 1 – Nominal chemical composition of the S31254 steel in weight percent.

	Fe	Cr	Ni	Mo	Cu	Mn	Si	N
S31254	bal.	19.68	17.77	6.08	0.67	0.65	0.20	0.19

The equilibrium liquidus and solidus temperatures of the S31254 steel were determined using a Setaram multi detector high temperature calorimeter apparatus. In this way, various heating rates were applied (2, 5, 10°C/min). The equilibrium liquidus and solidus temperatures were found to be 1413°C and 1376°C respectively.

2.2. Quenching during Directional Solidification

Experiments involving Quenching during Directional Solidification (QDS) were performed with a specially designed Bridgman type furnace as previously described by Lacaze and Lesoult [17]. These experiments are achieved under argon gas flow. The samples were electromachined into cylindrical rods, 5 mm in diameter and 200 mm in length. They were introduced in an alumina tube of 5-mm inner diameter. The thermal history of the sample during the experiment was recorded by a Pt/Rh B-type thermocouple, which is located inside the sample and protected by an alumina tube. The directional solidification is achieved by pulling the alumina tube and the sample along the vertical axis at a constant solidification velocity ($V = 5.10^{-5}$ m/s). The solidification velocity was chosen in order to have a directional solidification and not a longitudinal solidification. The thermal gradient, G, is controlled by the temperature difference between the inductive heater and the cooling box, which is filled with water. At constant solidification velocity, G controls the size of the different area on the sample, *i.e.* the mushy zone. Once the major part of the sample is solidified, the sample is quenched by pulling it instantaneously through the cooling box in order to freeze the microstructure and the microsegregations. The induced cooling rate is close to 100°C/s ($3.6.10^{5}$°C/h). This experiment enables to monitor separately the solidification velocity (V) from the thermal gradient (G). After the QDS experiments, the sample was polished along the longitudinal section, which is parallel to the growth direction. Each position in the sample corresponds to a given temperature of the steel before the quench at room temperature. In the present study, the average temperature gradient in the mushy zone was G= 10000°C/m. The average cooling rate of the sample was calculated to be about 0.5°C/s (1800°C/h).

2.3. Microstructure analysis

The QDS samples were prepared using standard metallographic polishing. Final sample preparation was accomplished with colloidal silica. The microstructural observations were performed using a scanning electron microscopy (SEM) Quanta FEG 650 in backscattered electron mode (BSE), operated at an acceleration voltage of 20Kv. Energy Dispersive X-ray (EDX) spectroscopy was used in order to analyse the elemental composition of the various phases.

3. Results and discussion

3.1. Experimental solidification path and phase transformations

Figure 1 shows the solidification microstructure at the interface between the quenched liquid and the dendrites tip and the thermal history of the sample during the experiment. A deviation of 15°C on the temperature determination is considered due to the thermocouple accuracy and the repeatability. As shown in Figure 1.a, the interface between the quenched liquid and the dendrites tip is represented by the red dotted line. The dendrites grow along a direction close to the thermal gradient. Figure 1.b illustrates a part of the sample thermal history that was recorded during the experiment. These results were used to determine specific temperatures *i.e.* the liquidus temperature assuming the tip undercooling is negligible and the end of solidification. These particular temperatures were obtained with two tangential lines. The first line was draw to fit the thermal signal in the liquid, and the other one was fitting to the thermal signal obtained from the mushy zone. The intersection of these two tangential lines indicates the dendrites tip temperature. The same procedure was used to determine the end of solidification. The solidification starts at 1430 ± 15°C and ends at 1345 ± 15°C.

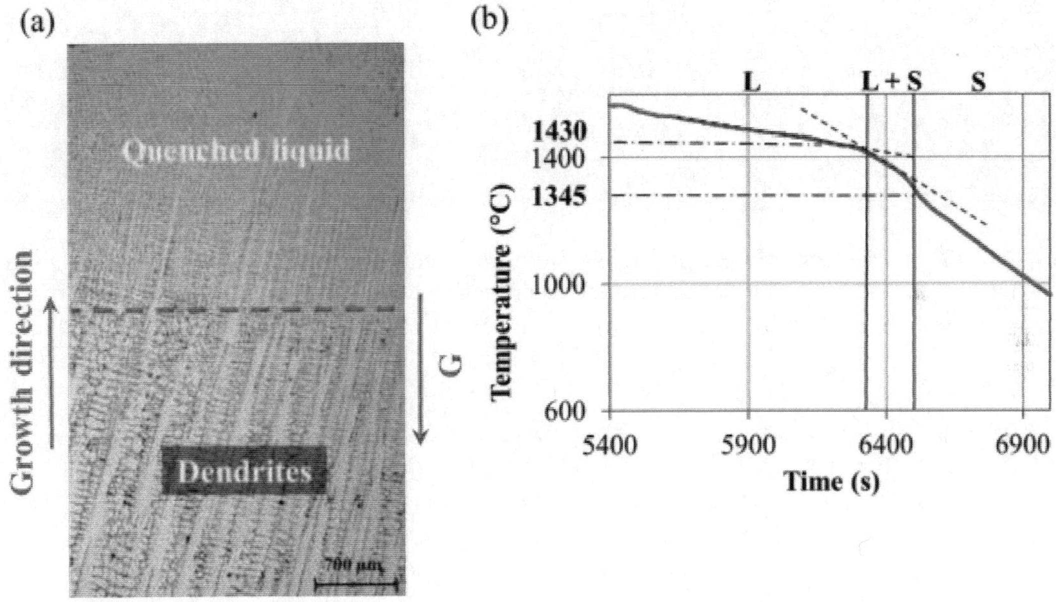

Figure 1 – QDS results (a) QDS micrograph after NaOH (20%) + H_2O (80%) electrolytic etching, 2.5 V, 5 seconds with the interface between the quenched liquid and the dendrites tip (red dotted line), (b) thermal history of the sample recorded by the thermocouple.

The microstructure evolutions during the solidification are reported with the corresponding temperatures on the Figure 2. The solidification starts at 1430°C ± 15°C with austenitic dendrites. At 1403°C ± 15°C, in the mushy zone, a new phase grow in the interdendritic spaces as shown in Figure 2.a. This phase should be δ-ferrite due to the segregation of Mo in the liquid during the solidification [16]. Moreover, with the Cr, Ni and Mo content, the primary solidification is austenite while δ-ferrite forms at a later stage[18]. The complete solidification occurs at 1345 ± 15°C. At this temperature, the microstructure consists of austenite dendrites and δ-ferrite in the interdendritic spaces. Thus, we can first conclude that no σ-phase was formed during the solidification period.

At lower temperature (1208°C ± 15°C), the solid-state transformation of interdendritic phase into two different phases occurs. As illustrated in Figure 2.b, the phase with a white grey level is enriched in Mo, Cr elements and depleted in Ni element, whereas the phase with a black grey level is enriched in Ni elements and depleted in Mo, Cr elements. These two phases seems to nucleate inside the δ-ferrite grain. Numerous studies in the literature exhibit a similar microstructure and solid-state transformation in different types of stainless steel (duplex, ferritic) [10–13,19]. According to the literature and the SEM/EDS analysis that were performed, these phases can be identified as σ-phase and γ_2 (eutectoid austenite). This transformation starts at 1208°C ± 15°C and ends at 1041°C ± 15°C. Below 1041°C ±15°C, the δ-ferrite is fully transformed into ($\sigma + \gamma_2$) and the microstructure is composed of austenite dendrites and ($\sigma + \gamma_2$) interdendritic precipitates, as shown in Figure 2.c.

Based on these assumptions, the solidification path and solid-state phase transformation of the S31254 steel can be expressed as following:

$$L \rightarrow L + \gamma\ (1430°C \pm 15°C) \rightarrow L + \gamma + \delta\ (1403°C \pm 15°C) \rightarrow \gamma + \delta\ (1345°C \pm 15°C) \rightarrow (1208°C \pm 15°C)\ \gamma + \delta + (\sigma + \gamma_2) \rightarrow (1041°C \pm 15°C)\ \gamma + (\sigma + \gamma_2)$$

These results should be confirmed by further crystallographic analysis.

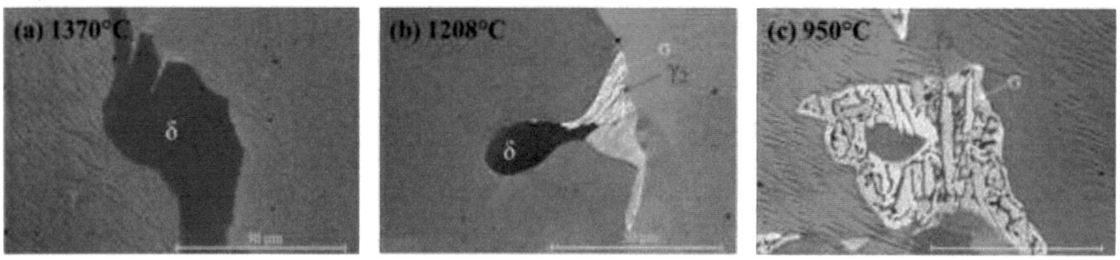

Figure 2 – BSE micrographs of three different representative zones of the QDS sample : (a) 1370°C ± 15°C, in the mushy zone composed of austenite dendrites and δ-ferrite, (b) 1208°C ± 15°C, partial eutectoid transformation of the δ-ferrite into $\sigma + \gamma_2$, (c) 950°C ± 15°C, δ-ferrite is fully transformed into $\sigma + \gamma_2$.

3.2. Solidification path: comparison between experimental and numerical results
Figure 3 compares the experimental and the numerical solidification path. It was chosen to report the difference between liquidus temperature and considered temperature, due to experimental error made on the temperature measurements in order to compare with ThermoCalc© calculations. The numerical path was calculated considering the Gulliver-Scheil model implemented in ThermoCalc© software. No diffusion for the substitutional elements and a complete diffusion for the faster elements that are carbon and nitrogen is assumed for the calculations. The numerical solidification interval (164°C) is larger than the experimental (85°C), this difference can be explained by the hypothesis considered in the Gulliver-Scheil model. There is no back-diffusion and infinite diffusion in the liquid, whereas in real conditions, the elements in the solid diffuse partially during the solidification process and the diffusion in the liquid is finite. An equilibrium calculation considering the lever rule showed that the solidus temperature is 1276°C with a fully austenite microstructure. A diffusion calculation for the Cr, Ni and Mo showed that diffusion length are low in austenite, which indicates that the QDS sample is not too far away from Scheil hypotheses. As a part of future work, the solidification should be described with a diffusion model to improve the comprehension of these results.

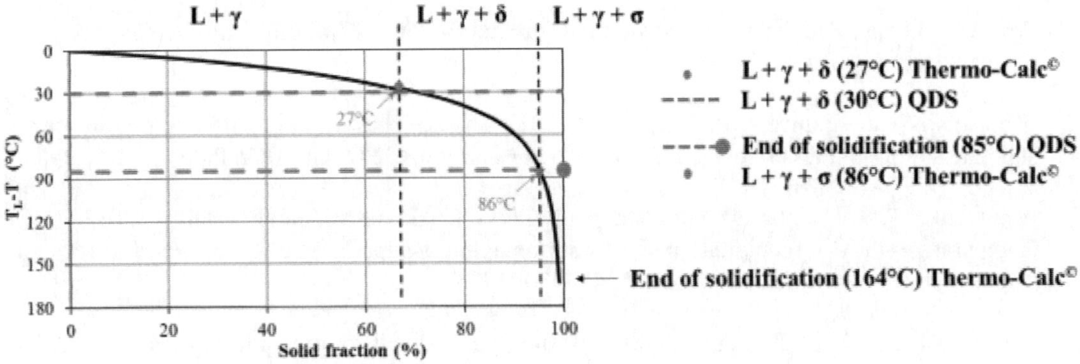

Figure 3 – Experimental and numerical solidification path

In both experimental results and numerical calculations, the solidification starts with austenite dendrites. 30°C below the liquidus temperature, δ-ferrite precipitates in the QDS sample. This value is close to that predicted with ThermoCalc© calculations (27°C). The solidification ends at 85°C below the liquidus temperature for the QDS sample. By calculation, the solidification progresses with the σ-phase formation at 86°C below the liquidus temperature and ends 164°C below the liquidus temperature. As a reminder, no σ-phase generated during the solidification process was observed in the QDS sample. This can be explained by the fact that the solidification range is smaller (85°C) in the QDS sample.

4. Conclusion

The solidification path and the solid σ-phase formed during the solidification process of the S31254 alloy were investigated using quenching direction solidification and SEM observations for a cooling rate of 0.5°C/s (1800°C/h). The experimental solidification path was reproduced with ThermoCalc© calculations with a good agreement. The solidification process in S31254 steel starts with austenite dendrites and progresses with δ-ferrite interdendritic formation due to Mo segregation in the liquid during solidification. The eutectoid decomposition $\delta \rightarrow \sigma + \gamma_2$ was showed by observations, EDX analysis and literature references. The transformation starts at 1208°C ± 15°C and ends at 1041°C ± 15°C when the δ-ferrite is completely transformed into $\sigma + \gamma_2$. As a part of future work, a QDS sample with a different cooling rate should be carried out on whether, the σ-phase nucleates during the solidification stage. This work should be also followed by crystallographic analysis in order to verify the assumptions made on the nature of the phases.

5. Acknowledgments

The authors would like to thank Industeel and ANRT, which support this work under a CIFRE Ph.D. fellowship (Grant Number 2016-0780).

References

[1] Lai J K L, Shek C H and Lo K H 2012 *Stainless steels: An introduction and their recent developments* (Bentham Science Publishers)

[2] Stein G, Hucklenbroich I and Feichtinger H 1999 Current and future applications of high nitrogen steels *Materials science forum* vol 318 (Trans Tech Publ) pp 151–160

[3] Olsson J and Minnich K 1999 Solid stainless steel for MSF once-through plants *Desalination* **124** 85–91

[4] Koutsoukis T, Redjaïmia A and Fourlaris G 2013 Phase transformations and mechanical properties in heat treated superaustenitic stainless steels *Mater. Sci. Eng. A* **561** 477–485

[5] Machado I F and Padilha A F 2000 Aging behaviour of 25Cr–17Mn high nitrogen duplex stainless steel *ISIJ Int.* **40** 719–724

[6] Koutsoukis T, Redjaïmia A and Fourlaris G 2011 Characterization of precipitation sequences in superaustenitic stainless steels *Solid State Phenomena* vol 172 (Trans Tech Publ) pp 493–498

[7] Koutsoukis T, Konstantinidis K, Papadopoulou E G, Kokkonidis P and Fourlaris G 2011 Comparative study of precipitation effects during aging in superaustenitic and superferritic stainless steels *Mater. Sci. Technol.* **27** 943–950

[8] Lee T-H, Kim S-J and Jung Y-C 2000 Crystallographic details of precipitates in Fe-22Cr-21Ni-6Mo-(N) superaustenitic stainless steels aged at 900 C *Metall. Mater. Trans. A* **31** 1713–1723

[9] Hsieh C-C and Wu W 2012 Overview of Intermetallic Sigma (σ) Phase Precipitation in Stainless Steels *ISRN Metall.* **2012**

[10] Pohl M, Storz O and Glogowski T 2007 Effect of intermetallic precipitations on the properties of duplex stainless steel *Mater. Charact.* **58** 65–71

[11] Tseng C C, Shen Y, Thompson S W, Mataya M C and Krauss G 1994 Fracture and the formation of sigma phase, M23C6, and austenite from delta-ferrite in an AlSl 304L stainless steel *Metall. Mater. Trans. A* **25** 1147–1158

[12] Llorca-Isern N, López-Luque H, López-Jiménez I and Biezma M V 2016 Identification of sigma and chi phases in duplex stainless steels *Mater. Charact.* **112** 20–29

[13] Chen T H and Yang J R 2001 Effects of solution treatment and continuous cooling on σ-phase precipitation in a 2205 duplex stainless steel *Mater. Sci. Eng. A* **311** 28–41

[14] Perricone M J and DuPont J N 2006 Effect of composition on the solidification behavior of several Ni-Cr-Mo and Fe-Ni-Cr-Mo alloys *Metall. Mater. Trans. A* **37** 1267–1280

[15] Hsieh C-C, Lin D-Y and Chang T-C 2008 Microstructural evolution during the δ/σ/γ phase transformation of the SUS 309LSi stainless steel after aging under various nitrogen atmospheric ratios *Mater. Sci. Eng. A* **475** 128–135

[16] Banovic S W, DuPont J N and Marder A R 2002 Dilution and microsegregation in dissimilar metal welds between super austenitic stainless steel and nickel base alloys *Sci. Technol. Weld. Join.* **7** 374–383

[17] Lacaze J and Lesoult G 1995 Modelling the development of microsegregation during solidification of an Al-Cu-Mg-Si alloy *ISIJ Int.* **35** 658–664

[18] Anderson T D, Dupont J N, Perricone M J and Marder A R 2007 Phase transformations and microstructural evolution of Mo-bearing stainless steels *Metall. Mater. Trans. A* **38** 86–99

[19] Kim Y-J, Chumbley L S and Gleeson B 2004 Determination of isothermal transformation diagrams for sigma-phase formation in cast duplex stainless steels CD3MN and CD3MWCuN *Metall. Mater. Trans. A* **35** 3377–3386

IOP Conf. Series: Materials Science and Engineering **529** (2019) 012009 doi:10.1088/1757-899X/529/1/012009

Prediction of Carbide Precipitation Using A Combined Partial Equilibrium- Para-Equilibrium - Lever Rule Approximation in Automobile Gear Materials

Hongwei Zhang[1,2], Keiji Nakajima[3], Mengmeng Su[1,2], Shaoshao Gao[1,2], Hong Lei[1,2], Qiang Wang[1,2], Jicheng He[1,2]

[1]Northeastern University, Key Laboratory of Electromagnetic Processing of Materials, Shenyang, 110004, P.R. China
[2] Northeastern University, School of Metallurgy, Shenyang, 110004, P.R. China
[3] KTH Royal Institute of Technology, Department of Materials Science and Engineering, SE-100 44 Stockholm, Sweden
E-mail: hongweizhang@epm.neu.edu.cn

Abstract. The carbide precipitation nature of an automobile gear material, i.e. Fe-C-Mn-Si-Cr-Mo alloy, during carburization and element addition process are investigated through PE+PA+LR prediction. Results show that, carburization process greatly increases the amount of cementite as well as the hardness index at the surface part. Besides, the addition of Ti and V helps to formation of TiC and (V,Mo)C carbides but suppress the precipitation of cementite, which contributes equivalent hardness at surface and lighter weight to the alloy.

1. Introduction

Automobile gear materials are well required for the wear-resistance around the surface and the toughness in the center. The basis composition of the gear material is 0.19wt%C, 1.20wt%Mn, 0.26wt%Si, 1.20wt%Cr, 0.021wt% Mo, 0.010wt%P, 0.020wt%S, 0.13wt%Ni, 0.021wt%Al, 12ppmO, 9ppmN. By using the gas carburization technique[1], it is possible to obtain a high carbon composition at the surface meanwhile maintain the original carbon composition over the bulk, which satisfies the dual industrial requirement for both surface and bulk of the alloy.

Recent years the demands in increasing of the gear hardness and at the same time reducing the weight of the automobiles have made it sound the addition of the substitutive elements into the traditional materials. Some VB group elements such as V and Nb etc, and IVB group elements like Ti and Zr etc are selected as the addition elements, for they promote the formation of MC type carbides with relatively lighter densities and much higher hardness[2,3] than usual M_3C carbide.

In order to clarify the effects of carbide composition and amount on the hardness of the automobile gear materials with the gas carburization and the element addition techniques, the solidification path of the automobile gear materials was studied through partial equilibrium (PE)- para-equilibrium (PA)-lever rule (LR) approximation. The PE-PA-LR approximation[4,5] is preferred to predict the solidification path from liquid down to the room temperature with considering the back diffusion of interstitial elements such as C, O etc in solids and also retrieving peritectic as well as eutectoid transformations.

The paper firstly investigated the precipitation nature of the carbides after carburization process and V, Ti element addition at different positions of the automobile gear materials by PE-PA-LR approximation, and then analyzed the effects of carburization and element addition on the hardness of the automobile gear materials.

2. Carbon concentration distribution in a carburized sample by Dictra prediction

The carbon concentration as a function of the distance from the surface of a sample is obtained by solving Fick's second law with the carbon flux prescribed at the carburized surface, $J_C = f_C(a_C^{gas} - a_C^{surface})$, where a_C^{gas} is the carbon activity at the steel in equilibrium with the gas. It is set as 0.64 corresponding to a carburizing carbon potential of 0.85 wt% according to the carburization condition [1]. $a_C^{surface}$ is the actual carbon activity in steel close to the surface. f_C is mass - transfer coefficient for carbon in steel, chosen as 8.25×10^{-9} mol/s[6]. The diffusion coefficient and thermodynamic data are retrieved from Thermo-Calc/MOB2 mobility database[7] and Thermo-Calc/TCFE6 steel database[8]. Then the distribution of carbon inside the sample can be predicted by Dictra [6].

Figure 1 shows the carbon concentration distribution for JIS SCr420 alloy (chemical composition: 0.21wt%C, 0.79wt%Mn, 0.25wt%Si, 1.21wt%Cr, 0.029wt%Al, 0.015wt%P, 0.015wt%S, 0.012wt%N) after carburization by experimental measurement[1] and by Dictra prediction. Reasonable agreement verifies the accuracy of Dictra prediction in carburization process.

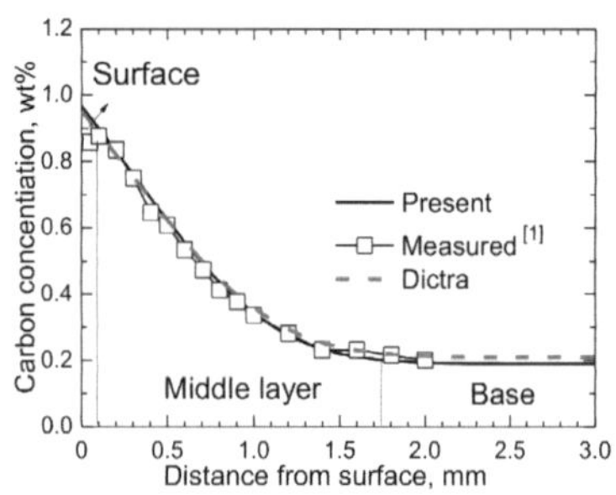

Figure 1 Predicted carbon concentration distribution after carburization for JIS SCr420 alloy and for the present Fe-C-Mn-Si-Cr-Mo alloy (0.19wt%C alloy) by Dictra. The measured data are from reference [1].

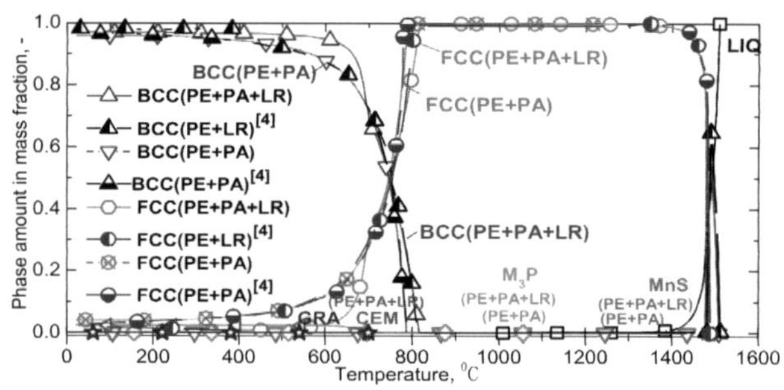

Figure 2 Predicted total range solidification path by several models, comparison with literature[4] is also shown.

Similar to JIS SCr420 alloy, a gear material Fe-C-Mn-Si-Cr-Mo alloy with 0.19wt%C base composition (0.19wt%C, 1.20wt%Mn, 0.26wt%Si, 1.20wt%Cr, 0.21wt% Mo) was adopted in the present carburization study. Through Dictra, its carbon composition distribution as a function of the distance from the surface is shown in figure 1, the carburization atmosphere, maintaining temperature and time are set same as those in reference[1].

Based on the carbon composition distribution, the sample is divided into three regime- surface (0~0.1mm distance from surface, with the averaged composition 0.85wt%C), middle part (0.1~1.7mm, with the averaged composition 0.5wt%C) and base part (>1.7mm, with the original base composition 0.19wt%C). Acting as a gear material in automobiles, it is expected a great deal of carbides existing at the surface which benefits to the hardness required there meanwhile almost no carbide precipitating at the base part to keep the good toughness. Through carburization, the gradient distribution of the carbon composition is achieved.

3. Prediction of solidification path to room temperature by PE+PA+LR approximation

There are several thermodynamic approximations for predicting the solidification path in a multi-component alloy system. The lever rule, LR, and Gulliver-Scheil approximation, GS, are the two extreme cases. The actual solidification path goes in between these two cases. The partial equilibrium approximation, PE[9], expands the GS model to distinguish the diffusivities of interstitual solutes as infinite and substitutionals as negligible respectively, but the prediction still terminate approaching to the solidus point. The para-equilibrium approximation, PA [4,10], which makes it possible to retrieve the solid-solid transformation, predicts not only the mushy zone peritectic transformation but also the eutectoid transformation at solid state. These approximations show potentials in predicting the solidification path at total solidification range, i.e. from liquid down to the final room temperature. For steels, the PE+PA approximation is a better choice to consider the back diffusion of interstitials in solids as well as the solid-solid paraequilibrium phase transformation.

3.1. PE+PA+LR approximation and its verification

Considering the case that only carbon is regarded as an interstitial element.

The partial equilibrium (PE) of carbon among all the phases which includes liquid and solid phases S_j, j=1, 2,…,m should satisfy the following conditions: the equal chemical potential of carbon among all the phases and the mass conservation of carbon during PE procedure, considering the partial equilibrium of an interstitial element; the unchanged composition in u-fraction of substitutional elements and the mass conservation of the substitutional element in each phase during PE procedure regarding the negligible diffusion of the substitutionals.

Para-equilibrium (PA) approximation focuses on the local equilibrium merely between the solids. The generation of a new solid as well as its variation in amount and composition from the mother solid is constrained by the interface paraequilibrium, i.e. no driving force, condition. It requires the interstitials, e.g. carbon, to follow the equal chemical potential and the mass conservation among mother and generating phases. The substitutionals behave as a unity during the paraequilibrium process due to their negligible diffusion, which obey the following constraints: the unchanged composition in u-fraction for mother phase during PA procedure, the equal composition in u-fraction for mother and generating phases; as well as the equal combined chemical potential (i.e. the summation of the products of the composition in u-fraction and the chemical potential for substitutionals) for two phases, and the mass conservation throughout the two phases during PA.

The procedure is programmed at Thermo-Calc/TQ programming interface[11] through coupling the thermodynamic equilibrium calculations. The temperature step adopted is -1K. The calculation finishes at room temperature.

In order to verify the calculation code, firstly the predicted solidification path by PE+PA and PE+PA+LR based on the composition of reference[4] (0.21wt%C, 1.64wt%Mn, 0.24wt%Si, 0.09wt%Al, 0.032wt%P and 0.007wt%S) was given in figure 2. In what we called 'PE+PA' model, both the peritectic and eutectoid transformation are predicted by PE+PA model, as it did in literature [4]. Meanwhile, in 'PE+PA+LR' model, the peritectic transformation is predicted by PE+PA model while the eutectoid transformation is predicted by a local equilibrium calculation between the matrix and the generation phase. This allows the austenite to transform to ferrite (or cementite) in a local equilibrium way. Comparatively, literature [4] predicted the peritectic and eutectoid transformation through PE+PA or PE+LR approximation, and literature [5] only made the PE+LR prediction around the peritectic transformation. As seen from figure 2, around the peritectic transformation regime, the predicted solidification path by PE+PA fits well with the results in literature[4]. Around 800^{0}C, both of them can retrieve the eutectoid transformation, although their starting points of transformation are a little bit different. As shown, the present prediction fits with the corresponding curves in reference[4], and the predicted curves by PE+PA+LR model fit with the ones by PE+PA model in solid-state transformation. Thus the PE+PA+LR model is adopted in the following prediction since it is prompt to deal with the solid-state phase transformations.

3.2. Solidification path for automobile gear materials after carburization

After carburization of a Fe-C-Mn-Si-Cr-Mo alloy sample, the carbide precipitation nature is checked according to the carbon concentration variation at surface, middle and base part through the prediction. As seen in figure 3, the existing carbides are cementite and M_7C_3. At 0.19wt%C content (figure 3(a)), the BCC-FCC phase transformation takes place and lasts from 1483^{0}C to 1475^{0}C. With the carbon content exceeding 0.5wt%, only FCC phase exists at this period (figure 3(b)). These three compositions all experience the FCC to BCC and FCC to cementite transformations at solid state. As seen in the figure, the carbide precipitation temperature rises up with the increase of the carbon content. At 0.19wt%C content, both M_7C_3 and cementite precipitate below the solidus (1164^{0}C), that is, 750^{0}C for M_7C_3 and 722^{0}C for cementite, respectively, companying with the FCC to BCC transformation taking place at 809^{0}C. With 0.85wt%C content, M_7C_3 forms at 1112^{0}C which is above the solidus temperature

1024^{0}C, and cementite precipitates at 854^{0}C which is also ahead of the FCC to BCC transformation point 738^{0}C. The uprise of the precipitation temperature certainly increases the amount of the corresponding carbides. As shown in figure 4, with 0.19wt%C content, both the M_7C_3 and cementite are little in amount; their total mass fraction is around 0.022(vf=0.0215). Then with the increase of carbon content, the amount of cementite rises up greatly and dominates the total amount of carbides, which are in mass fraction 0.064 (cementite, vf=0.067) and 0.071 (total, vf=0.074) at the middle part with 0.5wt%C content, and approaches in mass fraction 0.124 (cementite, vf=0.1306) and 0.125(total,vf=0.131) near the surface with 0.85wt%C content; meanwhile, the M_7C_3 almost vanishes.

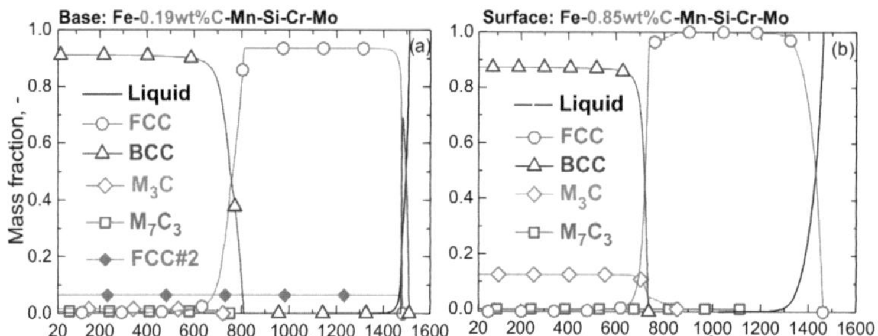

Figure 3 Solidification path predicted after carburization at (a) base, (b) surface part of the basis gear material. FCC#2 in (a) has the similar constitution as FCC phase.

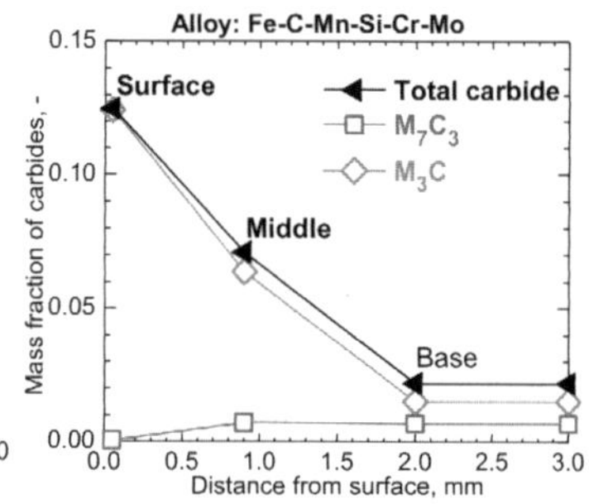

Figure 4 Predicted carbide amounts in basis gear material after carburization.

3.3. Effect of C on precipitating carbide composition and hardness

Through solidification path prediction, the carbide composition variation with the increase of the carbon content is determined. In the base sample, the constitution of cementite is $(Fe,Mn,Cr)_3C$, in which Mn dominantly occupies 12wt% and Cr is 9.8wt%, whose Vickers hardness is around 1360-1570[3]; while in middle part, it changes to $(Fe,Cr,Mn)_3C$ which contains 9.4wt%Cr and 6.7wt%Mn thus Vickers hardness increases to 1520-1890[2,3]. Up to the surface, the constitution alters to $(Fe,Cr)_3C$ where Cr occupies 8.1wt%, Mn decreases to 4.6wt%, instead, the Mo content increases up to 0.7wt%. In a word, with the increase of carbon content the constitution of cementite changes from $(Fe,Mn)_3C$ to $(Fe,Cr)_3C$, which increases its hardness.

Similarly, the M_7C_3 constitution in base part is $(Fe,Cr)_7C_3$ which contains 40wt%Cr and corresponding Vickers hardness is 2080-2570[3]; meanwhile, in the middle, M_7C_3 alters to $(Fe,Cr,Mo)_7C_3$ with 33.6wt%Cr and 7.4wt%Mo, its Vickers hardness changes to 2040 ~2570[3]; and at surface, its constitution changes to $(Fe,Mo,Mn,Cr)_7C_3$ where Mo content is up to 30.7wt%, Mn 20.4wt% and Cr down to 7.4wt%, Vickers hardness keeps 2040 ~2570[3]. Totally, the increase of carbon content changes the M_7C_3 from $(Fe,Cr)_7C_3$ to $(Fe,Mo,Mn,Cr)_7C_3$, which leads variations to its hardness.

4. Addition of V and Ti content in automobile gear materials

4.1. Effect of V on Carbide Precipitation

V helps to the formation of (V,Mo)C carbide (i.e. MC_eta). It precipitates approaching to the solidus temperature (1022^{0}C) at 0.1wt%V addition, as shown in figure 5, although keeps precipitating in solid state, its amount is not too much.

With the addition of V content increasing from 0.1wt% to 1.0wt%, the total amount of carbide decreases at all part of the sample, even in the base part. Figure 6(a) shows the effects at surface of the sample. As shown, with 0.1wt% to 1.0wt%V addition, the total carbide amount decreases from 0.131 to 0.104 in volume fraction, decreased by 0.027, which is due to the large decrease of cementite, although the M_7C_3 and (V,Mo)C do increase at the same time, their amounts are not enough to compensate the decrease.

On the other hand, the composition of the carbide is strengthened with the V addition. Comparing 0.1wt%V and further 1.0wt%V addition with the basis sample at surface part, cementite constitution evolves from $(Fe,Cr)_3C$ to $(Fe,Cr,Mn)_3C$. M_7C_3 constitution alters from $(Fe,Mo,Mn,Cr)_7C_3$ to

$(Fe,Mn,Cr,Mo)_7C_3$ and further to $(Fe,Cr,V)_7C_3$, in which Cr content increases from 7.4wt% in basis to 16.7wt% (+0.1wt%V) and further to 31.9wt% (+1.0wt%V), V content is increased up to 8.8 wt%. In (V,Mo)C, the V, Mo content occupies from 41.3wt%V and 43.7wt%Mo at 0.1wt%V addition up to 62.0wt%V and 20.8wt% Mo with 1.0wt%V addition.

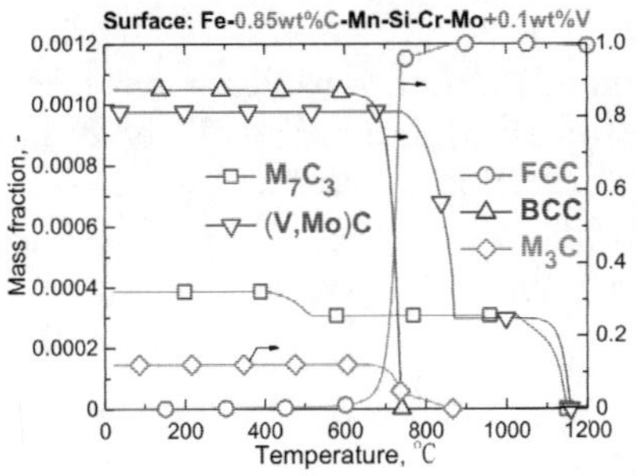

Figure 5 Predicted solidification path with 0.1%V addition at surface after carburization.

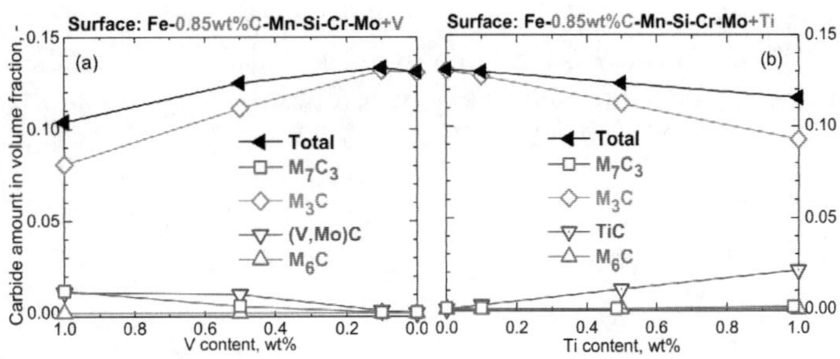

Figure 6 Predicted carbide amount at surface of the sample with the addition of (a)V and (b)Ti.

4.2. *Effect of Ti on Carbide Precipitation*

Ti addition is intended for the precipitation of TiC carbide. As seen from figure 7, TiC forms in the early stage of the solidification and precipitates both in liquid and solid state.

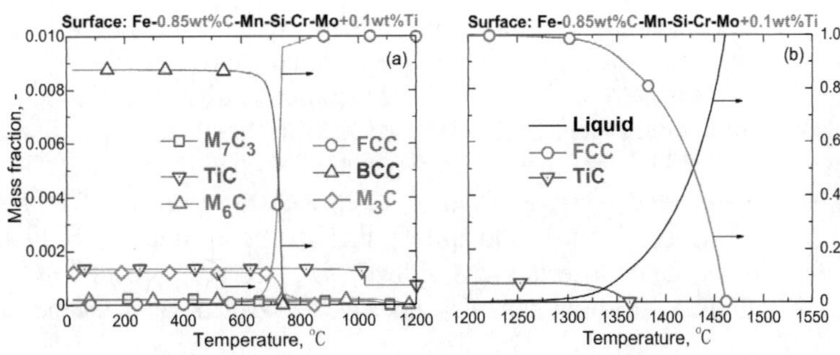

Figure 7 Predicted solidification path with 0.1%Ti addition at surface part after carburization.

From figure 6(b), Ti addition shows the similar changing tendency on carbide amount with V addition. However, TiC precipitation amount is larger than (V, Mo)C, thus at surface, the total carbide amount is decreased from 0.131 to 0.115 in volume fraction, decreased by 0.016, which is less than that in V addition. As cementite decreases by 0.038 in volume fraction, TiC forms up to 0.021 in volume fraction by 1.0wt%Ti addition; meanwhile, there is not much M_7C_3 left.

Ti doesn't dissolve in cementite and M_7C_3, but it does change their constitution. Comparing the composition at surface, with 0.1wt%Ti and further 1.0wt%Ti addition, cementite constitution evolves from $(Fe,Cr)_3C$ to $(Fe,Cr,Mn)_3C$. M_7C_3 constitution changes from $(Fe,Mo,Mn,Cr)_7C_3$ to $(Fe,Mn,Mo,Cr)_7C_3$ (+0.1wt%Ti) and further $(Fe,Cr,Mn,Mo)_7C_3$ (+1.0wt%Ti), in which Cr content changes from 7.4wt% in basis (no Ti) increasing to 8.0wt% (+0.1wt%Ti) and up to 29.1wt% (+1.0wt%Ti). Ti is the dominant element in TiC, where the Ti content occupies from 73.5wt% at 0.1wt%Ti addition up to 79.0wt% with 1.0wt%Ti addition. Since no more data of the variation of Vickers hardness with the composition is available, regardless of the constitution change, the Vickers hardness 1520-1890[3], 2040~2570[3] and 3200[2] are just adopted for cementite, M_7C_3 and TiC, respectively.

5. Hardness of automobile gear materials with C, V and Ti content variation

Considering the contributions of both the constitution and amount of the carbides to the alloy hardness, the hardness index of the alloy is plotted in figure 8. It is the summation of each carbide's hardness weighted by its volume fraction, as illustrated in figure 6.

Figure 8(a) shows the hardness index at different part of the sample as a function of carbon content, the measured actual hardness in reference [1] is also plotted in the figure. As seen, both the

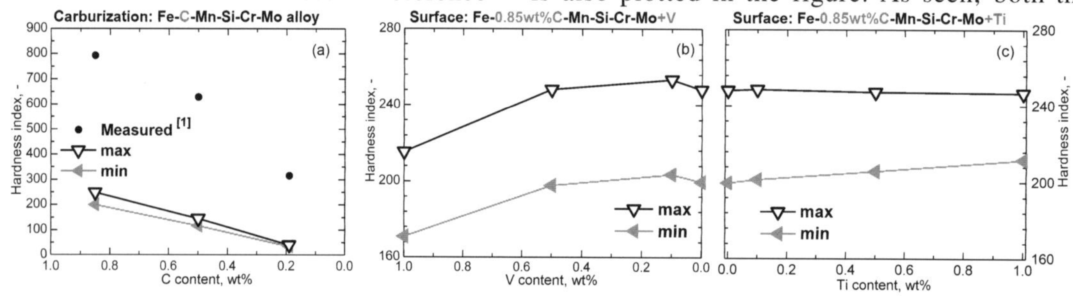

Figure 8 Effects of (a) C, (b) V and (c) Ti content on the surface hardness index of an automobile gear material.

measured data and the predicted hardness index have the same increasing tendency with the increase of the carbon content. However, the actual data increase in a larger slope. With carbon content ranging from 0.19wt% to 0.85wt%, the hardness index increases from 34~41 up to 200~250, increased by 166~209, while the actual hardness increases from 315 to 793, increased by 478. This hardness increase is mainly due to the great increase of the amount of the cementite, as shown in figure 4.

Figure 8(b) and 8(c) show the hardness index at surface with the change of V and Ti addition. As known, (V,Mo)C and TiC are high hardness carbide, whose Vickers hardness is 2100~2800[3] and 3200[2] respectively. Meanwhile, the density of (V,Mo)C, 5200~6200 kg/m^3, and TiC, 4600~4800 kg/m^3, is much less than that of cementite (7100~7500 kg/m^3), thus they contribute both the higher hardness and the lighter weight to the gear materials. As shown from figure, with the addition of 0.1wt% ~1.0wt%Ti or 0.1wt%~0.5wt%V, the total hardness index at the surface equalizes the original hardness index there.

6. Conclusion

The carbide precipitation nature of an automobile gear material, i.e. Fe-C-Mn-Si-Cr-Mo alloy, during carburization and element addition process is investigated through PE+PA+LR prediction. Results show that, carburization, with the carbon content increasing from 0.19wt% to 0.85wt%, greatly increases the amount of cementite at the surface and the hardness index correspondingly rise up by 166~209. Besides, the addition of Ti and V helps to the precipitation of TiC and (V, Mo)C carbides, which favour in higher hardness and lighter weight comparing to cementite. It is found through prediction that with 0.1wt%~0.5wt%V addition or 0.1wt%~1.0wt%Ti addition, the total hardness index at the surface equalizes the original one there; meanwhile, the addition of Ti and V contributes less weight to the automobile gear material through suppressing the formation of cementite.

References

[1] Koyama T, Kubota M and Yoshida S 2015 *Tetsu-to-Hagané* **101** 252-59
[2] Ogi K J 1994 *Jpn. Foundry Eng. Soc.*, **66** 764-71 (in Japanese)
[3] Sato T O, Nishizawa T, Ishiwara J 1959 *J. Japan Institute of Metals* **23** 403-07
[4] Koshikawa T, Gandin Ch -A, Bellet M, Yamamura H and Bobadilla M 2014 *ISIJ Inter.*, **54** 1274-82
[5] Chen Q, Engström A, Lu X G and Sundman B 2006 In: *Proc. of MCWASP* XI. ed Gandin Ch -A, Bellet M (Sophia Antipolis: The Minerals, Metals, and Materials Society) pp 529-36
[6] Borgenstam A, Engström A, Höglund L and Ågren J 2000 *J. Phase Equilib* **21** 269-80
[7] Helander T 1998 *MOB2-Mobility database* (Stockholm: Thermo-Calc software AB)

[8] Shi P 2008 *TCS steels/Fe-alloys database* V6.0 (Stockholm: Thermo-Calc software AB)
[9] Chen Q, Sundman B 2002 *Materials Transactions* **43** 551-59
[10] Hillert M 2008 *Phase Equilibria, Phase Diagrams and Phase Transformations*, 2nd (Cambridge: Cambridge University Press) pp 311-15
[11] Sundman B and Chen Q 2008 *Thermodynamic Calculation Interface TQ-Interface V7.0* (Stockholm: Thermo-Calc software AB)

Acknowledgements
The work is supported by National Natural Science Foundation of China [51574074] and [51425401], National Natural Science Foundation of China and Shanghai Baosteel [U1460108] and [U1560207], and Natural Science Foundation of Liaoning Province, China[L20150183]. The work is also supported by JERNKONTORET, Prytziska foden 2, Sweden.

IOP Conf. Series: Materials Science and Engineering **529** (2019) 012010 doi:10.1088/1757-899X/529/1/012010

Effects of interphase boundary anisotropy on the three-phase growth dynamics in the β(In) – In_2Bi – γ(Sn) ternary-eutectic system

S Mohagheghi[1], U Hecht[2], S Bottin-Rousseau[3], S Akamatsu[3], G Faivre[3], M Şerefoğlu[1]

[1] Koç University, Department of Mechanical Engineering, Rumeli Feneri Yolu, 34450, Sariyer, Istanbul, Turkey

[2] ACCESS e. V., Intzestrasse 5, D-52072 Aachen, Germany

[3] Sorbonne Université, CNRS UMR 7588, Institut des NanoSciences de Paris, Case Courrier 840, 4 Place Jussieu, 75252 Paris Cedex 05, France

mserefoglu@ku.edu.tr

Abstract. We present an experimental investigation on the effects of the interphase energy anisotropy on the formation of three-phase growth microstructures during directional solidification (DS) of the β(In)–In_2Bi–γ(Sn) ternary-eutectic system. Standard DS and rotating directional solidification (RDS) experiments were performed using thin alloy samples with real-time observation. We identified two main types of eutectic grains (EGs): (i) quasi-isotropic EGs within which the solidification dynamics do not exhibit any substantial anisotropy effect, and (ii) anisotropic EGs, within which RDS microstructures exhibit an alternation of locked and unlocked microstructures. EBSD analyses revealed (i) a strong tendency to an alignment of the In_2Bi and γ(Sn) crystals (both hexagonal) with respect to the thin-sample walls, and (ii) the existence of special crystal orientation relationships (ORs) between the three solid phases in both quasi-isotropic and anisotropic EGs. We initiate a discussion on the dominating locking effect of the In_2Bi–β(In) interphase boundary during quasi steady-state solidification, and the existence of strong crystal selection mechanisms during early nucleation and growth stages.

1. Introduction

In a directionally solidified eutectic material, the distinct crystal phases in the solid most often present several types of (special) orientation relationships (ORs). The free energy of the interphase boundaries can be substantially anisotropic, and that interfacial anisotropy, in particular, the number and the inclination of low-energy planes, depend on the OR. This has a great impact on the formation of both eutectic growth microstructures and eutectic grains (EGs) [1]. The interfacial-anisotropy effect is obviously absent from the reference, Jackson-Hunt theory of eutectic coupled-growth [2]. A major progress in this domain has been achieved thanks to the use of real-time observations in thin transparent systems [3], and, in particular, by performing real-time rotating directional solidification (RDS) experiments [4, 5]. During RDS, a thin alloy sample is slowly rotated with respect to the axis of the fixed unidirectional temperature gradient, in such a way that solidification occurs while the orientation of the crystals is continually varied. In a binary eutectic system, the trace of the interphase boundaries in the RDS microstructure was shown to be essentially homothetic to the equilibrium shape of the interphase boundary. As confirmed by numerical simulations [6], a weak interphase boundary anisotropy permits a fully "floating" dynamics of lamellar solidification front patterns, which allows uniformization processes that eventually lead to the formation of regular eutectic microstructures in the solid. In contrast, a strong anisotropy of the interphase boundaries can give rise to the so-called locked lamellar patterns. In the latter case, the lamellae follow a fixed direction, with the interphase boundaries aligning onto a plane that corresponds to a deep minimum in the Wulff-plot (*i.e.*, the interfacial energy-anisotropy function of the interphase boundary) [5]. Recently, an analysis, mostly by x-ray diffraction (XRD), of the crystal ORs in several eutectic grains in the In-In_2Bi alloy [7] has revealed, in brief, that (i) both quasi-isotropic and anisotropic eutectic grains exhibit special ORs, and (ii) lamellar locking mostly occurs onto atomically dense coincidence planes.

Inspiring observations and measurements have been recently made in three-phase systems, in particular the β(In)–In_2Bi–γ(Sn) alloy [8] under consideration here, and the well-known Al-Al_2Cu-Ag_2Al [9] alloy. However, a systematic study of the effects of the interfacial anisotropy and the ORs on the microstructure in such systems, which potentially present three kinds of interphase boundaries, still remains to be carried out.

In this work, we first show the microstructural evolution of the two main classes of eutectic grains, namely quasi-isotropic and anisotropic EGs, in the β(In)–In_2Bi–γ(Sn) system. In thin samples of that alloy, it is known that three-phase periodic solidification patterns exhibit an ABAC arrangement, where A = In_2Bi, B = β(In), and C = γ(Sn). In Figure 1, we show the evolution of the growth pattern of an anisotropic eutectic grain during an RDS experiment. It can be seen that the microstructure within the eutectic grain of interest is subject to substantial changes during the rotation. It successively exhibits unlocked parts with a floating behavior, and locked parts, separated by more disordered patterns. During the rotation, the position of the eutectic grain with respect to the rotation center remains globally constant. Therefore, the temperature gradient and the average growth velocity, V_{ave}, do not vary as well, and the morphological instabilities and changes of the growth pattern can be attributed to the sole variation of the orientation of the crystals with respect to the growth direction. In other words, on the light of the previous work of Refs. [4, 5], the RDS microstructure provides direct information on the Wulff-plot of the interphase boundary. More precisely, this kind of observation is similar, on a first sight, to previous observations in binary eutectic systems, but the situation for a three-phase system is more complex. It is first essential to determine the ORs between the three phases. For this purpose, we performed electron backscatter diffraction (EBSD) measurements in both quasi-isotropic and anisotropic grains. This led us to propose special ORs between the crystal phases, not only In_2Bi–β(In) and In_2Bi–γ(Sn) individually, but also with a coincidence plane common to all the three phases. Our measurements allowed us to determine that, in spite of the existence of clear ORs, the anisotropy of the In_2Bi–γ(Sn) interphase boundary is relatively weak, at least compared to that of the In_2Bi–β(In) interphase boundary which dominates the dynamics in anisotropic EGs.

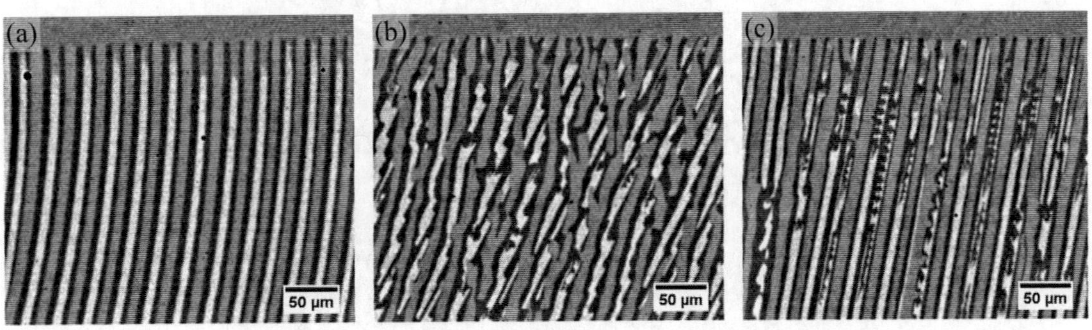

Figure 1. Evolution sequence of an anisotropic EG during an RDS experiment. (a) Unlocked, V_{ave} = 0.35 $\mu m/s$, (b) Transient, V_{ave} = 0.37 $\mu m/s$, and (c) Locked states, V_{ave} = 0.38 $\mu m/s$. G = 5.1 K/mm. RDS speed = 0.013°/s. Growth direction is counterclockwise.

IOP Conf. Series: Materials Science and Engineering **529** (2019) 012010 doi:10.1088/1757-899X/529/1/012010

2. Experimental procedure

A ternary eutectic alloy at nominal nonvariant composition, In–20.7Bi–19.1Sn (at.%) [10], was prepared from ultra-pure (99.999%) elements under Argon atmosphere. The thin-sample container, which was made out of two glass plates separated by a 13 μm-thick spacer, was filled with the molten alloy. The three phases in the eutectic solid are the body-centered tetragonal (space group number: 139) β(In), the hexagonal (space group number: 194) In_2Bi, and the hexagonal (space group number: 191) γ(Sn). Note that the indium-rich phase that is noted β here is the same phase as the one previously noted ε in Ref. [7], although the lattice parameters are different. Under optical microscope, these phases appear white, black, and gray, respectively, after numerical contrast enhancement. For gaining accuracy, we determined ourselves the crystal structures and lattice parameters of all the phases by powder x-ray diffraction. In our RDS experiments, the sample was continuously rotated (with the center of rotation being close to the solid-liquid interface) with respect to a fixed thermal gradient of 5.1 K/mm. Details of our RDS setup and experimental protocol can be found in Ref. [8]. For this study, both quasi-isotropic and anisotropic lamellar grains were processed by RDS, over several full turns, with real-time imaging. We recall that during RDS, solidification occurs one side of the sample, while the other side undergoes a directional melting. Additionally, there is a velocity ramp from the center to the edge of the sample at the solidification front. Due to this ramp, an average solidification velocity (V_{ave}) is reported in Figure 1. After a 180° rotation, the microstructure is fully renewed.

For ex-situ EBSD measurements, the glass plates were removed, without any apparent degradation of the metallic film which could then be directly exposed to the electron beam. Several high-quality crystal-orientation maps of the three phases were acquired. The area of each map was $\approx 35\times10^3$ μm^2 with a pixel size equal to 0.42 μm. The sample was aligned inside the scanning electron microscope (SEM) so that the axis of the temperature gradient during solidification was nearly parallel to the so-called rolling direction (RD) of the EBSD.

3. Results and discussions

In the RDS microstructure of Figure 2, two different regions can be distinguished according to their microstructures, namely, a central region with quasi-isotropic grain, and a distal region with anisotropic grain. The coexistence of the two types of growth dynamics in the same experiment again brings clear evidence of the influence of the interphase anisotropy.

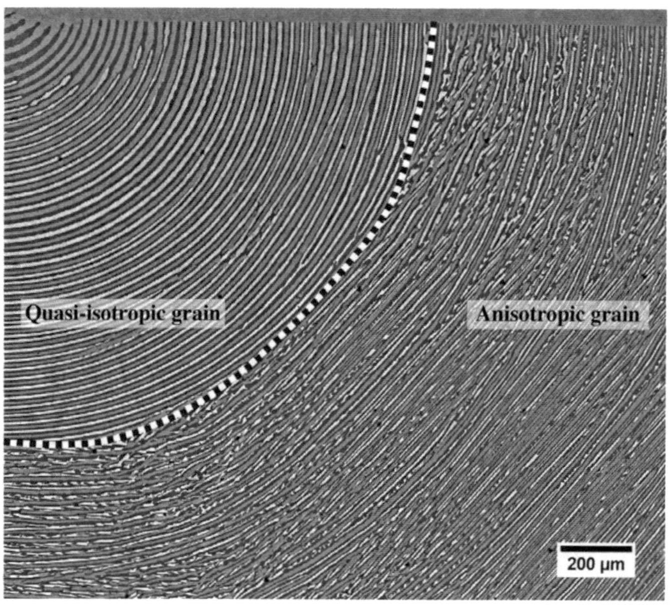

Figure 2. Quasi-isotropic and anisotropic EGs obtained in an RDS experiment. RDS speed = 0.013°/s. Growth direction is counterclockwise. Dashed line shows the grain boundary.

3.1. Identification of ORs in quasi-isotropic grains

We first focus on the floating region. In a quasi-isotropic grain, the lamellae grow essentially perpendicular to the average growth front. During RDS, the trajectories of the triple-junctions in the reference frame of the sample smoothly follow concentric circles, or nearly circular curves, within the possible effect of a small anisotropy and local deviations caused by spacing adjustment mechanisms due to the complex dynamics during RDS (Ex: spatial growth velocity ramp along the front, misalignment of the center of rotation, influence of grain boundary). In another sample, after a full rotation, the floating nature of the growth dynamics was further tested by standard DS without rotation (Figure 3). As expected, a regular ABAC lamellar structure formed in continuity with the RDS structure.

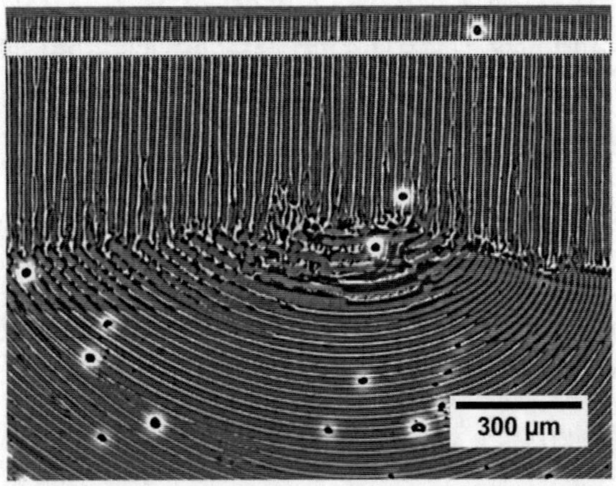

Figure 3. Evolution of the quasi-isotropic eutectic grain during RDS and DS experiments. The image on top shows the solid/liquid interface after 1500 μm growth with DS. RDS speed = 0.026°/s, DS speed = 0.5 μm/s.

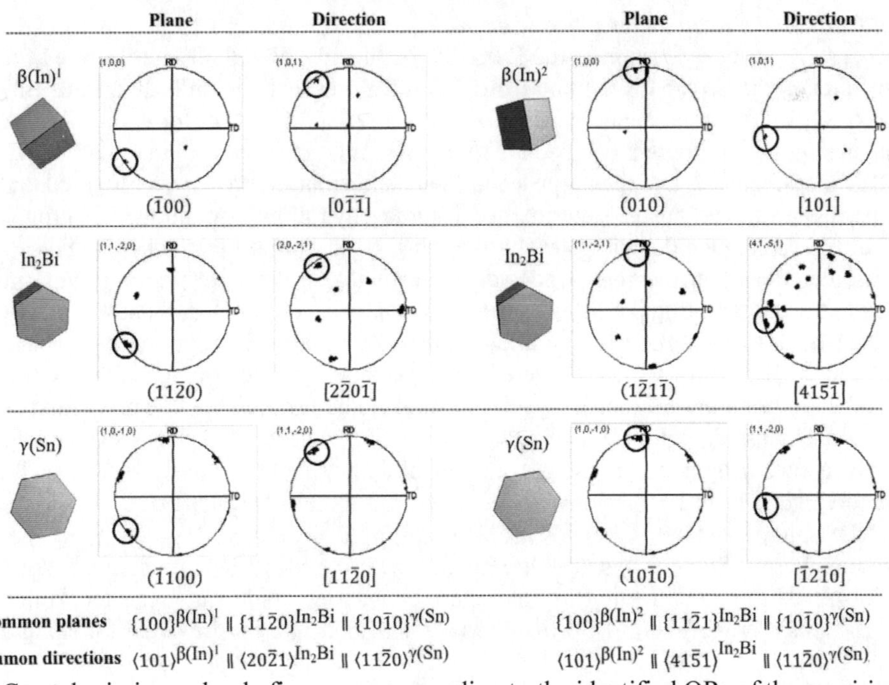

Common planes $\{100\}^{\beta(In)^1} \parallel \{11\bar{2}0\}^{In_2Bi} \parallel \{10\bar{1}0\}^{\gamma(Sn)}$ $\{100\}^{\beta(In)^2} \parallel \{11\bar{2}1\}^{In_2Bi} \parallel \{10\bar{1}0\}^{\gamma(Sn)}$

Common directions $\langle 101\rangle^{\beta(In)^1} \parallel \langle 20\bar{2}1\rangle^{In_2Bi} \parallel \langle 11\bar{2}0\rangle^{\gamma(Sn)}$ $\langle 101\rangle^{\beta(In)^2} \parallel \langle 41\bar{5}1\rangle^{In_2Bi} \parallel \langle 11\bar{2}0\rangle^{\gamma(Sn)}$

Figure 4. Crystal mimics and pole figures corresponding to the identified ORs of the quasi-isotropic pattern shown in Figure 3. EBSD maps are obtained from DS and RDS regions and both regions contain β(In)[1] and β(In)[2] phases. Common planes and directions are given below the pole figures.

We performed EBSD measurements in both the RDS and DS microstructures shown in Figure 3. We found that both In_2Bi and γ(Sn) phases keep one and the same orientation over the entire explored areas. In contrast, the β(In) phase presents two different orientations, named as $\beta(In)^1$ and $\beta(In)^2$, in both RDS and DS regions. In other words, the floating region is composed of two distinct types of eutectic grains, the difference between them being solely the orientation of the β(In) phase. The corresponding pole figures are shown in Figure 4 along with the crystal mimics of the three phases. In both $\beta(In)^1$–In_2Bi–γ(Sn) and $\beta(In)^2$–In_2Bi–γ(Sn) grains, the basal (0001) plane of the γ(Sn) crystals is parallel to the sample plane. The basal plane of the In_2Bi crystals is slightly inclined from the sample plane. From the pole figures extracted from the EBSD measurements, two special ORs between In_2Bi and γ(Sn) can be plausibly proposed, namely, $\{11\bar{2}0\}$ In_2Bi ∥ $\{10\bar{1}0\}$ γ(Sn), and $\{11\bar{2}1\}$ In_2Bi ∥ $\{10\bar{1}0\}$ γ(Sn), respectively, with angular departures between the involved In_2Bi and γ(Sn) lattice planes being less than about 2°. It is noticeable that those planes are perpendicular to the sample plane. We also identified ORs between In_2Bi and both $\beta(In)^1$ and $\beta(In)^2$. Both of them involve the {100} family planes of the β(In) crystals, namely: $(11\bar{2}0)$ In_2Bi ∥ $(\bar{1}00)$ $\beta(In)^1$, and $(1\bar{2}1\bar{1})$ In_2Bi ∥ (010) $\beta(In)^2$, respectively. Again, those planes are perpendicular to the sample plane.

A first conclusion from the above observations is that there is no clearly remarkable manifestation of the ORs in the quasi-isotropic EGs. In other words, in the quasi-isotropic grains, there is no measurable tendency of any preferred alignment of the interphase boundaries with the coincidence planes involved in the identified ORs, which indicates that these coincidence planes are associated to very shallow minima in the Wulff-plot. This phenomenon has also been previously found in the binary In-In_2Bi system [7]. In addition, our analysis shows that all the three phases have a common plane in coincidence as emphasized in Figure 4, which was not expected a priori. The generalization of this result has to be confirmed by more systematic observations in different samples. It would evidence the existence of particular, and strong crystal orientation selection mechanisms at play during the early nucleation and growth stages of eutectic solidification (including the alignment of the hexagonal crystals with respect to the sample walls).

3.2. Identification of ORs in an anisotropic grain

We now focus on the locked region of the RDS microstructure shown in Figure 2. We acquired EBSD maps in different regions with distinct microstructural features, namely unlocked, transient, and locked (Note that the EBSD measurements have been made after several RDS turns; at this final stage, the microstructure shown in Figure 2 had been melted, and replaced by new but similar ones). The results of the EBSD measurements and the proposed ORs are shown in Figure 5. We observed that the crystal orientations do not vary, within the experimental accuracy, by following continuously group of lamellae. In other words, even though the microstructures are highly dissimilar in unlocked, transient, and locked states, the same OR is preserved, as expected. Interestingly, in this eutectic grain, not only the γ(Sn) crystals (as in the quasi-isotropic EGs analyzed above) but also the In_2Bi crystals are oriented in such a way that their basal (0001) plane lies parallel to the sample plane. In addition, the prismatic planes of both γ(Sn) and In_2Bi crystals (which are therefore perpendicular to the sample plane) are strictly parallel with each other. In brief, the hexagonal crystals of the two phases have the same orientation. This defines a remarkable and unusual, high-symmetry OR between the two hexagonal phases even though the lattice parameters are substantially different. We also identified that {112} family planes of β(In) and the $\{10\bar{1}0\}$ family planes of In_2Bi, thus of γ(Sn) as well, are parallel with each other.

In the RDS microstructure of the anisotropic grain, we observed some facets along the In_2Bi-β(In) interphase boundaries. Some facets are well aligned with the identified coincidence planes. No clear facets are observed along the In_2Bi-γ(Sn) interphase boundaries. As it was suggested from the previous study [8], the anisotropy of the In_2Bi-β(In) interface is responsible for the overall locking dynamics of the eutectic grain.

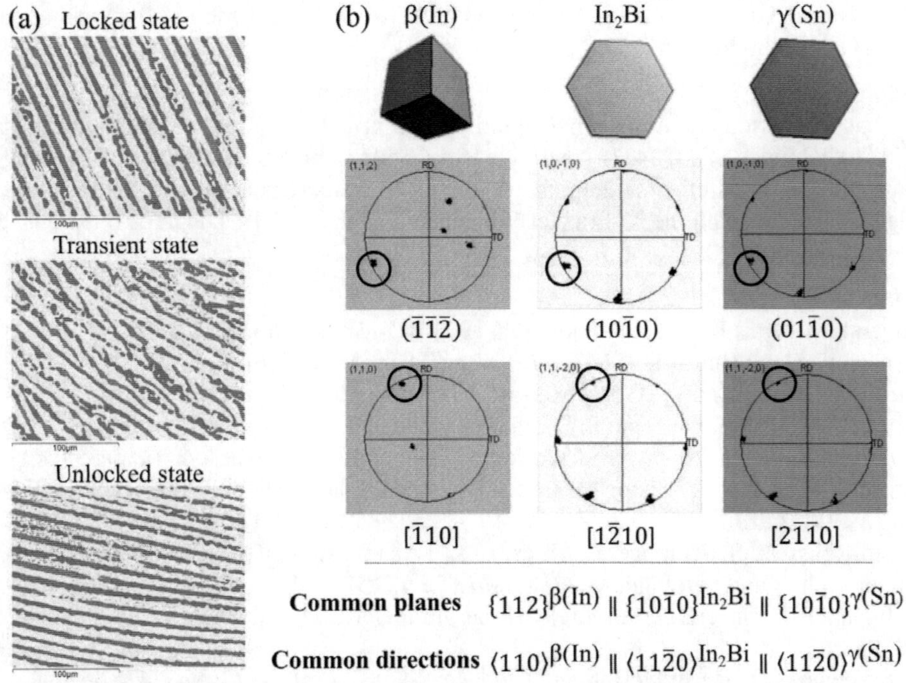

Figure 5. Crystal mimics and pole figures corresponding to the identified OR of unlocked, transient, locked states of the anisotropic grain. Common planes and directions are summarized as well.

Summary and conclusions

We used the RDS method for in-situ observation of the microstructure growth dynamics in thin samples of β(In)-In_2Bi-γ(Sn) three-phase system, and ex-situ EBSD analysis for determining the orientation of the crystals growing in a coupled manner. We have selected samples in which quasi-isotropic and anisotropic grains were clearly distinguishable, on a microstructural basis. We could identify special ORs all of which involve dense common planes in both cases between the three phases. In the quasi-isotropic grains, special ORs could be identified, but the interfacial anisotropies of both In_2Bi-β(In) and In_2Bi-γ(Sn) interphase boundaries remain weak. Due to this weakness, the floating growth dynamics are insensitive to the ORs. In the anisotropic grain, careful EBSD maps taken from three different states, namely, unlocked, transient, and locked, reveal that the crystal orientations stay constant during the evolution of the grain in the RDS experiment. More specifically, despite significant microstructural dissimilarities in these states, ORs are identical. In this grain, since the interphase boundaries of the In_2Bi-β(In) in the locked state are well-aligned with the identified common plane, this interphase is suggested to dominate the dynamics, and a locking effect is observed to occur in the vicinity of the coincidence family planes, namely, $\{112\}$β(In) ∥ $\{10\bar{1}0\}In_2Bi$. During solidification, the trajectory of the In_2Bi-γ(Sn) interphase boundaries is more or less slaved to the evolution dynamics imposed by the anisotropy of the In_2Bi-β(In) interphase boundaries. We can assume that the interfaces remain perpendicular to the sample walls as it is usually the case in a thin-sample geometry and in steady-state conditions. The anisotropy thus merely corresponds to a section of the Wulff-plot of the considered interface with the sample plane. For the In_2Bi and γ(Sn) crystals, which have basal planes nearly aligned with the sample plane, a weak anisotropy is thus apparently associated with the hexagonal symmetry of those lattice planes. The alignment of the hexagonal In_2Bi and γ(Sn) crystals in such a way that the basal planes are parallel to the sample plane is a remarkable feature. The selection mechanism at play is still unclear. It must obviously be determined during the early stages of the three-phase solid formation, including heterogeneous nucleation and competitive growth after the sample filling process. Further

EBSD and XRD analyses are needed to a more comprehensive identification of the locking planes and the associated ORs in this system.

Acknowledgments
This work was supported by European Commission Marie Curie Career Integration Grant FP7-PEOPLE-2012-CIG (NEUSOL 334216) and TÜBİTAK 3501 (Grant no: 212M013). The authors would like to thank Nicolas Casaretto for helping us for XRD measurements. Some of us would like to acknowledge funding through the M-Era.Net Project ANPHASES (UH: DFG HE 6938/2-1; SBR and SA: ANR-14-MERA-0004).

References

[1] Hogan L M, Kraft R W and Lemkey F D 1971 *Adv. Mater. Res.* **5** 83-126
[2] Jackson K A and Hunt J D 1966 *Trans. Metall. Soc. AIME* **236** 1129-42
[3] Caroli B, Caroli C, Faivre G and Mergy J 1992 *J. Cryst. Growth* **118** 135-50
[4] Akamatsu S, Bottin-Rousseau S, Serefoglu M and Faivre G 2012 *Acta Mater.* **60** 3199-205
[5] Akamatsu S, Bottin-Rousseau S, Şerefoğlu M and Faivre G 2012 *Acta Mater.* **60** 3206-14
[6] Ghosh S, Choudhury A, Plapp M, Bottin-Rousseau S, Faivre G and Akamatsu S 2015 *Phys. Rev. E* **91** 022407
[7] Bottin-Rousseau S, Senninger O, Faivre G and Akamatsu S 2018 *Acta Mater.* **150** 16-24
[8] Mohagheghi S and Şerefoğlu M 2018 *Acta Mater.* **151** 1-11
[9] Steinmetz P, Dennstedt A, Şerefoğlu M, Sargin I, Genau A and Hecht U 2018 *Acta Mater.* **157** 96-105
[10] Witusiewicz V T, Hecht U, Bottger B and Rex S 2007 *J. Alloys Compd.* **428** 115-24

The formation of Al_6 (Fe, Mn) phase in die-cast Al-Mg alloys

Xiangzhen Zhu and Shouxun Ji *

Brunel Centre for Advanced Solidification Technology (BCAST), Brunel University London, Uxbridge, Middlesex UB8 3PH, United Kingdom
* Corresponding author: Tel.: +44 1895 266663, Fax: +44 1895 269758, Email: shouxun.ji@brunel.ac.uk

Abstract. In aluminium alloys, iron is a common impurity as it is unavoidably picked up in practice. The excessive Fe is strongly prone to form various intermetallic phases. These Fe-rich intermetallics are generally brittle and act as stress raisers to weaken the coherence with Al matrix, therefore decreasing elongation. However, Fe addition in Al-Mg alloys may be beneficial because of the improvement in the yield strength with the scarification of ductility of die-cast aluminium alloys. The morphology of intermetallic phases has a vital effect on the properties of aluminium alloys. In the present work, the 3D morphology of Al_6 (Fe, Mn) in die-cast Al-Mg-Mn alloys with different levels of Fe contents were revealed. The formation of Al_6 (Fe, Mn) was also studied through crystal features and solidification behaviours.

1. Introduction

In die-cast aluminium alloys, Fe is a common deleterious impurity for elongation and is not economically removed from the melt [1-3]. Therefore the accumulation of iron has been a critical concern in recycled materials materials. Up to now, several methods have been invented to diminish the detrimental effect of Fe, such as melt-superheating treatment, chemical modification and rapid solidification [4-7]. Among them, Mn modification is a popular method to transform needle-like β-AlFeSi to blocky α-AlFeMnSi in Al-Si alloys. While, α-AlFeMnSi phase would forms large primary crystal and resulting in unacceptable mechanical properties when the total amount of Mn and Fe is more than 1.5 wt.%. Therefore, the application of recycled Al alloy with high Fe content is still an important concern in industry.

The obstacle of the application of recycled high-Fe Al alloy is that Fe-rich intermetallics would decrease elongation obviously. Therefore, a soft alloy matrix (such as ductile Al-Mg matrix) is necessary for high-Fe alloy. On the other side, it was found that Fe-rich phases in aluminium alloys are helpful in improving the yield strength with the scarification of ductility [8]. Besides, the addition of Mn in Al-Mg alloys can modify Al_3Fe to Al_6(Fe, Mn), and the increased cooling rate can significantly refine the Al_6(Fe, Mn) phase. It further increases the upper limit of the Fe level in aluminium alloys. Therefore, it is potential to use Fe as a strengthening element to design an Al-Mg-Mn-Fe alloy prepared by high pressure die casting (HPDC) which has a high cooling rate among various common casting methods.

The mechanical properties (especially the fracture behavior) of alloys are closed with the morphology of intermetallic phases. However, the Al_6(Fe, Mn) morphology and growth mechanism in Al-Mg alloys are still unclear. Therefore, the present work aims to reveal the 3D morphology and relative growth mechanism of Al_6(Fe, Mn) phase in Al-Mg-Mn-Fe alloys.

2. Experimental methods

Commercial pure Al and Mg ingots, Al-20% Mn and Al-45% Fe master alloys (all compositions quoted in this paper are in wt.% unless otherwise stated) were melted in an electric resistance furnace at 730 °C to prepare a series of Al-5Mg-xFe-0.6Mn alloys (x = 0.5, 1.5 and 2.5). After a homogenization process for about 30 min, the melt was manually dosed and subsequently released into the shot sleeve of a FRECH 4500 kN cold chamber HPDC machine. The HPDC die and shot sleeve were preheated at about 150 °C and 100 °C, respectively. The pouring temperature was 720 °C. Al_6(Fe, Mn) particles were collected after deep-etching or completely removing the matrix of samples by 15 vol.% HCl–distilled water solution, and then observed by a scanning electron microscope (SEM, Zeiss-Supra 35VP, Germany) equipped with an energy-dispersive X-ray spectroscope (EDS) and electron back-scattered diffractor (EBSD).

3. Results and discussion

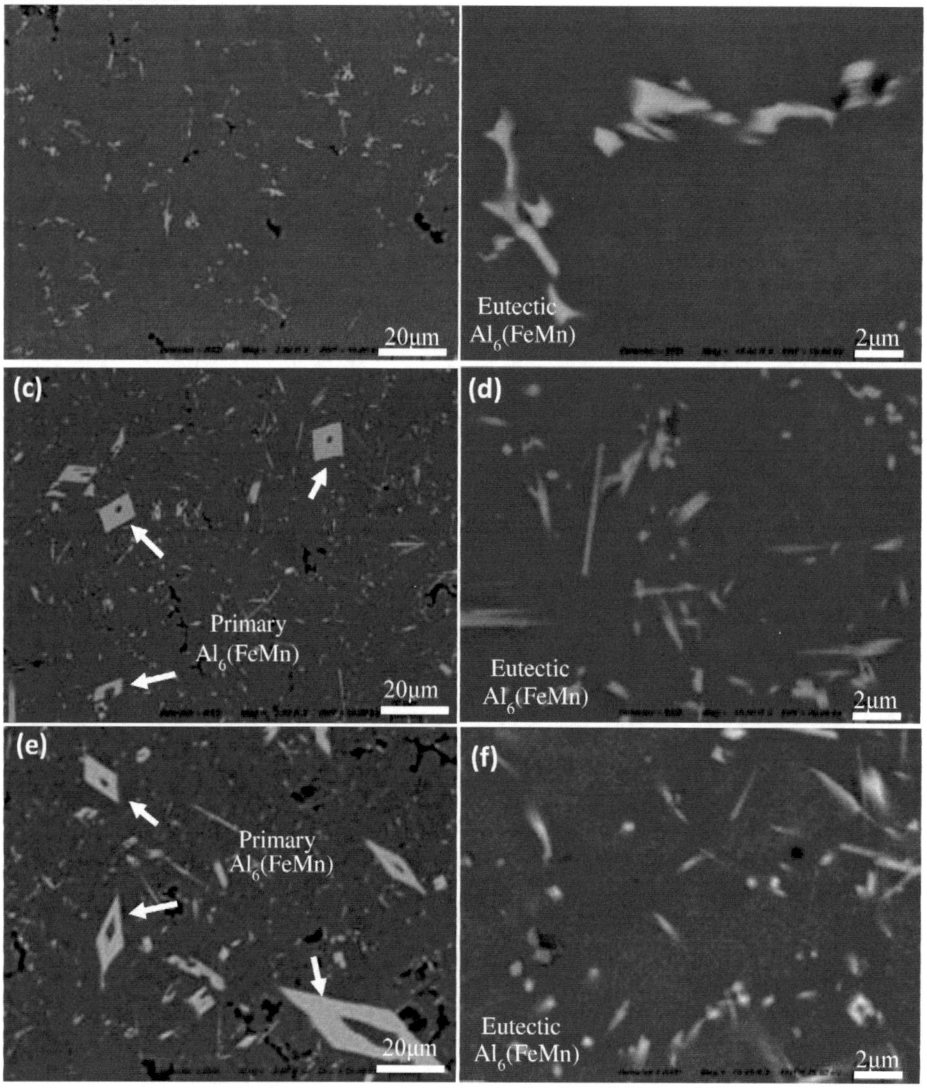

Figure 1. Backscattered SEM micrographs showing the microstructures of the Al-5Mg-0.6Mn-xFe alloys: (a, b) x=0.5, (c, d) x=1.5 and (e, f) x=2.5.

The as-cast microstructures of Al-5Mg-xFe-0.6Mn alloys (x = 0.5, 1.5 and 2.5) are shown in Figure 1. As observed, the morphology, size and amount of bright Al_6(Fe, Mn) phase varies obviously with the increase of the Fe level. In the alloy with 0.5 wt.% Fe, eutectic Al_6(Fe, Mn) were located at the grain boundaries and showed a lamellar morphology with some curved planes, as shown in Figure 1a and b. When the Fe level was at 1.5 wt.%, the rhombic or lath-like primary Al_6(Fe, Mn) phase was found (Figure 1c). Meanwhile, the eutectic Al_6(Fe, Mn) intermetallics were divorced from α-Al phase and exhibited as fine lath-like or rhombic morphologies with faceted surface (Figure 1d). When the Fe level was further increased to 2.5 wt%, both the amount and size of primary and eutectic Al_6(Fe, Mn) intermetallics were increased, while the morphology remains. It was noticed that the primary Al_6(Fe, Mn) intermetallics in the alloys with 2.5 wt.% Fe have two different size ranges. Similar with our previous result [3], the large and small primary Fe-rich intermetallics precipitated in the first solidification in shot sleeve and the secondary solidification in die cavity during cold-chamber HPDC process, respectively.

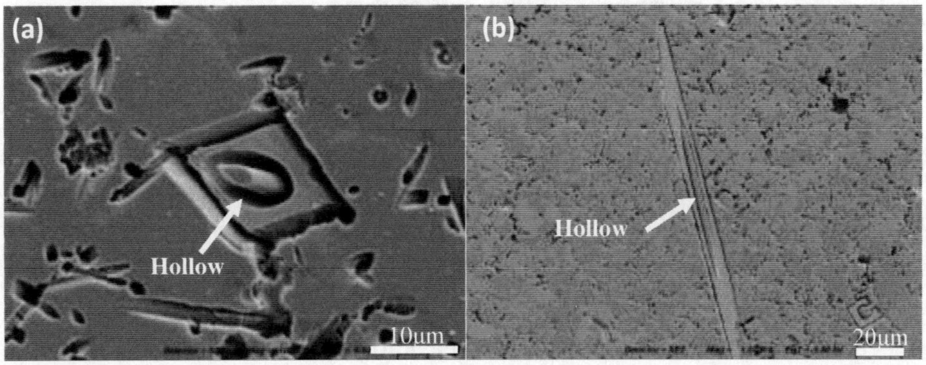

Figure 2. SEM micrographs showing (a) cross and (b) longitudinal sections of Al_6(Fe, Mn) phase.

It is interesting to note that both primary and eutectic phases have two morphologies: (1) lath / needle, (2) rhombus/dot. Besides, it was also found that almost all rhombic primary Al_6(Fe, Mn) crystals have a hollow. To further observe the morphology of the hollow primary Al_6(Fe,Mn) clearly, samples were deep-etched using a 15 vol% HCl-distilled water solution. Figure 2 indicates that the hollow is filled by α-Al phase and has a curved, smooth surface, although Al_6(Fe, Mn) is a strong faceted crystal. It means that these surface is not a certain exposed crystal surface, but the defects due to the incomplete crystal growth.

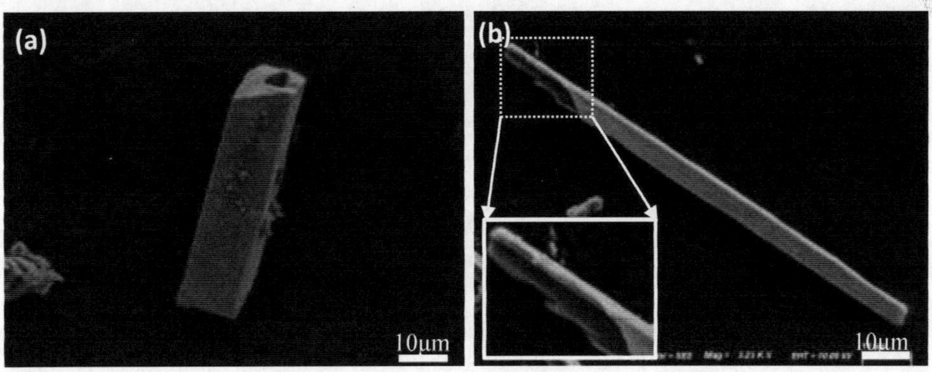

Figure 3. SEM micrographs showing 3D morphology of primary Al_6(Fe, Mn) phase: (a) a thick crystal and (b) a thin crystal with broken end.

To show the 3D morphologies of Al_6(Fe, Mn) crystals, a 15 vol% HCl water solution was applied to extract the intermetallic particles. The results shown in Figure 3 and 4 indicate that all primary and eutectic Al_6(Fe, Mn) crystals have one 3D morphology: quadrangular prism, and the lath/needle and rhombus/dot morphology in 2D sections are just the longitudinal and cross sections of Al_6(Fe, Mn) crystals, respectively. The difference is that the primary Al_6(Fe, Mn) crystals (shown in Figure 3) are

IOP Conf. Series: Materials Science and Engineering **529** (2019) 012011 doi:10.1088/1757-899X/529/1/012011

coarse and hollow, while the eutectic Al_6(Fe, Mn) crystals (shown in Figure 4) is thin (200-500 nm in cross section) and solid.

Figure 4. SEM micrographs showing 3D morphology of eutectic Al_6(Fe, Mn) phase.

The Bravais-Friedel-Donnay-Harker (BFDH) law is a well-accepted method to predict the possible exposed crystal faces. Al_6(Fe, Mn) has a Cmcm (63) space group and an orthorhombic structure (a = 0.7498 nm, b = 0.6495 nm, c = 0.8837 nm). According to the BFDH law, the 2 most possible exposed face are closed packed {100} and {002} faces. For orthorhombic crystal, the {110} only contains 4 (-110), (1-10) and (-1-10) faces, which form four side faces of an enclosed geometrical rhombic prism. While, {002} contains (112) and (00-2) faces, forming the top and bottom face of rhombic prism. Therefore, it can be speculated that the rhombic Al_6(Fe, Mn) is bounded by four {110} and two {002} faces.

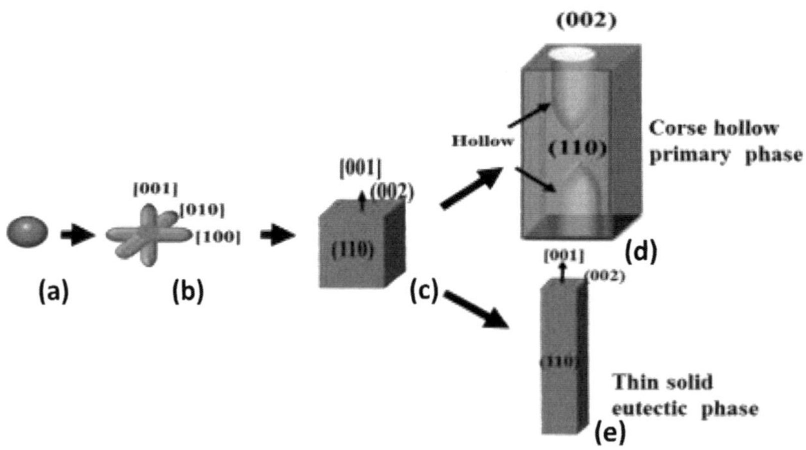

Figure 5. Schematic of the growth process of prism Al_6(Fe, Mn) crystal bounded by {110} and {002}: (a) spherical seed, (b) dentrite, (c) small solid rhombic prism, (d, e) coarse hollow primary phase and thin solid eutectic prism.

Figure 5 shows the growth process of prism Al_6(Fe, Mn) crystal in Al-Mg-Mn-Fe melt. As the decrease of melt temperature, the micro-segregation of atomic cluster becomes obvious and forms a spherical morphology with minimum free energy [10]. When the seed crystal grows and exceeds a critical size, it becomes unstable and grows rapidly along its first preferential growth directions ([001], [100] and [010]). At almost same time, new secondary branches generate on the first branches and grow between the first branches and finally fill these interspaces, making Al_6(Fe, Mn) crystal have a small prism morphology ({110}+{002}) with a strong intrinsic faceting feature. In the following growth stage, the rapid growth of {002}, which has a higher growth rate than {110} according to BFDH law, would elongate Al_6(Fe, Mn) prism, making it have a lath-like (primary phase) or needle-like (eutectic phase) morphology. The hollows inside Al_6(Fe, Mn) crystal also formed in the following process due to volume-diffusion. Compared with the diffusion in the side {110} faces, the diffusion in the central area of the faster growing {002} faces is relatively difficult. Therefore, the supplement of solute atoms and the ejection of impurities expelled from {002} faces are relatively slow. It will retard the growth in the central areas of the {002} faces, leading to the formation of hollows inside the primary Al_6(Fe, Mn) prism. While, the volume diffusion in small eutectic Al_6(Fe, Mn) crystal is not obvious, so eutectic Al (Fe,Mn) crystals remain solid.

4. Conclusions

In the present work, the 3D morphology and formation mechanism of Al_6(Fe, Mn) phase in HPDC Al-Mg-Mn-Fe alloys were revealed. It was found that Al_6(Fe, Mn) is the only eutectic Fe-rich phase and has an irregular morphology. As the Fe level increased to 1.5 wt.% and 2.5 wt.%, both primary and eutectic Al_6(Fe, Mn) have a rhombic prism morphology which is bonded by four {110} and two {002} faces. While, primary Al_6(Fe, Mn) phase has inside hollows, the eutectic phase is small and solid. In the late stage of crystal growth, volume-diffusion restrains the growth of the central areas of two {002} faces, leading to the formation of hollows inside the primary Al_6(Fe, Mn) phase. The small eutectic Al_6(Fe, Mn) phase is not affected by volume diffusion and remains solid.

5. References

[1] Shabestari S 2004 *Mater. Sci. Eng. A.* **383** 289
[2] Ashtari P, Tezuka H and Sato T 2005 *Scr. Mater.* **53** 937
[3] Ji S, Yang W, Gao F, Watson D and Fan Z *Mater. Sci. Eng. A.* **564** 130
[4] Fang X, Shao G, Liu Y and Fan Z 2007 *Mater. Sci. Eng. A.* **445** 65
[5] Kumari S, Pillai R, Rajan T and Pai B 2007 *Mater. Sci. Eng. A.* **460** 561
[6] Kumari S, Pillai R, Pai B, Nogita K and Dahle A 2006 *Metall. Mater. Trans. A.* **37** 2581.
[7] Shabestari S, Mahmudi M, Emamy M and Campbell J 2002 *Int. J. Cast. Met. Res.* **15** 17
[8] Yang H, Ji S and Fan Z 2015 *Mater. Des.* **85** 823
[9] Chen H, Zu F, Chen J, Zou L, Ding G and Huang Z 2008 *Sci.China, Ser. E: Technol. Sci.* **51** 1402

Optimization strategies for identifying the controlling mechanism for solid-state transformation in FeCrNi during rapid solidification

D M Matson, X Xiao, S Jeon

Department of Mechanical Engineering, Tufts University, Medford MA USA

douglas.matson@tufts.edu

Abstract. Retained free energy drives the transformation from metastable to stable phase during double recalescence of steel alloys. Statistical methods were used to identify the mechanism controlling cluster growth based on the principle of microstructural reversibility. Application of the coefficient of determination during optimization procedures showed that an extrinsic mechanism controls nucleation and that negligible healing occurs such that retained free energy from primary phase undercooling and melt convection enhance secondary nucleation.

1. Background

Double recalescence during rapid solidification involves nucleation of a metastable phase from the undercooled melt with subsequent transformation to the stable phase; both processes involve a rapid increase in temperature and the incubation time, or delay between these events, is a strong function of both undercooling and convection [1-3] for FeCrNi alloys. Previous work has shown that nucleation of the stable fcc austenite phase occurs along subgrain boundaries within the pre-existing metastable bcc ferrite phase and that the cluster geometry is of the form of two opposing hemispherical caps [4, 5] sharing a border that straddles the grain boundary (GB).

From classical nucleation theory the free energy change due to addition of an atom to the cluster of volume V and radius r is composed of positive terms hindering the transformation associated with the required creation of new surface, with a surface energy penalty of γ, and negative terms driving the transformation associated with reduction of volume free energy $\Delta G_v \equiv \Delta G_T$. The wetting angle θ is defined by a surface energy balance along the circular edge where the two caps meet such that

$$\cos\theta = \frac{\gamma_{M/M}}{2\gamma_{M/S}} \qquad f(\theta) = \frac{1}{2}(1-\cos\theta)^2(2+\cos\theta) \qquad r^* = \frac{2\gamma_{M/S}}{\Delta G_v} \tag{1}$$

The surface energy terms are strong functions of defect concentration [6] and thus the wetting angle seen during nucleation of the second phase may be influenced by conditions which occur during formation of the first. This effect can be quantified for solid-solid transformations where double recalescence is observed using the theory of retained free energy [7] where the system volume free energy is composed of two terms – the first driven by undercooling set by the equilibrium phase diagram ΔG_s and the second from damage retained from primary solidification ΔG_D. This second term is also composed of two terms – one describing retained free energy from primary undercooling ΔG_m and the other describing retained free energy due to convection during primary solidification ΔG_C.

$$\Delta G_T = \Delta G_s + f_x \Delta G_D \qquad \Delta G_s = \Delta T_s \frac{\Delta H_s}{T_{Ls}} \qquad \Delta G_D = \Delta G_m + \Delta G_c \tag{2}$$

$$\Delta G_m = \Delta T_m \frac{\Delta H_m}{T_{Lm}} \qquad \Delta G_c = \dot{\gamma}\left(\Delta_b - \Delta_m \dot{\gamma}\right) \tag{3}$$

The retained free energy may be reduced by some empirical fraction f_x representing healing of the retained damage through processes such as dynamic recrystallization or relaxation.

For the dual cap geometry the new surface area that must be formed is composed of two abutting hemispherical sections with area $A_{M/S}$ while the surface that is destroyed is the circular interface where they meet $A_{M/M}$. Subscripts refer to metastable (*M*) and stable (*S*) phase, respectively.

$$\Delta G = A_{M/S}\gamma_{M/S} - A_{M/M}\gamma_{M/M} - V\Delta G_T = 4\pi\left(r^2\gamma_{M/S} - \frac{r^3}{3}\Delta G_v\right) f(\theta) \tag{4}$$

The incubation time is evaluated based on the principle of microstructural reversibility as described by Feder *et al.* [8] where a cluster is assumed to be able to grow at the same rate as it would decompose:

$$\tau = -\frac{4 k_B T}{\beta\left(\dfrac{\partial^2 \Delta G}{\partial n^2}\right)} \tag{5}$$

Other authors have developed models which result in equations of similar form [9-15] with the only difference being the initial constant. In all these equations the key parameters are the atomic attachment rate from the parent phase to the cluster β and the curvature of the free energy curve with respect to addition of an atom to the cluster.

The purpose of this paper is to develop a strategy to investigate the sensitivity of potential statistical criteria to optimize the decision process and determine which mechanism controls cluster growth for a given alloy. The approach is to compare how well the observed data fits model predictions using two statistical quantities – the coefficient of determination [16] R^2 and the Pearson product-moment correlation coefficient [17] r:

$$R^2 \equiv 1 - \frac{\sum(y_i - f_i)^2}{\sum(y_i - \bar{y})^2} \tag{6}$$

$$r \equiv \frac{n\sum x_i y_i - \sum x_i \sum y_i}{\sqrt{\left[n\sum x_i^2 - \left(\sum x_i\right)^2\right]\left[n\sum y_i^2 - \left(\sum y_i\right)^2\right]}} \tag{7}$$

In these equations the subscript i on the dependent variable y and independent variable x represent observed individual events to be compared to corresponding model prediction behavior using the dependent variable f. The coefficient of determination is the proportion of the variance in y that is predicted from x. The Pearson correlation coefficient quantifies the strength of the linearity of a relationship between two sets of data. Note that the value obtained by linear regression using a widely-used commercial statistical package commonly employed by personal computer users labels the regression fit as an R^2-value but actually displays the correlation coefficient r. The goal of this paper is thus to identify a robust and unbiased method to optimize evaluation of the limiting cluster growth mechanism and from that gain insight on how cluster growth proceeds.

2. Transformation mechanisms

In order to develop an equation to predict the delay time given the melt condition, the principle of microstructural reversibility requires definition of the two key parameters identified in Equation 5.

From classical nucleation theory the curvature can readily be obtained by differentiation of Equation 4 twice. Since this relationship is a function of cluster radius and the differentiation is to be done relative to an incremental change in number of atoms the process requires a simple substitution based on:

$$V = \frac{4}{3}\pi r^3 f(\theta) = \frac{n\Omega}{N_A} \qquad n = \frac{4}{3}\pi r^3 f(\theta)\frac{N_A}{\Omega} \qquad \frac{dn}{dr} = \frac{8}{3}\pi r^2 f(\theta)\frac{N_A}{\Omega} \tag{8}$$

$$\frac{\partial \Delta G}{\partial n} = \frac{\partial \Delta G}{\partial r}\Big/\frac{\partial n}{\partial r} = \frac{\Omega_S}{N_A}\left(\frac{2\gamma_{M/S}}{r} - \Delta G_v\right) \qquad \frac{d}{dr}\left(\frac{\partial \Delta G}{\partial n}\right) = \frac{\Omega_S}{N_A}\left(-\frac{2\gamma_{M/S}}{r^2}\right) \tag{9}$$

$$\frac{\partial^2 \Delta G}{\partial n^2} = \frac{d}{dr}\left(\frac{\partial \Delta G}{\partial n}\right) \Big/ \frac{\partial n}{\partial r} = -\frac{\gamma_{M/S}}{2\pi r^4 f(\theta)} \frac{\Omega_S^{\ 2}}{N_A^{\ 2}} \quad \left.\frac{\partial^2 \Delta G}{\partial n^2}\right|_{r*} = -\frac{\Delta G_v^{\ 4}}{32\pi \gamma_{M/S}^{\ 3} f(\theta)} \frac{\Omega_S^{\ 2}}{N_A^{\ 2}} \qquad \textbf{(10)}$$

Note that this value is negative since the curvature is evaluated at the local maximum in the free energy curve such that the cluster is stable with addition of extra atoms. Thus the negative sign in Equation 5.

The result of this calculation is that the curvature is inversely proportional to the radius raised to the 4th power. With substitution for an evaluation of the critical cluster radius shown in Equation 1, where the radius is inversely proportional to free energy, the curvature becomes proportional to $(\Delta G_v)^4$.

The attachment rate β is a strong function of selection of the mechanism controlling cluster growth. If mass transfer through diffusion is the limiting step, the attachment rate will be proportional to the interfacial mass transfer area. Doubling the area results in twice the number of atoms attaching per unit time, halving the delay time. Since the interfacial area is some function of the cluster radius then the delay time will be a unique function of the transfer geometry. If mass transfer is not the rate limiting step for cluster growth and if some extrinsic influence, such as matrix strain or mobility of dislocations to the cluster to modify the wetting angle and augment interface potency [6], then the attachment rate is independent of r. These relationships are presented in Table I.

Table I – Relating delay time to free energy driving the transformation

Limiting mechanism	Attachment rate	Delay time
Bulk diffusion attachment	$\beta \propto A_S = 4\pi r^2 (1-\cos\theta)$	$\tau \propto \Delta G_v^{-2}$
GB diffusion attachment	$\beta \propto A_S = \delta (2\pi r \sin\theta)$	$\tau \propto \Delta G_v^{-3}$
Extrinsic factor limited	$\beta \propto$ constant	$\tau \propto \Delta G_v^{-4}$

If diffusion through the bulk controls, the transport area is the hemispherical cap surface which is proportional to r^2. If diffusion through the grain boundary controls, the area is the circumference of the basal plane multiplied by the grain boundary thickness δ which is proportional to r. If mass transfer is not the rate limiting step for cluster growth and if some extrinsic influence, such as matrix strain or mobility of dislocations to the cluster, then the attachment rate is independent of r. Combining this result with Equation 10 and eliminating r using Equation 1 yields the functional relationship between delay time and volume free energy. The extrinsic mechanism is functionally equivalent in form $\tau \propto \Delta G_v^{-4}$ to other previous models [9-15] and has been verified through simulation [18] and experiment [19] although the assumptions during development of the previous work were different than in the current study. The bulk mechanism is functionally similar to Shao and Tsakiropoulos [20] with the form $\tau \propto \Delta G_v^{-2}$

3. Experimental results

Results from levitation experiments have been presented previously for no-flow conditions using electrostatic levitation (ESL), laminar Marangoni flow using ESL with laser heating (ESL-L), transitional and turbulent flow using space electrostatic levitation (ISS-EML) and fully turbulent flow using ground-based electromagnetic levitation (EML).

Details of the experimental techniques are available in the literature [3, 7] based on an evaluation of the observed delay times for a 60Fe20Cr20Ni (wt%) alloy as a function of undercooling and melt convection. Melt shear was calculated in the same manner as presented by Lee *et al.* [21] adjusted for the thermophysical properties of the steel alloy using magnetohydrodynamic modelling.

4. Statistical optimization

The following analysis quantifies the performance of the three candidate attachment controlling mechanisms using the two statistical criteria described in Equations 6 and 7. Using these techniques, an evaluation is conducted to determine which mechanism is statistically most likely to control the delay behavior and determine the fraction of retained free energy f_x that contributes to the transformation total driving force ΔG_T. Alloy properties are documented in the literature supplemental material. [7]

A dimensionless technique [7] was used to compare statistical approaches. The reference state was for the thermodynamically-limiting condition defined by the equilibrium phase diagram. For double recalescence to occur the undercooling must be sufficient to access the metastable phase corresponding to a melt temperature at or below the metastable liquidus. This is the smallest undercooling that can result in formation of the metastable phase and defines the minimum total free energy possible, in the absence of melt stirring with $\Delta G_D = \Delta G_M = \Delta G_C = 0$, such that $\Delta G_R \equiv \Delta G_T = \Delta G_s$. This results in the longest possible delay time τ_R. From this condition the observed delay time defines a dimensionless delay time ratio N_τ and the corresponding undercooling defines a dimensionless driving force ratio N_M.

$$N_\tau = \frac{\tau_{EXP}}{\tau_R} \qquad N_M = \frac{\Delta G_T}{\Delta G_R} \tag{11}$$

These two independent quantities are shown in Figure 1 where the results for bulk diffusion attachment (gray line) and extrinsic limited mechanisms (black line) are displayed. The grain boundary attachment curve lies between these and is not shown for clarity. Filled triangles are ESL, open triangles are ESL-L, open diamonds are ISS-EML and filled diamonds are EML. The slopes agree well with theoretical values and a linear regression of the data results in slopes of $m_{bulk} = -1.994 \pm 0.032$, $m_{GB} = -3.008 \pm 0.029$, and $m_{ext} = -4.016 \pm 0.032$. Of particular interest is the inability of the D_{bulk} and D_{GB} mechanisms to describe behavior of highly stirred experiments using EML; the data does not follow the regression well and shows systematic deviation not observed for the extrinsic mechanism. This systematic deviation can easily be seen by looking at the trend for any given symbol type. In particular, the filled-diamond shape using EML at high dimensionless driving force for the D_{bulk} mechanism shows negative deviation at high driving force and positive deviation as the driving force is reduced. The behavior for the Extrinsic mechanism consistently follows the theoretical curve over the entire range. Similar behavior is seen for the filled triangles from ESL and open triangles from ESL-L testing.

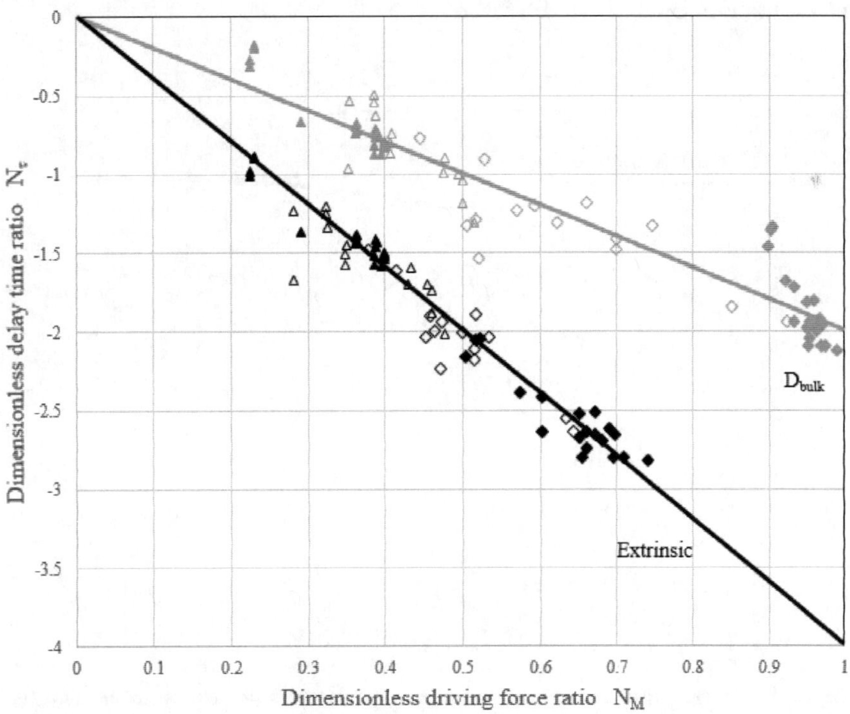

Figure 1 – Dimensionless predictions

This is further emphasized in Figure 2 where the predicted delay is compared to the observed delay. The solid black line with predictions from the extrinsic model shows a consistently better fit to both

ESL (zero melt shear) and EML (high shear) experimental data as compared to the dotted line D_{bulk} mechanism predictions.

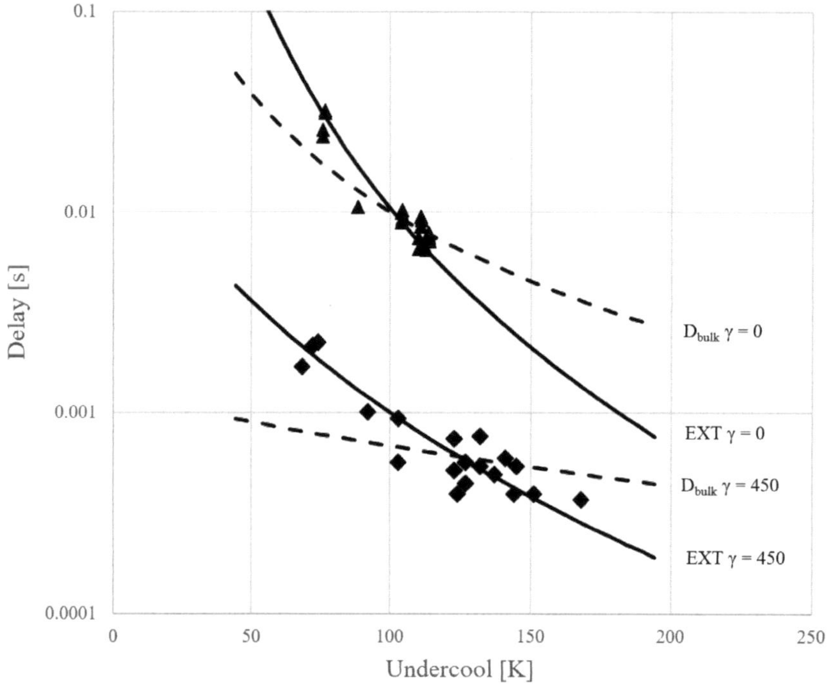

Figure 2 – Experimental data comparison

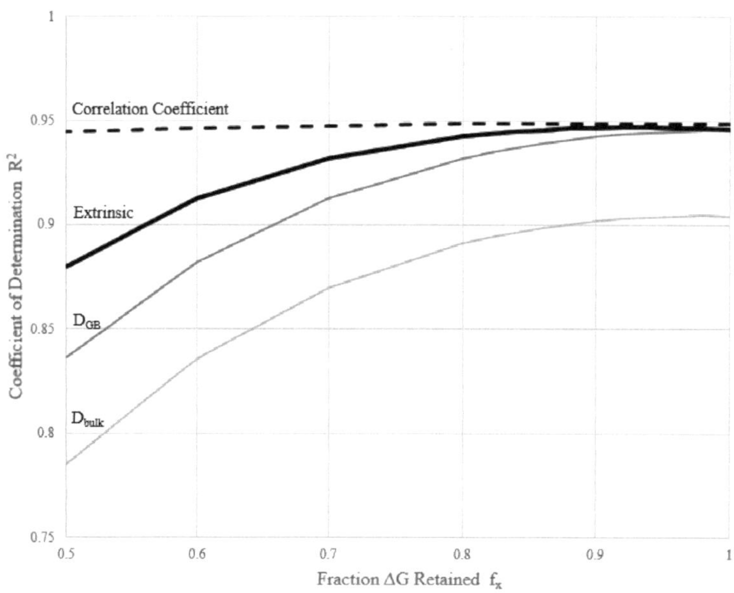

Figure 3 – Optimization of mechanism models as a function of fraction retained free energy

These trends are also evident when observing the statistical measures of fit. Figure 3 presents the optimized coefficient of determination as a function of the fraction of retained free energy over the range $0.5 < f_x < 1.0$. The best fit (highest R^2) occurs in all cases for $f_x > 0.9$ which implies that no healing occurs and all free energy from metastable solidification is retained to augment stable phase nucleation. Furthermore, the extrinsic mechanism outperforms both the grain boundary and bulk mechanisms as previously discussed – except for $f_x \rightarrow 1$ where R^2_{max} is 0.948 and 0.946 for extrinsic and grain boundary mechanisms respectively and thus differences are not statistically significant. The method is robust and trends are plainly evident because R^2 drops off significantly indicating that the measurement sensitivity is high. In comparison the dotted line represents the optimized correlation coefficient for the identical conditions. Not only is deviation nearly identical for all three mechanisms but the fidelity of the result limits the ability to define how f_x influences model performance. Use of the *r*-value is not recommended.

5. Conclusions

Optimization strategies were compared to define the mechanism that controls cluster growth during nucleation of the stable phase within a pre-existing metastable solid. The coefficient of determination outperforms the correlation coefficient. Extrinsic factors appear to control nucleation rather than either bulk or grain boundary diffusion. The empirical parameter describing the fraction of free energy that is retained has been evaluated for the FeCrNi system and found to indicate that free energy is fully retained following primary recalescence such that $f_x \rightarrow 1$ and the added energy promotes enhanced nucleation of the second phase. The dimensionless technique provides insight into the dynamics of this process.

Acknowledgements

This work was sponsored by NASA under grant NNX16AB59G. The authors wish to thank the NASA-MSFC Electrostatic Levitator team and the ESA/DLR operations support groups for valuable assistance during testing both on ground and in space using the ISS-EML facility.

References

[1] Koseki T and Flemings M C 1996 *Met. Trans.* **27A** 3226
[2] Hanlon A B *et al.* 2006 *Phil. Mag.* **86**(3) 165
[3] Matson D M *et al.* 2017 JOM **69**(8) 1311
[4] Koseki T and Flemings M C 1995 *Met. Trans.* **26A** 2991
[5] Matson D M 2012 *Solidification of Containerless Undercooled Melts* ed D Herlach and D M Matson (Wiley-VCH Weinheim Germany) chapter 10 p 213-238
[6] Matson D M and Xiao X 2018 Appl. Phys. Lett. 113(01) 011903-1
[7] Matson D M 2018 *npj Nat. Micrograv.* **4**(22) 1
[8] Feder J *et al.* 1966 *Advan. Phys.* **15**(57) 111
[9] Turnbull D 1948 *Trans. AIME* **175** 774
[10] Wakeshima H 1954 *J. Chem. Phys.* **22** 1614
[11] Collins F C 1955 *Z. Electrochem.* **59** 406
[12] Frisch H L 1957 *J. Chem. Phys.* **27** 90
[13] Chakraverty B K 1966 *Surf. Sci.* **4** 205
[14] Kashchiev D 1969 *Surf. Sci.* **14** 209
[15] Kantrowictz A 1951 *J. Chem. Phys.* **19**(9) 1097
[16] Kvalseth T O 1985 *Am. Stat.* **39**(4) 279
[17] Pearson K 1895 *Royal Society Proceedings* **58** 241
[18] Kelton *et al.* 1983 *J. Chem. Phys.* **79** 6262
[19] Kumomi H and Yonehara T 1994 *J. Appl. Phys.* **75** 2884
[20] Shao G and Tsakiropoulos P (1994) *Acta Metall. Mater.* **49**(9), 2937
[21] Lee J *et al.* 2015 *Met. Trans. B* **46**(1) 199

IOP Conf. Series: Materials Science and Engineering **529** (2019) 012012 doi:10.1088/1757-899X/529/1/012012

Nomenclature

Symbol	Units	Definition
A	(m^2)	area
D_{bulk}	(- -)	bulk diffusion-controlled mechanism
D_{GB}	(- -)	grain boundary diffusion-controlled mechanism
EXT	(- -)	extrinsically-controlled mechanism
f_i	(- -)	model prediction using the dependent variable
$f(\theta)$	(- -)	cluster geometry factor
f_x	(- -)	fraction of damage which is retained (not healed) before recalescence
ΔG	(J/m^3)	free energy
ΔG_m	(J/m^3)	free energy due to undercooling relative to the metastable phase
ΔG_s	(J/m^3)	free energy due to undercooling relative to the stable phase
ΔH	(J/mol)	enthalpy
k_B	(J/atom·K)	Boltzmann constant
m_{bulk}	(- -)	slope of the regression for bulk model
m_{GB}	(- -)	slope of the regression for grain boundary model
m_{ext}	(- -)	slope of the regression for extrinsic model
n	(atoms)	number of atoms in a cluster
n	(- -)	number of observed events
N_A	(atom/mole)	Avagadro's number
N_m	(- -)	dimensionless driving force ration obtained from observed undercooling
N_τ	(- -)	dimensionless delay time ratio obtained from observed incubation time
r	(m)	radius, with * for critical cluster radius
r	(- -)	Pearson product-moment correlation coefficient
R^2	(- -)	coefficient of determination
ΔT	(K)	undercooling
T_L	(K)	liquidus
V	(m^3)	volume
x	(-)	independent variable and abscissa value
y	(-)	dependent variable and ordinate value
$\overline{y}$	(-)	average dependent variable value
β	(atoms/s)	atomic attachment rate
γ	(J/m^2)	surface free energy
$\dot{\gamma}$	(s^{-1})	rate of shear in liquid due to forced convection
Δ_b	($J{\cdot}s/m^3$)	intercept for correlation of influence of convection on free energy
Δ_m	($J{\cdot}s/m^3$)	slope for correlation of influence of convection on free energy
τ	(s)	theoretical incubation delay time
τ_{EXP}	(s)	experimentally observed incubation delay time
θ	(radians)	wetting angle
Ω	(m^3/mol)	atomic volume

Subscripts

c	(- -)	convection damage
D	(- -)	total damage
i	(- -)	individual observed event
m	(- -)	metastable phase
M/M	(- -)	Metastable-metastable phase interface
M/S	(- -)	Metastable-stable phase interface
R	(- -)	reference value
s	(- -)	stable phase
T	(- -)	total
v	(- -)	volume

IOP Conf. Series: Materials Science and Engineering **529** (2019) 012013 doi:10.1088/1757-899X/529/1/012013

Investigation using 4D-CT of massive-like transformation from the δ to γ phase during and after δ-solidification in carbon steels

Hideyuki Yasuda[1], Takahiro Hashimoto[1], Naoki Sei[1], Kohei Morishita[1*], Masato Yoshiya[2]

[1]Department of Materials Science and Engineering, Kyoto University, Sakyo, Kyoto, 606-8501, Japan

[2]Department of Adaptive Machine Systems, Osaka University, Suita, Osaka, 565-0871, Japan

yasuda.hideyuki.6s@kyoto-u.ac.jp

Abstract. For a long time, the γ phase in metallic alloys has seemed to be produced through a peritectic reaction between the δ and liquid phases. However, direct observations have shown that a massive-like transformation, in which the δ phase transforms into the γ phase in the solid state, is dominant during or after solidification of the δ phase in carbon steels. To characterize such massive-like transformation, we use time-resolved tomography (4D-CT) to demonstrate the volume change during cooling from the melt and the crystallographic orientation relationship between the δ and γ phases. The volume changes from solidification and from the massive-like transformation from the δ to the γ phase were −3% and −0.5%, respectively. The transformation from the δ to the γ phase finished quickly, as demonstrated by the volume change. Fine γ grains were produced even in a single δ grain through the massive-like transformation. Also, the refined γ grains showed a wide crystallographic distribution.

1. Introduction

Recently, time-resolved in-situ tomography (4D-CT) has been used to observe the solidification of metallic alloys [1,2,3,4,5,6]. One key issue in 4D-CT is achieving sufficiently high time resolution in order to observe microstructure evolution. The solidification of Al–Cu alloys has been observed three-dimensionally by 4D-CT [1-5]. Solid grains isolated by a liquid phase, change in the solid fraction during cooling, shrinkage due to solidification, and the solid/liquid interfacial area (f_{ss}>0.7) have been observed or measured in Al–4mass%Cu alloys [1]. Segregation of Y_2O_3 particles during solidification of Al–Mg–Si alloys (AI-6082) with 0.54 vol.% of added Y_2O_3 particles (~500 nm) has also been demonstrated [2]. Dendrite growth calculated by a phase field model has been compared with the dendritic structure in Al–15mass% Cu alloys, obtained by 4D-CT [3]. The dendrite morphology, even early in solidification, was reconstructed by a 4D-CT technique using the interlaced view sampling (TIMBIR) method [4], which improved the time resolution. In addition to solidification, semisolid deformation in Al–15mass%Cu alloys has been observed by 4D-CT [5]. This study found that grains isolated by the remaining liquid phase experienced transgranular liquation cracking even at low stress (1–40 MPa) [5]. This finding gives new insight into controlling grain structure during solidification. Solidification of Ni–14mass%Hf, Fe–11mass%Hf, and Co–18mass%Hf alloys have also been observed [7]. Adding the Hf increased the contrast resolution between the solid and liquid phases, enabling solidification to be observed at high temperatures. Thus, 4D-CT techniques are expected to be useful in studying various metallic alloys.

An earlier study attempted to use 4D-CT to observe solidification in metallic alloys with low contrast resolution, such as Fe–C–Mn–Si alloys [8]. Although these reconstructed images had rather poor quality due to the low contrast resolution and low number of projection images, a filter method using a phase field model significantly improved the image quality. The development of the primary and secondary arms was observed, and the average curvature of the solid/liquid interface was obtained as a function of solid fraction. Characterizing dendritic growth, as done in Ref. [8], will be useful for developing physical models and numerical simulations.

Using 4D-CT with monochromatized X-rays has another advantage: X-ray diffraction spots can be measured in synchronization with rotation by using a 2D detector, and sequential analysis of the diffraction data can be used to obtain the distribution of crystallographic orientations over time. For example, it can find the relationship of crystallographic orientation between a mother phase and a product phase in a phase transformation. Additionally, the volume fractions of constituent phases can be evaluated by integrating the diffraction intensifies, and the volume change from thermal expansion and phase transformation can be found simply by counting voxels. Thus, this technique is attractive for studying phase transformation *in situ*.

For a long time, the γ phase has appeared to be produced though a peritectic reaction between the δ and liquid phases in carbon steel with a C content lower than 0.5 mass%. However, X-ray transmission imaging [9,10,11,12] has shown that a dominant massive-like transformation, in which the δ phase transforms into the γ phase in the solid state. As shown in Fig. 1, there is little nucleation of the γ phase, and the δ phase was undercooled below the peritectic reaction. Thus, the δ phase transformed into the γ phase in the solid state. Laser-scanning confocal microscopy has also shown that the δ phase massively transforms into the γ phase [13]. Moreover, the massive transformation was selected in Fe–Cr–Ni alloys [14]. Even so, the massive-like transformations in Fe–C and Fe–Cr–Ni alloys are not yet fully understood. Thus, to further understand massive-like transformations, we should measure the volume change during cooling and analyze the crystallographic orientation relationship between the δ and γ phases by using 4D-CT.

2. Experiments

We studied a specimen of Fe–0.18C– 0.6Mn–0.3Si (mass%) carbon steel, hereafter referred to as 0.18C steel. The specimen was 0.8 mm in diameter and 2–3 mm in height. Its height was sufficiently small to observe the entire specimen when evaluating specimen volume. The specimen was cooled from a melt at a cooling rate of 0.33 K/s in a vacuum (approximately < 1 Pa). The temperature gradient at the specimen was less than 1 K/mm.

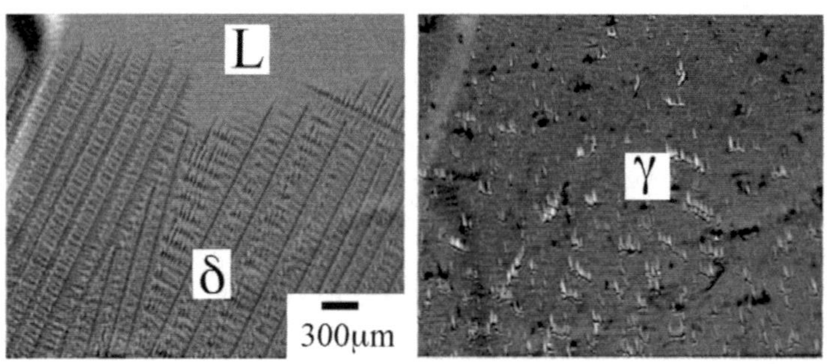

Figure 1. Example of the massive-like transformation observed in 0.45mass%C steel at a cooling rate of 0.83 K/min [9,10].

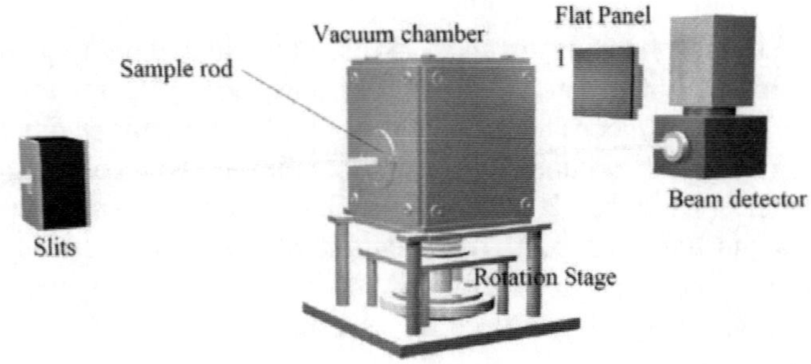

Figure 2. Setup of 4D-CT for observing transmission images and X-ray diffraction spots.

Figure 3. Reconstructed 3D image after solidification and the massive-like transformation. Concave shape at the top was caused by solidification shrinkage.

To measure the volume change and observe diffraction spots during cooling from the melt, we performed 4D-CT experiments at beamline BL20XU of the SPring-8 synchrotron radiation facility (Hyogo, Japan), as shown in Fig. 2. The size of the X-ray beam was adjusted using the slits, and the intensity was adjusted using the absorber, if needed. The specimen, inserted into a sintered Al_2O_3 pipe with inner and outer diameters of 0.8 mm and 2 mm, respectively, was placed in a graphite furnace. Transmission images were obtained by a beam monitor with a format of 1024 × 1024 pixels (pixel size: 6.5 μm × 6.5 μm). X-ray diffraction spots were observed with a flat-panel sensor with a format of 1008 × 682 pixels (pixel size: 100 μm × 100 μm, frame rate: 30 fps). Transmission images at a frame rate of 100 fps and diffraction images at a frame rate of 30 fps were continuously obtained by using monochromatized X-rays with an energy of 37.7 keV. The sample was rotated at 0.25 rps. Convolution back-projection (CBP) image reconstruction was performed using 200 transmission images over a 180° rotation every 4 s. Normal vectors of the $(110)_\delta$, $(200)_\delta$, $(211)_\delta$, and $(111)_\gamma$ planes in the specimen were calculated from the positions of X-ray diffraction spots on the flat-panel sensor and the rotation angle. The resolution of crystallographic orientation was 3° because 120 frames over 360° were obtained by the flat panel sensor. Because the flat panel had a small solid angle, the beam size was relatively large (4 mm horizontal and 3 mm vertical), and the diffraction spots changed rapidly after the massive-like transformation, we could not obtain a crystallographic orientation map of the specimen. The specimen volume was evaluated by counting specimen voxels (solid and liquid phases). Porosity detected in the specimen was excluded for this evaluation.

3. Results and Discussion

3.1. Volume change

Figure 3 shows a reconstructed 3D image of the 0.18C steel (solid and liquid phases) after the massive-like transformation following the ferrite solidification. The concave shape at the top was caused by solidification shrinkage. It was rather easy to track the boundary between specimen and vacuum because of the large difference in X-ray absorption coefficient. In addition, the reconstructed 3D images showed porosities with a diameter of more than 20 μm. The specimen volume was simply evaluated by counting specimen voxels (solid and liquid phases). The accuracy of the volume measurement depended on a threshold value to extract specimen voxels from the 3D images. However, the volume normalized by the volume of the δ phase immediately before the massive-like transformation has a relatively high sensitivity to volume change.

Figure 4 shows the change in volume during cooling from the melt at 0.33 K/s for the 0.18C steel. The volume of the liquid phase gradually decreased due to thermal shrinkage. The volume expansion coefficient estimated from the slope was 125×10^{-6} K^{-1}. Thus, the linear expansion coefficient, α, was estimated to be 42×10^{-6} K^{-1}. Jimbo and Cramb reviewed the reported values of liquid density (ρ) of Fe [15], giving the temperature dependence of density as shown in Eq. (1).

$$\rho = \rho_m + \Lambda\left(T - T_m\right) \quad (1)$$

Here, ρ_m is the density at the melting point, Λ is the coefficient, T is the temperature, and T_m is the melting point. The reported density at the melting point (ρ_m) ranges from 6.97×10^3 kg/m^3 to 7.28×10^3 kg/m^3. In addition, the reported coefficient, Λ, ranges from -0.8×10^{-3} kg/m^3K to -1.6×10^{-3} kg/m^3K. The linear expansion coefficient calculated from the reported values ranges from $38 \times 10^{-}$ K^{-1} to $74 \times 10^{-}$ K^{-1}. The measured value in the present study lies within the range of reported values. Thus, the present method has at least the same accuracy as the methods used in the previous works. Because the reported values were very scattered, further work was required to evaluate the accuracy of 4D-CT.

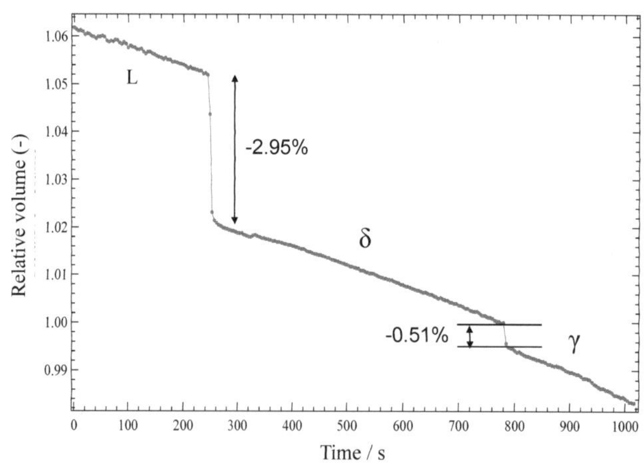

Figure 4 Volume change of 0.18mass%C steel during cooling from the melt. The volume was normalized by the volume immediately before the massive-like transformation.

Because solidification occurred in the undercooled melt, solidification completed within 20 s even at the low cooling rate. The volume change due to solidification was ~3%. After solidification, the volume of the δ phase decreased, and the linear expansion coefficient was estimated to be 39×10^{-6} K^{-1}. The massive-like transformation occurred at 780 s. The volume shrinkage due to the massive-like transformation was 0.5%. This shrinkage completed within the time resolution of 4D-CT (4 s). This fast volume change is consistent with reported in-situ observations with X-ray transmission imaging [9-12]. The observed shrinkage of 0.5% will help in understanding the strain and stress induced by the massive-like transformation. In addition, the linear expansion coefficient of the γ phase after the massive-like transformation was estimated to be 48×10^{-6} K^{-1}.

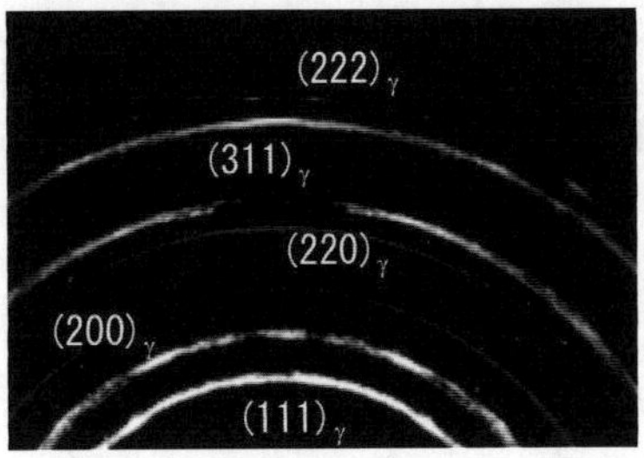

Figure 5 X-diffraction of γ phase after the massive-like transfixion, obtained by the flat panel sensor. The intensity of X-ray diffraction was integrated over 360-degree rotation.

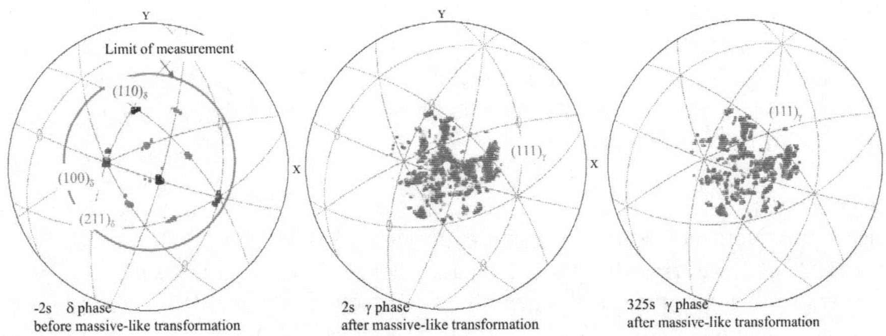

(a) (b) (c)

Figure 6 Distribution of crystallographic orientation of (a) δ phase at 2s before the massive-like transformation, (b) γ phase at 2 s after the massive-like transformation and (c) γ phase at 325 s after the massive-like transformation.

3.2. Crystallographic orientation

Figure 5 shows an example of X-ray diffraction pattern of the γ phase after the massive-like transformation. In this figure, 120 diffraction images over a 360° rotation were superimposed to show the diffraction rings. In the analysis, the positions of diffraction spots were measured on every diffraction image. The normal vectors of diffraction planes were geometrically calculated from the positions and the rotation angle of the specimen. Thus, we obtained a distribution of crystallographic orientations with an angle resolution of 3°, as shown Fig. 6. Each point on the stereographic projection corresponds to a normal vector of the diffraction plane. Normal directions outside the green circle in Fig. 6(a) were not measured because of the flat panel sensor had a limited solid angle. As shown in Fig. 6(a), the crystallographic orientation of the δ phase before the massive-like transformation showed that dendrites of the δ phase with a single orientation grew in the entire specimen. Namely, the massive-like transformation occurred in a single crystal of the δ phase.

Many diffraction spots of the γ phase appeared after the massive-like transformation (Fig. 5), indicating that it produced multiple γ grains. Figure 6(b) shows the distribution of the closest packed plane (111) at 2 s after the massive-like transformation. The color denotes the intensity of the diffraction spots. Ref. [9], using a white X-ray beam with a size of 50 μm × 50 μm, suggested that the closest packed plane (110) of the δ phase and the closest packed plane (111) of the γ phase tended to be adjoined. However, many (111) spots of γ grains appeared in the present study, so it was rather difficult to identify the crystallographic orientation relationship between the δ and γ phases in the specimen with a diameter of 0.8 mm and a height of 2–3 mm. In the present study, we found that the massive-like transformation produced fine γ grains with a wide crystallographic distribution, even though there was a certain crystallographic orientation relationship. Producing refined γ grains will be important to understanding microstructure evolution during and after solidification in carbon steel.

4. Summary

Using 4D-CT, we measured the change in volume during cooling from the melt. The distribution of crystallographic orientation was also measured for the δ and γ phases.

- The volume changes due to solidification and the massive-like transformation from the δ to the γ phase were 3% and 0.5%, respectively. This quantitative measurement will be useful for understanding the strain and stress induced by the massive-like transformation.
- The linear expansion coefficient of liquid was estimated to be 40×10^{-6} K^{-1}. The linear expansion coefficients of the δ and γ phases were estimated to be 39×10^{-6} K^{-1} and 48×10^{-6} K^{-1}, respectively.
- 4D-CT will be a useful tool for measuring volume changes from solidification phenomena.
- Fine γ grains were produced even in a single δ grain through the massive-like transformation.
- The refined γ grains showed a wide distribution of crystallographic orientations.
- The massive-like transformation was characterized by a volume change of 0.5% and the production of fine γ grains.

Acknowledgments

4D-CT experiments were performed as general projects (2016A1454, 2016B1458, 2017B1463) at BL20B2 and BL20XU at SPring-8 (JASRI), Japan. This study was initiated by the "visualization of solidification" research group in the 19th Committee on Steelmaking of JSPS. The study was developed in part by "Heterogeneous Structure Control: Towards Innovative Development of Metallic Structural Materials" in the Industry–Academia Collaborative R&D Program (JST). 4D-CT techniques developed by a Grant-in-Aid for Scientific Research (S) (No. 17H06155) were also used. We thank Joshua Yearsley, MS, from Edanz Group (www.edanzediting.com/ac) for editing a draft of this manuscript.

References

[1] Ludwig O, Dimichiel M, Salvo L, Suery M, Falus P 2005 Metall Mat Trans A **36A** 1515
[2] Daudin R, Terzi S, Lhuissier P, Salvo L, Boller E 2015 Mater. Design **87** 313
[3] Aagesen L K, Fife J L, Lauridsen E M, Voorhees P W 2011 Scr. Mater. **64** 394
[4] Gibbs J W, Mohan K A, Gulsoy E B, Shahani A J, Xiao X, Bouman C A, De Graef M, Voorhees P W 2015 Scientific Reports **5** 11824
[5] Cai B, Karagadde S, Yuan L, Marrow T J, Connolley T, Lee P D 2014 Acta Mater., **76** 371
[6] Cai B, Wang J, Kao A, Pericleous K, Phillion A B, Atwood R C, Lee P D 2016 Acta Mater. **117** 160
[7] Cai B, Wang J, Kao A, Pericleous K, Phillion A B, Atwood R C, Lee P D 2016 Acta Mater. **117** 160
[8] Yasuda H, Kawarasaki T, Tomiyori Y, Kato Y, Morishita K, submitted to IOP Conference Series: Materials Science and Engineering.
[9] Yasuda H, Nagira T, Yoshiya M, Sugiyama A, Nakatsuka N, Kiire M, Uesugi M, Uesugi K, Umetani K, Kajiwara K 2012 IOP Conf. Ser.: Mater. Sci. Eng. **33** 012036
[10] Yasuda H, Nagira T, Yoshiya M, Uesugi M, Nakatsuka N, Kiire M, Sugiyama A, Uesugi K, Umetani K 2012 IOP Conf. Ser.: Mater. Sci. Eng. **27** 012084

[11] Nishimura T, Morishita K, Nagira T, Yoshiya M, Yasuda H 2015 IOP Conf. Ser.: Mater. Sci. Eng. **84** 012062
[12] Nishimura T, Morishita K, Yoshiya M, Nagira T, Yasuda H 2019 Tetsu-to-Hagané **105** in press
[13] Griesser S, Reid M, Bernhard C, Dippenaar R 2014 Acta Mater. **67** 335
[14] Nishimura T, Matsubayashi R, Morishita K, Yoshiya M, Nagira T, Yasuda H 2019 ISIJ Int. **59** in press
[15] Jimbo I, Cramb A 1993 Metall. Trans. B **24B** 5

Numerical modelling of infiltration of Al into additive manufactured Fe preforms in order to obtain Fe-Al intermetallics

R. Berger[1,a], S. Jana[1], N. Focks[1], H. Michels[1], W. Kochanek[2], C. Charitidis[3]

1: Access e.V., Intzestraße 5, 52072 Aachen, Germany
2: Dr.Wolfgang Kochanek Ent., Freiheitstr. 57, 67434 Neustadt, Germany
3: NTUA, 9 Heroon Polytechniou St., Zographos, Athens, Greece GR-157 73
a: Corresponding Author: R. Berger - Email: r.berger@access-technology.de

Abstract. The current work, describes a numerical methodology to obtain deeper understanding of the kinetics of solidification and the dynamics of Al melt infiltration into porous iron preforms, in order to develop a near net shape process for a new class of highly advanced ductile and fine grained Fe-Al intermetallic. Investigations on macroscale simulations revealed that certain pressure and wetting conditions are beneficial for infiltration into finer structures. Microstructure showed a dependency of the solidification process on diffusion. Estimates for infiltration and solidification times are developed to determine which structures can be infiltrated before solidification stops melt flow.

1. Introduction and objectives of the work

Chromium and Nickel are listed in the table of Critical Raw Materials (CRMs) with a current projected lifetime of 25 – 100 Years [1-3]. In line with the strategy to find solutions to replace these CRMs in high volume end consumer products, a HORIZON-2020 project: EQUINOX [4] was funded to develop a novel process that allows to substitute Cr/Ni based (stainless) steel parts used in certain consumer products. The goals of this technological development is to develop a novel near net shape process for a new class of highly advanced ductile and fine grained Fe-Al based intermetallics. The basis of this development will be based on a reactive infiltration process of porous iron preforms with liquid Al or an Al-Fe-alloy, by using two different pressure assisted techniques: suction and centrifugal casting. Fine grain structures, which benefit tensile strength and ductility [5], are achieved by infiltrating preforms with fine structured porosity. Infiltration of Al in porous preforms can also be used for rapid prototyping [6]. Here the permeability of additive manufactured materials has a major impact on the infiltration process, but it is also important e.g. in biomedicine applications [7,8].

The first step is to effectively infiltrate melt into the preforms, which is directly related to the melt flow paths, or the structure of porosity in these preforms. Different preforms were manufactured by selective laser melting process (SLM), Metal injection Moulding (MIM), Powder metallurgy moulding and sintering (PMM) and Kochanek Process (KE) [9]. Preforms from SLM process consists of Fe-powder and characterized by thin long roughly square channels of approximately 400 -550 µm in size, representing around 50% of porosity. The unique process of manufacturing preforms developed by [9] uses powder moulding process of metal oxides rather than pure metal powders.

The flexibility of the KE process allows building preforms of different porosities and channels of different sizes. For all industrial purposes, the KE process is highly viable cost effective compared to

IOP Conf. Series: Materials Science and Engineering **529** (2019) 012014 doi:10.1088/1757-899X/529/1/012014

SLM and other processes to manufacture preforms. However, the preforms from well controlled SLM process can be used to develop the manufacturing of preforms through KE process.

Initial centrifugal casting experiments performed at Access e.V. indicated well designed channels in SLM preforms are very effective for infiltration into the depths of a preform body (see Fig. 1). This aspect of development is currently explored within the EQUINOX project, to yield similar channels as those of SLM and with that a substantial improvement in infiltration depth.

Due to the complexity of the process multiscale modeling tools are being employed to guide the experiments. In this work, a simulation methodology based on computational fluid dynamics macroscopic simulations using STAR-CCM+[10] and microstructure simulations using MICRESS [11] is presented to obtain deeper understanding of the kinetics of solidification, the formation of intermetallic phases and the dynamics of Al melt infiltration into porous iron preforms, that will aid in design of channels in preform.

The structure of the paper is as follows: Section 2 presents modelling of infiltration and solidification. Section 3 summarizes the results for infiltration and solidification times. Section 4: concludes and summarizes the work.

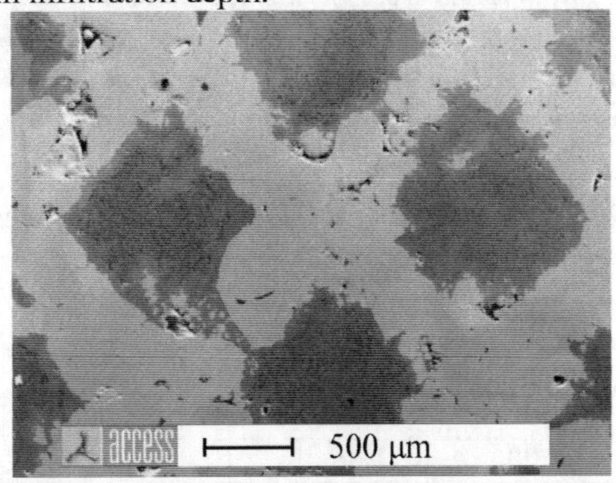

Figure 1: A metallographic image of a section of SLM sample after infiltration with Al.

2. Modelling and Simulation

2.1. Macrosimulation

Macrosimulations under isothermal conditions were performed to understand infiltration dynamics in these preforms. In the simulations two phases were considered, starting with gas filled preforms and variations considering surface tension, and contact angle effects, viscosity, pressure etc., were included in order to understand how these parameters affect filling of these preforms consisting of inlay channels.

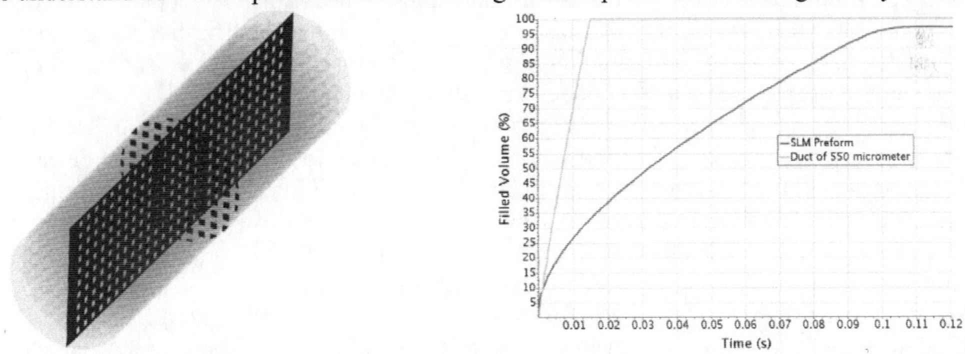

Figure 2: Left: Snapshot of CAD model of SLM manufactured preform (transparent view shows the entire geometry with internal structure, and cross-sections are shown in blue). **Right**: Comparison of filled volume of SLM preform (shown in Fig 1) and a simplified duct of 550 µm with inlet pressure of 20000 Pa, Surface tension coefficient of 0.9 N/m and wetting angle of 110°.

At first fluid flow in the centrifugal casting machine was simulated to obtain the typical melt pressure acting on the preform. This was obtained to be around 20000 Pa, which was used as inlet boundary conditions for subsequent investigations. Next filling simulations were then performed on an SLM preform (see Fig. 1) with an inlay channels of size of 550 µm and 40mm in length and a comparison of filling times are shown in Figure 2.

IOP Conf. Series: Materials Science and Engineering **529** (2019) 012014 doi:10.1088/1757-899X/529/1/012014

Simulations were performed on different square cross-section channel sizes, with a constant length of 4 cm. Results are shown in Table 1. A typical simulation setup consists of a square channel with inlet and outlet. At the inlet a boundary condition with a definite pressure is applied. All other boundaries are considered as walls with no slip boundary condition. The channel is assumed to be filled with air initially. Starting at simulation time t= 0 s, the melt is allowed to flow in with a pressure P = 20000 Pa. The simulations were performed using Al with 2700 Kg/m^3 density, 0.00125 Pa-s viscosity, 0.9 N/m Surface tension coefficient and the second medium was considered as air with 1.18415 Kg/m^3 density and 1.85x10^{-5} Pa·s viscosity. Initially the melt was assumed non-wetting (110° wetting angle) with the channel walls.

Table 1: Infiltration times for different square cross-section channel sizes and wetting angles.

No.	Channel Size (µm)	Wetting Angle (°)	Time for infiltration (s)	No.	Channel Size (µm)	Wetting Angle (°)	Time for infiltration (s)
1	550	110	0.01492	9	30	90	1.32
2	400	110	0.01842	10	30	80	0.655
3	200	110	0.056	11	10	90	8.8
4	100	110	0.354	12	10	80	2.51
5	100	90	0.145	13	3	90	Partial Infiltration
6	50	110	No Infiltration	14	3	80	6.8
7	50	100	1.4	15	1	90	Partial Infiltration
8	50	90	0.55	16	1	80	12

2.2. Thermodynamics
Prior to solidification simulations the thermodynamics of the Al-Fe system was investigated using the CALPHAD- (CALculation of PHAse Diagrams) software Thermocalc [12] and the High Entropy Alloys database [13]. Several solid phases occur in the Al-Fe system: The FCC aluminium phase, the intermetallic phases $Al_{13}Fe_4$, Al_5Fe_2, Al_2Fe and Al_5Fe_4, for low Al concentrations between 900°C and 1400°C FCC-iron (austenite) and a BCC-iron solid solution phase that can contain up to 50mol% Al. To account for the thermodynamics of phase transformations in the solidification simulations the phase field simulation is linked to the same thermodynamic database. Further details about the multiphase field model and the required material model parameters can be found in [14,15].

Because of its material properties the aim of the process is a material mainly consisting of the BCC-phase. In accordance with experimental work of Plevanchuk et al. [16] the average density of the Fe preform and the infiltrated Al melt is lower than the density of the intermetallic phase formed due to the reaction between the melt and the preform. This leads to volume shrinkage and thus porosity during the reaction. By infiltrating with an Al-Fe alloy instead of pure Al the above mentioned effects together with release of reaction heat are reduced or completely eliminated.

For this reason, and also to reduce formation of other intermetallic phases than the BCC phase infiltration with Al-Fe alloys of various compositions (25mol%, 33mol% and 50mol% Fe) was investigated in the solidification simulations.

2.3. Microsimulation
Phase field

Microstructure simulations were performed for solidification of melt filled SLM, PMM, MIM and KE preforms using the phase field method detailed in [17] and implemented in the software MICRESS [11]. Initial structures for the simulations were generated from CT-scans of the preforms. From these Representative Volume Elements (RVEs) were cut out and discretized on a regular grid. For SLM-structures an RVE of 400×400×250µm representing a quarter of a channel with surrounding material was discretized with a grid spacing of 2µm. For PMM-structures a region of 660×660×660µm was discretized with 3mm grid spacing and for KE structures 10×10×10μm were discretized with 0.05μm spacing. Simulations for MIM-structures 2D simulations were done on a 25×50 μm grid with 0.1 μm spacing. In addition a series of 2D simulations was performed for channels with a square cross section and channel width varying from 1 μm to 400 μm. In these simulations a quarter of the channel and surrounding preform material (representing an initial solid fraction of 75% of the RVE) was discretized on a 200×200 grid. The initial structures for 3d simulations are shown in figure 3. To avoid the whole preform being represented by a single grain an artificial grain structure was superimposed on the solid region as shown in Figure 3.

At the start of the solidification simulations the pores were filled with melt, assuming completed infiltration. For KE- and MIM- structures this assumption is not realistic, since solidification blocks infiltration paths very fast for these fine structures and applicable infiltration pressure is limited to avoid structural damage of the preform. Nevertheless from simulations for these hypothetical conditions general trends for reaction times and permeabilities can be gained that are transferable to other structures and length scales. The starting temperatures were chosen to be about 20K above the liquidus temperature,

Initial melt composition and starting temperatures were: 25mol% Fe, 1450K (SLM 2D, SLM 3D, MIM, KE), 33mol% Fe, 1470K (PMM, KE) and 50mol% Fe 1610K (SLM 2D, square channel, MIM, KE). As thermal conditions for SLM- and PMM-structures heat extraction rates of 5, 10, 20 and 50 W/cm^3 were investigated. For the finer KE- and MIM-structures the reaction times are much shorter and exothermic heat release would drive temperatures up by 300K under the above process conditions. Under such conditions the initial microstructure of the preform would become quickly irrelevant due to coarsening, which would determine the features of the resulting structure. Thus for KE- and MIM-structures isothermal conditions were investigated, assuming coupling to a heat reservoir at constant temperature. For better comparability these conditions were also chosen for the square channel structures.

Figure 3: Initial RVEs. From left to right: SLM, PMM, KE, KE with superimposed grain structure.

Diffusivity

An important (and under isothermal conditions the only) contribution to the solidification in this process is the exchange of Fe for Al between BCC phase (i.e. preform) and melt, so that the Fe-enrichment raises the liquidus temperature. Thus the diffusivity in the BCC-phase plays an important role, since it determines how fast Fe is transported to and Al from the solid liquid interface. In all microstructure simulations the temperature dependency of the diffusivity in the BCC-Fe-phase is described by an Arrhenius law. Based on literature data [18] for various alloy compositions Q = 240 kJ/mol and D_0 = $2\cdot10^{-3}$ m^2/s were chosen as Arrhenius parameters. The resulting diffusivities are $D_{1450}=4.52\times10^{-12}$ m^2/s, $D_{1470}=5.92\times10^{-12}$ m^2/s and $D_{1610}=3.27\times10^{-11}$ m^2/s.

Results

The SLM, PMM, MIM and KE microstructures are very different, specifically with respect to the length scales. While in the SLM structures the length scale of the porosity is in the range of several hundred µm, in the PMM structures the diameters of pores and connecting channels are initially some 10 µm and in KE structures some 0.1 µm. The specific surface S_V (or its inverse) is chosen to quantify the length scale. It is defined as interfacial area (of the melt solid interface) per volume. The value of S_V was determined for the initial structures and tracked during the solidification.

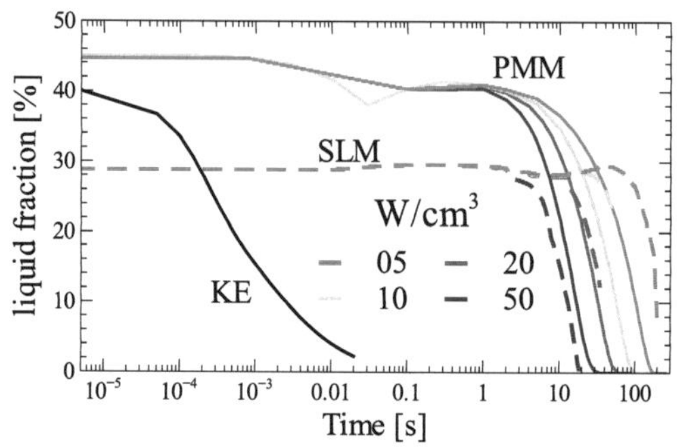

Figure 4: Solidification over time

The differences in length scale have a large impact on the time scale of the diffusion driven solidification and on the permeability as shown in figure 4. Compared are

- KE, infiltrated with Al33at%Fe with constant temperature condition (dotted, black)
- PMM, infiltrated with Al33at%Fe (solid lines) and
- SLM infiltrated with Al25at%Fe (dashed) at four cooling rates (see legend).

Figure 4 shows that in the fine KE structures it takes only several milliseconds for diffusion to increase the Fe-content in the melt, thereby solidifying it. For PMM structures solidification times vary from 30 to 190 s depending on the cooling rate, for SLM structures solidification takes over 200 s at the lowest cooling rate.

2.4. Analytic

To infiltrate an amount V_L of melt into a structure the energy $J_\sigma=\Delta\sigma\cdot S_L\cdot V_L$ is necessary in the non-wetting case to overcome surface tension. $\Delta\sigma$ is the difference between the interfacial energy of the free surface of the solid and the interfacial energy of the solid liquid interface and $S_L=S_V/f_L$ the surface per liquid volume. So a pressure $\Delta P_\sigma= J_\sigma/V_L =\Delta\sigma\cdot S_L$ is necessary to overcome surface tension during infiltration.

Additional pressure is needed to force the melt through the porous structure. For laminar flows this is described by the Darcy law that describes the dependency of the flux v_f (melt volume passing through an area per time and area) on the viscosity µ and a structure dependent permeability k. For the permeability k a simplified (approximating the tortuosity with 1) Carman-Kozeny rule can be applied:

$$k = \frac{1}{C_K}\cdot\frac{f_l^3}{S_V^2}$$

The Carman-Kozeny constant C_K depends on the structure. The pressure difference due to the Darcy permeability along the infiltration depth x is $\Delta P_{perm}=g_p\cdot x$. Inserting k given by Carman Kozeny in the Darcy law yields:

$$\Delta p_{perm} = g_p\cdot x = \frac{\mu}{k} v_f\cdot x = C_K\cdot\mu\cdot S_L^2\cdot v\cdot x$$

Here $v = \dot{x} = v_f/f_L$ is the velocity of the advancing liquid in the pore volume. With $\Delta p=\Delta p_s+\Delta p_{perm}$ and $2x\dot{x} = {}^{d}/_{dt}\,(x(t)^2)$ integration yields for an infiltration depth $x(t_l)=l$:

$$t_{kc} = \frac{C_K}{2}\cdot\frac{\mu\cdot S_L^2\cdot l^2}{\Delta p-\Delta\sigma\cdot S_L} \qquad (1)$$

For most isotropic structures a Kozeny-constant $C_K=5$ is a good choice [19]. For channels with round cross section one finds $S_L=U_{circ}/A_{circ}=4/d$, applying the Hagen-Poiseuille-equation results in $C_{K,circ}=2$ which reflects a higher permeability along the direction of the channels (but zero in perpendicular directions). For square channels $S_L=4/a$ and $C_{K,sq}\approx 1.78$ [20]. The denominator in equation (1) indicates, that under nonwetting conditions a pressure $\Delta p_{min}=\Delta\sigma\cdot S_L$ has to be exceeded to infiltrate the structure.

3. Discussion

The macrosimulations show, that for finer structures infiltration becomes harder and takes longer, while the phase field simulations show a trend of shorter solidification times for finer structures. For a given set of process and material parameters this results a lower limit for the length scale of structures that can be infiltrated before melt flow is stopped by solidification.

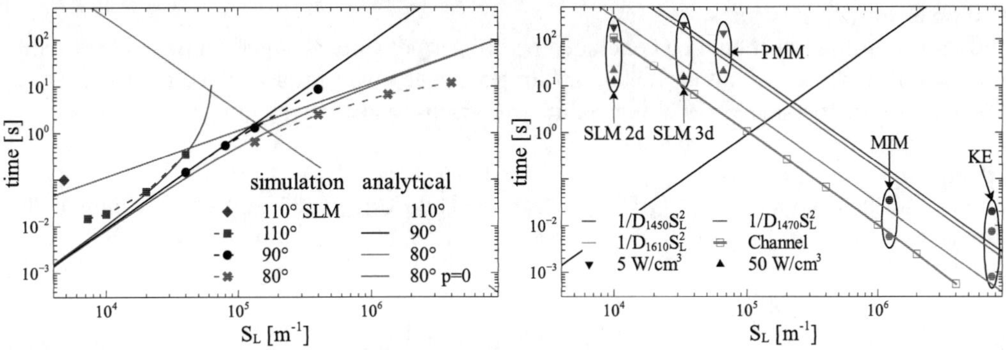

Figure 5: Left: infiltration times, **Right:** solidification times. For better comparability the $1/D_{1610}$-S_L^2 graph is shown with the infiltration times (red, dotted) and the analytical curve for 90° wetting angle with the solidification times (black).

Figure 5 shows infiltration and solidification times side by side, as a measure of (inverse) length scale the specific surface per liquid volume is used. On the left infiltration times for square channels resulting from the macrosimulations are compared to the analytical results for laminar flow through square channels from the previous section for the process and material parameters as discussed above. The numerical results show, that $t \sim S_L^2$ for $\Delta\sigma = 0$ (90° wetting angle), wetting conditions (green) result in faster infiltration and for non wetting conditions (blue) infiltration of finer structures becomes impossible for a given pressure.

The right hand side of figure 5 shows solidification times. As solidification time $t(f_s{=}95\%)$ is given, since solidification slows down when only small pockets of liquid remain, also a sharp decrease of permeability was observed between 90% and 95% solid fraction. Different melt compositions and starting temperatures are indicated by colour; blue for 25mol%Fe and 1450K, green for 33mol%Fe and 1470K and red for 50mol%Fe and 1610K. Triangles indicate heat extraction of 5 W/cm^3 (up) and 50 W/cm^3 (down), circles indicate isothermal conditions as do squares for the simulations of square channels.

As discussed previously, solidification depends strongly on diffusion in the BCC phase. A dimensional analysis (i.e. application of the Buckingham π theorem) shows that for a process only dependent on diffusion a scaling behaviour $t \sim (D \cdot S_L^2)^{-1}$ should be expected, so the functions $(D_T \cdot S_L^2)^{-1}$ are given as reference as dotted lines. For the coarser SLM and PMM structures the solidification time strongly depends on the heat extraction rate, but for the finer MIM- and KE-structures and isothermal conditions solidification times are governed by structure size. The times for the square channel simulations, which compare the same structure scaled to different sizes, follows the scaling rigidly, the results match $t_{95\%}=(3 \cdot D_{1610} \cdot S_L^2)^{-1}$. It should be noted, that for these simulations due to the high initial Fe- content in the melt no additional intermetallic phases nucleate, so during the solidification the channel shrinks in diameter while the inner melt phase is transformed into the outer BCC phase. At lower initial Fe concentrations in the melt intermetallic phases that form at the solid-liquid interface affect transport of Fe into the melt and hence solidification times.

4. Conclusions

- For non-wetting conditions a certain pressure must be overcome to infiltrate the structures, this pressure is larger for finer structures.

• The simulations show, that wetting conditions (wetting angles < 90°) are beneficial for infiltration. This effect is stronger in finer structures, and can predominate an applied pressure, as was shown for thinner channels or when comparing SLM, to finer PMM and much finer KE structures.
• The diffusivity in the BCC-phase determines how fast Fe can be transported to the solid liquid interface. This strongly affects solidification times, especially for finer structures and leads to short solidification times for fine structures.
• The condition that infiltration should be concluded before permeability is significantly decreased due to solidification limits which structures can be infiltrated under given process conditions (pressure, temperature) and material parameters (wetting conditions, viscosity, diffusivity).

Acknowledgement
This work was performed under project EQUINOX supported by Horizon 2020-European Union, under Grant Agreement No. 689510 and patent application by Kochanek, and co-workers, on EQUINOX manufacturing path is in progress.

References

[1] https://ec.europa.eu/growth/sectors/raw-materials/policy-strategy/sustainable-supply-global_en
[2] https://ec.europa.eu/growth/tools-databases/eip-raw-materials/
[3] https://ec.europa.eu/growth/tools-databases/eip-raw-materials/en/content/strategic-implementation-plan-sip-0
[4] T.B. Sercombe and G. Schaffer: Science 29 Aug 2003, vol. 301, Issue 5637, pp. 1225-1227, DOI: 10.1126/science.1086989
[5] C.T. Liu, E.P. George, P.J. Maziasz, J.H. Schneibel: Mat. Sci. and Eng. A, Vol- 258, Issues 1–2, (1998) pp 84-98, DOI: 10.1016/S0921-5093(98)00921-6
[6] Z. Zhang, D. Jones, S. Yue, P. D. Lee, J. R. Jones, C. J. Sutcliffe and E. Jones: Mat. Sci. Eng. C 33 (7) pp 4055-62 (2013), DOI: 10.1016/j.msec.2013.05.050
[7] Z. Zhang, L. Yuan, P. D. Lee, E. Jones and J. R. Jones: J. Biomed Mater Res B, Nov 2014, VOL 102B (8) pp. 1689-1699, DOI: 10.1002/jbm.b.31804
[8] www.equinox-project.eu
[9] W. Kochanek, US 6939509 METHOD FOR MANUFACTURING METAL PARTS.
[10] https://mdx.plm.automation.siemens.com/star-ccm-plus
[11] MICRESS®, the microstructure evolution simulation software, http://web.micress.de/
[12] Thermocalc Software, www.thermocalc.com, (accessed 1 August 2017).
[13] https://www.thermocalc.com/solutions/by-material/high-entropy-alloys/
[14] B. Böttger, J. Eiken, and M. Apel: Comput. Mater. Sci., 2015, vol. 108, pp. 283–92.
[15] A. Carré, B. Böttger, and M. Apel: J. Cryst. Growth, 2013, vol. 380, pp. 5–13.
[16] Y. Plevachuk, I. Egry, J. Brillo, D. Holland-Moritz and I. Kaban: Int. J. Mater. Res. 2007, 98, 107–111
[17] J. Eiken, B. Böttger, and I. Steinbach: Phys. Rev. E. Stat. Nonlin. Soft Matter Phys., 2006, vol. 73, p. 66122.
[18] M. Eggersmann and H. Mehrer: Philosophical Magazine A, 80:5, (2000) 1219-1244 DOI: 10.1080/01418610008212112
[19] T. Ozgumus, M. Mobedi and U. Ozkol: Engineering Applications of Computational Fluid Mechanics Vol. 8, No. 2, pp. 308–318 (2014)
[20] F. M. White, Fluid Mechanics (McGraw-Hill, 2003)

Recent advances in the understanding of the role of vanadium carbonitride precipitation to improve surface edge cracking on continuous casting of blooms

E Erdem Hornauer[1], U Grafe[1], U Plociennik[1], W Klos[1], M Reifferscheid[1] and M Luettenberg[2]

[1] SMS group GmbH, Eduard-Schloemann-Str. 4, 40237 Duesseldorf, Germany
[2] ArcelorMittal Duisburg, Vohwinkelstrasse 107, 47137 Duisburg, Germany

E-mail: esra.erdem.hornauer@sms-group.com

Abstract. This study aims in combining the material properties and numerical modelling techniques through practical application to provide understanding of surface edge cracking caused by V(C,N) precipitation to enhance the yield of the continuous bloom caster at ArcelorMittal Duisburg. The investigation for this work is carried out on three different micro-alloyed steel grades; one of them being the most crack sensitive 20MnV6. A process model is used which calculates the solidification process of the strand to design an optimum cooling strategy. Therefore, two cooling patterns are employed for the steel grade 20MnV6. Reduction of area values of the steel grades 20MnV6, 27MnSiVS6 and 38MnSiVS5 evaluated from hot tensile tests using a Gleeble simulator have been used as an indication in assessing steel's cracking behaviour. Further precipitates have been analysed by SEM at ArcelorMittal. These laboratory results suggest that precipitation kinetics of V(C,N) influences the crack sensitivity of the micro-alloyed steel. The software MatCalc® is used to simulate precipitation for process parameters of continuously cast blooms at Arcelor Mittal Duisburg as well as the parameters of the Gleeble experiments. From these simulations the Zener pinning force (ZPF), resulting from V(C,N) particles on grain boundaries, is evaluated which can be used as a measure for the crack sensitivity. Using the values of the ZPF it is possible to identify the set of casting parameters of the steel grade 20MnV6 which lead to the minimum crack intensity on the surface of the blooms. Moreover it is possible to make a ranking list with respect to the ductility drop occurring in the Gleeble experiments for the analysed steel grades. The proposed statement is an issue for a continuing investigation of precipitation modelling.

1. Introduction

Transverse edge surface cracking is a common long-standing problem in continuous casting which has been studied for decades and understanding is a still challenging task for steel producers. This review describes the control of the kinetics of V(C, N) precipitation resulting in the second ductility trough in the austenitic temperature range from 800 °C to 1200 °C of the investigated steels and eliminating the edge cracking during straightening operations in the bloom casting. The investigation for this work is carried out on three different vanadium containing micro-alloyed steel grades.

The SMS group developed a process model X-Pact® Solid Control (former DSC® Dynamic Solidification Control) which calculates the solidification process of the strand and controls the various zones of the secondary cooling system [1]. An optimum cooling strategy depending on the temperature range of low ductility of each steel has to be designed to prevent transverse crack formation. Therefore, according to the experience of the casting practice at ArcelorMittal Duisburg two cooling patterns are employed for the steel grade 20MnV6. Reduction of area values of the steel grades 20MnV6, 27MnSiVS6 and 38MnSiVS5 from the hot tensile tests using the thermo-mechanical simulator Gleeble HDS V-40 has been used as an indication in assessing steel's cracking sensitivity. Further precipitates have been analysed by SEM at ArcelorMittal. These laboratory results suggested that precipitation kinetics of V(C,N) influences the crack sensitivity of the micro-alloyed steel.

Grain boundary precipitates are generally detrimental to ductility because they promote void formation during grain boundary sliding. When grain boundary sliding in austenite is active as a process of plastic deformation, micro-alloy particles at the austenite grain boundaries will encourage the development of intergranular failure, leading to increased steel brittleness. Large volume fractions of fine precipitates are likely to increase strength and refine the grain size, but they also reduce the ductility of the steel owing to a more effective pinning action of grain boundaries, allowing more time for crack nucleation [2-5].

The thermo-kinetic software MatCalc® is used for simulating the different casting process parameters as well as the parameters of the Gleeble experiments. From these simulations the Zener pinning force (ZPF) resulting from V(CN) particles on grain boundaries, is evaluated which can be used as a measure for the crack sensitivity.

The hindrance of grain boundary migration by precipitates due to a back driving force on the grain boundary is known as Zener pinning [6]. The Zener pinning force characterises the particle drag to inhibit the grain boundary migration by taking into account the volume fraction and also the size of the precipitates. The respective retarding force (ZPF) is expressed by

$$ZPF\ (k,i) = \frac{3}{2}\gamma \sum_i \sum_k \frac{f_{k,i}}{r_{k,i}}$$

where f is the volume fraction of the particles, r the mean precipitate radius, γ the grain boundary energy, i is to account for the different precipitate types and k the number of size classes [7]. In technical applications, the presence of precipitates resulting in Zener pinning is used to control the grain growth during heat treatment and thermo-mechanical processing to achieve better mechanical properties [7-8].

In the temperature regime of continuous casting, creep is the predominant effect of plastic deformation reducing stresses arising from mechanical loading and shrinkage. In case that the volume fraction is large and the size of the precipitates is small the ZPF is maximised. These conditions lead to a reduction of creep deformation (grain boundary sliding) and thus reduce the ability to compensate external stress by plastic deformation. In other words the material becomes brittle and the crack susceptibility increases. Therefore, we propose the Zener pinning force to characterise the crack susceptibility in the second ductility trough which is caused by the precipitates.

Using the values of the ZPF it is possible to identify the set of casting parameters of the steel grade 20MnV6 which lead to the minimum crack formation on the surface of the blooms. Moreover it is possible to make a ranking list with respect to the ductility drop occurring in the Gleeble experiments employing the values of the ZPF for the analysed steel grades. The proposed statement is an issue for a continuing investigation of precipitation modelling.

2. Simulation data

The thermo-kinetic simulations were performed with the MatCalc software, version 6.00 release 0.256 using the thermodynamic database mc_fe_2.059.tdb and kinetic database mc_fe_v2.011.ddb. In the simulations, the choice of the type of nucleation sites determines the number of potential nucleation sites in the evaluation of the precipitate nucleation rate. The nucleation site for the V(C,N) precipitate phase at austenite grain boundary was defined according to the SEM fracture surface observations and EPMA investigations at ArcelorMittal previously presented in the paper from Leuschke et. al. [9]. Since the aim of the present paper is not the calculation of grain growth, the grain boundary energy was set to 1 J/m^2 in the simulations.

The simulations of precipitation kinetics for the as-cast material should not be carried out with the nominal compositions of steels due to micro-segregation. Scheil-Gulliver simulations are performed to estimate the segregation of the alloying elements in the interdendritic spaces and the phase fractions of primary precipitates during solidification. The nominal compositions of the steels are given in Table 1. Back diffusion of the fast diffusing elements carbon and nitrogen as well as the peritectic transformation are taken into account. The interdendritic compositions for which the precipitation simulations were conducted are evaluated from the micro-segregation simulation at a fraction of solid of 0.97 [10].

Table 1. Chemical composition (wt.-%) of the investigated steel grades

	C	Si	Mn	P	S	Al	N	Ti	V	Nb
20MnV6	0.220	0.440	1.580	0.015	0.030	0.020	0.021	0.003	0.180	-
27MnSiVS6	0.289	0.060	1.430	0.014	0.024	0.014	0.018	0.013	0.088	-
38MnSiVS5	0.375	0.565	1.355	0.013	0.068	0.026	0.015	0.002	0.099	0.003

The precipitation kinetics which are based on the temperature histories are calculated for two cases. For the first case, the simulation of the precipitation kinetics of V(C,N) on grain boundaries (gb) is performed for the temperature profile of the Gleeble experiments. For the second case of simulations, simplified surface edge temperature profiles for two cooling patterns of the continuously cast blooms at ArcelorMittal Duisburg via the thermal process model are applied. The two cooling strategies have been adapted to the temperature-dependent ductility drops especially during unbending to prevent transverse edge cracking on the bloom surface. Thermal simulations and temperature measurements have shown that the austenite-ferrite transformation does not take place at the edges of the bloom within the strand guide. The calculated A_{e3} temperature of the interdendritic composition of the three steel grades is lower than 790 °C which is below unbending temperature. Therefore, the kinetic simulations are performed by only accounting for the V(C,N)gb precipitation within the austenitic domain without consideration of the austenite-ferrite transformation which can be detrimental to ductility (third ductility trough).

3. Hot ductility curves

The plots of experimentally determined reduction of area (RoA) values against test temperatures for all steels examined in the present work are shown in figure 1. Over the whole temperature range the hot ductility is slightly decreased to values about 70 % which might be attributed to the formation of manganese sulphide in the solid. MnS precipitation in the solidification interval will produce coarse precipitates which should not significantly affect hot ductility. Of all the steels, 38MnSiVS5 exhibits the best ductility in the temperature range of 1000 - 950 °C followed by 27MnSiVS6 and 20MnV6. Due to the falling temperature a decrease in the RoA values is observed between 900 and 800 °C for steel 38MnSiVS5 but it exhibits higher values compared to the other investigated steel grades. It is therefore assumed to be less susceptible to cracking during unbending. The steels 20MnV6 and 27MnSiVS6 show a similar hot ductility behaviour with little variations in RoA values especially in the temperature range of the unbending process.

IOP Conf. Series: Materials Science and Engineering **529** (2019) 012015 doi:10.1088/1757-899X/529/1/012015

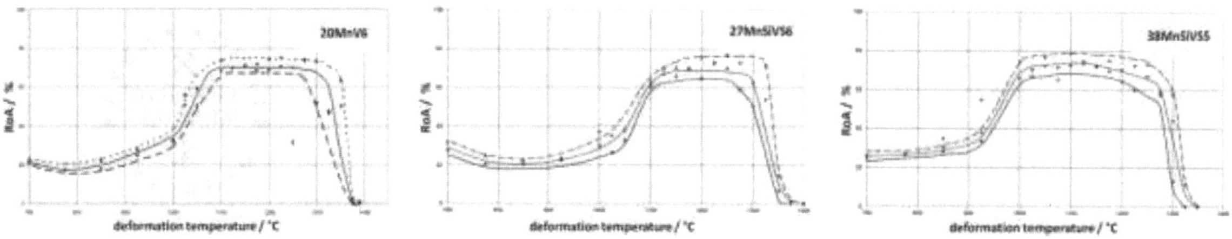

Figure 1. Hot ductility curves of the examined steels.

4. Results of thermo-kinetic simulations

4.1. Thermo-kinetic simulation of Gleeble curves

Results of thermo-kinetic simulations for V(C,N)gb precipitates in steel 20MnV6 are summarized in figure 2, showing the volume fraction, number density, size and the resulting Zener pinning force for a Gleeble temperature profile as used in the simulations. A particle volume fraction of 10^{-7} at the grain boundaries was suggested to be enough to obtain the predicted precipitation start temperatures. At 963 °C, precipitation of V(C,N)gb starts in steel 20MnV6 and at the same time the Zener pinning force starts to increase due to continuous nucleation of new fine precipitates.

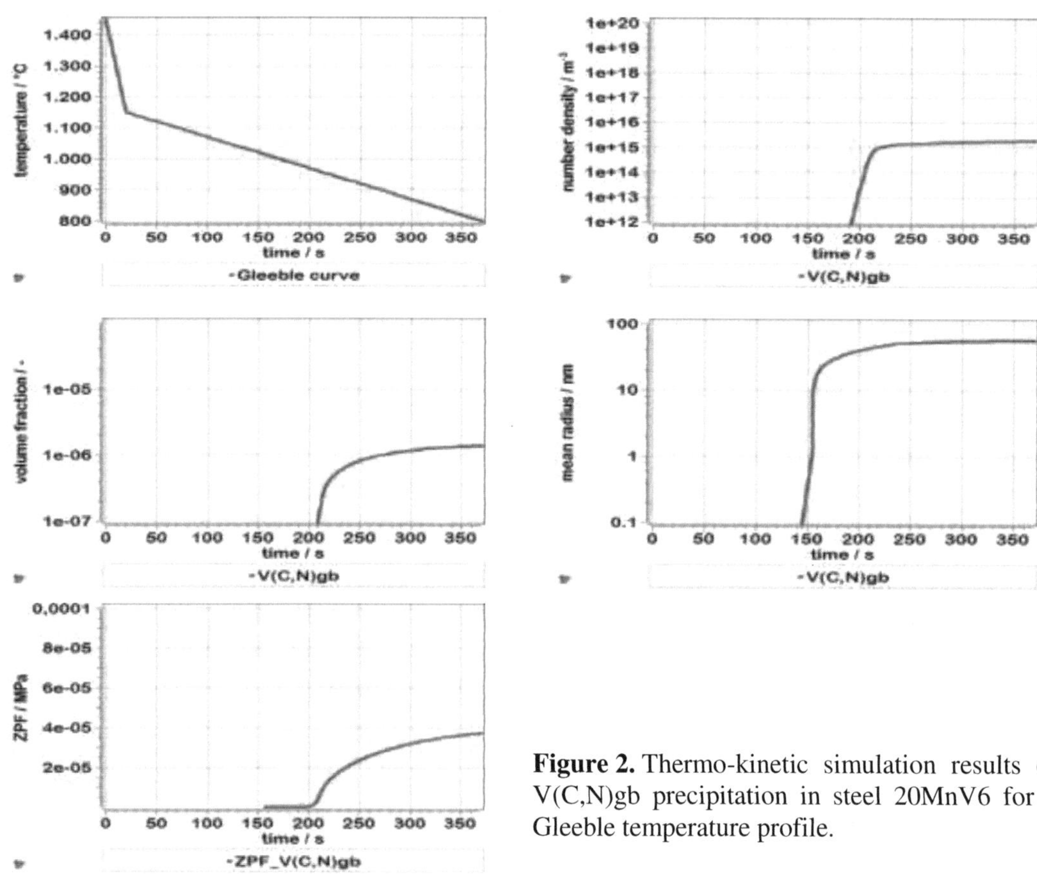

Figure 2. Thermo-kinetic simulation results of V(C,N)gb precipitation in steel 20MnV6 for a Gleeble temperature profile.

Figure 3 shows the comparison of RoA values for three steel grades related to figure 1. Mintz [3] evaluated a critical value of 40 % RoA from his experimental data (valid for a certain strain rate), which is necessary to prevent transverse surface cracking during continuous casting, which is drawn as dashed line in the comparison diagram. It should be noted that the measured temperature-dependent hot ductility of the steels shows mainly the general trend of the loss in ductility influenced by combined effects of such as MnS, V(C,N) and (AlN) formation. Lowest RoA values indicate a high cracking sensitivity and poor ductility of the steel. Since the steels 20MnV6 and 27MnSiVS6 show a similar hot ductility loss in the temperature range of unbending operations during continuous casting, it cannot be estimated which of the steels 20MnV6 and 27MnSiVS6 has a higher cracking risk. According to the casting practice it is known that 20MnV6 exhibits higher crack intensity.

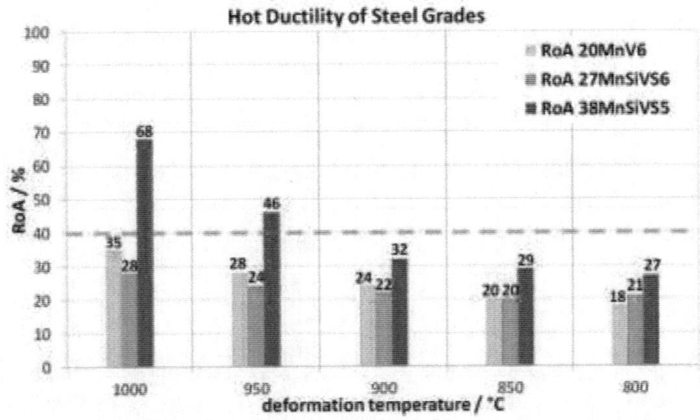

Figure 3. Gleeble experiments: Temperature-dependent hot ductility values for the examined steel grades related to figure 1.

Zener pinning force simulation results for V(C,N)gb precipitation obtained for the Gleeble temperature profiles can be seen in figure 4. The results show that values of the ZPF for V(C,N)gb correlate with the RoA values: Higher ZPF values have an adverse effect on the ductility of the material, meaning that reduction of area values decrease as the ZPF increases. Figure 4 shows that 20MnV6 is most crack sensitive.

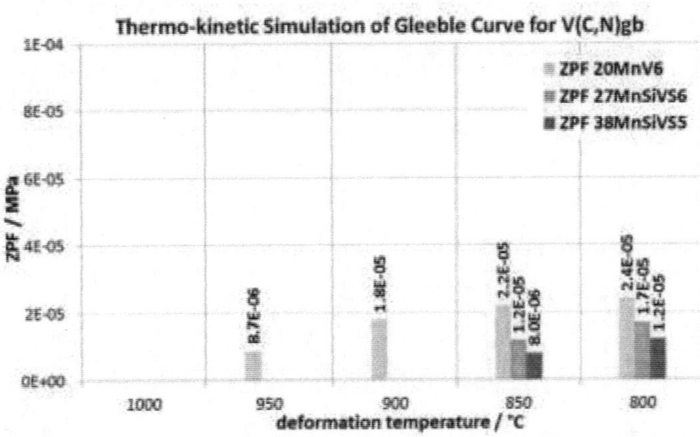

Figure 4. Gleeble experiments: Simulation results of the Zener pinning force for V(C,N)gb precipitation in the examined steel grades.

4.2. Thermo-kinetic simulation of continuously cast bloom edge temperature profiles

Figure 5 presents the results of kinetic simulations of V(C,N)gb precipitation based on the two simplified surface edge temperature profiles. According to the experience of the casting practice at ArcelorMittal Duisburg the best results have been achieved for the prevention of edge cracks with a soft cooling strategy in steel grade 20MnV6. The surface target temperature for the edges is increased by 70 K during unbending (at a time of 1300 sec.). With this cooling strategy a significant increase in the ductility as a consequence of reduced V(C,N) precipitation can be observed. According to the simulation results the precipitation of V(C,N)gb starts earlier applying a hard cooling than soft cooling. When comparing two cooling patterns, it can be seen that soft cooling profile induces lower Zener pinning forces at the entry of unbending zone and reaches to a similar level of hard cooling at the end of unbending. The kinetic simulations carried out confirm that by applying a soft cooling for steel 20MnV6 the surface edge cracking risk can be minimised.

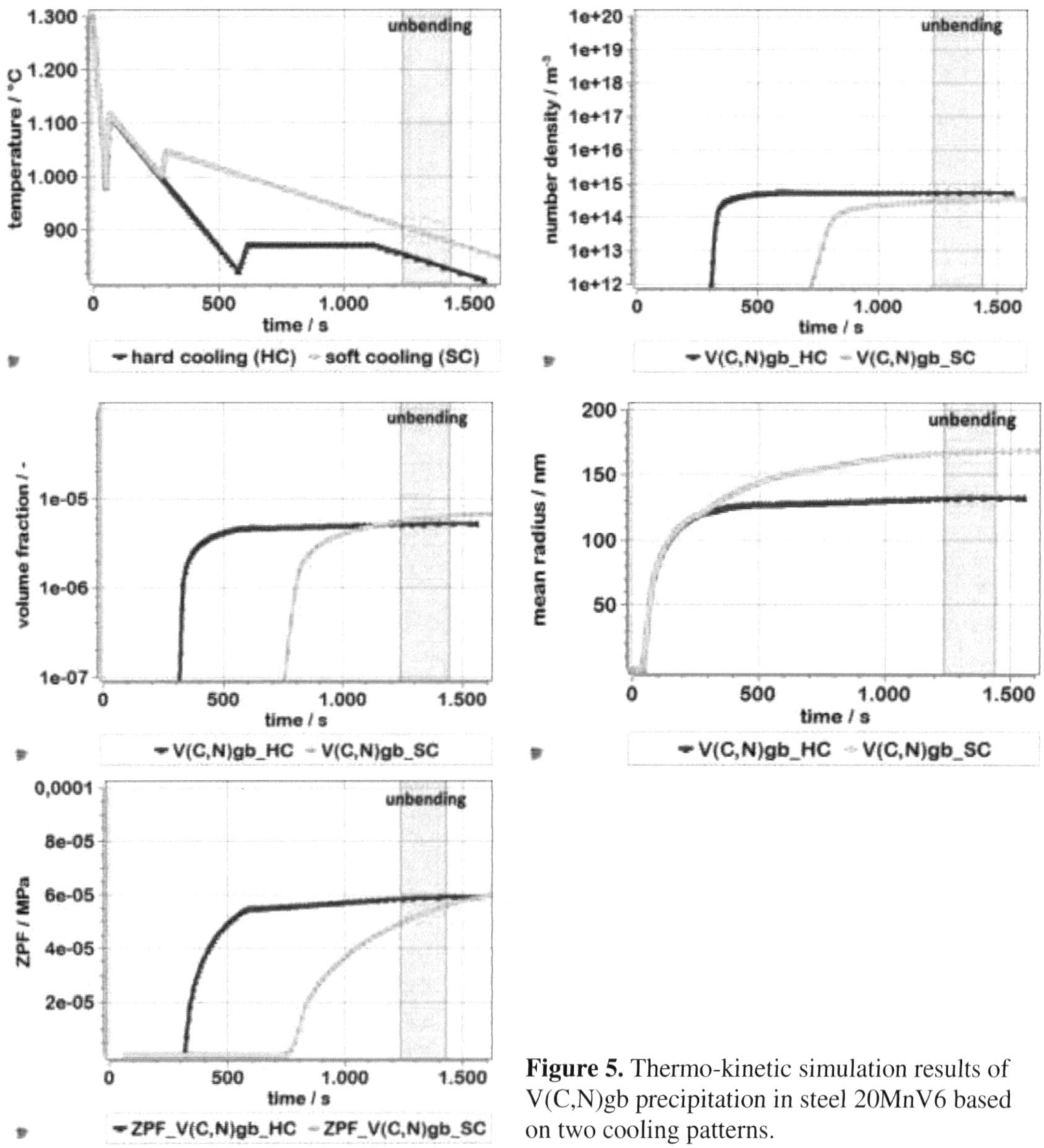

Figure 5. Thermo-kinetic simulation results of V(C,N)gb precipitation in steel 20MnV6 based on two cooling patterns.

IOP Conf. Series: Materials Science and Engineering **529** (2019) 012015 doi:10.1088/1757-899X/529/1/012015

5. Conclusion

In this paper, a new approach is introduced to characterise the transverse surface edge cracking susceptibility below the upper temperature limit of the second ductility trough which is caused by precipitates by means of thermo-kinetic modelling. It is not sufficient to assess the precipitate radius and volume fraction separately to estimate the crack sensitivity of the steel. The new approach combines the pinning effect of second phase particles on grain boundaries during continuous casting of blooms with ductility values resulting from tensile tests directly after melting and re-solidifying of the samples using a Gleeble machine. The Zener pinning force (ZPF) effect is particularly important when the volume fraction of precipitates becomes large and the mean size becomes small by generating ductility loss in the critical temperature range where the strand unbending takes place.

From the simulations it is shown that an increase in Zener pinning forces resulting from V(C,N) particles on grain boundaries can be used as an indicator for ductility in the studied vanadium containing micro-alloyed steel grades. Using the values of ZPF, it is possible to predict the edge cracking risk on the surface of the blooms for different process parameters of continuously cast blooms. On the basis of thermo-kinetic calculations, improved process parameters can lead to significant reduction in crack formation and thus to an increase in casting throughput.

Further simulations for precipitation under casting conditions are planned to be conducted to study the effect of Zener pinning force on the crack sensitivity of vanadium micro-alloyed grades 38MnSiVS5 and 27MnSiVS6.

References

[1] Reip T, Barani A, Stevermuer F, Plociennik U, Grafe U, Reifferscheid M and Wessely A 2015 *European Stainless Steel & Duplex Stainless Steel Conf., 8th,* Graz, AT, April 28-30, pp 157-165

[2] Vedani M, Dellasega D and Mannuccii A 2009 *ISIJ International* **49** No.3 446-52

[3] Mintz B 1999 *ISIJ International.* **39** No.9 833-55

[4] Mintz B, Yue S and Jonas J J 1991 *Int. Materials Reviews* **36** No.5 187

[5] Mohamed Z 2002 *Materials Science and Engineering* **A326** 255-60

[6] Gottstein G 2003 *Physical Foundations of Material Science* (Berlin Heidelberg: Springer)

[7] Buken H and Kozeschnik E 2017 *Metallurgical and Materials Transactions A* **48** No.6 2812-18

[8] Apel M, Böttger B, Rudnizki J, Schaffnit P and Steinbach I 2009 *ISIJ International* **49** No.7 1024-29

[9] Leuschke U, Kirsch D, Grafe U, Plociennik U and Reifferscheid M 2017 *stahl und* eisen **137** No.5 63-69

[10] Six J, Wojcik T, Ilie S and Kozeschnik E 2013 *ISIJ-VDEh-Jernkontoret Joint Symposium, 14th VDEh-ISIJ Seminar on Science and Technology of Process Metallurgy, 8th Japan-Nordic Countries Joint Symposium on Science and Technology of Process Metallurgy (Osaka,* pp 173-181

Control of Solidification Pattern of Cast Irons by Simultaneous Thermal and Contraction / Expansion Analysis

I Riposan, S Stan*, M Chisamera, L Neacsu, A M Cojocaru, E Stefan, I Stan

Materials Processing & Ecometallurgy, POLITEHNICA University of Bucharest, 313, Independentei Spl., 6th District, 060042 Bucharest, ROMANIA

*Corresponding author: constantin.stan@upb.ro

Abstract. An experimental device conducts thermal analysis and volume change measurements in a single ceramic cup with cast iron quality as the variable. The recorded data are processed using specialized software. Experiments compare solidification patterns for white [WI], grey [GI] and ductile [DI] irons, to correlate the most important events between the cooling curves and contraction curves, to evaluate the sensitivity to shrinkage formation. All of the irons have similar values for initial expansion up to the start of eutectic freezing [0.437 – 0.443%]. Graphite formation promotes expansion [WI-0.002%, GI-0.109%, DI-0.596%], resulting a difference in the reached maximum expansion [WI-0.465%, GI-0.552%, DI-1.032%], placed between the end of eutectic recalescence and the end of solidification. Higher graphite expansion, greater the shrinkage sensitivity: open shrinkage increased while the density of a casting when considering total shrinkage and micro-shrinkage formation decreased. Prolonged graphitization at the beginning of eutectic reaction increased the expansion and, consequently, shrinkage sensitiveness. More graphite formation at the end of this stage also increased expansion, but this phenomena contributed to reduce of shrinkage level, due to the better access of liquid iron to compensate contraction holes. Special metallurgical treatments can favour a strong graphitization process at the end of solidification, with beneficial effects on the castings soundness.

Keywords: white, grey and ductile cast irons; solidification; thermal analysis; cooling & contraction curves; graphite expansion; shrinkage

1. Introduction

Contraction defects, such as macro and micro-shrinkage usually characterize iron castings, depending on the solidification pattern, specifically for carbides or graphite precipitation during solidification and graphite morphology type, too. Micro-shrinkage formation in inter-eutectic cells region is also favoured by a strong segregation of carbides-promoting elements in these regions.

Mould deformation and enlargement tendency, volume-to-surface area ratio (cooling modulus) of casting, chemical composition of cast iron [especially expressed by carbon equivalent], graphite characteristics [morphology, count, size], carbides [amount and distribution] and eutectic cells [size, count, distribution] characteristics, pouring temperature, metallurgical treatments [spheroidising and inoculation] act as the most important factors influencing the contraction defects. [1-9]

Thermal analysis testing, involving the cooling curve itself, as well as its derivatives and related parameters becomes an indicator of cast iron quality. It is important to measure the energy signatures of different phases and structures in cast iron and interpret them into usable chemistry and microstructure readings. Solidification thermal [cooling curve] analysis by thermocouple positioned in the mould was combined with the measurement of linear displacement during solidification by different techniques to obtain more information on solidification pattern and iron castings characteristics. Using specialized software, both cooling and contraction / expansion curves and their pertinent readings are able to be displayed simultaneously. [9 - 20]

The main objective of the present paper is to summarize the performance of this equipment illustrated by representative shrinkage tendency results in ductile iron as affected by mould rigidity (green sand and furan resin sand moulds), and its adaption for commercial foundry use, by comparing the solidification patterns for white, grey and ductile cast irons, focusing on the early stages of solidification, previously published in part. New data added, to deeply correlate the most important events between the cooling and contraction curves, to evaluate the sensitivity to shrinkage formation.

2. Background

In previous laboratory experiments [9 – 12, 19], a technique incorporating two moulds was used to test different inoculants (FeSi-based alloys, including in different variants Ca, Ba, Zr, Al, Ce, S, O), different inoculant additions, different mould media (green sand and furan resin sand moulds), and different iron chemistries (hypo and hyper-eutectic compositions). The ductile iron shrinkage sensitiveness is mainly depending on the mould rigidity. A much higher level of eutectic expansion $(\varepsilon_{di})_{max}$ was obtained in green sand mould, as compared to furan resin mould, and consequently, the highest level of both concentrated and total shrinkage was recorded (Fig. 1). With higher maximum initial expansion (shortly before the end of solidification) and/or greater under-cooling (more negative) at the end of solidification, there is an increase in concentrated and dispersed shrinkage volumes, which lowers the apparent density especially of ductile iron castings. [19]

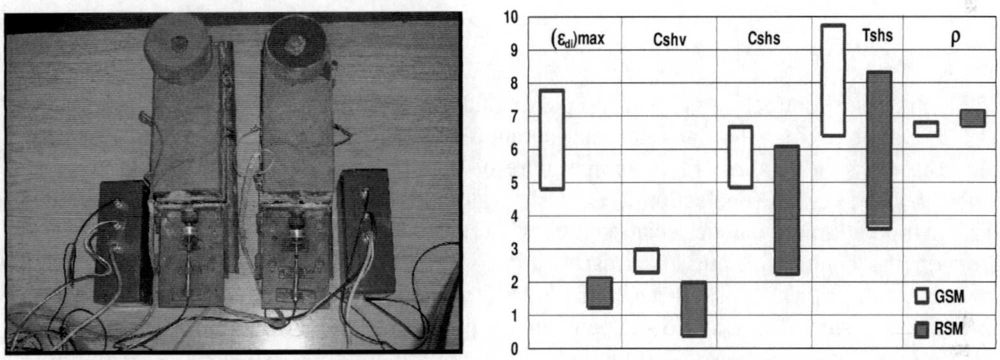

(a) (b)

Fig. 1. Arrangement of twin mould cooling and contraction / expansion curves analyser (CCCA) (a) and maximum initial expansion [(ε_{di})max x10, %], concentrated shrinkage volume [Cshv, vol.%], concentrated [Cshs, cm^2] and total [Tshs, cm^2] shrinkage surface and apparent density [ρ_1, g/cm^3] of inoculated ductile irons solidified in green sand moulds (GSM) and resin sand moulds (RSM) (b). [11]

3. Experimental procedure

The dual purpose technique was adapted to apply in the melt shop, similar to current thermal analysis systems (Fig. 2a). [11] The intent was to provide a simple-to-use system for comprehensive evaluation of treated iron quality, as the variable, under the same mould media solidification conditions. A special ceramic cup was designed, to include a thermocouple (for cooling curve analysis) and a device for contraction and expansion curve analysis. A shell mould was formed using phenol-formaldehyde resin (Novolak) coated sand (Croning process), with a high rigidity (cca. 95 Dieter Hardness number). This sample (750g) is characterized by a cooling modulus (CM) of 0.70cm, close to the standard ceramic cup – Quik-cup™ (0.75cm). This equipment was upgraded for more accuracy data registration, including a high speed type PCI-6284 interface to record simultaneous temperature and linear displacement data. [13] Recently [13, 20] this equipment was used to compare solidification pattern of hypo-eutectic grey [GI] and ductile [DI] irons with white [WI] iron as reference.

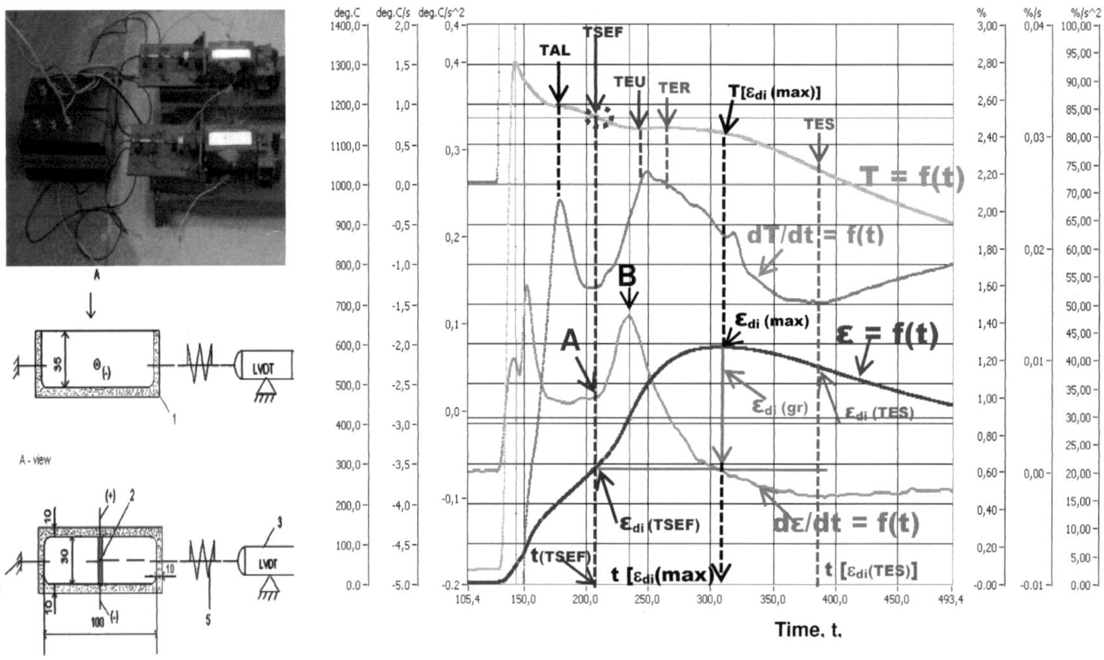

(a) (b)

Fig. 2 Cooling and contraction / expansion curve analyser (a) and typical Cooling Curve [T = f(t)] and its first derivative [dT/dt = f(t)] and contraction curve [ε = f(t)] and its first derivative [dε/dt = f(t)] for ductile cast iron solidification. [TAL-temperature of austenitic liquidus, ^{0}C; TSEF-temperature of the start of eutectic freezing (nucleation), ^{0}C; TEU-temperature of the maximum eutectic undercooling, ^{0}C; TER-temperature of eutectic recalescence, ^{0}C; TES temperature of the end of solidification, ^{0}C; A-position on the contraction curve corresponding to TSEF moment (the beginning of the intensive eutectic graphitization); ε_{di}(TSEF)–expansion before the start of eutectic freezing, %; ε_{di}(max)–maximum expansion, %; ε_{di}(TES)–expansion at the end of solidification, %; ε_{di}(gr)=ε_{di} (max)-ε_{di}(TSEF), graphitic expansion, %; B-maximum value of the graphitic expansion acceleration].

Experimental grey and ductile cast irons were obtained by standard procedures, including inoculation for grey irons (0.15wt.% Sr-FeSi alloy, ladle treatment) and double treatment for ductile irons (a tundish cover nodularization technique, 2.0wt.% FeSICaMgRE alloy followed by 0.5wt.% Ca,Ce,S,O-FeSi alloy, ladle inoculation). By Te-addition to the grey iron melt into the ceramic cup of cooling-contraction analysis equipment, before solidification, a non-graphitic (carbidic) structure resulted, or white (WI) cast iron, respectively. Hypo-eutectic cast irons were considered for these experiments, according to obtained carbon equivalent (CE) level: CE = 3.56 – 3.77% for grey cast irons, CE = 3.98 – 4.11% for ductile cast irons and CE = 3.62 - 3.64% for white cast irons.

The experiments were repeated, [2 samples for each heat] with 8 heats tested for grey irons, 5 heats for ductile irons and 2 heats for white irons. Each heat was controlled to ensure similar thermal and chemistry histories. The cast iron chemistries for the base chemical composition and level of minor elements were evaluated in the base irons and treated irons by the use of a performance spectrometer (SPECTROLAB M 10, Hybrid Optic) capable of determining very low levels of minor elements.

4. Results and Discussion

Figure 2b shows a typical assembly of simultaneous recorded solidification cooling and contraction curves, including their first derivatives. Previous preliminary reported results (Fig. 3) [13, 20] illustrate specific behaviour of the tested cast irons, as cooling and contraction / expansion parameters.

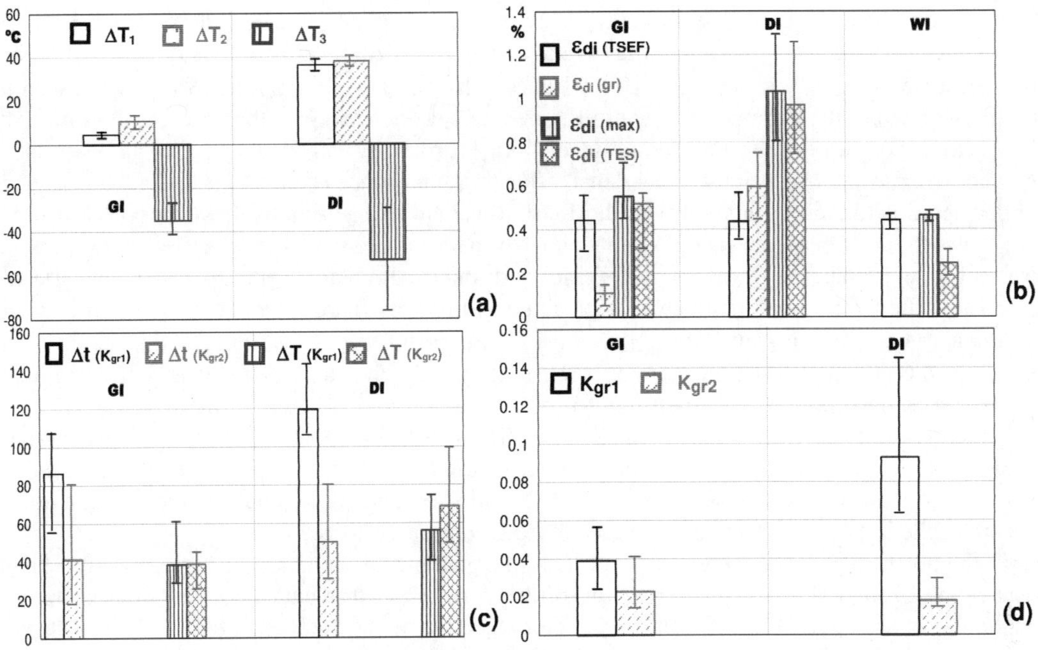

Fig. 3 Undercooling comparing to metastable eutectic temperature (a), specific expansion stages (b), time (Δt) and temperature (ΔT) variation (c) and acceleration (K_{gr1}) and deceleration (K_{gr2}) of graphitic expansion (d), before and after reaching the maximum expansion values.

Grey cast iron castings [GI] are characterized by greater undercooling during the eutectic reaction referring to the metastable (Tmst) equilibrium eutectic temperature, in iron-carbide system diagram [lower ΔT_1 = TEU – Tmst and ΔT_2 = TER - Tmst], while ductile iron castings [DI] shown a greater (more negative) undercooling at the end of solidification [ΔT_3 = TES - Tmst] (Fig. 3a).

All of the tested cast irons have similar values for initial expansion up to the start of eutectic freezing [$\varepsilon_{di\ (TSEF)}$ = 0.44%] due to the ferro-static pressure, silica sand mould expansion, mould movement etc. A-point (Fig. 2b) illustrates the separation between the mould expansion and graphitic expansion. The maximum expansion [$\varepsilon_{di\ (max)}$], reached between the end of the eutectic recalescence and the end of solidification, depends on the carbides / graphite ratio and graphite morphology: WI-0.465%, GI-0.552%, DI-1.032%, as averages. Graphitic expansion [$\varepsilon_{di(gr)}$], absent for WI, increased to 0.109% [GI] and up to 0.596% [DI]. For both GI and DI the expansion at the end of solidification [$\varepsilon_{di\ (TES)}$] is only 6% lower compared to the maximum level, while for WI it decreased more than 50% because of matrix contraction (Fig. 3b).

Specifically the acceleration rate of the graphite expansion up to the maximum level [K_{gr1}] is different compared to its deceleration to the end of solidification [K_{gr2}], and also are different for irons being tested, GI versus DI. These parameters were defined in a previous paper [13] by the difference in expansion [$\Delta\varepsilon_{di}$], in decreasing temperature [ΔT] and increasing time [Δt] values in these two sections [$Kgr = \Delta\varepsilon_{di} / (\Delta T \times \Delta t)$, μm/s.^{0}C]. Nodular graphite led to the [K_{gr1}] being 2.5 times higher, compared to GI, whereas only a slight difference was observed between GI and DI for the K_{gr2} (Fig. 3c,d).

There were found good relationships between graphite expansion [ε_{di}(gr)] and both specific factors K_{gr1} and K_{gr2} for grey cast irons: higher ε_{di}(gr), higher K_{gr1}, but lower K_{gr2}. For ductile irons, the same relationships present some irregularities: higher expansions, higher K_{gr1} factor, or the acceleration of expansion up to the maximum values, in the same way as in grey irons, but with less dependency. The K_{gr2} factor does not appear to be in a clear relationship with both ε_{di}(gr) and ε_{di}(max) expansions.

Higher graphite expansion in DI led to higher shrinkage sensitiveness than in GI, as measured in furan resin mould test castings: open shrinkage increased [Sh = 1.99 cm^3/kg (GI) versus 4.33cm^3/kg (DI)] while the real density of a casting when considering total shrinkage and micro-shrinkage formation decreased [ρ = 7.18 kg/cm^3 (GI) versus 6.73 kg/cm^3 (DI)] (Fig. 4a). More analysis on the obtained results led to the summarized data in Figures 4b,c,d. Increasing of the acceleration rate of the graphitic expansion up to the maximum level [K_{gr1}] promoted an increasing of the shrinkage sensitiveness, expressed by decreased real density due to the presence of contraction defects. Contrary, higher deceleration to the end of solidification [K_{gr2}], lower shrinkage sensitiveness for both irons.

Not only the graphite expansion $\varepsilon_{di}(gr)$ and resulted maximum expansion level $\varepsilon_{di}(max)$ are important to determine the level of shrinkage and micro-shrinkage, but also the solidification characteristics during the maximum expansion landing process. Increasing the length (time) of the expansion at the maximum level $\Delta t_{[\varepsilon di(max)]}$ favours the shrinkage formation, as the result of the increasing of the graphitic force on the mould walls (Fig. 4c). The same effect was registered as the effect of the specific temperature drop during the maximum expansion landing time $K_{[\varepsilon di(max)]}$ (Fig. 4d). Experiments pointed out that the graphite expansion has two contrary effects during graphite formation process. Increased graphitic expansion (force) led to higher shrinkage sensitiveness during the first part of eutectic reaction, but to the decreasing of shrinkage at the end of solidification, due to the forcing of the last liquid iron to occupy the previous formed cavities.

From the practical point of view, it is important to apply special metallurgical treatments to favour a strong graphitization process at the end of solidification. The test data acquired offer a possible explanation of other reported results [2, 3, 9-11] on reduced shrinkage tendency in ductile iron castings when using Ce,Ca,S,O-FeSi inoculation or La-MgFeSi nodularisation treatment. This is attributed to a bimodal or skewed nodule size distribution (fewer large nodules formed at the start of solidification and a high number of much smaller nodules formed later, towards the end of solidification).

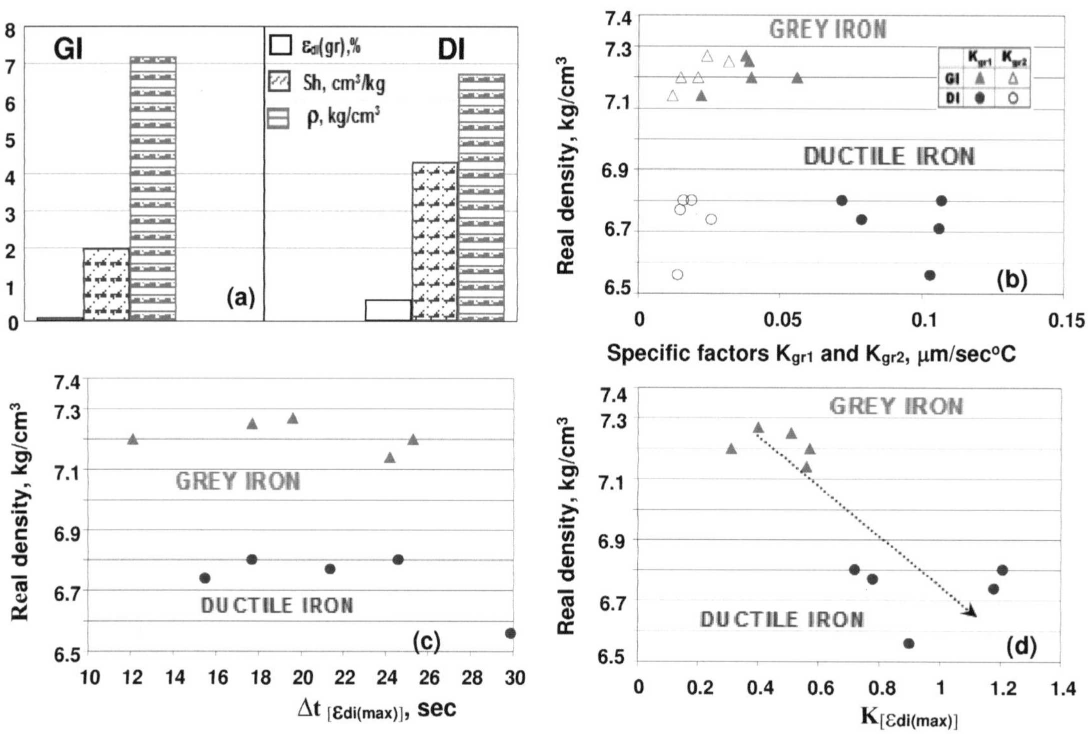

Fig. 4 Influence of graphite expansion value (a), specific factors $\mathbf{K_{gr1}}$ and $\mathbf{K_{gr2}}$ (b), maximum expansion landing time $\mathbf{\Delta t[\varepsilon_{di(max)}]}$ (c) and the temperature drop during the expansion landing time $\mathbf{K\ [\varepsilon_{di(max)}]}$ (d) on the shrinkage sensitiveness expressed by open shrinkage **(Sh)** and real density **(ρ)**.

5. Conclusions

*A ceramic cup variant of the dual technique [simultaneous cooling / contraction analysis], with constant mould media and with recorded data processed using specialized software was obtained to apply in the melt shop similar to current thermal analysis systems;
*All of the irons have similar values for initial expansion up to the start of eutectic freezing [cca 0.44%]; graphite formation promoted expansion [WI-0.002%, GI-0.109%, DI-0.596%], and, consequently, different maximum expansion [WI-0.465%, GI-0.552%, DI-1.032%], and higher shrinkage sensitivity for ductile irons, respectively;
*Increasing of the acceleration rate of the graphitic expansion up to the maximum level [K_{gr1}] promoted an increasing of the shrinkage sensitiveness (decreased real density). Contrary, higher deceleration to the end of solidification [K_{gr2}], lower shrinkage sensitiveness for both irons;
*Increasing the length (time) of the expansion at the maximum level $\Delta t_{[\varepsilon di(max)]}$ favours the shrinkage formation, as the result of the increasing of the graphitic force on the mould walls;
*Experiments pointed out that the graphitic expansion has two contrary effects. Increased graphitic expansion (force) led to higher shrinkage sensitiveness during the first part of eutectic reaction, but to the decreasing of shrinkage at the end of solidification, due to the forcing of the last liquid iron to occupy the previous formed cavities;
*Strong graphitization process promotion at the end of solidification favours the castings soundness.

Acknowledgments

This work was financed by a grant of the Romanian National Authority for Scientific Research & Innovation, CNCS/CCCDI - UEFISCDI, project PN-III-P2-2.1-PED-2016-1793, within PNCDI III.

References

[1] Wallace J F, Samal P K, Voss J D, 1984 *AFS Trans.,* vol. 92, pp. 765-784.
[2] Skaland T, 2003 *Int. J. Cast Metals Res.,* vol. 16(1-3), pp. 11-16.
[3] White D, 2011 *Proc. 115th AFS Casting Congress*, Paper 11-141.
[4] Li W Z, Liu B C, Li J R, 2001 *J. Mater. Sci. Techn.* vol. 17(6), pp 610-614.
[5] Han Z Q, Su J Y and Cai K K, 2000 *J. Univ. Sci. Technol. Beijing*, vol. 17(1), pp. 24-29.
[6] Ohnaka I, Iwane J, Yasuda H, Zhu J D, 2003 *Int. J. Cast Metals Res.,* vol. 16(1-3), pp. 293-299.
[7] Stefanescu D M, 2005 *Int. J. Cast Metals Res.,* vol. 18(3), pp. 129-143.
[8] Bates C E, Oliver G L, McSwain R H, 1977 *AFS Trans.,* vol. 85, pp. 289-298.
[9] Riposan I, Chisamera M, Stan S, Gadarauteanu C, Skaland T, 2003 *Proc. of Keith Millis Symposium on Ductile Cast Iron*, Hilton Head Island, USA, pp. 125-135.
[10] Chisamera M, Riposan I, Stan S, Toboc P, Skaland T, White D, 2011 *Int. J. Cast Metals Res.,* vol. 24(1), pp. 28-36.
[11] Chisamera M, Riposan I, Stan S, Barstow M, 2012 *AFS Trans.*, vol. 120, pp. 375-388.
[12] Stan S, Chisamera M, Riposan I, Neacsu, L, Cojocaru AM, Stan I, 2018 *J. Mater. Eng. Performances,* vol. 27(10), pp. 5187-5196.
[13] Stan S, Chisamera M, Riposan I, Stefan E, Neacsu L, Cojocaru AM, Stan I, 2018 *Keith Millis Symp.*, Hilton Head, SC, USA; Int J Metalcast, 2019, DOI: 10.1007/s40962-018-0285-5
[14] Tadesse A, Fredriksson H, 2017 *Int. J. Cast Metals Res.,* vol. 30(3), pp 159-170.
[15] Stefanescu DM, Moran M, Boonmee S, Guesser WL, 2012 *AFS Trans.*, vol.120, pp. 365-374.
[16] Tadesse A, Fredriksson H, 2018 *Int. J. Cast Metals Res.,* vol. 31(2), pp. 108 - 117.
[17] Svidro P, Dioszegi A, 2014 *Int. J. Cast Metals Res.,* vol. 27(1), pp. 26-37.
[18] Alonso G, Loizaga A, Zarrabeitia G, Stefanescu DM, Suarez R, 2014 *119th AFS Congr.*, 14–006.
[19] Stan S, Chisamera M, Riposan I, Neacsu L, Cojocaru AM, Stan I, 2017 *IOP Conf. Ser.: Mater. Sci. Eng.,* vol. 209, 012017 (10pp).
[20] Stan S, Chisamera M, Riposan I, Stefan E, Neacsu L, Cojocaru AM, Stan I, 2018 *European Symposium on Thermal Analysis and Calorimetry (ESTAC12)*, August 2018, Brasov, Romania.

Chunky graphite in spheroidal graphite iron: review of recent results and definition of an predicting index

Jon Sertucha[1], Jacques Lacaze[2] and Rodolfo González-Martínez[1]

[1] Investigación y Desarrollo de Procesos Metalúrgicos, IK4-Azterlan, Aliendalde Auzunea 6, E-48200 Durango (Bizkaia), Spain.
[2] CIRIMAT, Université de Toulouse, BP 44362, F-31030 Toulouse, France

Jacques.lacaze@ensiacet.fr

Abstract. Graphite degeneracy in heavy-section spheroidal graphite cast irons is mostly associated with the formation of chunky graphite which consists of large eutectic cells with interconnected graphite strings. At low level, appearance of chunky graphite is limited to its non-aesthetic effect on machined surfaces, while at higher level it is detrimental for mechanical properties of the components. Chunky graphite is often related to high silicon levels and too high cerium additions during the spheroidization treatment. The appearance of this defect may be limited by controlled additions of antimony that is thought to tight the excess of cerium, but other impurities and low level elements may have to be considered during melt preparation. This contribution proposes a review of recent results and approaches on chunky graphite appearance, primarily but not exclusively in the case of heavy-section cast irons. Based on this literature review and series of experimental data, a predictive index for evaluating the risk of chunky graphite appearance is proposed. Lines for further research work aimed at a better understanding of graphite degeneracy are finally suggested.

Introduction

The need for melt control before casting of spheroidal graphite cast irons have led for long to look for graphite degeneracy as resulting from melt chemistry. Thielemann developed a trace element evaluation number S_b for assessing the ability of the charge to give well-formed nodules [1]:

$$S_b = 4.4 \cdot w_{Ti} + 2.0 \cdot w_{As} + 2.3 \cdot w_{Sn} + 5.0 \cdot w_{Sb} + 290 \cdot w_{Pb} + 370 \cdot w_{Bi} + 1.6 \cdot w_{Al} \quad (1)$$

where w_i stands for the content in "i" element, expressed in wt.%. When S_b is close to one, the casting should be free from graphite degeneracy while when it is higher than this value addition of rare earth (RE) should be made. This equation is often referenced in the technical literature and sometimes applied to the final cast alloy composition and not to the charge. In his study, Thielemann considered melts cast into samples with thicknesses 8–45 mm. Equation (1) is valid for residual Mg content in the range 0.04–0.08 wt.% while the maximum content in other elements was previously listed [2].

Using a series of casting results from literature, Javaid and Loper [3] suggested to correlate the capability of RE to counteract the detrimental effect of Bi, Pb and Sb to the possible formation of compounds RE_aX_b (with RE=Ce or La and X=Bi, Pb or Sb) such as CeBi, Ce_2Bi or La_3Sb_2 for example. One may write the following balance for the formation of such compound in an iron melt:

$$w_{RE}^{total} - w_{RE}^{limit} = \frac{b \cdot M_X}{a \cdot M_{RE}} \cdot \left(w_X^{total} - w_X^{limit}\right) \quad (2)$$

or

$$w_{RE}^{total} = \frac{b \cdot M_X}{a \cdot M_{RE}} \cdot w_X^{total} + \left(w_{RE}^{limit} - \frac{b \cdot M_X}{a \cdot M_{RE}} \cdot w_X^{limit} \right) \quad (3)$$

where w_i^{total} is the amount of element "i" (RE or X) in the liquid cast iron at the time of pouring, w_i^{limit} is the solubility limit of element "i" in liquid iron with respect to precipitation of compound RE_aX_b, and M_i is the molar mass of element "i". It should be noted that w_i^{limit} may in principle depend on temperature and on the whole iron composition. The statistical analysis performed by Javaid and Loper [3] gave the ratio $\frac{w_{RE}}{w_X}$ – which should be the term $\frac{b \cdot M_X}{a \cdot M_{RE}}$ - equal to 1.1206, 0.840 and 0.914 for Bi, Pb and Sb respectively. However, these authors noted that the scatter of the data was quite large, and the experimental ratio was found to vary in between about 0.5 and 1.5 for all three elements. This scatter is such that any of the compounds of the RE-X systems could have precipitated depending on other melt parameters.

Javaid and Loper [3] then extended the above correlations by expressing that the weight ratio of RE content to the sum of subversive elements should be higher than a critical value for avoiding graphite degeneracy. Considering equation (3), this is certainly an oversimplified approach which could however be efficient. They found that this ratio depends on the effect of other elements, namely P, Si, Mg and Ni, and also on the casting section size. After this seminal work, many authors considered similar ratios for analyzing their results, in particular, in the case of studies on the formation a peculiar graphite degeneracy, the so-called chunky graphite illustrated in figure 1.

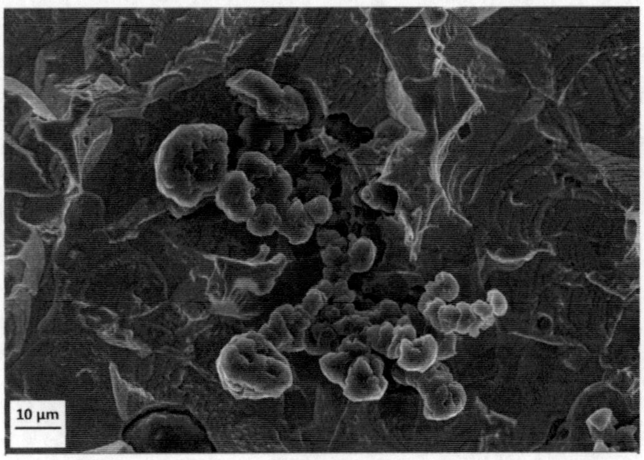

Figure 1 – Chunky graphite strings in a deep-etched ductile iron sample.

While Ce and other rare earths are used to balance the effect of trace elements including Sb (equation 1), Sb is used to counteract the detrimental effect of Ce in heavy-section castings. According to Javaid and Loper [3], the optimum weight ratio of RE to Sb should be 0.916 for RE to counteract the deleterious effect of Sb. This suggests that the optimum ratio of Sb to RE for Sb to counteract the effect of RE in chunky appearance should be 1.095. In fact, Tsumura et al. [4] found that Sb counteracts the effect of Ce when added at a level so that the ratio w_{Sb}/w_{Ce} is higher than 0.8 and Larrañaga et al. [5] confirmed this value.

Following the above lines, Löblich [6] emphasized that elements leading to chunky graphite (Si, Cu, Ni, Ca and Ce) in heavy-section castings may be counteracted by elements known to lead to intercellular graphite (Bi, Pb, Sb and As). Analysing a series of castings containing 2.25 wt.% to 2.70 wt.% silicon, Löblich found that the formation of chunky graphite depends on the silicon content and on the ratio $\frac{w_{Ce}}{w_{Pb}+w_{Sb}+w_{As}}$. For castings solidifying in about 50 min, Löblich suggested an empirical equation giving the critical silicon content, $w_{Si,limit}$, above which chunky graphite appears. This critical content may be written as:

$$w_{Si,limit} = 2.344 \cdot \left(1 + \frac{0.056}{1.563 - \frac{w_{Ce}}{w_{Pb}+w_{Sb}+w_{As}}}\right) \tag{4}$$

In the absence of cerium and whatever the content in other elements, this equation predicts that no chunky graphite would appear for silicon content lower than 2.43 wt.%. Similarly, a value of 1.563 for the ratio $\frac{w_{Ce}}{w_{Pb}+w_{Sb}+w_{As}}$ would ensure no chunky graphite appears even at high silicon content. Using values indicated by Löblich, namely w_{Mg} at 0.045 wt.%, w_{Si} at 2.7 wt.% and a casting thickness of 20 cm, equation (4) in Javaid and Loper [3] predicts a critical weight ratio of w_{Ce} over the sum of subversive elements of 1.05 which is reasonably close to Löblich's estimate. Recently, Stets et al. [7] extended the work by Löblich by considering the ratio (w_{Ce}+w_B+w_{Ca}+w_{Al}) to (w_{Bi}+w_{Pb}+w_{Sb}+w_{As}) for silicon contents lower than 3 wt.%.

Sertucha et al. [8] have conducted a statistical analysis leading to a relation between the extent of the casting section affected by chunky graphite and the composition of the cast irons. The study was based on about 60 melts cast in cubic blocks 300 mm in size with a limited range of silicon contents. The analysis gave a very good correlation coefficient and showed simple and coupled effects of many elements, namely Ce, Cu, La, P, Sb and Sn. However, it appeared difficult to give a physical meaning to these coefficients and this certainly demonstrates the limits of a statistical approach when the number of experiments is not large enough or when hidden factors enter to play.

In a more recent study, González-Martínez et al. [9] investigated the effect of Si on the microstructure of near-eutectic cast irons with various additions of Ce and Sb on a series of 31 alloys cast in Y2 blocks. The silicon content varied in the range 2.29 wt.% to 9.12 wt.%. The area fraction of chunky graphite was measured at the centre of the castings and the ratio of this quantity to the total area fraction of graphite was used as an output parameter. It was first confirmed that silicon promotes chunky graphite as cerium does, and it was also observed that magnesium acts as cerium. Following the type of analysis conducted by Javaid and Loper [3] and expressed with equation (2), it was considered that part of Ce could combine with Sb and analysis of the results showed this happens probably as $CeSb_2$ compound [9]. Limiting the analysis to Si, Ce, Sb and Mg, the following index was proposed:

$$\Omega_{Si} = w_{Si} + 800 \cdot \left(w_{Ce} \cdot \frac{55}{140.1} - 2 \cdot w_{Sb} \cdot \frac{55}{121.8}\right) + 50 \cdot w_{Mg} \cdot \frac{55}{24.3} \tag{5}$$

where M=55 stands for the molar mass of the cast iron.

When plotting the area fraction of chunky graphite as function of Ω_{Si}, it was found that this degeneracy appears when the index is higher than about 7 wt.% [9]. In a companion paper [10], a second series of 24 alloys were cast that all contained some Sb to limit chunky graphite appearance. In this second series, the change in silicon content was limited to the domain where maximum mechanical properties are observed, namely 4.84 wt.% to 5.42 wt.% Si. The results of both series are plotted in figure 2-a where it is seen they agree well between each other, indicating again a critical index value of about 7 wt.%.

It appeared of interest to check if this index could apply to previously published results obtained on cubic blocks 300 mm in size [5, 11, 12]. The silicon content of these alloys varied in between 1.92 wt.% and 2.41 wt.% and castings were either or not post-inoculated. The local area fraction of chunky graphite listed in these works is plotted in figure 2-b as a function of Ω_{Si}. It is seen again that chunky graphite appears above a critical value of the index. This critical value is decreased to about 4 wt.% in agreement with the larger size of the cubic blocks than that of the Y2 keel-blocks corresponding to the case shown in figure 2-a.

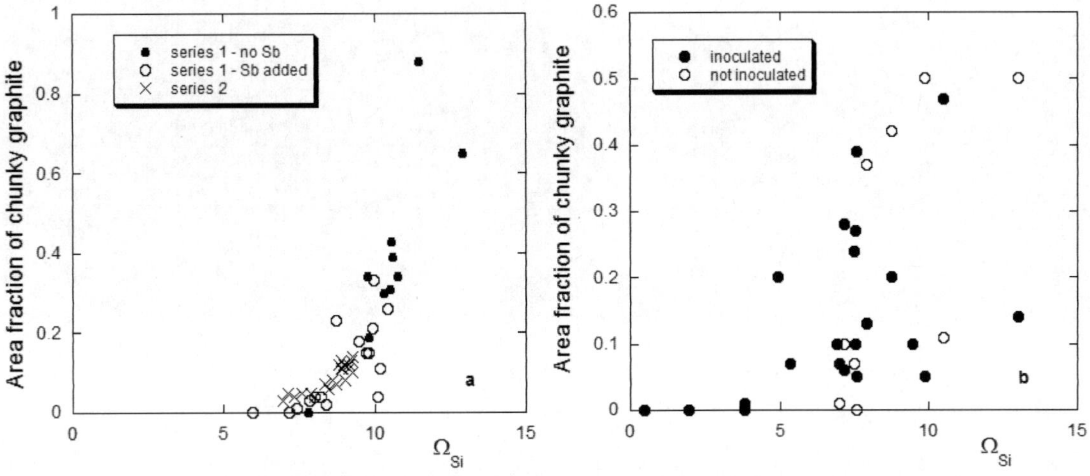

Figure 2 – Fraction of chunky graphite as function of Ω_{Si} for Y2 keel-blocks (a) and 300 mm in size cubic blocks (b).

Discussion

As already mentioned, other elements should be considered and enter in equation (5). This is the case of bismuth whose role on graphite shape in heavy-section ductile iron castings has been recently reviewed by Pokopec et al. [13]. This however needs appropriate experimental data which are not yet available.

Quite recently, Labrecque and Cabanne [14, 15] have reviewed various quality indexes proposed in the literature over several decades. These indexes relate to graphite degeneracy, pearlite promoters or, else, low temperature brittleness. However, none of them are able to account for the fact that there may be an optimum addition of counteracting elements as was stressed by Gagné and Argo [16] when studying the role of Sn and Sb on the formation of chunky graphite. From the present study, it appears that such an optimum should be considered also for Mg and Ce (or more generally RE). It would thus be highly helpful to develop a basic knowledge on the thermodynamic interactions in real cast irons as well as a large database dedicated to a statistical analysis of the cross-effects of low level elements.

Finally, it is worth stressing that the growth mechanism of chunky graphite is far from being understood. As a matter of fact, it was observed that the alloy with 9 wt.% Si in the studied series [9] presented little chunky graphite, at most a fraction of 0.17. This graphite is illustrated in figure 3. This low graphite degeneracy is in contrast with the high Ω_{Si} value for this alloy. This certainly has to do with the fact that this alloy solidified with ferrite as iron-rich phase and not austenite in agreement with the assessed Fe-C-Si phase diagram [17]. As chunky graphite grows as a coupled eutectic with the iron-rich matrix, a possible schematic of the solidification front has been proposed previously for the case when the iron-rich phase is austenite [18]. This schematic is shown in figure 4 where growth of graphite at the interface with the liquid is assumed to follow a 2D nucleation growth model [19]. In figure 4, the value of the surface tension between liquid and graphite, $\gamma^{l/G}$, is such that it counter-balances exactly the sum S of the austenite/graphite, $\gamma^{\gamma/G}$, and austenite/liquid, $\gamma^{\gamma/l}$, surface tensions. As it is admitted that the surface tension between ferrite and liquid, $\gamma^{\alpha/l}$, is lower than $\gamma^{\gamma/l}$ [20], the forces will not be any more balanced at the three-phase junction making chunky graphite growth more difficult when the iron-rich phase is ferrite.

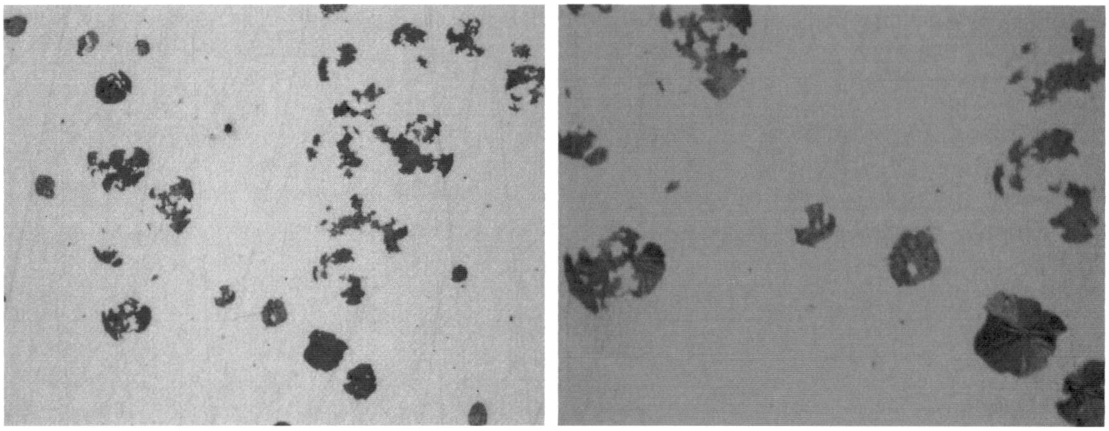

Figure 3 – Light optical micrograph of graphite in the 9 wt.% Si alloy: bright field image (a) and polarized light image of part of the area in a (b)

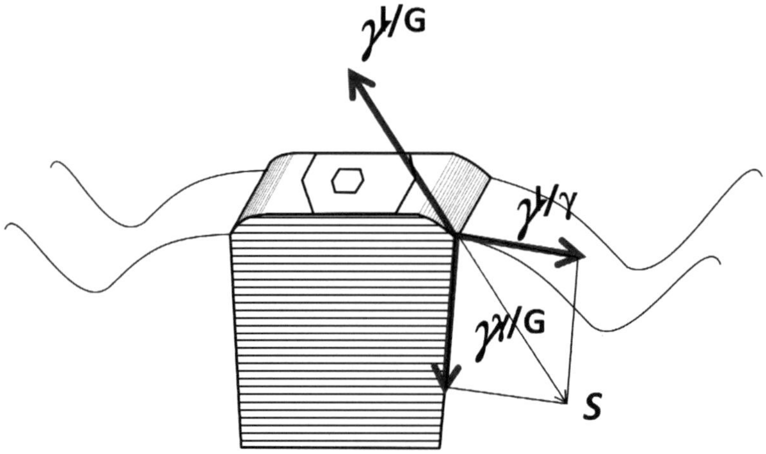

Figure 4 – Schematic of the coupled eutectic solidification front of chunky graphite and austenite

Conclusion

While it has been known for long that chunky graphite in spheroidal graphite cast irons depends also on melt composition, very few quantitative studies are available. More important than that, it appears impossible to compare the results of these rare studies as full details for such a comparison are not available. The index proposed here to characterize the propensity of a melt to chunky graphite degeneracy has been developed on the basis of more than 50 melts cast in Y2-blocks. A tentative extension to other cooling conditions has been shown which would need being better substantiated.

Interestingly enough, growth of chunky graphite is certainly not definitely understood though figure 4 suggests one mechanism. It is evident that clarifying this mechanism would help finding ways to avoid its appearance.

References

[1] Thielemann T 1970 *Giessereitechnik* **16** 16-24
[2] Magnusson Åberg L, Hartung C and Lacaze J 2014 Proc. 10th Int. Symp. on the Science and Processing of Cast Iron SPCI-10
[3] Javaid A and Loper C R 1995 *AFS Trans.* **103** 135-150
[4] Tsumura O, Ichinomiya Y, Narita H, Miyamoto T and Takenouchi T 1995 *Imono* **67** 540–545

[5] Larrañaga P, Asenjo I, Sertucha J, Suárez R, Ferrer I and Lacaze J 2009 *Metall. Mater. Trans. A*. **40** 654-661
[6] Löblich H 2006 *Giesserei* **93** 28–41
[7] Stets W, Petzschmann U, Hentsch R and Lötschert A 2014 *Giesserei* **101** 36–47
[8] Sertucha J, Lacaze J, Armendariz S and Larrañaga P 2013 *Metall. Mater. Trans. A* 2013 **44** 1159–1162
[9] González-Martínez R, de la Torre U, Lacaze J and Sertucha J 2018 *Mater. Sci. Eng. A* **712** 794–802
[10] González-Martínez R, de la Torre U, Ebel A, Lacaze J and Sertucha J 2018 *Mater. Sci. Eng. A* **712** 803–811
[11] Sertucha J, Suárez R, Asenjo I, Larrañaga P, Lacaze J, Ferrer I and Armendariz S 2009 *ISIJ Int.* **49** 220–228
[12] Lacaze J, Armendariz S, Larrañaga P, Asenjo I, Sertucha J and Suárez R 2010 *Mater. Sci. Forum* **636-637** 523–530
[13] Pokopec I M, Mrvar P and Bauer B 2017 *Mater. Technol.* **51** 275-281
[14] Labrecque C and Cabanne P M 2010 Proc. World Foundry Congress, pp. 411–418
[15] Labrecque C and Cabanne P M 2017 ASM Handbook, vol 1A, Cast iron science and technology, ed D M Stefanescu pp 629–644.
[16] Gagné M and Argo D 1987 Proc. Int. Conf. Advanced casting technology, ed J. Easwaran, ASM Int. pp. 231–244
[17] J. Lacaze, B. Sundman 1991 *Metall. Trans. A* **22A** 2211-2223
[18] de la Torre U, Lacaze J and Sertucha J 2016 *Int. J. Mater. Res.* **107** 1041-1050
[19] Lacaze J, Bourdie J and Castro Roman M J 2017 *Acta mater* **134** 230-235,
[20] Bobadilla M, Lacaze J and Lesoult G 1988 *J. Crystal Growth* **89** 531-544

IOP Conf. Series: Materials Science and Engineering **529** (2019) 012018 doi:10.1088/1757-899X/529/1/012018

Cellular-to-Dendritic and Dendritic-to-Cellular Morphological Transitions in a Ternary Al-Mg-Si Alloy

C. Brito[1], H. Nguyen-Thi[2], N. Mangelinck-Noël[2*], N. Cheung[3], J.E. Spinelli[4], A Garcia[3]

[1] Campus of São João da Boa Vista, São Paulo State University, UNESP, 13876-750 São João da Boa Vista/SP Brazil.
[2] Aix-Marseille Université, CNRS, IM2NP UMR CNRS 7334, Campus Saint Jérôme, case 142, 13397 Marseille Cedex 20, France.
[3]Department of Manufacturing and Materials Engineering, University of Campinas, UNICAMP, 13083-970 Campinas/SP, Brazil.
[4]Department of Materials Engineering, Federal University of São Carlos – UFSCar, 13565-905 São Carlos/SP, Brazil.
*E-mail: nathalie.mangelinck@im2np.fr

Abstract. The study is focused on the influence of solidification thermal parameters upon the evolution of the microstructure (either cells or dendrites) of an Al-3wt%Mg-1wt%Si ternary alloy. It is well known that the application properties of metallic alloys will greatly depend on the final morphology of the microstructure. As a consequence, various studies have been carried out in order to determine the ranges of cooling rates associated with dendritic-cellular transitions in multicomponent alloys. In the present research work, directional solidification experiments were conducted using either a Bridgman (steady-state) device or another device that allows the solidification under transient conditions (unsteady-state). Thus, a broad range of cooling rates ($\dot{T}$), varying from 0.003K/s to 40K/s could be achieved. This led to the identification of a complete series of cellular/dendritic/cellular transitions. For low cooling rate experiments, low cooling rate cells to dendrites transition happens. Moreover, at a high cooling rate, a novel transition from dendrites to high cooling rate cells could be observed for the Al-3wt%Mg-1wt%Si alloy. Additionally, cell spacing λ_C and primary dendritic spacing λ_1 are related to the cooling rate by power function growth laws characterized by the same exponent (-0.55) for both steady-state and unsteady-state solidification conditions.

1. Introduction

The microstructural morphology in metallic alloys is of prime importance. Properties such as mechanical strength, corrosion resistance and wear resistance were proved to be affected by the morphology and scale of the α-Al matrix. Then, due to the direct relationship observed between the solidification microstructure and the final properties, several experimental studies and models were devoted to the conditions in which the morphology of the solid-liquid (S/L) interface changes. The solid-liquid interface instability depends on the growth rate, the thermal gradient and the distribution of solute ahead of the S/L interface leading to the constitutional undercooling [1-10].

The constitutional undercooling criterion is largely used to predict whether a S/L planar interface is stable or not, despite showing some important limitations. Firstly, the constitutional undercooling is based on thermodynamic equilibrium considerations, whereas the evolution of the S/L interface during solidification is a non-equilibrium process. Secondly, the local curvature of the interface is not taken into account although it can impact the free energy of the system. Thirdly, the evolutionary trend of the interface morphology is not described. This criterion is not valid for rapid solidification or processes involving non-equilibrium solidification.

Hunt and Lu [1] developed a numerical model for the growth of cells and dendrites that predicts the transition between these two types of microstructures. This transition was considered diffuse. It ends when the preferred direction of growth is reached and the secondary dendritic arms are already defined.

Although the specialized literature shows investigations about the transitions regarding the sequence planar - cellular - dendritic - cellular - planar, few studies were devoted to the reverse dendritic to cellular microstructural transition in multicomponent metallic alloys. Most of the studies dealt with the cellular to dendritic transition that occurs combining dilute alloys, low growth rate and high temperature gradients. The cellular to dendritic transition with cells known as high-speed cells was only supposed to occur at very high growth rates [2]. However, no consensus on which thermal and compositional parameters are the determinant factors leading to this type of transition has been established to date.

In the present investigation, an Al-3wt%Mg-1wt%Si alloy is directionally solidified (DS) under transient and stationary solidification conditions with a wide range of cooling rates. It is shown that the investigated alloy transitioned from fully cellular to dendritic and from dendritic to cellular for high cooling rates. The microstructural transitions are characterized, and the evolution laws for cellular and the dendritic length scales with the solidification cooling rate are also established.

2. Experimental procedure

In order to obtain a broad range of cooling rates, two solidification systems were used to perform the experiments. A water-cooled directional solidification (DS) setup permitted unsteady-state conditions and high cooling rates to be achieved, whereas a Bridgman type device guaranteed solidification under constant temperature gradient and cooling rate. During the unsteady-state solidification, the alloy was melted *in situ* by radial electrical wiring positioned around a cylindrical stainless steel split mold with inner diameter of 60 mm; height of 150 mm and wall thickness of 5mm. More details about this system can be found elsewhere [8, 11]. The Bridgman setup is built to operate with a cold and a hot part separated by an adiabatic zone. The hot part is constituted by two heating elements wounded around a 15 mm diameter cylindrical alumina piece. In this set-up, the controlled displacement of the crucible inside the furnace is allowed. During the Bridgman furnace experiments, the temperature gradient was fixed at a value of 11K/cm while the solidification rates (v) were varied in the range from 0.3 μm/s to 70 μm/s [11].

The solidification thermal parameters, growth rate (V_L) and cooling rate ($\dot{T}$), during unsteady-state growth were determined from seven type K thermocouples (with 0.2 mm diameter wire) positioned along the length of the DS casting at the positions 3, 6, 12, 20, 30, 50 and 70mm from the metal/mold interface. These sensors were connected by coaxial cables to a data logger interfaced with a computer, capable of automatically record temperature data at a frequency of 5 Hz.

Optical microscopy analyses were performed using an Olympus inverted metallurgical microscope (model 41GX). The cellular (λ_C), primary dendritic arm (λ_1) and secondary dendritic arm (λ_2) spacings were measured from the optical images of transverse and longitudinal section metallographies. The triangle method [12] was used in order to determine λ_C and λ_1 measurements. The intercept method was adopted for measuring λ_2. At least 50 measurements were performed for each sample related to a single cooling rate in both steady and unsteady state regimes.

3. Results and discussion

Figure 1 shows the variations of cellular and primary dendritic spacings with the cooling rate. The range of experimental cooling rates associated with the unsteady-state solidified samples was

$0.2<\dot{T}<45$ K/s. Cooling rates higher than 2 K/s were characterized by the growth of α-Al cells and cooling rates lower than 0.8 K/s are related to the prevalence of α-Al dendrites. The intermediate microstructures, constituted of both cells and dendrites, remained associated with $0.8<\dot{T}<2.0$ K/s.

Additionally, the experiments carried out following steady state conditions have allowed to monitor the influence of the solidification thermal parameters (here synthesized by the cooling rate for easier comparison with the unsteady state experiments) on the final microstructure. From these experiments, the cellular-to-dendritic transition for low cooling rates could be studied. Moreover, a complete description of the cellular→dendritic→cellular transitions for the ternary Al-3wt%Mg-1wt%Si alloy could be deduced. The range of experimental cooling rates during the steady state experiments was $0.003<\dot{T}<0.17$ K/s. For cooling rates between 0.03K/s and 0.17 K/s only typical α-Al dendritic morphologies could be observed. A gradual cellular-to-dendritic transition started to occur for cooling rates lower than 0.03 K/s and ended at about 0.005 K/s. For $\dot{T}<0.005$K/s, no evidence of dendrites could be observed. The range where the aforementioned morphologies prevailed can be seen in Figure 1. The present experimental approach appears to be very practical with a view to determining microstructural morphological transitions in other Al-based multicomponent alloys of interest.

Contrary to the expectations from the literature for metallic systems [13,14], the Al-3wt%Mg-1wt%Si alloy shows high-velocity cells occurring even for moderate $\dot{T}$ values. The length-scale of the cells was observed to be relatively coarse, that is, λ_C varied from 10 μm to 40 μm. Typical dendritic morphologies were observed with λ_1 varying in the range 115 - 270 μm. The length scale of the microstructure is reduced to about 1/3 when the dendritic morphology changes into high-velocity cells.

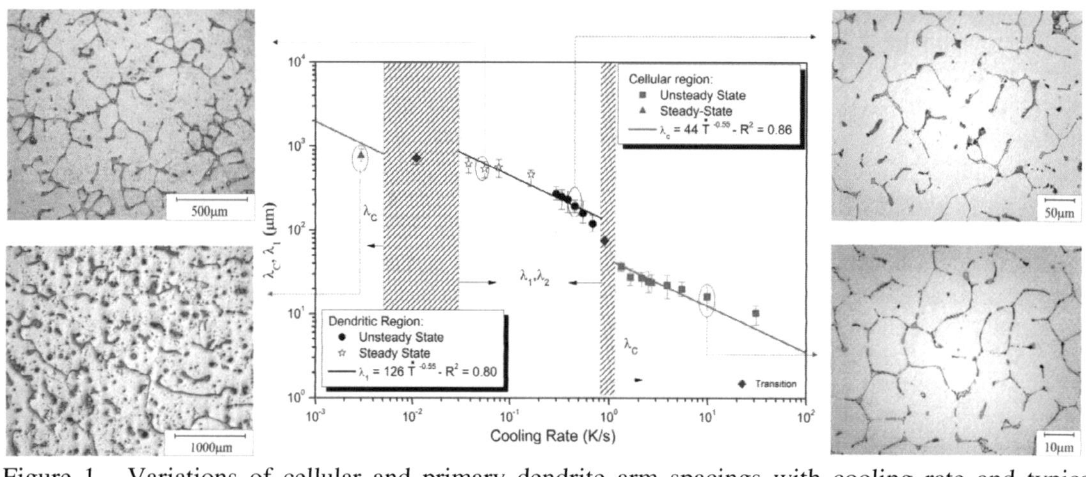

Figure 1 - Variations of cellular and primary dendrite arm spacings with cooling rate and typical transverse microstructures of each α-Al morphological region of directionally solidified Al-3wt%Mg-1wt%Si alloy samples.

The present results allowed comparisons of the experimental scaling laws relating cellular and primary dendrite arm spacing to the cooling rate. The λ_C and λ_1 were related to $\dot{T}$ by power function growth laws characterized by a same exponent of -0.55, that is, $\boldsymbol{\lambda_1 = 44}\dot{T}^{-0.55}$ and $\boldsymbol{\lambda_1 = 126}\dot{T}^{-0.55}$ for cellular and dendritic regions, respectively. Such experimental scaling laws confirm the application of the exponent −0.55 in order to correlate the λ_C and λ_1 evolution with the cooling rate. This exponent has been largely applied for the dendritic growth of a number of binary and ternary metallic alloys [9, 11, 14, 15, 16, 17].

Figure 2 shows the evolution of the secondary dendrite arm spacings against the cooling rate. The experimental scatter is represented by average spacing values with their standard deviations for both samples solidified under steady state and unsteady state regimes. λ_2 decreases with the increase in cooling rate according to the experimental growth law $\lambda_2 = 22\dot{T}^{-0.55}$, obtained from a fit on the experimental scatters. In a recent study [15], the secondary dendritic growth of a binary Al-3wt.%Mg alloy was shown to be represented by a -1/3 exponent power function growth law. The addition of 1wt%Si seems to change the intensity of lateral solute segregation inside the region corresponding to the secondary arm regions. The addition of Si appears to make λ_2 more sensitive to variations in the solidification cooling rate. Recent research works [14, 16] stated that the effects of lateral solute segregation on the wavelengths of instabilities along the sides of primary dendritic stems for multicomponent alloys are still not well understood. The same exponent, -0.55, was also used to correlate the experimental growth of the secondary dendrite arm spacing against the cooling rate for ternary Al-Cu-Ni alloys [9].

Figure 3 shows a schematic representation of the complete cycle of morphological transitions obtained in the present investigation. An unusual form of transition reported in the literature as the reverse dendritic to cellular shift for metallic systems was observed. For binary systems, the well-known theories from the literature have shown that the cellular-to-dendritic transition is typical of dilute alloys and for lower cooling rates (4, 18). The morphology known as high-speed cells was only supposed to occur at very high growth rates [2, 5]. Further investigations regarding to alloy cooling rate and a variety of solute contents appear to be decisive for the understanding of the occurrence of the reverse transition in multicomponent alloys.

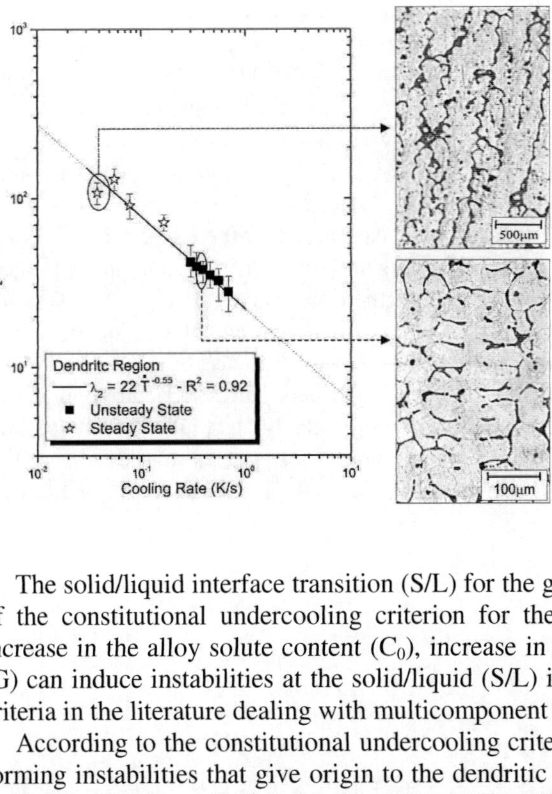

Figure 2 –Variations of secondary dendritic spacing *vs.* cooling rate and typical longitudinal microstructures of each morphological region of the directionally solidified Al-3wt%Mg-1wt%Si alloy samples.

The solid/liquid interface transition (S/L) for the growth of cells and dendrites is explained in terms of the constitutional undercooling criterion for the growth of binary alloys. In this criterion the increase in the alloy solute content (C_0), increase in growth rate (v) and decrease in thermal gradient (G) can induce instabilities at the solid/liquid (S/L) interface [1, 4]. However, there are few available criteria in the literature dealing with multicomponent alloys.

According to the constitutional undercooling criterion, increase in C_0 destabilizes the S/L interface forming instabilities that give origin to the dendritic growth. For multicomponent alloys, the increase in C_0 (or, ΣC_0, sum of all solutes) seems to affect differently. In the case of the examined non-dilute alloy, the sum of the solute chemistries is equal to 4wt.%, which would lead to the growth of a totally dendritic morphology since it does not characterize a dilute alloy. This is not so considering the present experimental results. Cells grew for either $\dot{T}$<0.005 K/s or $\dot{T}$> 2.0 K/s. In contrast, another study showed that the ternary Al-1wt%Fe-1wt%Ni alloy solidified under transient solidification conditions [17], developed only cells along the entire range of examined cooling rates. In contrast, dendrites prevailed for the binary Al-1.0wt%Ni alloy solidified under similar conditions [20]. This reinforces the need for detailed studies mapping the morphologies and the factors affecting their growth for multicomponent Al-based alloys.

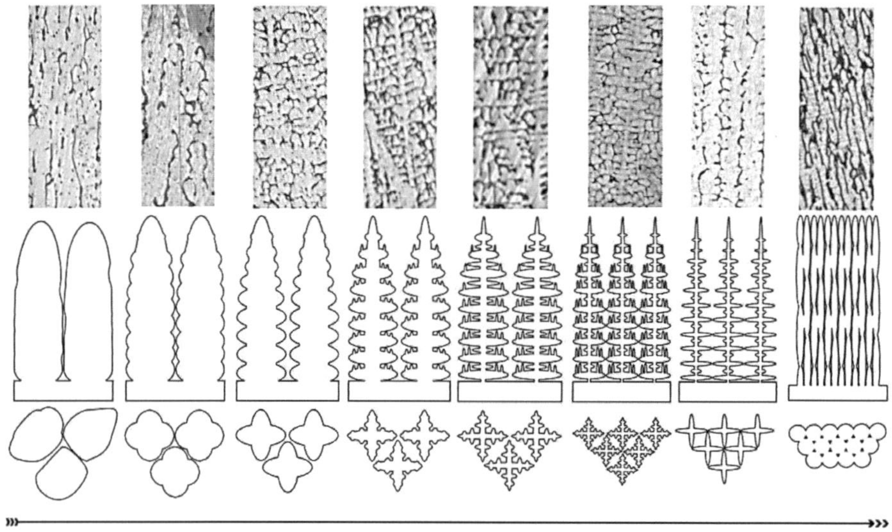

Figure 3 - Schematic progression of α-Al morphologies as a function of cooling rate and representative microstructures of the directionally solidified Al-3wt%Mg-1wt%Si alloy.

In order to compare the experimental growth law $\lambda_1 = 126\dot{T}^{-0.55}$ proposed in the present study, experimental λ_1 data from the literature [21] have been included as open circles in Figure 4. The λ_1 scatter from the literature were related to stationary cooling rates calculated by the expression $\dot{T} = G \times v$, where the growth rate, v, ranged from 0.0172mm/s to 0.1 mm/s and the thermal gradient, G, ranged from 0.45 K/mm to 17 K/mm. These λ_1 data are compared with the experimental growth law (Figure 1) for the dendritic region derived in the present work. An excellent agreement can be observed between the present results and the results from the literature, despite differences in the alloys compositions. The good fit observed demonstrated that the -0.55 exponent can also be applied in the case of another Al-Mg-Si alloy composition, being valid for both steady-state and unsteady-state solidification conditions. This gives indications that the experimentally derived scaling laws using the present results could be representative of the dendritic growth for other Al-Mg-Si alloy compositions.

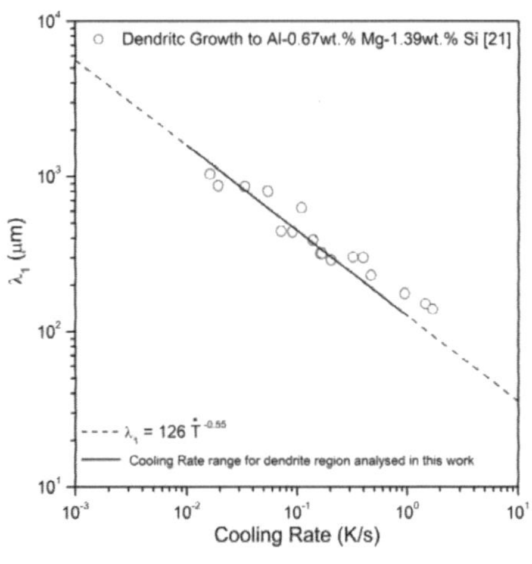

Figure 4 - Comparison of experimental steady-state results [21] showing λ_1 as a function of $\dot{T}$ in the present study and for another ternary Al-Mg-Si alloy.

4. Conclusions

The effect of the cooling rate on the morphology of the α-Al matrix was examined for Al-3wt%Mg-1wt%Si alloy samples solidified upon a wide range of solidification conditions. For this purpose, directional solidification experiments were conducted using a Bridgman (steady state) device and another device that allows the solidification under transient conditions (unsteady state). Consequently, a broad range of solidification cooling rates ($\dot{T}$), varying from 0.003K/s to 45K/s could be assessed as well as the corresponding microstructures. This combination of techniques led to the identification of a complete cycle of α-Al morphologies from cells to dendrites and reverting from dendrites to cells. The range of experimental cooling rates associated with the steady-state experiments was 0.003K/s < $\dot{T}$ <0.17K/s. For cooling rates between 0.03K/s and 0.17K/s only dendritic morphologies prevailed. For $\dot{T}$ values lower than 0.005K/s a degenerate growth of α-Al cells was observed. For unsteady state conditions, the range of experimental cooling rates was 0.2K/s < $\dot{T}$ < 45K/s. The observed growth of high cooling rate cells was related to $\dot{T}$ > 2K/s whereas the dendrites prevailed for $\dot{T}$ < 0.8K/s. Values in between refer to a zone in which both cells and dendrites could be found. Experimental scaling laws were proposed correlating both λ_C and $\lambda_1$with$\dot{T}$. It is worth noting that the same power function law could be found for all laws related to cell, primary and secondary dendrite spacings.

ACKNOWLEDGMENTS

The authors acknowledge the financial support provided by FAPESP (Grants 2012/08494-0; 2012/16328-2, 2013/23396-7 and 2014/25809-0), CNPq and CAPES/COFECUB (Grant 23038.000069/2015-04).

References

[1] Hunt JD and Lu SZ. 1996. *Metall Mater Trans A*. 27(A) 611-623.
[2] Kurz W and Fisher DJ. 1992. *Fundamentals of Solidification*. (Switzerland: Trans. Tech. Public.) p 87.
[3] Trivedi R, Sekhar JA, Seetharaman V. 1989. *Metall Trans A*. 20(A) 769-777.
[4] Mullins WW and Sekerka RF. 1964. *J Appl Phys*. 35(2) 444-451.
[5] Fu JW, Yang YS, Guo J J, & Tong WH. 2008. *Mater Sci Tech*. 24(8) 941-944.
[6] Lee JH, Kim HC, Jo CY, Kim SK, Shin JH, Liu S, & Trivedi R. 2005. *Mat Sci Eng A*. 413 306-311.
[7] Chikama H, Shibata H, Emi T, & Suzuki M. 1996. *Mater Trans*. 37(4) 620-626.
[8] Brito C, Vida T, Freitas E, Cheung N, Spinelli JE, & Garcia A. 2016. *J Alloys Compd*. 673 220-230.
[9] Rodrigues AV, Lima TS, Vida TA, Brito C, Garcia A, & Cheung N. 2018. *Met Mater-Int*. 24 1058–1076.
[10] Bertelli F, Freitas ES, Cheung N, Arenas MA, Conde A, Damborenea J, & Garcia A. 2017. *J Alloys Compd*. 695 3621-3631.
[11] Jung H, Mangelinck-Noël N, Nguyen-Thi H, Billia B. 2009. *J Alloys Compd*. 484(1-2), 739-746.
[12] Gündüz M, and Çadırlı E. 2002. *Mater Sci Eng A*. 327(2), 167-185.
[13] Goulart PR, Spinelli JE, Cheung N, Mangelinck-Nöel N, & Garcia A. 2010. *J Alloys Compd*. 504(1) 205-210.
[14] Canté, MV, Brito, C, Spinelli, JE, Garcia, A. 2013. *Mater Design*. 51, 342-346.
[15] Brito C, Costa TA, Vida TA, Bertelli F, Cheung N, Spinelli JE, & Garcia A. 2015. *Metall Mater Trans A*. 46(8) 3342-3355.
[16] Peres MD, Siqueira CA, & Garcia A. 2004. *J Alloys Compd*. 381(1) 168-181.
[17] Bertelli F, Brito C, Ferreira IL, Reinhart G, Nguyen-Thi H, Mangelinck-Noël N, Cheung N & Garcia A. 2015. *Mater Design*. 72 31-42.
[18] Canté MV, Brito C, Spinelli JE, & Garcia A. 2013. *Mater Design*. 51 342-346.
[19] Kurz W, and Fisher DJ. 1981. Dendrite growth at the limit of stability: tip radius and spacing. *Acta Metall*. 29(1) 11-20.
[20] Canté MV, Spinelli JE, Ferreira IL, Cheung N, & Garcia A. 2008. *Metall Mater Trans A*. 39(7) 1712.
[21] McCartney DG and Hunt JD. 1981. *Acta Metall*. 29(11) 1851-1863.

Impact of gravity-related phenomena on the grain structure formation: comparative study between horizontal and vertical solidification of a refined Al-20wt.%Cu alloy

H Soltani[1,2], G Reinhart[1], M C Benoudia[3], M Zahzouh[2] and H Nguyen-Thi[1]

[1] Aix Marseille Univ, CNRS, IM2NP, UMR 7334, Campus scientifique de St-Jérôme, case 142, 13397 Marseille Cedex 20, France
[2] Badji Mokhtar University, P.O. Box 12, 2300, Annaba, Algeria
[3] L3M, Ecole Nationale Supérieure des Mines et de la Métallurgie, Annaba, Algeria

Abstract. On earth, gravity-related phenomena are unavoidable, such as thermo-solutal convection caused by density gradients in the melt and buoyancy when the liquid phase is denser than the solid phase. Such phenomena can drastically affect both the grain density and their morphology during equiaxed solidification processes. For these reasons, fundamental studies comparing the influence of solidification parameters with and without gravity effects are important to obtain benchmark data, which are useful to understand and then control the final structure of materials in industrial processes. In the present work, the impact of the solidification parameters on the dendritic grain structure formation and on the final grain size and shape was investigated *in situ* by using X-radiography for different growth orientations with respect to gravity. In a first step, experiments were carried out with various solidification parameters and with the furnace in horizontal position, with the main surface of the sample being perpendicular to gravity to limit gravity-related phenomena. In a second step, experiments were carried out with identical solidification parameters but with the furnace in a vertical position, and for two solidification directions (upward and downward). A comparative study between horizontal and vertical experiments was carried out. Phenomena related to gravity have been highlighted and their respective impact on the solidification front propagation was analysed.

1. Introduction

The mechanical properties of metal alloys strongly depend on the grain structure forming during the solidification step, so a precise control of the growth process is crucial in engineering. Two types of grain structures are commonly obtained during metal alloy solidification: a columnar grain structure with anisotropic properties, or an equiaxed grain structure with more uniform and isotropic properties. Equiaxed grain structure is required in most applications of aluminum-based alloys. The occurrence of equiaxed grains structures is often promoted by adding refining particles acting as preferential sites for heterogeneous nucleation. These particles are activated in liquid regions that are undercooled [1] and their efficiency depends on several characteristics: chemistry, amount, size distribution.

The control of the grain refinement involves the understanding of additional complex mechanisms, from nucleation to dendritic growth and solutal interactions. Grain structure prediction is made even more difficult when phenomena such as convection and buoyancy must be considered. Any density gradient in the melt that is not suitably oriented with respect to the gravity vector can cause fluid flow. Even when solidification is performed in both thermal and solutal stabilizing configuration (i.e. upward solidification and increased liquid density with solute segregation), horizontal residual thermal gradients can initiate strong convective flow [2]. Convection in the melt can yield to significant inhomogeneity in solute distribution and thus liquid undercooling variations, and ultimately micro- and macro-segregation. Furthermore, newly formed solid grains can be carried away by fluid flow [3]. Upward motion of grains can also occur due to buoyancy forces when the liquid phase is denser than the solid phase [4], and downward motion (sedimentation) when the solid phase is denser than the liquid phase [5]. Experimental investigations of gravity effects in controlled environments are therefore of great interest for the validation of theoretical and numerical models.

In situ and time-resolved X-ray imaging is a method of choice for unveiling the dynamical evolution of dendritic microstructures and grain structure formation during the solidification of metal alloys [6]. Investigations taking advantage of the high brilliance of synchrotron sources began with the application of radiography [7], and a combination of radiography and diffraction imaging [8]. Then, the application of tomography enabled three-dimensional observations to be carried out [9]. Although radiography only provides a two-dimensional projected image, it remains an attractive imaging technique due to its ease of implementation. Recently, new opportunities arose with the improvement of compact micro-focus sources and X-ray sensitive detectors that enable *in situ* and time-resolved radiography to be carried out in laboratories, with a sufficient spatial resolution to distinguish the microstructure features [10], enabling different orientations of the sample with respect to gravity [11] and compatible with microgravity platforms [12].

The present investigation reports on the solidification of refined Al-20wt.%Cu samples observed *in situ* using the SFINX (Solidification Furnace with IN situ X-radiography) laboratory device. Experiments were carried out with identical applied parameters but with different orientations of the the furnace with respect to the gravity vector. A comparative study between horizontal and vertical experiments was thus carried out to analyze and highlight phenomena related to gravity and their impact on the solidification front propagation.

2. Experimental details

2.1. SFINX facility

The SFINX facility is a duplicate of the device used during the MASER-12 sounding rocket mission [12] and parabolic flight campaigns [13]. These facilities were developed within the framework of the ESA MAP named XRMON (In-situ X-ray monitoring of advanced metallurgical processes under micro gravity and terrestrial conditions), devoted to the application of X-radiography during microgravity experiments [14].

The X-radiography system figures a micro-focus X-ray source with a molybdenum target (3 μm focal spot), which provides enough photon flux with two peaks of energy 17.4 keV and 19.6 keV that ensure a good image contrast to study Al–Cu based alloys. The camera system is made of a scintillator plate and a digital camera with a CCD-sensor. Due to the X-ray beam divergence, a geometric magnification of the object is observed at the detector, which is the ratio of the source-detector by source-sample distances (figure 1). In this work a magnification of approximately 5 was used, for a Field-of-View (FoV) of about 5 × 5 mm^2 leading to an effective pixel size of ~4 μm. The acquisition rate was set to 2 frames per second. The solidification furnace consists of two heaters, separated by a gap, that applies a longitudinal temperature gradient G_{app}. The gradient furnace enables directional solidification with applied temperature gradient within the range of 5–15 K/mm and cooling rates R within the range of 0.01–1.5 K/s.

In the present work, the studied alloy was Al-20wt.%Cu inoculated with 0.1wt.% AlTiB. Sheet-like rectangular samples with dimension of 5 mm × 50 mm in area and about 250 μm in thickness were used. The sample was placed in the middle of stainless-steel spacers sandwiched between two flexible glassy carbon sheets sewn together with a silica thread. The crucible was enclosed inside the furnace and meets both sides to achieve the expected thermal profile.

The furnace was set in two positions allowing either horizontal solidification of the sample with its main surface perpendicular to the gravity vector $\boldsymbol{g}$ (figure 1a), or vertical solidification of the sample with its main surface parallel to the gravity vector $\boldsymbol{g}$ (figure 1b and 1c). The horizontal configuration is chosen to minimize gravity-related phenomena such as buoyancy and thermo-solutal convection [15]. The vertical configuration is chosen to highlight gravity related-phenomena for two directions of the solidification: bottom-up ($\boldsymbol{G_{app}}$ anti-parallel to $\boldsymbol{g}$, figure 1b) and top-down ($\boldsymbol{G_{app}}$ parallel to $\boldsymbol{g}$, figure 1c) as previously termed by R. H. Mathiesen *et al.* [16].

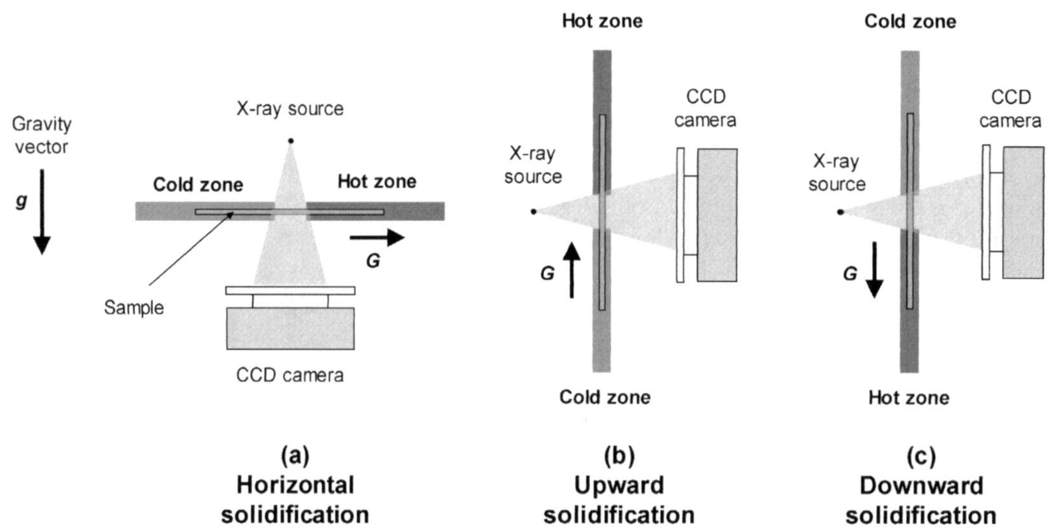

Figure 1. Schematic layout of the three configurations used to study the impact of the growth direction with respect to the gravity direction.

2.2. *Image processing*

During the sample solidification, the X-radiography system recorded a stack of raw images. The grey level variations in the acquired images are related to the attenuation coefficient of each different parts of the sample. Image legibility is enhanced by applying a "flat-field" correction, which consists in dividing each frame by a reference image recorded just before solidification. This procedure reduces the noise and removes defects related to the detector and crucible [17]. After applying the image processing, the bright regions in the processed radiograph are solid Al-grains whereas the dark regions are the Cu-rich liquid regions.

3. Results and discussion

In the present study, we will consider experiments with two different cooling rates, R_1 = 0.1 K/s and R_2 = 0.9 K/s, with the same applied temperature gradient G_{app} = 7.5 K/mm. These parameters were chosen to compare the evolution of the microstructure formation for relatively low and rapid solidification front velocities with respect to the achievable parameters of the furnace, and thus investigate the impact of gravity-related phenomena at different growth velocities.

3.1. *Impact of gravity related-phenomena on the solidification front at low growth rate*

For each experiment, the directional solidification started with fully melted samples and the solidification front was observed passing through the field of view after the cooling rates were applied (figure 2a, 2b and 2c). During the horizontal solidification experiment, the equiaxed front propagated in a similar way to the observations reported by H. Nguyen-Thi *et al.* [18], A. Prasad *et al.* [19] or E. Liotti *et al.* [20] using synchrotron X-radiography, and recently by Xu *et al.* [21] by using a laboratory device similar to the present one. Several dendritic grains grew concomitantly towards the top of the field of view. This grain network formed an effective front that progressively created a constitutionally undercooled zone ahead. Then, a new series of grains nucleated and developed in the undercooled zone. The new grains eventually blocked the previous ones by solutal interactions and were incorporated to the effective front.

A significantly different solidification front was observed for the vertical experiment during upward solidification (figure 2b). Aluminum grains are buoyant for the present alloy composition [22]. Consequently, after nucleation they floated from the bottom to the top due to buoyancy force. The moving grains appear blurred on the images during their upward motion due to the too long acquisition time with respect to their floatation velocity (≈ 100 μm/s). The dendritic grains rapidly melted in the hot region of the sample until their complete disappearance. A similar behavior was reported by L. Abou-Khalil *et al.* [13] and G. Zimmermann *et al.* [23] for dendrite fragments detached from columnar structures. Only a few grains remained stuck in the thickness of the sample when their size was too large compared to the sample thickness. The solidification front was then composed of much larger grains than during horizontal solidification, with large liquid zones that ultimately lead to highly segregated areas.

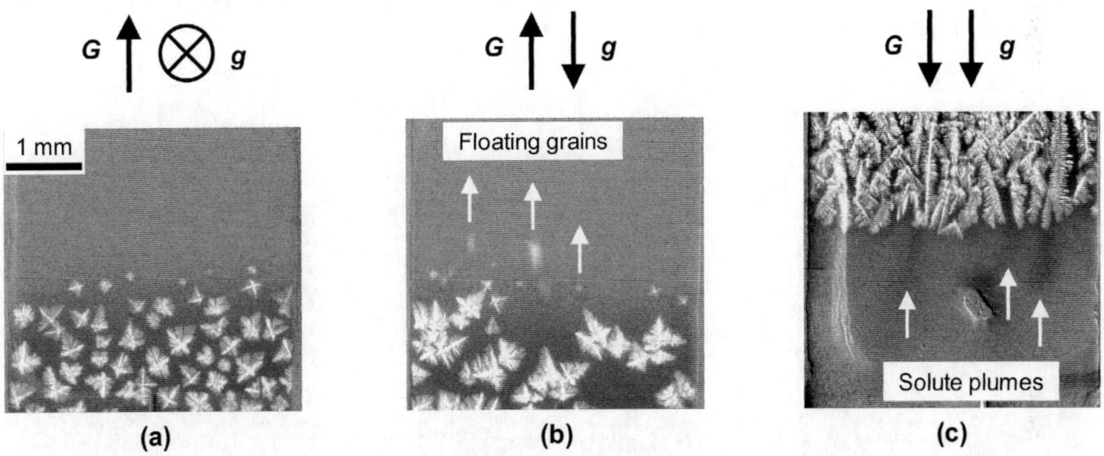

Figure 2. Radiographs of the solidification front for the low cooling rate (R_l = 0.1 K/s and G_{app} = 7.5 K/mm) during (a) horizontal solidification, (b) upward solidification and (c) downward solidification. The grey rounded shape surrounded by a white and black line in the bottom of the field of view in (c) corresponds to an artefact of the image processing following the formation of a porosity in this region of the sample (cf. figure 3c).

Downward solidification also led to the observation of a very different solidification front compared to both horizontal and upward solidification. Regarding the development of convective flow this configuration is both thermally unstable with the hot zone in the bottom part of the sample, and solutally unstable due to the rejection of heavy solute. No obvious effect of convective flow of thermal origin was detected, but large solute plumes made of Cu-rich liquid were seen developing toward the bottom part of the sample (figure 2c). It is worth noting that no segregated liquid channel was obtained for the presently investigated parameters, when this was the case in Ga-In alloys solidified upward (thermally stable and solutally unstable configuration) as reported by N. Shevchenko *et al.* [24]. In addition to the development of solute plumes, a striking difference with the previous configurations is the growth of very thin and elongated grains, even columnar grains in some areas. This observation suggests that many refining particles were inhibited by the propagation of the plumes that locally bring solute-rich liquid. As consequence, the intensity of the undercooling in the liquid is significantly decreased in these regions, which prevents the nucleation of new grains on the refiners.

3.2. Solidification fronts at high growth rate

It has been shown that if the velocity of the solidification front propagation is higher than the characteristic velocity of gravity related-phenomena such as convective flow, then the microstructure formation is close to a solidification in a diffusive regime [25]. Experiments have been carried out with a high cooling rate to investigate such an effect.

The solidification front during horizontal solidification at high growth velocity (figure 3a) is analogous to the solidification front at low growth velocity (figure 2a) but composed of smaller and more compact equiaxed grains. It is worth noting that it is also comparable to the solidification front observed during reduced gravity periods aboard parabolic flights [13].

By comparing figure 3a for horizontal solidification and figure 3b for upward solidification, it is visible that both growth front are very similar. This can be attributed to the fact that the nucleation front in the upward experiment has been propagating at a faster rate ($\approx$ 250 µm/s) than the average grain floatation velocity ($\approx$ 100 µm/s). After nucleating, grains only have a short time to move upward before being stop by a layer of new grains that nucleated just above. In a similar way, the development of solute plumes was not observed at high growth rate during downward solidification, when the solidification front velocity is faster than solute sedimentation.

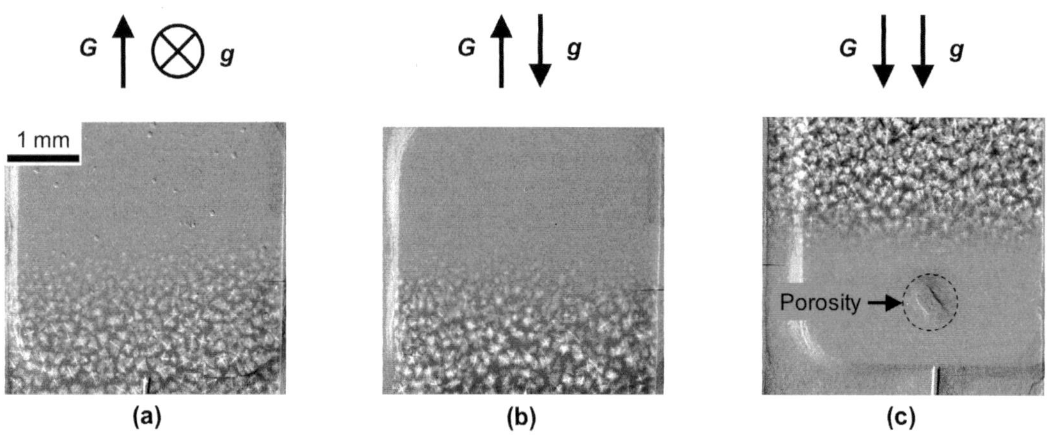

Figure 3. Radiographs of the solidification front for the high cooling rate (R_2 = 0.9 K/s and G_{app} = 7.5 K/mm) during (a) horizontal solidification, (b) upward solidification and (c) downward solidification.

4. Conclusion

A study of the dendritic grain growth during directional solidification of a refined Al-20wt.%Cu alloy has been carried out for various solidification directions with respect to the gravity vector orientation. The *in situ* observation of the development of the solidification front was made possible at low and high growth rate by using the SFINX facility. Gravity-related phenomena strongly disturb the solidification front development at low growth rate, such as grain buoyancy during upward solidification and solute plumes during downward solidification. The impact of these effects becomes significantly reduced when the velocity of the solidification front is higher than both the characteristic grain floatation velocity and solute sedimentation. More detailed analyses of these phenomena are in progress to be compared quantitatively with analytical and numerical models in the future.

Acknowledgement
This work is supported by the XRMON project (AO-2004-046) of the MAP programme of the European Space Agency (ESA), by the French National Space Agency (CNES) and the French-Algerian doctoral fellowship program PROFAS B+. The authors would also like to thank the Swedish Space Corporation (SSC) for the development of the SFINX facility and the technical support.

References

[1] Quested T E and Greer A L 2004 *Acta Mater.* **52** 3859–68
[2] Burden M H, Hebditch D J and Hunt J D 1973 *J. Cryst. Growth* **20** 121–4
[3] Quested T E 2004 *Mater. Sci. Technol.* **20** 1357–69
[4] Rerko R S, de Groh III H C and Beckermann C 2003 *Mater. Sci. Eng. A* **347** 186–97
[5] Nguyen-Thi H, Bogno A, Reinhart G, Billia B, Mathiesen R H, Zimmermann G, Houltz Y, Löth K, Voss D, Verga A and Pascale F de 2011 *J. Phys. Conf. Ser.* **327** 012012
[6] Akamatsu S and Nguyen-Thi H 2016 *Acta Mater.* **108** 325–46
[7] Mathiesen R H, Arnberg L, Ramsøskar K, Student T W P D, Rau C and Snigirev A 2002 *Metall. Mater. Trans. B* **33** 613–23
[8] Reinhart G, Buffet A, Nguyen-Thi H, Billia B, Jung H, Mangelinck-Noël N, Bergeon N, Schenk T, Härtwig J and Baruchel J 2008 *Metall. Mater. Trans. A* **39** 865–74
[9] Ludwig O, Dimichiel M, Salvo L, Suéry M and Falus P 2005 *Metall. Mater. Trans. A* **36** 1515–23
[10] Rakete C, Baumbach C, Goldschmidt A, Samberg D, Schroer C G, Breede F, Stenzel C, Zimmermann G, Pickmann C, Houltz Y, Lockowandt C, Svenonius O, Wiklund P and Mathiesen R H 2011 *Rev. Sci. Instrum.* **82** 105108
[11] Klein S, Braeuer D, Becker M, Knipstein A, Meckel S, Sondermann E and Kargl F 2016 *Int. J. Microgravity Sci. Appl.* **33** 330405
[12] Nguyen-Thi H, Reinhart G, Salloum-Abou-Jaoude G, Browne D J, Murphy A G, Houltz Y, Li J, Voss D, Verga A, Mathiesen R H and Zimmermann G 2014 *Microgravity Sci. Technol.* **26** 37–50
[13] Abou-Khalil L, Salloum-Abou-Jaoude G, Reinhart G, Pickmann C, Zimmermann G and Nguyen-Thi H 2016 *Acta Mater.* **110** 44–52
[14] Browne D J, Garcia-Moreno F, Nguyen-Thi H, Zimmermann G, Kargl F, Mathiesen R H, Griesche A and Minster O 2017 *Magnesium Technology 2017* ed K N Solanki, D Orlov, A Singh and N R Neelameggham (Cham: Springer International Publishing Ag) pp 581–90
[15] Xu Y, Casari D, Du Q, Mathiesen R H, Arnberg L and Li Y 2017 *Acta Mater.* **140** 224–39
[16] Mathiesen R H, Arnberg L, Bleuet P and Somogyi A 2006 *Metall. Mater. Trans. -Phys. Metall. Mater. Sci.* **37A** 2515–24
[17] Buffet A, Nguyen-Thi H, Bogno A, Schenk T, Mangelinck-Noël N, Reinhart G, Bergeon N, Billia B and Baruchel J 2010 *Mater. Sci. Forum* **649** 331–6
[18] Nguyen-Thi H, Reinhart G, Mangelinck-Noël N, Jung H, Billia B, Schenk T, Gastaldi J, Härtwig J and Baruchel J 2007 *Metall. Mater. Trans. A* **38** 1458–64
[19] Nogita K, Yasuda H, Prasad A, McDonald S D, Nagira T, Nakatsuka N, Uesugi K and StJohn D H 2013 *Mater. Charact.* **85** 134–40
[20] Liotti E, Arteta C, Zisserman A, Lui A, Lempitsky V and Grant P S 2018 *Sci. Adv.* **4** eaar4004
[21] Xu Y, Casari D, Mathiesen R H and Li Y 2018 *Acta Mater.* **149** 312–25
[22] Ganesan S and Poirier D 1987 *Metall. Trans. -Phys. Metall. Mater. Sci.* **18** 721–3
[23] Zimmermann G, Pickmann C, Hamacher M, Schaberger-Zimmermann E, Neumann-Heyme H, Eckert K and Eckert S 2017 *Acta Mater.* **126** 236–50
[24] Shevchenko N, Boden S, Gerbeth G and Eckert S 2013 *Metall. Mater. Trans. -Phys. Metall. Mater. Sci.* **44A** 3797–808
[25] Dupouy M . and Camel D 1998 *J. Cryst. Growth* **183** 469–89

Direct observation of dendrite fragmentation in the solidification of undercooled melts

Andrew M Mullis[1] & Nafisul Haque[1]

[1]School of Chemical and Process Engineering, University of Leeds, Leeds, UK

E-Mail: A.M.Mullis@leeds.ac.uk

Abstract. The fragmentation of dendrites immediately following the recalescence phase of growth during the solidification of undercooled melts has been invoked to explain various rapid solidification microstructures. Despite this, little direct evidence of such a fragmentation process usually survives in the as-solidified material. We report on the rapid solidification of the single phase, congruently melting intermetallic β-Ni_3Ge. During equilibrium solidification this material solidifies to the chemically ordered $L1_2$ crystal structure. Conversely, during rapid solidification, disorder trapping results in solidification to a random fcc solid solution, thereby providing a means to distinguish the rapidly solidified structures. We present results which show a range of microstructures in which the dendrite fragmentation process has been captured in progress. Results from EBSD Euler mapping reveal that dendrite fragmentation is a potential, but not particularly efficient, route to grain refinement.

1. Introduction

Spontaneous grain refinement (SGR) during the solidification of deeply undercooled melts has been a subject of enduring interest within rapid solidification. Defined as an abrupt reduction in grain size, typically of at least an order of magnitude, above a critical undercooling, ΔT^*, the first such observation was made by Walker in pure Ni in 1959 [1]. Since then many other pure metals [2, 3, 4] and alloys [5, 6, 7] have been observed to undergo similar transitions, with alloy systems often displaying a more complex behavior involving two distinct grain refinement transitions. For a typical alloy system undergoing SGR, a fine grained structure is observed above a lower SGR transition undercooling ΔT_1^*, reverting to a coarse grain structure at higher undercooling. In most systems a second episode of grain refinement is then observed at yet high undercooling, ΔT_2^*.

Where growth velocity measurements have been made, the onset of SGR above ΔT^* (ΔT_2^* in alloys) is often accompanied by a discontinuous break in the velocity-undercooling curve [2]. Below ΔT^*, the growth velocity, V, can generally be described by a power-law relationship of the form $V \propto (\Delta T^*)^\beta$, with β typically being in the range 2-3, while above ΔT^*, V is often linear in ΔT. High speed imaging of the solidification front has also suggested that there is a change in the front morphology from an angular euhederal morphology below ΔT^* to a smooth morphology above ΔT^* [8]. However, in alloy systems there is generally no signature in the velocity undercooling curve corresponding to the lower SGR transition at ΔT_1^*. This has prompted suggestions that the upper and lower SGR transitions have different origins [9, 10], an idea that ties in with some of the earliest observation of SGR in which it was noted that there were subtle difference in the solute segregation patterns observed at high and low undercoolings [11]. At high undercooling the solute segregation pattern is typically spheroidal, whereas at low undercooling small, well-developed dendritic fragments are observed.

Numerous mechanisms have been proposed to account for SGR. Early theories included; copious nucleation ahead of the solidification front resulting from a shrinkage induced pressure pulse [1], the action of minor solute additions [12], and recrystallization [13]. However, perhaps the most enduring model for SGR is that of remelting and fragmentation of dendrites shortly after the recalescence phase of solidification, during which time the solid and liquid co-exist. Early proponents of fragmentation based models included Horvay [14] and Kattamis & Flemings [11], although perhaps the best known of these fragmentation models is that due to Schwarz *et al.* [15]. They argue, based upon a mathematical analysis due to Karma [16] for the growth of Rayleigh instabilities along the dendrite, that there are two key timescales to consider. The first is that for the breakup, τ_{br}. This is the timescale over which fragmentation and subsequent spheroidisation of secondary dendrite arms will occur during co-existence of the solid and liquid following recalescence. The second is the plateau timescale, τ_{pl}, which is the time for which the solid and liquid co-exist. If $\tau_{br} < \tau_{pl}$ then breakup is assumed to occur with the fragments providing nuclei for the subsequent growth of refined grains. τ_{pl} is simply a function of the heat extraction rate from the sample while τ_{br} is a function primarily of the size of the dendrite, which in turn, due to the self-similarity of dendrites, is equivalent to it being a function of the dendrite tip radius, ρ, with smaller radii giving shorter breakup times. Consequently, for a given cooling rate there is a critical tip radius, ρ^*, below which fragmentation occurs.

An attractive feature of the Schwarz model [15] is that it appears to give a natural explanation for there being only one SGR transition in pure metals and two in alloys; the tip radius decreasing monotonically in a pure metal but passing through a local minimum at intermediate undercoolings in alloy systems. There are however problems with the Schwarz model. It is for instance difficult to see how remelting, which by definition must occur post-recalescence, can give rise to either a break in the velocity-undercooling curve or a change in the morphology of the solidification front. That both occur concurrently with the onset of SGR must therefore be ascribed to coincidence. We have previously argued [17] that a change in the primary solidification morphology from dendritic to seaweed is therefore a more likely precursor to the onset of SGR, as this may also be expected to have a signature in both the morphology of the solidification front and the velocity-undercooling behavior. However, there is also a further objection to the Schwarz model for SGR. There is an implicit assumption that remelting and fragmentation of the dendrites formed during recalescence will lead to grain refinement. In fact, it is not at all clear that even if dendrite fragmentation and spheroidisation occurs as described by the model of Karma [16] that such fragments will subsequently nucleate the copious number of new grains required for SGR.

Unfortunately, direct experimental evidence of dendrite fragmentation in rapidly solidified metals has been extremely difficult to obtain as any indication of such fragmentation is usually obscured in the final, as-solidified sample. Such remelting has been predicted by [18] for 250 μm droplets of Al-10 wt% Cu alloy, wherein it was found that the fraction of the recalescence solid subject to remelting was a strong function of the nucleation undercooling, with high undercooling leading to greater remelting. Subsequent EBSD analysis of Al-Fe alloys [19] showed evidence of such remelting, with detached and slightly misoriented primary and secondary dendrite arms being observed.

Recently we have identified that remelting and fragmentation of the primary dendrites formed during recalescence may be preserved during rapid cooling of the congruently melting intermetallic compound β-Ni_3Ge. As such, β-Ni_3Ge is an interesting vehicle to probe whether remelting is a likely origin of SGR. Here we report on drop-tube experiments upon the β-Ni_3Ge compound.

2. Experimental Methods

β-Ni_3Ge is a congruently melting intermetallic with the $L1_2$ ordered fcc crystal structure and a homogeneity range of 22.5 to 25.0 at. % Ge [20]. The congruent melting temperature is 1405 K and upon slow cooling the compound will grow in the ordered form directly from the liquid. However, disorder trapping occurs for undercoolings in excess of 168 K [21].

The starting alloy for the drop-tube experiments was produced by arc-melting together Ni and Ge (both 99.99% purity) in the correct proportions under a protective Ar atmosphere with the resulting ingot being checked to ensure it was single phase β-Ni_3Ge using XRD (see below). For rapid solidification processing a 9.6 g ingot was melted by induction heating at the top of the Leeds 6.5 m drop-tube, which is maintained at a pressure of 50 kPa throughout processing. The sample is contained within an alumina crucible with three 300 μm laser drilled holes in the base. Ejection of the sample is occasioned by pressurizing the crucible to 400 kPa, wherein a fine spray of droplets is produced which solidify in flight down the tube. The melt ejection occurred at a temperature of 1480 K, corresponding to a melt superheat of 75 K. More details of the drop-tube method are given in [22].

Following drop-tube processing the powders are sieved into 9 standard size fraction between 850 μm and 38 μm. Each size fraction is subject to XRD analysis using a PANalytical Xpert Pro before being mounted in thermosetting resin, ground and polished to a 0.25 μm finish for microstructural analysis. In order to reveal the underlying microstructure samples were etched for 20 s using a highly aggressive mixture comprising equal parts of undiluted HF, HCl and HNO_3. Microstructural analysis was undertaken using a Carl Zeiss EVO MA15 Scanning Electron Microscope (SEM) equipped with an Oxford Instrument X-Max Energy-Dispersive X-Ray (EDX) detector. Electron Backscatter Diffraction (EBSD) was undertaken using a FEI Quanta 650 FEGSEM with Oxford/HKL Nordlys EBSD system. For EBSD unetched samples were used that had been polished to a 0.1 μm surface finishing using a colloidal silica suspension.

3. Results and Discussion

Figure 1a shows a sample from the 150-106 μm sieve fraction displaying a typical, fully dendritic morphology. The cooling rate for this sample, estimated using the methodology described in [23], is 7800 - 13000 K s^{-1}. In [24] the principal growth direction, for both the recalescence and post-recalescence material, was determined as <100> for solidification close to equilibrium. This is as expected for a material with the cubic $L1_2$ crystal structure. Normal <100> growth was observed to persist for cooling rates up to 26,000 K s^{-1}, (75 μm diameter droplets) wherein a transition to a mixed <100> and <110> growth mode was observed. It is clear that within the microstructure a large number of differently directed dendrites are present, which is highly suggestive of multiple nucleation events having occurred. The average grain size in these samples has been measured at 10.2 μm [24] and, with no evidence of dendritic fragmentation, it seems reasonable to assume that each grain originates from an independent nucleation event, giving an estimate nucleation density of 9.45×10^{14} m^{-3}. Even at the very highest cooling rates, ≥ 62,000 K s^{-1}, for droplets ≤ 38 μm diameter, we still observe large numbers of grains per droplet. This is highly unusual as most other materials display only 1 grain per droplet at these very high cooling rates. Moreover, there is little evidence of porosity, either at the grain boundaries or in the interdendritic space, suggesting minimal solidification shrinkage.

Figure 1b shows an EDX line scan across one of the dendrite trunks. It is clear that, to within the normal errors associated with EDX, there is no difference in composition between the dendrite and the surrounding interdendritic material, as would be expected for a congruently melting compound. In addition, XRD analysis confirms that all of sieve fractions comprise single phase β-Ni_3Ge, with all of the diffraction peaks being unambiguously matched to ICCD reference pattern 04-004-3112 for β-Ni_3Ge. Etching normally produces topographic contrast resulting from differential dissolution due to differences in either phase or composition. Accordingly, the question here is what produces that topographic contrast in a single phase sample of homogeneous composition. In fact, we show in [25] that the contrast arises due to the degree of local chemical ordering present, with the dendrites being disordered and the surrounding material being ordered. Such a situation would arise due to disorder trapping in the rapidly solidified material growing during recalescence, with the interdendritic material growing more slowly, and hence as the ordered phase, post-recalescence. This would be consistent with the known resistance of intermetallic compounds to chemical attack, with the disordered, and hence more metallic, phase preferentially dissolving in the aggressive etch used here. The upshot is the rare situation in which etching will reveal only the solid formed during recalescence, leaving the post-recalescence solid unetched. Consequently this is, as far as we are aware, a unique material insomuch as it permits direct imaging of the solid formed during recalescence in the as-solidified sample.

As described above, droplets in the 150-106 μm sieve fraction display an as-solidified microstructure which is typically of dendritic solidification in a cubic material. For the smaller sieve fractions, and hence higher cooling rates, a number of microstructural changes are observed including a change in the dendrite growth direction and the occurrence of seaweed structures. However, it is the large sieve fractions, and hence lower cooling rates, that are of interest here.

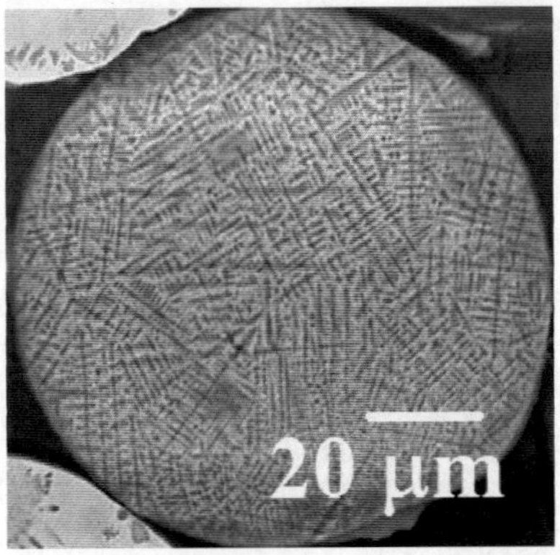

Figure 1. Typical microstructure as observed in a sample of drop-tube processed β-Ni_3Ge from the 150-106 μm sieve fraction following etching in a mixture of HF + HCl + HNO_3. Numerous, multiply branched dendrites are observable. Contrast arises due to preferential etching of the ordered material formed during the recalescence phase of solidification.

Figure 2a shows a typical sample from the 300-212 μm sieve fraction (cooling rate 2800 – 4600 K s^{-1}). Numerous linear arrays of near spherical fragments are evident. We believe that such linear arrays of particles could only arise by the melting, and subsequent spheroidisation of primary dendrites. Typical fragment size in these arrays is 5-10 μm Moreover, in Figure 2b, now taken from the 212-150 μm sieve fraction, the higher cooling rate, estimated as 4600 - 7800 K s^{-1}, appears to have resulted in a dendrite being captured in the process of fragmenting. Specifically, the uppermost secondary arm on the left-hand side of the micrograph has detached from the primary trunk and appears to have undergone almost complete fragmentation into 6 more-or-less distinct spherical fragments while the secondary arm below this displays the modulation characteristic of the onset of the Rayleigh instability. The mean fragment diameter in the fully fragmented arm in the top left of the image is ≈ 2.4 μm. As the nature of the contrast in these samples is between the rapidly solidified (chemically disordered) material and that solidified close to equilibrium (chemically ordered) material, there is unfortunately no means to distinguish between material that grew post-recalescence and material that grew during the recalescence phase of solidification but subsequently remelted. Both ultimately solidified close to equilibrium and hence will be the chemically ordered phase.

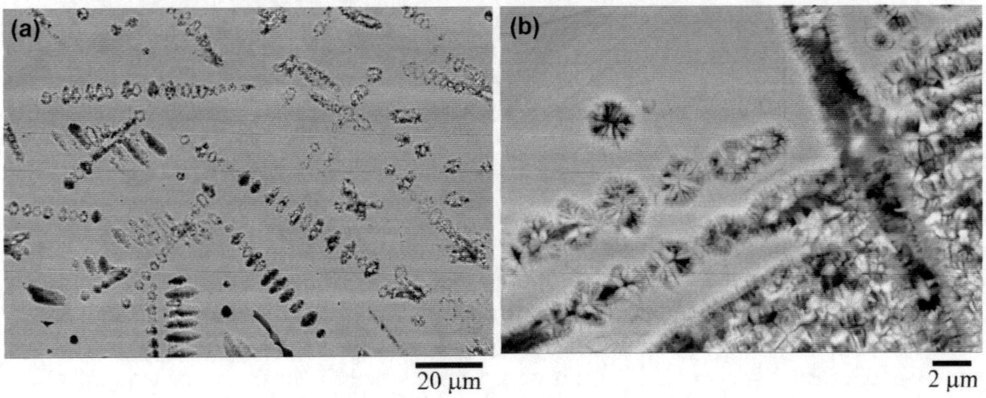

Figure 2. SEM micrograph of fragmented dendrites β-Ni_3Ge from (a) the 300-212 μm sieve fraction and (b) the 212-150 μm sieve fraction.

Based on the evidence present above we contend that in rapidly cooled β-Ni_3Ge there exist a range of cooling rates, and therefore plateau times, τ_{pl}, in which the fragmentation of dendrites may be observed directly in the as-solidified microstructure. Such fragmented dendrites appears to be the dominant morphology in the 300-212 μm sieve fraction, for which τ_{pl} is estimated to be in the range 0.22-0.12 s (corresponding to the upper and lower sieve limits respectively). In the next sieve size down, 212-150 μm, wherein τ_{pl} = 0.12-0.06 s respectively, partial fragmentation, particularly now of the secondary dendrite arm may still be observed, while for all smaller sieve fractions such fragmentation appears to be wholly inhibited by the high cooling rate. Given that we can now, with reasonable certainty, conclude that dendritic fragmentation is occurring, the pertinent question becomes are these samples also grain refined. The polished and etched samples do not display obvious grain boundaries, so we have used EBSD to investigate the grain size distribution.

Figure 3a shows an EBSD Euler orientation map from a sample in the 300-212 μm sieve fraction, i.e. in which fully fragmented dendrites are expected to be the dominant morphology. The EBSD map reveals there are a small number of large grains, typically > 50 μm, with a much larger number of small grains in the size range 5-10 μm. Interestingly, the large grains all appear to dappled with small regions of material displaying different crystallographic orientations to the main grain within which they are embedded. From this it would appear that some of the dendritic fragments give rise to small grains of < 10 μm diameter, but that many do not, instead being incorporated into much larger grains despite an obvious mismatch in orientation. The observed structure, comprising mixed large and small grains is does not resemble the classic SGR microstructure of uniformly distributed fine grains. For completeness we also show, in Figure 3b, an EBSD map characteristic of the 212-150 μm sieve fraction. Given that, from the SEM micrographs, dendrite fragmentation is unlikely to be complete, with primary trunks and some secondary arms intact, we may not expect to observe a grain refined structure. As with the larger sieve fraction the structure remains a mixture of large and small grains, although without the dappling of the large grains evident in the 300-212 μm sieve fraction.

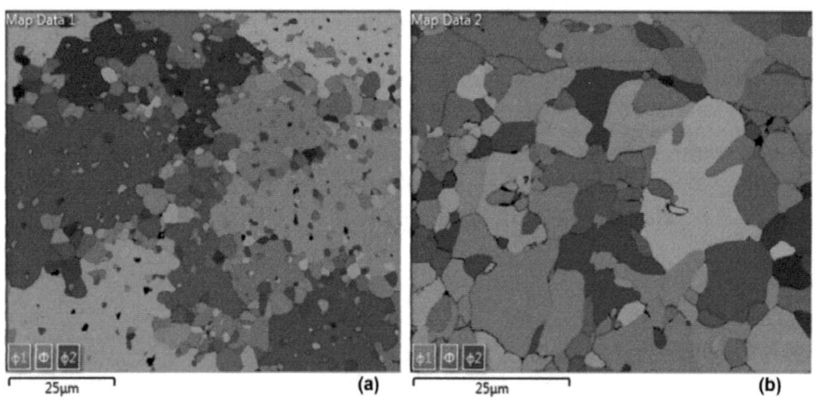

Figure 3. EBSD Euler maps obtained from β-Ni_3Ge in (a) the 300-212 μm sieve fraction and (b) the 212-150 μm sieve fraction. Note in (a) the large number of small fragments that appear to have been incorporated into much larger grains.

We have then a system in which we can establish that dendritic fragmentation occurs over a certain well defined range of cooling rates. However, such fragmentation does not appear to be a particularly effective mechanism for grain refinement. It appears that some fragments are likely to nucleate fine grains, as evidenced by the large number of small grain in the EBSD map for the 300-212 μm sieve fraction, but this does not seem to be pervasive, with many large grains remaining. Indeed, > 90% of the field of view in the EBSD map (≈8750 μm^2) is occupied by just 12 grains. Consequently, the resulting structure is not the one typically associated with SGR. We note that for droplets in free-fall in the drop-tube flow in the melt is likely to be suppressed and we believe it is possible that if flow were present to disperse the fragments, grain refinement by dendritic melting may be more effective. In this regard we note that electromagnetic levitation induces strong flow within the melt and that this has been shown to influence the results obtained by this technique [26]. However, in such a case it is questionable whether SGR by dendrite fragmentation should be considered an inherent property of the material or a consequence of the processing technique. This would also not account for the many samples observed to display SGR processed by techniques which do not induce extensive flow in the melt (melt fluxing, drop-tube, electrostatic levitation). We conclude the dendrite remelting and fragmentation is not an efficient cause of SGR.

4. Summary and Conclusions

The rapid solidification of single-phase, congruently melting β-Ni_3Ge has been studied in by drop-tube processing. Rapid growth during the recalescence phase of solidification results in a disordered solid which may be etched preferentially to the ordered solid that grows post-recalescence. Over a range of cooling rates from 2800 - 7800 K s^{-1} dendrites fragmented due to remelting may be observed directly in the as-solidified samples. However, despite containing many fully fragmented dendrites these samples do not display the grain size distribution normally associated with spontaneous grain refinement. We therefore conclude that the remelting and fragmentation of dendrites is not an efficient route to spontaneous grain refinement in deeply undercooled melts.

References

[1] Walker J L 1959 *The Physical Chemistry of Process Metallurgy Part 2* ed G L St. Pierre (New York: Interscience) 845
[2] Wilnecker R, Herlach D M and Feuerbacher B 1989 *Phys. Rev. Lett.* **62** 2707
[3] Battersby S E, Cochrane R F and Mullis A M 1997 *Mater. Sci. Eng. A* **226** 443
[4] Dragnevski K I, Cochrane R F and Mullis A M 2004 *Mater. Sci. Eng. A* **375-377** 479
[5] Battersby S E, Cochrane R F and Mullis A M 2000 *J. Mater. Sci.* **35** 1365
[6] Lu S Y, Li J F and Zhou Y H 2007 *J. Crystal Growth* **309** 103
[7] Zhang T, Liu F, Wang H F and Yang G C 2010 *Scripta Mater.* **63** 43
[8] Matson D 1998 *Solidification 1998* eds S P Marsh, J A Dantzig, R Trivedi, W. Hofmeister, M G Chu, E J Lavernia and J-H Chun (Warrendale PA: TMS) 233
[9] Castle E G, Mullis A M and Cochrane R F 2014 *Acta Mater.* **66** 378
[10] Castle E G, Mullis A M and Cochrane R F 2014 *Acta Mater.* **77** 76
[11] Kattamis T Z and Flemings M C 1967 *Mod. Casting* **52** 97
[12] Powell G L F and Hogan L M 1969 *Trans. AIME* **245** 407
[13] Southin R T and Weston G M 1973 *J. Aust. Inst. Met.* **18**
[14] Horvay G 1962 *Proc. of the 4 US National Congress of Appl. Mech.* (Univ. California) 1315
[15] Schwarz M, Karma A, Eckler K and Herlach D M 1994 *Phys. Rev. Lett.* **73** 1380
[16] Karma A 1998 *Int. J. Non-equilibrium Process.* **11** 201
[17] Mullis A M, Dragnevski K and Cochrane R F 2004 *Mater. Sci. Eng. A* **375-377** 157
[18] Heringer R, Gandin Ch-A, Lesoult G, Henien H 2006 *Acta Mater.* **54** 4427.
[19] Chen J, Dahlborg U, Bao C M, Calvo-Dahlborg M and Henien H 2011 *Metall. Mater. Trans. B* **42** 557.
[20] Nash A and Nash P 1976 *US National Bureau of Standards Monograph Series 25*, (Ohio:Elsevier, ASM) 35
[21] Ahmad R, Cochrane R F and Mullis A M 2012 *J. Mater. Sci.* **47** 2411
[22] Cao L, Cochrane R F and Mullis A M 2015 *Metall. Mater. Trans. A* **46** 4705
[23] Oloyede O, Bigg T D, Cochrane R F and Mullis A M 2016 *Mater. Sci. Eng. A* **654** 143
[24] Haque N, Disorder Trapping in Rapidly Solidified Intermetallic Compounds, PhD Thesis submitted to the University of Leeds, July 2018.
[25] Haque N, Cochrane R F and Mullis A M 2016 *Intermetallics* **76** 70
[26] Binder S, Galenko P and Herlach D M 2014 *J. Appl. Phys.* **115** 053511

IOP Conf. Series: Materials Science and Engineering **529** (2019) 012021 doi:10.1088/1757-899X/529/1/012021

Rapid solidification of Al-Cu droplets of near eutectic composition

A-A. Bogno[1], J. Valloton[1], D.D. Jimenez[1], M. Rappaz[2], H. Henein[1]

[1] Department of Chemical and Materials Engineering
University of Alberta, Edmonton, AB T6G 2G6 Canada

[2] Institute of Materials, Ecole Polytechnique Fédérale de Lausanne
CH-1015 Lausanne Switzerland

Email: bogno@ualberta.ca / hhenein@ualberta.ca

Abstract. Near eutectic Al-Cu droplets were rapidly solidified by Impulse Atomization. A wide range of microstructural scales was obtained at different cooling rates and undercoolings. The micrographs of the investigated samples revealed two distinct zones of different structural morphologies: An undulated eutectic morphology developed during recalescence following the single grain nucleation and a regular lamellar eutectic morphology resulting from the solidification of the remaining liquid post recalescence. The volume fraction of each zone was measured as a function of the droplet diameter, and the nucleation undercooling was deduced using the hypercooling limit equation. Scanning Electronic Microscopy imaging and microhardness measurements were used to evaluate the microstructural scale, and mechanical properties.

1. Introduction

High undercooling / cooling rate associated with rapid solidification of metallic alloys results in various beneficial microstructural features. These include minimal segregation, grain refinement, solubility extension and the formation of metastable phases [[1]][[2]][[3]][[4]]. Thus, in recent decades, rapid solidification techniques have been extensively used to develop new advanced materials. However, it is sometimes difficult to relate the observed microstructure with process parameters due to experimental difficulties. These include, measurements of variables such as cooling rate, undercooling and solidification rate, which are very well known to control the final solidification microstructure.

Rapid solidification of alloys of eutectic composition is important from both technological and scientific points of view as it can result in very fine microstructures and metastable phases. The Al-Cu eutectic system has been extensively studied and various types of microstructures have been reported to form under rapid solidification conditions, depending on the solidification rate [[5]][[6]][[7]][[8]]. At low cooling rates, a regular lamellar eutectic structure, consisting of alternate α-Al and θ-Al_2Cu lamellae, is formed. After nucleation the two phases grow cooperatively with a fairly well-established relationship between the solidification rate V and the interlamellar spacing λ, given by the equation $V\lambda^2$ = constant [[9]]. At higher cooling rates, Al-Cu eutectic solidifies into degenerate / anomalous eutectic structure consisting of irregular θ-Al_2Cu precipitates embedded in an α-Al matrix or into undulated eutectic

microstructures, where α-Al and θ-Al_2Cu lamellae become wavy [[5]][[6]]. At even higher cooling rates, a banded eutectic [[5]][[10]] or a fully supersaturated single α-Al solid solution were observed [[11]][[12]].
This paper present the observations made on rapidly solidified Al-Cu droplets of eutectic composition, obtained by Impulse Atomization (IA). This single fluid atomization technique yields solidification microstructures formed over a wide range of undercooling, from the onset to the end of recalescence and during further cooling.

2. Materials and methods

Near-eutectic Al-Cu droplets were generated using IA. Aluminum and copper were melted in a graphite crucible by heating up to 1200 °C. Atomization was achieved under a stagnant argon atmosphere with oxygen level less than 20ppm. The melt was superheated to 750°C, i.e., about 200°C above the equilibrium eutectic temperature. A wide range of droplet sizes (< 200 μm to 1000 μm) were obtained to ensure microstructural analysis over a wide range of cooling rates and undercoolings.
The composition of the eutectic droplets (given in the table below) were obtained by chemical analysis performed according to ASTM E1097-12 and ASTM E1479-16. As can be noted, both alloys are slightly hypoeutectic, hence the near-eutectic designation.

Sample ID	Al	Cu	Si	Fe	Zn
Alloy 1	Balance	31.9	0.01	<0.01	None
Alloy 2	Balance	32.2	None	0.02	0.04

The as-atomized microstructures were characterized by Scanning Electron Microscopy (SEM) using both Secondary Electrons (SE) and Back Scattered Electrons (BSE) modes, on both ground and polished sections of the droplets as well as on their as-atomized external surface. The instrument used was a Zeiss Sigma Field Emission SEM equipped with a Bruker energy dispersive X-ray spectroscopy (EDS) system with dual silicon drift detectors, each with an area of 60 mm^2 and a resolution of 123 eV.
The scale of the microstructure was determined by eutectic spacing measurement using line intercept method on the micrographs, according to ASTM E112-13. The strength of the samples was evaluated through microhardness measurement using a Buehler VH 3100 microhardness machine. The device was calibrated using a manufacturer provided steel block. Five indentations were randomly applied on each sample with a load of 100gf held for 10s.

3. Results

3.1. Nucleation and microstructure formation

Figures 1a through 1f show typical micrographs of as-atomized droplets surface and cross-sections of the two investigated alloys. The BSE imaging providing an atomic number (Z) contrast, α-Al appears dark while the Cu rich θ-Al_2Cu phase is in light grey. As shown in Figure 1, the majority of the investigated droplets have a single nucleation point from where the microstructure propagates. Furthermore, the nucleation point is most often located at the surface of the atomized droplets, but in some cases it can be within the volume, as is for example the case for the Alloy 2 shown in (e). Nucleation usually occurs in two steps, i.e., the nucleation of one phase helps that of the second one, as shown in (e) and (f). Once both phases are formed, the coupled eutectic grows with a very undulated morphology corresponding to Zone A in the figure, followed by a regular eutectic in Zone B.

Zone A, formed around the nucleation point, which in most cases is an impurity, is characterized by an undulated eutectic morphology. The solidification speed of the eutectic front in this region is assumed to be very high, since it forms during the recalescence period. In some of the droplets, as shown in Figure 1e, primary nucleation occurs with a Cu-rich region (Figure 1f), which then grows as a dendrite before coupled eutectic growth occurs. The elemental composition of this dendrite, confirmed by EDX, corresponds to that of the θ-Al_2Cu phase. As the melt surrounding this growing dendrite is depleted of Cu, there is a driving force to form α-Al. This is clearly shown in the zoom of Figure 1f (dark layer surrounding the θ-dendrite). Once both phases are formed, the undulated eutectic forms.

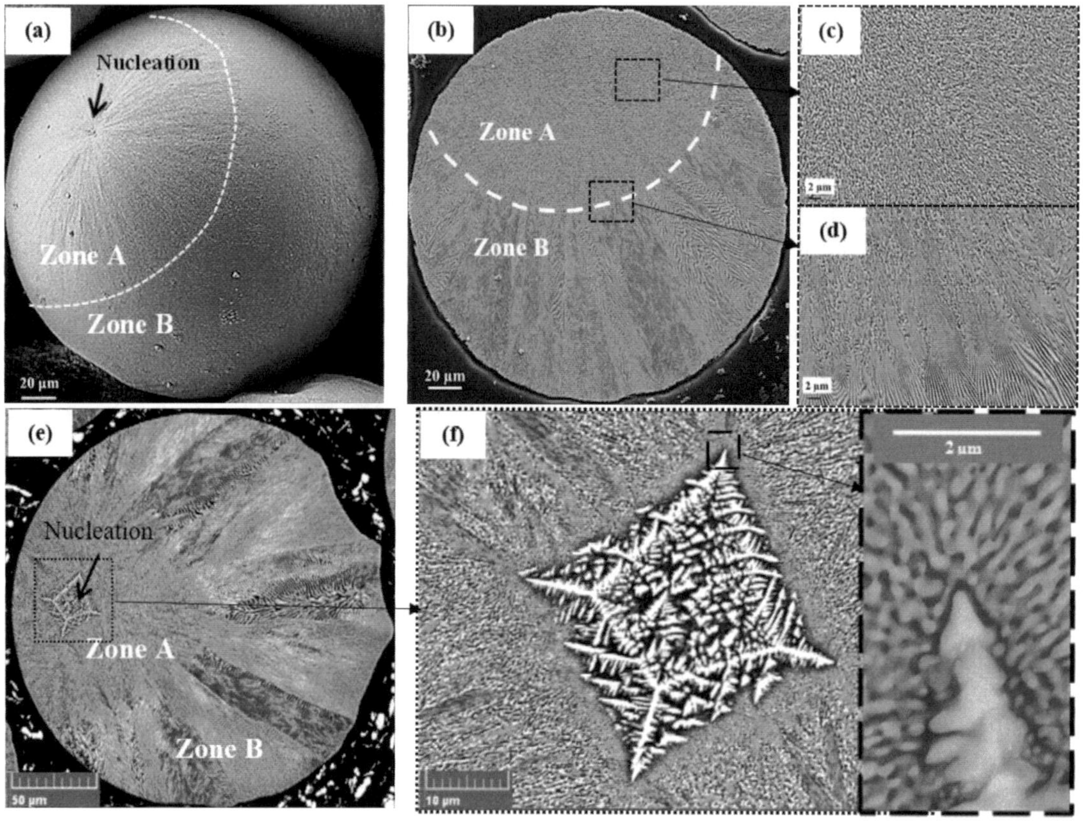

Figure 1: SEM (BSE) micrographs of the two investigated eutectic alloys. (a) Surface of a 215µm diameter droplet from Alloy 1, the nucleation point is indicated on the surface by an arrow. (b) The corresponding cross section of the droplet showing the undulated microstructure (Zone A) and the regular lamellar structure (Zone B). Magnifications of Zone A and of the transition between Zones A and B are shown in (c) and (d), respectively. (e) Cross section of a 230 µm droplet from Alloy 2. The microstructure propagated from the Cu-rich dendrite (nucleation point) to a Zone A (undulated eutectic) which then becomes a regular lamellar structure (Zone B). (f) Magnification of the transition between the Cu-rich dendrite (light grey) and the undulated eutectic, with formation of an α-Al layer in between.

These observations are similar to those of laser remelting of the irregular eutectic Al-Si, i.e, formation of Si first, with an α-Al layer around it that grows before coupled growth occurs [[13]]. However, in this last case, this sequence of phase formation could be expected since this is an irregular eutectic and the composition was hypereutectic. It is well known that irregular eutectics have a coupled zone skewed toward the facetted phase, i.e., silicon [[14]]. But Al-Cu is a regular eutectic, i.e., both α-Al and θ-Al_2Cu are non-faceted (this is quite evident for the θ-phase in Figure 1(f)). Since the coupled zone should not be skewed toward either phase, it is hard to predict which phase has a growth advantage at large undercooling. Considering Jackson's criterion for the formation of a faceted phase, the entropy of fusion ΔS_f for the θ-phase has been calculated with Thermo-Calc [[15]] and the thermodynamic database TTAl7. It is equal to 17.5 J $mol^{-1}K^{-1}$ at 869 K, thus giving $\Delta S_{f,} \approx 2R$, where R is the gas constant. This means that the stable θ-phase is close to being a faceted phase according to this criterion and the coupled zone might be skewed toward θ. For a eutectic composition, this would rather predict the formation of α and not θ. It should be recalled that the concept of coupled zone is solely based on a growth competition undercooling advantage, and not at all on nucleation. Further analyses are required to see if primary nucleation of Al_2Cu is induced by trace elements (in this case Fe and Zn).

Once both phases are present, they grow in a coupled but very unstable way in Zone A, thus producing an undulated microstructure at large undercooling. Once the recalescence ends and the undercooling is reduced, there is a transition to the expected regular lamellar eutectic structure of Al-Al_2Cu (Zone B). It is worth noting that Zone B consists of eutectic colonies, characteristic of eutectic structures resulting from the rejection of impurities (Si, Fe or Zn in this case) during solidification [[16]]. The colonies could also have their origin from a thermal instability as the growth, at least during recalescence, occurs in an undercooled melt (i.e., negative thermal gradient).

3.2. Nucleation undercooling

Zone A is assumed to have formed under conditions dominated by internal heat flow under adiabatic conditions during recalescence. Numerical simulations indeed show that recalescence is very fast (less than 5% of the total solidification time account for primary phase nucleation and growth during recalescence [[14]]) and that heat losses of the droplet can be neglected. Under such conditions, the nucleation undercooling, ΔT_N, can be defined by Eq. 1, i.e., by the product of the solid fraction f_R formed during recalescence and the characteristic hypercooling limit of the melt, ΔT_{hyp}.

$$\Delta T_N = f_R \cdot \Delta T_{hyp} \quad \text{(Eq.1)}$$

$$\Delta T_{hyp} = \frac{\Delta H_f}{C_p^l} \quad \text{(Eq.2)}$$

where C_p^l (2.86×10^6 $Jm^{-3}K^{-1}$) and ΔH_f (1.23×10^9 Jm^{-3}) are the melt heat capacity and the latent heat of fusion, respectively (values are taken from [[17]]).

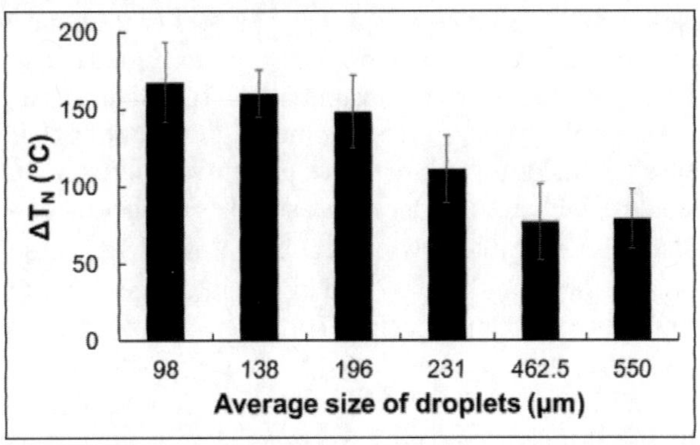

Figure 2: Variation of nucleation undercooling with average droplet size for Al-Cu of eutectic composition, sample ID: Alloy 1

Figure 2 shows the variation of nucleation undercooling with respect to the average droplet size. The results are obtained based of the measurement of the volume fraction of Zone A of the samples for Alloy 1 where nucleation occurred on the droplet surface. Figure 2 shows that the nucleation undercooling in each droplet is fairly large, which explains the formation of the undulated microstructure in Zone A, promoted by high growth rate (0.2 to 0.5 ms^{-1}) as observed in rapid solidification microstructure of Al-Cu eutectic alloy obtained by laser remelting [[5]].

3.3. Scale of the microstructures and mechanical properties

The eutectic spacing, λ, was measured for each droplet as a function of the normalized position (*Z*/*D*) on each investigated droplet of diameter *D*, where *Z* is the distance from the nucleation point along the growth direction within the two zones. In Zone B, the measurement of the average lamellar spacing was carried out on the finer eutectic colonies only, in order to minimize stereological effects of the lamellae orientation in 2-dimensional cross sections. In Zone A, the interphase spacing λ was measured from a mean intercept. Figure 3a shows the variation of the average eutectic spacing with the normalized position (*Z*/*D*) on each investigated droplet of alloy 1. Similar results were obtained for Alloy 2.

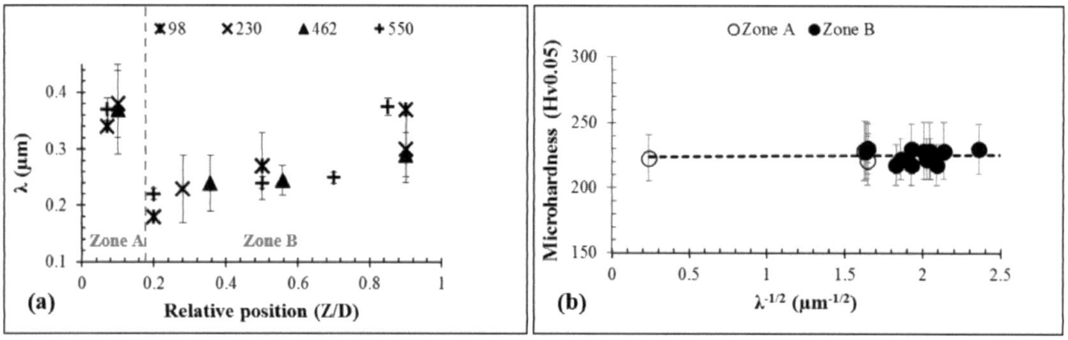

Figure 3. (a) Spatial variation of interphase (Zone A) and lamellar (Zone B) spacing λ as a function of the normalized distance Z/D from the nucleation point.(b) Hall-Petch relationship between microhardness and $\lambda^{-1/2}$.

In Zone B, the average lamellar spacing increases with *Z* as expected. After recalescence, the eutectic front interface velocity *V* is controlled by the heat flow extracted from the droplet during the nearly isothermal eutectic reaction. As the solid-liquid interface increases, *V* has to decrease to keep the latent heat release about constant, thus explaining the increase in λ with the relationship λ^2V = constant. Surprisingly, the interphase spacing λ in Zone A is nearly constant (within the error bars), regardless of the droplet size, and larger than the lamellae spacing measured in Zone B. This should be the opposite since the undercooling during recalescence is substantially larger than during the nearly isothermal plateau. This observation, also reported during laser remelting (but not explained) in [[5]], is not well understood but clearly shows that the growth kinetics of these two microstructures is different.

Finally, Figure 3b shows the average microhardness measured as a function of $\lambda^{-1/2}$, describing the Hall-Petch relationship associated with one of the two eutectic phases [[18]]. Assuming that it is the α-Al that deforms plastically, dislocations could be stopped by the interfaces with the θ-Al_2Cu phase, hence microhardness could depend on the eutectic spacing. However, microhardness is found to be constant around 220±10 HV within both Zones A and B, independently of the investigated droplet size. It is worth noting that there is ~50% in volume of α-Al and θ-Al_2Cu in both Zone A and Zone B, based on measurement performed on the 2-dimensional cross sections of all the investigated samples.

4. Summary

Impulse Atomization was used to investigate microstructure development in rapidly solidified Al-Cu droplets of near-eutectic composition as a function of cooling rate (droplet size). From microscopy analysis, evidence of a single nucleation point on the surface was found for all the investigated samples. Two types of eutectic morphologies have been observed: an undulated eutectic during recalescence and a regular lamellae structure during nearly isothermal solidification after recalescence. From the volume of the undulated eutectic zone, assumed to occur during adiabatic recalescence, the nucleation undercooling could be determined and was found to vary between 80 K and 170 K, depending on the droplet size. While the eutectic spacing of Zone B seems to follow the λ^2V relationship typical of

IOP Conf. Series: Materials Science and Engineering **529** (2019) 012021 doi:10.1088/1757-899X/529/1/012021

eutectic, the interphase spacing of Zone A is surprising. It is almost independent of the droplet size (and thus undercooling) and larger than the lamellae spacing of Zone B. Further investigations need to be done to understand the growth kinetics of this morphology, which is definitely not like 1-λ or 2-λ oscillations reported in the literature [[5]][[19]]. Also, 3-dimensional characterization of Zone A is an ongoing effort.

Acknowledgements
The authors are grateful to the Natural Sciences and Engineering Research Council of Canada (NSERC) and the European Space Agency (ESA) for their financial support. Dr. Mark Gallerneault of Alcereco Inc. is also acknowledged for providing the aluminum raw materials. Finally, thanks are due to Dr. Jose Eduardo Spinelli for fruitful discussions.

References
[1] Kear, B.H., and Giessen, B.C.,1982, *MRS Bull.*, **7**,(02), 5.
[2] Bogno, A.-A., Riveros, C., Henein, H., and Li, D.,2016, *Metall. Mater. Trans. B - Process Metall. Mater. Process. Sci.*, **47** , (6), 3257.
[3] Banerjee, S., and Dey, G.K., 2004, *J. Mater. Sci.*, **39**, (12), 3985.
[4] Liu, R.P., Herlach, D.M., Vandyoussefi, M., and Greer, A.L., 2004, *Metall. Mater. Trans. A*, **35**, (2), 607.
[5] Zimmermann, M., Carrard, M., andKurz, W., 1989, *Acta Metall.*, **37**, (12),3305.
[6] Mueller, B.A., and Perepezko, J.H.,1988, *Mater. Sci. Eng.*, 98, 153.
[7] Lee, S.M., and Hong, C.P., 1998, *ISIJ Int.*, **38**, (2),157.
[8] Sahoo, S., and Ghosh, S.,", 2015, *Appl. Phys. A Mater. Sci. Process.*, **121,** (1), 45.
[9] Jackson, K.A., and Hunt, J.D., 1966, *Trans. Am. Inst. Min.Engrs*, **236**, 1129.
[10] Zimmermann, M., Carrard, M., Gremaud, M., and Kurz, W., 1991, *Mater. Sci. Eng. A*, **134**, 1278.
[11] Williams, D.B., and Edington, J.W., 1976, *Philos. Mag. A J. Theor. Exp. Appl. Phys.*, **34**, (2), 235.
[12] Scott, M.G., and Leake,J.A., 1975, *Acta Metall.*, **23**, (4), 503.
[13] Gremaud,M., Allen, D.R., Rappaz,M., and Perepezko, J.H., 1996, *Acta Mater.*,**44**, (7), 2669.
[14] Bogno,A.-A., Khatibi,P.D., Henein,H., and Gandin, Ch-A., 2016, *Metall. Mater. Trans. A Phys. Metall. Mater. Sci.*, **47A,** (9), 4606
[15] Andersson, J.O., Helander,T., Höglund,L., Shi,P., and Sundman,B., 2002, *Calphad Comput. Coupling Phase Diagrams Thermochem.*, **26**, (2), 273.
[16] Rutter, J.W., 1977, *J. Cryst. Growth*, **42**, 515.
[17] Zimmermann, M., 1990, "Solidification rapide de l'eutectique Al-Al_2Cu par refusion laser," *Institute of Materials, Ecole Polytechnique Fédérale de Lausanne*.
[18] Mason, D.P., and Van Aken, D.C., 1993, *Scr. Metall. Mater.*, **28**, (2), 185.
[19] Karma, A., 1987, *Phys. Rev. Lett.*, **59**, (1), 71.

IOP Conf. Series: Materials Science and Engineering **529** (2019) 012022 doi:10.1088/1757-899X/529/1/012022

Primary Dendrite Trunk Diameter in Al-7wt% Si Alloy Directionally Solidified Aboard the International Space Station

SR Upadhyay[1], SN Tewari[2], M Ghodes[3], RN Grugel[4], DR Poirier[5] and M Lauer[6]

[1] Engineer-Metallurgical, RioTinto Kennecott Copper, 11600 W. 2100 South, Magna (UT) USA.
[2] Professor, Chemical and Biomedical Engineering Department, Cleveland State Uniersity, Cleveland, OH (44115) USA.
[3] Assistant Professor, Middle East Technical University, Northern Cyprus Campus, Güzelyurt, Mersin 10, Turkey
[4] Scientist, NASA-Marshall Space Flight Center, Huntsville, AL (35812) USA
[5] Professor, Department of Materials Science and Engineering, The University of Arizona, Tucson, AZ (85721) USA
[6] Engineer, ME Elecmetal Inc., Duluth, MN (55808) USA

E-mail: S.Tewari@csuohio.edu

Abstract. Under a NASA (National Aeronautics and Space Agency)-ESA (European Space Agency) collaborative research project, MICAST (Microstructure formation in casting of technical alloys under a diffusive and magnetically controlled convection conditions), three Al-7wt% Si samples (MICAST-6, MICAST-7 and MICAST2-12) were directionally solidified at growth speeds varying from 10 to 50 μm s^{-1} aboard the International Space Station to determine the effect of mitigating convection on the primary dendrite array. The observed primary dendrite trunk diameters during steady-state growth of MICAST samples show a good agreement with predictions from a coarsening based model developed by the authors. The trunk diameters in the terrestrial-grown equivalent samples were larger than those predicted from the model. This suggest that thermosolutal convection increases the trunk diameter of primary dendrites, perhaps by increasing their tip radius due to compositional changes.

1. Introduction

During directional solidification (DS) of alloys in a positive thermal gradient (G_L) a mushy-zone consisting of array of primary dendrite arms and interdendritic liquids in between, develops between the melt above and the solid below. Analytical and numerical theories have been developed to describe various aspects of the dendrite network, e.g., cell to dendrite transition, dendrite tip radius and primary dendrite spacing [1-8] under diffusive growth conditions. Given that mechanical properties depend upon the dendrite spacing and distribution, numerous experiments have been carried out to measure primary spacing as a function of solute content (C_o,), growth speed (R), thermal gradient (G_L) and the physical properties of the alloy to compare with model predictions [9-27]. Invariably the discrepancy between the experimental observations and model predictions has been attributed to convection in the melt. Convection itself has been treated by a pseudo-analytical approach of defining an overall mushy-zone Rayleigh Number for the growth conditions [28-40], and correlating, for example, the increasing propensity for convection with observed reduction in mean spacing in terrestrial-grown samples as

IOP Conf. Series: Materials Science and Engineering **529** (2019) 012022 doi:10.1088/1757-899X/529/1/012022

compared with those solidified in microgravity [30-33, 40]. While primary spacing has been extensively evaluated as the morphology parameter for the dendritic arrays, another feature, the primary dendrite trunk-dimeter [41-44] has remained relatively unexplored. The trunk-diameter responds to an abrupt speed change more quickly because it is more reflective of the dendrite tip radius as compared with the primary spacing which achieve the new steady-state value only after further DS for two to three mushy-zone length [45]. Here we present primary dendrite trunk diameter measurements from three Al-7Si alloy samples (MICAST6, MICAST-7 and MICAST2-12), which were directionally solidified on the Space Station at growth speeds varying from 5 to 50 μm s^{-1}. It is seen that the observed primary dendrite trunk-diameters during steady-state growth of MICAST samples are in a good agreement with predictions from a coarsening based model developed by some of the authors [44].

2. Experiments

Cylinders with a diameter of 7.8 mm and a length of ~25-cm were machined from [100] oriented terrestrially grown dendritic Al-7Si samples and inserted into alumina ampoules kept within the Sample Cartridge Assembly (SCA) inserts of the Low Gradient Furnace (LGF) for MICAST-6 and MICAST-7 and in the Specimen Quench Furnace (SQF) for MICAST2-12. The A-7Si cylinders were partially re-melted in space and directionally solidified by withdrawing the furnace with respect to the SCA; their unmelted solid portions at cold end acted as [100] oriented seed. Fig. 1 shows a schematic representation of the solidification processing conditions. The unmelted portion is left of T_E (eutectic isotherm) with the melted and directionally solidified portion on the right. The mushy-zone lengths at the onset of DS are indicated in the figure. For MICAST-6 the withdrawal speed was 5 μm s^1 for 3.75 cm, and then 50 μm s^{-1} for rest of the 11.2 cm long melt. For MICAST-7 it was 20 μm s^{-1} for 8.5 cm and then 10 μm s^{-1} for the rest 10.5 cm. For MICAST2-12 it was 40 μm s^{-1} along its entire 12 cm long melt column.

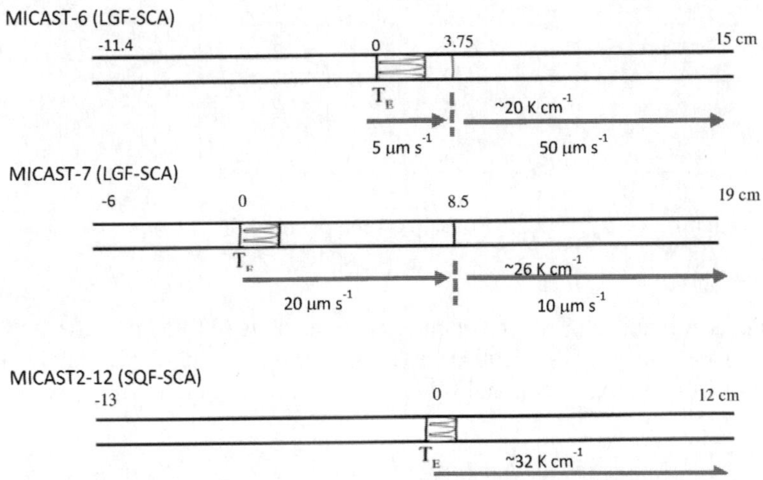

Figure 1. Solidification processing conditions of the MICAST samples. The unmelted portion is left-of eutectic temperature (T_E) with the melted and directionally solidified portion on the right. The mushy-zone lengths at the onset of directional solidification are indicated near T_E.

Temperatures along the sample length during DS were recorded by eleven thermocouples which were attached at regular intervals on the outer surface of the alumina ampoules along their length. Growth speeds and thermal gradients extracted from such thermal profiles are plotted in Fig. 2 for the MICAST-6, MICAST-7 and MICAST2-12 samples. Fig. 2(a), on the left, plots the thermal gradients at the liquidus (G_{TL}) (solid lines) and at the eutectic (G_{TE}) (dashed lines) temperature as a function of distance from the cold end of the samples. Fig. 2(b), on the right, plots the liquidus (V_{TL}) and eutectic (V_{TE}) isotherm velocities along the sample lengths. The thermal gradients were observed to vary during

IOP Conf. Series: Materials Science and Engineering **529** (2019) 012022 doi:10.1088/1757-899X/529/1/012022

directional solidification. For example G_{TL} varied from 22 to 14 K cm^{-1} during growth of MICAST-6, from 26 to 24 K cm^{-1} for MICAST-7, and from 33 to 31 K cm^{-1} for MICAST2-12.

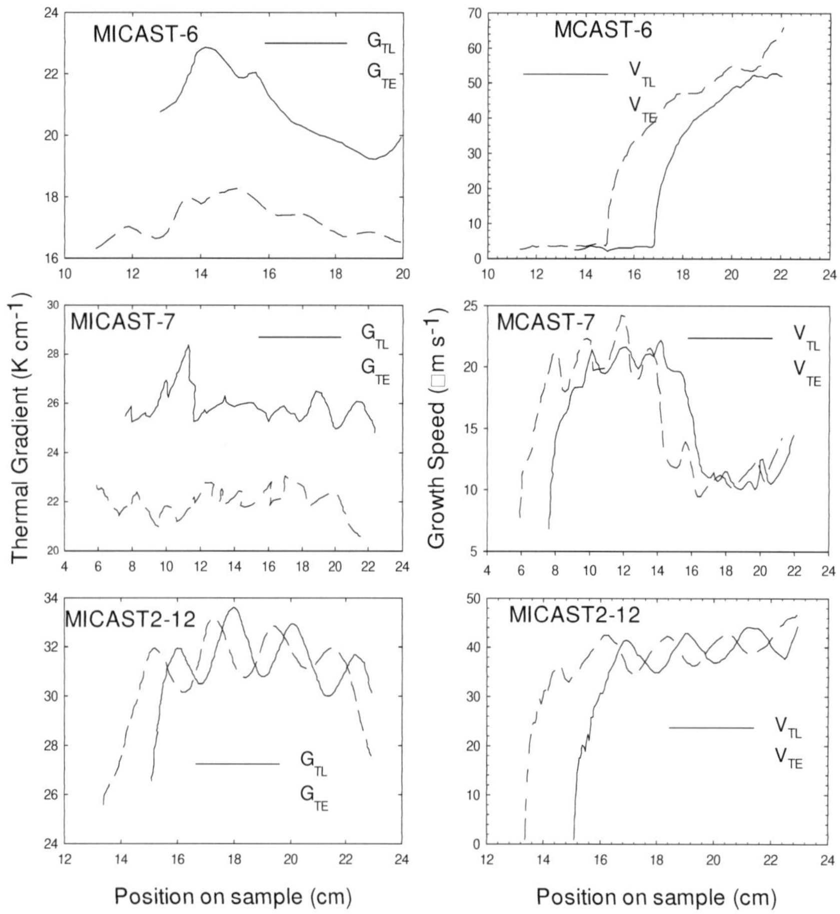

Fig. 2(a) Fig. 2(b)

Figure 2. Thermal gradients at liquidus (G_{TL}) and at eutectic (G_{TE}) (Figs. 2(a) on the left), and velocities of liquidus (V_{TL}) and eutectic (V_{TE}) isotherms (Figs. 2(b) on the right) as a function of distance from the sample bottom during directional solidification of MICAST-6, MICAST-7 and MICAST2-12 samples.

3. Results

Fig.3 shows typical transverse microstructures along the lengths of the MICAST samples. Trunk diameters of primary dendrites on several such cross-sections along the length of the three MICAST samples were measured in the manner describes in Ref. 44. Figure 4 shows the variation in the primary dendrite-trunk diameter along the length of the three samples. Circles are average primary dendrite trunk diameter and error bars are the ± standard deviation; the dotted lines are predictions from a model proposed by the authors [44]. Recall that for the MICAST-6 samples there was a step increase in the growth speed from 5 to 50 μm s^{-1} at the distance of 3.75 cm (Fig. 1(a)) and for the MICAST-7 sample an abrupt decrease from 20 to 10 μm s^{-1} at 8.5 cm (Fig. 1(b)). The entire 12 cm long melted portion of MICAST2-12 sample was grown at 30 μm s^{-1} (Fig. 1(c)). The trunk-diameters respond to the speed changes almost immediately. Also, the predictions from the simple coarsening-based trunk-diameter model [44] are in a good agreement with the experimental observations during the transient and the steady-state DS.

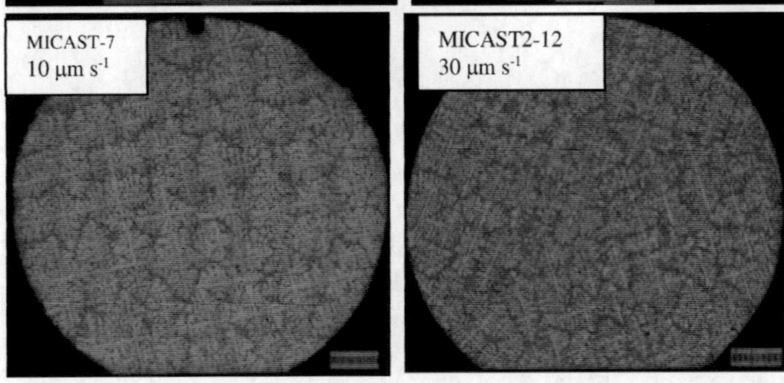

Figure 3. Primary dendrites on transverse sections of directionally solidified MICAST-6, MICAST-7 and MICAST2-12 samples. Markers correspond to 1-mm.

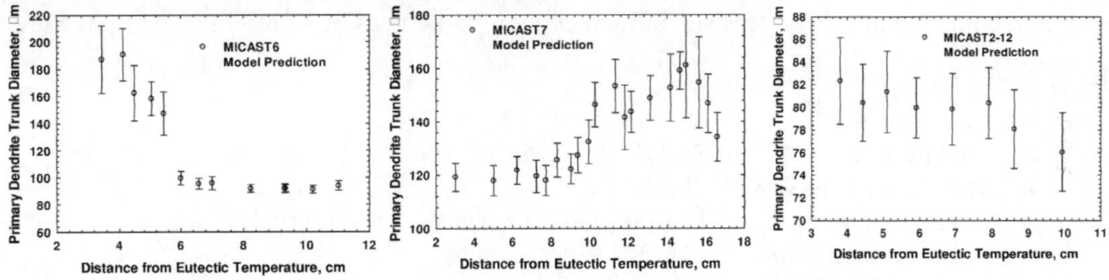

Figure 4. Primary dendrite trunk diameters along the DS length of MICAST-6, MICAST-7 and MICAST2-12 samples. Symbols are average values and error bars are ± one standard deviation; the dotted lines are predictions from a model proposed by the authors [44].

If we include data only from the steady-stare DS portion of the MICAST samples, the agreement appears to be even better (Fig. 5). For each growth speed the trunk-diameter measurements from only those transverse sections were pooled together which met the following conditions, (i) at least, 2-mushy zone lengths of directional solidification has occurred at one growth speed, (ii) at least, 2-mushy zone lengths of solid exists below T_E, and, also at least 2-mushy zone lengths of liquid exists above T_E. Fig.5 shows that predictions from the model (the dotted line) are in a very good agreement with the observed trunk diameters (symbols and ± standard deviation).

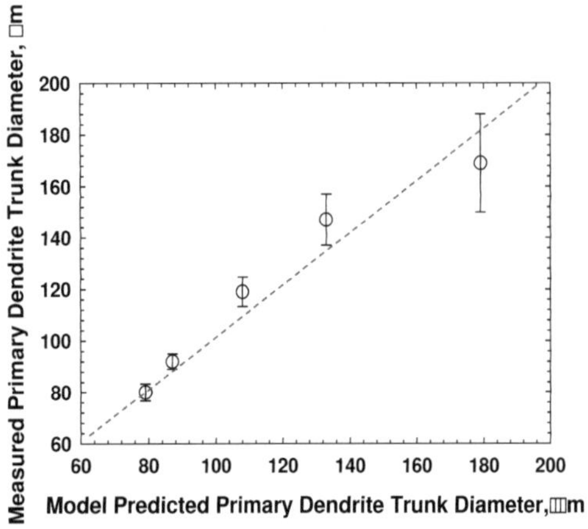

Figure 5. Comparison of the experimentally observed primary trunk diameters during steady-state directional solidification conditions in the MICAST-6, MOCAST-7, and MICAST2-12 samples with those predicted from a model proposed by the authors [44].

4. Discussion and Conclusion

Following additional observations were made during this study, details of which are not included in this paper because of the lack of space. The MICAST samples grown at 10, 20, 30 and 50 μm s^{-1} did not show the radial macrosegregation which invariably occurs during terrestrial DS of Al-7Si alloy [46] and is attributed to the "steepling" type thermosolutal convection [28-32, 46]. This strongly suggests that the three MICAST samples were indeed directionally solidified under convection-free diffusive transport conditions. We have also seen that primary dendrite trunk diameters in the Al-7Si samples, which were directionally solidified terrestrially under approximately similar growth conditions as the low-g processed MICAST samples, are larger than those predicted from the model. These observations strongly suggest that the thermosolutal convection during directional solidification not only reduces the spacing between primary dendrites [30-33] it also increases their trunk diameter as compared with growth under diffusive transport conditions, perhaps by increasing their tip radius.

5. References

[1] Burden MH and Hunt JD 1974 *J. Crystal Growth* **22** 99-108.
[2] Kurz W and Fisher DJ 1979 *Int. Metall. Rev.* **5(6)** 177-204.
[3] Trivedi R 1980 *J. Cryst.Growth* **49** 219-232.
[4] Hunt JD 1979 *Solidification and Casting of Metals*, The Metals Society, London, 3.
[5] Trivedi R 1984 *Metall. Trans. A* **15A** 977-984.
[6] Lu SZ and Hunt JD 1992 *J. Cryst. Growth* **123** 17-34.
[7] Hunt JD and Lu SZ 1996 *Metall. Mater. Trans.* **27A** 611-23.
[8] Takaki T, Sakane S, Ohno M, Shibuta Y, Shimokawabe T, Aoki T 2016 *Acta Materi.* **118** 230-243.
[9] Klaren CM, Verhoeven JD and Trivedi R 1980 *Metall. Trans.* **11A** 1853-1861.
[10] McCartney DG and Hunt JD 1981 *Acta Met.* **29** 1851-1863.
[11] Mason JT, Verhoeven JD and Trivedi R 1982 *J. Cryst. Growth* **59** 516-524.
[12] Somboonsuk K, Mason JT and Trivedi R 1984. *Metall. Trans.* **15A** 967-975.
[13] Miyata Y, Suzuki T and Uno J 1985 *Metall. Trans.* **16A** 1779-1814.
[14] Chopra MA and Tewari SN 1991 *Metall. Trans.* **22A** 2467-2474.
[15] Billia B, Jamgotchian H, Thi HN 1991 *Metall. Trans.* **22A** 3041-3050.
[16] Noel N, Jamgotchian H, and Billia B 1997 *J. Crystal Growth.* **181** 117-132.
[17] Odell SP, Ding GL and Tewari SN 1999 *Metall. Mater. Trans.* **30A** 2159-2165.
[18] Hui J, Chen R, Tiwari R, Wu X, Tewari SN and Trivedi R 2002 *Metall. Mater. Trans.* **33A** 3499-3510.
[19] Trivedi R, Mazumder P and Tewari SN 2002 *Metall. Mater. Trans.* **33A** 3763-3775.

[20] Tschopp MA, Miller JD, Oppedal AL and Solanki KN 2014 *Metall. Mater.Trans.* **45A** 426-437.
[21] Chopra MA and Tewari SN 1991 *Metall. Trans.* **22A** 2467-2474.
[22] Han SH and Trivedi R 1994 *Acta Metall. Mater.* **42** 25-41.
[23] Tewari SN, Weng YH, Ding GL and Trivedi R 2002 *Metall. Mater. Trans.* **33A** 1229-1243.
[24] Gunduz M and Cardili E 2002 *Mater. Sci. Eng. A* **3127** 167-185.
[25] Tewari SN and Laxmanan V 1987 *Acta Met.* **35** 175-183.
[26] Trivedi R and Somboonsuk K 1984 *Matls. Sci. Engg.* **65** 65-74.
[27] Esaka H and Kurz W 1984 *J. Cryst. Growth* **69** 362-366.
[28] Dupouy MD, Camel D and Favier JJ 1989 *Acta. Metall. Mater.* **37[4]** 1143-1157.
[29] Dupouy MD, Camel D and Favier JJ 1992 *Acta Metall Mater.* **40[7]** 1791-1801.
[30] Dupov MD and Camel D 1998 *J. Crystal Growth.* **183** 469-489.
[31] Thi HN, Dabo Y, Drevet B, Dupouy MD, Camel D, Billia B, Hunt JD and Chilton A 2005 *J. Cryst. Growth* **281** 654-668.
[32] Drevet B, Thi HN, Camel D, Billia B and Dupouy MD 2000 *J. Cryst. Growth.* **218** 419-433.
[33] Tewari SN and Shah R 1996 *Metall.l Trans.* **27A** 1353-1361.
[34] Beckermann C, Gu JP, and Boettinger WJ 2000 *Metall. Mater. Trans.* **31A** 2545-52.
[35] Tewari SN, Tiwari R and Magadi G 2004 *Metall. Materi. Trans.* **35A** 2927-2934.
[36] Tewari SN and Tiwari R 2003 *Metall. Mater. Trans.* **34A** 2365-2376.
[37] Ojha SN, Ding G, Lu Y, Rye J and Tewari SN 1999 *Metall. Mater. Trans.* **30A** 2167-2171.
[38] TewariSN, Shah R and Chopra MA 1993 *Metall. Mater. Trans.* **24A** 1661-1669.
[39] Ojha SN and Tewari SN 2000 *Int. J. Cast Metals Res.* **13** 207-213.
[40] Zimmerman G and Weiss A 2005 *Microg. Sci. Techn.* **XVI** 143-147.
[41] Grugel RN 1992 *Materials Characterization* **28** 213-219.
[42] Grugel RN 1995 *Metall. Mater. Trans.* **26A** 496-499.
[43] Pratt RA and Grugel RN 1993 *Materials Characterization* **30** 225-231.
[44] Tewari SN, Grugel RN and Poirier DR 2014 *Metall. Materials. Trans.* **45A** 4758-4761.
[45] Hui J, Chen YS, Wu X and Tewari SN 2003 *J. Cryst. Growth,* **253** 413-423.
[46] Ghods M, Johnson L, Lauer M, Tewari SN, Grugel RN and Poirier DR 2016 *J. Cryst. Growth,* **441** 107-116.

Acknowledgements
This research was supported by NASA-Grants (NNX08AN49G and NNX14AJ73G). Appreciation is expressed to all the scientists and engineers from NASA and ESA whose combined effort made this research possible, and to Dr Men G. Chu (ALCOA) for providing the Al-7Si alloy feed-stock.

Characterization of dendritic growth in Fe–C system using time-resolved X-ray tomography and physics-based filtering

H Yasuda, T Kawarasaki, Y Tomiyori, Y Kato and K Morishita*

Department of Materials Science and Engineering, Kyoto University, Sakyo, Kyoto, 606-8501, Japan
* Present address: Department of Materials Science and Engineering, Kyushu University, Nishi-ku, Fukuoka 819-0395, Japan

E-mail: yasuda.hideyuki.6s@kyoto-u.ac.jp

Abstract. Time-resolved *in situ* tomography of dendritic growth in Fe–0.45 mass% C carbon steel was performed using synchrotron radiation X-rays at SPring-8 synchrotron radiation facility (Japan) with improvement of the image quality using a physics-based filter. The voxel size of the reconstructed image was approximately 6.5 μm × 6.5 μm × 6.5 μm, and the time resolution (duration of 360° rotation) was 4 s (0.25 rps). Three-dimensional images of the dendrites were reconstructed even without image processing; however, the low contrast resolution in Fe–C alloys led to poor image quality. Consequently, it was impossible to precisely track the solid/liquid interface or evaluate the average curvature. To improve the image quality, a physics-based filter (a PF filter) was developed using a phase-field model. In the PF filter, images were retrieved in terms of interface curvature. The PF filter significantly improved the computed tomography image quality. As a result, dendritic growth was clearly observed even in Fe–C alloys. Moreover, the average curvature of the solid/liquid interface was evaluated as a function of solidification time (solid fraction). The ability to systematically characterize growing dendrites will be beneficial for modeling and simulation of solidification phenomena.

1. Introduction

Time-resolved *in situ* observations using synchrotron radiation X-rays with high brilliance and high coherency have allowed us to observe solidification and related phenomena such as the solidification of low-melting-temperature alloys (Sn, Zn, and Al alloys) [1-5] and high-melting-temperature alloys (cast iron, carbon steel, and stainless steel) [6-11]. These direct observations are beneficial for understanding microstructural evolution and casting defect formation as well as for constructing physical models, enabling, for example, the direct observation of the massive transformation in Fe–C alloys [9]. The γ phase (austenite, face-centered cubic) has long been considered to be produced via a peritectic reaction between the δ phase (ferrite, body-centered cubic) and liquid phase during solidification of carbon steel with a carbon content of less than 0.5 mass%. However, time-resolved *in situ* observations of Fe–C alloys [9, 10, 12] revealed that a massive transformation, in which the δ phase transforms into the γ phase in the solid state, is dominantly selected. In addition to radiography, laser-scanning confocal microscopy has also revealed this massive transformation of the δ phase into the γ phase [13]. This massive transformation was also observed in Fe–Cr–Ni alloys [14].

All these observations of solidification and massive transformation in Fe-based alloys were performed using X-ray transmission imaging of a thin specimen (typically 100 μm), which has some inherent limitations. Solute diffusion and the development of dendrite arms are constrained in thin specimens. The limited space also affects the curvature of the solid/liquid interface and dendrite arm spacings. Time-resolved computed tomography (referred to as 4D-CT), which allows *in situ* observation of the microstructural evolution, can be applied to overcome these issues for characterization of the 3D solidification structure.

A critical issue for the 3D observation of dendrites during solidification is the time resolution. Ultrafast 4D-CT was performed using synchrotron radiation X-rays at the European Synchrotron Radiation Facility (ESRF) to observe the microstructural evolution in Al–4 mass% Cu alloys (1 mm in diameter, cooling rate of 0.1 K/s) [15]. A pink beam (peak energy of 40 keV) was used to increase the transmission X-ray intensity. For the reconstruction, 500 projections over 180° rotation (10 s/rotation) were used. The 4D-CT observations revealed the formation of solid grains isolated by a liquid phase, a change in the solid fraction during cooling, shrinkage due to solidification, and the solid/liquid interfacial area (f_s>0.7). In addition, 4D-CT was used to observe the melting and solidification of a metal-matrix nanocomposite in Al–Mg–Si alloys (Al-6082) with the addition of 0.54 vol.% Y_2O_3 particles (~500 nm) [16]. In total, 1440 projections over 180° (8.6 s/rotation) were obtained using a pink beam (peak energy of 17.6 keV). Coarsening of the dendrite arms in Al–15 mass% Cu alloys was also observed using 4D-CT (monochromatized X-ray: 30 keV, 721 projections over 180°, voxel edge length: 1.4 μm), and the results were compared with phase-field simulations [17]. More recently, the dendrite morphology during the early stage of solidification in Al–24 mass% Cu alloys (1 mm in diameter) was also observed using 4D-CT by applying interlaced view sampling (interval between reconstruction: 1.8 s, voxel edge length: 0.65 μm) [18]. The morphologies of the primary and secondary dendrite arms were well reconstructed using the time-interlaced model-based iterative reconstruction (TIMBIR) method [18]. Semi-solid deformation in Al–15 mass% Cu alloys was also observed using 4D-CT (3 mm in diameter, 720 projections over 180°, interval between reconstruction: 4 s) [19]. The study revealed that transgranular liquation cracking of grains isolated by the remaining liquid phase could occur even at low stress (1–40 MPa) [19].

In addition, 4D-CT has been applied to observe the solidification in Ni–14 mass% Hf, Fe–11 mass% Hf, and Co–18 mass% Hf alloys (1.8 mm in diameter) at high temperatures [20]. Higher contrast resolution between the solid and liquid phases in the alloys containing Hf (an absorption edge of 65.351 keV) was achieved using high-energy X-rays (65–80 keV). The 4D-CT technique clearly revealed the solid/liquid interface even in Ni, Fe, and Co alloys.

To the best of our knowledge, however, observation of the evolution of the 3D structure in alloys with low contrast resolution such as Fe–C steels (Fe–C–Mn–Si alloys) remains challenging. For carbon steels (Fe–C–Mn–Si alloys), X-rays with relatively high energy are required because of the large X-ray absorption coefficient of iron; however, X-rays with lower energy are preferable to increase the contrast resolution. Our previous study [21] showed that 3D images of dendrites, even in Fe–C alloys, could be obtained using 4D-CT. This study demonstrates the use of 4D-CT combined with image processing using a phase-field model, with the curvature of the solid/liquid interface evaluated from the processed images.

2. Experiments

Time-resolved tomography experiments were performed at beamline BL20XU of SPring-8 (synchrotron radiation facility, Hyogo, Japan) [21]. The X-rays were monochromatized using a Si (111) double-crystal monochromator. An X-ray energy of 37.7 keV was selected. Figure 1 shows the 4D-CT apparatus and configuration of the rod specimen (0.8 mm in diameter), X-ray beam (approximately 3 mm in width and 3 mm in height), and beam monitor (pixel size 6.5 μm × 6.5 μm). The specimen was rotated at 0.25 rps (4 s/rotation), and transmission images were obtained at 100 fps. Namely, convolution back-projection (CBP) image reconstruction was performed using 200 transmission images (approximately 200 × 200 pixels) over 180° rotation. It should be noted that 200 projections were not sufficient to reconstruct a cross-section (specimen area of 200 × 200 pixels) using normal procedures.

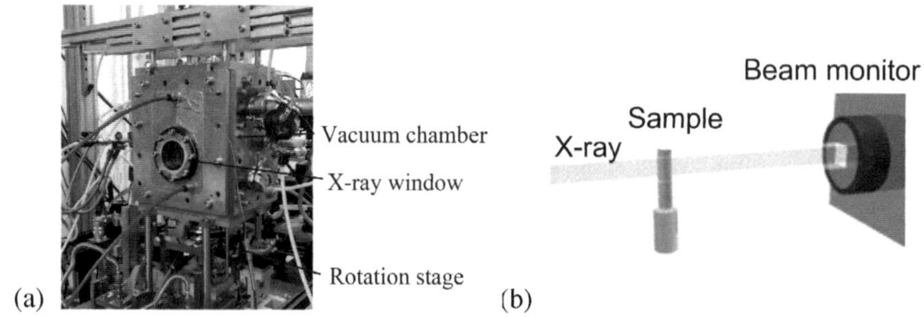

Figure 1. (a) Time-resolved tomography apparatus and (b) configuration of X-ray beam, specimen, and beam monitor.

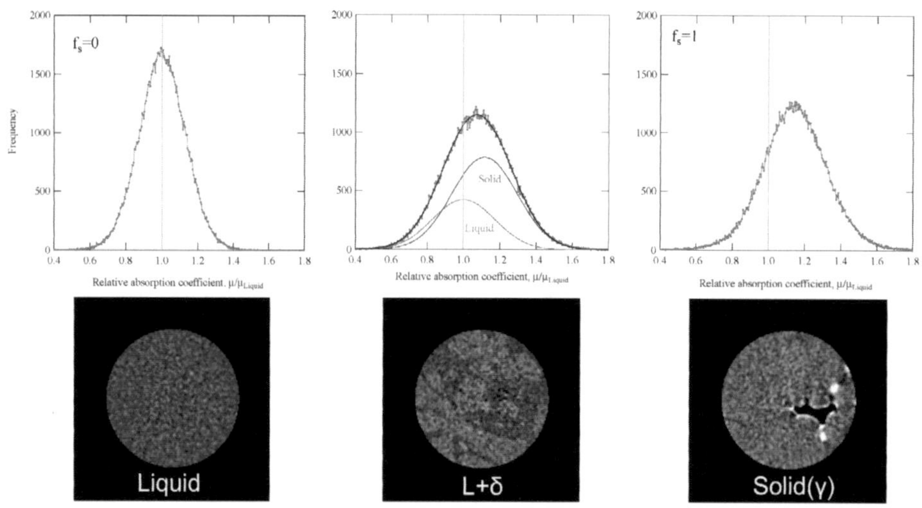

(a) (b) (c)

Figure 2. Reconstructed images (bottom) and distribution of the relative absorption coefficient (top). The relative absorption coefficient was normalized by that of the liquid phase before solidification.

Conventional carbon steel, Fe–0.45C–0.6Mn–2Si (mass%), was used as the specimen. The specimen rod was inserted into a sintered Al_2O_3 pipe (inner and outer diameters of 0.8 and 2 mm, respectively). The specimen was melted and then cooled at a cooling rate of 0.17 K/s in a vacuum atmosphere (approximately 1 Pa). The temperature gradient at the specimen was less than 1 K/mm. Projected images were continuously recorded during the cooling procedure.

3. Results and Discussion

3.1. Solid fraction

Figure 2 presents the reconstructed images without image processing and plots of the distribution of the relative absorption coefficient. The relative absorption coefficient was obtained by normalizing the absorption coefficient to that of the liquid phase. Thus, the average value in the distribution at $f_s = 0$ (liquid phase) was 0, as shown in Fig. 2(a). The full width at half maximum was as large as 40% because of the low signal-to-noise (S/N) ratio, which arose from the low contrast resolution and the small number of projected images (200 projection images). Although dendrite arms were visible in the reconstructed images, as shown in Fig. 2(b), it was impossible to smoothly track the solid/liquid interface.

As observed in Fig. 2(c), the average value of the absorption coefficient was 1.08 at $f_s = 1$ (austenite phase). Thus, the change in the absorption coefficient during solidification was measured even though the S/N ratio was low. The distribution of the absorption coefficient was well fitted by two Gaussian curves for the liquid and solid phases, as shown in Fig. 2(b). Figure 3 shows the solid fraction, f_s, obtained by the fitting. The solid fraction roughly followed a square-root law and rapidly increased at 740 s owing to the transformation from the δ to γ phase. Because the transformation occurred below the peritectic temperature, solidification was immediately completed after the transformation.

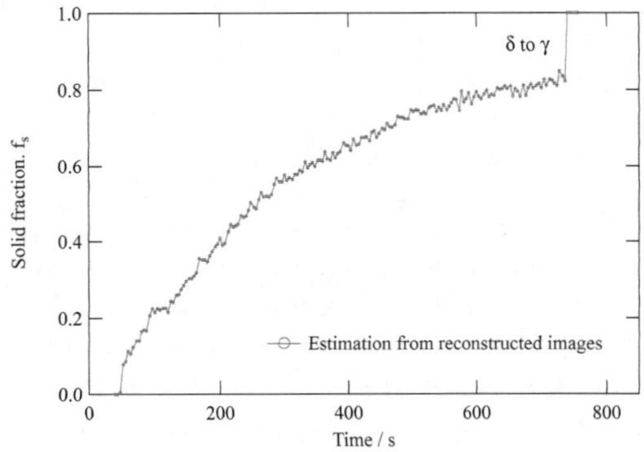

Figure 3. Solid fraction estimated from the distribution of the relative absorption coefficient.

The present results indicate that a 3D solidification structure can be obtained using 4D-CT. However, considerable noise was introduced because of the low contrast resolution of Fe–C alloys in the reconstructed images. The change in the solid fraction during solidification was evaluated despite the considerable noise.

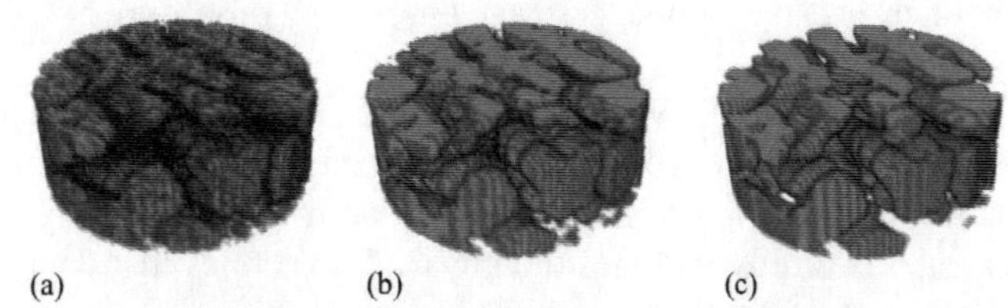

Figure 4. Reconstructed images: (a) raw data and date after (b) Gaussian filtering and (c) PF filtering.

3.2. Filtering using a phase-field model

The considerable noise introduced during the reconstruction process by the low contrast resolution needed to be removed to track the solid/liquid interface and evaluate the interfacial area and curvature. In conventional image processing, the reconstructed images are improved by blurring/smoothing, which reduces detail, and unsharp masking, which increases acutance. These image procedures are widely used for various applications. As shown in Fig. 4, the smoothing (Gaussian filter) improved the reconstructed image, and consequently, dendritic arms were clearly observed in the 3D reconstructed image. Thus, conventional image processing was sufficient to observe dendrites. Although the surface became smooth,

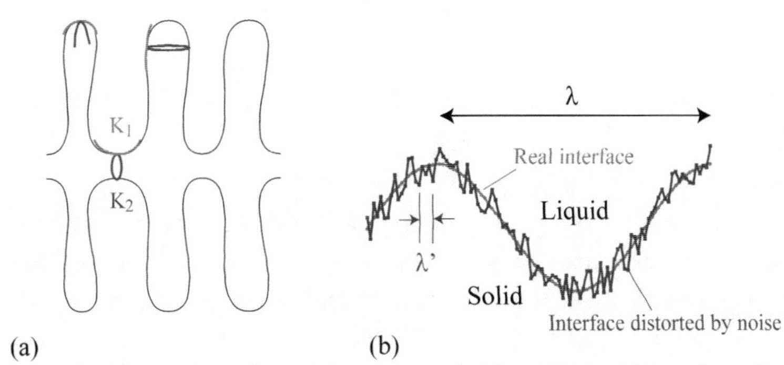

Figure 5. (a) Schematic illustration of dendrite arms and (b) solid/liquid interface distorted by noise.

the noise (on the order of the voxel size) still remained and degraded the evaluation of the curvature. Furthermore, the filtering conditions are often determined by sensory experience.

We attempted to develop an image processing procedure that overcomes these disadvantages. Figure 5(a) presents a schematic illustration of the dendrite shape. The average curvature, K, is defined by

$$\frac{1}{K} = \frac{1}{K_1} + \frac{1}{K_2}. \quad (1)$$

Here, K_1 and K_2 are defined as shown in Fig. 5(a). If a pure substance is placed under isothermal conditions, the average curvature should be equal at any solid/liquid interface point in equilibrium. For alloy dendrites under isothermal conditions, the presence of a non-uniform distribution of solutes modifies the average curvature. However, the distributions of solute concentrations will be rather small between the dendrite arms (the typical scale ranges from 10 to 100 μm).

Figure 5(b) shows the solid/liquid interface distorted by noise induced during the reconstruction process. The typical period of the noise induced in the reconstruction procedure is on the order of the voxel size (λ'). When the typical period (λ) of the dendrite arms is larger than that of the noise, as shown in Fig. 5(b), the real interface can be tracked by removing the short-wavelength component in the reconstructed 3D images. As the average curvature is significantly distorted by the short-wavelength component, we developed an image processing procedure based on the average curvature as one possible approach.

Phase-field models were developed to simulate microstructural evolution such as dendritic growth [22-25]. These phase-field models have certain advantages. For instance, the model does not explicitly consider the solid/liquid interface, and 3D images are easily converted into a phase field. In addition, the model naturally includes the physical principles that control the microstructural evolution, such as the curvature effect. In this study, a phase-field model was used for filtering the reconstructed images (referred to as PF filtering). In the first step, 3D images, such as those in Fig. 2, were converted into a phase field using eq. (1). In the present case, each CT voxel was divided into eight phase-field voxels (2 × 2 × 2 voxels).

$$\phi = \begin{cases} 0 & (\bar{\mu} < \mu_L) \\ \dfrac{\bar{\mu} - \mu_L}{\mu_S - \mu_L} & (\mu_L \le \bar{\mu} \le \mu_S) \\ 1 & (\bar{\mu} > \mu_S) \end{cases} \quad (2)$$

In the second step, the phase field was sequentially calculated using eqs. (2)–(4).

$$\Delta\phi = -(M\Delta t)\left[6\phi(1-\phi)\Delta f + 2\phi(1-\phi)(1-2\phi)W - \varepsilon^2\nabla^2\phi\right] \quad (3)$$

$$\Delta f = -\sigma K = -\sigma\frac{2}{r} \quad (4)$$

$$W = \frac{3\sqrt{2}\sigma}{\delta} \quad (5)$$

$$\varepsilon^2 = 3\sqrt{2}\sigma\delta \quad (6)$$

Here, ϕ is the phase field, t is the time, M is the interface mobility, Δf is the driving force of solidification expressed by eq. (3), σ is the solid/liquid interfacial energy (0.2 J/m^2), and δ is the thickness of the solid/liquid interface. The thickness δ used in this study was 2–4 times larger than the edge length of the phase-field voxel. W and ε^2 were determined from the solid/liquid interfacial energy and thickness of the solid/liquid interface. In the PF filter, the value of $M\Delta t$ was selected such that the change in ϕ was less than 0.01 at each step.

Equation (4) expresses the driving force for the interface motion. In the third step, an appropriate value for r in eq. (4) is determined. Once the parameters are given, the PF filtering evolves the solid/liquid interface and the solid fraction converges to a certain value. In this study, the parameter *r* was selected such that the solid fraction estimated from the phase field was equal to that estimated from the distribution of the relative absorption coefficient, as shown in Fig. 2(b). In other words, the solid fraction estimated from 4D-CT was used as a restraint condition in the PF filtering. Figure 2(c) shows an example of the reconstructed dendrites after the PF filtering. Although there is no significant difference between Figs. 4(b) and (c), the solid/liquid interface was modified by the PF filtering. The details are provided in the next section.

3.3. Characterization of dendrite shape

Figure 6 presents the cross-sectional images and the distributions of the average curvature. The numbers in the figure represent the number of iterations of the PF filtering. As the number of iterations increased, the noise was remarkably reduced; the shapes of the dendrite arms were clearly identifiable in the cross-sectional images after 1000 and 1500 steps. It was thus possible to track the solid/liquid interface in the cross-sectional images after 1000 and 1500 steps.

As shown in the histogram of the average curvature in Fig. 6, the distribution of the average curvature was remarkably modified by the PF filter. In the reconstructed images without the PF filtering (0 steps), an intense artifact induced by noise in the rectangular coordinate system produced 5–6 sharp peaks. The average curvature of 10^{6} m (curvature radius: 2 μm) is much larger than the real curvature, indicating that the noise contributed to the large curvatures. After the PF filter was applied, the distribution became Gaussian-like and the full width at half maximum gradually decreased with increasing number of iterations. In principle, as expressed by eq. (2), the PF filtering eliminated distortion with short wavelength. Thus, the PF filter was effective in reducing noise induced during the reconstruction process.

Figure 7 presents 3D images of the growing dendrites and a plot of the average curvature as a function of solidification time (the solid fraction was also determined from Fig. 2). Even the secondary dendrite

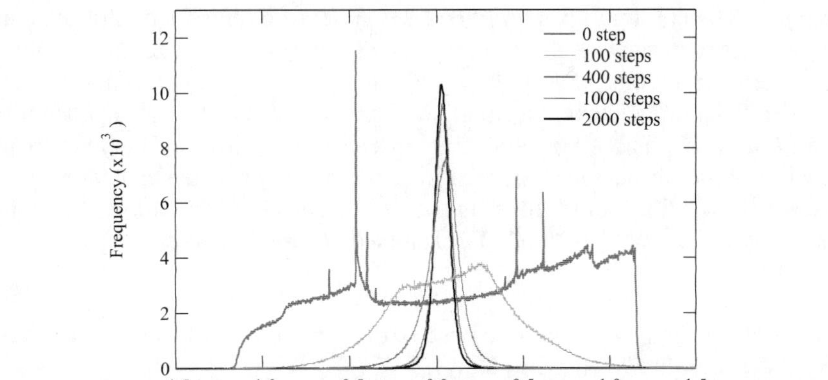

Figure 6. Filtered images (bottom) and distribution of solid/liquid interface curvature (top) as a function of number of iterations.

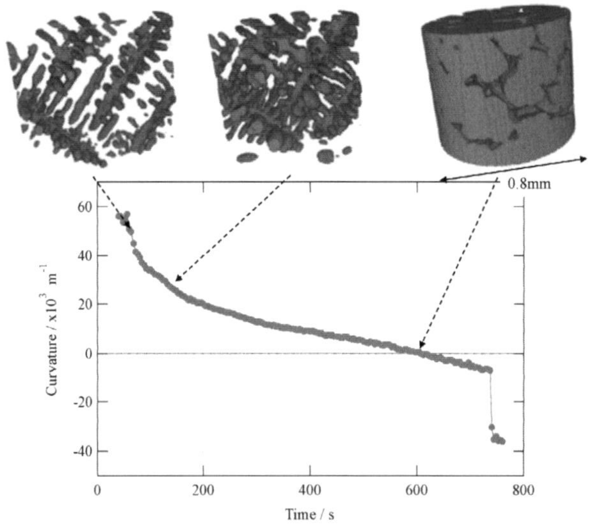

Figure 7. 3D images of dendrites after filtering (top) and plot of average curvature as a function of solidification time (bottom).

arms were clearly identified in each time step because of the elimination of noise. In addition, the average curvature of the solid/liquid interface as a function of solidification time (solid fraction) was successfully evaluated, as shown in the top images in Fig. 7. The average curvature rapidly decreased at the beginning of solidification, and then, the decreasing rate was gradually reduced. The average curvature became negative at a solid fraction of 0.75. The rapid decrease at 750 s was caused by the massive transformation from the δ to γ phase [9]. It should be noted that the filtered 3D images were uniquely obtained by the PF filtering in which the solid fraction determined by 4D-CT was used as a constraint parameter. The constraint condition enabled systematic reproduction of the dendrite shape. Further studies are needed to evaluate the accuracy of this process and to discuss its limitations.

These results demonstrate that 4D-CT with a PF filter can be used for time-resolved *in situ* 3D observation of dendritic growth and quantitative characterization of growing dendrites (i.e., the curvature, interfacial area, and secondary arm spacing) even in Fe–C alloys. This improved understanding of dendrite shapes will be beneficial for modeling and simulation of solidification phenomena. Moreover, 4D-CT with PF filtering enables unique characterization of growing dendrites. The validity of the use of the PF filter should be examined in a future study.

4. Summary

Time-resolved *in situ* tomography was used to observe the dendritic growth and to characterize growing dendrites in Fe–0.45C–0.6Mn–2Si (mass%) alloys.

- In this work, 4D-CT was performed using monochromatized X-rays with an energy of 37.7 keV. The specimen was rotated at 0.25 rps, and 200 projections over 180° rotation were used for the reconstruction.
- Dendrite arms were recognized in the reconstructed images without the filter (pixel edge length: 6.5 μm). However, it was impossible to track the solid/liquid interface or evaluate the average curvature of the solid/liquid interface because of the poor image quality.
- A PF filtering procedure was developed to improve the quality of the reconstructed images. For the PF filtering, the solid fraction determined from the 4D-CT was used as a constraint condition. The driving force of solidification in the PF filtering was a key parameter for the filtering. Use

of this PF filter with an appropriate value of this driving force significantly improved the image quality.

- The primary and secondary dendrite arms were clearly identified at each time step. Time evolution of three-dimensional dendrites was well reproduced from the filtered images.
- The distribution of the average curvature was Gaussian-like with a relatively low full width at half maximum. In addition, the average curvature was determined as a function of solidification time (solid fraction).
- 4D-CT with the PF filtering is expected to improve the characterization of growing dendrites even in alloys with relatively low contrast resolution. The resulting quantitative data can be used for modeling and simulation of solidification phenomena.

Acknowledgments
The synchrotron radiation experiments were performed as general projects at beamlines BL20B2 and BL20XU at SPring-8 (JASRI), Japan. The study was developed by "Heterogeneous Structure Control: Towards Innovative Development of Metallic Structural Materials" in Industry–Academia Collaborative R & D Program (JST). An observation technique including the PF filter was developed by a Grant-in-Aid for Scientific Research (S) (No. 17H06155). We thank Tiffany Jain, M.S., from Edanz Group (www.edanzediting.com/ac) for editing a draft of this manuscript.

References
[1] Mathiesen R H, Arnberg K, Mo F, Weitkamp T, Snigirev A 1999 *Phys. Rev. Lett.* **83** 5062
[2] Yasuda H, Ohnaka I, Kawasaki K, Sugiyama A, Ohmichi T, Iwane J, Umetani K 2004 *J. Cryst. Growth* **262** 645
[3] Mangelinck-Noel N, Nguyen-Thi H, Reinhart G, Schenk T, Cristiglio V, Dupouy M D, Gastaldi J, Billia B, Hartwig J 2005 *J. Phys. D,* **38** A28
[4] Li B, Brody H D, Black D R, Burdette H E, Rau C 2006 *J. Phys. D* **39** 4450
[5] Wang T M, Xu J J, Xiao T Q, Xie H L , Li J, Li T J, Cao Z Q 2010 *Phys. Rev. E* **81** 042601
[6] Yasuda H, Yamamoto Y, Nakatsuka N, Nagira T, Yoshiya M, Sugiyama A, Ohnaka I, Umetani K, Uesugi K 2008 *Int. J. Cast Met. Res.* **22** 15
[7] Yasuda H, Nagira T, Yoshiya M,, N. Nakatsuka, A. Sugiyama, K. Uesugi, K. Umetani, ISIJ Int. 51 (2011) 402.
[8] Nagira T, Gourlay C M, Sugiyama A, Uesugi M, Kanazawa M, Yoshiya M, Uesugi K, Umetani K, Yasuda H 2011 *Scripta Mater.*, **64** 1129
[9] Yasuda H, Nagira T, Yoshiya M, Sugiyama A, Nakatsuka N, Kiire M, Uesugi M, Uesugi K, Umetani K, Kajiwara K 2012 *IOP Conf. Ser.: Mater. Sci. Eng.* **33** 012036
[10] Nishimura T, Morishita K, Nagira T, Yoshiya M, Yasuda H 2015 *IOP Conf. Ser.: Mater. Sci. Eng.* **84** 012062
[11] Yamane K, Yasuda H, Sugiyama A, Nagira T, Yoshiya M, Morishita K, Uesugi K, Takeuchi A, Suzuki Y 2015 *Metall. Mater. Trans, A* **46A** 4937
[12] Nishimura T, Morishita K, Yoshiya M, Nagira T, Yasuda H **2019** *Tetsu-to-Hagané* **105** in press.
[13] Griesser S, Reid M, Bernhard C, Dippenaar R 2014 Acta Mater. **67** 335
[14] Nishimura T, Matsubayashi R, Morishita K, Yoshiya M, Nagira T, Yasuda H 2019 *ISIJ International* **59** in press.
[15] Ludwig O, Dimichiel M, Salvo L, Suery M, Falus P 2005 *Metall Mat Trans A* **36A** 1515
[16] Daudin R, Terzi S, Lhuissier P, Salvo L, Boller E 2015 *Mater. Design* **87** 313
[17] Aagesen L K, Fife J L, Lauridsen E M, Voorhees P W 2011 *Scripta Mater.* **64** 394
[18] Gibbs J W, Mohan K A, Gulsoy E B, Shahani A J, Xiao X, Bouman C A, De Graef M, Voorhees P W 2015 *Scientific Reports* **5** 11824
[19] Cai B, Karagadde S, Yuan L, Marrow T J, Connolley T, Lee P D 2014 *Acta Mater.* **76** 371
[20] Cai B, Wang J, Kao A, Pericleous K, Phillion A B, Atwood R C, Lee P D 2016 *Acta Mater.* **117** 160

[21] Yasuda H, Tomiyori Y, Kawarazaki T, Kato K, Morishita K 2018 *Proc. Solidification and Gravity VII* 75
[22] Kobayashi R 1993 *Physica D* **63** 410
[23] Wheeler A A, Murray B T, Schaefer R 1993 *Physica D* **66** 243
[24] Wang S-L, Sekerka R F. 1996 Phys. Rev. E **53** 3760
[25] Karma A, Rappel W-J 1996 *Phys. Rev. E* **53** R3017

IOP Conf. Series: Materials Science and Engineering **529** (2019) 012024 doi:10.1088/1757-899X/529/1/012024

Microstructure in A356/AA6xxx after bimetallic casting with flux coating

Aina Opsal Bakke[1], Lars Arnberg[1] and Yanjun Li[1]

[1]Department of Materials Science and Engineering, Norwegian University of Science and Technology, Trondheim, Norway

Corresponding author: aina.o.bakke@ntnu.no

Abstract. This work focuses on bimetallic casting between A356 alloy melt and profiles of AA6xxx wrought aluminum alloys through a gravity casting process, with the aim to improve the component's mechanical properties. However, combining two aluminum alloys is difficult due to the stable aluminum oxide present on the surface of the aluminum inserts and the advancing liquid melt front. The oxide layer strongly reduces the wettability between aluminum melt and solid metal. It will also prevent diffusion and formation of a metallurgical bond. In order to obtain sound metallic bonding between the two alloys, different surface treatments, including flux coating and chemical treatments of the profiles have been tested. The influences of preheating temperature and melt flow modes on the quality of the bimetallic casting have been addressed. Based on a detailed microstructure characterization of the bonding layer in the casting, by using Optical Microscopy (OM), Scanning Electron Microscopy (SEM) and Energy Dispersive X-ray Spectroscopy (EDS), the solidification structure development at the interface has been discussed. Results showed that when flux coating was applied, magnesium diffused to the insert surface and prevented formation of a metallurgical bond. Without flux coating, a metallurgical bond was achieved due to slight melting of the insert surface.

1. Introduction

Joining of two aluminum alloys have gained interest, especially in the automotive industry, as aluminum with its high strength-to-weight ratio [1] can replace heavier materials to achieve weight reduction, thus also reduce emission from the vehicles. Such a multi-material is also beneficial, as it is able to meet specific requirements that one material alone often cannot.

Bimetallic casting, also known as compound casting, is a process in which two materials, one solid and one liquid are joined together. The solid material will be placed in the mold prior to casting, where the liquid material is cast around, creating a diffusion zone between the two materials which leads to a metallurgical bond [2]. Although several other processes have been used to join aluminum, such as friction stir welding [3], explosive welding [4], gas metal arc welding [5] and direct chill casting [6], bimetallic casting has in general higher process efficiency and less geometric limitations.

The difficulty in joining aluminum alloys is mainly due to the surface oxide layer. Aluminum oxides form spontaneously and are thermodynamically stable [7], while their melting points are significantly higher than typical process temperatures [8]. Due to this stability, the oxides reduce wettability of the aluminum surface, which hinders formation of a metallurgical bond [9]. To remove surface oxides, aluminum is often pickled or etched and then coated with a material to prevent reoxidation. Use of zinc as coating has showed good potential, as it melts at a low temperature and thus allows for bonding with an aluminum surface free of oxides [9] [10] [11]. Coating with Ni and Cu has also shown formation of a metallurgical bond between a cast and a wrought aluminum alloy [12].

In this research, a wrought and a cast aluminum alloy has been joined through a bimetallic casting process. Various flux coatings have been used to remove oxides from the aluminum surface and thus enhance wettability. The flux coatings are typically used for brazing, and therefore the parameters in the bimetallic casting will resemble those of a brazing process. Microstructure in the interface has been investigated and discussed.

2. Experimental procedure

6060 aluminum bolts with a diameter of 10mm, were cut to lengths of 10cm. Prior to casting, the bolts were ground with 1200grit sandpaper and cleaned with ethanol, before being immersed into a bath of 10% HCl for 30sec followed by further cleaning in distilled water and then ethanol. This was to remove some of the surface oxides. A third of the bolts were then coated with a liquid flux consisting of $K_{1-3}AlF_{4-6}$, $KZnF_3$, Li_3AlF_6 and metallic silicon particles [13]. Another third was coated with a mixture of $K_{1-3}AlF_{3-6}$ and $CsAlF_4$ [14]. The Cs containing flux was mixed with distilled water to a concentration of approximately 51%. The mixture was then painted on the bolts. The remaining third was left without surface treatment.

The treated 6060 aluminum bolt was then inserted into a graphite mold. A sketch of the mold is shown in Figure 1. Both the mold and the insert were heated to 600°C prior to casting. An A356 alloy was melted in an induction furnace to 730°C. Composition of both aluminum alloys are given in Table 1. Casting was performed at a pouring temperature of 720°C. After pouring, the mold remained in the furnace at 600°C for five minutes before being removed and air-cooled.

Table 1 Chemical composition of aluminum alloys 6060 and A356

Alloy	Composition [wt.%]								
	Mg	Si	Fe	Zn	Cu	Mn	Ti	Cr	Others	Al
6060 [13]	0.45-0.6	0.3-0.6	0.1-0.3	0.15	0.1	0.1	0.1	0.05	0.05-0.15	Bal.
	Si	Mg	Ti	Fe	Sr	Ga	Zn	Others	Al	
A356	7.0	0.41	0.11	0.082	0.013	0.0089	0.0042	0.0029	Bal.	

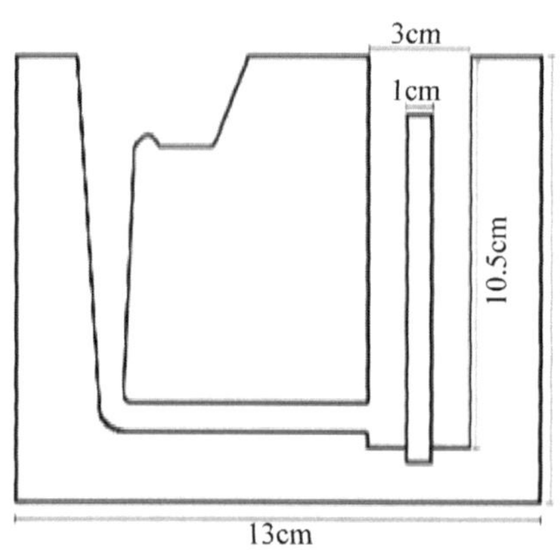

Figure 1 Sketch and dimensions of the graphite mold

2.1. Materials characterization

Samples of 1cm thickness were cut perpendicular to the bolt from each casting. These were ground to 4000 grits using a Struers LaboPol-21, then polished with diamond suspension of 3μm and 1μm on a Struers Tegramin-20. A Leica MEF4M optical microscope was used to examine the interface and the bonding properties between the aluminum alloys. To characterize and detect the elements in the interface, a Zeiss Supra 55VP Field Emission Scanning Electron Microscope (FESEM) with Energy Dispersive X-ray Spectroscopy (EDS) was used. A working distance of 10mm and an acceleration voltage of 15kV were applied in the FESEM observation.

3. Results

3.1. *Effect of flux coating*

The samples from the two different flux coatings will in the results be mentioned as SF and SN, where SF is the sample coated with the liquid flux, while SN is the sample coated with Cs flux. Figure 2 shows optical micrographs of the 6060/A356 interface in the two flux coated samples.

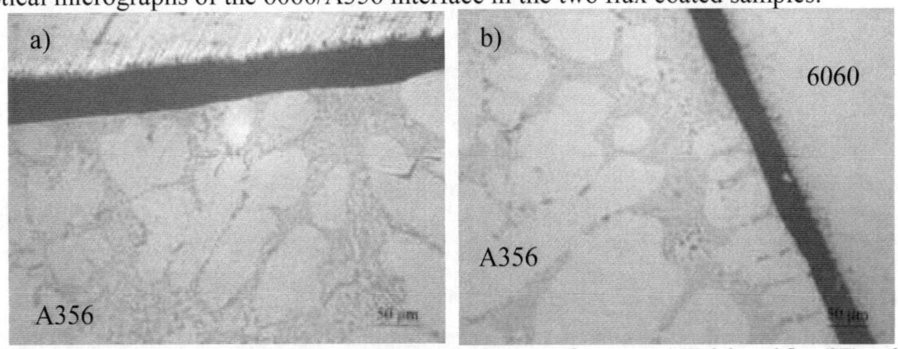

Figure 2 Optical micrographs of the 6060/A356 interface in the flux-coated castings a) With liquid flux (SF) and b) With Cs flux (SN)

As seen in Figure 2, there is a significant gap between the 6060 insert and the cast A356 for both flux-coated castings. This gap was seen throughout the interfaces and shows that no metallic bonding was achieved for either casting. For casting SF the gap width average was measured to approximately 40μm, while it was slightly thinner for SN measuring to approximately 30μm. Also visible in the micrographs, are the relatively large dendrites in the cast alloy. The mold remaining at 600°C for five minutes significantly decreases the cooling rate and allows the dendrites to grow. Average secondary dendrite arm spacing (SDAS) was measured to approximately 65μm for SF and 73μm for SN.

To further examine the interface, an EDS analysis was performed at areas close to the gap on the 6060 insert. The areas analyzed are shown in Figure 3, where backscattered electron (BSE) imaging is used to see the different atom contrasts. By observing the micrographs, there appears to be some flux remaining on the 6060 insert in casting SF (Figure 3a). However, for casting SN (Figure 3b), there is definitely something remaining on the insert, as a porous material can be observed in the interface. This was observed across the whole interface from the edge of the 6060 insert to the cast A356, indicating that indeed something has remained in the interface.

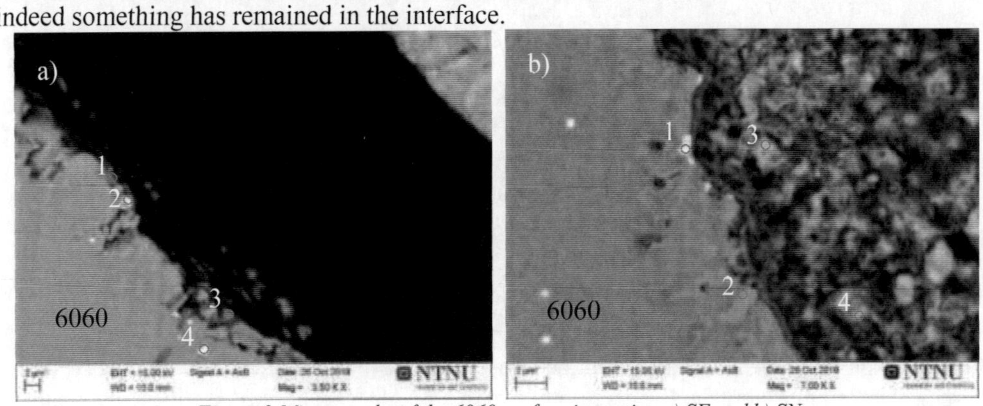

Figure 3 Micrographs of the 6060 surface in casting a) SF and b) SN

Table 2 Compositions detected through EDS analysis at the surface of the 6060 insert in castings SF and SN

Area	Composition [at%]							
	C	O	F	Mg	Al	Si	K	Fe
A-1	41.55	19.70	4.38	1.04	31.31	1.61	0.40	-
A-2	39.33	24.22	-	-	31.85	2.88	-	1.73
A-3	53.73	29.08	4.89	1.26	10.02	0.43	0.60	-
A-4	-	-	28.59	12.00	49.27	-	10.14	-
B-1	-	16.83	17.52	6.47	44.71	3.40	5.19	5.88
B-2	-	5.36	27.38	9.56	49.23	-	8.47	-
B-3	13.97	22.14	33.01	-	13.54	-	17.34	-
B-4	19.92	24.24	28.65	-	13.03	-	14.17	-

The EDS results in Table 2 show that potassium and fluorine are detected in nearly all analyzed areas, which are two of the main elements in the two flux coatings. This implies that the flux coating still remains on the insert after casting. Interestingly, high concentrations of magnesium have also been detected at the insert surface. 6060 has a magnesium content of 0.4-0.6wt%, which is below all the concentrations detected in the EDS analysis. In combination with the potassium and fluorine concentrations found, this could suggest diffusion of magnesium to the surface, and formation of K-F-Mg of F-Mg phases. Especially Area A-4 is believed to be a ternary K-F-Mg phase.

In Area A-2 and B-1, iron is detected. Considering that both these areas are white particles, it is likely that these particles are similar. Although fluorine, magnesium and potassium are also detected in Area B-1, it is suggested that these white particles are Al-Fe-Si ternary particles as they are observed further in the insert as well. The high concentration of oxygen shows that there might be oxides at the surface. Comparing the EDS analysis of the two different flux coatings, it is evident that a higher concentration of flux coating has remained in the interface of casting SN.

3.2. Bimetallic casting without flux coating

When analyzing the sample without flux coating, it was possible to see that the outer part of the insert had melted. This was also observed in the optical microscope, as the insert no longer displayed a circular form. In higher magnification, it was possible to see that due to the melting, a metallurgical bond had formed between the insert and the cast alloy. Figure 4 shows an optical micrograph of the 6060/A356 interface, where the transition from cast to wrought alloy can be distinguished by the characteristic eutectic silicon in A356.

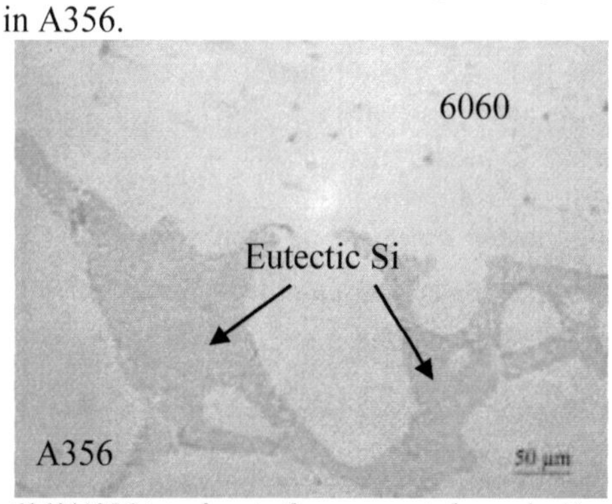

Figure 4 Optical micrograph of the 6060/A356 interface in the non-coated sample. No gap can be observed between the two alloys

Although a metallurgical bond formed in the non-coated casting, cracks were still observed in certain areas. As seen in Figure 5a, the crack propagates through the cast A356, following a path through the interdendritic eutectic silicon. Figure 5b shows that the crack continues through the eutectic silicon at the 6060/A356 interface, as silicon can be observed on both sides. This shows that the crack has formed mainly in the A356 alloy side upon solidification, likely due to built-up stress as aluminum shrinks upon cooling and solidification.

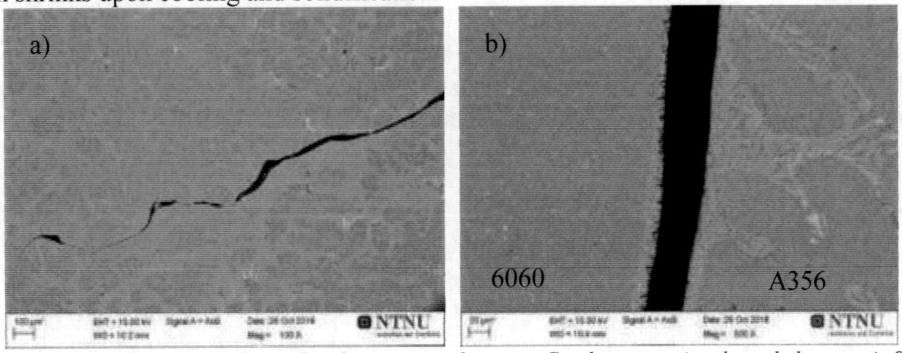

Figure 5 Micrographs of the crack observed in the non-coated casting. Crack propagating through the eutectic Si a) In the cast A356. b) Adjacent to the 6060/A365 interface

4. Discussion

4.1. Formation of a metallurgical bond

According to the manufacturer's specifications, the 6060 aluminum inserts have a melting range of 555-660°C [13]. This means that a preheating temperature of 600°C is well above the solidus temperature of the 6060 alloy. Upon casting, the insert is further heated due to the pouring temperature of 720°C. The outer layer will then remelt, causing a liquid-liquid interface and the formation of a metallurgical bond. It has been suggested that the bond forms due to interdiffusion of the alloying elements, and that through a liquid-liquid reaction no treatment to remove the oxide layers was necessary [6] [14]. The interface between the two alloys can be recognized from the eutectic silicon structure observed in the A356 alloy. In the non-coated casting, the only defect observed was the crack propagating through the eutectic silicon. As aluminum cools and solidifies, it shrinks and thus creates stresses at the interface that in this case has resulted in a crack.

4.2. Flux coating reactions

According to the product description of the fluxes, the liquid flux has a melting range of 564-572°C [15], while it is 558-566°C for the Cs flux [16]. This implies that under the conditions used in the experiment, the flux should be molten and ready to react with the surface oxides. As mentioned for the non-coated casting, the preheating temperature was sufficient to melt the outer layer of the insert. However, for both flux coatings, no change in the insert shape was observed, suggesting that no melting had occurred. It has previously been reported that K_xAlF_y fluxes are highly reactive with magnesium. Therefore, solute magnesium atoms in the 6060 insert will diffuse towards the flux layer and form compounds such as MgF_2 and $KMgF_3$, which due to their high melting points will reduce the flux efficiency [17]. The EDS analysis in Table 2 shows magnesium at the insert surface exceeding that of the 6060 alloy composition provided in Table 1. As the mold and insert are kept in the furnace until the preheating temperature is reached, it allows for solid-state diffusion of magnesium in the insert. Since such magnesium compounds were found, they have probably decreased the flux' efficiency and wettability so that upon casting no bonding will occur.

Comparing the two fluxes, the Cs flux shows in general a slightly thinner gap. However, this could be due to the flux drying upon heating allowing it to shrink slightly, while the liquid flux remains as a paste. Based on the element detection on the insert surface and in the interface, and that the gap had approximately similar width throughout the interface, it is believed that the flux coating remained on the interface after casting for both fluxes. Overall, lower concentrations of magnesium is detected on the insert surface using the liquid flux. This could be due to the lower concentration of K_xAlF_y in the liquid flux compared to the Cs flux.

5. Conclusions

Metallurgical bonding between 6060/A356 aluminum could be achieved without flux coating. The preheating temperature of the insert allowed for a slight melting of the insert surface, promoting formation of a metallic bond.

Using flux coating containing K_xAlF_y prevented a metallurgical bond due to diffusion of magnesium to the insert surface and formation of magnesium compounds with high melting points. It also prevented remelting of the 6060 insert surface.

6. Acknowledgements

The authors would like to thank the Research Council of Norway for financial support through the IPN project "AluLean" (90141902).

References

[1] Miller, W., Zhuang, L., Bottema, J., Wittebrood, A., De Smet, P., Haszler, A., & Vieregge, A. (2000). *Materials Science and Engineering: A*, **280**(1), 37-49.

[2] Tayal, R. K., Singh, V., Kumar, S., & Garg, R. (2012). *Proceedings of the National Conference on Trends and Advances in Mechanical Engineering,* 501-510.

[3] Prime, M. B., Gnäupel-Herold, T., Baumann, J. A., Lederich, R. J., Bowden, D. M., & Sebring, R. J. (2006). *Acta Materialia*, **54**(15), 4013-4021.

[4] Findik, F. (2011). *Materials & Design*, **32**(3), 1081-1093.

[5] Praveen, P., & Yarlagadda, P. K. (2005). *Journal of Materials Processing Technology,* **164-165**, 1106-1112.

[6] Fu, Y., Jie, J., Wu, L., Park, J., Sun, J., Kim, J., & Li, T. (2013). *Materials Science and Engineering: A*, **561**, 239-244.

[7] Papis, K. J., Loeffler, J. F., & Uggowitzer, P. J. (2009). *Science in China, Series E: Technological Sciences*, **52**(1), 46-51.

[8] Aylward, G., & Findlay, T. (2002). *SI Chemical Data* (5th ed.). Milton: John Wiley & Sons Australia.

[9] Papis, K. J., Hallstedt, B., Löffler, J. F., & Uggowitzer, P. J. (2008). *Acta Materialia*, **56**(13), 3036-3043.

[10] Liu, T., Wang, Q.-d., Liu, P., Sun, J.-w., Yin, X.-l., & Wang, Q.-g. (2015). *Transactions of Nonferrous Metals Society of China*, **25**, 1064-1072.

[11] Koerner, C., Schwankl, M., & Himmler, D. (2014). *Journal of Materials Processing Technology*, 1094-1101.

[12] Feng, J., Ye, B., Zuo, L., Wang, Q., Wang, Q., Jiang, H., & Ding, W. (2017). *Metallurgical and Materials Transactions A*, **48A**, 4632-4644.

[13] Solvay. (2017, December 22). Safety Data Sheet - Nocolok® Sil Flux - Fine Grade. Retrieved from https://www.solvay.us/en/binaries/P02-19419-USA_CN_EN-237549.pdf

[14] Solvay. (2017, March 3). Safety Data Sheet - NOCOLOK®Cs FLUX (TM) BW. Retrieved from https://www.solvay.us/en/binaries/P36966-USA_CN_EN3-238040.pdf

[15] Smith Stål. Smith Stål - Storage catalogue. Retrieved from https://www.smithstal.no/SmithStaal/Produkter/lagerkatalog-smithstal.no.pdf

[16] Sun, J., Song, X., Wang, T., Yu, Y., Sun, M., Cao, Z., & Li, T. (2012). *Materials Letters*, **67**(1), 21-23.

[17] Takemoto, T., Matsunawa, A., & Shibutani, T. (1997). *Welding International*, **11**(11), 845-851.

IOP Conf. Series: Materials Science and Engineering **529** (2019) 012025 doi:10.1088/1757-899X/529/1/012025

Formation of micro-plumes at a planar solid/liquid interface in a temperature gradient

J P Mogeritsch, T Peifer and A Ludwig

Department of Metallurgy, Chair for Simulation and Modelling Metallurgical Processes, Montanuniversitaet Leoben, Austria
*Corresponding author: johann.mogeritsch@unileoben.ac.at

Abstract: For directional solidification experiments with low withdrawal rates near and even below the constitutional undercooling limit of the corresponding alloy, a sample is needed which reveals a homogeneous concentration along the sample axis. As stirring in samples of a certain length is difficult, homogenization by rapid solidification is very common. Hereby, the liquid sample is rapidly moved from hot to cold and thus fine and long dendrites with segregated interdendritic liquid are formed. By subsequently melting and holding the sample in a resting position within a given temperature gradient, diffusion in the liquid and enhanced diffusion in the hot solid diminishes local concentration differences. In this paper, we report on homogenization and solidification experiences with transparent organic model compounds of TRIS-NPG. During one hour of homogenization we observed (i) formation of a coarse grain structure, (ii) liquid film and droplet migration by temperature gradient zone melting, and (iii) formation and eventually disappearance of liquid channels. During solidification experiments we occasionally found liquid channels, which revealed a connection to the solid/liquid interface; liquid that rose in the channel and thus formed micro-plumes at the planar solid/liquid interface. This upwards motion is explained by solutal buoyancy of NPG-rich liquid and a possible feeding through an interconnected network of channels.

Keywords: organic model system, peritectic layered structures, Bridgman-furnace, plumes, thermos-solutal convection

1. Introduction

The interesting characteristic in peritectic systems is the possibility of forming peritectic layered structures around the peritectic temperature T_P. For a peritectic reaction, a solid primary phase α and a liquid phase L reacts during cooling to form a second solid peritectic phase β ($\alpha + L \rightarrow \beta$). Close to or below the limit of constitutional undercooling peritectic layers are possible as a result of a competitive growth of both phases directly from the melt [1, 2]. Since the formation of such peritectic layered structures is highly influenced by convection in the melt, sample thicknesses in the range of µm are often used as those flat containments suppress fluid flow to a large extent. Using the Bridgman technique, the organic model system TRIS-NPG [3] has been used by the authors for in-situ observation of the dynamics of the solid/liquid (s/l) interface which leads to layered peritectic solidification morphologies [4-13].

Often concentration homogenization in such samples is performed by fast melting and rapid solidification which in turn leads to a fine dendritic structure. Afterwards when the sample is stopped, the temperature gradient leads to local remelting and solidification phenomena [14-19]. These processes involved morphological changes and coarsening, as well as macroscopic mass transport, migration of liquid droplets and liquid film migration through the solid matrix, known as temperature gradient zone melting (TGZM) [14].

To deepen our understanding, investigations were performed to visualize the influence of convection in the melt and migration movements in the solid on the formation of layered peritectic structures by using seeding particles as tracers [20].

2. Experimental set-up

Our Bridgman-furnace for in-situ observation consists of a hot zone of 10 mm and a cold zone of 40 mm departed by an adiabatic observation gap of 7 mm. The moving slide to pull the glass sample through the furnace is located about 40 mm below the cold zone and the sample stands out up to 100 mm above the outer rim of the hot zone after fixing. The two organic substances TRIS and NPG were obtained as a powder with a purity of 99% for NPG and 99.9^{+} % for TRIS. Both substances exhibit a facetted low temperature phase and an orientationally disordered crystalline high temperature phase, called "plastic phase". In order to the visualize the flow pattern tracers in the form of hollow glass spheres seeding particles [21] with an average distribution size of 11 ± 9 µm were added to the organic substances. We produced the organic components by melting and cooling together both components in the appropriate composition together with the particles. A glass tube sample (length ~ 200 mm) was filled with the seeding-enriched component by capillary force. Within the Bridgman-furnace, the sample was illuminated through the adiabatic zone to observe the forming solidification morphology with a light transmission microscope. Images were recorded by a CCD camera for a subsequent evaluation. More details concerning the sample preparation and the experimental set-up are given in [6, 8].

In-situ investigations were carried out by placing the sample within the preheated Bridgman-furnace. The temperature gradient G_T within the adiabatic area was controlled by the selected temperature of the hot and cold zone of the furnace. For the experiments presented in this paper the gradient was measured with a fine thermocouple within a constantly moving, filled sample to be $G_T = 5.6$ K/mm. Once the sample was connected with the pulling device it was drawn through the furnace from hot to cold with a relatively fast velocity of $V = 15.7$ µm·s-1 for 50 minutes. It should be noted that in this time period approximately 47 mm sample length are pulled through the 10 mm width hot zone. The organic component melts when entering the hot zone and solidify within the adiabatic gap again. Finally, the sample remains still in the furnace for 60 minutes before the solidification experiments with a pulling velocity rate of $V = 0.174$ µm·s^{-1} for several hours was carried out.

Here, we have recorded the processes which are activated by the selected homogenization method and observed the corresponding influence on the formation of layered structures for a hyper-peritectic alloy of $x = 0.52$ mol fraction NPG.

3. Results

At room temperature, the organic component only displayed the facetted face. By placing the sample within the Bridgman-furnace the faceted phase rapidly molt or transformed partly into the plastic phase, see figure 1a, depending on the position within the adiabatic zone. The s/l interface showed a convex surface towards the two sideward glass walls during the melting process - an effect caused by the small variation in the heat conductivity of the glass sample and the organic component.

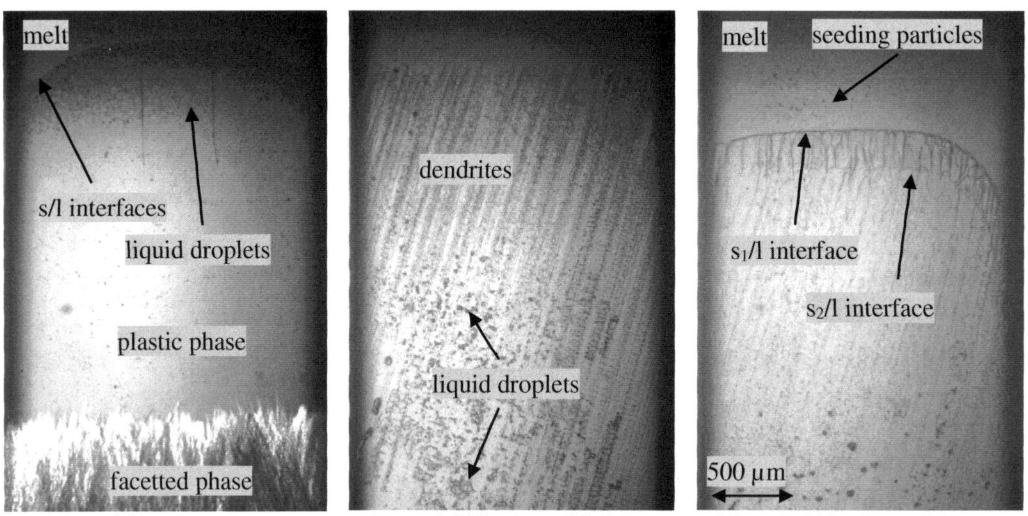

(a) $t = 300$ s / $x = 0$ mm (b) $t = 2{,}820$ s / $x = 39.5$ mm (c) $t = 3{,}300$ s / $x = 47.1$ mm

Figure 1: (a) Partial melting of the organic substance while the sample remains at rest. (b) Dendritic solidification structure while the sample is pulled through the temperature gradient with $V = 15.7$ µm·s^{-1}. (c) Solidification morphology after one hour at rest. Two interfaces at different temperature levels are formed. The dark shadow on the pictures left and right side originates by the side glass walls. Below the pictures the experimental time and the solidification length are given.

The transparent plastic phase displayed a bubble-like structure close to the s/l interface which disappeared within minutes by further melting. The remaining faceted phase showed a filamentous structure. As soon as the sample was fixed to the pulling device (Δt ~ 5 minutes) and before the thermodynamic equilibrium was reached, it was moved through the temperature gradient at a velocity of V = 15.7 μm·s^{-1}.

Due to the enforced solidification, the s/l interface changed into dendritic solidification morphology, see figure 1b. After around 50 minutes, which corresponded to a solidification length of 47.1 mm, the process was stopped and the sample was kept at the same position for one hour to achieve thermal stability. During this period, the morphological structure transformed from dendritic to planar and two boundary layers were formed at different temperature levels - one planar s/l interface, labelled s_1/l, and a second one, terms s_2/l, at a lower temperature level compare to s_1/l (figure 1c).

After 1 hour at rest, the sample was pulled (specified solidification time t = 0 s) with V = 0.174 μm·s^{-1} and both interfaces recoiled to a lower temperature level. Simultaneously, movement of the particles in the melt exhibited a descending flow in the middle of the sample and an upward motion along the two lateral glass walls. Particles from liquid films connected to the s_1/l interface rose up into the melt and formed several temporary micro-plumes.

After Δt = 3,600 s, as the s_1/l interface passed the initial position of the s_2/l interface, the growth of a new phase was observed. This happened on the sample's right side where the new phase in form of a band grew laterally towards the center of the sample. Subsequently, a new band began to grow laterally towards the center of the sample, coming in from the left side (see figure 2a).

During the formation of the bands, an enduring strong upward flow compare to the downwards flow in form of a micro-plume near the center of the sample was observed. It strength was sufficient to change the flow direction in the sample (figure 2b) to an upward movement slightly offset from the center and a downward movement along the sidewall of the glass tube.

Shortly before the bands merged the initial phase is the new preferred phase and start to grow dominantly again. It spreads out in both lateral directions which can be recognized in the form of a growing triangle. Whereby, the lateral growth was hampered on the left side by the above-mentioned micro-plume. Simultaneously, a new band was formed on the right side of the sample. In contrast to the first band, the new phase grew in between the front and back of the existing phase and the glass wall (figure 2c). It expands toward the left side until it was stopped at the micro-plume. The growth of these bands constrained the existing phase in such a way that this led to peritectic coupled growth (PCG) on the right side of the sample. In contrast, the s/l interface at the center became unstable and cellular growth occurred. The left side was unaffected and the s/l interface grew planar (see figure 2d).

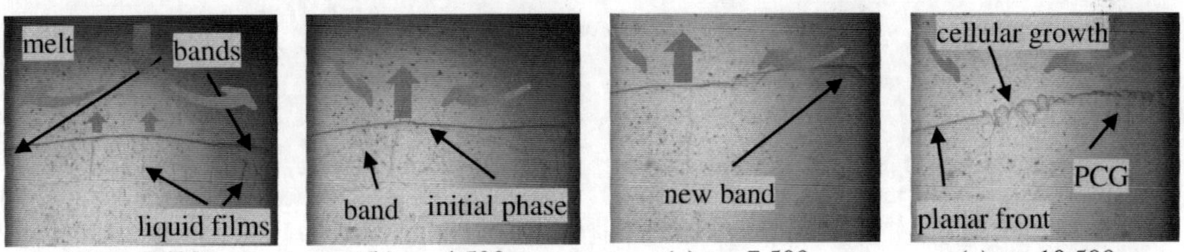

(a) t = 3,600 s (b) t = 4,500 s (c) t = 7,500 s (c) t = 10,500 s

Figure 2: (a) Two bands form close to the sample side walls. (b) The original phase is again preferred. (c) A new band forms at the right side and the planar front becomes unstable at the center. (d) Transition from banded structure to peritectic coupled growth (PCG). Convection: The arrows show the flow in the melt and the micro-plumes. The width of the picture is 2,000 μm.

4. Discussion

Since the mushy zone was formed via directed rapid solidification, elongated fine dendrites and corresponding interdendritic liquid developed. Particles were frozen in the solid structure of the initial dendrites or trapped within the interdendritic liquid (figure 3a). As soon as the sample became stable, a series of complex processes occurred. The dendrites widened and grew together and a planar s/l interface formed. Since the mushy zone was within a temperature gradient, liquid inclusions migrated in the solid phase toward higher temperatures up to the melt (figure 3b and c). Careful observation of the migration showed that the initial dendrite tips have grown at two different temperature levels and in different depths of the sample. This fact was not observable during the rapid solidification process due to the transparent nature of the organic components. It seems to be that due to the fast solidification rate, dendrites consisting of the primary α phase in the center of the sample depth and the peritectic β phase close to the front glass wall were formed. As a consequence, two planar interfaces at different temperature levels were generated (figure 3c).

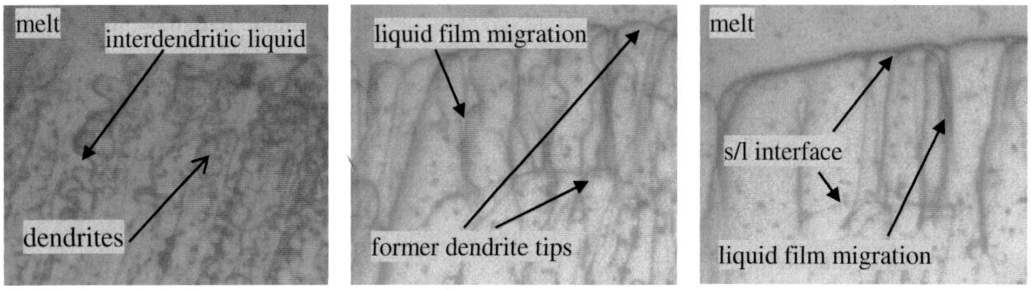

(a) t = 60 s (b) t = 1,140 s (c) t = 2,340 s

Figure 3: Sample in rest (a-c): migration of liquid inclusions and formation of two s/l interfaces. The width of the picture is 500 µm.

After one hour in rest, the solid was still full of liquid films. According to the phase diagram, the enclosed liquid must be enriched in NPG. Whereby, the density of the liquid decrease with the amount of NPG. Note, that the lower the temperature level, the higher the NPG content in the trapped liquid. Due to the migration of the liquid film within the temperature gradient, temporarily, liquid channels were formed by connecting various liquid films at different temperature levels up to the melt. Whereby, feeding though a liquid film network is postulate, high NPG-enriched liquid rises upwards through the liquid channels towards the melt forming micro-plumes. Just such a channel leads to the formation of the special plume as described in the previous chapter.

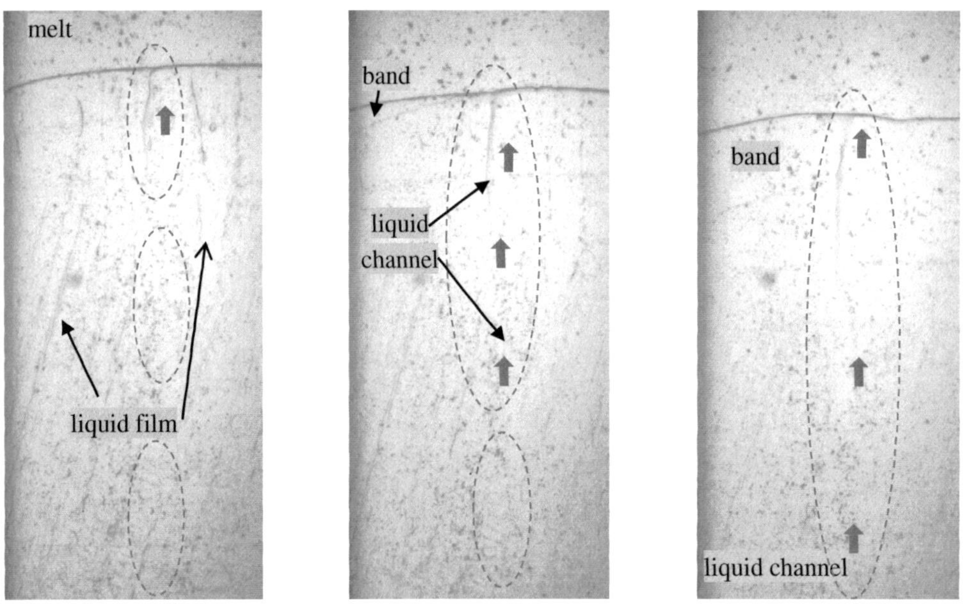

(a) t = 2,670 s (b) t = 3,870 s (c) t = 5,970 s

Figure 4: Formation of a long channel by connecting several liquid films (circles). Particle within the liquid channel are moving upwards (arrows) into the melt. The width of the picture is 1000 µm.

Figure 4 shows the formation of such a channel and the corresponding movement of particles within a time period of Δt = 3,300 s. This liquid channel formed a length of approx. 1,500 µm deep in the solid phase. The fluid channels were particularly recognizable by the particles moving therein. Whereby, at this temperature level only concentrations close to pure NPG are still liquid. In contrast to particles in the bulk melt, the particles within the liquid channel kind of shoot out into the bulk melt. This upward flow was sufficient to alter the flow pattern in the melt as shown in figure 2a and b.

The position of the observed micro-plumes, the liquid flow direction, and the interpretation of the micro structure formation is shown in figure 5. Initially, there were two solid phases present, recognized as the α and β phases, respectively. Ahead of the interface, thermosolutal convection initiated a downward movement in the center and a corresponding upward motion along the sample side walls. For the given solidification rate, both interfaces tried to reach stable growth conditions at the corresponding solidus temperature; therefore, both s/l interfaces started to recoil. As soon as the s/l interface of the α phase (s_1/l) had passed the initial temperature level of the β phase (s_2/l), the peritectic phase started to grow in form of two band (figure 5a). Additionally, the liquid channel described in figure 4 was formed and the strong micro-plume was established. Whereby, it was not possible to detect a direct influence of the micro-plume on the peritectic band. However, as a consequence of the micro-plume, a new convection pattern in the melt was induced. During the growth of the peritectic phase on both sides, the micro-plume is still active (figure 5b). Shortly prior to the two β phase bands merging at the center, the α phase became again the preferred phase. This can be seen by the expanded laterally growing triangle. The growth of the initial primary phase from the center to the left edge was halted as the α-phase came into contact with the micro-plume. Unlike the spread of the peritectic phase, here, the flow prevented the lateral growth of the α phase. Simultaniously, a new α phase band appeared on the right side of the sample (figure 5c). The phase seems to be formed by nucleation at the β/l interface close to the wall. Since the α phase had grown between the glass wall and the existing β phase, the peritectic phase became enveloped by the primary phase except at the top. This circumstance led to a competing growth between the two phases in form of PCG, cellular growth in the center, and planar growth on the left side.

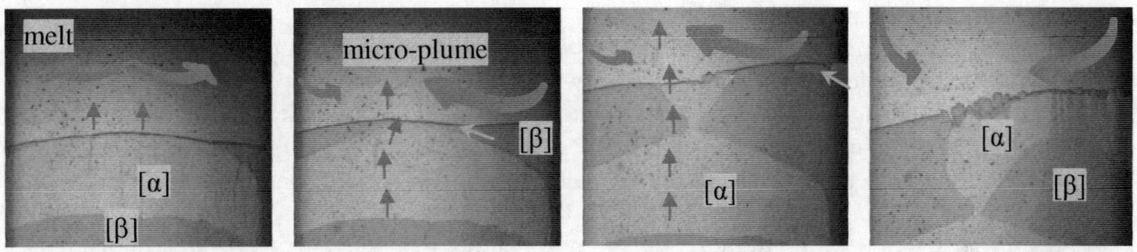

(a) $t = 0$ s (b) $t = 4{,}500$ s (c) $t = 7{,}500$ s (d) $t = 10{,}500$ s

Figure 5: (a) Formation of two β island bands unaffected by the liquid flow of NPG-enriched plumes (arrows). The dots in the pictures are the seeding particles and the arrows indicate the thermosolutal convection at the front of the s/l interface. (b) Shortly before both β-island bands was going to merge the α phase became the new preferred phase. (c) The α phase formed a new band coming from the right side. The new band solidifies in a planar manner. The lateral growth of the α phase growing at the center is stopped on the left side by the micro plume. (d) A planar, cellular, and peritectic coupled growth front was formed. The width of the picture is 2,000 μm.

5. Conclusion

Solidification experiments to investigate peritectic layered structures were carried out with the transparent peritectic modelling system TRIS-NPG. In order to visualize the thermosolutal convection flow, seeding particles were added to the organic substances as tracers. We have managed to demonstrate that the rapid solidification led to a simultaneous growth of α and β phases in form of fine dendrites. The corresponding interdendritic fluid was, accordingly, NPG-enriched. Once the sample was at rest, the dendrites broadened to form a solid matrix and the interdendritic fluid became trapped in the form of liquid films and few droplets. Since there was a temperature gradient, the trapped liquid began to migrate by local melting and remelting. This migration remained at the beginning of the solidification experiment. As a result, several liquid films merged to one liquid channel connecting to the s/l interface. Due to the density difference, the liquid enclosed in this channel flowed upwards towards the bulk melt and formed a special micro plume there. This flow phenomenon was sufficiently strong to change the convection pattern in the melt. For the observed formation of peritectic layered structures, first, no interaction of the liquid channel with the corresponding special micro plume was observed. In the further course of peritectic layered solidification, however, a direct influence of the liquid channel and micro plume on the solidification pattern was found.

Acknowledgements

This research has been supported by the Austrian Research Promotion Agency (FFG) within the frame-work of the METTRANS projects and by the European Space Agency (ESA) a part of the METCOMP project.

References

[1] Boettinger W J 1974 The structure of directionally solidified two-phase Sc-Cd peritectic alloys, *Met. Trans.* ***5*** 2023–2031
[2] Trivedi R 1995 Theory of layered-structure formation in peritectic systems, *Metal. Mater. Trans.* ***26*** *(6)* 1583-90
[3] Barrio M, Lopez D O, Tamarit JL, Negrier P and Haget Y 1995 Degree of miscibility between non-isomorphous plastic phases: Binary system NPG -TRIS *J. Mater. Chem.* ***5*** 431-39
[4] Ludwig A, Mogeritsch J 2011 In-situ observation of coupled peritectic growth, *Solidification Sci. Tech.: John Hunt Int. Symp.* 233-42
[5] Mogeritsch J, Ludwig A 2011 In-situ observation of coupled growth morphologies in organic peritectics, *IOP Conf. Ser. Mater. Sci. Eng.* ***7*** 12028
[6] Ludwig A, Mogeritsch J, Grasser M 2012 In-situ observation of unsteady peritectic growth modes, *Trans. Indian Inst. Metals* ***62*** 433-436
[7] Mogeritsch J, Ludwig A 2012 Microstructure formation in the two phase region of the binary peritectic organic system TRIS-NPG, TMS Annual Meeting Symposium, *Materials Res. Micro-gravity*, Orlando, Florida, USA 48-56
[8] Ludwig A, Mogeritsch J 2014 Recurring instability of cellular growth in a near peritectic trans-parent NPG-TRIS alloy system, *Mater. Sci. Forum* **790** 317-22
[9] Mogeritsch J, Ludwig A 2015 In-situ observation of the dynamic of peritectic coupled growth us-ing the binary organic system TRIS-NPG, *IOP Conf. Ser. Mater. Sci. Eng.* **84** 12055
[10] Ludwig A, Mogeritsch J 2016 Compact seaweed growth of peritectic phase on confined, flat pro-peritectic dendrites, *J. Crystal Growth* **455** 99-104
[11] Ludwig A, Mogeritsch J, Pfeifer 2017 In-situ observation of coupled peritectic growth in a binary organic model alloy *Acta. Mat.* ***126***, 329-335
[12] Mogeritsch J, Ludwig A 2018 Investigation on Peritectic Layered Structures by using the Binary Organic Components TRIS-NPG as Model Substances for Metal-Like Solidification, *Crimson Publishers, Res. Dev. Mater. Sci.* ***4****(1)* 1-3
[13] Mogeritsch J, Ludwig A 2018 In-situ observation of growth morphologies in organic peritectics, ICTIMESH-18, Dubai, submitted
[14] Pfann WG 1955 Temperature gradient zone melting, *J. Minerals, Metals & Materials Society 7(9)* 961–964
[15] Tiller W A 1963 Migration of a liquid zone through a solid: Part I, *J. Appl. Phys.* ***34*** 2757–2762
[16] Rettenmayr M 2009 Melting and remelting phenomena, *Inter. Mater. Rev.* **54**(11) 1-17
[17] Jabbareh M A, Assadi H 2009 Modelling of microstructure evolution in transient-liquid phase diffusion bonding under temperature gradient , *Scripta Mater.* ***60*** 780–782
[18] Liu D, Li X, Su Y, Rettenmayr M, Guo J, Fu H 2014 Local melting/solidification during peritectic solidification in a steep temperature gradient: analysis of a directionally solidified Al–25at%Ni, *Appl. Phys. A* ***116****(4)* 1821–1831
[19] Li Y j, Wang L, Ni P, Tan Y 2017 Growth of bulk Si from Si-Al alloy by temperature gradient zone melting, *Mater. Sci. Semicond. Process* ***66*** 170–175
[20] Mogeritsch J P, Pfeiler T, Ludwig A 2018 Investigation on the Liquid Flow ahead of the Solidifi-cation Front During the Formation of Peritectic Layered Solidification Structure, *7th Int. Conf. on Solidification and Gravity*, University of Miskolc 319-324
[21] Information on https://www.dantecdynamics.com

Solidification cracking during welding of steel: *In situ* X-ray observation

L. Aucott[a], J. Li[a], A. Rack[b], H. B. Dong[a *]
[a] Department of Engineering, University of Leicester, UK
[b] European Synchrotron Radiation Facility, Grenoble, France
Corresponding author: hd38@le.ac.uk

Abstract. Solidification cracking is an important issue during welding, casting and some of the additive manufacturing process. In order to illuminate the failure mechanisms, solidification cracking during arc welding of steel are investigated in situ with high-speed, high energy, synchrotron X-ray radiography approach. Analysis of the *in situ* radiography sequence revealed the solidification cracking initiates in the weld sub-surface trailing the welding electrode at relatively low true strain of about 3.1% in the form of micro-cavities. Although both material type and bending speed influence solidification cracking, cracks propagate from the core of the weld towards the free surface along the solidifying grain boundaries was found at a speed of between 1.7 -2.6 × 10^{-3} m s^{-1} for three different steels. In addition, a three-stage mechanistic model for solidification cracking during welding of steel is proposed.

1. Introduction

Solidification cracking, also known as hot cracking, is a key phenomenon associated with fusion welding, casting and some of the additive manufacturing processes. If undetected, the cracking defects can act as stress concentration sites which lead to premature failure via fatigue, as well as offer favourable sites for hydrogen assisted cracking [1, 2] and stress corrosion cracking [3]. Therefore, the solidification cracking is widely studied and have been the subject of investigation for several decades [4-6].

Early experimental efforts to observe the phenomenon were focussed on favourable organic alloys because of the technical problems involved with metallic alloys. Farup *et al* [7] successfully observed solidification cracking *in situ* in a directionally solidified succinonitrile-acetone alloy using a Bridgman type experiment under a microscope. Recent *in situ* observations of semi-solid deformation and solidification cracking in metallic alloys has been born from the interesting new fields of experimental research created by the development of third generation synchrotron sources. Gourlay *et al* [8] demonstrated the existence and importance of granular behaviour in solidification using *in situ* radiography experiments. Vernede, Rappaz, and co-workers then adopted the granular simulation method for discrete element simulations of solidification cracking [9-11]. The studies show that solidification cracking is a highly complex phenomenon, which involves heat flow, fluid flow under capillary action and various other factors such as material chemistry and processing parameters. Phillion *et al* [12] used a synchrotron micro-tomography approach to image the three dimensional morphology of solidification cracks in aluminium alloy AA5182. A similar approach was then employed by Aveson *et al* [13], although at a much smaller scale, to observe the initiation and propagation of solidification cracks at the dendritic scale in an Al-Sn binary alloy. Kareh *et al* [14] utilised time-resolved tomography to reveal that grains in semi-solid equiaxed Al-Cu alloys deform and rotate as independent bodies under compressive load.

IOP Conf. Series: Materials Science and Engineering **529** (2019) 012026 doi:10.1088/1757-899X/529/1/012026

In this research, a novel small-scale Varestraint test rig (ADW test rig) has been developed and embedded into the European Synchrotron Radiation Facilities (ESRF) high resolution imaging beamline, ID19. The beamline is installed on a low-beta section of the storage ring. ID19 has a small source size (30 µm vertical x 120 µm horizontal) and long 145 m straight section from the source to experimental hutch. This allows for the coherence properties of the beam to be exploited [15]. Solidification cracking during welding of steel is observed *in situ* using high-speed, high-energy radiography.

2. Materials and methodology

Figure 1 shows the experimental setup and schematic illustration of the test apparatus components. The equipment is designed to carry out stand-alone small-scale Varestraint weldability tests, while also allowing for integration to ESRF ID19 experimental hutch. The welding was carried out using a tungsten inert gas (TIG) weld method at 10 V and 98-108 A with a non-consumable tungsten electrode in DC-ve polarity.

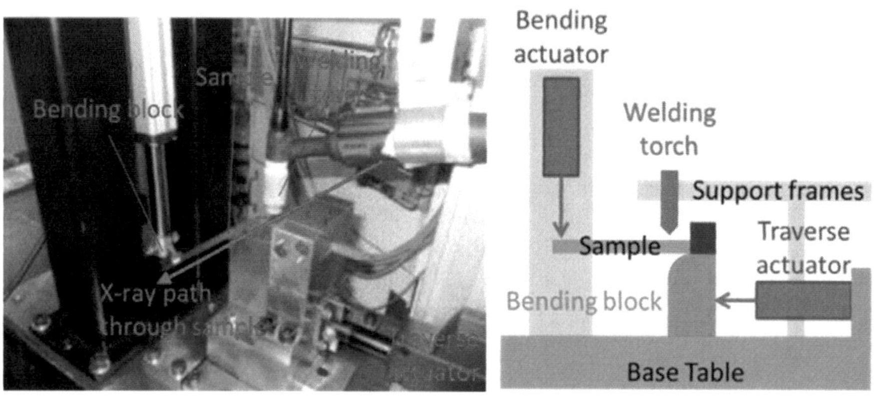

Figure 1: a) Varestraint test rig embedded into ID19, b) Schematic illustration of the test apparatus components.

8 mm square samples were created 300 mm long from three steels with poor weldabilty. The chemistry of the materials is defined in Table 1. To improve X-ray image quality, it is imperative to obtain the smallest path length through the sample. The 8 mm sample thickness was selected as it was the smallest feasible to weld upon. The samples were locked into the clamping device on top of the bending radius with a 30 mm overhang to attach the earth clamp for welding.

Table 1: Nominal chemistries of sample materials.

	C	Mn	Si	P	S	Cr	Ni
Mild Steel-EN1A	0.15	1	0.35	0.06	0.6	--	--
Stainless Steel-S303	0.15	2	1	0.2	0.15	19	10
Carbon Steel	0.4	0.9	0.2	0.06	0.6	--	--

The bending block assembly (with the sample clamped in) is attached to one of two electric actuators powered by a 12 V battery. The horizontally orientated actuator in Figure 1 supplies a load at speeds of up to 14 mm/s across a 40 mm stroke length. This actuator is connected to the bending block assembly and traverses it along the table to create the motion to weld a 40 mm length along the sample. Guide rails located on the surface of the base table interlink with the bending block assembly to ensure a straight travel path, normal to the beam. The opposite side of the test sample is then attached to a second actuator with a vertical orientation. This provides a load at speeds of up to 135 mm/s across a 150 mm stroke length. This actuator bends the sample during welding. The bending causes the sample to deform around the radius (60 mm) of the bending block and induces an augmented strain to the sample to stimulate solidification cracking. The weld pool solidification temperature was measured by tungsten-rhenium Type C thermocouples.

In order to reach a sufficiently high photon flux density at the desired X-ray energy of around 110 keV, ID19 was operated in white beam mode. The light of the beamline's wiggler (gap 45) was filtered by a diamond window, 5.6 mm Al and 6 mm Cu. The resulting bandwidth is comparably large (delta_E/E = 90%, peak around 117 keV) but allows for radiography imaging at frame rates up to 1 kHz. The resultant photon flux density was approximately 7 x 10^{11} photons $mm^{-1}s^{-1}$. For the imaging, a view window of 1872 x 1000 pixels was employed with a 10 µm/pixel resolution. The camera recorded continuously images at a rate of 1000 fps to satisfy the temporal demands in observing the solidification cracking *in situ*. All post-processing was performed with freely available ImageJ software. A simple processing routine was implemented for all analysis.

3.Results and discussion

3.1Solidification cracks at macro-scale

The novel small-scale Varestraint test rig commissioned for the tests successfully produces

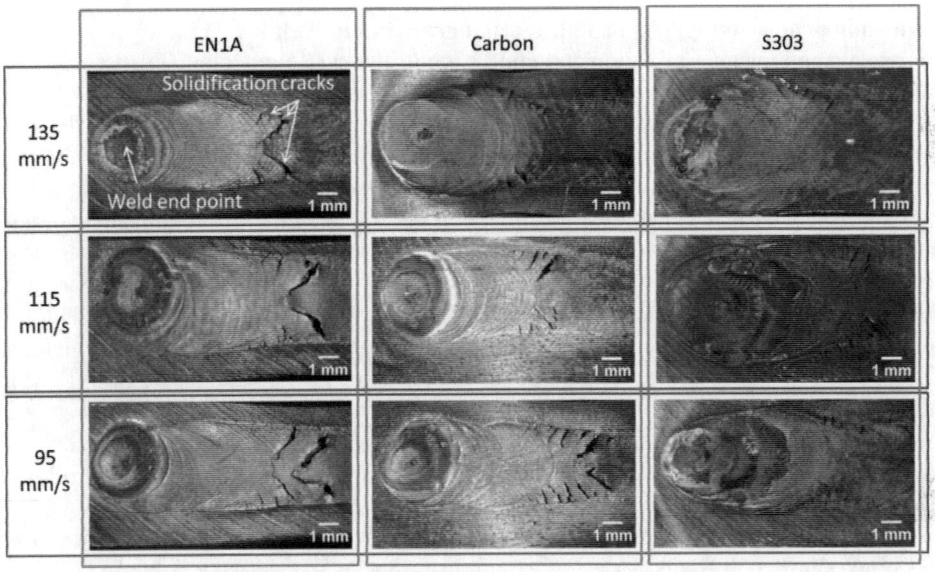

Figure 2: The array of weld solidification cracking observed in EN1A, Carbon, and S303 materials tested at 135, 115, and 95 mm/s bending speeds.

consistent and repeatable solidification cracks. Figure 2 displays macro images of the surface cracking observed in the samples selected for analysis. Each of the nine tests depicted in Figure 2 was repeated three times to ensure that the results obtained were consistent. It is evident from initial inspection of Figure 2 that both the material type and bending speed employed during Varestraint testing has significant impact on solidification cracking.

ImageJ software was used to quantify the surface cracking observed in the samples in terms of total crack length criterion (TCL) with the results plotted in Figure 3. TCL is derived from:

$$TCL = \Sigma Crack1 + Crack2 + \cdots CrackN \quad (1)$$

Analysis of Figure 3 confirms that both material type and bending speed influence solidification cracking. There is clearly a 3 tier susceptibility based on material type, with the EN1A sample being the most susceptible, and the S303 being the least susceptible. It is also evident from Figure 3 that the TCL of solidification cracks is also heavily dependent upon the bending speed utilised during Varestraint testing.

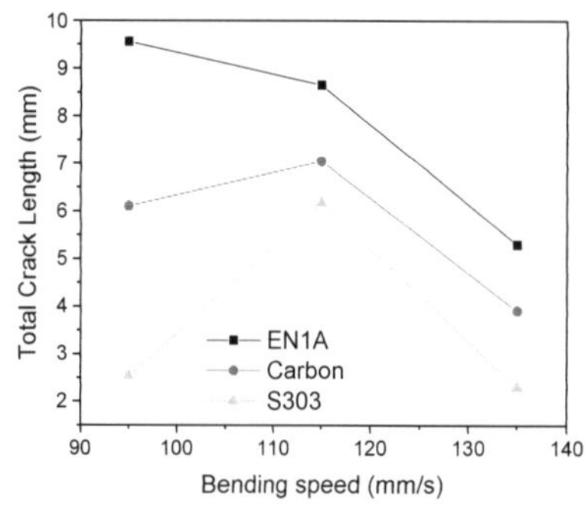

Figure 3: Total length of surface solidification cracking observed in the weld after Varestraint testing at a range of bending speeds.

3.2 *In situ* radiography imaging of solidification cracking in welding

Figure 4 displays the selected initiation and end stages of the cracking from the processed radiographs. The white tip in the top left of the images is the tungsten welding electrode. The large white part in the centre of the images is an EN1A test sample. Figure 4a) highlights the point at which weld solidification cracks were observed to initiate in the radiography sequence. The cracks are easily distinguishable, being darker than the bulk sample. This is due in part to a lower attenuation through the cracking interface. However, the main source of the interface clarity can be attributed to the refraction based coherent imaging technique and wide propagation distance between the sample and detector.

The cracks appear after 0.134 s of bending under a loading rate of 135 mm/s. The approximate strain at cracking initiation is 3.25%. The solidification cracks appear to initiate trailing the welding electrode by 2.69 mm at 1.45 mm sub-surface. The velocity of the fracture is measured at approximately 2.2 mm/s in the 2D plane of imaging, which could equate to speeds of up to 3.2mm/s to the 3D nature of the cracking when considering propagation at angles up to 45° in the z plane.

After the initial cracks have penetrated the surface, further cracking then begins to initiate sub-surface in the region to the left of the initial nucleation site (closer to the heat source) and follow the same vertical propagation path. This time however, it appears that the cracks also begin to converge laterally with the initial crack. Cracking continues to develop in this manner until 0.832s of bending (Figure *4*4b). The cracks then cease to develop at 10.53% strain. From the analysis it is possible to conclude that the weld solidification cracking occurred in the EN1A sample (with a 135 mm/s bending rate) over a 3.25-10.53% strain range and 0.698s time period.

The same analysis was completed for all samples referenced in Figure 2. The results are presented in Table 2. All solidification cracks were found to initiate sub-surface during analysis of micro-radiographs. It is clear that the least susceptible material – the S303 alloy – initiates at a shallower (0.47 – 0.98 mm) location when compared to the more susceptible EN1A and Carbon alloys. Furthermore, it also demonstrates that solidification cracks initiate at lower strains (2.42 – 3.42 %) in the more susceptible materials - the EN1A and carbon alloys. In the less susceptible S303 alloy, the cracks do not appear until strains have reached 4.00 – 5.06 %. The most susceptible material in terms of the commonly applied TCL criterion – the EN1A sample – exhibits the widest strain range in which solidification cracking propagates. The least susceptible alloy – the S303 sample – exhibits the shortest strain range. Again, it is apparent that all of the features are again heavily dependent upon the Varestraint bending speed. However, the measured fracture velocity remains fairly consistent throughout the experiments with an average velocity of 2.33 mm/s calculated.

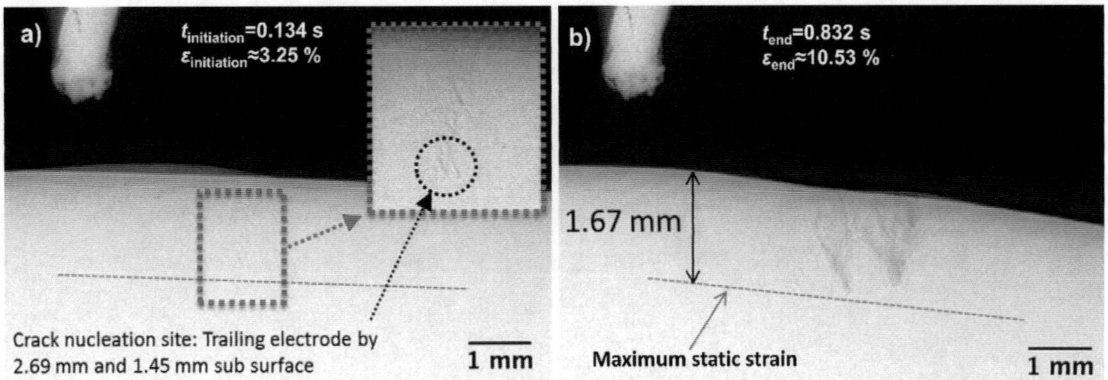

Figure 4: Results of synchrotron radiography analysis on the EN1A sample tested with a 135 mm bending speed, showing a) the initiation of weld solidification cracking at 0.134s of bending with an approximate strain of 3.25 %, the broken blue line depicts the fusion boundary measured from post-test microscopic observations b) the end of solidification cracking at 0.832s of bending and an approximate strain of 10.53 % (adapted from [6])

Table 2: Summary of properties observed from in situ micro-radiography.

	EN1A			S303			Carbon		
	95mm/s	115mm/s	135mm/s	95mm/s	115mm/s	135mm/s	95mm/s	115mm/s	135mm/s
Crack initiation time (s)	0.15	0.11	0.13	0.52	0.26	0.22	0.19	0.16	0.18
Crack end time (s)	1.33	0.88	0.83	1.15	0.87	1.00	1.22	0.87	0.71
Cracking period (s)	1.19	0.77	0.70	0.63	0.61	0.78	1.03	0.71	0.52
Crack initiation strain	2.47	2.58	3.25	5.06	4.00	4.35	2.42	2.56	3.42
Cracking end strain	8.33	8.89	10.53	9.52	8.70	10.13	7.69	8.16	9.52
Cracking strain range	5.86	6.31	7.27	4.46	4.70	5.78	5.27	5.60	6.11
Crack nucleation depth (mm)	1.18	1.63	1.45	0.47	0.86	0.98	0.82	1.05	1.73
Fracture velocity (mm/s)	1.66	2.46	2.20	2.48	2.54	2.53	2.33	2.18	2.59

3.3. Mechanisms proposed for cracking during welding of low-alloy high-impurity steels

Figure 5 schematic the proposed formation mechanism of hot cracking during weld pool solidification. After the transition from the dendrite to the grain structure when the volume fraction of liquid is <0.1 and temperature is ~1430 °C, solute elements (Mn, Si, Al) and impurities (S) segregate to grain boundaries. As a result, low melting point films with eutectic chemistries form in the inter-granular regions. Under tensile strains of ~3.25%, hot cracks form sub-surface at the inter-granular sites where the eutectic liquid films are concentrated, as illustrated in Figure 5(a). The cavities initially form sub-surface in the weld centre, as illustrated in Figure 5(a) by the red globules.

As the loading increases, the hot cracks then coalesce, predominantly in the "y" direction through the fracture of network separating bridges. Sometimes the hot cracks remain isolated and sometimes they coalesce with each other but not the bulk crack, as evidenced in which circle in Figure 5(b). Eventually, it is proposed that the coalescence of the hot cracks propagates through the grain structure to the solidifying dendritic structure.

Once the coalescence of the inter-granular hot cracks has propagated through to the solidifying dendrite structure, hot tearing begins. At this stage, the volume fraction of liquid is higher and inter-dendritic fracture is observed from the separation of primary dendrite arms as a result of tensile deformation as described in Figure 5(c). As the propagation is linked to the solidifying dendrites, the propagation path is related to that of the growing dendrites in terms of direction (towards the heat source) and velocity (~2.2-2.8 mm/s. However, as the load is applied along the length of the weld (z-axis), the tearing of the dendrites is also induced parallel to the x-axis, resulting in the somewhat angular cracking path towards the heat source as illustrated in Figure 5(c) and evidenced in Figure 2.

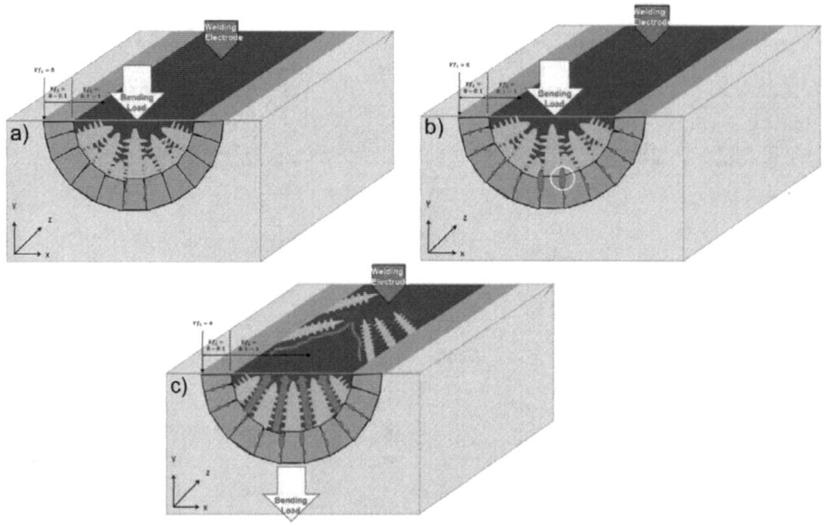

Figure 5: 3D schematic of hot crack formation mechanism in respect to weld pool: (a) inter-granular hot crack initiation; (b) coalescence of hot cracks; (c) hot tearing propagation.

Concluding Remarks

Weld solidification cracking in steel has been successfully observed *in situ* using a high speed, high energy, synchrotron X-ray radiography approach. Analysis of the *in situ* radiography sequence revealed the solidification cracking initiates in the weld sub-surface trailing the welding electrode. The results reveal that both material type and bending speed influence solidification cracking. However, the measured fracture velocity remains fairly consistent throughout the experiments with an average velocity of 2.33 mm/s calculated. In addition, a three-stage mechanistic model for solidification cracking during welding of steel is proposed.

References

[1] V. Shankar, J. Devletian, Science and Technology of Welding and Joining, 10 (2005) 236-243.
[2] S. Park, Ohio State University, 1989.
[3] J.R. Davis, Corrosion of Weldments, ASM International, Materials Park, OH, USA, 2006.
[4] I. Medovar, Avtomatich. , 7 (1954) 12–28.
[5] Aucott, L., Huang, D., Dong, H.B. et al. Metall and Mat Trans A (2018) 49: 1674.
[6] L. Aucott, D. Huang, H. B. Dong, S. W. Wen, J. A. Marsden, A. Rack & A. C. F. Cocks Scientific Reports (2017) volume 7, Article number: 40255
[7] I. Farup, J.M. Drezet, M. Rappaz, Acta Materialia, 49 (2001) 1261-1269.
[8] C.M. Gourlay, A.K. Dahle, T. Nagira, N. Nakatsuka, K. Nogita, K. Uesugi, H. Yasuda, Acta Materialia, 59 (2011) 4933-4943.
[9] S. Vernède, J.A. Dantzig, M. Rappaz, Acta Materialia, 57 (2009) 1554-1569.
[10] S. Vernède, M. Rappaz, Acta Materialia, 55 (2007) 1703-1710.
[11] S. Vernède, P. Jarry, M. Rappaz, Acta Materialia, 54 (2006) 4023-4034.
[12] A.B. Phillion, S.L. Cockcroft, P.D. Lee, Materials Science and Engineering: A, 491 (2008) 237-247.
[13] J. Aveson, IOP Conference Series: Materials Science and Engineering, 2012.
[14] K.M. Kareh, P.D. Lee, R.C. Atwood, T. Connolley, C.M. Gourlay, Nat Commun, 5 (2014).
[15] T. Weitkamp, P. Tafforeau, E. Boller, P. Cloetens, J.-P. Valade, P. Bernard, F.c.o. Peyrin, W. Ludwig, L. Helfen, J.e. Baruchel, AIP Conf. Proc. (ICXOM20), 1221 (2010) 35-38.

IOP Conf. Series: Materials Science and Engineering **529** (2019) 012027 doi:10.1088/1757-899X/529/1/012027

Capillary-mediated solid-liquid energy fields: their detection with phase-field method

M Glicksman[1] **and K Ankit**[2]

[1] Florida Institute of Technology, 150 West University Blvd., Melbourne, FL 32901-6975, United States

[2] School for Engineering of Matter, Transport and Energy, Arizona State University, 551 E. Tyler Mall, ERC 265, Tempe, AZ 85287, United States

E-mail: [1]mglicksman@fit.edu, [2]kumar.ankit@asu.edu

Abstract. Observations of melting crystallites in microgravity showed unusual shape changes as melting proceeded toward extinction. When re-analyzed in 2011, shape evolution data showed needle-like crystallites becoming spheroids as they melted toward extinction, suggesting that some type of capillary phenomenon at solid-liquid interfaces was responsible for an energy release capable of spherodising particles on melting, and stimulating pattern formation during unstable crystal growth. The presence of these previously undetected energy fields was recently uncovered using phase-field simulations that employ an entropy density functional. Simulations allow measurement of interfacial energy distributions on equilibrated solid-liquid interfaces configured as *stationary* grain boundary grooves (GBGs). Interfacial energy source fields—related to gradients in the Gibbs-Thomson temperature—entail persistent cooling along GBG profiles, a new result that fully confirms earlier predictions based on sharp-interface thermodynamics. This study also provides new insights to improve microstructure control at reduced scales by explaining the thermodynamic fields responsible for pattern formation in castings.

1. Introduction

Solidification microstructures remain subjects of importance, as they affect the mechanical and chemical properties of primary metals, cast alloys, solder joints, and fusion welds. The origins of complex pattern formation in general [1, 2], including those observed throughout nature, such as snow flakes and crystalline minerals, remain as scientific issues with important technological applications. To achieve better control of microstructure formation in materials one must determine: 1) If pattern-forming "instructions" are somehow encoded on solid-liquid interfaces? If so, 2) Do these instructions devolve from some deterministic thermodynamic mechanism, or are they imposed extrinsically by noise and other random disturbances?

2. Grain Boundary Grooves

Grain boundary grooves (GBGs) are naturally occurring crystalline microstructures found along polycrystalline solid-liquid interfaces. GBGs affect interfacial dynamic behavior by reducing an interface's morphological stability through enhanced local heat and solute rejection into the melt. Their interactions with both stationary and moving solid-liquid interfaces were first reported experimentally in the late 1960s using hot-stage optical and electron microscopy.

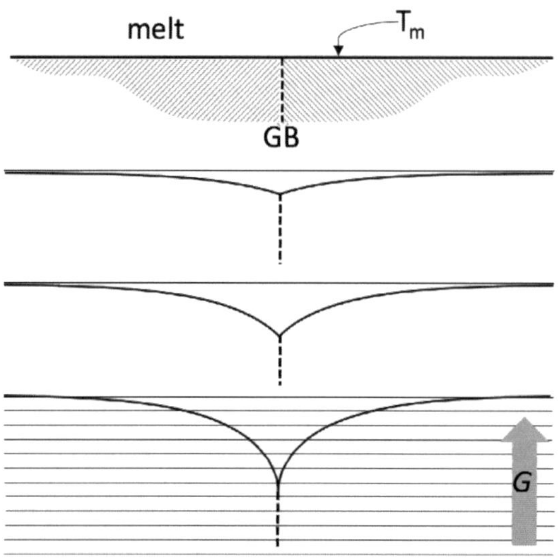

Figure 1. Sequence of virtual work steps to equilibrate an isolated GBG with a small dihedral angle. **Top**: Initial state, grain boundary (GB) intersects a flat solid-liquid interface. This interface locates the melting point isotherm, $T = T_m$. **Mid-panels**: Intermediate states, where the interface curves and lengthens, while the grain boundary shortens and the triple junction is drawn downward. **Bottom**: Final state, consisting of undercooled melt within the cusp depression in a vertical, constant temperature gradient, G, suggested by the stack of equally spaced isotherms.

Early studies on dilute alloys demonstrated that persistent defects, including GBGs, initiate morphological instabilities that stimulate complex patterns on solid-liquid interfaces [3, 4, 5]. Recently, dynamic effects induced by GBGs on solid-liquid interfaces in alloys were probed with phase-field simulation to follow morphological changes associated with interfacial instability [6].

2.1. Variational GBGs

The equilibrium profile for an isolated GBG, separating phases with equal thermal conductivity, was solved using variational calculus by Bolling and Tiller [7]. The profile, $\mu(\eta;\Psi)$, for a mirror-symmetric variational GBG, between phases of equal thermal conductivity, with dihedral angle, $\Psi \in [0,\pi)$, is described by the following pair of equations, plotted in figure 2 for $\Psi = 0$,

$$\begin{aligned}\mu(\eta;\Psi) &= -\frac{1}{2}\left[\ln\left(\frac{\sqrt{1-\eta^2}-1}{\eta}\right)-\ln\left(\cot\frac{3\pi+\Psi}{8}\right)\right]-\sqrt{1-\eta^2}+\sin\left(\frac{\pi+\Psi}{4}\right), \quad \mu\geq 0\\ \mu(\eta;\Psi) &= +\frac{1}{2}\left[\ln\left(\frac{\sqrt{1-\eta^2}-1}{\eta}\right)-\ln\left(\tan\frac{\pi-\Psi}{8}\right)\right]+\sqrt{1-\eta^2}-\cos\left(\frac{\pi-\Psi}{4}\right), \quad \mu\leq 0\\ &\qquad \eta\in[\eta^\star(\Psi),0).\end{aligned} \tag{1}$$

The leading minus sign in equation (1) denotes a GBG's right-side profile, $0 \leq \mu \leq \infty$, whereas the leading plus sign denotes its left-side profile, $-\infty \leq \mu \leq 0$. Also, the limited range of negative η-values allowed for equations (1) is $\eta^\star(\Psi) \leq \eta < 0$, where $\eta^\star(\Psi) = -\sqrt{\frac{1}{2}(1-\cos\frac{\pi-\Psi}{2})}$.

The system's thermo-capillary length, $\Lambda \equiv \sqrt{\frac{\gamma_{s\ell}\,\Omega}{G\Delta S_f}}$, determines the equilibrated (physical) size for variational GBGs derived from calculus. The materials constants defining Λ are $\gamma_{s\ell}$ [J/m^2], the solid-liquid interfacial energy density; Ω [m^3/mole] the molar volume of both phases; and ΔS_f [J/mole·K], the solid-liquid system's molar entropy change on melting. G, however, is the system parameter setting the magnitude of the temperature gradient and the GBG's size.

2.2. Equilibria on curved solid-liquid interfaces

The scaling law implicit in the definition of Λ also defines the appropriate temperature interval needed to non-dimensionalize the Gibbs-Thomson equilibrium interface temperature into a useful steady-state thermo-potential distribution along curved solid-liquid interfaces in the presence of a constant temperature gradient. The Gibbs-Thomson temperature distribution provides several important thermodynamic effects, largely ignored to date, that are intimately related to pattern formation during solidification.

At fixed pressure in the melt, the Gibbs-Thomson temperature distribution connects interface curvature, $\kappa(y(x))$ [m^{-1}], of a solid-liquid interface with its local equilibrium temperature, or chemical potential. For the case of a stationary GBG separating pure solid and liquid with equal thermal conductivities, held in a uniform temperature gradient, G, the interface temperature, $T_{int}(y(x))$ [K], decreases *linearly* with depth coordinate, y. Local undercooling within the melt cusp, viz., $(T_{int}(y(x)) - T_m)$ [K], increases linearly as $G \times y$ [K]. Thus, for equilibrated GBGs, portrayed in figures 1 and 2, their undercooling [K] and interface curvatures [m^{-1}] remain proportionate per the well-known Gibbs-Thomson relationship,

$$T_{int}(y(x)) - T_m = -\left(\frac{\gamma_{s\ell}\Omega}{\Delta S_f}\right)\kappa(y(x)). \tag{2}$$

Melt undercooling [K] can be transformed to a non-dimensional thermo-potential, ϑ, by dividing equation (2) by a suitable temperature interval. The temperature interval characteristic of a GBG is $2G \times \Lambda$ [K] that approximates the maximum undercooling developed at a GBG's triple junction. This choice defines a dimensionless thermo-potential, $\vartheta(y(x))$, namely,

$$\vartheta(y(x)) \equiv \frac{(T_{int}(y(x)) - T_m)}{2G\Lambda} = -\left(\frac{\gamma_{s\ell}\Omega}{2G\Lambda\Delta S_f}\right)\kappa(y(x)). \tag{3}$$

A non-dimensional curvature, $\hat{\kappa}$, defined consistently with the system's capillary length, Λ, is $\hat{\kappa}(\eta(\mu)) \equiv \kappa(y(x)) \times 2\Lambda$. Introducing that curvature into the foregoing Gibbs-Thomson relation, equation (3), allows a GBG's thermo-potential to be expressed in non-dimensional (μ, η) coordinates as,

$$\vartheta(\eta(\mu)) = -\left(\frac{\gamma_{s\ell}\Omega}{2G\Lambda\Delta S_f}\right)\hat{\kappa}(\eta(\mu)) \times \left(\frac{1}{2\Lambda}\right). \tag{4}$$

Gathering all the material and system constants that appear in equation (4), and inserting the lumped parameters that define the GBG's length scale Λ, shows that $\frac{\gamma_{s\ell}\Omega}{G\Lambda^2\Delta S_f} = 1$. Thus, judicious selections of the length and temperature scales allow the Gibbs-Thomson temperature for a variational GBG to assume the simple form, $\vartheta(\eta(\mu)) = -\frac{1}{4}\hat{\kappa}(\eta(\mu))$. The in-plane curvature distribution for any variational groove profile, equations (1), can be evaluated using the well known differential calculus formula [8], $\hat{\kappa}(\mu(\eta)) \equiv \frac{d^2\eta/d\mu^2}{(1+(d\eta/d\mu)^2)^{2/3}} = -4\eta$, $(0 \geq \eta \geq \eta^\star(\Psi))$. Note, that under the constrained equilibrium provided by the applied temperature gradient,

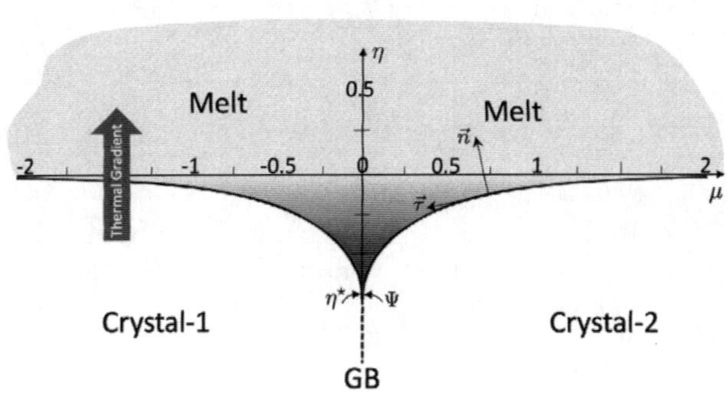

Figure 2. Variational GBG profile, $\mu(\eta; \Psi)$, plotted with equation (1), and its scaled coordinates $\mu = x/2\Lambda, \eta = y/2\Lambda$. The interfacial slope and curvature both vanish as $\mu \to \pm\infty$. The length scale, Λ, depends inversely on the magnitude of the gradient, G, in the $+\eta$ direction. This gradient constrains the entire profile to lie between the μ-axis and the groove's triple junction at $\mu = 0$, $\eta = \eta^\star(\Psi)$.

G, the interface curvature and the thermo-potential both approach zero as $\eta \to 0$, and $\mu \to \pm\infty$. Moreover, when the curvature, $\hat{\kappa} = -4\eta$, is substituted into the thermo-potential, $\vartheta(\eta(\mu)) = -\frac{1}{4}\hat{\kappa}(\eta(\mu))$, one finds that this interface thermo-potential equals the value of the GBG's depth coordinate, η,

$$\vartheta(\eta(\mu)) = \eta(\mu), \quad (0 \geq \eta \geq \eta^{\star}(\Psi)). \tag{5}$$

As shown next, this linear distribution of scalar Gibbs-Thomson thermo-potentials along the depth coordinate, η, implies much more than merely matching chemical potentials at thermodynamic equilibrium. They also provide a vector gradient field along the interface.

2.3. Interface gradients and fluxes

A directional gradient field from the Gibbs-Thomson scalar thermo-potential distribution, equation (5), exists tangentially along the solid-liquid interface. This field may be calculated from the following vector operational sequence,

$$\vec{\nabla}_{\tau}[\vartheta(\eta(\mu))] = \frac{d}{d\eta}[\vartheta(\eta(\mu))]\left(\frac{d\eta}{d\hat{s}}\right)\vec{\tau}, \quad \eta(\mu) \in \mathbb{R}^2 \tag{6}$$

where $\hat{s}$ is the GBG's dimensionless arc length of its profile, scaled similarly by twice the thermo-capillary length, 2Λ. The Frenet frame vector in equation (6), $\vec{\tau}$, is the unit tangent along the interface. The arc-length derivatives, $d\eta/d\hat{s}$, on the left and right profiles are, respectively,

$$d\eta/d\hat{s} = \pm 1/\sqrt{1 + (d\mu/d\eta)^2} = \mp 2\eta\sqrt{1-\eta^2}, \quad (\eta < 0). \tag{7}$$

When equations (5) and (7) are substituted into the chain rule, equation (6), one obtains the explicit expressions for the *vector* arc-length gradients of the Gibbs-Thomson thermo-potential along the right and left GBG profiles, respectively,

$$\vec{\nabla}_{\tau}[\vartheta(\mu(\eta);\Psi)] = \pm 2\eta\sqrt{1-\eta^2}\cdot\vec{\tau}, \ (0 \geq \eta \geq \eta^{\star}(\Psi)), \tag{8}$$

The presence of vector gradients in the thermo-potential requires an *irreversible* interfacial response according to Fourier's law of heat conduction in 2-D. Specifically, an energy flux [W/m] is spontaneously stimulated along the solid-liquid interface in directions opposite to those of the interface's local thermo-potential gradient vectors, equations (8). Application of Fourier's law introduces an interfacial thermal conductivity, k_{int} [W/K]—the value for which is the *bulk* thermal conductivity [W/m-K] of the solid and liquid phases multiplied by a thickness [m], of the order of the effective width of the solid-liquid interface. This interfacial transport number is well defined and has been analyzed theoretically [9]. Unfortunately, estimates for its values have received little attention from experimental heat transfer research for solid-liquid systems. Nevertheless, the dimensional forms of these tangentially stimulated Gibbs-Thomson energy fluxes on the right and left GBG profiles may be expressed as,

$$\vec{\phi}_{\tau}(y(x);\Psi) = \mp\left(\frac{2Gk_{int}}{\Lambda}\right) y\sqrt{1 - \frac{y^2}{4\Lambda^2}}\cdot\vec{\tau} \quad (y < 0) \quad [\mathrm{W/m}]. \tag{9}$$

The non-dimensional form of this capillary-mediated heat flux, $\hat{\Phi}(\eta(\mu);\Psi)$, is found by dividing equations (9) by the lumped constant Gk_{int} [W/m], to yield expressions identical to equations (8) for the tangential gradients, but of opposite sign.

$$\hat{\Phi}(\eta(\mu);\Psi) = \mp 2\eta\sqrt{1-\eta^2}\cdot\vec{\tau}. \tag{10}$$

2.4. Capillary bias fields

The presence of tangential interface fluxes, equations (10), has several interesting inferences. First, as *tangential* energy fluxes, these capillary-mediated energy streams, caused by gradients in the distribution of the Gibbs-Thomson thermo-potentials along the solid-liquid interface, are not directly interactive with the interface. The reason for their indirect interaction is that the energy balance controlling interface speed responds only to fluxes directed *normal* to the interface, not tangentially. Second, the quasi-static thermal fields, $T(\mu,\eta)_i$, $i{=}s,\ell$, within the bulk phases forming a GBG, which are established by the applied temperature gradient, G, each obey Laplace's equation. Laplace temperature distributions *within* the bulk phases are "harmonic" fields, which preclude divergence of their heat fluxes. This harmonic restriction, however, does not apply to the capillary-mediated fluxes moving along the interface itself, i.e., heat flowing *between* the phases! This peculiar and, to date, unnoticed distinction is of critical importance to understanding how perturbing energy fields arise from capillarity and modify the local speed of solid-liquid interfaces. Tangential fluxes obey Poisson's equation, and are divergent, i.e., they result in small releases or removals of energy along the interface, which act as Poisson sources. This qualitative difference among the system's interacting thermal fields accounts for capillary-induced contributions to the local energy budget of curved solid-liquid interfaces, both moving and stationary.

Specifically, capillary-mediated thermo-potential gradients along interfaces obey the following Poisson equation [10],

$$\nabla^2_\tau \vartheta(\mu(\eta);\Psi) = \mathfrak{B}(\eta(\mu)), \tag{11}$$

where $\mathfrak{B}(\eta(\mu))$, the bias field, acts as the interfacial source/sink function.

Equation (11) shows that bias fields, mathematically, are also the *surface Laplacian* of the Gibbs-Thomson potential, or equivalently, they equal (minus) the arc-length divergence of the capillary-mediated tangential vector flux, $-\vec{\nabla}_\tau \cdot [\hat{\Phi}(\eta(\mu);\Psi)]$. The key issue to be resolved in this paper is to demonstrate how interfaces, such as stationary GBGs, also support 'active' Poisson sources or sinks that provide intrinsic interface perturbations. Stationary microstructures, open to energy exchange, also interact with their own Poisson bias fields and provide an unique opportunity to generate a *persistent perturbations* once the microstructure and its source field are equilibrated. This occurs for GBGs without suffering the additional complications of complex pattern generation or other significant microstructure changes associated with morphologically unstable moving solid-liquid interfaces.

The Poisson source, or bias field, for a grain boundary groove may be calculated analytically from its Gibbs-Thomson potential distribution by twice arc-length differentiating equation (5). This is accomplished applying the following nested set of elementary operations in $\mathbb{R}^2$,

$$\nabla^2_\tau \vartheta(\eta(\mu)) = \frac{d}{d\eta}\left[\frac{d}{d\eta}[\vartheta(\eta(\mu);\Psi)]\left(\frac{d\eta}{d\hat{s}}\right)\vec{\tau}\right]\cdot\left(\frac{d\eta}{d\hat{s}}\right)\vec{\tau}. \tag{12}$$

The operational sequence shown in equation (12) defines the surface Laplacian of the Gibbs-Thomson equilibrium thermo-potential specified in equation (5). The bias field, $\mathfrak{B}(\eta(\mu))$, which in 2-D is $d^2\vartheta/d\hat{s}^2$, describes the dimensionless distribution of rates at which divergent capillary-mediated fluxes steadily remove thermal energy along a GBG's solid-liquid interface. The result for any GBG is the (negative) cubic distribution of energy rates,

$$\mathfrak{B}(\eta(\mu);\Psi) = 16\eta(1-2\eta^2),\ \ (0 \geq \eta \geq \eta^\star(\Psi)). \tag{13}$$

Equation (13), provides a close approximation to the actual Poisson perturbation field resident on a fully equilibrated GBG. This occurs, somewhat fortuitously, because the shape modifications of a stationary GBG induced by capillary perturbation fields are extremely small, so variational

GBG profiles accurately portray the actual curvature distributions along an equilibrated grain boundary groove supporting a basis field.

Equation (13), plotted as smooth curves in figure 3, shows that the bias field along any stationary GBG is everywhere negative and steady state. Negative bias fields cause cooling, i.e., energy removal, and correspond to positive vector divergences of the tangential capillary-mediated flux. Thus, the tangential heat flux, by virtue of its (positive) divergence, steadily *removes* energy from a GBG's solid-liquid interface *everywhere*, especially deep within the cusp. This self-induced energy removal cools the interface slightly, causing its curvature to increase (responding to the Gibbs-Thomson effect), and spontaneously "biases" the interface toward local equilibrium, as predicted by the Le Châteliér-Braun effect.

Bias fields on *moving* solid-liquid interfaces can behave in more complex ways, and may heat and cool different interfacial regions. Added or diminished energy rates both affect a moving interface's local energy budget. The interface speed responds immediately to conserve energy at all points by slowing or accelerating latent heat production, respectively. In this sense, the capillary field biases or modulates the motion of a moving interface, and can stimulate complex microstructure pattern development if the interface is morphologically unstable.

3. Phase-field Simulations and Results

We formulated an entropy density functional, $\mathcal{S}(e,\boldsymbol{\phi})=\int_V(s(e,\boldsymbol{\phi})-(\epsilon a(\boldsymbol{\phi},\nabla\boldsymbol{\phi})+\frac{1}{\epsilon}w(\boldsymbol{\phi})))dV$, for phase-field computations of thermal grain boundary grooving, a model that ensures consistency with classical irreversible thermodynamics. The bulk entropy density, s, depends on the internal energy density, e, and phase indicator, $\boldsymbol{\phi}=(\phi_\alpha)_{\alpha=1}^N$, the latter being a vector of phase-field variables that lies in the $N-1$ dimensional plane. Two grains and their melt ($N=3$) are the phases needed to model a GBG microstructure. $a(\boldsymbol{\phi},\nabla\boldsymbol{\phi})$ and $w(\boldsymbol{\phi})$ represent the gradient and obstacle potential type energy density, respectively; ϵ is a small length-scale parameter related to the thickness of the diffuse interfaces; and V represents the domain volume. Details of this simulation method are available in [11, 12]. Interface residual measurements to detect bias fields are fully disclosed in [13, 14].

Phase-field simulations, as very briefly described here, were used to probe stationary GBGs, which, upon equilibration in an applied linear temperature gradient, spontaneously develop interfacial Poisson sources (capillary-mediated perturbation fields). The opportunity to equilibrate a stationary microstructure, and then 'uncover' and measure its resident bias-field, presents a novel extension of phase-field simulation. Importantly. it permits independent verification of the existence of steady-state interfacial bias-fields.

Precision interface potential measurements conducted in our study reached levels of detection well below that which could possibly be accomplished experimentally, which allowed successful extraction of small "residuals" caused by non-linear interactions between the interface and its capillary-mediated fields. Residuals were the differences of the measure isoline potentials and the linear potentials imposed by the applied constant temperature gradient, G.

In order to obtain residuals data from simulations, GBG microstructures must achieve equilibrated stationary configurations in an applied temperature gradient, so that their input shapes become slightly modified by interaction with their self-generated capillary bias fields.

4. Conclusions

(i) Steady-state distributions of interfacial GBG thermo-potentials are the sum of the applied gradient, in which the GBGs are equilibrated, plus small non-linear contributions that arise from autogenous interfacial energy sources. The key issue divulged in this study that led to independent verification of these higher-order thermodynamic fields was spontaneous energy absorption from GBG interfaces. The energy removal occurred because tangential

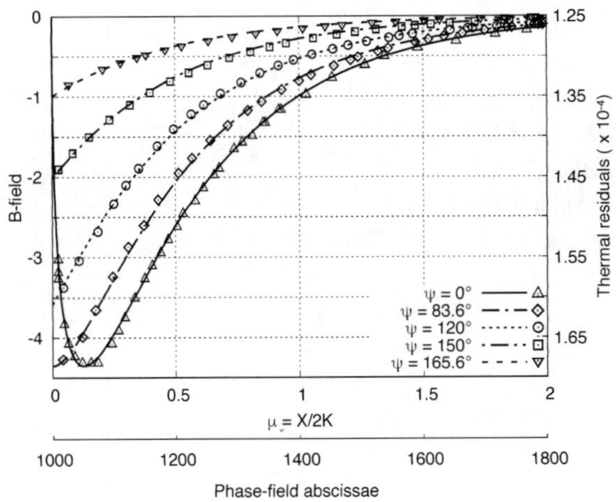

Figure 3. Comparison of phase-field residuals (data) from simulated GBGs equilibrated at steady-state with predicted 𝔅-fields (curves) obtained from sharp-interface theory. Residuals data measured on simulated diffuse interfaces agree well with theoretical estimates based on sharp interface thermodynamics. These results show that capillary-mediated bias fields are present at equilibrium on equilibrated GBG interfaces.

gradients and heat fluxes of the Gibbs-Thomson thermo-potential are divergent vector fields. Positive divergence of this capillary-mediated tangential energy flux causes energy loss at the solid-liquid interface, i.e., cooling, which results in small decreases of the Gibbs-Thomson thermo-potential, and consequent small increases in the interface curvature required by the system's shift toward local equilibrium.

(ii) The distribution of residuals is the result of—and, as explained in [13, 14], proportionate to—the capillary source distribution affecting the local thermal balance. Residuals are plotted as the data points in figure 3, all of which closely mimic the distributions of cooling perturbations, 𝔅-fields, plotted as smooth curves predicted from sharp-interface thermodynamics and field theory for variational GBGs.

(iii) When simulated, GBG profiles achieve thermal equilibration, and we observed no significant changes from their analytical shapes. We detected only "cooler" interfaces, i.e., by detecting negative residuals of the potential, caused by energy removal through the GBG's (negative) capillary bias field, equation (13). Again, the quantitative agreement demonstrated in figure 3 between residuals, i.e., non-linear potential depressions, and their calculated causal cooling rates, $\mathfrak{B}(\mu(\eta))$, reflects the predicted proportionality expected between interfacial cooling rates and measured decreases in thermo-potential.

(iv) These data provide indirect, but firm, support for our previous claim that it is capillary-mediated interfacial energy fields, previously unrecognized as required for *omnimetric* interfacial energy balances, that are causal for microstructure formation during solidification. Future microstructure control strategies during casting should recognize their presence for further improvement and control of microstructure patterns.

Acknowledgments

Author MG thanks the Allen S. Henry endowed Chair of Engineering, which provided support for this research at the Florida Institute of Technology, Melbourne, Florida. Both authors acknowledge financial support from the National Aeronautics and Space Administration (NASA) through Grant Number 80NSSC18K1440.

References

[1] Thompson D 1948 *On Growth and Form,* 2nd *Edn.* (Cambridge, UK: Cambridge University Press)

[2] Kassner K 1996 *Pattern Formation in Diffusion-Limited Crystal Growth* Directions In Condensed Matter Physics (Singapore: World Scientific Publishing Co. Pte. Ltd.)

[3] Morris L and Winegard W 1969 *J. Cryst. Growth* **5** 361–375
[4] Schaefer R and Glicksman M 1970 *Met. Trans.* **1** 1973–1978
[5] Sharp R and Hellawell A 1970 *J. Cryst. Growth* **6** 334–340
[6] Yeh S, Chen C and Lan C 2011 *J. Cryst. Growth* **324** 296–303
[7] Bolling G and Tiller W 1960 *J. Appl. Phys.* **31** 1345
[8] Kells L 1946 *Calculus* 7th Ed. (New York: Prentice-Hall, Inc.)
[9] Gurtin M 1988 *Archive for Rational Mechanics and Analysis* **104** 195–221
[10] Glicksman M 2016 *J. Cryst. Growth* **450** 119–139
[11] Garcke H, Nestler B and Stinner B 2004 *SIAM J. Appl. Math.* **64** 775
[12] Nestler B, Garcke H and Stinner B 2005 *Phys. Rev. E* **71** 041609
[13] Glicksman M and Ankit K 2017 *Metals* **7**(12) 547
[14] Glicksman M and Ankit K 2018 *J. Mater. Sci.* **53** 10955–10978

IOP Conf. Series: Materials Science and Engineering **529** (2019) 012028 doi:10.1088/1757-899X/529/1/012028

Efficient model for the prediction of dendritic grain growth using the lattice Boltzmann method coupled with a cellular automaton algorithm

Stephan Jäger[1] and Andreas Ludwig[2]

[1] Light Metals Technologies Ranshofen, Center for Low-Emission Transport, Austrian Institute of Technology, 5282 Braunau am Inn - Ranshofen, Austria

[2] Simulation and Modelling of Metallurgical Processes, Department of Metallurgy, University of Leoben, 8700 Leoben, Austria

E-mail: stephan.jaeger@ait.ac.at

Abstract. An efficient model for the prediction of dendritic grain growth is developed coupling the lattice Boltzmann method for solving the transport of solute and a cellular automaton algorithm for determining the evolution of grains' envelope and the release of solute during phase change. In contrast to solving equations from the field of continuum mechanics the new model is more related to particular occasions what is more similar to the behaviour of cellular automaton algorithms. The resulting dendritic grain growth shows qualitative correctness, although the consideration of solute conservation is still missing. It is shown that neglecting proposed conditions regarding the choice of time step size can destabilize the solid-liquid interface resulting in secondary and ternary dendrite arms.

1. Introduction

There are several approaches of simulating dendritic grain growth. Besides phase field and front tracking simulations, cellular automaton (CA) is quite often used based on the work presented by Gandin and Rappaz [1, 2]. Herein, the evolution of the grain's envelope is related to the thermal undercooling using the model proposed by Kurz, Giovanola and Trivedi (KGT)[3, 4] for rapid solidification. This approach is extended in [5] considering solute and curvature undercooling and continued for three dimensions [6]. It is coupled to the lattice Boltzmann method (LBM) considering fluid flow in [7, 8] and compared with results from phase field simulations in [9]. A different approach is chosen in [10] employing thermal LBM with an enthalpy-porosity technique for the simulation of dendritic grains with fluid flow. These all have in common the calculation of the curvature of the interface between the solid and liquid phase what is demanding on the resolution and the computational effort calculating spatial gradients.

The objective of the present work is to more efficiently simulate solidification offering dendritic structures using a CA for determining the evolution of growing grains and the LBM for solving transport of solute. The proposed model aims to be more accurate with respect to the grain shape and microsegregation than the original CA of [1, 2] by considering solutal undercooling. In contrast to [11, 7, 10] undercooling due to the curvature of the solid-liquid interface is neglected. Thus, no gradients using the finite difference method have to be computed. The new approach should not act as an alternative for solving continuum mechanical equations in its equilibrium state but combining the state driven behaviour of cellular automaton [12] and analytical equations. The increased efficiency should enable the integration of grain growth models in full scale industrial application in the future.

2. Model description

LBM is coupled with a CA for the simulation of the dendritic grain growth. These are not weakly but fully coupled following the categorization of [13]. Compared to previously mentioned work the release of solute is not done in the way of continuum mechanics but following the idea of discrete occasions which is explained at the end of this section.

2.1. Lattice Boltzmann method for solute transport

LBM is applied for the computation of the diffusion of solute c neglecting any advection

$$\frac{\partial c}{\partial t} = D\frac{\partial^2 c}{\partial x^2} \tag{1}$$

assuming constant diffusivity D. LBM is based on the relaxation of non-equilibrium states derived from a statistical distribution of particles. There are basically two operatrions which define the computational sequence: there are the *collision* and the *streaming* operation. The *collision* of the discrete distribution of particles $g_i(\vec{x}, t+1)$ at the spatial position $\vec{x}$ and the time $t+1$ is determined as

$$g_i(\vec{x}, t+1) = g_i(\vec{x}, t) - \frac{1}{\tau}(g_i(\vec{x}, t) - g_i^{eq}(\vec{x}, t)). \tag{2}$$

using the Bathnagar-Gross-Krook dynamics with its linear collision operator [14, 15] and the relaxation time τ. The equilibrium state $g_i^{eq} = cw_i$ is defined using the lattice weights w_i following the multiscale Chapman-Enskog analysis [16]. The solute is reconstructed by

$$c = \sum_{i=0}^{n-1} g_i. \tag{3}$$

The sum is limited by the amount n of discrete velocities $\vec{e_i}$ and corresponding weights w_i which depend on the chosen lattice. On the two-dimensional cartesian grid, the D2Q5 lattice with $n = 5$ is chosen. This reduces the computational effort and the requirements regarding memory capacity compared to the commonly used D2Q9 lattice. The D2Q5 lattice guarantees second order rotational invariance [16]. This is sufficient for the reconstruction of a diffusion equation. The diffusivity D in equation 1) is related to the relaxation time τ by

$$D = c_s^2(\tau - \frac{1}{2})\frac{\Delta x^2}{\Delta t} \tag{4}$$

with the lattice speed of sound $c_s = 1/\sqrt{3}$ for the D2Q5 lattice. Δx and Δt are the spatial and the temporal discretizations which have to be chosen in a way that $\tau > 1/2$. Thus, the relaxation time τ is defined for the liquid and the solid phase depending on the diffusivity in the liquid D_l and in the solid phase D_s, respectively.

The *collision* is fully local which requires no communication to neighbouring voxels (discretization points) since the left hand side and the right hand side of equation 2 are function of $\vec{x}$. Whereas, the *streaming* operation streams information from $\vec{x}$ to the next neighbour $\vec{x}+\vec{e_i}$ which is defined as

$$g_i(\vec{x}+\vec{e_i}, t+1) = g_i(\vec{x}, t+1). \tag{5}$$

2.2. Grain growth algorithm

The CA algorithm for simulating the evolution of growing grains is derived from the original work [1, 2] and the decentred squares [17]. The state of a cell - it can be either liquid, interface or solid - and its evolution depend on the state of the neighbouring cells, the local temperature and the local solute. Thus, it is not fully local because there are eight neighbours (and the cell itself) in two dimensions using a cartesian grid. In the original CA grain growth algorithm, the spatial discretization Δx is recommended to be about the secondary dendrite arm spacing λ_2 [2]. For the binary alloy Al-Cu and moderate cooling rates, the secondary dendrite arm spacing is estimated to $\lambda_2 \approx 10 - 40$ µm [18] for a parameter study. In the modified CA algorithms the spatial resolution has to be much smaller since the models fully resolve the interface and the secondary dendrite arms as well. For this, the spatial resolution is chosen $\Delta x \approx 1$ µm or even smaller [11, 7, 9]. In the presented model, the spatial discretization should be 1 µm $< \Delta x < \lambda_2$ to become more reliable for large-scale applications with respect to the computational effort. Since the discretization can be chosen in a way the secondary dendrite arms can not be resolved, different models have to be applied. Therefore, the grain growth algorithm is defined for two regions. The inner region Ω_{in} is enclosed by the envelope region Ω_{env}. The envelope region Ω_{env} describes the interface between the fully liquid and the partially solid region with the solid fraction $0 < f_s \ll 1$ whereas Ω_{in} can be fully solid or partially solid ($0 < f_s \leq 1$). Basically, the models are taken from the work presented in [19] and adapted in a way to get more precise information about the grain shape. In all cases the growth velocity is related to the supersaturation

$$\Omega = \frac{c_l^* - c_l}{c_l^* - c_s^*} = \frac{c_l^* - c_l}{c_l^*(1-k)} \tag{6}$$

with the partition coefficient $k = c_s^*/c_l^*$, solid c_s^* and liquid c_l^* solutal equilibrium concentration and local liquid concentration c_l. It is assumed that for Ω_{env} the model proposed by Lipton, Glicksman and Kurz (LGK) [20, 4] is valid although it is derived assuming a paraboloid primary dendrite tip in steady state condition without any neighbours. The growth velocity v_{env} is determined as [20]

$$v_{env} = v_{tip} = v_{LGK} = \frac{D_l\, m_l\, c_l(k-1)}{\Gamma \pi^2} (Iv^{-1}(\Omega))^2 \tag{7}$$

with the solute diffusivity D_l, the liquidus slope m_l, the solute concentration of the liquid c_l, Gibbs-Thomson coefficient Γ and the inverse of the Ivantsov equation $Iv^{-1}(\Omega)$ for the two-dimensional space [21]. The initialization of the solid fraction for the inner region $f_{s,init}$ after going through the envelope region Ω_{env} has to be assumed at the current state of this work. For the inner region, the growth velocity is defined as

$$v_{in} = \frac{D_l}{l_d} \frac{c_l^* - c_l}{c_l^*(1-k)} \tag{8}$$

substituting the interdendritic (liquid) concentration c_d from [19] by the liquid concentration c_l. The characteristic diffusion length l_d can be estimated according to [22] by $l_d = \beta_2(\lambda_2 - d_2)/2$ with the diameter of the secondary dendrite arm d_2, and the constant $\beta_2 = 1.0$. It can be approximated by [19]

$$l_d = \beta_2 \frac{\lambda_2 f_l}{2}. \tag{9}$$

substituting the interdendritic liquid f_d^e in [19] by the local liquid fraction f_l within Ω_{in}. The equation for the growth algorithm is applied as long as the the cell is still partially liquid $f_s < 1$. If $f_s = f_{s,limit}$ the cell becomes fully solid and the diffusivity is changed to $D = D_s$. A surface area concentration S is missing in combination with the growth velocity v_{in} for determining the change of the solid fraction $df_s/dt = v_{in}\, S$. The envelope liquid fraction f_e is substituted by the liquid fraction f_l and the Avrami factor in [19] is neglected due to the increased resolution compared to [19]. This results in the surface area concentration

$$S = \frac{2}{\lambda_2} f_l. \tag{10}$$

Following this approach, results in a non-discrete interface between the liquid and the solid region and the co-existence of solid solute and liquid solute at each cell are allowed. It is an averaged field of solute and solid fraction. Thus, if one would determine the solute only depending on the solid fraction [7] the solute at the envelope region would be underestimated and the high gradients, especially close to the primary dendrite tip, would not be represented. This would lead to an underestimated concentration field and a wrong tip velocity using the LGK model. Therefore, the solute release is defined as a particular occasion setting the concentration of the liquid phase $c_l = c_l^*$ and the concentration for the solid phase $c_s = c_s^*$. These are defined each time the integrated growth velocity over time t

$$l = \int_{t_{j-1}}^{t_j} v \, dt \tag{11}$$

exceeds a certain characteristic length l starting at the last time an occasion occured t_{j-1} until the current time t_j. The characteristic length is determined as the spatial resolution $l = \Delta x$. The growth velocity v is either the one of the envelope region v_{env} or the inner region v_{in}. Subsequently, the released solute is diffused in the liquid phase using LBM. The time step for the simulation basically depends on several parameters. Thus, it is restricted by the stability of LBM for the transport of solute (see eq. 4) and by the growth velocity of CA. The supersaturation can restrict the time step for the CA depending on the cooling rate. The time step is defined depending on the maximum of velocity v_{env} with the scale factor γ - which is in general bounded by $0 < \gamma \leq 1$ for a stable algorithm and proposed to be chosen as $\gamma = 1/10$ in [23]:

$$\Delta t = \gamma \frac{\Delta x}{\max\{v_{env}(c_l)\}}. \tag{12}$$

3. Simulation setup

The evolution of a single equiaxed grain with the crystallographic orientation of 20° in an Al-4.7wt.%Cu alloy is simulated in a two-dimensional domain of the size 1 mm x 1 mm. The alloy related parameters are listed in table 1 taken from [18]. The simulation starts at the liquidus temperature T_l and is isotropically reduced applying a constant cooling rate $\dot{T} = 30$ K/s. The free model parameters λ_2, $f_{s,init}$ and $f_{s,limit}$ are assumed to be constant (see table 1). Two spatial resolutions are distinguished: (1) $\Delta x = 5$ µm and (2) $\Delta x = 2.5$ µm. The time step is defined equally for both spatial resolutions as $\Delta t = 1.5 \cdot 10^{-4}$ s. Thus, for the higher resolved simulation $\gamma \leq 1$ is not fulfilled (see eq. 12) which results in a destabilized behaviour.

Table 1 Simulation parameters for Al-4.7wt.%Cu

Thermophysical properties	Thermodynamic properties	Model parameters
$D_l = 3 \cdot 10^{-9}$ m²/s	$T_m = 933.5$ K	$\lambda_2 = 10$ µm
$D_s = 8 \cdot 10^{-13}$ m²/s	$m_l = -344$ K	$f_{s,init} = 0.1$
$L = 3.97 \cdot 10^5$ J/kg	$k = 0.145$	$f_{s,limit} = 0.85$
$\rho_s \approx \rho_l = 2743$ kg/m³	$\Gamma = 2.41 \cdot 10^{-7}$ m K	

4. Results

At time $t = 0.87$ s and $t = 1.02$ s the extent of the dendrite tips and the solute concentration is different with respect to the spatial resolution (see figure 1). Herein, the solute concentration is determined as $c = c_l(1 - f_s) + c_s f_s$. Focusing on the solid fraction f_s, the lower resolution (left half of the pictures) shows a more averaged field whereas the higher resolution (right half of the pictures) shows some detailed resolved secondary dendrite arms ($f_s > 0$). The solute concentration at $t = 1.02$ s shows a characteristic behaviour of this algorithm: resolution (1) with the stable CA shows a very diffusive behaviour. In contrast, resolution (2) with the unstable CA shows less dominant diffusion. There is a very high gradient of the solute concentration close to the liquid-solid interface especially in the region of secondary dendrite tips which seem to overtake the diffusive process. Further refining as well as coarsening shows that the results are strongly dependent on the spatial resolution. In general the solute is not conserved using this approach. The change of solute over time depends on the growth velocity and in most cases the total solute strongly increases.

The destabilized simulation shows secondary dendrite arms and even tertiary dendrite arms in a later stage. It seems to be physical in a way, but the destabilization is not related to the interface anisotropy. In the present work it is related to the ratio of user-defined time step and growth velocity which again depend on the solute and thermal undercooling.

The simulation time for the stable and the destabilized setup is 25 s and 96 s, respectivley, running in parallel on a quadcore notebook processor.

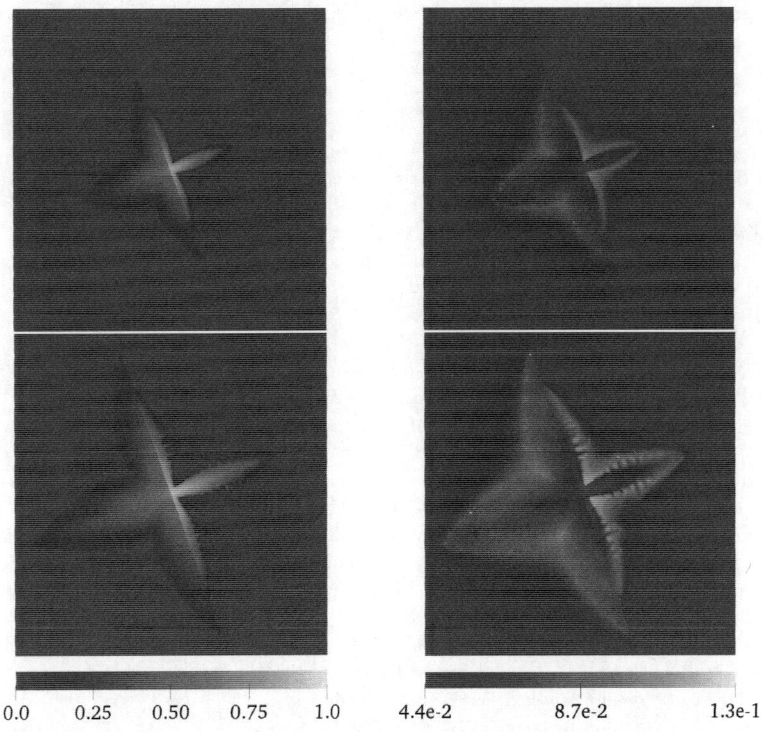

(a) Solid fraction f_s (b) Solute concentration c

Figure 1 Solid fraction f_s and solute concentration c during equiaxed dendritic solidification with the crystallographic orientation of 20° in an isotropically undercooled ($\dot{T} = 30$ K/s) Al-4.7wt.%Cu melt at two different times $t = 0.87$ s (top) and $t = 1.02$ s (bottom) for resolution (1) (left half in each figure) and resolution (2) (right half)

5. Discussion

The presented model meets the expecations regarding the determination of grain growth showing secondary dendrite arms or an averaged field depending on the user-defined resolution. On the contrary, the solute is not conserved and the accuracy depends on the spatial and temporal resolution. This also related to the issue that there is no energy or force at the envelope which pushes the solute concentration which has to be improved in future work. There is no sharp transition of phases. Future modification has to strongly focus on the conservation of species.

Furthermore, it has been shown that growing secondary dendrite arms can be simulated using a destabilized simulation setup. For physically correct destabilzation the condition regarding the growth veocity and the influence of spatial and temporal resolution has to be further investigated.

6. Outlook

This work should constitute the basis for more sophisticated simulations considering the interaction of multiple grains, microsegregation as well as the transport of grains in 3D. Therefore, the future work will heavily focus on the efficiency of the newly developed model considering the conservation of species. Furthermore, its application in the three-dimensional space must be approved.

Acknowledgments

The Author would like to thank the Austrian Research Promotion Agency (FFG), the Federal Ministry for Transport, Innovation and Technology (BMVIT) and the State of Upper Austria for funding this work. This work was conducted in the framework of COMET project Amoree (Grant: 843537) and the framework of FTI Struktur Land Oberösterreich of Upper Austria project PSHeRo.

References

[1] Rappaz M and Gandin C A 1993 *Acta Metallurgica et Materialia* **41** 345–360
[2] Gandin C A and Rappaz M 1994 *Acta Metallurgica et Materialia* **42** 2233–2246
[3] Kurz W, Giovanola B and Trivedi R 1986 *Acta Metallurgica* **34** 823 – 830
[4] Kurz W and Fisher D 1986 *Fundamentals of solidification* Bd. 1 (Trans Tech Publications)
[5] Zhu M F and Hong C P 2001 *ISIJ International* **41** 436–445
[6] Zhu M F and Hong C P 2002 *ISIJ International* **42** 520–526
[7] Sun D, Zhu M, Pan S and Raabe D 2009 *Acta Materialia* **57** 1755 – 1767
[8] Sun D, Zhu M, Wang J and Sun B 2016 *International Journal of Heat and Mass Transfer* **94** 474–487
[9] Zaeem M A, Yin H and Felicelli S D 2013 *Applied Mathematical Modelling* **37** 3495 – 3503
[10] Chatterjee D and Chakraborty S 2006 *Physics Letters A* **351** 359–367
[11] Luo S and Zhu M Y 2013 *Computational Materials Science* **71** 10 – 18
[12] Chopard B and Droz M 1998 *Cellular Automata Modeling of Physical Systems* Collection Alea-Saclay: Monographs and Texts in Statistical Physics (Cambridge University Press)
[13] Gandin C A, Desbiolles J L, Rappaz M and Thevoz P 1999 *Metallurgical and Materials Transactions A* **30** 3153–3165
[14] Qian Y H, D'Humières D and Lallemand P 1992 *EPL (Europhysics Letters)* **17** 479
[15] Lallemand P and Luo L S 2000 *ICASE Report* 1–33
[16] Latt J 2007 *Hydrodynamic Limit of Lattice Boltzmann Equations* Ph.D. thesis University of Geneva
[17] Gandin C A and Rappaz M 1997 *Acta Materialia* **45** 2187–2195
[18] Wu M and Ludwig A 2009 *Acta Materialia* **57** 5632 – 5644
[19] Wu M and Ludwig A 2009 *Acta Materialia* **57** 5621 – 5631
[20] Lipton J, Glicksman M and Kurz W 1984 *Materials Science and Engineering* **65** 57 – 63
[21] Dantzig J and Rappaz M 2016 *Solidification: 2nd Edition* Engineering Sciences (EPFL Press)
[22] Wang C Y and Beckermann C 1993 *Metallurgical and Materials Transactions A* **24** 2787–2802
[23] Rai A, Markl M and Körner C 2016 *Computational Materials Science* **124** 37–48

IOP Conf. Series: Materials Science and Engineering **529** (2019) 012029 doi:10.1088/1757-899X/529/1/012029

Finite diffusion microsegregation model applied to multicomponent alloys

V. Maguin[1,2], G. Guillemot[1], V. Jaquet[2], N. T. Niane[2], L. Rougier[2], D. Daloz[3], J. Zollinger[3] and Ch.-A. Gandin[1]

[1] MINES ParisTech, PSL Research University, CEMEF – Centre de mise en forme des matériaux, CNRS UMR 7635, CS 10207, 1 rue Clause Daunesse, F-06904 Sophia Antipolis Cedex, France

[2] Safran, site de Paris-Gennevilliers, 171 Boulevard de Valmy, F-92700 Colombes, France

[3] Université de Lorraine, Institut Jean Lamour, CNRS UMR 7195, F-54011 Nancy, France

E-mail: `vincent.maguin@mines-paristech.fr`, `vincent.maguin@safrangroup.com`

Abstract. During casting, diffusion phenomena are often simplified by Lever-Rule or Gulliver-Scheil hypotheses, thus simplifying chemical diffusion to extreme configurations. The present work proposes an extension of the Tong-Beckermann microsegregation model to multi-component alloys while considering finite diffusion in liquid and solid phases plus the effect of tip undercooling of the columnar front. The behaviour of this model is studied according the solidification conditions (growth velocity and thermal gradient) and comparison with microprobe measurements is proposed for a seven components nickel-base superalloy.

1. Introduction

Upon solidification of metallic alloys, complex segregation and diffusion phenomena occur during the growth of the dendritic microstructure. Several microsegregation models can be used to determine the solidification path. The Lever-Rule (LR) solidification path assumes full equilibrium in the liquid and solid phases, equivalent to complete diffusion of all chemical species. The solid fraction is evaluated according the tie line for the nominal composition. However, diffusion in solid and liquid phases differs by several orders of magnitude. Typical cooling rates during casting do not allow reaching the thermodynamic equilibrium. The Gulliver-Scheil (GS) solidification path only considers thermodynamic equilibrium between a small portion of solid at the interface with liquid phase, thus incrementally accumulating solid with various compositions as no solid diffusion. Both Lever-Rule and Gulliver-Scheil models are approximations mainly because they do not consider the microstructure undercooling and finite diffusion in all phases. Brody and Flemings [1] and Clyne and Kurz [2] proposed models is accounted for finite (and non-zero) diffusion in the solid phase. Wang and Beckerman [3] simplified the dendritic network to three domains (1) solid, (2) interdendritic liquid and (3) extradendritic liquid. Tong and Beckerman (TB) [4] proposed a modification where the Fourier number in the liquid can be estimated thanks to a function of the dendrite tip supersaturation. Their model was applied to columnar solidification without diffusion in the solid phase for a binary Ag - 15 wt.% Cu alloy. Linearisation of the phase diagram was assumed.

In the present work, the equations of the TB model are written for whatever multicomponent

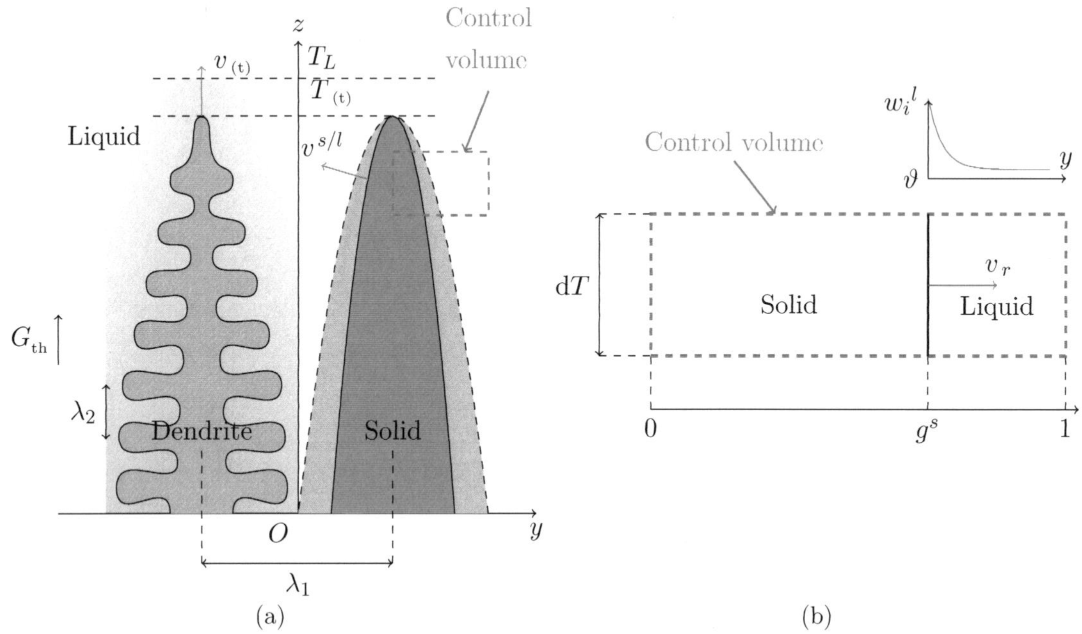

Figure 1: Schematics of (a, left) a columnar dendrite and (a, right) its simplification with (b) the control volume for mass balances. (Inspired from [4])

alloy including diffusion in solid phase. This microsegregation model is directly coupled with a dendrite tip model which considers the solidification conditions.

2. General equations

The present model describes the growth of a columnar dendritic network in the same direction as the temperature gradient, G_{th}, with a primary arm spacing, λ_1, as schematized in fig. 1a. The dendrite tips grow with velocity $v_{(\text{t})}$ at temperature $T_{(\text{t})}$. The complex shape of the primary trunk (fig. 1a, left) is simplified by a dendrite envelope (fig. 1a, right). The equations are written for a control volume (fig. 1b) defined by the green rectangle in fig. 1a, *i.e.* in which solidification proceeds directionally over a distance $\lambda_1/2$ at a temperature evolving with time according to $T = T_{(\text{t})} + G_{\text{th}} \cdot v_{(\text{t})} \cdot t$. Velocity $v^{s/l}$ is then approximated bye a radial contribution v_r.

2.1. Dimensionless quantities

This model is written with a dimensionless form. Chemical diffusion is described by Fourier numbers for each chemical specie i according the ratios:

$$F_i^l = \frac{D_i^l\, t_s}{(\lambda_1/2)^2} \quad \text{(1a)} \qquad\qquad F_i^s = \frac{D_i^s\, t_s}{(\lambda_2/2)^2} \quad \text{(1b)}$$

where D_i^l and D_i^s are the chemical diffusion coefficients for the liquid and solid phases, respectively. The model associates the Fourier number in the liquid phase with the primary dendrite arm spacing, λ_1, and the Fourier number in the solid phase with the secondary dendrite arm, λ_2. These two different characteristic lengths were already used by Wang and Beckermann [3] for a dendritic microstructure. The solidification time, t_s, is firstly estimated according a Lever-Rule solidification interval ΔT_0 (taking into account the possible eutectic temperature) and the cooling rate $\dot{T}$, $t_s = \Delta T_0 / \dot{T}$. However another Fourier number $\widetilde{F_i^l}$ is defined for the liquid phase as:

$$\widetilde{F_i^l} = \tilde{\sigma}_i F_i^l = \tilde{\sigma}_i \frac{D_i^l \, t_s}{(\lambda_1/2)^2}. \tag{1c}$$

The estimation of this modified Fourier number is determined later according to the working point of the dendrite tip. The constant $\tilde{\sigma}_i$ is evaluated for each chemical specie for a given value of the primary dentrite arm spacing.
For the dendrite tip the chemical Péclet number, Pe_{c_i}, is defined for each solute i as:

$$Pe_{c_i} = \frac{r_{\,(\mathrm{t})} \; v_{\,(\mathrm{t})}}{2 \; D_i^l}, \tag{2}$$

with $r_{\,(\mathrm{t})}$ and $v_{\,(\mathrm{t})}$ the radius and the growth velocity of the dendrite tip, respectively. It is worth noticing that a stationary growth regime is assumed so $v_{\,(\mathrm{t})} = v_L$, where v_L is the velocity of the liquidus isotherm T_L.

2.2. Solute balances

Three mass balances for each solute i describe the system for liquid and solid phases and for the control volume [4]. In the liquid phase, figure 1b gives a schematic representation of the concentration profile in the lateral growth direction $\overrightarrow{Oy}$ by assuming no diffusion of solute in the main solidification direction $\overrightarrow{Oz}$. This profile is expressed as :

$$\frac{{w_i}^l - w_{i0}}{w_i^{l(s/l)} - w_{i0}} = \exp\left(- \frac{1}{\widetilde{F_i^l} g^s} \frac{y}{\lambda_1} \right), \tag{3}$$

where ${w_i}^l$ represents the liquid composition for a y position in the liquid phase, w_{i0} is the nominal composition of the alloy and $w_i^{l(s/l)}$ is the composition of liquid at the solid/liquid interface. By integration of eq. (3) in the liquid phase, the mean composition $\langle w_i \rangle^l$ for each solute i takes the form:

$$\frac{\langle w_i \rangle^l - w_{i0}}{w_i^{l(s/l)} - w_{i0}} = \frac{2\widetilde{F_i^l} g^s}{(1 - g^s)} \left[1 - \mathcal{F}_i \left(g^s \right) \right] \tag{4}$$

with for each solute: $\mathcal{F}_i\left(g^s\right) = \exp\left(- \left(1 - g^s\right) / \left(2\widetilde{F_i^l} g^s\right) \right)$.
For the solid phase the solute mass balance in the solid phase for each solute i is written [3]:

$$g^s \frac{\mathrm{d} \langle w_i \rangle^s}{\mathrm{d} g^s} = \left(1 + 6F_i^s\right) \left(w_i^{s(s/l)} - \langle w_i \rangle^s \right). \tag{5}$$

The last solute balance for each solute i is expressed for the whole control volume:

$$g^s \langle w_i \rangle^s + g^l \langle w_i \rangle^l = w_{i0} \qquad \text{with} \qquad g^s + g^l = 1, \tag{6}$$

2.3. Phase diagram for multicomponent alloy

The present microsegregation model is coupled with a phase diagram. For a working point $(T, w_i^{l(s/l)})$ the liquidus surface follows the relationship:

$$T = T_{L0} + \sum_i m_{Li} \left(w_i^{l(s/l)} - w_{i0} \right) \tag{7}$$

with T_{L0} is the liquidus temperature for a given reference composition w_{i0}. The solid interfacial composition $w_i^{s(s/l)}$ is defined by the tie line relation according to the segregation coefficient k_i:

$$k_i = \frac{w_i^{s(s/l)}}{w_i^{l(s/l)}}, \tag{8}$$

3. Numerical Solution

3.1. General solution

This section presents the solution methods in order to solve the equations of this microsegregation model. With combination of equations (4), (5) and (6), a single equation is obtained for each solute i:

$$\mathcal{G}_{1_i}(g^s)\frac{\mathrm{d}\, w_i^{l(s/l)}}{\mathrm{d}g^s} = \mathcal{G}_{2_i}\left(g^s, w_i^{l(s/l)}, w_i^{s(s/l)}\right), \tag{9}$$

with

$$\left\{\begin{array}{rcl} \mathcal{G}_{1_i}(g^s) & = & 2\widetilde{F_i^l} g^s (1-\mathcal{F}_i(g^s)) \\ \mathcal{G}_{2_i}\left(g^s, w_i^{l(s/l)}, w_i^{s(s/l)}\right) & = & (1+6F_i^s)\left(w_{i0} - w_i^{s(s/l)}\right) \\ & + & \left(w_i^{l(s/l)} - w_{i0}\right)\left(\dfrac{\mathcal{F}_i(g^s)}{g^s} - 2\widetilde{F_i^l}(1+6F_i^s)(1-\mathcal{F}_i(g^s))\right) \end{array}\right. \tag{10}$$

By the Newton integration method of this 1st order differential equation, the increments of the interfacial compositions in liquid phase $w_i^{l(s/l)}$ are estimated for a unique solid fraction step. The new interfacial composition in the solid $w_i^{s(s/l)}$ is deduced from the tie line equation (7). The temperature is evaluated *a posteriori* using the phase diagram and relation (7).

3.2. Initialisations

The modified Fourier number, $\widetilde{F_i^l}$, has to be evaluated in the microsegregation model. This is done by considering that Equation 9 must be consistant with the operating point of the dendrite tip. In this way the differential equation is evaluated for $g^s = 0$, at the beginning of the solidification. By remarking that $\mathcal{G}_{1_i}$-functions reaches the zero-value and by solving the systems $\lim\limits_{g^s \to 0} \mathcal{G}_{2_i}\left(g^s, w_i^{l(s/l)}, w_i^{s(s/l)}\right) = 0$, one can show that:

$$\widetilde{F_i^l} = \frac{w_{i0} - w_{i(\mathrm{t})}^{s(s/l)}}{2\left(w_{i(\mathrm{t})}^{l(s/l)} - w_{i0}\right)} \quad \text{or} \quad \tilde{\sigma}_i = \frac{1 - \mathrm{Iv}(\mathrm{Iv}(Pe_i))}{2\,\mathrm{Iv}(Pe_i)}, \tag{11}$$

where $w_{i(\mathrm{t})}^{s(s/l)}$ and $w_{i(\mathrm{t})}^{l(s/l)}$ represent respectively the interfacial composition at the tip in solid and liquid phases. The working point of the dentrite tip and also $\left(w_{i(\mathrm{t})}^{s(s/l)}, w_{i(\mathrm{t})}^{s(s/l)}\right)$ quantities are estimated with a KGT dendrite tip model [5] applied to multicomponent alloys [6]:

$$w_{i(\mathrm{t})}^{l(s/l)} = \frac{w_{i0}}{1-(1-k_i)\,\mathrm{Iv}(Pe_{c_i})}, \tag{12}$$

where $\mathrm{Iv}(x)$ represents the Ivantsov function of quantity x.

The tip radius computed using the KGT model is also used for evaluating the first interdendritic dendrite arm spacing according to Kurz's expression [7]:

$$\lambda_1 = \sqrt{\frac{3\Delta T_0\, r_{(\mathrm{t})}}{G_{\mathrm{th}}}} \tag{13}$$

Ni	Cr	Co	Mo	W	Al	Ti	Ta
Base	7.8	6.5	2.0	5.7	5.2	1.1	7.9

Table 1: Nominal composition of AM1 [12]

Case	G_{th} ($\text{K} \cdot \text{cm}^{-1}$)	v_L ($\text{mm} \cdot \text{min}^{-1}$)	Γ ($\text{K} \cdot \text{m}$)	$\propto D_i^s$	λ_1 (μm)	λ_2 (μm)
#T1	40	6	10^{-11}	$\times 1$	196	196
#T2	40	6	10^{-11}	$\times 10^3$	196	196
#E1	40	6	10^{-7}	$\times 1$	196	100
#T3	1	6	10^{-7}	$\times 1$	1240	100
#T4	40	100	10^{-7}	$\times 1$	97	100

Table 2: Cooling configurations for AM1 solidification paths

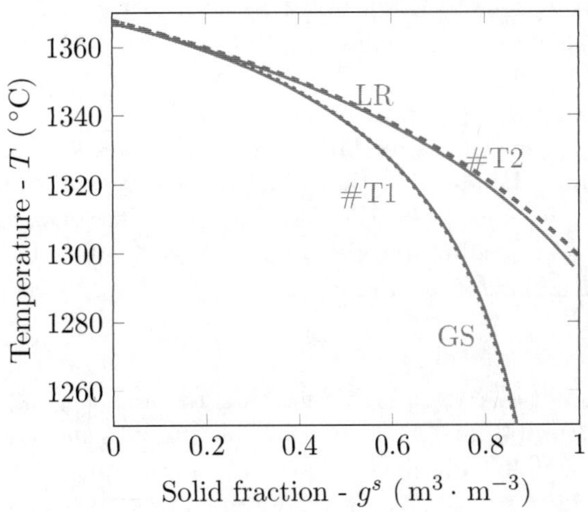

(a)

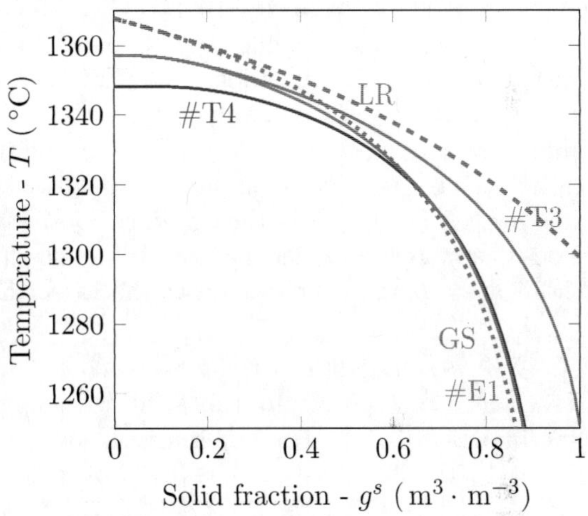

(b)

Figure 2: Solidification path for the AM1 superalloy, (a) validation of the model for cases (—) #T1 and (—) #T2 (b) effect of thermal gradient and growth velocity for cases (—) #E1, (—) #T3 and (—) #T4. For both graphs grey lines represent (- - -) Lever-Rule and (·····) Gulliver-Scheil.

4. Results and discussion

Firstly, a multicomponent nickel based superalloy (AM1), with nominal composition reported in Table 1, is considered. The model is validated for Lever-Rule and Gulliver-Scheil solidification paths. Then the effects of solidification conditions (growth velocity and thermal gradient) are studied. One comparison with experimental data is proposed. All thermodynamic and mobility properties are extracted from the NI25 [8] and MOBNI4 [9] databases. The linearization of the phase diagram and diffusion coefficients for each chemical species were computed with the Thermo-Calc interface.

4.1. Model's behaviour

AM1 is a 1st generation nickel based superalloy, daily used for the production of single crystal turbine blades of jet engines. Several solidification conditions are presented in table 2 in addition

Case	G_{th} ($\text{K}\cdot\text{cm}^{-1}$)	v ($\text{mm}\cdot\text{min}^{-1}$)	λ_1 (μm)	λ_2 (μm)
#E1	40	6	196	100
#E2	40	3	233	100
#E3	40	0.5	364	110

Table 3: Cooling configurations for AM1 solidification experiments

with the primary and secondary dendrite arm spacing.
For the test cases #T1 and #T2, the Gibbs-Thomson coefficient Γ is fixed to an artificially small value, equal to $10^{-11}\,\text{K}\cdot\text{m}$. In this way the tip undercooling is close to zero (fig. 2a). Due to the relative large arm spacing, the Fourier number is very small and thus diffusion in solid phase is limited. The case #T1 is close to the Gulliver-Scheil solidification path. For case #T2, the diffusion coefficients in solid phase are increased by 3 orders of magnitude, leading to a result that approaches the Lever-Rule approximation.
For cases #E1, #T3 and #T4 the Gibbs-Thomson coefficient is fixed to a consistent value of $10^{-7}\,\text{K}\cdot\text{m}$. The secondary arm spacing is now set to a value of 100 μm. Case #E1 is computed with a solidification condition of $G_{\text{th}} = 40\ \text{K}\cdot\text{cm}^{-1}$ and $v_L = 6\ \text{mm}\cdot\text{min}^{-1}$. The tip undercooling reaches a value of 10.7 K. When the temperature gradient is decreased to a value of $G_{\text{th}} = 1\ \text{K}\cdot\text{cm}^{-1}$ (case #T3), the modification on the working point of the tip remains limited. Due the increase of the solidification time the effect of the diffusion in solid is more pronounced. For $g^s > 0.4$ the solidification path comes closer to LR. For case #T4 the growth velocity is increased to $100\ \text{mm}\cdot\text{min}^{-1}$. The tip undercooling reaches 19.6 K. For $g^s > 0.6$ this solidification path is superposed to the case #E1 and close to GS .

4.2. Comparison with experimental results
Case #E1 corresponds to the solidification conditions examined by Zollinger and Daloz's experiments [10, 11]. The experimental points were obtained by micro-probe analyses in four cuts at 1350, 1325, 1300 and 1275 °C of a directionally solidified cylindrical sample quenched during processing. Two other solidification conditions, #E2 and #E3 were investigated as presented in table 3. The evolutions of the solid fraction are compared to experimental points as well as the GS and LR solidification paths in fig. 3 while using $\Gamma = 10^{-7}\ \text{K}\cdot\text{m}$.
Simulations are very different from experimental measurements. These differences can be firstly explained by the model assumptions. The model uses a linearization of the phase diagram at the liquidus temperature. One can evaluate its effect be considering the black line in fig. 3 providing the LR solidification path with full coupling with the CALPHAD database [8]. The corresponding Gulliver-Scheil path shows almost no difference with its counterpart using linearization and thus is not plotted. Direct coupling of the present microsegregation model with thermodynamic equilibrium may improve the results. Because the time cost for CALPHAD equilibrium is heavy, a direct CALPHAD-coupling is not compatible for coupling at macroscopic scale. The thermodynamic computation might be externalized and the tie lines gathered in tabulations [13]. For information purposes, the solidification path computation with the linearization hypothesis for the AM1 alloy needs only few seconds for a solid fraction integration step of 10^{-4}. It makes quite possible to use the present microsegregation model for coupling at macroscopic scale. The phase diagram storage has yet to balance the accuracy and the computation time. The tabulations also prevent convergence difficulties which are encountered for heavy thermodynamic databases during CALPHAD computations.
This model describes a columnar front growth for a constant nominal concentration. Due to relative low cooling rate (from $0.4\,\text{K}\cdot\text{s}^{-1}$ for case #E1 to $0.03\,\text{K}\cdot\text{s}^{-1}$ for case #E3), solute

IOP Conf. Series: Materials Science and Engineering **529** (2019) 012029 doi:10.1088/1757-899X/529/1/012029

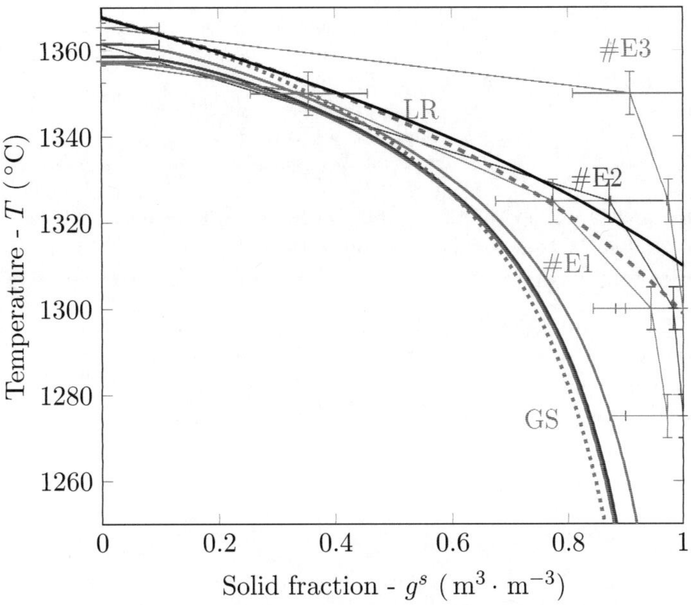

Figure 3: Comparison of solidification paths for the AM1 superalloy between (continuous lines, —) the microsegregation model and (—+—) experimental data. The color is associated to case presented in table 2 (—) #E1, (—) #E2 and (—) #E3 configurations. Grey lines represent (- - -) Lever-Rule and (·····) Gulliver-Scheil. The black line (—) represents Lever-Rule with full CALPHAD coupling.

segregation phenomena through liquid flow is conceivable. This kind of effect cannot be described by a microsegregation model alone. A macrosegregation framework including solute transport and liquid flow is required. Diffusion coefficient are computed for the nominal composition at liquidus temperature and constant in the present model.

A constant λ_2 value for the entire solidification interval. However, λ_2 has been described by Zollinger and Daloz as a function $\propto t_s^{1/3}$. In this model diffusion in the solid is underestimated at the beginning of the solidification path. In addition, measurement report that λ_2 measurement is independent of cooling rate even thought the ratio between the lowest and highest cooling rates is 12. Due to coarsening, the measurement of the secondary arm spacing also is over-evaluated by the *post mortem* analysis. A time-dependant secondary arm spacing could increase the effect of diffusion in the solid phase. More broadly, the confidence of the thermodynamic and diffusion database could also be discussed.

5. Conclusions

The present work is an extension of the Tong-Beckermann's microsegregation model for a multicompoment alloy with consideration of diffusion in solid and liquid phases. Solidification paths similar to LR and GS were obtained by artificially changing the Gibbs-Thomson and diffusion coefficients. The comparison with experimental data gives good prediction of the tip undercooling and coherent evolutions when changing the processing parameters. In order to improve this model, direct coupling with tabulated thermodynamic equilibrium is being conducted.

References

[1] H. D. Brody and M. C. Flemings 1966 *Trans. TMS-AIME* 236 615

IOP Conf. Series: Materials Science and Engineering **529** (2019) 012029 doi:10.1088/1757-899X/529/1/012029

[2] T. W. Clyne and W. Kurz 1981 *Metallurgical Transactions A* 12 **6** 965-71
[3] C.Y. Wang and C. Beckermann 1993 *Materials Science and Engineering* A **171** 199-211
[4] X. Tong and C. Beckermann 1998 *Journal of crystal growth* 2 **187** 289-302
[5] W. Kurz, B. Giovanola and R. Trivedi 1986 *Acta metallurgica* 34 **5** 823-30
[6] W. J. Boettinger 2016 *Journal of phase equilibria and diffusion* 37 **1** 4-18
[7] W. Kurz and D. J. Fisher 1998 *Fundamentals of Solidification, Fourth Revised Edition* ed Trans Tech Publications p255-6
[8] NI25 : Ni-based alloys and superalloy solutions database Thermo-Calc Software AB, 2017
[9] MOBNI4 - TCS Nickel Mobility Database [Version 4.0 : May 2015] 2015
[10] H. Ben Hamouda 2012 *Ecole Nationale Supérieure des Mines de Paris* https://pastel.archives-ouvertes.fr/tel-01202342
[11] J. Zollinger and D. Daloz, 2010 Essais de trempe en cours de solidification sur AM1 : réalisation, expertises métallurgiques, synthèse
[12] P. Caron and T. Khan 1999 it Aerospace Science and Technology 3 **8** 513–523
[13] A. Saad, Ch.-A. Gandin and M. Bellet 2015 *Comp. Mat. Sc.* 99 221-31

IOP Conf. Series: Materials Science and Engineering **529** (2019) 012030 doi:10.1088/1757-899X/529/1/012030

Three-phase numerical modeling for equiaxed solidification of Sn–10 wt.%Pb alloy under forced convection driven by electromagnetic force

T Wang[1,2], E Wang[1], Y Delannoy[2], Y Fautrelle[2] and O Budenkova[2]

[1] Key Laboratory of Electromagnetic Processing of Materials (Ministry of Education), Northeastern University, P. R. China
[2] Univ. Grenoble Alpes, CNRS, Grenoble INP, SIMAP, F-38000 Grenoble, France
tao.wang1@grenoble-inp.fr

Abstract. A three-phase equiaxed solidification model where macroscale heat transfer and fluid flow are coupled with microscale nucleation and dendrite growth, is applied to the simulation of the macrosegregation in binary alloy solidification subjected to the electromagnetic stirring. The investigated experimental solidification case is conducted in a cavity which has a good control of the thermal boundary conditions. The proposed model uses a double time step scheme to accelerate the solution. Electromagnetic force is introduced as a source term into momentum equation in analytical form. To account for the friction from the side walls, a 2D½ flow model is applied to a three-dimensional experimental configuration. A comparison between the results of simulation and experimental ones is made.

1. Introduction

Macrosegregation defects often appear in ingot especially during a long-time solidification. These defects cannot be removed by the subsequent heat treatment or rolling process, and will affect the product properties. Magnetic fields and electromagnetic force have been widely applied to control solute distribution and grain structure [1–3]. It was shown that using the electromagnetic stirring during solidification, a larger equiaxed grain zone and finer grain size can be obtained and macrosegregation can be reduced [4]. However, sometimes heavier macrosegregation will be generated by the strengthened convection if parameters of electromagnetic stirring (EMS) are not appropriately chosen [3]. It is of high significance to have deep understanding of the mechanism how EMS affects the solute distribution during solidification that can be obtained through dedicated experiment and numerical modelling.

Volume/Statistical average method is a probable way to anticipate segregation formation since it can relate macroscopic variables to the microscopic ones. Various volume-averaged models have been developed aiming predict macrosegregation and solidification structure, including continuum model [5,6], columnar model [7], equiaxed model [8–12] and mixed model [13–16]. In each of them, phases attribution and growth kinetic vary based on different physical assumption.

In order to provide an experimental basis for the validation of numerical simulation, SIMAP/EPM laboratory in France designed a benchmark solidification setup, referred hereafter AFRODITE. The main characteristics of this experiment setup are: (1) controlled thermal boundary conditions (2) real-time whole-field temperature measurements (3) possibility to realize various solidification condition and (4) possibility to use different modes of EMS during solidification. During solidification, a temperature map is recorded by a lattice of 50 thermocouples. After solidification, the post-mortem analysis of the sample is made regarding the segregation pattern (with X-ray and/or using probes); the grain structures and dendrite parameters. Several experiments have been performed already. First, Wang and Fautrelle [17] carried out solidification experiment on pure tin with natural convection. Then, Hachani and co-workers studied the effect of different initial concentration in solidification of a tin-led alloy with natural convection [18]. The latest results obtained with the AFRODITE set-up were related to the effect of the various modes of the electromagnetic stirring during the solidification of a tin-lead alloy on the segregation and final structure of ingot [19]. Numerical calculations were performed for AFRODITE solidification without EMS using volume and statistical average method in two-dimensional and three-dimensional configurations, using columnar solidification model [20] or mixed columnar-equiaxed solidification model [21]. It was shown that the solidification phenomena have been well repeated through volume averaged method, although some boundary conditions and crucial parameters should be further confirmed.

The objective of current work is to simulate the influence of EMS on the AFRODITE benchmark solidification through a new volume averaged equiaxed solidification model, and compare numerical results with experimental ones. This work focus on the grains' growth and phases' movement and formation of macrosegregation, we do not consider the dendrite break-off phenomenon although it could happen under strong convection.

2. Experimental setup

The concerning experiment is a solidification of Sn–10 wt.%Pb alloy in a rectangular cavity with inside size of 100 × 60 × 10 mm. The temperature of left and right boundary are controlled to keep a difference of 40 K between the thermocouples placed outside of the liquid phase at the heaters, and to cool them down with same rate which for the case under consideration was CR = 0.03 K/s [20], other walls can be regarded as non-slip and adiabatic. In such configuration a natural thermal convection develops in the liquid which drives the latter upward along the hot wall, downward along the cold wall, and consequently, from the hot wall toward the cold wall along the top of the cavity and in the opposite direction along the bottom [20]. To introduce the EMS in the liquid phase, a travelling magnetic field was created with a linear motor placed horizontally along the cavity and below it. The Lorentz force acts in the liquid mostly along the bottom and decreases exponentially in the vertical direction. An analytical expression for the resulting Lorentz force can be found elsewhere [22] and was used in the numerical simulations. Hereafter we consider the experiment in which the Lorentz force created the motion of the melt opposite to the one generated by the buoyancy force, the frequency and the intensity of the coil current was $f = 50$ Hz and I = 8.2 A, respectively.

More details about the experiment and related results can been found in the reference [19] for the experimental case number 3. In the work, most material properties are taken from the reference [23] and some information is not provided here for the sake of communication brevity. However, liquid/solid density and electrical conductivity that taken elsewhere are provided in Table 1 with the indication of the source in the literature.

Table 1. Some material properties Sn–10 wt.%Pb alloy.

	Units		Ref.
Reference density(liquid)	kg m^{-3}	7246	[24]
Solid density	kg m^{-3}	7199	[24]
Electrical conductivity (liquid)	$Ohm^{-1}cm^{-1}$	2×10^4	[25]
Electrical conductivity (solid)	$Ohm^{-1}cm^{-1}$	2.5×10^4	[25]

3. Model description and numerical approach

In the numerical simulations, only the equiaxed model is employed, because the post-mortem analysis of the sample solidified under described conditions revealed mostly equiaxed grain structure (see figure 10c in [19]). Similar to [26], [14] and [27], in our model we distinguish three phases : the liquid outside the envelope of equiaxed dendrite (l-phase), the interdendritic liquid (d-phase) and the solid dendrite (s-phase) that gives the corresponding fractions $f_d + f_s + f_l = 1$. Lipton–Glicksman–Kurz (LGK) tip growth kinetics is applied to estimate the evolution of the grain envelope. The solidification of interdendritic melt is driven by the difference between equilibrium liquid concentration and interdendritic liquid concentration. The solute diffusion equation around the growing grains' envelope is calculated according to [12], in which the diffusion length l_{ld} considered the influence of growth status, grain's size and fluid flow as:

$$l_{ld} = \min\left(\frac{D_l}{v_{en}}, \frac{d_{eq}}{Sh}\right)$$

Where v_{en} is envelope growth velocity, d_{eq} is the average grains' diameter, Sh is Sherwood number accounting for the effect of fluid flow on grains's movement.

Similar to [8], grain shape factors $\Phi_{env} = 0.6827$ and sphericity of grain envelope $\Phi_{sph} = 0.283$ are applied to correlates the growth velocity and surface area concentration of volume equivalent sphere respectively.

Considering that temperature varies in the cavity, a simple melting model [28] was included. Other details about the grain growth model and macroscopic conservation omitted here because of brevity are similar to the ones in [27].

In terms of momentum transport, s-phase and d-phase are treated as union, which is literally the dendritic structure of an equiaxed grain (e-phase), whose fraction is $f_e = f_d + f_s$, respectively. This means also that the interdendritic liquid and solid dendrite share the same velocity field while the velocity of the extradendrite liquid is different. The momentum exchange between the grain and the liquid phase is accounted for with a drag model proposed in [29] (equation (26) and Table I) The flow partitioning effect between the interdendritic and extradendritic liquid is neglected for simplicity. When $f_e \geq 0.637$, the grain phase is packed, i.e. it becomes immovable while the permeability of the grain fraction is changed and is calculated with Carman-Kozeny equation based on a constant characteristic length number related to grains' density. This characteristic length number is given as 2×10^{-4} m, according to [22]. Nucleation happens in extradendritic liquid immediately when the alloy temperature is lower than liquidus temperature. Nuclei density is defined as a scalar and move with grain phase.

In our approach the processes which occur at macroscopic and microscopic scale are modelled with different time steps. A larger time step is used for the calculation of the transport phenomena at the macro-sale and a smaller time sub-steps is used for the calculation of the processes within the grain (diffusion, solidification) and the grain growth. The macroscopic calculation, including multiphase flow field and transport of energy, solute and nuclei density, are solved by ANSYS FLUENT® software in which the modeling of the solidification at the microscale is implemented with user-defined functions.

4. Results and discussion

The AFRODITE experiment is supposed to be a quasi-2D benchmark although the thickness of the cavity is only 10mm that affects strongly the flow. Yet, the two-dimensional simulation of the experiment is still possible if an additional source term is introduced into the momentum equations to take into account the frictions from the large lateral walls which are absent in two-dimensional problem statement [30]. In the present simulation this term was implemented in the momentum equation for the liquid phase as well as the analytical expression for the Lorentz force. Similar to the previous simulations, it is supposed that the temperature difference between the cold and the hot boundaries is less than those seen on the external thermocouples because of a thermal resistance. Based on the temperature distributions in liquid [21] we estimated this difference around 15 degrees which was used in the simulations.

The preliminary simulations of the melt flow performed with laminar approach did not converge. It was found that the motion caused by the EMS force overcomes the buoyancy convection at the bottom and near the cold wall of the cavity. However, the upward flow near the hot wall remains and interaction of two vortices near the cavity center causes flow instabilities. Consequently, for further simulations a RNG k-e turbulent model was used. It should be noted that the calculated turbulent viscosity was rather weak.

As shown in table 1, for Sn–10 wt.%Pb alloy electrical conductivity of the solid is larger than the liquid one [25], therefore, the Lorentz force acting on grain phase has to be taken into account proportionally to the solid fraction. At the beginning of solidification, the Lorentz force acting on grains does not have crucial effect because solid fraction is still low. This force will be important when the grain phase is going to be packed. The force on solid could strength the melt flow via drag force and retard the grain packing phenomenon.

The temperature distributions after 50s of solidification (figure 1b) is quite similar to the one observed in the experiment and two vortices described above exist in the liquid. The maximum liquid velocity is 3.48 cm/s which corresponds to experimental estimations [19] and to the Reynolds number Re=4307, i.e. to a transition regime or a weakly turbulent flow. At this stage the grain phase prefers to nuclei and grow first at upper right corner, where has lowest temperature in cavity because the liquid flow upwards at this cold side. Some grains remain at the corner and further grow up while others flow into bulk liquid and then remelt gradually.

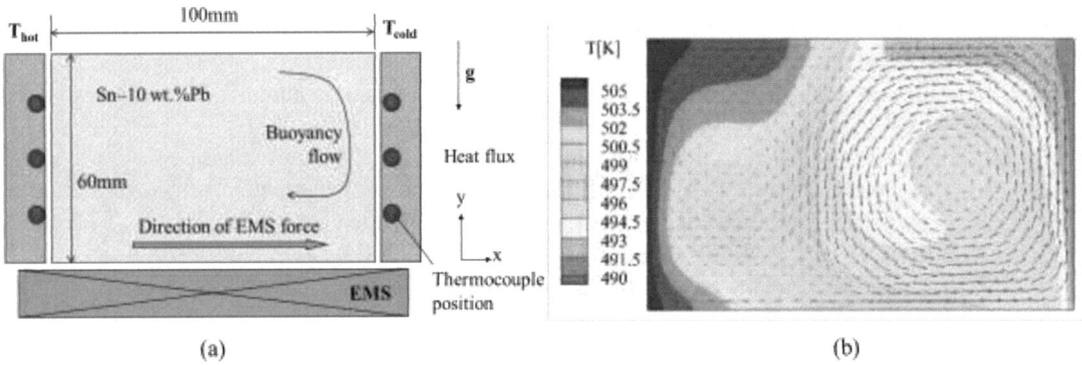

Figure 1. Experiment schedule (a); Temperature distribution and averaged velocity distribution at 200s ($\left|\langle\vec{v}_{average}\rangle\right|_{max}$ = 34.8 mm/s) (b).

At 540s, two vortexes with different direction still persist in the cavity, yet, the vortex generated by EMS force is located only at a lower part of the cavity. Another vortex forms because of the thermal-solute effect, which become pronounced with rejection of Pb during solidification. These two flows collide near the center of hot wall and then separate. Both vortices bring some unmelted grains back to the cold region and attend solidification again. However, another factor which is important to the grain distribution is the density. According to the chosen properties, the density of the solid phase is constant and larger than the density of the liquid if the concentration of Pb in liquid is less than about 8.7wt%, depending also on temperature. The upper flow will take the grains to the top region, some grains can remain at top because they are lighter than liquid thus forms negative segregation region. At grain faction $f_e = 0.637$, the grain phase velocity vanish while liquid flow become weak in this region due to rapidly increasing flow resistance. The packed grains layer which is poorer in Pb first form at right wall then gradually extend leftwards. Because of the EMS force driving flow, there is a thicker equiaxed grain layer at top rather than bottom.

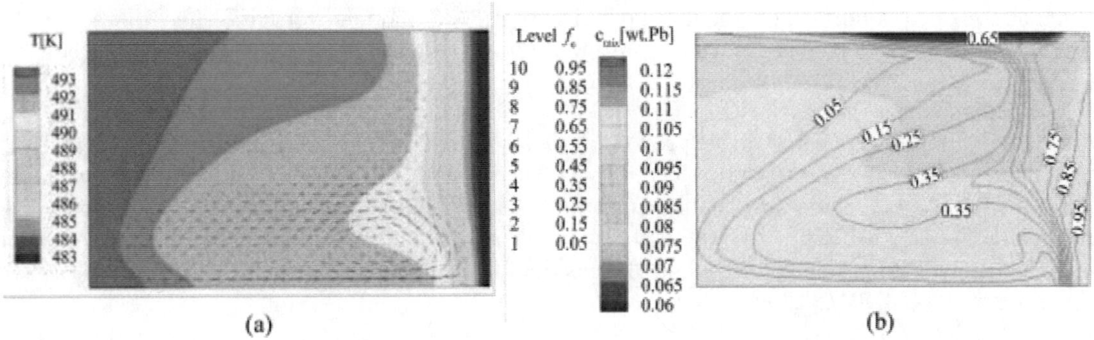

Figure 2. Temperature distribution and averaged velocity distribution at 540s ($\left|\langle\vec{v}_{average}\rangle\right|_{max}$ = 24.1 mm/s) (a); concentration and grain fraction distribution at 540s (b).

After 900s, almost whole the cavity except the bottom layer is filled by the packed grain and become rigid. Minor flow still exists at bottom layer because the liquid is highly enriched and will solidify later when the temperature becomes lower than local liquidus temperature, or eutectic temperature. The maximum value of liquid velocity in the layer is only 1.41 mm/s, and has negligible influence on concentration transport.

Final macrosegregation map is shown in figure 3. A positive segregation zone forms from the central bottom and seems flow towards upper rightwards, which is quite similar to the experimental results (see figure 12c in [19]). This happens because the flow driven by EMS brings enriched liquid rightwards at the bottom of cavity. The enriched flow turns upwards when being blocked by packed grains layer at right side. A slightly negative segregated layer at the top is also predicted, and is formed because of two reasons. Firstly, some unmelted grains stay at the top because they have lighter density than liquid, they will continue to grow if temperature is lower than liquidus temperature. Secondly, Pb rejected from solid during solidification will enrich liquid phase thus give it higher density. The enriched liquid tends to flow downwards and will be replenish by fresh alloy from bulk liquid. However, the segregation channel at right lower corner is not repeated maybe it involves the columnar grain structure, which are not included in current research.

Figure 3. Macrosegregation map obtained from calculation at 1440s.

5. Summary

A 3-phase equiaxed solidification model is applied in predicting the AFRODITE experiment for Sn–10 wt.%Pb alloy under the forced convective flow driven by EMS. The grain fraction distribution, liquid flow field and macrosegregation map have been presented and analyzed. The result shows that the applied EMS force can change the flow motion at bottom of cavity. Strong EMS driving flow bring more unpacked grains from solidification frontier to the bulk liquid. The stirring has effect of enlarging the isothermal region. Calculated macrosegregation map shows similar negative segregated layer at top and positive segregation zone at bottom as experimental one. Future work corresponds to develop mixed columnar/equiaxed solidification model under different EMS modules. Additionally, the complete model can be applied to predict grain structure and macrosegregation in industrial casting processes.

Acknowledgements
This work is a joint cooperation between SIMAP laboratory of Grenoble INP (France) and Key Laboratory of EPM of Northeastern University (P. R. China). The authors gratefully acknowledge financial support from China Scholarship Council (No. 201706080074), National Key R&D Program of China (Grant No. 2017YFE0107900) and National Nature Science Foundation of China (Grant No. U1760206). The SIMAP laboratory acknowledges the financial support through ESA-MAP MICAST project contract 14347/01/NL/SH.

References
[1] Spitzer K-H, Reiter G and Schwerdtfeger K 1996 ISIJ Int. 36 487–92
[2] Tzavaras A A and Brody H D 1984 JOM J. Miner. Met. Mater. Soc. 36 31–7
[3] Medina M, Du Terrail Y, Durand F and Fautrelle Y 2004 Metall. Mater. Trans. B Process Metall. Mater. Process. Sci. 35 743–54
[4] Campanella T, Charbon C and Rappaz M 2004 Metall. Mater. Trans. A Phys. Metall. Mater. Sci. 35 A 3201–10
[5] Plotkowski A and Krane M J M 2016 Comput. Mater. Sci. 124 238–48
[6] Bennon W D and Incropera F P 1988 Numer. Heat Transf. 13 277–96
[7] Bellet M, Combeau H, Fautrelle Y, Gobin D, Rady M, Arquis E, Budenkova O, Dussoubs B, Duterrail Y, Kumar A, Gandin C A, Goyeau B, Mosbah S and Založnik M 2009 Int. J. Therm. Sci. 48 2013–6
[8] Wu M and Ludwig A 2009 Acta Mater. 57 5621–31
[9] Ni J and Beckermann C 1993 J. Mater. Process. Manuf. Sci. 2 217–31
[10] Založnik M and Combeau H 2010 Comput. Mater. Sci. 48 1–10
[11] Wang C Y and Beckermann C 1996 Metall. Mater. Trans. A 27 2754–64
[12] Appolaire B, Combeau H and Lesoult G 2008 Mater. Sci. Eng. A 487 33–45
[13] Wu M, Fjeld A and Ludwig A 2010 Comput. Mater. Sci. 50 32–42
[14] Ciobanas A I and Fautrelle Y 2007 J. Phys. D. Appl. Phys. 40 3733–62
[15] Ciobanas A I and Fautrelle Y 2007 J. Phys. D. Appl. Phys. 40 4310
[16] Silva J N, Moutinho D J, Moreira A L, Ferreira I L and Rocha O L 2009 J. Alloys Compd. 478 358–66
[17] Wang X and Fautrelle Y 2009 Int. J. Heat Mass Transf. 52 5624–33
[18] Hachani L, Zaidat K, Saadi B, Wang X D and Fautrelle Y 2015 Int. J. Therm. Sci. 91 34–48
[19] Hachani L, Zaidat K and Fautrelle Y 2015 Int. J. Heat Mass Transf. 85 438–54
[20] Boussaa R, Hachani L, Budenkova O, Botton V, Henry D, Zaidat K, Ben Hadid H and Fautrelle Y 2016 Int. J. Heat Mass Transf. 100 680–90
[21] Zheng Y, Wu M, Karimi-Sibaki E, Kharicha A and Ludwig A 2018 Int. J. Heat Mass Transf. 122 939–53
[22] Wang X, Moreau R, Etay J and Fautrelle Y 2009 Metall. Mater. Trans. B 40 104–13
[23] Založnik M, Kumar A and Combeau H 2010 Comput. Mater. Sci. 48 11–21
[24] Stankus S V. and Khairulin R A 2006 High Temp. 44 389–95
[25] Plevachuk Y, Sklyarchuk V, Yakymovych A, Willers B and Eckert S 2005 J. Alloys Compd. 394 63–8
[26] Wang C Y and Beckermann C 1993 Metall. Trans. A 24 2787–802
[27] Wu M and Ludwig A 2009 Acta Mater. 57 5632–44
[28] Ludwig A and Wu M 2002 Metall. Mater. Trans. A 33 3673–83
[29] Wang C Y, Ahuja S, Beckermann C and de Groh H C 1995 Metall. Mater. Trans. B 26 111–9
[30] Botton V, Boussaa R, Debacque R, Hachani L, Zaidat K, Ben Hadid H, Fautrelle Y and Henry D 2013 Int. J. Therm. Sci. 71 53–60

IOP Conf. Series: Materials Science and Engineering **529** (2019) 012031 doi:10.1088/1757-899X/529/1/012031

A 3D discrete-element model for simulating liquid feeding during dendritic solidification of steel

Yi Feng[1], Miha Založnik[2], Brian Thomas[3], André Phillion[1]
[1]Department of Materials Science and Engineering, McMaster University, Hamilton, Canada
[2]Université de Lorraine, CNRS, IJL, F-54000 Nancy, France
[3]Department of Mechanical Engineering, Colorado School of Mines, Golden, USA
E-mail: andre.phillion@mcmaster.ca

Abstract. A 3D meso-scale discrete-element model has been developed to simulate fluid flow during dendritic solidification of steel. The model domain is a representative volume element consisting of a set of equiaxed dendritic grain envelopes along with extra-dendritic liquid channels, where the final grain shape is given by a Voronoi tessellation. Solidification of each grain is simulated via a volume average approach. The output of the solidification simulation at a given solid fraction is used as the input mesh for the fluid flow simulation. A single domain Darcy-Brinkman model is used to calculate the pressure field within the liquid channels, with Poiseuille flow assumed to occur in the extra-dendritic region, and Darcy flow assumed to occur within the dendrite envelope. Mass conservation over each element is then used to derive a flow equation that is solved via the finite element method. The results of this new model are first compared with a previously-developed granular model [1] where fluid flow only occurs between the grains, and then compared with different forms of the Carman-Kozeny equation. It is shown that the intra-dendritic liquid fluid flow plays a major role in the semi-solid pressure field, and thus needs to be included when investigating hot tearing susceptibility in engineering alloys undergoing dendritic solidification.

1. Introduction

Crack formation during solidification is a serious problem in many casting processes including the continuous casting of steel slabs. During the casting process, the volume change of the solidifying metal due to the liquid to solid phase transformation as well as deformation of the semisolid mush needs to be compensated by the convection of liquid metal through the dendritic network [2]. The friction on the liquid when going through the dendritic channels leads to a pressure drop that can form a hot tear if the loss of pressure is too great [2, 3]. Transverse surface cracks observed in the cast slab section may start with the occurrence of hot tears near the meniscus [4], which are often found along the boundaries of "blown grains" [5]. After they initiate near the meniscus region in the mold, these cracks grow lower in the caster when the temperature drops into the lower-temperature ductility trough at ~700-900 °C, especially on the inner radius during unbending, where strain is large. Hot tearing is an important initial step in the formation of these detrimental cracks, which deserves more study.

Whether the microstructure is dendritic, globular, or cellular, understanding fluid flow within liquid channels of a semisolid is critical for predicting hot tearing formation. During steel solidification, the microstructure evolves from equiaxed cellular to columnar cellular to columnar dendritic during the initial solidification. Recently a set of discrete-element or granular models of fluid flow [1, 6, 7] were proposed to investigate flow localization at the grain boundaries as part of multi-physics models of hot tearing. These models assumed that the microstructure consisted of equiaxed-globular grains, considered only fluid flow between the grains, and modeled the fluid as Poiseuille flow. For the dendritic structure found in steel, fluid flow occurs both in the extra-dendritic region between grains as well as intra-dendritic zone. The friction encountered when fluid flow occurs through dendrites may accelerate hot tear formation.

In this present work, a 3D discrete-element model is presented to simulate fluid flow within the equiaxed dendritic structure of steel. The severe temperature gradient near the meniscus, and the transformation from equiaxed grains at the surface to columnar grains are neglected. The steel selected for analysis is an alloy steel having an equivalent carbon content of 0.07 wt.%. This is a low carbon alloy that solidifies between 1532-1503 °C. The concept of equivalent carbon simplifies the solidification simulation from a five element problem to a binary one, reducing significantly computational cost. First, the methodology for creating semi-solid equiaxed-dendritic grains is reviewed along with the corresponding solidification model. Second, the new dendritic liquid feeding model is described. Third, the pressure contours and velocity within the liquid channel predicted by the new dendritic fluid flow model are compared with the previous model for equiaxed-globular grains. Finally, the results are discussed in the context of hot tearing.

2. Meso-Scale Solidification Model for Dendritic Alloys

As a first step in modeling the initial solidification of steel, a meso-scale model domain of 1000 grains was chosen, shown in Fig. 1(a). The empty space within the domain represents the liquid outside the envelope. To create this geometry, a Voronoi tessellation is applied to approximate the final grain morphology based on randomly-placed seeds acting as nuclei for equiaxed-dendritic grains. Each grain, Fig. 1(b), is made up of polyhedral structures, Fig. 1(c), that are further divided into tetrahedral elements.

Given that the microstructure of steels is dendritic, the evolution in solid fraction in the interior of the envelope within each tetrahedral element is predicted using a volume average approach [8]. The main idea of this approach is that each semisolid grain contains not two phases (solid and liquid) but instead three phases, delta-ferrite g_δ, intra-dendritic liquid g_l^{id}, and extra-dendritic liquid g_l^{ed}. The sum of the volume fraction of each phase is equal to unity, $g_\delta + g_l^{id} + g_l^{ed} = 1$. As proposed by Rappaz and Thevoz [9] the dendrite envelope represents the region enclosed by linking the dendrite tips, and thus the intra-dendritic liquid consists of all the liquid between the dendrite arms. The volume fraction of the dendrite envelope is the sum of the solid and the intra-dendritic liquid phases, $g_g = g_\delta + g_l^{id}$. Temperature within each grain is assumed to be uniform and the evolution of each phase is predicted by solving explicitly the master equations given in Ref. [8]. Note that solidification within each tetrahedral element is simulated independently, and solute exchange between the neighbouring elements is neglected. All of the tetrahedral elements are then combined to create the semisolid geometry shown in Fig. 1(a) which will be used for the fluid flow calculation.

3. Meso-scale Fluid Flow Model for Dendritic Alloys

Fluid flow within a semisolid occurs between all the grains that have not yet coalesced. In the case of equiaxed-dendritic structures, flow occurs both in the extra-dendritic and intra-dendritic regions as can be seen in Fig. 1(d). This fluid velocity through this composite can be described by the average form of the Navier-Stokes momentum equation over a small control volume V that is large enough to smooth the morphological complexities but small enough to capture the global transport properties. Detailed derivations of the master equation and information regarding the averaging concepts can be found in Ref. [10].

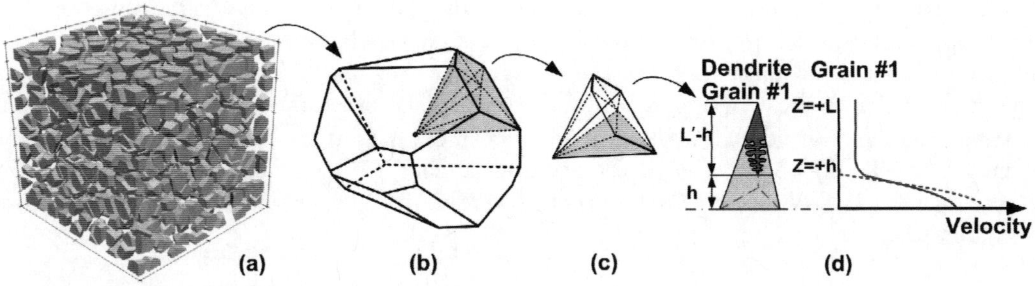

Figure 1. (a) Meso-scale simulation domain with 1000 grains; (b) Single Voronoi grain; (c) Polyhedral structure; (d) Tetrahedral element with dendrite, and the corresponding fluid velocity profile (blue line). In (d), the fluid velocity profile from the prior equiaxed-globular case (red line) is also shown for comparison purposes.

In the extra-dendritic region, fluid flow is similar to the case of equiaxed-globular gains, however, the no-slip boundary condition at the dendrite envelope interface (when z=+h in Fig. 1(d)) is no longer valid. Assuming quasi-steady-state, the flow is described as Poisuille flow,

$$-\nabla p + \mu_l \frac{d^2 \vec{v}^{ed}}{dz^2} = 0, \text{and } - g_l' \nabla p + \mu_l \frac{d^2 \vec{v}^{id}}{dz^2} - \frac{\mu_l g_l'}{K\left(g_l'\right)} \vec{v}^{id} = 0, \tag{1}$$

where μ_l represents the dynamic viscosity, $\vec{v}^{ed}$ is the fluid velocity in the extra-dendritic region, p is the pressure. g_l' represents the local volume fraction of liquid, $\vec{v}^{id}$ is the fluid velocity vector within the dendrite envelope and $K\left(g_l'\right)$ represents the individual hydrodynamic permeability as a function of liquid fraction. The Carman-Kozeny equation [11] is used to determine the permeability within each tetrahedral element. Assuming that $\frac{\partial}{\partial z}\vec{v}|_{z=0} = 0$ and $\vec{v}$ is finite when $z \to \infty$, Eq. 1 can be solved analytically,

$$\vec{v}^{ed} = \left(\frac{z^2}{2\mu_l} + C_1\right) \nabla p, \text{and } \vec{v}^{id} = \left(C_2 e^{\frac{z}{\delta}} - C_3\right) \nabla p. \tag{2}$$

In Eq. 2, C_1 and C_2 represent two unknown constants, $C_3 = \frac{g_l'\delta^2}{\mu_l}$ and $\delta = \sqrt{\frac{K(g_l')}{g_l'}}$. The unknown constants can be further solved with additional constraints. Since all of the fields are continuous within the whole domain, the velocity and the viscous stress at the interface between the intra-and extra-dendritic regions must be continuous. Then,

$$C_1 = -\frac{\delta h}{\mu_l} - \frac{g_l'\delta^2}{\mu_l} - \frac{h^2}{2\mu_l}, \text{and } C_2 = \frac{-\frac{\delta h}{\mu_l}}{e^{-\frac{h}{\delta}}}. \tag{3}$$

By integrating the above equations over two facing tetrahedral elements, with symmetry plane shown by the dotted line in Fig. 1(d), one then obtains the flux; then using the divergence theorem and the Green's theorem, the equation for dendritic flow can be obtained:

$$\left[\left(\frac{h^3}{3\mu_l} + 2C_1 h\right) S_1 + 2\left(C_2\delta e^{-\frac{h}{\delta}} + C_3 h - C_2\delta e^{-\frac{L_e'}{\delta}} - C_3 L_e'\right) S_2\right] \nabla^2 p = 0, \tag{4}$$

where S_1 and S_2 are dendrite envelope area and solid/liquid interface area based on the equivalent solid radius, respectively, and L' is the total length of grain.

A schematic of the resulting fluid flow in one dendrite is given in Fig. 1(d) (blue line). For comparison purposes, the flow prediction from the equiaxed-globular case is also given (red line). In this second curve, flow only occurs in the intergranular region, and thus for a given fluid flux the maximum velocity must be greater in order to displace an equivalent amount of fluid in a given time. For the entire domain, the fluid flow mesh consists of triangular prismatic elements representing the liquid phase of each tetrahedral element with connectivity provided by the solidification model. The finite element solution of Eq. 4 is achieved following the procedure outlined Ref. [1].

4. Results and Discussion

Below, a set of simulations are presented to compare the influence of microstructure on the fluid flow within the dendritic and globular structures at different solid fractions where the extra-dendritic liquid channel widths for the dendritic structure are approaching zero due to the high cooling rate of 50 °C·s^{-1}, and secondary armspacing is selected to be 33 μm. The domain was 5 mm×5 mm× 5 mm consisting of 1000 grains, and $\mu_l = 7.0 \times 10^{-3}$ Pa·s. For the bulk permeability assessment, an isothermal flow simulation is conducted using the following boundary conditions: a constant pressure on the top surface, $p_t = 0$ Pa, a constant non-zero flux on the bottom surface, $q_b = 50$ μm·s^{-1} and closed lateral surfaces, $q_s = 0$ μm·s^{-1}.

4.1. Fluid Flow

Pressure contours within the liquid channels obtained from the two models are shown in Fig. 2. For better visibility, the grains are shaded with the same colour of the pressure in the liquid channel. First, it can be seen that a lower pressure is obtained at a further distance away from the top of the domain. Second, a higher pressure drop was found in the dendritic microstructure as compared with the globular structure. This is because the increased fluid friction in the dendritic network will require a larger pressure drop to achieve the same fluid flow. Although not shown, the localization of the liquid feeding along several preferential channels can be observed in the velocity contours. Since the permeability of the dendritic structure is lower than that of globular structure at the same solid fraction, the maximum velocity within the liquid channels is also lower, being 162 μm·s^{-1}, in the dendritic structure as compared to 5949 μm·s^{-1} in the globular structure.

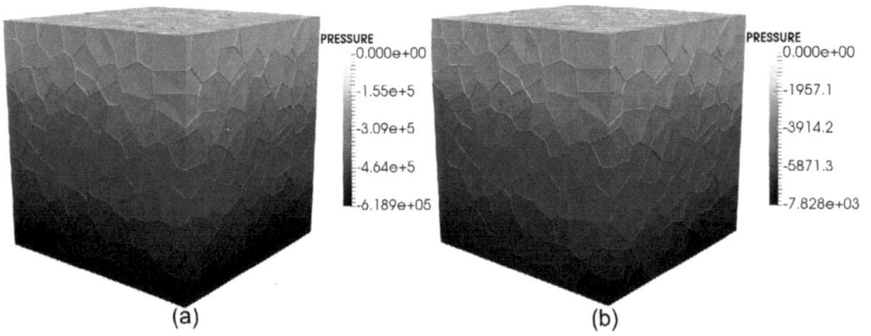

Figure 2. Pressure distribution at a solid fraction of 0.95: (a) dendritic and (b) globular.

4.2. Hot Tearing Sensitivity Coefficient

It is well known that pressure gradient within a semisolid domain is a contributing factor to hot tearing sensitivity [1]. The pressure gradient itself can be utilized as a sensitivity coefficient.

Fig. 3(a) compares the hot tearing sensitivity coefficients for both the dendritic and globular structures. As can be seen, the values are rather low at low solid fraction due to the adequate liquid feeding; hot tearing is unlikely to occur. At high solid fraction, there is an increase in the sensitivity coefficient however the sharp rise occurs both earlier in the dendritic structure and to a much greater extent as compared to the globular structure. Thus, the model shows that hot tearing is more prevalent in dendritic structures because the increased difficulty in fluid flow increases the pressure drop within the semisolid and increases hot tearing susceptibility at a given solid fraction.

4.3. Model Verification

Finally, the dendritic fluid flow model can be verified by comparing its predictions of permeability against the analytical Carman-Kozeny equation. For this comparison, a brief review of the Carman-Kozeny equation is useful. The local permeability shown in Eq. 1 can be calculated from the local fraction liquid and the solid/liquid interfacial area concentration S_v,

$$K = \frac{(g_l')^3}{5S_v^2}. \tag{5}$$

S_v is influenced by grain growth, impingement and coarsening processes [12], and can be approximated in one of two ways, by assuming a one-dimensional platelike geometry of the secondary dendrite arm [13], or through an empirical link with solid fraction [12]). These are shown below, where λ_2 is the cooling-rate dependent [14] SDAS, and c, p and q are constants [12, 15],

$$S_v = \frac{2}{\lambda_2}, \tag{6}$$

$$S_v = c{g_s}^p(1 - g_s)^q. \tag{7}$$

For the verification, simulations at different solid fraction were performed using the same conditions. K is then calculated as,

$$K = \mu_l \frac{q_b}{\left(\frac{p_t - p_b}{L}\right)}, \tag{8}$$

where L is the distance between the two surfaces, and p_b is the bottom surface pressure.

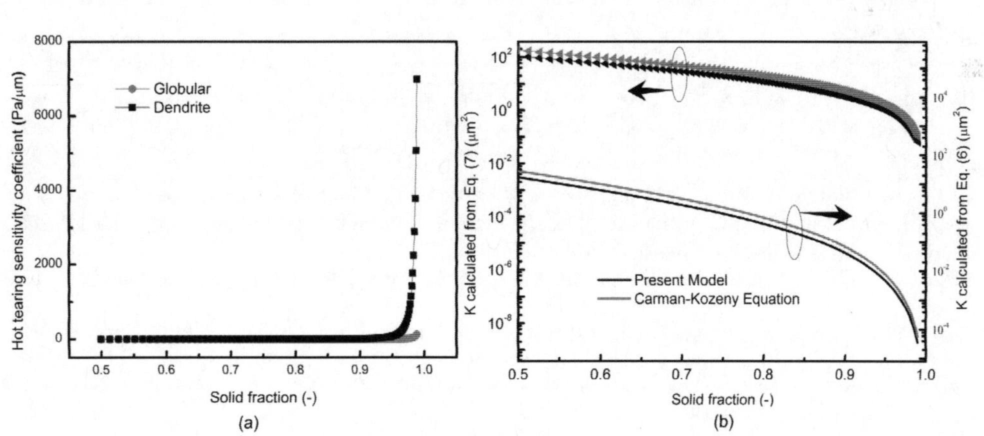

Figure 3. (a) Variation in hot tearing sensitivity coefficient with solid fraction in both dendritic and globular structure. (b) Permeability comparisons for the two S_v terms.

Fig. 3(b) shows the evolution in bulk permeability predicted by the model and Carman-Kozeny equation for both S_v equations, a good match has been achieved. All the curves display similar trends where the permeability decreases with increasing solid fraction, and the gap between any two curves narrows near the end of solidification as the liquid fraction within the dendrite envelope $\rightarrow 0$ and thus the remaining liquid must flow around the grains. Further, a higher value in permeability is always predicted by Carman-Kozeny, irrespective of S_v. This might due to the assumption that the fluid flow direction is always parallel to the bottom surface of tetrahedral element, resulting in lower permeability due to the increased friction forces. The different S_v equations also predict significantly different values of permeability, with the method based on SDAS being nearly three orders of magnitude smaller at the end of solidification.

5. Conclusions

Preliminary work to model grain structure effects related to hot tearing during steel solidification was carried out, where the fluid flow behaviour within semisolid steel having a composition of 0.07 wt.%C with dendritic microstructure has been predicted at the meso-scale. First, the semisolid microstructure in an ensemble of grains was created using a volume average approach to model solidification of individual grain. Each grain consists of three phases, delta-ferrite, intra-dendritic liquid, and extra-dendritic liquid in order to simulate the dendritic structure. Second, at a prescribed solid fraction, the Darcy-Brinkman form of the Navier-Stokes equation was used to predict fluid flow between and through the grains. The extra-dendritic position was treated as a pure liquid zone, while the region within the dendrite envelope was treated as a porous medium. The results of the new dendritic fluid flow model have been compared to a previous model consisting of globular microstructure [1], as well as the well-known Carman-Kozeny equation. It is shown that, for a given solid fraction, dendritic microstructure undergoes a higher pressure drop than globular microstructure. This difference is due to the fluid friction occurring within the dendrite envelope; this friction is not present in globular microstructure. The Carman-Kozeny also over-estimates the permeability for the same reasons. A hot tearing sensitivity coefficient was developed, demonstrating that the increased hot tearing susceptibility in dendritic structures is a result of an increase in the difficulty for fluid flow to occur.

Future work would focus on the application of present model to the meniscus of casting steel by simulating the fluid flow behavior within a domain containing a few grains, and predict the initiation of hot tearing by imposing the strain on the domain.

Acknowledgements

The authors wish to thank the Natural Sciences and Engineering Research Council of Canada (NSERC) and the McMaster Steel Research Centre for financial support.

References

[1] Sistaninia M, Phillion A, Drezet J M and Rappaz M 2012 *Acta Materialia* **60** 3902–3911
[2] Rappaz M, Drezet J M and Gremaud M 1999 *Metallurgical and materials transactions A* **30** 449–455
[3] Flemings M C 2006 *Materials Science and Technology*
[4] Mišičko R, Fujda M, Fedáková S, Fedáková D and Szekeres S 2012 Further study of the "blown grains" condition
[5] Szekeres E 2002 A review of strand casting factors affecting transverse cracking *Proceedings of 6th International Conference on Clean Steel* pp 324–338
[6] Vernede S, Jarry P and Rappaz M 2006 *Acta Materialia* **54** 4023–4034
[7] Rajani H Z and Phillion A 2018 *Materials & Design* **144** 45–54
[8] Tourret D and Gandin C A 2009 *Acta Materialia* **57** 2066–2079
[9] Rappaz M and Thevoz P 1987 *Acta Metallurgica* **35** 1487–1497
[10] Le Bars M and Worster M G 2006 *Journal of Fluid Mechanics* **550** 149–173
[11] Bear J 2013 *Dynamics of fluids in porous media* (Courier Corporation)
[12] Neumann-Heyme H, Eckert K and Beckermann C 2017 *Acta Materialia* **140** 87–96
[13] Wang C, Ahuja S, Beckermann C and De Groh H 1995 *Metallurgical and Materials Transactions B* **26** 111–119
[14] El-Bealy M and Thomas B G 1996 *Metallurgical and materials transactions B* **27** 689–693
[15] Domitner J, Kharicha A, Wu M and Ludwig A 2013 *Metallurgical and Materials Transactions A* **44** 607–616

IOP Conf. Series: Materials Science and Engineering **529** (2019) 012032 doi:10.1088/1757-899X/529/1/012032

A novel route to the coupling of molecular dynamics and phase-field simulations of crystal growth

P C Bollada[1], H Men[2], C Fang[2], P K Jimack[3], Z Fan[2], A M Mullis[1]

[1] School of Chemical and Process Engineering, University of Leeds, Leeds LS2 9JT
[2] BCAST, Brunel University, Kingston Lane, Uxbridge, Middlesex, UB8 3PH
[3] School of Computing, University of Leeds, Leeds LS2 9JT
Email: p.c.bollada@leeds.ac.uk

Abstract

Molecular dynamic simulations, *ab initio* (DFT) calculations and experimental evidence suggests that there is a liquid-solid transition region which may be characterised by an order parameter. In this interface region the order parameter is not observed to be symmetrical, rather it tends to be steep on the solid side and exponentially decreasing on the liquid side. The order parameter in phase field computations is, to date, always assumed to give a symmetrical interface region. Hence, we ask how to extend the phase field model to give a profile that fits this data, and how such a model affects the simulation.

Keywords: crystal formation; diffuse interface; alloy solidification;

1 Introduction

The integration of modelling techniques that operate on disparate length and timescales is one of the greatest challenges facing the computational simulation of materials. At the atomistic scale techniques such as density functional theory (DFT) and molecular dynamics (MD) track every atomic degree of freedom but are severely limited with regard to the size of domain that can be considered. The largest such MD simulations reported [1, 2], facilitated by massively parallel supercomputers, contain $O(10^{12})$ atoms (in MD, very significantly less in DFT) wherein, for an example material such as metallic Ni at standard temperature and pressure, we have a simulation volume of $\approx 10^{-17}\mathrm{m}^3$. Conversely, continuum models cover length scales from the mesoscale (10^{-6} to 10^{-4}m), with techniques such as phase-field modelling [3] and level sets [4]) to that of large engineering structures and beyond.

Linking these length scales is critically important to the quantitative understanding of material behaviour. However, to date most such linking is of the 'weak' type, that is, parameters or constitutive equations derived from models at small length or timescales are incorporated into models operating at larger scales. For instance, a cascade of such weak coupling from the atomistic to the component scale, is described by [5], in which results obtained from DFT models of dislocation core structures are used to describe solute strengthening in FE models of the tensile deformation of aluminium alloys. Strong coupling, in which models operating at disparate length scales are directly linked, is much rarer, particularly with regard to linking from the atomistic to the continuum scale (but see [16]). With regard to fluid dynamics Neumann *et al.* [6] stated "only one framework-like piece of software has been developed so far for ... coupling molecular-continuum code", while in the solid-state [7] present a rare example of an embedded atom model for a Co-Ni-Al particulate being incorporated into an FE model of an aluminium matrix.

Solidification is a phenomena that intrinsically bridges multiple length scales with crystallisation morphologies being determined both by macroscale diffusive-convective fluxes of heat and solute and atomistic scale capillary and kinetic effects [8]. Despite this there are remarkably few examples of models coupling across the atomistic-continuum divide. There are a few examples of weak coupling (e.g. [9]) in which capillary and kinetic parameters derived from MD simulations are fed into mesoscale phase-field simulations but, as far as we are aware, this is the extent of coupling across the atomistic-continuum divide in the field of crystallisation. Note that we explicitly exclude from this definition models of the phase-field crystal (PFC) type as, while these model discrete atom positions by means of a continuous density function, the length scales modelled are comparable to large MD simulations.

Within solidification modelling, phase-field simulation is one of the most popular and successful continuum techniques for modelling microstructure formation. In fact there exists a very natural route to linking phase-field and atomistic simulations such as MD, via the structure of the diffuse interface. The interface between the solid and the liquid is modelled via an order parameter, ϕ, which varies continuously between 1 in the solid and 0 in the liquid over some finite interface width. When we look at MD simulations of a solid-liquid interface we see just such a progressive transition, with crystalline order being lost over an interface width of several atom diameters. There are various ways in which this loss of order can be parameterised, for instance using a in-plane order parameter or via the locus of the density maxima across the interface. The in-plane order parameter, calculated via MD for the growth of Al from its parent melt along the $< 100 >, < 110 >$ and $< 111 >$ directions, is shown in Figure 1. It is clear from the figure that the effective interface width (i.e. the number of atom planes over which long-range crystalline ordering is lost) varies with growth direction, with the width of the $< 111 >$ interface being significantly smaller than for growth in either the $< 110 >$ or $< 100 >$ directions.

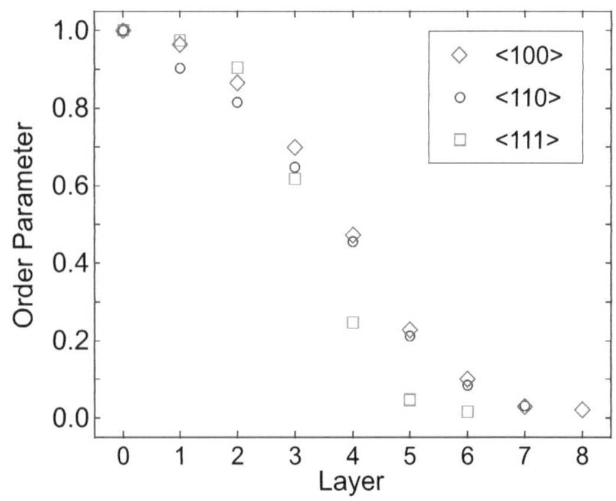

Figure 1. In plane order parameter for the growth of Al in the $< 100 >, < 110 >$ and $< 111 >$ direction at 880K (53K undercooling) as calculated by molecular dynamics.

The atomic structure of the proximal liquid to a solid crystalline interface has been the subject of considerable recent interest [14], with liquid atoms becoming layered due to the crystalline order within the solid. However, the asymmetry in the interface, i.e. that order is lost rapidly in the first 1-2 liquid layers but that vestiges of ordering persist several layers out, has only recently become apparent. [see e.g. Fig. 2 in Ref. [15]. This asymmetry appears to be due to a large number of defects being introduced into the first layer adjacent to the solid, with a more gradual increase in defect density thereafter, although phenomenon has been far from fully explored.

In most cases the phase-field diffuse interface is taken as a tanh profile, the form of which is $\phi = \frac{1}{2} - \frac{1}{2}\tanh((x - x_0)/\delta)$ where x_0 is the interface position and δ the interface width. However, this is generally no more than a computational expedient, with no attempt made to match either the shape or length scale to what might be considered realistic values from MD simulation. Over the last several years we have shown how the application of advanced numerical techniques such as adaptive mesh refinement, implicit time-stepping, non-linear multigrid solvers and parallelisation allows realistic interface widths to be used in phase-field simulation [3, 10, 11, 12]. We now turn our attention to the shape of the interface. It can be seen from Figure 1 that, irrespective of the growth direction, the in-plane order parameter decreases more steeply on the solid side of the solid-liquid interface than on the liquid side. In fact, the MD simulation interface can be parameterised by a tanh curve only if we take d to be different on the solid and liquid sides of the interface. As an example, the profile for growth along the $< 110 >$ direction can be parameterised by a tanh curve if $\delta = 2.65(0.5 \leq f < 1)$ and $\delta = 3.54(0 < \phi \leq 0.5)$. This is in stark contrast to phase-field, where a symmetry around $\phi = 0.5$ is assumed. Our contention here is that additional physics can be captured by matching the phase-field interface to the shape and length-scale predicted by atomistic simulations of the solid-liquid interface. We present below a phase-field formalism that allows an arbitrary monotonic order parameter to be matched in the interface region.

IOP Conf. Series: Materials Science and Engineering **529** (2019) 012032 doi:10.1088/1757-899X/529/1/012032

2 Phase field fitting to MD data

The question we wish to address is: given a set a data from *ab-initio* or molecular dynamic simulation how can we model this within phase field? We begin by assuming we have a given phase field profile, $\phi(x)$, which is determined in advance by measurements from MD data. Within a phase field model, this profile is determined by a double well potential, $f(\phi)$, as part of the equilibrium equations:

$$\frac{\delta F}{\delta \phi} = 0 \tag{1}$$

where

$$F = \int \delta^2 \nabla\phi \cdot \nabla\phi + f. \tag{2}$$

Computing the variational derivative gives

$$\frac{\delta F}{\delta \phi} = -\delta^2 \nabla^2 \phi + \frac{\partial f}{\partial \phi} = 0 \tag{3}$$

which, in 1D reduces to

$$\delta^2 \phi''(x) = \frac{\partial f}{\partial \phi}. \tag{4}$$

This can be written

$$\delta^2 \frac{\partial}{\partial x}\frac{\partial \phi}{\partial x} = \frac{\partial f}{\partial \phi} \tag{5}$$

Now $\frac{\partial \phi}{\partial x} = 1/\frac{\partial x}{\partial \phi}$ and so by the chain rule

$$\begin{aligned} \frac{1}{\delta^2}\frac{\partial f}{\partial \phi} = \frac{\partial}{\partial x}\frac{\partial \phi}{\partial x} = \frac{\partial \phi}{\partial x}\frac{\partial}{\partial \phi}\left(\frac{\partial \phi}{\partial x}\right) &= \frac{\partial \phi}{\partial x}\frac{\partial}{\partial \phi}\left(\frac{\partial x}{\partial \phi}\right)^{-1} \\ &= -\frac{\partial \phi}{\partial x}\left(\frac{\partial x}{\partial \phi}\right)^{-2}\frac{\partial^2 x}{\partial \phi^2} \\ &= -\left(\frac{\partial x}{\partial \phi}\right)^{-3}\frac{\partial^2 x}{\partial \phi^2} \\ &= \frac{\partial}{\partial \phi}\left[\frac{1}{2}\left(\frac{\partial x}{\partial \phi}\right)^{-2}\right] \end{aligned} \tag{6}$$

and thus

$$f = \frac{\delta^2}{2}\left(\frac{\partial x}{\partial \phi}\right)^{-2}. \tag{7}$$

For example, the standard, symmetrical, 1D phase field solution

$$\phi = 1/2\left[1 + \tanh\left(\frac{x}{\sqrt{2}\delta}\right)\right] \equiv \frac{1}{1 + \exp(-\sqrt{2}x/\delta)} \tag{8}$$

which has inverse

$$x = \frac{\delta}{\sqrt{2}}[\ln(\phi) - \ln(1-\phi)] \tag{9}$$

determines the double well potential:

$$f = \frac{\delta^2}{2}\left(\frac{\partial x}{\partial \phi}\right)^{-2} = \phi^2(1-\phi)^2. \tag{10}$$

Notice in the above, we required the inverse function $x = x(\phi)$, which is easily obtainable by fitting to data of the form

$$[[x_0, \phi_0], [x_1, \phi_1], ..., [x_n, \phi_n]]. \tag{11}$$

By choosing a function to fit the data carefully we will be able to retain some desirable properties of the potential. We impose the property of $x(\phi)$ that

$$1/x'(0) = 1/x'(1) = 0 \tag{12}$$

Keeping in mind the function for the symmetric profile, Eq. 9, we choose the simplest ansatz for an arbitrary position

$$x(\phi) = a\ln(\phi) + b\ln(1-\phi) + c \tag{13}$$

where the constants a, b, c are determined from the data by least squares. Clearly if $a = -b$ we return the symmetrical tanh phase profile with some interface width (determined by a) and position along the x-axis (determined by c). Our fit to one of the sets of data from Figure 1 is shown together with the corresponding potential wells in Figure 2. Also shown, to demonstrate that profiles with a more extreme skew can be accommodated, is a fit to the interface for Al nucleated on a static substrate calculated via ab-initio simulation. Note, that the data here is more usefully presented as x as a function of phase, ϕ.

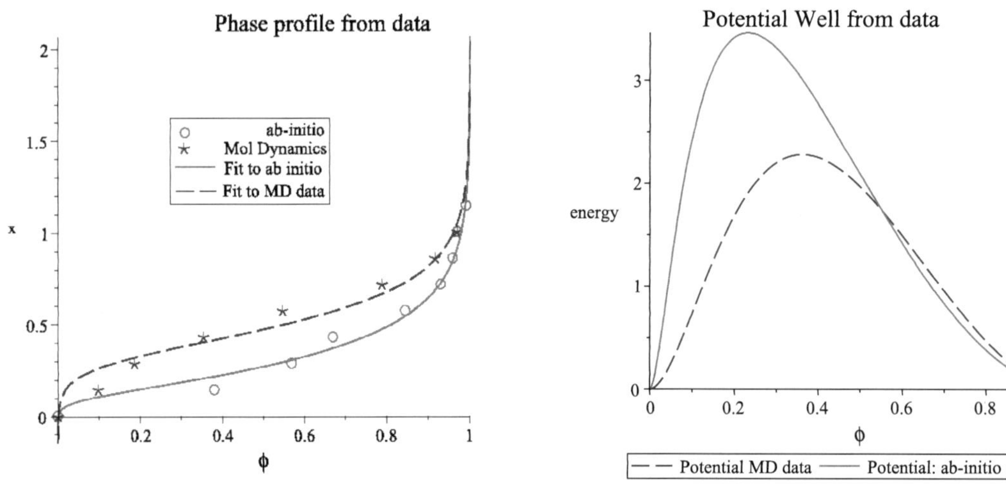

Figure 2. Fitted data (crosses and circles) using Eq. 13, determines the potential wells on the right in corresponding colour and line style. In line with phase field convention, we choose $\phi = 0$ to be solid (in contrast to the convention in Figure 1), and we fit the $< 110 >$ data from that figure (blue stars) along with some heavily skewed DFT data (red circles). The more non-symmetrical the data, the more skewed the double well potential, $f(\phi)$. Notice that the the less symmetry in the ab initio data (red circle) gives the more skewed potential well (red solid line).

3 Interpolation function

We now have a method by which we can reproduce a fitted curve that approximates given experimental or *ab initio* computations, by determining the appropriate double well potential for an equilibrium solution. However, for a non-equilibrium solution it is not sufficient to only modify the double well potential. We find that it is necessary to modify the forcing term also. The key function for this is the interpolation function, $g(\phi)$, which arises in the energy,

$$F = \int \delta^2 \nabla\phi \cdot \nabla\phi + f(\phi) + g(\phi) f_L(c, T) + (1 - g(\phi)) f_S(c, T), \tag{14}$$

IOP Conf. Series: Materials Science and Engineering **529** (2019) 012032 doi:10.1088/1757-899X/529/1/012032

where f_L, f_S are the free energies of the pure liquid and pure solid respectively obtained, for example, from a thermodynamic data base. Typically, the interpolation function, $g(\phi)$, used in phase field is constructed from the double well potential (e.g. Wheeler)

$$g'(\phi) = C \int f(\phi)\, \mathrm{d}\phi \tag{15}$$

where C is some constant and $g(0) = 0, g(1) = 1$. This would make g a 5th order polynomial if the double well potential, $f(\phi)$, is the standard 4th order form. We do not adopt this and instead relate g to a moving tanh profile,

$$\phi = \tfrac{1}{2}\left[1 + \tanh\left(\frac{2(x - Vt)}{\delta}\right)\right] \tag{16}$$

where V is the speed in which this profile is moving to the right. We require

$$\dot{\phi} = -Cg'(\phi) \tag{17}$$

for some positive constant C. We find

$$\dot{\phi} = -4V\phi(1 - \phi)/\delta \tag{18}$$

which gives the (normalised) cubic polynomial

$$g = \phi^2(3 - 2\phi), \tag{19}$$

and so

$$\dot{\phi} = -4V\phi(1 - \phi)/\delta = -\frac{2V}{3\delta} g'(\phi). \tag{20}$$

We may note that, in this case, it appears to suggest the rule

$$g'(\phi) \propto \sqrt{f(\phi)}. \tag{21}$$

If this is generally true we can apply this rule to obtain $g(\phi)$ from any $f(\phi)$. So, assume both

$$\frac{\mathrm{d}^2\phi}{\mathrm{d}x^2} = \alpha(\phi), \frac{\mathrm{d}\phi}{\mathrm{d}x} = \beta(\phi) \tag{22}$$

where clearly $\alpha(\phi) = f(\phi)/\delta^2$ and $\beta(\phi) \propto g'(\phi)$. It follows that

$$\alpha(\phi) = \frac{\mathrm{d}^2\phi}{\mathrm{d}x^2} = \frac{\mathrm{d}\beta}{\mathrm{d}x} = \frac{\mathrm{d}\beta}{\mathrm{d}\phi}\frac{\mathrm{d}\phi}{\mathrm{d}x} = \frac{\mathrm{d}\beta}{\mathrm{d}\phi}\beta = \tfrac{1}{2}\frac{\mathrm{d}(\beta^2)}{\mathrm{d}\phi} \tag{23}$$

Hence, in general, $f \propto [g'(\phi)]^2$. and so

$$g = C \int \sqrt{f(\phi)}\, \mathrm{d}\phi \tag{24}$$

where C is chosen so that $g(0) = 0, g(1) = 1$.

4 The effect of phase profile on solute partitioning

Figure 3a illustrates a phase field example of a skewed profile. The use of skewed profiles necessarily changes the dynamics of the model, for example, using a series of increasingly skewed profiles affects the solute partitioning at the interface- see Figure 3b.

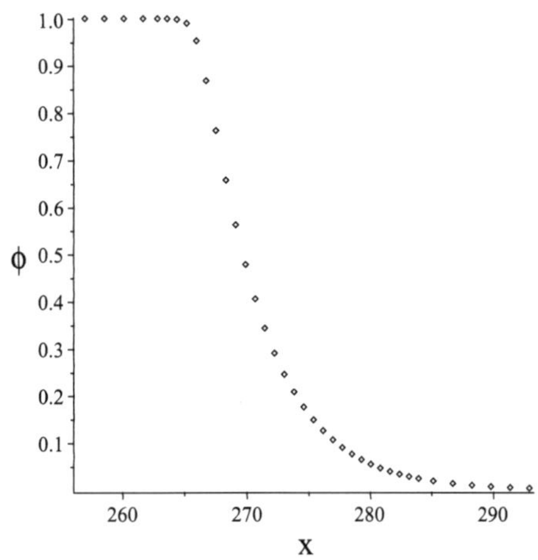

Figure 3a A phase field simulation using a large skew double well potential

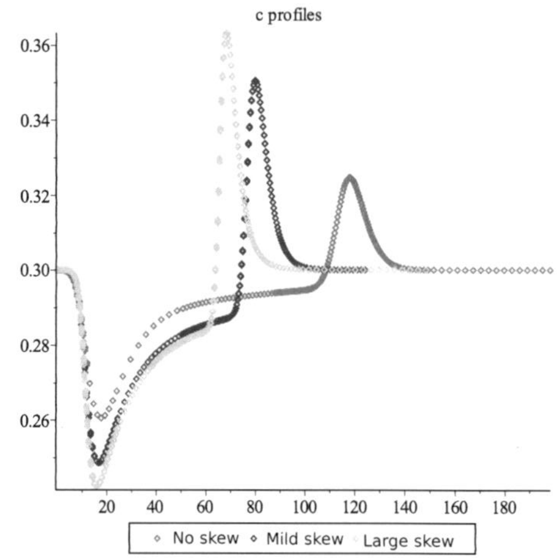

Figure 3b A series of increasingly skewed phase profiles from red (symmetrical) to very skewed blue to green.

5 Summary

We have described a procedure whereby we can use external data to dictate the profile shape in a phase field model. This involves using a least squares fit to a limited function with only three free coefficients. Once this function is known we describe how to obtain the interpolation function so that the profile shape is maintained even under non equilibrium conditions. There is some evidence that profile shape affects the dynamics although it is hard to isolate these from effects due to reduction in interface width.

6 Acknowledgements

This research was funded by EPSRC Innovative Manufacturing Research Hub in Liquid Metal Engineering (LiME), Grant No. EP/N007638/1.

References

[1] Germann T C and Kadau K 2008 Int. J. Mod. Phys. C 19 1315.
[2] Niethammer C et al. 2014 J. Chem. Theory Comput. 10 4455.
[3] Bollada P C, Goodyer C E, Jimack P K, Mullis A M and Yang F W 2015 J. Comput. Phys. 287 130.
[4] Moghadam M M and Voorhees P W 2018 Comp. Mater. Sci. 143 454.
[5] Krajewski PE, Hector LG, Qi Y, Mishra RK, Sachdev AK, Bower AF and Curtin WA 2011 JOM 63 24.
[6] Neumann P, Eckhardt W and Bungartz H-J 2014 Comput. Math. Appl. 67 272.
[7] Yamakov V, Hochhalter J D, Leser W P, Warner J E, Newman J A, Purja Pun G P and Mishin Y 2016 J Mater. Sci. 51 1204.
[8] Hoyt J J, Asta M and Karma A 2003 Mater. Sci. Eng. R 41 121.
[9] Bragard J, Karma A, Lee Y H 2002 Interf. Sci. 10 121.
[10] Rosam J, Jimack P K and Mullis A M 2007 J. Comput. Phys. 225 1271.
[11] Rosam J, Jimack P K and Mullis A M 2008 Acta Mater. 56 4559.
[12] Bollada P C, Goodyer C E, Jimack P K and Mullis A M 2015 Appl. Phys. Lett. 107 053108.
[13] Bollada P C, Jimack P K and Mullis A M 2017 Comp. Mat. Sci. 126 426-437.
[14] W.D. Kaplan and Y. Kauffmann, Annu. Rev. Mater. Res. 36 (2006) 1–48
[15] B. Jiang, H. Men, Z. Fan, Comp. Mater. Sci. 153 (2018) 73–81.
[16] E. Miyoshi et al, Comp. Mater. Sci. 152 (2018) 118

IOP Conf. Series: Materials Science and Engineering **529** (2019) 012033 doi:10.1088/1757-899X/529/1/012033

Effect of the Coriolis force on the macrosegregation of aluminum in the centrifugal casting of Ti-Al alloys

M. Cisternas Fernández[1], M. Založnik[1], H. Combeau[1], C. Huang[2], J. Zollinger[1] and U. Hecht[2]

[1] Université de Lorraine, CNRS, IJL, F-54000 Nancy, France
[2] Access e.V., Intzestr. 5, D-52072 Aachen, Germany

E-mail: martin.cisternas-fernandez@univ-lorraine.fr

Abstract. Within the framework of the ESA GRADECET project, experiments of directional solidification of cylindrical Ti-Al samples were conducted in hypergravity. The experiments were performed in a centrifuge with the apparent gravity (sum of centrifugal and terrestrial gravity) aligned along the cylinder centerline. 3D numerical simulations of aluminum macrosegregation in these samples are presented. A volume-averaging solidification model is used that accounts for centrifugal and Coriolis accelerations in a non-inertial rotating reference system. We compare the melt flow pattern and the macrosegregation formation under terrestrial gravity and under $20\vec{g}$ centrifugation. The results show that the Coriolis acceleration, although very weak, breaks the symmetry of the thermosolutal convection, having an important impact on the final macrosegregation pattern. The macrosegregation is entirely modified in comparison with a sample solidified under terrestrial gravity conditions. Besides the aluminum segregation intensity increases with the centrifugation level.

1. Introduction

Titanium aluminum alloys have been studied during the last 30 years due to their low density and high resistance in high temperature environments, making them a very good candidates for automotive and aerospace applications. However, their high reactivity in the liquid state makes them difficult to cast as they need to be cast at very low superheats and at fast filling rates. Centrifugal casting is a possibility to enhance the mold filling speed by combining the effect of terrestrial gravity and centrifugal acceleration, preventing undesired reactions and porosity defects.

In spite of the benefits of centrifugal casting, it can increase the magnitude of the liquid thermosolutal convection during solidification and consequently increase the segregation of alloying elements. In the case of Ti-Al alloys, controlling the macrosegregation of centrifugally cast ingots is important since the microstructure formation is very sensitive to the aluminum composition [1]. Solidification during centrifugation has been studied before. Ramachandran et al. [2] performed numerical simulations attemping to emulate the liquid thermodriven convection that takes place in a centrifuge-based Bridgman crystal growth configuration, reporting that the Coriolis acceleration played a stabilizing role in cases where the centrifugal acceleration was parallel to the thermal gradient. Experimental studies of solidification in centrifuges, such as the one of Rodot et al. [3] or Muller et al. [4] concluded that the Coriolis acceleration and centrifugal acceleration gradients can play an important role on liquid convection. Although

IOP Conf. Series: Materials Science and Engineering **529** (2019) 012033 doi:10.1088/1757-899X/529/1/012033

Coriolis acceleration is known to have an impact on the liquid convection during solidification, its effect on macrosegregation has not been studied yet.

The European Space Agency (ESA) launched the GRADECET ("GRAvity DEpendence of Columnar to Equiaxed Transition") project, whose main objective is to investigate the influence of gravity on the Columnar-to-Equiaxed Transition - *CET* - and on microstructure formation in peritectic Ti-Al alloys. Within the framework of this project, several experiments of directional solidification in hypergravity were carried out on the ESA "Large Diameter Centrifuge" (LDC). These experiments gave relevant information on the solidification microstructure formed under different gravity levels [1]. To fully understand the experimental observations, characterization of the liquid thermosolutal convection during solidification is needed. This work aims to support the understanding of the liquid convection and of macrosegregation formation by means of numerical simulations.

We implemented a solidification model for columnar growth that takes into account the liquid themosolutal convection considering the non-inertial accelerations - centrifugal and Coriolis - into the finite-volume framework OpenFOAM. We perform 3D numerical simulations of the GRADECET experiments for the alloy Ti-48Al-2Cr-2Nb and we show that the Coriolis acceleration has a large impact on the macrosegregation because it breaks the symmetry of the liquid convection, leading to an entirely three-dimensional aluminum segregation pattern.

2. The model

The volume-average solidification model is similar to the one used by Combeau et al. [5] and accounts for columnar growth with a fixed solid phase in a binary alloy. The momentum equation is defined in terms of the intrinsic average liquid velocity $\langle \vec{v}_l \rangle^l$ and takes into account the centrifugal and Coriolis accelerations. The buoyancy term is modeled using a Boussinesq approximation which considers the thermal and solutal contributions. In addition, the energy equation is solved for the averaged enthalpy $\langle h \rangle = c_p T + g_l L_f$, where c_p is the specific heat, g_l the liquid fraction and L_f the latent heat. It is assumed that the temperature in the liquid is the same as in the solid $T = T_l = T_s$. Finally, a model assuming infinitely fast diffusion at the microscopic scale is used for the solid growth (lever rule). The equations of the model are:

- *Mass conservation:* Only the liquid movement is considered.

$$\nabla \cdot \left(g_l \langle \vec{v}_l \rangle^l \right) = 0 \tag{1}$$

- *Liquid momentum conservation:*

$$\begin{aligned} \frac{\partial}{\partial t}\left(g_l \langle \vec{v}_l \rangle^l \right) + \nabla \cdot \left(g_l \langle \vec{v}_l \rangle^l \langle \vec{v}_l \rangle^l \right) + 2 g_l \left(\vec{\omega} \times \langle \vec{v}_l \rangle^l \right) = \\ - g_l \nabla p + \nabla \cdot \left(g_l \nu_l \nabla \langle \vec{v}_l \rangle^l \right) - \frac{\nu_l \rho g_l^2}{K} \langle \vec{v}_l \rangle^l + g_l \rho_l^b \left[\vec{g} - \vec{\omega} \times (\vec{\omega} \times \vec{x}_c) \right] \end{aligned} \tag{2}$$

Where $\rho_l^b = \rho \left(1 - \beta_T (T - T_{ref}) - \beta_C (C_l - C_{ref})\right)$ and $K = \frac{\lambda_2^2 g_l^3}{180(1-g_l)^2}$

- *Energy conservation:*

$$\frac{\partial \langle h \rangle}{\partial t} + \nabla \cdot \left(g_l \langle \vec{v}_l \rangle^l h_l \right) - \frac{1}{\rho} \nabla \cdot (k \nabla T) = 0 \tag{3}$$

Where $h_l = c_p T + L_f$

- *Solute conservation:* Macroscopic solute diffusion is neglected.

$$\frac{\partial \langle C \rangle}{\partial t} + \nabla \cdot \left(g_l \langle \vec{v}_l \rangle^l C_l \right) = 0 \tag{4}$$

- *Supplementary relations:*
  - Averaged solute concentration:

$$\langle C \rangle = g_l C_l + g_s C_s \tag{5}$$

  - Microsegregation model (infinitely fast microscopic diffusion):

$$T = T_f + m_l C_l \tag{6}$$

$$C_s = k_p C_l \tag{7}$$

The alloy with which the GRADECET experiments were performed undergoes a peritectic transformation in its solidification path near 1480 [°C] [6]. A simulation of the solidification path was performed using Thermo-Calc Software [7] considering full thermodynamic equilibrium between the phases. Figure 1 presents the averaged aluminum concentration in solid and liquid phases obtained with Thermo-Calc and the approximations for the *liquidus* and *solidus* lines used in this work. In addition, Figure 2 presents the corresponding solidification path that assumes thermodynamic equilibrium between all phases along with the solidification path used in this work. The simplified solidification path considers a pseudo-eutectic transformation at T = 1479.7[°C]. Table 1 presents the list of thermophysical parameters that are used in the numerical simulations.

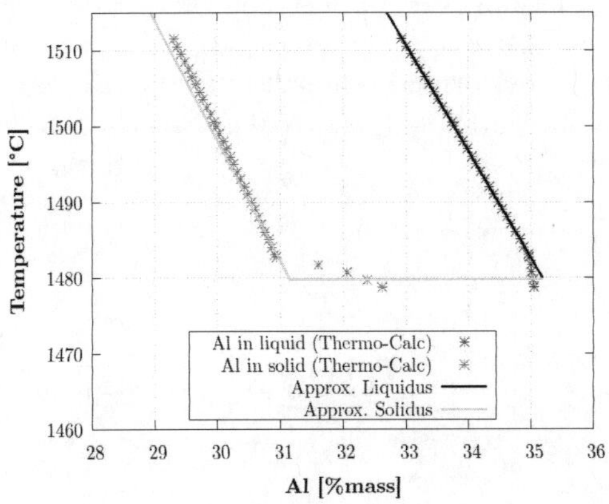

Figure 1. Averaged concentration of aluminum in the different phases vs temperature.

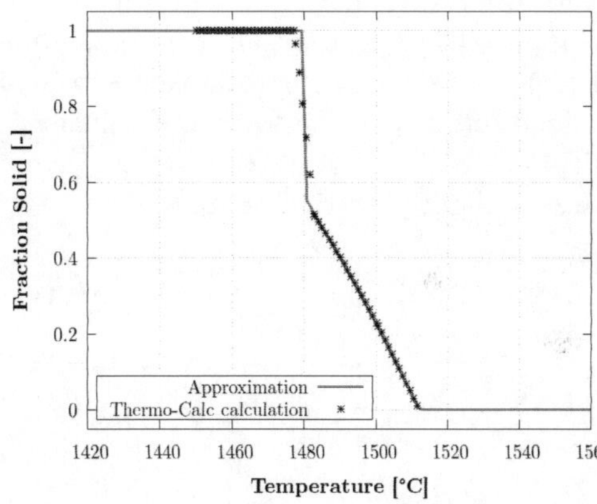

Figure 2. Solidification path obtained with Thermo-Calc Software and approximation used in the modeling.

3. Case study

The GRADECET experiments consisted of cylindrical Ti-Al samples (8 [mm] diameter, 165[mm] length) which were first remelted and then directionally solidified by means of a furnace that followed a specific thermal treatment to induce the *CET*. The furnace was mounted in the LDC and its thermal treatment was performed using three heaters located along the sample. The LDC centrifuge arm is 4[m] long, so that large centrifugal accelerations can be produced at relatively low rotational velocity. The furnace was free to tilt, such that the total apparent gravity $\vec{g}_{tot}$ was aligned with the cylinder centerline.

Figure 3 shows a schematic of the furnace in the centrifuge with the three heaters and the Ti-Al sample. The main plane of study is the one formed by the total apparent gravity $\vec{g}_{tot}$

Table 1. Thermophysical properties used in the simulations.

Property	Symbol	Units	Value
Nominal aluminum concentration	C_0	$[\%mass]$	32.96
Partition coefficient	k_p	$[-]$	0.885
Melting point of the pure substance	T_f	$[^\circ C]$	1975.66
Liquidus slope	m_l	$[^\circ C/\%mass]$	-14.089
Pseudo-eutectic concentration	C_{eut}	$[\%mass]$	35.2
Reference density	ρ	$[kg/m^3]$	3877.8
Kinematic viscosity	ν_l	$[m^2/s]$	1.65×10^{-6}
Thermal conductivity	k	$[W/m^\circ C]$	19.9
Solutal expansion coeff.	β_C	$[\%mass^{-1}]$	8.465×10^{-3}
Thermal expansion coeff.	β_T	$[^\circ C^{-1}]$	1.1785×10^{-4}
Latent heat	L_f	$[J/kg]$	3.62×10^5
Secondary arm spacing	λ_2	$[m]$	6.50×10^{-5}

and flight velocity $\vec{F}_{vel}$ vectors (as shown in Fig. 3) since the effect of the Coriolis force is more comprehensive on it. Thermal boundary conditions at the surface of the Ti-Al sample were obtained by heat transfer simulations of the full furnace by a dedicated furnace model. The temperature evolution extracted from the thermal simulations was imposed as Dirichlet b.c. at the sample surface in the solidification simulations.

Numerical simulations of macrosegregation were performed considering one level of centrifugation, where the total apparent gravity (sum of centrifugal acceleration and terrestrial gravity) was $20\vec{g}$ ($\vec{g}$: normal terrestrial gravity) and one simulation accounting only for terrestrial gravity (without centrifugation).

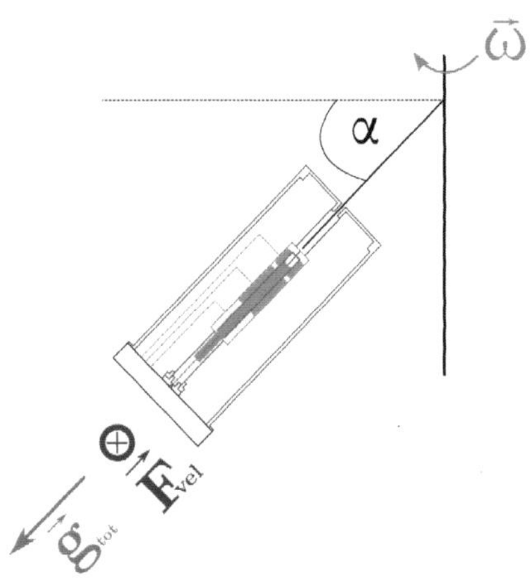

Figure 3. Schematic of the furnace and the sample in the centrifuge. Ti-Al sample (grey), heaters (yellow and orange) and thermocouples (red).

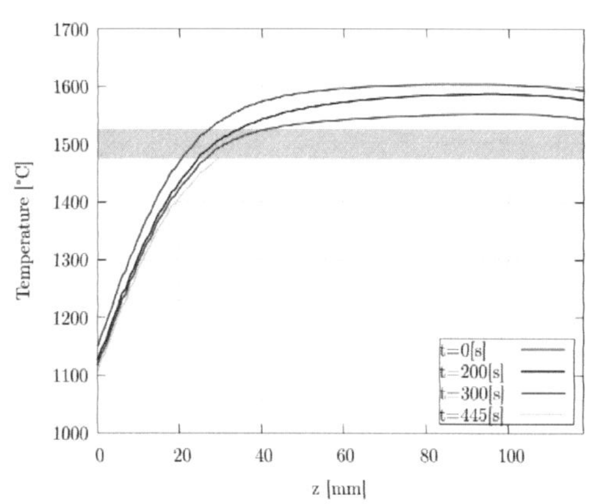

Figure 4. Temperature boundary conditions on the cylinder surface at different times.

IOP Conf. Series: Materials Science and Engineering **529** (2019) 012033 doi:10.1088/1757-899X/529/1/012033

4. Results and discussion

During solidification, the temperature gradient along the cylinder was anti-parallel to the total apparent gravity. With the onset of solidification, the cylinder side is slightly colder than its center due to the change of temperature in the heaters, so that, the first solid is formed at the cylinder side. Aluminum is rejected from the growing solid enriching the liquid in the mushy zone and thus decreasing its liquid density.

Figure 5 presents a comparison of the liquid flow between the cases 1G ($\vec{g}_{tot} = 1\vec{g}$) and 20G ($\vec{g}_{tot} = 20\vec{g}$) at t = 300[s] in the main plane of study. It can be observed that in the 1G case, the liquid flow is completely axisymmetric while in the 20G case, the Coriolis force, although very weak ($\frac{2(\vec{\omega}\times\langle\vec{v}_l\rangle^l)}{\vec{g}_{tot}} \sim 10^{-4}$), entirely modifies the convection producing a three-dimensional fluid flow pattern. In the 1G case, the aluminum enriched liquid, that at the beginnig is located in the cylinder sides, moves upward from the bottom of the mushy zone to the top entering to the fully liquid region. This aluminum enriched liquid advances along the cylinder by the cylinder side arriving to a point where the superheated liquid, that is lighter due to the thermal effect, blocks its advance. Finally, the liquid returns to the mushy zone by the cylinder centerline. On the other hand, in the 20G case, the Coriolis acceleration pushes the aluminum enriched liquid, that moves upward from the bottom of the mushy zone, to the flight velocity side of the cylinder. The lighter liquid flows from the mushy zone to the top of the cylinder by the flight velocity side and returns to the mushy zone by the opposite to flight velocity side. We found that the maximum liquid velocity in the 1G case is around 0.1[mm/s] while in the 20G case is 5[mm/s].

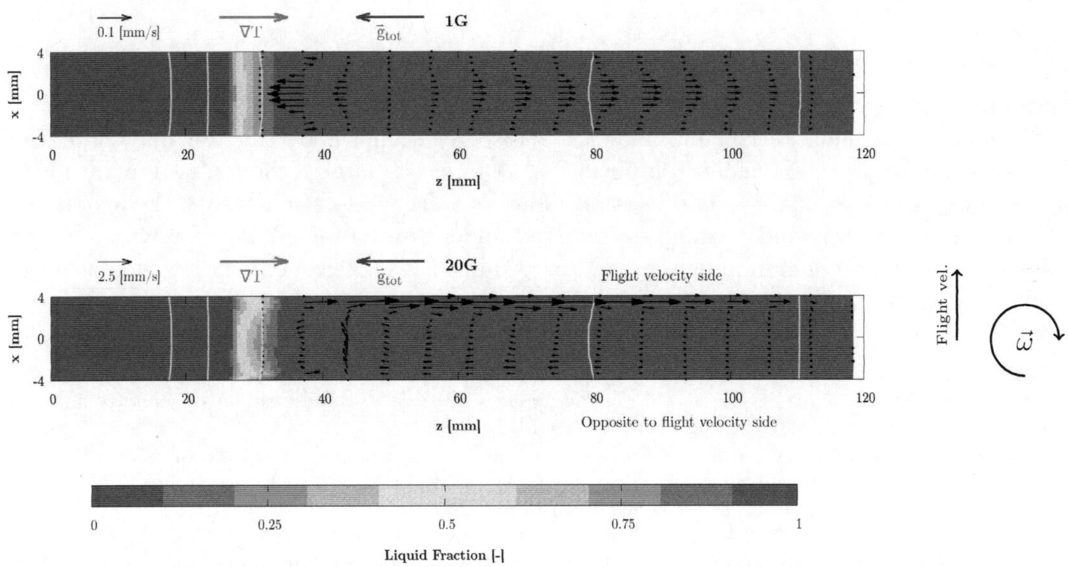

Figure 5. Comparison of the liquid convection between the 1G and 20G cases at $t = 300[s]$. Temperature contours in yellow from left to right: 1400,1450,1500,1550 [° C].

The three-dimensional thermosolutal driven flow advects the aluminum throughout the domain leading to macrosegregation. Figure 6 presents the aluminum segregation map in the main plane of study for the 1G and 20G cases. We can see that increasing the total apparent gravity, the segregation intensity is increased due to the stronger thermosolutal convection. The aluminum segregation in the 1G case is axisymmetric as well as its liquid motion while in the case with centrifugation, the flight velocity side of the sample is aluminum enriched. This segregation pattern follows the criterion $\frac{\partial\langle C\rangle}{\partial t} \sim \langle\vec{v}_l\rangle^l \cdot \nabla T$ that can be obtained by combining Eqs. 1, 4 and

6.

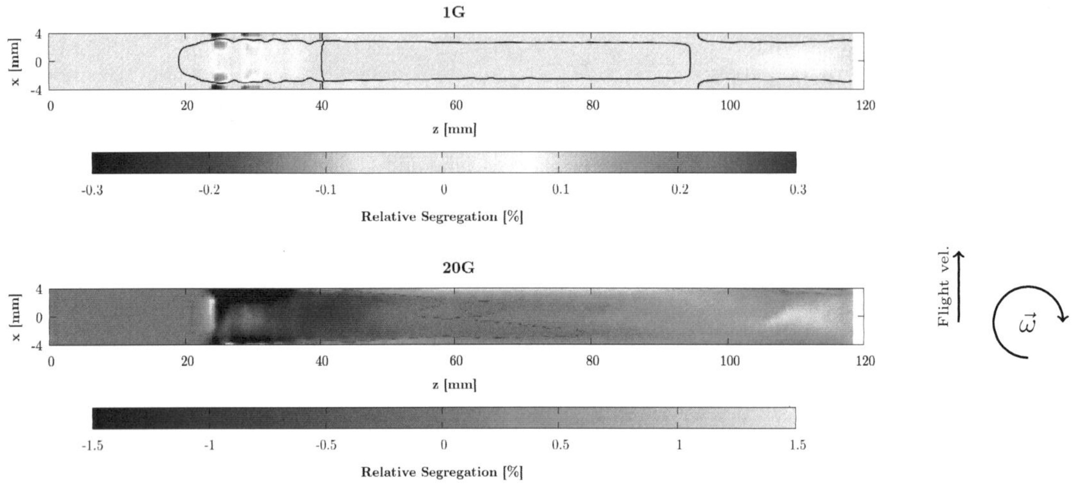

Figure 6. Aluminum segregation in the main plane at the end of solidification for the 1G and 20G cases. Relative Segregation: $\frac{\langle C\rangle - C_0}{C_0} \times 100\%$. Contours at $\langle C\rangle = C_0$ (red line) separate the positively and negatively segregated zones.

5. Conclusions

The 3D numerical simulations of aluminum macrosegregation show that the impact of the Coriolis acceleration in centrifugally solidified Ti-Al samples is large although its magnitude is small compared to the centrifugal acceleration. The Coriolis acceleration breaks the symmetry of the flow structure compared to samples solidified under normal terrestrial gravity conditions and changes the pattern of aluminum segregation. Additionally, the centrifugation increases the aluminum segregation intensity.

References

[1] Reilly N 2016 *Hétérogénéités de fabrication des aluminiures de titane: caractérisation et maîtrise de leurs formations en coulée centrifuge* Ph.D. thesis Université de Lorraine
[2] Ramachandran N, Downey J, Curreri P and Jones J 1993 *Journal of crystal growth* **126** 655–674
[3] Rodot H, Regel L and Turtchaninov A 1990 *Journal of crystal growth* **104** 280–284
[4] Müller G, Neumann G and Weber W 1992 *Journal of crystal growth* **119** 8–23
[5] Combeau H, Bellet M, Fautrelle Y, Gobin D, Arquis E, Budenkova O, Dussoubs B, Du Terrail Y, Kumar A, Gandin C A *et al.* 2012 Analysis of a numerical benchmark for columnar solidification of binary alloys *IOP Conference Series: Materials Science and Engineering* vol 33 (IOP Publishing)
[6] Witusiewicz V, Bondar A, Hecht U, Rex S and Velikanova T Y 2008 *Journal of Alloys and Compounds* **1** 185–194
[7] Andersson J O, Helander T, Höglund L, Shi P and Sundman B 2002 *Calphad* **26** 273–312

Acknowledgments

This work was carried out as part of the GRADECET (GRAvity DEpendence of Columnar to Equiaxed Transition in TiAl Alloys) research project and was funded by the European Space Agency (contract 4000114221/15/NL/PG). The view expressed herein can in no way be taken to reflect the official opinion of the European Space Agency.

IOP Conf. Series: Materials Science and Engineering **529** (2019) 012034 doi:10.1088/1757-899X/529/1/012034

A Study of the Mass Transfer Kinetics during the Dissolution of Ti-N Particles in Liquid Titanium

L Yao, S L Cockcroft and D M Maijer

Department of Materials Engineering, University of British Columbia, 309-6350 Stores Road, Vancouver, BC, Canada, V6T 1Z4

yaolu@mtrl.ubc.ca

Abstract. This paper summarizes the development of a computational fluid dynamics (CFD) model based on the commercial software package ANSYS-CFX, which has been used to examine the kinetics of dissolution of a single stationary Ti-N particle surrounded by moving liquid titanium. The model accounts for diffusional mass transport in the solid particle, the formation of a two-phase solid/liquid boundary layer and both advective and diffusional transport to the bulk liquid. The results have been used to estimate an effective mass transfer coefficient that may be applied in models that only consider solid-state mass transport of nitrogen. The results show that a correction is needed to the conventional Ranz-Marshall correlation to more accurately calculate the mass transfer coefficient during the Ti-N particle dissolution process.

1. Introduction

The Electron Beam Cold Hearth Remelting (EBCHR) and Plasma Arc Melting (PAM) processes have been developed to produce premium quality titanium for applications in the aerospace industry. These processes are designed to remove a variety of inclusions including hard-alpha (Ti-N) particles using a combination of dissolution and density separation. Dissolution, the focus of this work, relies on solute mass transport to drop the melting temperature of the particle to below or at the local temperature of the fluid. Therefore, understanding the dissolution kinetics of Ti-N particle inclusions in liquid titanium is important for the purpose of process design.

Previous studies have shown that the dissolution of Ti-N particles in liquid titanium alloys is strongly affected by the particle composition (N content), temperature, residence time and the velocity of the particle relative to the fluid [1-3].

In this work, a CFD model was developed to examine the kinetics of dissolution of a single stationary Ti-N particle within flowing liquid titanium. The model was developed using the commercial CFD software package ANSYS-CFX.

2. Model description

The CFD particle dissolution model was developed using the multicomponent flow solver available in ANSYS-CFX (v15.0). The model is assumed isothermal with a constant temperature of 1700°C. The multicomponent model consists of two components: nitrogen (solute) and liquid titanium (solvent).

2.1 Governing equations

Since the model is isothermal the energy conservation equation is not solved. The governing partial differential equations describing mass conservation and momentum conservation are presented in Equations (1) and (2):

$$\frac{\partial \rho}{\partial t} + \nabla \cdot (\rho \mathbf{u}) = 0 \quad (1)$$

$$\frac{\partial \rho \mathbf{u}}{\partial t} + (\mathbf{u} \cdot \nabla)\rho \mathbf{u} = \nabla \cdot \boldsymbol{\tau} - \nabla \mathbf{p} + \boldsymbol{S}_M \quad (2)$$

where ρ (kg·m^{-3}) is the density, t(s) is time, **u** (m·s^{-1}) is the velocity vector, μ(Pa·s) is the viscosity, τ (N·m^{-2}) is the shear stress tensor, **p**(N·m^{-2}) is the pressure force vector.

ANSYS-CFX [5] solves Equations (1) and (2) for the entire domain including the "solid" Ti-N particle. In order to suppress flow within the solid material, the dynamic viscosity is ramped linearly with increasing solid fraction from 0.003 Pa·s in the liquid [6] to 3 Pa·s in the solid. In addition, a flow dampening term $\boldsymbol{S}_M$ was added based on Darcy's law. Both approaches were found to be necessary to fully curtail flow in the solid.

The compositional transport equation that calculates mass transport of solute (nitrogen) is shown in Equation (3):

$$\frac{\partial \rho m_N}{\partial t} + \nabla \cdot \rho m_N \mathbf{u} = D_N \nabla^2 \cdot \rho m_N \quad (3)$$

where m_N is the mass fraction of nitrogen, and D_N (m·s^{-2}) is the diffusivity of nitrogen in titanium. The material properties, such as liquid viscosity, diffusivity and density, used in the model are summarized in Table 1[3,6].

2.2 Mesh and geometry

The dimensions of the domain used in the model are 6mm × 6mm × 18mm as shown in Figure 1(a). Liquid titanium flows into the domain from the left side surface (Inlet in Figure 1(a)) and out of the domain from the right surface (Outlet in Figure 1(a)). A cross-sectional view of the mesh, normal to the flow direction, is shown in Figure 1 (b). The initial radius of the particle is assumed to be 1mm.

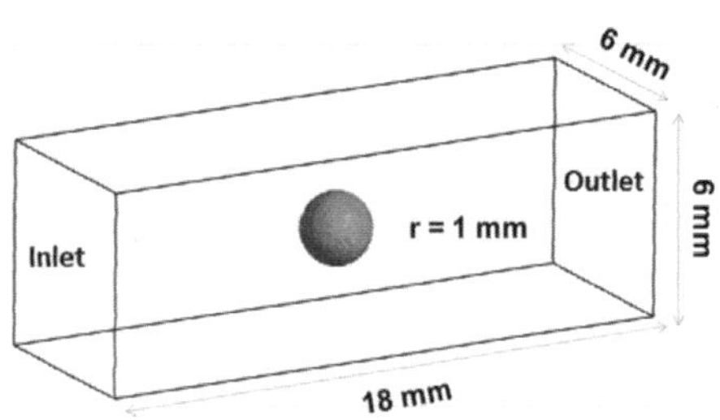

(a)

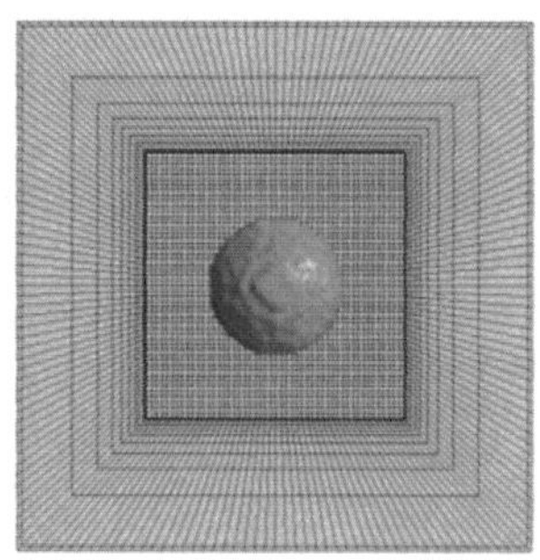

(b)

Figure 1. Geometry of the computational domain (a) and cross-sectional view of the mesh (b) used in the model.

2.3 Initial and boundary conditions

As would be expected from the Ti-N phase diagram – see Figure 2 – α-Ti (HCP) and β-Ti (BCC) may be expected to be present within the particle at 1700°C dependent on the nitrogen content. Assuming an initial nitrogen content of 5wt% for the bulk of the particle, the solid particle would start as pure α–Ti due to the stabilizing effect of the N. As the nitrogen is removed from the particle surface, the α-Ti would be expected to transform to β-Ti. In the model a small layer of β-Ti, 10μm in thickness is assumed to be present initially in order to setup the appropriate boundary conditions. The various phase boundary interfaces are assumed to be at equilibrium. The interface compositions and diffusivity for the various phases are given in Table 1, see also Figure 2 (b). The bulk fluid is assumed to initially contain no nitrogen. At the inlet boundary, the fluid velocity is set equal to the slip velocity of the particle. A free slip adiabatic wall boundary condition is applied to all the other surfaces of the domain.

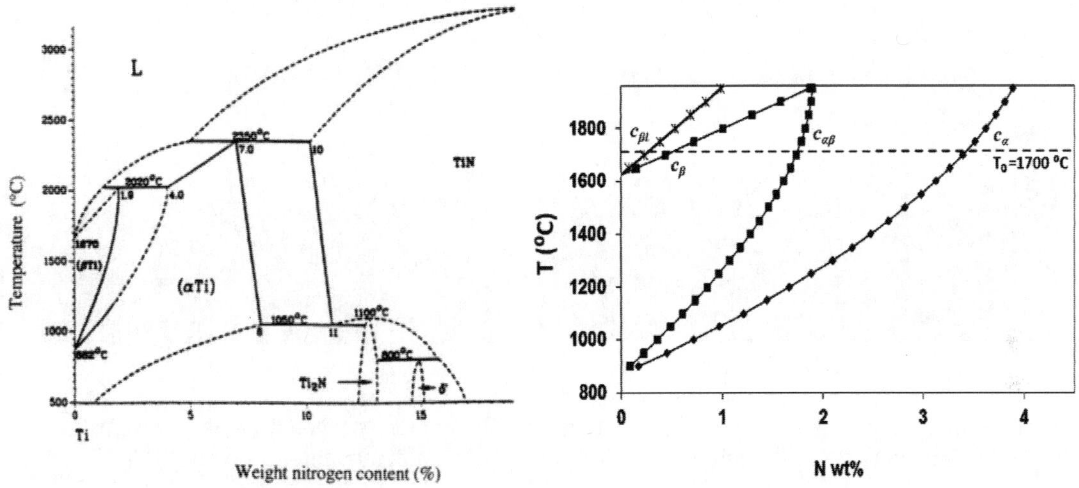

(a) (b)

Figure 2. Thermodynamic phase diagram of Ti-N (a) and replot (b) of the zoom-in area around the temperature 1700°C used in the model [7].

Table 1. Calculation parameters used in the model.

	Phase α	Phase β	Liquid
Initial radius/thickness (m)	10^{-3}	10^{-5}	--
Equilibrium N mass fraction at phase boundaries	0.035@ α/αβ	0.017@ αβ/β 0.0025@β/βl	0.0015@βl /liquid
Initial N mass fraction	0.05	0.015	0
Viscosity (Pa·s)	3	3 in β; 0.003-3 in βl	0.003[6]
Diffusivity ($m^2 \cdot s^{-1}$) [2]	$3.1\times10^{-7}\exp\left(-\frac{168\times10^3}{RT}\right)$	$4.6\times10^{-5}\exp\left(-\frac{185\times10^3}{RT}\right)$	5.13×10^{-9}
Density ($kg \cdot m^{-3}$) [3]	4520	4370	4100
Slip velocity ($m \cdot s^{-1}$)		0.005 - 0.1	
Temperature (°C)		1700	

3. Results and discussions

3.1 Nitrogen concentration profile

Figure 3 shows the calculated nitrogen concentration profile in the liquid and particle, and the fluid velocity vectors on the cross-sectional plane at various simulation times during particle dissolution. The black outline highlights the solid particle areas surrounded by the β/βl interface. It is shown that the particle does not keep its spherical shape as it dissolves. The surface facing upstream is preferentially eroded by the incoming fluid due to the enhanced convective mass transfer.

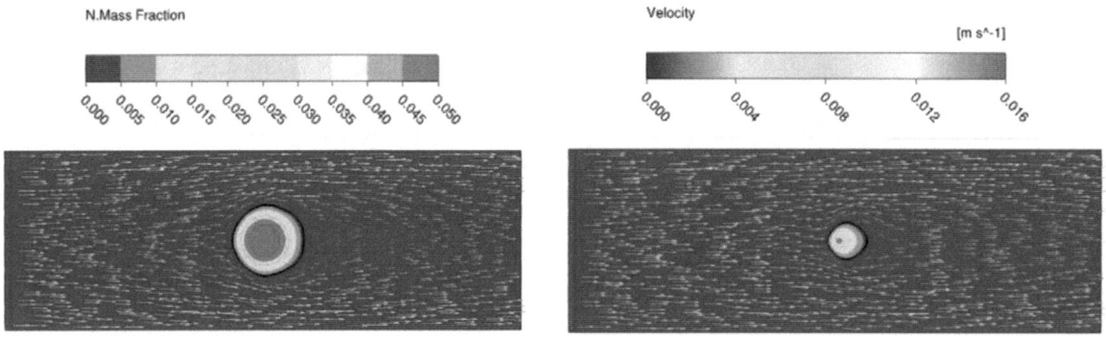

(a) (b)

Figure 3. Nitrogen concentration profile (contour) and fluid velocity (vectors) at the cross-sectional plane at (a) 30s, and (b) 120s predicted by the particle dissolution model (slip velocity = 0.01 m·s^{-1}).

3.2 Evolution of particle size during dissolution

Figure 4 shows the evolution of the radius of the particle estimated with the model for a series of different slip velocities. Initially, the diffusion rate to the particle/liquid interface is higher than the transport rate from the particle/liquid interface, causing the particle to grow, which is observed in all the cases examined.

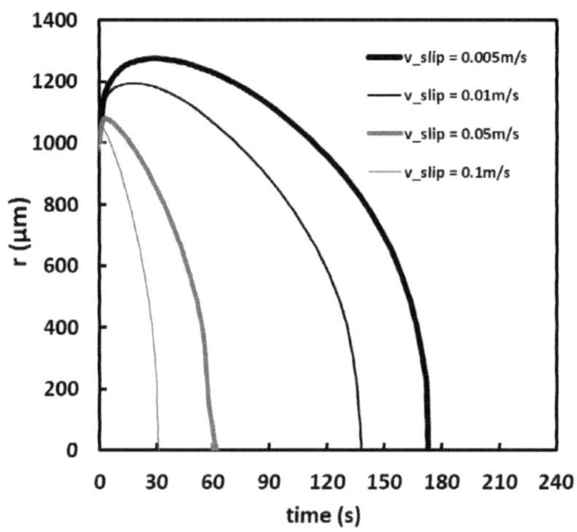

Figure 4. Evolution of the particle radius predicted at four different slip velocities.

The result shows a range of dissolution times from 30s, for a slip velocity of 0.1 m·s^{-1}, to in excess of a 160s, for a slip velocity of 0.005 m·s^{-1}. Obviously, prediction of dissolution times in an industrial setting will hinge on a good understanding of slip velocities experienced by a typical particle in an industrial furnace. Note: the various equilibrium phase boundaries within the particle and their evolution with time have not been plotted for the sake of brevity.

3.3 Effective mass transfer coefficient

An effective mass transfer coefficient, *MTC*, has been calculated based on the result of the particle dissolution model, using the following expression [8]:

$$MTC_e = \frac{(m_N^t - m_N^{t-\Delta t})}{A \cdot \Delta t \cdot (C_{\beta l}\rho_\beta - C_l\rho_l)} \tag{4}$$

where m_N^t (kg) is the total mass of nitrogen within the particle at the current time step t; Δt(s) is the time step size; A(m^2) is the surface area of the particle; $C_{\beta l}$is the equilibrium nitrogen mass fraction at the β/β+l interface and C_l is the nitrogen mass fraction in the bulk liquid; and ρ_β(kg·m^{-3}) and ρ_l(kg·m^{-3}) are the densities of β and bulk liquid, respectively. The effective mass transfer coefficient was calculated at the four slip velocities and has been plotted as solid lines in Figure 5(a), together with the *MTC* estimated using the Ranz-Marshall correlation for the various slip velocities, plotted as the dashed lines.

The results show that the effective MTC_e calculated from the CFD dissolution model follows the same trends as the *MTC* predicted by the Ranz-Marshall correlation – i.e. the variation in mass transfer coefficient with slip velocity and particle radius. However, there is an appreciable difference, with the Ranz-Marshall correlation tending to over predict the mass transfer coefficient, particularly for the lower slip velocities examined. Potential reasons for this have been examined, but not presented here for sake of brevity.

To offset the difference, a "tuning" coefficient, k, has been proposed for the Ranz-Marshall correlation to bring it into alignment with the effective MTC_e predicted by the CFD model – i.e. $MTC_e = k{\cdot}MTC$. Figure 5 (b) shows a comparison between the CFD-based MTC_e and the result for the tuned Ranz-Marshall relationship. The fitting coefficients are presented as a function of the slip velocities in Figure 6, together with a linear equation describing the dependence of MTC_e on v_{slip}. At the highest slip velocity studied in this work, v_{slip} = 0.1 m·s^{-1}, the fitting coefficient is one, meaning the unmodified Ranz-Marshall correlation is applicable.

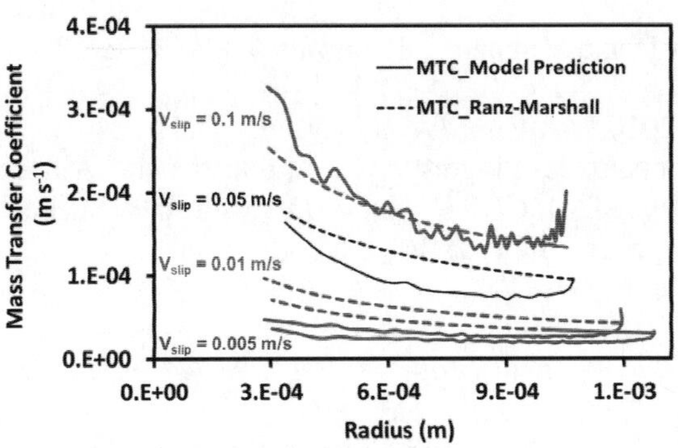

(a)

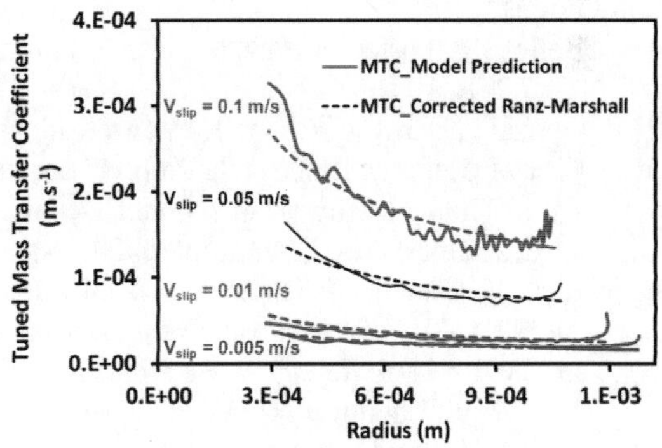

(b)

Figure 5. Comparison of the effective mass transfer coefficient calculated by the CFD particle dissolution model and the original Ranz-Marshall correlation (a) and modified Ranz-Marshall correlation (b).

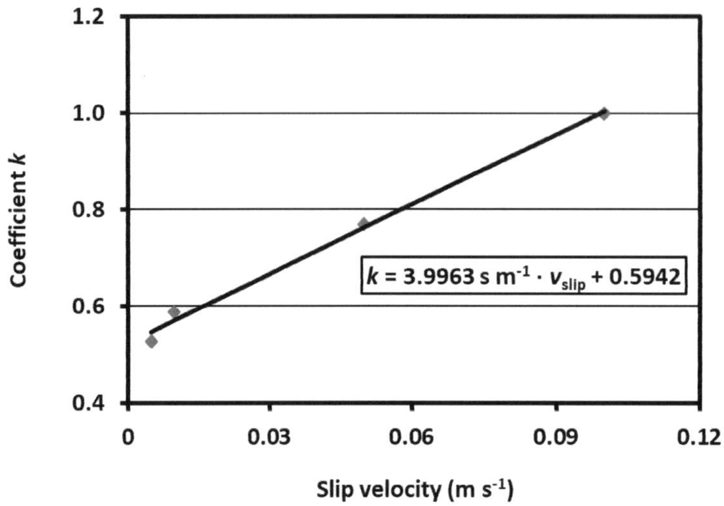

Figure 6. Coefficients used to correct Ranz-Marshall correlation for the effective mass transfer coefficient.

4. Conclusions

A CFD particle dissolution model has been developed to predict the dissolution time of a Ti-N particle in liquid titanium. The effect of particle/fluid slip velocity was assessed. The results show that the dissolution time decreases as the slip velocity increases. The effective mass transfer coefficient was estimated based on the CFD model's predictions and compared with the conventional Ranz-Marshall correlation. The result shows that a correction is needed to the Ranz-Marshall correlation to more accurately describe the mass transfer kinetics during the Ti-N particle dissolution process, particularly at lower slip velocities.

References

[1] Bellot J P, Foster B, Hans S, Hess E, Ablitzer D and Mitchell A 1997 *Metall. and Mater. Trans. B* **28B** 1001
[2] Ghazal G, Jardy A, Chapelle P, and Millet Y 2010 *Metall. and Mater. Trans. B* **41B** 646
[3] Ghazal G, thesis: Etude de la Propreté Inclusionnaire des Lingots VAR - Application aux Alliages de Titane, Institut Jean Lamour,Departement SI2M, CNRS-Nancy-Universite-UPVM, Ecole des Mines, Parc de Saurupt, 54042 Nancy Cedex, France,2010
[4] Ranz W E and Marshall Jr. W R 1952 *Chem. Eng. Prog.* **48** 141.
[5] *ANSYS CFX-Solver Theory Guide*, release 15.0, ANSYS Inc., PA November 2013
[6] I.C. Mills 2002 *Recommended Values of Thermophysical Properties for Selected Commercial Alloys*, Woodhead, Cambridge, England
[7] Wriedt H A and Murray J L 1986 *Binary Alloy Phase Diagrams* vol 3, eds Massalski T B(ASM) p 2705
[8] Ou J, Chatterjee A, Cockcroft S, Maijer D, Reilly C, and Yao L 2015 *Int. J. Heat Mass Transf.* **80** 386

IOP Conf. Series: Materials Science and Engineering **529** (2019) 012035 doi:10.1088/1757-899X/529/1/012035

Kinetics of rapid crystal growth: phase field theory *versus* atomistic simulations

P K Galenko[1,2,5], **A Salhoumi**[2,3] **and V Ankudinov**[2,4]

[1] Friedrich-Schiller-Universität-Jena, Faculty of Physics and Astronomy, Otto Schott Institute of Materials Research, 07743 Jena, Germany
[2] Ural Federal University, Theoretical and Mathematical Physics Department, Laboratory of Multi-Scale Mathematical Modeling, 620000 Ekaterinburg, Russia
[3] University of Hassan II Casablanca, Faculty of Sciences Ben M'Sik, Department of Physics, Laboratory of Condensed Matter Physics (LPMC), BP 7955 Casablanca, Morocco
[4] Udmurt State University, Department of Physics and Energetics, Laboratory of Condensed Matter Physics, 426034 Izhevsk, Russia

E-mail: peter.galenko@uni-jena.de, ahmedsalhoumi@gmail.com, vladimir@ankudinov.org

Abstract. Kinetics of crystal growth in undercooled melts is analyzed by methods of theoretical modeling. Special attention is paid to rapid growth regimes occurring at deep undercoolings at which non-linearity in crystal velocity appears. A traveling wave solution of the phase field model (PFM) derived from the fast transitions theory is used for a quantitative description of the crystal growth kinetics. The "velocity – undercooling" relationship predicted by the traveling wave solution is compared with the data of molecular dynamics simulation (MDS) which were obtained for the crystal-liquid interfaces growing in the ⟨100⟩-direction in the $Ni_{50}Al_{50}$ alloy melt.

1. Introduction

Interface kinetics at high driving forces of growth/melting of crystals have received special attention due to advanced experimental techniques [1, 2, 3]. In fact, the rapid solidification front may undergo diffusionless (chemically partitionless) transformation which may proceed in wide or narrow intervals of driving forces [1, 3, 4]. In such case, the crystal growth kinetics often exhibit the non-linear behavior [5] which is not predicted by the traditional kinetic theories or phase field models (PFM) based on local thermodynamic equilibrium [6]. Recent attempts to describe non-linearity in the crystal growth velocity by the local equilibrium phase field model with the mobility extracted from molecular dynamics simulation (MDS) are limited so far by a small interval of driving forces [7] in comparison with a wide range of undercooling attained in atomistic simulations for the crystal growth kinetics [8, 9, 10, 11, 12].

The present work extends the kinetic equation formulation presented for description of growth kinetics of Ni-crystal in Refs. [13, 14] to investigation of crystal growth kinetics from the undercooled $Ni_{50}Al_{50}$ melt. Tang and Harrowell [12] published the data of MDS for the growth of $Ni_{50}Al_{50}$ crystals growing in the ⟨100⟩-direction which are used by us as a reference to check consistency of our kinetic equation and consistency of the well-known Wilson-Frenkel kinetic equation formulated as an outcome of the diffusion limited theory (DLT).

[5] Corresponding author

2. Diffusion Limited Theory

In application to the melting/crystallization, the kinetic rate theory (or thermally activated growth theory [15]) compares two atomic fluxes at the moving crystal-liquid interface: the first one goes from the liquid over to crystal per unit time at a single kink or atomic micro-roughness, and the second one comes from crystal to liquid if the atomic configuration in liquid favors to solid atoms detachment. This results in the non-zero interface velocity [4] which is proportional to the temperature dependent kinetic growth coefficient, $\beta_k(T)$, having various atomistic theories adopted to the concrete growth mechanism of crystals depending of the absolute temperature T and the Gibbs free energy change on transformation ΔG given by

$$\Delta G = G_s(T,C) - G_l(T,C) \left\{ \begin{array}{ll} < 0, & \text{solidification}, \\ > 0, & \text{melting}, \end{array} \right. \tag{1}$$

where $G_l(T,C,\phi)$ and $G_s(T,C,\phi)$ are the Gibbs free energies of the liquid and solid phases, respectively.

The diffusion-limited mechanism of growth as was analyzed by Wilson and Frenkel [16, 17], using the approximation $\Delta G \ll k_B T$ and a simplest expression for the driving force [18],

$$\Delta G = \Delta H_f(-\Delta T)/T_m, \quad \Delta T = T_m - T, \tag{2}$$

yields the Wilson-Frenkel equation taking into account the liquid diffusivity and expressing the solid-liquid interface velocity V as

$$V = \beta_k^{(DLT)} \Delta T = \frac{a}{\lambda^2} f_0 D(T) \frac{\Delta H_f}{k_B T_m^2} \Delta T, \tag{3}$$

with $\beta_k^{(DLT)}$ stands for the kinetic coefficient in the framework of DLT, a is the spacing between crystalline layers, λ is the mean free path of atoms, f_0 is the fraction of collisions with the crystal that contributes to the growth of the crystal, ΔH_f is the melting enthalpy, $D(T)$ is the diffusion coefficient taken as [4] $D(T) = D_0 \exp(-\Delta E_B/k_B T)$ with the diffusion prefactor $D_0 = \Lambda^2 \widetilde{\nu}/6$, given by [19], where $\widetilde{\nu}$ is the frequency of thermal vibrations of an atom in the crystal and the liquid, Λ is an elementary diffusive jump distance of particles in the liquid, E_B is the activation energy for diffusion, T_m is the melting (crystallization) point temperature, k_B is the Boltzmann constant and ΔT is the undercooling which is necessary for non-zero attachment/detachment of atoms in the case of solidification/melting. This undercooling is defined in the dendrite growth models as the "kinetic undercooling" [2, 3, 20, 21].

In Eq. (3), DLT introduces the diffusion transport coefficient $D(T)$ of the supercooled liquid and, therefore, the kinetic coefficient exhibits the strong temperature dependence associated with an activated process. Quantitative estimations of kinetic growth coefficients for various crystal growth models can be found in Refs. [8, 11] as obtained from atomistic simulations and in Ref. [20] as obtained from experimental measurements. However, DLT has difficulties to quantitatively describe the growth kinetics for a wide range of temperatures [8, 9]. The results of calculations by Eq. (3) for pure Ni in comparison with data of MDS [5, 11] are given in Ref. [14] from which follows that (*i*) in the relatively small range of undercooling and overheating, the MDS-data exhibits linear behavior for the interface velocity, both in melting and in crystallization that is well described by DLT; (*ii*) for the large range of undercooling, the predictions of DLT behaves inconsistently in comparison with MDS-data qualitatively and, as a consequence, quantitatively.

3. Phase Field Model

Consider a binary mixture consisting of solvent and solute undergoing phase transition, solidification/melting, from the undercooled/overheated state for which the free energy is described by

$$\mathcal{G} = \int_{v_0} \left[\frac{\varepsilon_\phi^2}{2} |\vec{\nabla}\phi|^2 + G(T, C, \phi, \partial\phi/\partial t) \right] dv_0, \tag{4}$$

where v_0 is the volume of the mixture, ε_ϕ is the gradient energy coefficient related to the interface energy γ, ϕ is the phase field variable defined as $\phi = 0$ in the liquid phase and $\phi = 1$ in the solid phase, $\partial\phi/\partial t$ is the gradient flow for the phase field, and C is the solute concentration.

Introducing the phase field ϕ and gradient flow $\partial\phi/\partial t$ as independent thermodynamic variables in the Gibbs potential $G(T, C, \phi, \partial\phi/\partial t)$ can be considered in the full analogy with Newtonian mechanics, where the initial position and velocity of a particle must be specified to determine their evolution and velocity. Indeed, if inertial effects are sufficiently low in comparison with dissipative effects during phase field propagation, $\partial\phi/\partial t$ will be determined directly by a dynamical equation in terms of ϕ and its gradient. Otherwise, ϕ and $\partial\phi/\partial t$ will be independent and an equation for $\partial^2\phi/\partial t^2$ must be found [22, 23]. Therefore, the Gibbs potential $G(T, C, \phi, \partial\phi/\partial t)$ which is a combination of local equilibrium contribution, $G_{eq}(T, C, \phi)$, and local non-equilibrium contribution, $G_{neq}(T, \partial\phi/\partial t)$, takes the following form,

$$G(T, C, \phi, \partial\phi/\partial t) = [1-p(\phi)]G_l(T, C)+p(\phi)G_s(T, C)+W_\phi(T, C)g(\phi)+(\alpha_\phi(T)/2)\left(\partial\phi/\partial t\right)^2, \tag{5}$$

where $p(\phi)$ is the interpolation function $p(\phi) = (3 - 2\phi)\phi^2$, $g(\phi)$ is the double-well function $g(\phi) = (1 - \phi)^2\phi^2$ [24], $W_\phi(T, C)$ is the barrier between phases, $\alpha_\phi(T)$ is phenomenological coefficient being proportional to the relaxation time τ_ϕ of the gradient flow $\partial\phi/\partial t$ which is introduced as independent thermodynamic variable [25]. The latter term in Eq. (5), which is proportional to $(\partial\phi/\partial t)^2$, is considered as the kinetic energy term added to Gibbs potential [26].

A stable evolution of the entire system is given by the Lyapunov condition of non-positive change of the total Gibbs free energy in time [27]. Application of this condition to the functional (4) yields the following phase field equation [13]

$$\tau_\phi \frac{\partial^2\phi}{\partial t^2} + \frac{\partial\phi}{\partial t} = D_\phi \nabla^2\phi - M_\phi \left[\Delta G \frac{dp(\phi)}{d\phi} + W_\phi(T, C) \frac{dg(\phi)}{d\phi} \right], \tag{6}$$

where the Gibbs free energy difference ΔG is defined by Eq. (1) and the diffusion coefficient of the phase field $D_\phi(T) = \varepsilon_\phi^2 M_\phi(T)$ is essentially depends on the temperature if the phase transition is considered in a wide temperature range [28].

The hyperbolic equation (6) describes relaxation of two variables: relaxation of the slow ϕ-field by the first time derivative and relaxation of the gradient flow, $\partial\phi/\partial t$, by the second time derivative. In this sense, due to introducing relaxation of $\partial\phi/\partial t$, Eq. (6) describes the evolution of the local non-equilibrium system. Further we show how it is important to use of the gradient flow relaxation and, consequently, what is the role the local non-equilibrium in fast crystal growth kinetics. On one hand, Eq. (6) admits, in the equilibrium state $\Delta G = 0$, $G_s(T, C) = G_l(T, C)$, one dimensional steady solution $\phi(x) = (1/2)\left[1 - \tanh\left(x/\delta_I\right)\right]$, at $\partial\phi/\partial t = 0$ along the spatial x-axis with the stationary width of diffuse interface $\delta_I = \varepsilon_\phi\sqrt{2/W_\phi(T, C)}$ and the surface energy of the crystal-liquid interface $\gamma = \delta_I W_\phi(T, C)/6$. On the other hand, Eq. (6) has, in the dynamic state $\Delta G = G_s(T, C) - G_l(T, C) \neq 0$, one dimensional traveling-wave solution $\phi(x, t) = (1/2)\left[1 - \tanh\left((x - Vt)/\ell\right)\right]$ [13, 29]. This particular solution with the hyperbolic tangent function follows from the general set of analytical solutions of Allen-Cahn-type equations [30] which is given in the present model by Eq. (6). In this solution, the crystal growth velocity V is limited by a maximum speed of phase field propagation, V_ϕ because the phase field itself dictates the interface shape and its velocity, i.e. $V < V_\phi$, [13]. In the simplest approximation $\Delta G \ll k_B T$ (see Eq. (2)) the velocity has the following form [31]

$$V = \beta_k^{(PFM)} \Delta T = \frac{D_\phi(\Delta T)\Delta H_f}{\gamma T_m \sqrt{1 + \left[\frac{D_\phi(\Delta T)\Delta H_f}{\gamma T_m V_\phi(\Delta T)} \Delta T\right]^2}} \Delta T, \tag{7}$$

which is consistent with the approximation of kinetic equation (3) and for which the kinetic coefficient, $\beta_k^{(PFM)}$, depends on the undercooling ΔT. The maximum speed V_ϕ of phase field propagation in Eq. (7) is defined by the diffusion of the phase field as

$$V_\phi(\Delta T) = \sqrt{D_\phi(\Delta T)/\tau_\phi}, \tag{8}$$

where the relaxation time τ_ϕ is taken as a constant independent from the temperature in the present analysis.

The diffusion coefficient of phase field in Eqs. (7) and (8) is given by

$$D_\phi(\Delta T) = D_\phi^0 \exp\left(-\frac{E_A}{T_m - \Delta T - T_A}\right), \tag{9}$$

where the diffusion factor D_ϕ^0, the energetic barrier E_A and the pseudo-glass transition temperature T_A are parameters of the phase field propagation. The form of Eq. (9) is similar to the form of phase field mobility $M_\phi(T)$ given in Ref. [28] and it shows that, as soon as the undercooling approaches its critical value, the phase field diffusion begins its steep decrease down to its zero value.

Table 1: Material parameters for $Ni_{50}Al_{50}$ used in calculations

Parameter	Set 1	Set 2	Reference
Melting temperature, T_m (K)	1520	1520	[9, 12]
Enthalpy of melting, ΔH_f ($J \cdot m^{-3}$)	1.74×10^9	1.74×10^9	[21]
Interface energy, γ ($J \cdot m^{-2}$)	0.24	0.24	[21]
Pseudo-glass transition temperature, T_A (K)	950	950	present work
Relaxation time, τ_ϕ (s)	8.22×10^{-11}	$\to 0$	present work
Diffusion factor, D_ϕ^0 ($m^2 \cdot s^{-1}$)	1.10×10^{-8}	1.06×10^{-8}	present work
Energetic barrier, E_A (K)	368.19	403.07	present work

4. Predictions of DLT and PFM in comparison with MDS-data

Previously, the nonlinearity in the growth kinetics having a form of the velocity with saturation has been described for pure Ni [13, 14]. However, there are other type of non-linearity in the "velocity – undercooling" relationship which is presented by the crystal growth velocity curve having a maximum at fixed undercooling [8, 9, 10, 12]. Such curves are typical for glass forming metals and alloys. PFM-solutions summarized by Eqs. (7)-(9) shall now be tested against MDS-data found in growth kinetics of $Ni_{50}Al_{50}$ crystals [12]. Note that $Ni_{50}Al_{50}$ belongs to congruently melting alloys which are solidifying without chemical segregation. Therefore we do not need to introduce constitutional effects in consideration and Eqs. (7)-(9) can be used to describe the growth of $Ni_{50}Al_{50}$ crystals without chemical contributions. All calculations have

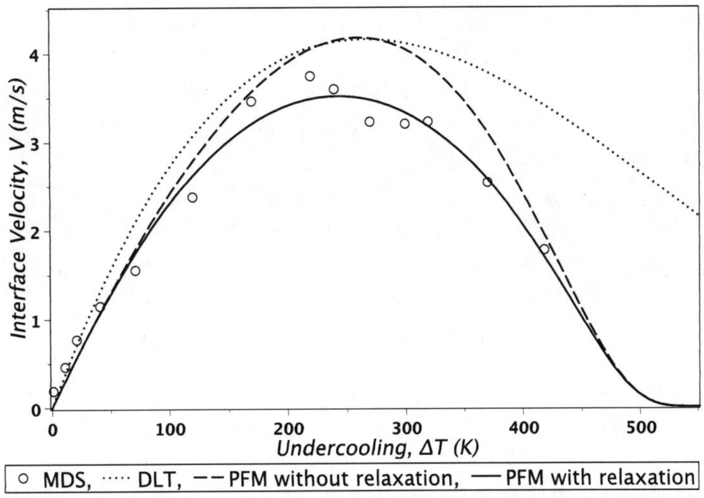

Figure 1: Comparison of growth kinetic equations based on Diffusion Limited Theory (DLT)- and Phase Field Model (PFM) (curves) with data kinetics of molecular dynamics simulations (MDS) of Tang and Harrowell [12] for $Ni_{50}Al_{50}$-crystals in the $\langle 100 \rangle$ direction (○). Calculations are given by: (*i*) DLT (······), Eq. (3) with $\lambda \approx 2.25a$, $\Lambda \approx 4a$ and the other constants from [9, 12]; (*ii*) PFM without relaxation (- - - -), Eqs. (7)-(9) with $\tau_\phi \to 0$, $V_\phi \to \infty$ and all other parameters are given by Set 2 from Table 1; (*iii*) PFM with relaxation (——), Eqs. (7)-(9) with $\tau_\phi \neq 0$, finite V_ϕ and other parameters given by Set 1 from Table 1.

been made using material parameters for $Ni_{50}Al_{50}$ from Table 1, where the relaxation time, τ_ϕ, the diffusion factor, D^0_ϕ, and the energetic barrier, E_A, are considered as free parameters at a fixed pseudo-glass transition temperature, T_A. They can be obtained from molecular dynamics simulation (e.g., τ_ϕ) and from phase field simulations (e.g., D^0_ϕ and E_A).

Figure 1 shows that Wilson-Frenkel equation (3) based on DLT (dotted curve) describes well the molecular dynamics data for the growth of $Ni_{50}Al_{50}$ crystals (open circle) only for smallest values of undercoolings. Beyond the small undercooling, Wilson-Frenkel equation (3) disagrees with MDS-data despite its tendency to reach the maximum value of growth velocity.

Two cases for solution of Eqs. (7)-(9) have been considered and plotted in Figure 1: with the local non-equilibrium effect ($\tau_\phi \neq 0$, V_ϕ is finite) and without it ($\tau_\phi \to 0$, $V_\phi \to \infty$). With no local non-equilibrium effects, the predicted velocity well describes data of atomistic modeling only at very small and very high undercooling (see dashed curve which is obtained with the new parameters of E_A and D^0_ϕ from Table 1 providing the better fit to data of simulations). If, however, the local non-equilibrium effect is included, we recover perfectly the atomistic simulation data in the entire undercooling range and the growth rate (solid curve). The solidification kinetics of glass forming alloys is well described by the theory which includes local non-equilibrium effects in the form of relaxation of the gradient flow in the phase field. Therefore, good comparison with MSD-data confirms our initial theoretical assumption in PFM about the predominant influence of local non-equilibrium effects in crystal growth under large driving forces.

5. Conclusions

- The non-linearity in the "interface velocity – kinetic undercooling" behavior was described by the traveling wave solution of the PFM for $Ni_{50}Al_{50}$ crystals growing in the $\langle 100 \rangle$-direction. This non-linearity was found for the velocity with maximum at the fixed value of undercooling.

- Predictions of the developed PFM and DLT have been compared with the data of MDS for $Ni_{50}Al_{50}$-crystals growing in a wide range of undercoolings.
- The Wilson-Frenkel equation based on DLT describes the data of MDS only for limited range of undercoolings and disagrees elsewhere despite its tendency to reach the maximum velocity at fixed undercooling.
- Using the traveling wave solution of PFM the MDS-data are predicted perfectly if the deviation from local thermodynamic equilibrium is taken into account in the form of local relaxation of the gradient flow of the phase field. Such good prediction of MSD-data confirms an existence of the local non-equilibrium effect on the growth kinetics of crystals under large driving forces.

Acknowledgments

This work was supported by the Russian Science Foundation [grant number 16-11-10095] and the German Space Center Space Management under contract number 50WM1541. A.S. thanks Prof. M. Bennai for incorporating the present work within the research activities of LPMC.

References

[1] Miroshnichenko I S 1982 *Quenching from the Liquid State* (Metallurgia, Moscow)
[2] Herlach D M and Matson D M 2012 *Solidification of Containerless Undercooled Melts* (Wiley-VCH,Weinheim)
[3] Herlach D M, Galenko P K and Holland-Moritz D 2007 *Metastable Solids from Undercooled Melts* (Elsevier, Amsterdam)
[4] Chernov A A 1984 *Modern Crystallography III. Crystal Growth* (Springer, Berlin)
[5] Hoyt J J, Sadigh B, Asta M and Foiles S M 1999 *Acta Mater.* **47** 3181–7
[6] Berghoff M, Selzer M and Nestler B 2013 *Scientific World J.* **2013** 564272
[7] Guerdane M and Berghoff M 2018 *Phys. Rev.* B **97** 1441051–10
[8] Ashkenazy Y and Averback R S 2007 *Europhys. Lett.* **79** 26005
[9] Kerrache A, Horbach J and Binder K 2008 *Europhys. Lett.* **81** 58001
[10] Chan W L, Averback R S, Cahill D G and Ashkenazy Y 2009 *Phys. Rev. Lett.* **102** 095701
[11] Mendelev M I, Rahman M J, Hoyt J J and Asta M 2010 *Mod. Simul. Mat. Sc. Eng.* **18** 074002
[12] Tang C and Harrowell P 2013 *Nat. Mat.* **12** 507–11
[13] Salhoumi A and Galenko P K 2016 *Physica* A **447** 161–71
[14] Salhoumi A and Galenko P K 2017 *IOP Conf. Series: Mat. Sci. Eng.* **192** 12014
[15] Christian J W 1975 *The Theory of Transformations in Metals and Alloys* (Pergamon, Oxford)
[16] Wilson H A 1900 *Phil. Mag.* **50** 238–50
[17] Frenkel J 1946 *Kinetic Theory of Liquids* (Oxford University Press, New York)
[18] Thompson C V and Spaepen F 1979 *Acta Metal.* **27** 1855–9
[19] Jackson K A 2002 *Interface Sci.* **10** 159–169
[20] Herlach D M, Simons D and Pichon P Y 2018 *Phil. Trans. Roy. Soc.* A **376** 201702051–13
[21] Hartmann H, Galenko P K, Holland-Moritz D, Kolbe M, Herlach D M and Shuleshova O 2008 *J. Appl. Phys.* **103** 073509
[22] Galenko P and Jou D 2005 *Phys. Rev.* E **71** 046125
[23] Galenko P, Danilov D and Lebedev V 2009 *Phys. Rev.* E **79** 051110
[24] Wang S L, Sekerka R F, Wheeler A A, Murray B T, Coriell S R, Braun R J and McFadden G B 1993 *Physica* D **69** 189–200
[25] Danilov D A, Lebedev V G and Galenko P K 2014 *J. Non-Equilibrium Therm.* **39** 93–111
[26] Jou D, Casas-Vazquez J and Lebon G 2010 *Extended Irreversible Thermodynamics* 4th ed (Springer, Berlin)
[27] Lebedev V, Sysoeva A and Galenko P 2011 *Phys. Rev.* E **83** 026705
[28] Novokreshchenova A A and Lebedev V G 2017 *Tec. Phys.* **62** 642–4
[29] Galenko P K, Abramova E V, Jou D, Danilov D A, Lebedev V G and Herlach D M 2011 *Phys. Rev.* E **84** 041143
[30] Nizovtseva I G and Galenko P K 2018 *Phil. Trans. Roy. Soc.* A **376** 20170202
[31] Galenko P K, Ankudinov V 2019 *Acta Mater.* (2019) doi: 10.1016/j.actamat.2019.02.018

Prediction of solidification structures in a 9.8 t steel ingot

Benjamin Gerin[1], Hervé Combeau[1], Miha Založnik[1], Isabelle Poitrault[2] and Maya Cherif[2]

[1] Université de Lorraine, CNRS, IJL, F-54000 Nancy, France
[2] ArcelorMittal Industeel, 56 rue Clemenceau, F-71201 Le Creusot, France

E-mail: benjamin.gerin@univ-lorraine.fr, herve.combeau@univ-lorraine.fr, miha.zaloznik@univ-lorraine.fr, isabelle.poitrault@arcelormittal.com, maya.cherif@arcelormittal.com

Abstract.
The control of the carbon macrosegregation level in steel ingots is important for the structural integrity of the final component. During solidification, the fragmentation of the columnar dendrites is an important source of equiaxed grains, and has a large influence on the macrosegregation and the grain structure. The goal of this study is to show that a numerical model that takes into account fragmentation can describe the formation of the structures and the macrosegregation during solidification of a large steel ingot. The present article describes the multiphase numerical model used in the simulations. The simulation results are compared to experimental data from a 9.8 t ingot cast in A5/6 steel by ArcelorMittal Industeel. The model can then be used to explain the formation of the observed structures. For example, we show that the structural transitions, first from columnar to equiaxed globular and then to equiaxed dendritic at the bottom of the ingot are a consequence of the concurrent transport and growth of the dendrite fragments from the columnar zone. Furthermore, we show that most of the structures are formed very early on during solidification, whereas macrosegregation develops much more gradually.

1. Introduction

During solidification of large steel ingots, the average concentration of alloying elements does not stay homogeneous throughout the ingot. Most alloying elements are more soluble in the liquid than in the solid phase, leading to an enrichment of the liquid as the solidification progresses. This inhomogeneity, called macrosegregation, can lead to variations of local mechanical properties, most notably when the carbon macrosegregation is high. During the forging of the ingot, the macrosegregation is carried through to the final component, which will therefore exhibit inhomogeneous mechanical properties. This is especially important for very large forged components such as pipes and reactor pressure vessels, where the large ingot size can lead to severe macrosegregation. Being able to predict and control macrosegregation is thus crucial for the structural integrity of the components.

The solidification structures (columnar, equiaxed, grain morphology) also influence segregation and mechanical properties. Thus accurately predicting both the macrosegregation and the solidification structures is an important issue for steel ingot producers. The prediction of the structures requires taking into account the fragmentation of the columnar dendrites in the model [1]. The present study uses the SOLID® multiscale and multiphysics model to predict the structure and macrosegregation of a 1.44 m diameter, 9.8 t steel ingot by taking into account the fragmentation during solidification. The originality of this study lies in prediction and description of the formation of the structures and the macrosegregation, which is not commonly found in literature for such large products.

Fragmentation of the columnar dendrites is an important source of equiaxed grains, especially in castings with no inoculation. Some secondary arms detach and can be swept into the bulk liquid, where they can grow and sediment, leading to a columnar to equiaxed transition. The detachment of secondary arms does not occur because of mechanical breakage [2] but rather due to local dissolution at the root of the secondary arms [3] or remelting of the arms due to temperature increase [4]. During normal coarsening of the columnar structure, secondary arm detachment is quite slow. However, certain situations massively increase fragmentation, such as unconstrained growth of the columnar front [5] or slowdown of the columnar tip velocity [6]. In addition, in steels, the peritectic transformation can also promote the increase in size of the equiaxed region [7].

2. Model used in the SOLID® software

The SOLID® software uses a 2D finite volume method to discretise the set of partial differential equations of a multiphase model (liquid, columnar, equiaxed and interdendritic liquid for both solid phases). The discretised volumetric elements will be referred to as "cells". A cell can contain both columnar and equiaxed solid, in addition to liquid. Cells containing both columnar and equiaxed grains are called mixed cells. Solidification begins with a columnar front starting on the mould walls which can grow and fragment. Fragmentation generates equiaxed grains in the undercooled region ahead of the columnar front. These equiaxed grains are not individually simulated, but are instead represented by average grain size and grain density values in each cell. Fragmentation is implemented using the model developed by Nicolas Leriche in his PhD [8]. Fragmentation uses two parameters: the fragmentation flux density (number of generated grains per surface area of columnar front per second) and the thermal gradient activation threshold. The threshold means that the fragmentation occurs only when the temperature gradient at the columnar front is locally smaller than a given threshold value.

The model calculates the movement of the liquid phases (intergranular and extragranular for both the equiaxed and columnar phases) and of the equiaxed grains. The grains are thus swept up by the liquid flow and grow or remelt depending on local undercooling, and sediment when they reach a large enough size. Grain movement in a cell is blocked when the extragranular liquid fraction is below 0.6. The columnar front in a cell is blocked when the envelope fraction of equiaxed grains in that cell reaches 0.5, leading to a columnar to equiaxed transition (CET). Unlike the columnar structure, equiaxed cells cannot progress from one cell to the next. However, they can generate an equiaxed to columnar transition (ECT), producing new columnar cells. In the model, ECT is set to occur only if there are no equiaxed grains in the surrounding liquid. When ECT starts, a new columnar front is started in the adjacent liquid cells, which can then grow, fragment and can itself be blocked. Further information about the model can be found in [1, 8, 9].

3. Structures and macrosegregation experimentally observed in the 9.8 t ingot

The focus of the present article is a 9.8 t experimental research ingot cast in A5/6 steel by Industeel. It has a diameter of 1.44 m and a height of 0.74 m, which is much smaller than typical commercial Industeel ingots. This ingot was cast specifically for research purposes and was not destined to produce a component. Ingots of this shape (large diameter compared to height) are used to make flat component such as thick plates and disks. The chemical composition of the alloy, given in table 1, was determined by spectroscopy just before casting. The mould is made

IOP Conf. Series: Materials Science and Engineering **529** (2019) 012036 doi:10.1088/1757-899X/529/1/012036

Table 1. Chemical composition in wt% of the A5/6 steel used in the 9.8 t ingot.

C	S	P	Si	Mn	Ni	Cr	Mo	Cu	Sn	Al	Fe
0.184	0.026	0.008	0.245	0.59	0.17	0.23	0.035	0.250	0.025	0.018	Balance

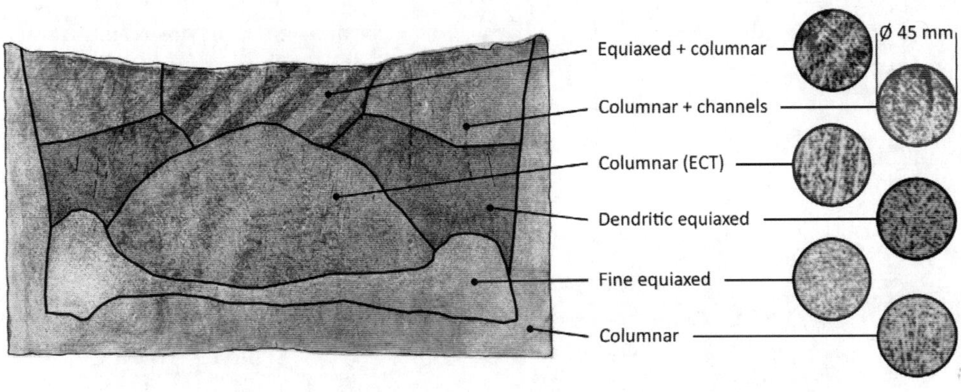

Figure 1. Experimentally observed structures of the 9.8 t ingot (Baumann print).

of cast iron and there is no hot top, however the liquid metal is covered by two layers of powder: an exothermal powder with an insulating powder on top. No inoculation was used during the casting. The total solidification time was less than 6 hours. After solidification, the ingot was cut in half along a diameter in order to analyse its macrostructure and segregation. Fig. 1 shows a sulphur print (Baumann method) of the ingot, which detects the sulphur distribution using photo-sensitive paper. This reveals the solidification structures such as dendrites and equiaxed grains. The different macrostructures were then visually identified and manually colour-coded. A columnar region (green) can be seen along the mould walls and the bottom part of the ingot. This bottom columnar layer is covered by an equiaxed structure with fine grains (light blue). In the centre of the ingot is a large columnar zone (orange) with an equiaxed structure of larger dendritic grains (dark blue) on each side. The top of the ingot is mostly columnar, with the centre region being a mix of columnar and equiaxed grains.

Spectroscopy analysis was performed on the open face of the ingot cross-section to measure carbon segregation. Shallow holes of diameter 6 to 8 mm were drilled and the analysis was performed on the resulting chips. The experimental measurements of carbon concentration are presented in fig. 2. Each data point is located by a black dot and its value is given as a percentage variation of the nominal concentration ($\Delta C/C_0$). Contour lines were manually drawn to show carbon macrosegregation. The bottom part of the ingot is negatively segregated, reaching -15% of the nominal composition. The top centre region, which is the last to solidify, features a large positive segregation, with a maximum of $+331\%$ relative to the nominal composition.

4. Numerical simulation

The solidification of the 9.8 t ingot was simulated using the SOLID® software. The mould geometry is simplified to a cylindrical shape, assuming an axial symmetry. In order to simplify the calculation, a binary Fe–C alloy is used, with 0.184 wt% carbon. The fragmentation parameters were determined by performing a sensitivity analysis on these parameters (not

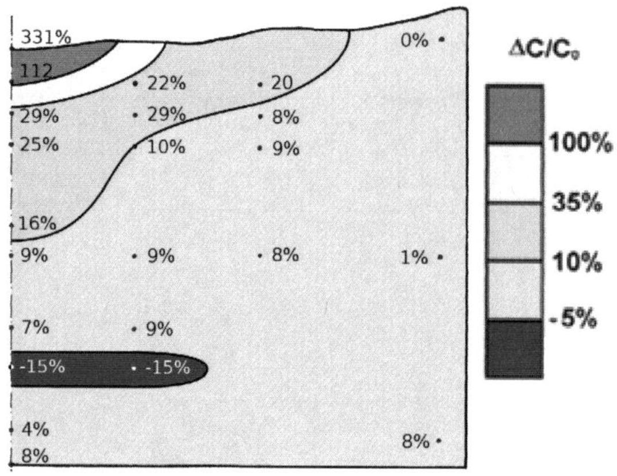

Figure 2. Experimental measurements of carbon concentration in a cross-section of the 9.8 t ingot, represented as percentage variation of nominal composition ($\Delta C/C_0$). Contour lines were manually drawn according to the values. Only half of the ingot is represented, the centreline is on the left.

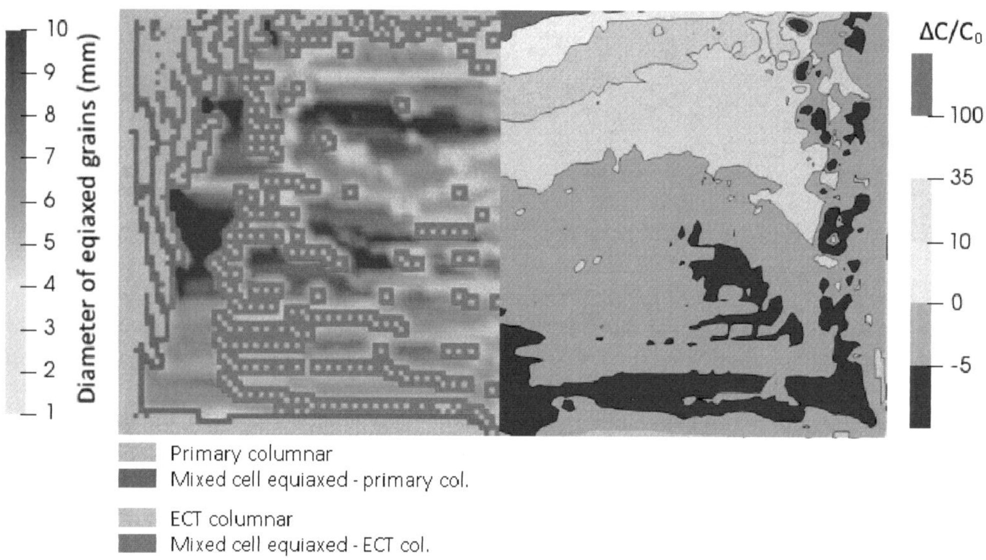

Figure 3. Simulation results for the 9.8 t ingot. The structure map is given on the left side, showing the columnar structure (either primary or from ECT) and the grain diameter for the equiaxed structure. On the right side, the carbon composition is given as a percentage of nominal composition.

discussed in this article), and then choosing the values that gave the results closest to the experimental ingot. The chosen fragmentation parameters are: fragmentation flux density of 2000 frag.m^{-2}.s^{-1} and activation of fragmentation at a thermal gradient below 100 K.m^{-1}. The formation of columnar grains is disabled on the top surface of the ingot, where the liquid metal is in contact with the exothermal powder. Filling was not simulated, the initial state is the mould entirely filled with liquid at a uniform 2°C superheat. With a mesh size of 1 cm and a time step of 0.1 s, the total computation time on one CPU core was 40 hours.

The final structure map and carbon segregation pattern at the end of the simulation are given in fig. 3. The same structure colour code was used as in fig. 1: green and orange for columnar, blue for equiaxed. In addition, mixed columnar-equiaxed cells are distinguished by a darker columnar colour (dark green or dark orange). Coarser mesh sizes give similar results, with a roughly equivalent columnar thickness. Mixed cells contain both columnar and equiaxed grains, and the equiaxed grains can block the growth of the columnar front if their envelope fraction reaches 0.5. The equiaxed grains in a blocked mixed cell can then lead to an ECT with a new columnar front growing in an adjacent cell. This is why along the mould wall and near the bottom, the columnar changes from primary (green) to ECT (orange) without any visible equiaxed (blue) cells in between. The simulated macrosegregation is similar to the experimental results (fig. 2), with a negative segregation at the bottom and a large positive segregation with the maximum value along the centre axis. The simulated structures also show similarities to the Baumann print (fig. 1). The top part of the ingot features a large ECT region, which is similar to the columnar structure observed in fig. 1. Additionally, a transition in equiaxed grain size can be seen near the middle of the ingot. The lower part of the equiaxed structure is comprised of grains with a size between 2 and 5 mm in diameter (light blue), but this quickly gives way to grains which are closer to 10 mm in diameter (dark blue). This transition in grain size corresponds to the transition from fine equiaxed structure to a larger dendritic structure with larger grains as observed in the real ingot. However, compared to the experiment, the negative segregation in the simulation is too close to the bottom of the ingot, and the 10% to 35% range for the positive segregation does not extend as low. For the macrostructure, two major differences are immediately apparent: the columnar region along the bottom mould wall is not as thick as expected, and the large columnar region resulting from ECT (orange) in the centre of the ingot is absent. Instead, the ECT is limited to a few discrete cells in the central part of the ingot.

5. Analysis

The simulation shows the entire solidification sequence, explaining when and how the various structures were formed. The solidification sequence can be divided into the following stages:

- Solidification starts immediately with a columnar front initiating on the mould walls (except for the top surface, as this was deliberately prevented). Fragmentation starts after a few seconds, generating fragments in the liquid ahead of the columnar front.
- The equiaxed grains produced by fragmentation sediment and start blocking the columnar front on the bottom surface. At $t = 200$ seconds, the equiaxed structure starts forming at the bottom from the accumulation of sedimented grains.
- At $t = 290$ s, the equiaxed structure covers the entire bottom of the ingot and has the flattened bowl shape seen in fig. 1. The first ECT cells appear on top of this equiaxed structure, with the ECT staring in the centre of the "bowl". This can be seen in fig. 4.a, which shows the simulated structure at $t = 290$ s, with the columnar front, the sedimented equiaxed grains and the ECT starting at the centre. The bowl shape is formed because most of the sedimenting grains are coming down from the columnar front growing along the mould wall, leading to a higher number of grains on the outer area. The lower grain density in the centre is also why the ECT starts there. These structures have a shape similar to the experimental results, however the thickness of both the columnar and the equiaxed structures is much smaller than expected. In addition, the ECT does not grow into a large columnar structure and is instead limited to a few discrete cells. This is because the newly formed columnar cells are very quickly blocked by both the continuous flow of sedimenting grains, and by their own fragments. A columnar front from an ECT can therefore only grow unhindered only in the upper corner of the ingot, where the downward movement of the grains prevents them from blocking the columnar front.
- At $t \approx 1340$ s, the transition from small equiaxed grains to larger grains starts. Most of the columnar front growing along the mould wall has been blocked by equiaxed grains, leaving only the top part to fragment. This reduces the equiaxed grain density along the columnar

IOP Conf. Series: Materials Science and Engineering **529** (2019) 012036 doi:10.1088/1757-899X/529/1/012036

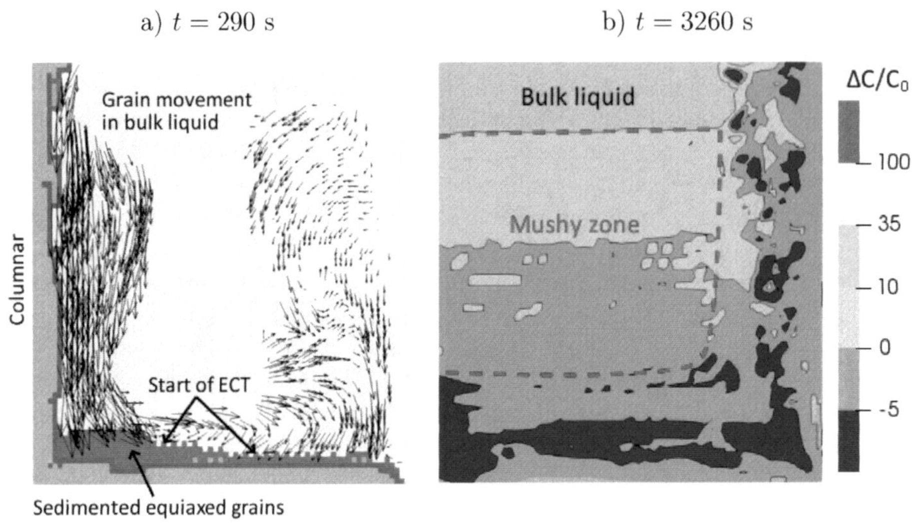

Figure 4. a) Simulated structures at $t = 290$ s, showing the columnar structure (green), the sedimented equiaxed grains (blue) and the grain movement in the bulk liquid. b) Simulated carbon segregation at $t = 3260$ s, the mushy zone is outlined in red.

front, allowing the grains to grow to larger sizes. In addition, this gives the grains more time to grow as they sediment from the top of the ingot to the equiaxed structure at the bottom. These two combined effects lead to an increase in the size of the sedimenting grains and to a rapid increase in the height of the equiaxed region.

- As the equiaxed zone progresses upward, the distance between the upper fragmenting columnar cells and the equiaxed region is reduced, sedimenting grains therefore have less time to grow. The sedimentation of smaller grains slows down the upward progress of the equiaxed region. This slowdown happens at $t \approx 3260$ s, when the top of the equiaxed region is at a distance of 15 cm from the top of the ingot. At that time, less than 20% of the total volume of the ingot is still fully liquid, however there is a large mushy zone formed by the equiaxed structure. The macrosegregation of carbon at $t = 3260$ s is shown in fig. 4.b. At this stage, the macrosegregation is still low: the carbon content in the bulk liquid is +15% relative to nominal composition.
- For $t > 3260$ s, most of the macrostructures are already in place and there are no major changes. Along the top, the columnar front slowly grows inward from the mould wall, competing with the equiaxed region to fill the remaining liquid volume. This results in the columnar and equiaxed structures that are visible in the top part of the ingot in fig. 3. The end of solidification happens at $t \approx 21800$ s, finishing in the upper region at the centre of the ingot. This stage of the solidification is by far the longest (18540 s, 85% of total solidification time), during which the macrosegregation continues to evolve slowly due to liquid flow through the equiaxed mushy zone.

6. Conclusion

The simulation results show that by using only fragmentation as a source of equiaxed grains, it is possible to reasonably predict both the macrosegregation of carbon and the macrostructure of a large steel ingot. The simulation shows that most of the structures are formed early on during solidification, 3260 s in our case, 15% of the total 21800 s. Conversely, the large positive

IOP Conf. Series: Materials Science and Engineering **529** (2019) 012036 doi:10.1088/1757-899X/529/1/012036

segregation forms slowly over a much longer period.

However, the results are not entirely predictive, as the centre ECT columnar region is much smaller in the simulations, and the thickness of the columnar and equiaxed regions below the ECT are much thinner than observed experimentally. The simulation presented here was without filling and using a binary alloy. In order to improve the simulation results, the next step of this study will be to account for the filling stage of the ingot and multiple alloying elements. In addition, further work must be done in order to accurately predict the centre ECT structure.

Acknowledgments

The authors would like to thank the partners of this study for their support: Industeel, EDF, Framatome and Sciences Computers Consultants.

References

[1] Leriche N, Combeau H, Gandin Ch-A, and Založnik M (2015). Modelling of Columnar-to-Equiaxed and Equiaxed-to-Columnar Transitions in Ingots Using a Multiphase Model. *IOP Conf. Series: Mat. Science and Engineering* 84.

[2] Pilling J and Hellawell A (1996). Mechanical deformation of dendrites by fluid flow. *Met. and Mat. Trans.* A, **27A** 229–32.

[3] Neumann-Heyme H, Shevchenko N, Lei Z, Eckert K, Keplinger O, Grenzer J, Beckermann C, and Eckert S (2018). Coarsening evolution of dendritic sidearms : from synchrotron experiments to quantitative modeling. *Acta Materialia*, **146** 176–86.

[4] Sato T, Kurz W, and Ikawa K (1987). Experiments on dendrite branch detachment in the succinonitrile-camphor alloy. *Trans. of the Japan Institute of Metals*, **28**(12) 1012–21.

[5] Gandin Ch-A (2000). From constrained to unconstrained growth during directional solidification. *Acta Materialia*, **48**(10) 2483–2501.

[6] Liu S, Lu S-Z and Hellawell A (2002). Dendritic array growth in the systems NH_4Cl–H_2O and $[CH_2\ CN]_2$–H_2O: the detachment of dendrite side arms induced by deceleration. *J. of Crystal Growth*, **234**(4) 740–50.

[7] Tiwari S N and Beech J (1981). The role of the peritectic reaction in the formation of the equiaxes zone in iron base alloys. *Trans. ISIJ*, **21** 554–8.

[8] Leriche N (2015). *Etude de la Transition Colonnaire-Equiaxe dans les Lingots et en Coulée Continue d'Acier et Influence du Mouvement des Grains.* Phd Thesis, Université de Lorraine, Nancy.

[9] Combeau H, Založnik M, Hans S, and Richy P E (2009). Prediction of Macrosegregation in Steel Ingots: Influence of the Motion and the Morphology of Equiaxed Grains. *Met. and Mat. Trans.* B, **40B** 289–304.

IOP Conf. Series: Materials Science and Engineering **529** (2019) 012037 doi:10.1088/1757-899X/529/1/012037

Semi-grand canonical Monte Carlo simulation for derivation of thermodynamic properties of binary alloy

Kensho Ueno and Yasushi Shibuta

Department of Materials Engineering, The University of Tokyo, 7-3-1, Hongo, Bunkyo-ku, Tokyo 113-8656, Japan

E-mail; ueno@mse.mm.t.u-tokyo.ac.jp

Abstract. Semi-grand canonical Monte Carlo (SGCMC) simulations are performed to derive thermodynamic properties of binary alloy from atomistic-based simulations. Particularly, solidus and liquidus compositions are directly derived for Fe-Cr alloy described by two different EAM potentials. Although the SGCMC simulation can derive relationship between the free energy and composition at any temperature straightforwardly, partial phase diagram obtained from SGCMC simulations strongly depends on the choice of interatomic potential.

1. Introduction

Solid-liquid interfacial properties have a great influence on the solidification microstructure [1-3]. Therefore, it is important to know interfacial properties such as the interfacial energy and mobility. However, it is not easy to experimentally obtain such interfacial properties with high accuracy. Therefore, in recent years, derivation of these properties by numerical analysis method such as molecular dynamics method (MD) is becoming popular. At present, there are many reports on interfacial properties of pure metals based on MD simulations [2,3]. However, there are not so many reports on those of alloy systems since derivation techniques on the estimation of interfacial properties of the alloy system are not established yet. For example, it is difficult to handle the solute distribution at the solid-liquid interface in the MD simulation since time scale of diffusion is much longer than that of a conventional MD simulation. Moreover, it is not straightforward to deterministically define the entropy of a system from the coordinate and velocity of atoms obtained by MD simulations since the number of states cannot

IOP Conf. Series: Materials Science and Engineering **529** (2019) 012037 doi:10.1088/1757-899X/529/1/012037

be defined in principle for a system governed by the classical equation of motion [4]. This makes it difficult to validate the accuracy of the interatomic potential of alloy systems. Therefore, a Monte Carlo (MC) simulation is often employed as an alternative technique to derive properties of alloys although it is not straightforward for MC simulations to treat dynamics of the interface. For example, a Metropolis Monte Carlo (MMC) simulation in conjunction with an atoms swap technique is often used to estimate equilibrium concentrations of a solid-liquid biphasic system of a binary alloy at a finite temperature [5].

Moreover, the Gibbs free energy of mixing can be derived from a semi-grand canonical MC (SGCMC) simulation [6-8], in which an equilibrium concentration for given chemical potential difference is obtained. In this paper, we performed SGCMC simulations to derive solidus-liquidus composition of binary alloy, which is an important thermodynamic factor determining solid-liquid interfacial properties. Here, Fe-Cr alloy described by for several interatomic potentials is employed as a model system since the partition coefficient changes significantly between Fe-rich and Cr-rich compositions.

2. Methodology

In this study, SGCMC and MD simulations are performed using LAMMPS (Large-scale Atomic/Molecular Massive Parallel Simulator) [9]. The EAM (embedded atom method) potentials fitted by Bonny et al. [10] (abbreviated as EAM(BPTM)) and by Stukowski et al. [11] (abbreviated as EAM(SSEC)) are employed as the interatomic potential of Fe-Cr alloy. Parameter files for the EAM potentials from the database of NIST-IPR project [12] are utilized for simulations by the LAMMPS. The Nose-Hoover thermostat and barostat [13,14] are employed to control temperature and pressure in MD simulations. We follow literatures [6-8] for practical procedure of SGCMC simulations.

3. Results and discussion

3.1 Melting point, enthalpy, and difference in Gibbs free energy between solid and liquid

Prior to SGCMC simulations of derivation of solidus and liquidus compositions for Fe-Cr alloy, melting points of pure Fe and difference in Gibbs free energy between solid and liquid for EAM potentials employed in this work are estimated by MD simulations. For the estimation of melting point, the convergence temperature technique [15,16] is performed as follows. Two bcc crystals consisting of 75 $\times$ 25 $\times$ 25 unit cells (93750 atoms) for the EAM(BPTM) potential and 100 $\times$ 25 $\times$ 25 unit cells (250000 atoms) for the EAM(SSEC) potential are annealed at 500 and 3000 K for 10 ps with the number of atoms, volume and temperature (NVT)-constant ensemble to obtain solid and liquid structures, respectively. Then, these structures are connected with the (100) plane appearing on the interface and the energy minimization is performed to avoid the unexpected proximity of atoms at the solid-liquid interface. The obtained these structures are used as the initial configuration of the following simulations.

The solid-liquid biphasic system of pure Fe is relaxed with the number of atoms, pressure and temperature (NPT)-constant ensemble for 25 ps at 1700, 1750 and 1800 K. Then, the systems are relaxed with the number of atoms, pressure and enthalpy (NPH)-constant ensemble for subsequent 1975 ps. The pressure is maintained at 0 Pa in all processes. In this technique, the convergence temperature is regarded as the melting point of the target material [15,16]. Figure 1 shows temperature as a function of time for the solid-liquid biphasic systems of pure Fe described by EAM (BPTM) and EAM(SSEC) potentials, respectively. Melting points are estimated to 1770 ± 10 K for EAM(BPTM) and 1770 ± 10 K for EAM(SSEC), respectively, which are closed to the experimental value (1811 K).

Next, the difference in Gibbs free energy between solid and liquid ΔG_{SL} (= $G_L - G_S$) for pure Fe is estimated using the following Gibbs-Helmholtz relation:

$$\left[\frac{\partial}{\partial T}\left(\frac{\Delta G_{SL}}{T}\right)\right]_P = -\frac{\Delta H_{SL}}{T^2}. \qquad (1)$$

where ΔH_{SL} (= $H_L - H_S$) is the difference in enthalpy between solid and liquid and T is the temperature. Since ΔG_{SL} becomes zero at the melting point T_M, the following relation is obtained:

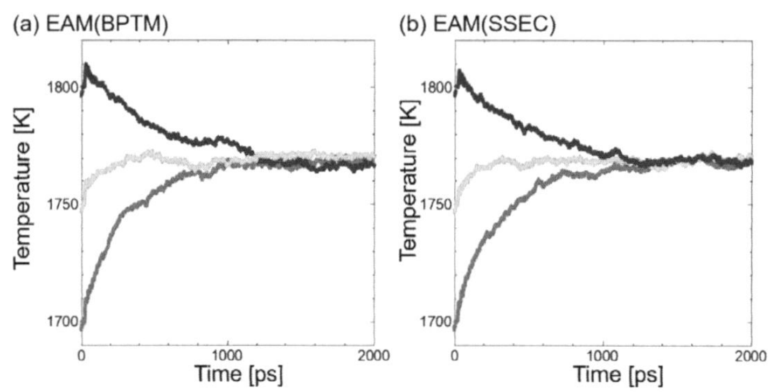

Figure 1. Time change of temperature for solid-liquid biphasic system of pure Fe described by (a) EAM(BPTM) and (b) EAM(SSEC) potentials.

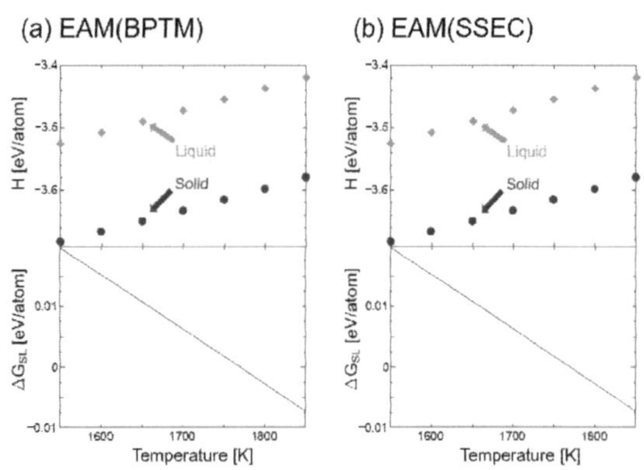

Figure 2. (Top) Enthalpy of solid and liquid systems and (bottom) difference in Gibbs free energy between solid and liquid for pure Fe described by (a) EAM (BPTM) and (b) EAM (SSEC) potentials.

$$\Delta G_{\mathrm{SL}}(T) = T\int_{T}^{T_{\mathrm{M}}} \frac{\Delta H_{\mathrm{SL}}(T')}{T'^2} dT' . \tag{2}$$

Since the enthalpy is directly calculated from MD simulation, ΔG_{SL} can be estimated from the difference in enthalpy between solid and liquid, which is calculated as follows. For a solid system, a bcc crystal consisting of 20 × 20 × 20 unit cells (16000 atoms) is annealed at various temperatures between 1550 and 1850 K at intervals of 50 K with the NPT-constant ensemble to calculate enthalpy. A liquid structure obtained by pre-anneal of the bcc crystal at 5000K for 10 ps is annealed at same temperatures in the solid system. The enthalpy of solid and liquid system obtained by MD simulations and the difference in Gibbs free energy between solid and liquid for pure Fe derived from Equation (2) are shown in Figure 2.

3.2 Solidus and liquidus compositions from SGCMC simulations

Then, SGCMC simulations are performed using a bcc crystal consisting of 20 × 20 × 20 unit cells (16000 atoms). A liquid structure obtained by pre-anneal of the bcc crystal at 5000K for 10 ps is used for calculations of liquid system. In a SGCMC simulation, an equilibrium concentration is obtained for the given chemical potential difference $\Delta\mu$ at given temperature. Here, SGCMC simulations between $\Delta\mu = -1.0$ and 1.0 eV/atom at intervals of 0.1 eV/atom are performed at various temperatures. Figure 3(a) and 3(b) show the relationship between $\Delta\mu$ and equilibrium Cr composition X_{Cr} at 1800 and 1900 K for EAM(BPTM) potential and 1900 and 2000 K for EAM(SSEC) potential, respectively. Plots obtained from SGCMC simulations are fitted to the following equation to obtain the fitting parameters A_i (i = 0,1,…n).

$$\Delta\mu = kT \ln\left(\frac{x_{\mathrm{Cr}}}{1 - x_{\mathrm{Cr}}}\right) + \sum_{i=0}^{n} A_i x_{\mathrm{Cr}}^i , \tag{3}$$

where k is the Boltzmann constant. Here, n = 5 is employed. Once fitting parameters are obtained, a Gibbs free energy of mixing is described as

$$\Delta G_{\mathrm{mix}} = kT\left[x_{\mathrm{Cr}} \ln x_{\mathrm{Cr}} + (1 - x_{\mathrm{Cr}})\ln(1 - x_{\mathrm{Cr}})\right] + \sum_{i=0}^{n} \frac{A_i x_{\mathrm{Cr}}^{i+1}}{(i+1)} . \tag{4}$$

Figure 3(c) and 3(d) shows the Gibbs free energy of mixing for solid and liquid phases of the Fe-Cr alloy for corresponding temperatures. Here, Gibbs free energy of mixing for solid at $X_{Cr} = 0$ is defined as a reference value (i.e. zero). Therefore, Gibbs free energy of mixing for liquid at $X_{Cr} = 0$ takes a value of ΔG_{SL}, which is derived from Equation (2). Finally, solidus and liquidus compositions are derived from the contact points of the common tangent line as shown in the figure.

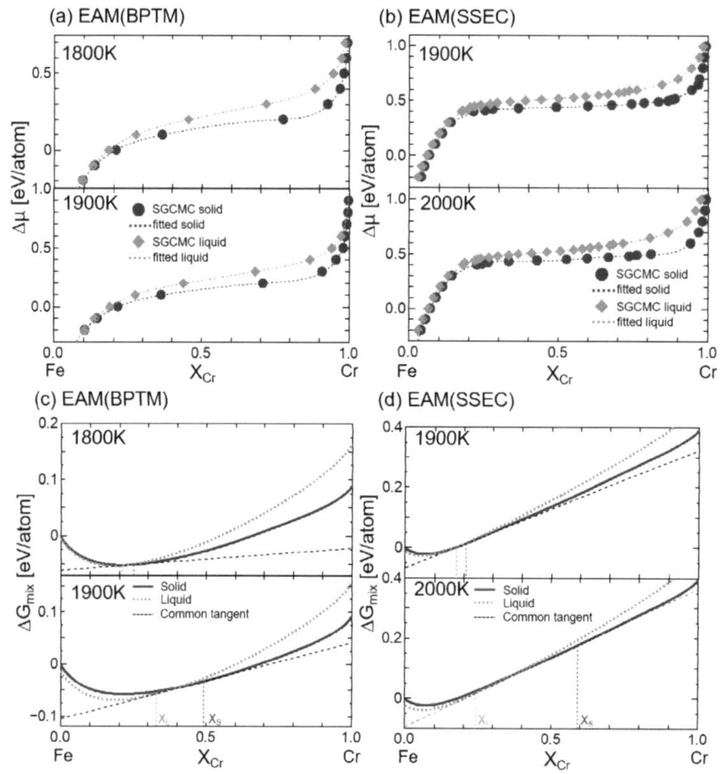

Figure 3. (a)(b) Relationship between $\Delta\mu$ and equilibrium Cr composition and (c)(d) Gibbs free energy of mixing for solid and liquid phases of the Fe-Cr alloy at 1800 and 1900 K for EAM(BPTM) potential and 1900 and 2000 K for EAM(SSEC) potential.

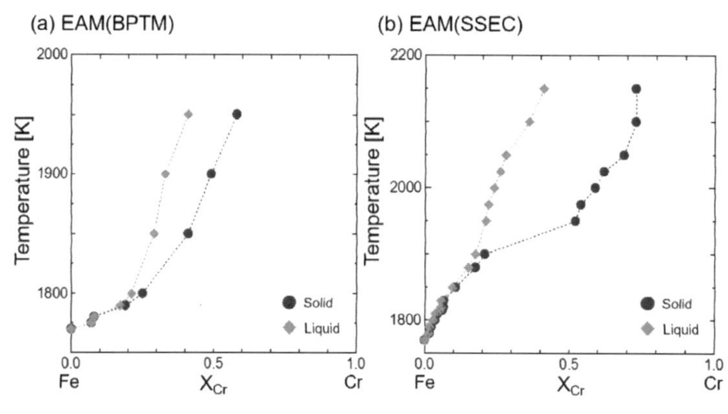

Figure 4. Partial phase diagram from SGCMC simulations: (a) EAM(BPTM) potential and (b) EAM(SSEC) potential.

Solidus and liquids composition at various temperatures can be obtained by repeating SGCMC simulation with different temperatures. Figure 4 shows the partial phase diagram obtained by SGCMC simulations. In general, there is no significant difference between solidus and liquidus composition at

IOP Conf. Series: Materials Science and Engineering **529** (2019) 012037 doi:10.1088/1757-899X/529/1/012037

the Fe-rich compositions, whereas there is large difference in it at the Cr-rich compositions for both interatomic potentials. However, there is a large difference in shape of solidus and liquidus lines between EAM(BPTM) and EMA(SSEC) potentials. Especially, solidus and liquidus compositions at high temperature are not well defined for the EAM(BPTM) potential. It is not straightforward to develop the interatomic potential which reproduce solidus and liquidus compositions at high temperature since these properties are not explicit in the interatomic potential curve itself.

4. Conclusions

In this study, SGCMC simulations are performed to derive solidus and liquidus composition of binary alloys described by two different EAM potentials for Fe-Cr alloy. Although the SGCMC simulation derives relationship between the free energy and composition at any temperature straightforwardly, obtained results strongly depend on the choice of interatomic potential. Moreover, this technique is not used straightforwardly when the relationship between $\Delta\mu$ and equilibrium is not monotonic, which causes a miscibility gap in the phase diagram. Therefore, it is essential to establish methodology for derivation of phase diagram from interatomic potential to discuss its reliability. Further investigation is needed to fill the gap between atomistic-based simulations and macroscopic thermodynamics for alloy systems.

References

[1] Dantzig J A and Rappaz M, *Solidification* 2009 (Lausanne: EPFL Press) pp. 287-344.
[2] Shibuta Y, Ohno M and T. Takaki 2015 *JOM* **67** 1793.
[3] Shibuta Y, Ohno M and T. Takaki 2018 *Adv. Theo. Sim.* **1** 1800065.
[4] Shibuta Y 2019 *Mater. Trans.* submitted.
[5] Ueno K and Shibuta Y 2018 *Materialia* **4** 553.
[6] Ramalingam H, Asta M, van de Walle A and Hoyt J J 2002 *Interface Sci.* **10** 149.
[7] Becker C A, Asta M, Hoyt J J and Foiles S M 2006 J. *Chem. Phys.* **124** 164708.
[8] Becker C A, Olmsted D L, Asta M, Hoyt J J and Foiles S M 2009 *Phys. Rev. B* **79** 054109.
[9] Plimpton S J 1995 *J. Comput. Phys.* **117** 1, http://lammps.sandia.gov.
[10] Bonny G, Pasianot RC, Terentyev D and Malerba L 2011 *Philo. Mag.* **91** 1724.
[11] Stukowski A, Sadigh B, Erhart P and Caro A 2009 *Modeling Simul. Mater. Sci. Eng.* **17** 075005.
[12] Becker C A, Tavazza F, Trautt Z T and Buarque de Macedo R A 2013 *Curr. Opin. Solid State Mater. Sci.* **17** 277, http://www.ctcms.nist.gov/potentials.
[13] Nose S 1984 *J. Chem. Phys.* **81** 511.
[14] Hoover W G 1985 *Phys. Rev. A* **31** 1695.
[15] Shibuta Y and Suzuki T 2007 *Chem. Phys. Lett.* **445** 265.
[16] Shibuta Y, Takamoto S and Suzuki T 2008 *ISIJ Int.* **48** 1582.

Numerical simulations of solidification structures and macrosegregation by a cellular automaton model coupled with flow calculations

Y Natsume

Department of Materials Science, Akita University, 1-1 Tegata-Gakuenmachi, Akita 010-8502, Japan

natsume@gipc.akita-u.ac.jp

Abstract. In this study, a numerical model was developed, and direct simulations were performed to predict solidification grain structures and macrosegregation based on a three-dimensional cellular automaton finite difference (CAFD) method coupled with flow calculation of natural convection and shrinkage flow. First, to evaluate the model coupled with natural convention, simulations of unidirectional solidification for Al-10wt% Mg alloy were performed. Mg-rich plumes rising in the melt were seen due to subsequent upward flow, and Mg-rich channels forming in the mushy zone were observed. Columnar grains were then formed, and they became coarse afterwards. Next, to evaluate the model coupled with shrinkage flow, simulations of casting Al-10wt% Cu alloy in a unique mold, which can form macrosegregation in the central region of the small ingot, were performed. The bridging of columnar grains formed during solidification, and the positive segregation was generated in the region below the bridging. Thus, the main factor for this macrosegregation is the shrinkage flow with bridging. From the comparison of simulation results with and without the chill for the unique mold, it was established that the shrinkage flow and the bridging of solidification structures play an important role in macrosegregation.

1. Introduction

The cellular automaton (CA) model is one of the tools used to efficiently predict solidification structures. This model coupled with a finite element (FE) heat flow calculation (CAFÉ model) has been shown to predict grain structures successfully in various solidification processes [1-3]. Recently, microscopic scale CA models have been developed to simulate microstructures [4-8]. Like phase-field simulations, they can also simulate dendritic morphologies. However, the microscopic scale CA model has lower computational costs. On the other hand, predicting the dendritic grain structures for solidification of samples such as large ingots, billets, and slabs will incur high computational costs. Thus, the macroscopic scale CA model is the most practical and efficient tool to predict solidification grain structures.

According to the conventional model developed by Gandin and Rappaz [1-3], the macroscopic scale CA model coupled with FE heat flow calculations consists of nucleation and growth of the dendritic grain envelope. In this model, the distribution of solute concentration is not considered, thus, the initial solute concentration is given for the whole region. The conditions of nucleation and grain growth are determined by local undercooling obtained from heat flow calculations. To simulate solute segregation, the distribution of solute concentration must follow the models. Moreover, to simulate and predict macrosegregation, the flow of liquid during solidification needs to be considered [9, 10]. Recently, the CAFÉ model coupled with macrosegregation has been developed and validated by comparing *in situ* real-time X-ray radiography observations [11-13]. A microscopic scale CA simulation coupled with flow calculation was carried out to have a new mechanism of freckle formation [14]. Natsume and Ohsasa also proposed a macroscopic scale CA model coupled with a finite difference (FD) heat flow calculation (CAFD) model to predict microsegregation in solidification grain structures [15]. In this CAFD model, the effect of fluid flow is considered by an effective partition coefficient. However, this has not been coupled with flow calculation. In the previous work, Natsume developed a numerical model to predict solidification grain structures and macrosegregation based on a macroscopic scale CAFD method coupled with flow calculation of natural convection and shrinkage flow [16, 17]. In the present work, direct simulations of grain growth and macrosegregation using this model were performed to determine the influence of bridging and shrinkage flow on macrosegregation. Also, simulation results of grain structures and macrosegregation under natural convection were demonstrated.

2. Model

In this macroscopic scale CAFD model, nucleation, grain growth, and solute concentration are calculated by the mesoscopic scale model using a cubic CA cell, and heat flow is calculated by the macroscopic scale model using a cubic FD grid. The nucleation adopted the original model proposed by Rappaz and Gandin [18], while a grain growth model based on the octahedral decentered growth algorithm [2, 3] was used. A new method to calculate solute concentration was developed in the previous work [16] while the enthalpy method was used in the heat flow calculation. The flow of liquid is performed in the macroscopic scale model by solving equations (1) and (2).

$$\frac{\partial \boldsymbol{u}}{\partial t}+\boldsymbol{u}\cdot\nabla\boldsymbol{u}=-\frac{\nabla p}{\rho}+\nu\nabla^2\boldsymbol{u}-\frac{\nu}{K}\boldsymbol{u}-\left\{\beta_T(T-T_{reff})+\beta_c(c-c_{reff})\right\}f_L\boldsymbol{g} \tag{1}$$

$$\nabla\cdot\boldsymbol{u}+\beta\frac{\partial f_S}{\partial t}=0 \tag{2}$$

where $\boldsymbol{u}$ is velocity, t is time, p is pressure, ρ is density, ν is kinematic viscosity, K is permeability, β_T and β_c are coefficients of thermal and solutal expansion, respectively, T is temperature, c is solute concentration, f_L and f_S are fraction of liquid and solid, respectively, $\boldsymbol{g}$ is gravity, and β is solidification shrinkage rate. In the permeability in equation (1), an anisotropic permeability model proposed by Natsume *et al.* [19, 20] was used. In this model, the anisotropic permeability of parallel and normal direction for primary dendrite arms, K_P and K_N, are given as follows:

$$K_P=0.0194\left[\frac{d_2}{1+(d_2/d_1)}\right]^2\frac{f_L^3}{(1-f_L)^2} \tag{3}$$

$$K_N=0.0097\left[\frac{d_2}{1+(d_2/d_1)+(d_2/d_1)^2}\right]^2\frac{f_L^3}{(1-f_L)^2} \tag{4}$$

where d_1 and d_2 is the primary and secondary dendrite arm spacing, respectively. In the simulations, the growth of columnar grains was only considered because it was not introduced in the movement of equiaxed grains due to fluid flow. The anisotropic permeability, K_P and K_N, was set to K_x, K_y, and K_z for simplicity.

3. Results and discussion

3.1. Simulations of grain structures and macrosegregation under natural convection

To confirm the effect of natural convection, simulations of unidirectional solidification cooled on the bottom surface of the mold were performed for an Al-10wt% Mg alloy. The size of the calculation domain was 10 x 100 x 80 mm, and the sizes of CA cell and FD grid were 0.25 and 1.25 mm, respectively.

The initial temperature in the entire domain was assumed to be homogeneous at a constant temperature of 640 °C, and the chill was assumed to be set on the bottom of the calculation domain. The temperature and heat transfer coefficient of the chill are always 25 °C and 3000 W/K/m^2, respectively. Zero-flux boundary conditions were used for temperature and solute concentration on four sides and top surfaces. The anisotropic permeability used was as follows: $K_x = K_y = K_N$ and $K_z = K_P$. In this simulation, the solidification shrinkage was not considered.

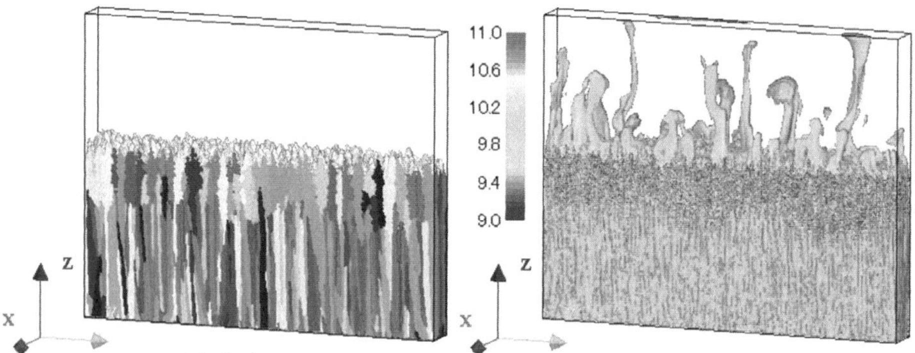

(*a*) Grain structure (*b*) Mg concentration

Figure 1. Grain structure and Mg concentration distribution during solidification.

Figure 1 shows the snapshots of grain structure and distribution of Mg concentration during solidification. Columnar grain structures have formed from bottom to top in alloy melt while Mg-rich plumes rising in the melt were seen due to the subsequent upward flow caused by the thermosolutal buoyant force. Once the plume occurred in the melt above the mushy zone, the morphology of columnar grains varied, and the grains became coarse due to branching of primary and secondary dendrite arms, but it cannot be seen in detail because the grain growth followed the dendritic grain envelope model. Moreover, regions like Mg-rich channels forming in the mushy zone can be observed. Such regions could delay solidification inside the liquid-rich channels and might result in a freckle defect. Similarly, the plume and freckles were observed in the directional solidification of Ga-In alloys under the effect of thermosolutal convection by means of X-ray radioscopy by Shevchenko *et al.* [21].

Figure 2 shows the average Mg concentration profile of the cross section from bottom to top. The negative segregation occurred in the middle of solidification and the positive segregation occurred at the end of solidification. From these results, the potential of the model to predict macrosegregation and grain structures under natural convection were demonstrated, and it was confirmed that the proposed model was effective in predicting the macrosegregation coupling with the grain structure formation.

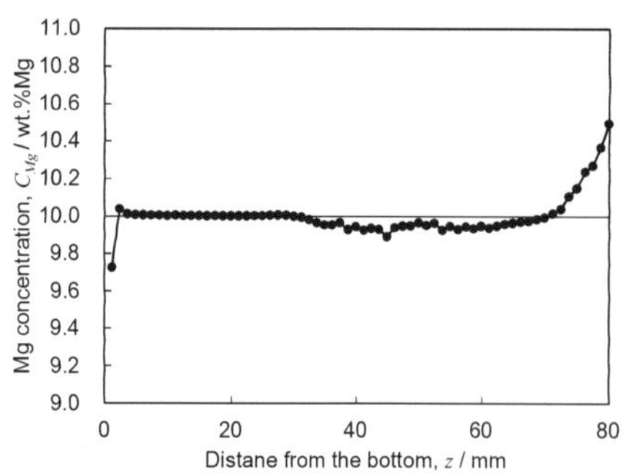

Figure 2. The profile of average Mg concentration of cross section at the end of solidification obtained by simulation of unidirectional solidification.

3.2. Simulations of grain structures and macrosegregation with shrinkage flow

To confirm the effect of shrinkage flow, the simulations were performed in the condition wherein macrosegregation formation is possible like the center segregation observed in a continuously cast slab. The original concept of the condition used in the simulation was that of the model experiment of Al-10wt% Cu alloy carried out by Sato *et al.* [22]. Sato *et al.* developed a unique mold which can form macrosegregation in the central region of the small ingot, known to be "Sato's mold" in Japanese research groups. Figure 3 shows the solidification system used in this simulation. The domain size was 50 x 30 x 120 mm, and the sizes of CA cell and FD grid were 0.5 and 2.0 mm, respectively. In a system with the chill shown in Figure 3 (a), the chill was set on the *xz* surfaces, 70-80 mm from the bottom of the calculation domain, and the *xz* surfaces other than the chill were set on as the mold surfaces wherein the different heat transfer coefficient were given. The initial temperature in the entire domain was assumed to be homogeneous at a constant temperature of 650 °C. The values of heat transfer coefficient of the chill, mold-1, and mold-2 are 1000 W/K/m^2, 400 W/K/m^2, and 200 W/K/m^2, respectively. Zero-flux boundary conditions were used for solute concentration. The anisotropic permeability used was as follows: $K_x = K_z = K_N$ and $K_y = K_P$. For the conditions of simulation, two types of simulation were performed: with and without the chill. In this simulation, natural convection was not considered to determine the influence of shrinkage flow on macrosegregation.

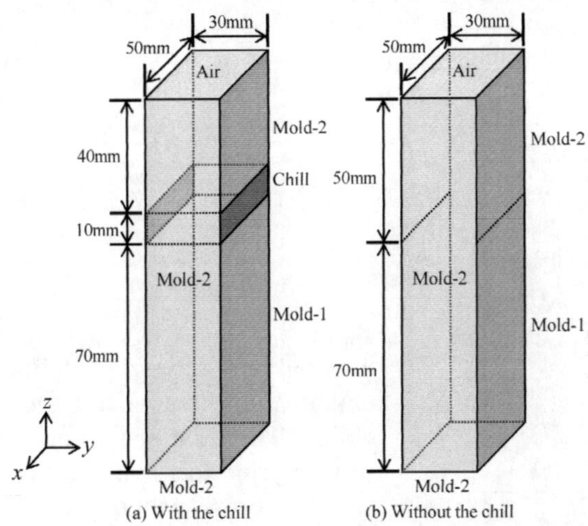

Figure 3. Solidification system used in the simulations. (b) is the same system as that of the experiment reported by *Sato et al.* [22]

Figure 4 and Figure 5 shows the snapshots of grain structures and distributions of Cu concentration during solidification for a system with the chill, respectively. The bridging of columnar grains formed near the center of the ingot during solidification due to the chill. Molten alloy flowed down through the columnar grains before bridging because of shrinkage flow. In a system without the chill, this phenomenon was not observed. Figure 6 shows the line profile of Cu concentration ratio on centerline at the end of solidification. As shown in Figure 6 (a), the positive segregation was generated in the region below the chill, and the negative segregation was generated in the region above the chill. In a system without the chill, the negative and positive segregations were observed on the upper and lower region, respectively. However, the degree of segregation was weak because the bridging was not formed. Therefore, the primary factor for this macrosegregation was the shrinkage flow with the formation of bridging. From these results, it was confirmed that the shrinkage flow and the bridging of solidification structures played an important role in macrosegregation. Also, in the previous study, it was confirmed that the degree of positive and negative segregation is enlarged by the presence of natural convection. Although the simulated results could not be quantitatively compared with the experimental results by Sato *et al.* due to the difference in solidification condition, the simulated result could explain the mechanism forming this macrosegregation.

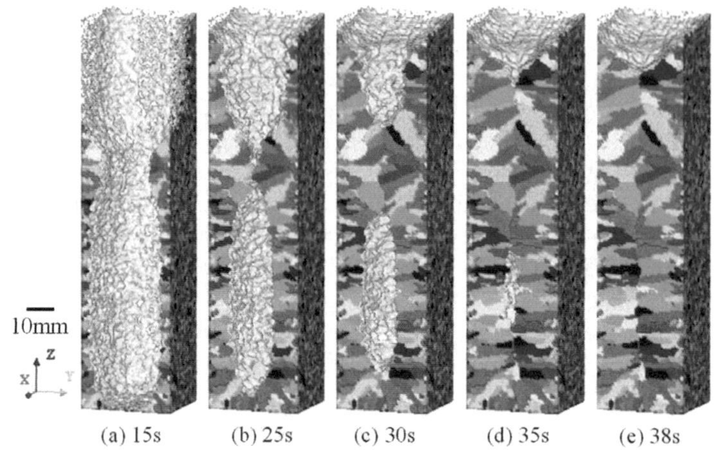

Figure 4. Grain structures during solidification obtained by the simulation with the chill. The elapsed time since solidification start is (a) 15s, (b) 25s, (c) 30s, (d) 35s, and (e) 38s, respectively.

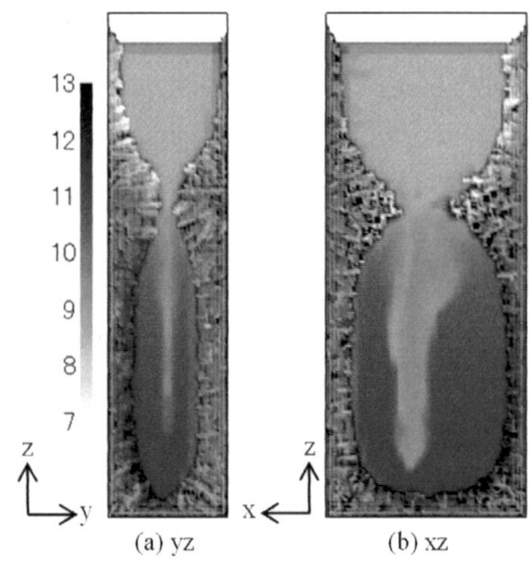

Figure 5. Cu concentration distribution during solidification obtained by the simulation with the chill. The elapsed time since solidification start is 25 sec. (a) shows yz cross section of x = 25 mm, and (b) shows xz cross section of y = 15 mm.

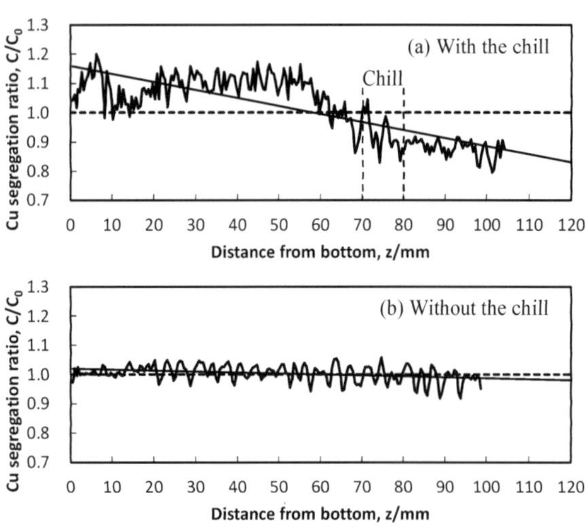

Figure 6. Line profiles of Cu segregation ratio on centerline (x = 25 mm, y = 15 mm) at the end of solidification obtained by the simulations (a) without and (b) with the chill.

4. Conclusions

Two types of direct simulations of grain growth and macrosegregation were performed using a numerical model based on a 3D CAFD method coupled with flow calculation of natural convection and shrinkage flow: (1) the simulation of unidirectional solidification to evaluate the model coupled with natural convection, and (2) simulation for a unique mold proposed by Sato *et al.*[22] to evaluate the model coupled with shrinkage flow. For the first type, simulations of unidirectional solidification cooled on the bottom surface of the mold were performed for a Al-10wt% Mg alloy. Mg-rich plumes rising in the melt were seen due to the subsequent upward flow caused by the thermosolutal buoyant force, and Mg-rich channels forming in the mushy zone were observed. The columnar grain structures have formed from the bottom to the top in molten alloy, and some grains became coarse after Mg-rich plumes were observed. For the second type, simulations of casting of Al-10wt% Cu alloy for Sato's mold were performed. The bridging of columnar grains formed during solidification upon cooling, and in the region below it, the positive segregation was generated. Analyzing the simulated results, it was found that the main factor for this macrosegregation is the shrinkage flow with the bridging. In the previous work [16, 17], the influence of the bridging on the macrosegregation was unclear because the simulation without the chill was not performed. In the present work, it has been established that the shrinkage flow and bridging of solidification structures play an important role in the formation of macrosegregation. For future research, the simulation results will be reported in comparing with the experimental data using modified Sato's mold to evaluate quantitatively the validity of the model in the near future. In addition, a huge amount of computational time will be required to perform 3D flow calculations, especially in large-scale simulations such as continuous casting. A new macroscopic scale CA model coupled with a Lattice Boltzmann method optimized for parallel computing to perform larger-scale simulations will also be reported.

References

[1] Gandin C A and Rappaz M 1994 *Acta Metall. Mater.* **42** 2233
[2] Gandin C A and Rappaz M 1996 *Acta Metall. Mater.* **44** 3339
[3] Gandin C A and Rappaz M 1997 *Acta Metall. Mater.* **45** 2187
[4] Wang W, Lee P D and McLean M 2003 *Acta Mater.* **51** 2971
[5] Zhu M F and Stefanescu D M 2007 *Acta Mater.* **55** 1741
[6] Pan S and Zhu M F 2010 *Acta Mater.* **58** 340
[7] Natsume Y and Ohsasa K 2015 *Mater. Sci. Eng.* **84** 012050
[8] Natsume Y 2015 *J. JFS* **87** 636
[9] Beckermann C 2002 *Int. Mater. Rev.* **47** 243
[10] Wu M, Ludwig A and Kharicha A 2017 *Appl. Math. Model.* **41** 102
[11] Guillemot G, Gandin C A and Bellet M 2007 *J. Crystal Growth* **303** 58
[12] Carozzani T, Gandin C A, Digonnet H, Bellet M, Zaidat K and Fautrelle Y 2013 *Metall. Mater. Trans A* **44** 873
[13] Saad A, Gandin C A, Bellet M, Shevchenko N and Eckert S 2015 *Metall. Mater. Trans A* **46** 4886
[14] Yuan L and Lee P D 2012 *Acta Mater.* **60** 4917
[15] Natsume Y and Ohsasa K 2014 *ISIJ Int.* **54** 415
[16] Natsume Y 2017 *Tetsu-to-Hagane* **103** 723
[17] Natsume Y 2017 *Tetsu-to-Hagane* **103** 738
[18] Rappaz M and Gandin C A 1993 *Acta Metall. Mater.* **41** 345
[19] Natsume Y, Takahashi D, Kawashima K, Tanigawa E and Ohsasa K 2013 *ISIJ Int.* **53** 838
[20] Natsume Y, Takahashi D, Kawashima K, Tanigawa E and Ohsasa K 2014 *ISIJ Int.* **54** 366
[21] Shevchenko N, Boden S, Gerbeth G and Eckert S 2013 *Metall. Mater. Trans. A* **44** 3797.
[22] Sato F, Esaka H and Shinozuka K 2013 *Tetsu-to-Hagane* **99** 101

IOP Conf. Series: Materials Science and Engineering **529** (2019) 012039 doi:10.1088/1757-899X/529/1/012039

Analysis of columnar-to-equiaxed transition experiment in lab scale steel casting by a multiphase model

S. Sachi[1], M. Založnik[1], H. Combeau[1], Ch.-A. Gandin[2], M. Gennesson[3], J. Demurger[3], M. Stoltz[3] and I. Poitrault[4]

[1] Université de Lorraine, CNRS, IJL, F-54000 Nancy, France
[2] MINES ParisTech, CEMEF UMR CNRS 7635, F-06904 Sophia, Antipolis, France
[3] Ascométal CREAS, Avenue de France, BP 70045, 57301, Hagondange, France
[4] ArcelorMittal Industeel, 71201 Le Creusot, France

E-mail: savya.sachi@univ-lorraine.fr

Abstract.
Correct prediction of composition heterogeneities and grain structure across a steel ingot is critical in optimizing the industrial processing parameters for enhanced performance. The columnar to equiaxed transtion (CET) is a microstructural transition which is strictly controlled as it affects the mechanical properties of the final product along with the macrosegregation patterns. Larger equiaxed regions are preferred for most industrial applications. CET is significantly affected by the number density of equiaxed grains and by the nucleation undercooling. 8 kg 42CrMo4 alloy steel ingots (240 mm x 60 mm x 60 mm) were cast. The cast structure was characterized by ASCOMETAL. The experiments were simulated with a process-scale model of solidification that incorporates a multiscale description of the microstructure formation. The goal of the present study is to show the capabilities of such a process-scale solidification model to explain the observed structure distributions (extent of the columnar and equiaxed zones, equiaxed-to-columnar transition).

1. Introduction

Solidification processing is one of the most important routes to produce metals and alloys. During solidification, parameters such as the temperature gradient and the growth rate which are dependent on the casting practice and on alloy properties, result in a wide variety of microstructures and material behaviour. Solidification results in two types of grain morphologies: columnar and equiaxed. Nucleation of crystals in an undercooled melt results in equiaxed grains whereas columnar grains form when the heat flux is unidirectionally oriented. Transition from columnar to equiaxed growth is called Columnar-to-equiaxed (CET) transition and occurs when nucleation of equiaxed grains occurs ahead of the columnar front. CET is significantly affected by the number density of the equiaxed grains. The nuclei which subsequently evolve into equiaxed grains are introduced through heterogeneous nucleation or dendrite fragmentation. Heterogeneous nucleation can occur during any time of solidification due to constitutional supercooling. Dendrite fragmentation can occur due to convection or fallout from the upper surface of the ingot.

Numerical simulations have demonstrated that density of grains in the melt can affect the intensity of macrosegregation [1, 2, 3]. Deposition of the solute depleted grains along with the rejection of solute into the liquid is the primary cause of macrosegregation. The grain size and the morphology (dendritic or globular) determine the settling dynamics of the equiaxed grains. The size and the morphology of the deposited equiaxed grains further affects the permeability of the deposited layer, thus influencing the permeation of solute rich liquid through the packed layers. Globular grains lead to more pronounced macrosegregation than dendritic grains. It has further been observed that segregation can also result in "solutal blocking" of the columnar front and result in CET [4]. Thus, the final microstructure obtained after casting is a result of multitude of physical phenomena occurring simultaneously. In this work, the influence of origins of the equiaxed grains on the CET in a small steel casting has been investigated.

2. Model description

The multiscale multiphase model of solidification SOLID is used for this study. It is based on the volume averaging method with three hydrodynamic phases: the solid phase and two liquid phases (interdendritic and extradendritic) [5]. The present version is an extension of the two phase model [1, 6] and models the growth of the columnar grains along with the development of moving equiaxed grains [2, 7]. The morphology of the equiaxed grains in the volume averaged model is based on the idea of dendrite envelope as proposed by Rappaz and Thevoz [8]. The model distinguishes between the liquid outside the envelope (extradendritic liquid) and the liquid inside the envelope (interdendritic liquid) which is assumed to be homogeneous. Same envelope shape and morphological parameters are used to model the columnar and equiaxed structure. A columnar grain is schematized as an assembly of equiaxed grains arranged on a cubic sublattice with a volume density of substructures distinct from that of equiaxed grains [2]. Columnar grains originate at the mold wall and the growth direction is aligned with the local thermal gradient ahead of the front. The columnar front is tracked using a method developed by Ludwig and Wu [9]. Equiaxed grains originate from dendrite fragments or heterogeneous nuclei. The growth of the columnar and equiaxed grains envelopes are calculated according to the LGK model [10]. In the present model, a mixed zone containing columnar and equiaxed grains is considered in the vicinity of the columnar tips. In this zone, the system contains six phases, three phases associated with the equiaxed grains and three phases associated with the columnar grains.

The model contains two parts: a) macroscopic transport model which accounts for the momentum, mass, heat, solute mass and grain populations conservation b) microscopic growth model which accounts for the nucleation and growth of the grains. The two parts of the model are coupled using an operator splitting scheme [6]. The motion of the liquid melt and the solid grains at the macroscopic scale are modeled by taking into account: inertial forces, drag forces at the solid-liquid interface and the buoyancy forces. The buoyancy forces are induced by liquid density variation (thermosolutal convection modeled with Boussinesq approximation) and density difference between liquid and solid. Heterogeneous nucleation of equiaxed grains is modeled using an instantaneous nucleation law. Multiple sets of heterogeneous nucleation sites (nuclei) are used, each set is described by a distinct nucleation undercooling and a population density. Columnar dendrite fragments are a second source of new equiaxed grains. They are introduced ahead of the columnar front at a predefined rate when the local thermal gradient drops below a critical value. Once nucleation occurs, equiaxed grains grow as dispersed solid phases and freely move. They are introduced as globular grains with a diameter of 1 μm. The population balance for grains and nuclei are accounted for. While the grains move with the solid velocity, the nuclei move along with the liquid. When the envelope fraction of the equiaxed grains is above the packing fraction, solid phase is considered stationary and the flow regime changes from a slurry flow to a porous flow. Two phenomena responsible for CET are

IOP Conf. Series: Materials Science and Engineering **529** (2019) 012039 doi:10.1088/1757-899X/529/1/012039

considered: "mechanical blocking" and "solutal blocking". "Mechanical blocking" occurs when the equiaxed grain fraction in the mixed zone reaches a critical value which can range from 0.2 [11] to 1. In the present work, the mechanical blocking fraction is set to 0.5 according to the criterion proposed by Hunt [12]. "Solutal blocking" occurs when the chemical undercooling reduces sufficiently to prevent the growth of the columnar front, effectively blocking it [4]. "Equiaxed-to-Columnar" (ECT) transition, which is observed at the top of industrial steel ingots [3], is also modeled.

3. Experimental setup, observations and simulation setup

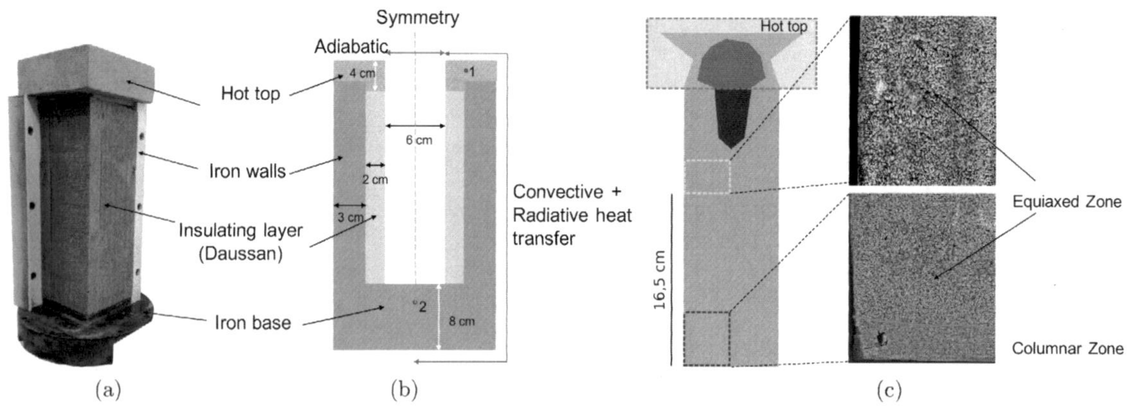

Figure 1. (a) Experimental setup used for the ingot casting with two of the iron walls have been disassembled to reveal the added insulation layer in the mold and (b) geometry used for the simulations. The red dots show the positions of the thermocouples 1 and 2 respectively (c) Diagram showing the etched locations along with the associated microstructure

8 kg Experimental trials were conducted by Ascométal to see the effect of inoculants on the cast structure. The composition of the 42CrMo4 steel is 0.4 wt% C, 0.2 wt% Si, 0.7 wt% Mn, 0.3 wt% Mo and 1.1 wt% Cr. The ingots have a square cross section as shown in Fig 1a. The hot top is a composite material with a chemical composition of 45% Al_2O_3 - 44 % SiO_2 - 5 % Fe_2O_3 - 6% binders. The walls of the mold have been insulated from the inside with a fibrous composite produced by Daussan France (Woippy, France). Its chemical composition is 83% SiO_2 - 2% Al_2O_3 - 2% CaO - 13 % binders. This setup promotes directional solidification from the bottom. The metal to be cast is heated to 1560 °C after which it is poured into the preheated mold. The experiments were instrumented using thermocouples. Macroetching were performed on the entire surface of the ingot. Fig. 1c shows a ≈1.5 cm thick columnar zone at the bottom for a cast ingot with no inoculating agent additions. No columnar zone was observed along the vertical walls of the casting. A structural gradient can be observed between the edge and the center of the sample. A coarser structure is observed in the top and a much finer structure is observed in the bottom.

We consider a simplified 2D axis-symmetric geometry as shown in Fig 1b. The numerical domain was discretized into approximately 1700 cells with an average cell of 4 mm. The time step varies from 5×10^{-3} to 0.01 seconds. The thermo-physical parameters used for the simulations are listed in Table 1. To model the steel, we consider a binary iron-carbon alloy with a nominal composition of C_0=0.4 wt.% and neglecting the other alloying additions. This is because carbon has the strongest effect on the solidification path and the solutal buoyancy. The experimental temperature measurements in the mold were used to fit the thermal boundary conditions and to set the initial temperature of the mold components. An external temperature of 200 °C was used. The filling stage has been neglected and the initial temperature and carbon concentration fields in the steel are assumed to be uniform.

Table 1. Thermophysical properties used in the simulations.

Property	Symbol	Units	Value
Nominal carbon concentration	C_0	$[\%mass]$	0.4
Partition coefficient	k_p	$[-]$	0.18
Melting point of the pure substance	T_f	$[°C]$	1538
Liquidus slope	m_l	$[°C/\%mass]$	-85
Liquidus temp. at C_0	$T_{liq}(C_0)$	$[°C]$	1504
Reference density	ρ	$[kg/m^3]$	7040
Solutal expansion coeff.	β_C	$[\%mass^{-1}]$	1.4×10^{-2}
Thermal expansion coeff.	β_T	$[K^{-1}]$	1.7×10^{-4}
Latent heat	L_f	$[J/kg]$	3.09×10^5
Thermal conductivity	k	$[W/(mK)]$	30
Heat capacity	C_p	$[J/(Kkg)]$	500
Kinematic viscosity	ν_l	$[kg/(ms)]$	4×10^{-3}
Diffusion coefficient	L_f	$[m^2s]$	3×10^{-9}

4. Results and discussion

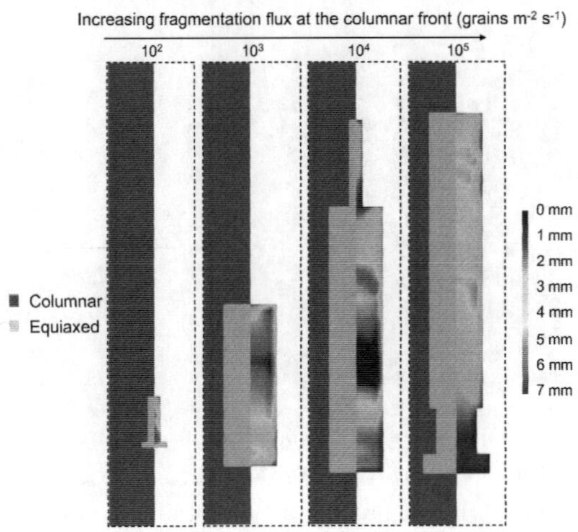

Figure 2. Predicted structure and grain size distribution for fragmentation rates 10^2, 10^3, 10^4 and 10^5 $m^{-2}s^{-1}$. Left hand section shows the grain structure (equiaxed and columnar) and the right half shows the grain size only for the equiaxed regions

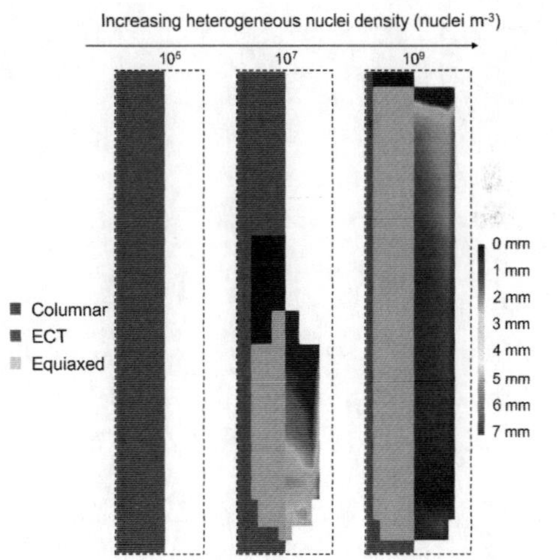

Figure 3. Predicted structure and grain size distribution for heterogeneous nuclei density of 10^5, 10^7 and 10^9 m^{-3}

The simulation results matched with the experimental thermocouple measurements which validated the thermal boundary conditions being used. We use the software SOLID to investigate the structure obtained for the above experimental setup. The simulations investigated the two scenarios of equiaxed grain formation and the impact on CET:

(i) Only fragmentation of columnar dendrites with no heterogeneous nucleation

(ii) Only heterogeneous nuclei with no fragmentation of the columnar dendrites

In all cases, solidification starts with the formation of the columnar dendrites. Simulations with only fragmentation are shown in Fig. 2. Fluxes of fragments across the columnar front, ϕ_0 (nuclei. $m^{-2}s^{-1}$), of 10^2, 10^3, 10^4 and 10^5 are used. The fragments are introduced in the liquid adjacent to the columnar front. Equiaxed grains start to form from the fragments in front of the columnar zone. Initially the equiaxed grains are small and move along with the liquid. At around 40-50 seconds, the grains start sedimenting. The columnar front is blocked first at the bottom and then further up in the ingot. Lower fragmentation rates results in delayed blocking of the columnar front. As the equiaxed grains descend, they are blocked when the grain envelope fraction (g_{env}) reaches 0.3. The first grains that settle to the bottom are small and globular. As solidification progresses upwards, the equiaxed morphology changes to dendritic and the grain size increases. Lower fragmentation rates results in larger grain size towards the top part of the equiaxed region. Fragmentation rate of 10^5 nuclei. $m^{-2}s^{-1}$ gives a satisfactory fit with the experimentally observed observed grain size at the bottom. However, it fails to produce the observed thickness of the columnar zone at the bottom and the size evolution of the equiaxed grains along the height of the cast structure. It can be concluded that fragmentation alone cannot reproduce the observations.

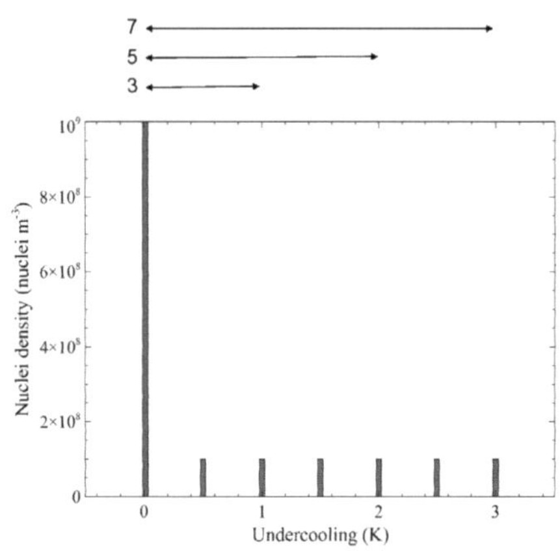

Figure 4. Discretization of nucleation sites with respect to critical undercooling

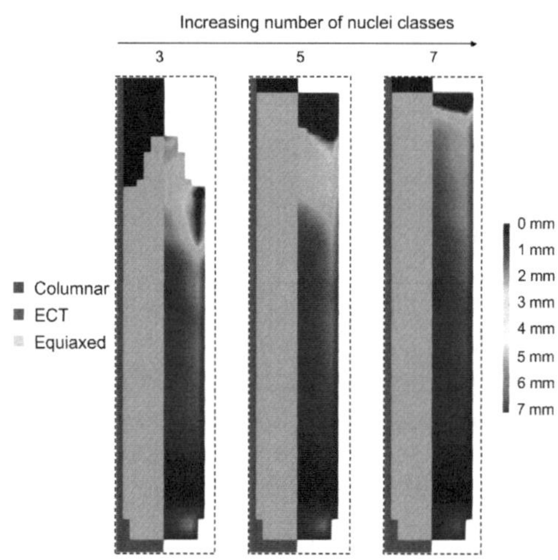

Figure 5. Predicted structure and grain size distribution for heterogeneous nuclei density of 10^9 m^{-3} with varying number of nuclei classes

Fig. 3 shows the results for the heterogeneous nucleation of equiaxed grains without fragmentation. The heterogeneous nucleation simulates the effect of inoculants as well as the possible grains detached from the mold surface during filling. Multiple classes of nuclei as shown in Fig. 4 are used for the next shown simulations [13]. Solidification starts with the formation of the columnar front on the bottom and the development of fluid flow due to thermal convection. The liquid convection results in the ingot being undercooled and all the nuclei are activated. The highest undercooling is obtained in the vicinity of the columnar front at the bottom and the liquid flow drives the liquid to the bottom. This favours the growth of the equiaxed grains which sediment to the bottom of the ingot and block the growth of the columnar front. No equiaxed zone is obtained for a nuclei density of 10^5 nuclei. m^{-3}. Nuclei density of 10^9 nuclei.

m^{-3} produces a completely equiaxed structure with fine columnar zone of 0.4 mm thickness along the vertical walls. The thickness of the columnar zone on the bottom is 0.8 cm and the grain size is in the range of 1-2 mm. Experimental observations are reproduced except for the coarse grain size at the top. Once all the equiaxed grains have sedimented, the columnar structure is free to develop in the remaining liquid which results in an ECT occurring at the top of the sedimented equiaxed grains. Further,the columnar structure originating from ECT can meet the columnar front originating from the mold as can be seen for volumetric injection rate of 10^7 nuclei. m^{-3} in Fig. 3. Fig. 5 shows the influence of number of nuclei classes on the structure and the grain size. For simulations with 3 classes, nuclei were activated at an undercooling of 0, 0.5 and 1 K while for 5 classes, nuclei were activated at an undercooling of 0, 0.5, 1, 1.5 and 2 K. Using multiple classes of heterogeneous nuclei delays the formation of ECT as can be seen in Fig. 5. The size classes used for the simulations is shown in Fig. 4. Nucleation sites requiring large undercooling for activation are introduced towards the later stages of solidification when the liquid fraction is small. The equiaxed grains obtained from these nuclei prevent the growth of the columnar grains unless they sediment and allow an ECT transition. As these nuclei are injected in a highly undercooled liquid, coarse equiaxed grains are obtained at the top as is observed in Fig. 5.

5. Conclusions

Simulations were carried out for 8 kg steel ingot cast by Ascométal. Different origins of equiaxed grains: injection of fragments into the liquid ahead of the columnar front and classical heterogeneous nucleation were simulated using SOLID. A parametric study of the fragmentation shows that this mechanism cannot accurately predict the experimental CET. The top of the structure exhibits a purely columnar structure which is not observed experimentally and the grain size obtained was larger than what was experimentally observed. Heterogeneous nucleations were incorporated. A single class of nuclei resulted in the top region being depleted of the grains and resulted in CET transition. Thus, multiple nuclei classes were incorporated which seems to match better with the experimental observations. Further work is required to understand the mechanism better and the effect of solidification path on this phenomena.

Acknowledgements

This work was supported by Ascométal Creas, ArcelorMittal Industeel Creusot, Aubert & Duval, Aperam Alloys, Transvalor, Sciences Computers Consultants, Affival and IXTREM in the framework of the FUI SOFT-DEFIS project.

References

[1] Combeau H, Založnik M, Hans S and Richy P E 2009 *Metall. Mater. Trans. B* **40** 289–304
[2] Leriche N 2015 *Etude de la Transition Colonnaire-Equiaxe dans les lingots et en coulée continue d'acier et influence du mouvement des grains* Ph.D. thesis University of Lorraine
[3] Kumar A, Založnik M, Combeau H, Demurger J and Wendenbaum J 2012 *International conference on ingot casting, rolling and forging, Aachen, Germany*
[4] Martorano M A, Beckermann C and Gandin C A 2003 *Metall. Mater. Trans. A* **34** 1657–1674
[5] Tveito K O, Pakanati A, MHamdi M, Combeau H and Založnik M 2018 *Metall. Mater. Trans. A* **49** 2778–2794
[6] Založnik M and Combeau H 2010 *Comput. Mater. Sci.* **48** 1 – 10
[7] Leriche N, Combeau H, Gandin C A and Založnik M 2015 *IOP Conference Series: Mater. Sci. Eng.* vol 84
[8] Rappaz M and Thevoz P 1987 *Acta Metallurgica* **35** 1487–1497
[9] Ludwig A and Wu M 2005 *Mater. Sci. Eng. A* **413** 109–114
[10] Lipton J, Glicksman M and Kurz W 1984 *Mater. Sci. Eng.* **65** 57–63
[11] Biscuola V and Martorano M 2008 *Metall. Mater. Trans. A* **39** 2885–2895
[12] Hunt J 1984 *Mater. Sci. Eng.* **65** 75–83
[13] Heyvaert L, Bedel M, Založnik M and Combeau H 2017 *Metallurgical and Materials Transactions A* **48** 4713–4734

IOP Conf. Series: Materials Science and Engineering **529** (2019) 012040 doi:10.1088/1757-899X/529/1/012040

Three-dimensional mesoscopic modeling of equiaxed dendritic solidification in a thin sample: effect of convection flow

A. Olmedilla[1], **M. Založnik**[1], **M. Cisternas Fernández**[1], **A. Viardin**[2], **H. Combeau**[1]

[1] Université de Lorraine, CNRS, IJL, F-54000 Nancy, France
[2] Access e.V., Intzestr. 5, D-52072 Aachen, Germany

E-mail: antonio.olmedilla.aero@gmail.com

Abstract. A 3D mesoscopic envelope model is used to numerically simulate the experimental X-ray observations of the equiaxed dendritic isothermal solidification of a thin sample of Al-20 wt%Cu alloy including the natural convection flow. Several four-grain simulations are run to investigate the effect of the convection, of the grain position, and of the grain rotation on the tip growth kinetics of one of the grains. We show that the effect of convection flow – consequence of the presence of gravity parallel to the sample thickness direction, z – on the growth kinetics of the reference grain depends significantly on the position of the grain along the sample thickness.

1. Introduction

Different approaches can be used in numerical modeling of solidification microstructures, depending on the desired simulation scale. At the microscopic scale, approaches such as the phase-field modeling [1], the pseudo front tracking [2], the enthalpy method [3], or the microscopic CA [4] can be used, whereas at a mesoscopic scale, the needle network (DNN) [5], the cellular-automaton finite-element (CAFE) model [6] and the envelope model [7, 8, 9] are available. The mesoscopic approaches achieve a less detailed description of the dendritic grain morphology compared to the microscopic approaches, but have a much smaller computational cost, i.e. larger domains can be easily simulated.

Herein we focus on the mesoscopic envelope approach, which is convenient to simulate solidification phenomena for which interactions at the scale of a group of grains are important [8]. We show that the three-dimensional mesoscopic envelope model is a useful numerical tool to describe the grain morphology in case of equiaxed dendritic solidification in thin samples used for in-situ observations of metallic alloys [10]. Additionally, we show the effect of convective flow on the growth kinetics. This effect depends on the initial grain position along the sample thickness direction, z, which is parallel to gravity, and the grain orientation, more precisely, the misalignment of the grain preferential growth directions $\langle 100 \rangle$ with respect to the direction z.

2. Mesoscopic Envelope Model

We use the in-house mesoscopic envelope computer code CrystalFOAM. In the mesoscopic envelope model the complex dendritic grain morphology is approximated as an envelope [7].

For an alloy under the assumption of a homogeneous temperature field, the growth of the dendrite tips is governed by the solute flux that the tips eject to the liquid. The Ivantsov analytical tip growth solution [11] is employed in a stagnant film formulation [12] to relate the tip growth Peclet number, Pe_{tip}, to the supersaturation, Ω_δ, at a finite distance, δ, from the tip. The Peclet number is defined by $Pe_{tip} = R_{tip}V_{tip}/(2D_l)$ where R_{tip} is the tip radius, V_{tip} is tip growth velocity, and D_l is the solute diffusion coefficient in the melt. The supersaturation is defined by $\Omega_\delta = (C_l^* - C_l^\delta)/((1-k_p)C_l^*)$ where C_l^* is the liquidus equilibrium solute concentration, C_l^δ is the solute concentration in the liquid at the distance δ from the tip, and k_p is the alloy partition coefficient.

The tip velocity is obtained from Pe_{tip} by a tip selection criterion that reads $V_{tip} = 4\sigma^* D_l Pe_{tip}^2/d_o$, where $d_o = \Gamma/(m_l C_l^*(k_p - 1))$ is the capillary length, m_l is the liquidus slope, Γ is the Gibbs-Thomson coefficient, and σ^* is the selection parameter. We consider $\sigma^* = (2\pi)^{-2}$ [13].

The concentration field is obtained by means of the solute convection-diffusion equation. The phase-field sharp-interface capturing method [14] is used to track the envelope contour once the growth velocity field of the grain envelope is computed. The complete description of CrystalFOAM for purely diffusive dendritic growth (no melt flow) is available in [8, 9].

In order to account for the solutal buoyancy driven flow in presence of gravity, $\vec{g}$, a two phase solid-liquid averaging approach is adopted [15]. The interior of the envelopes is considered as a two-phase medium. The distribution of the liquid fraction in the envelopes is given by a Scheil solidification model [8] and the drag on the liquid is modeled by a permeability of the porous medium. The equations of liquid mass conservation (Eq. 1), and liquid momentum conservation (Eq. 2) are included. Herein, the solid phase is assumed immobile, therefore no equations are needed for the solid.

$$\nabla \cdot \left(g_l \langle \vec{v}_l \rangle^l\right) = 0 \tag{1}$$

$$\frac{\partial}{\partial t}\left(g_l \langle \vec{v}_l \rangle^l\right) + \nabla \cdot \left(g_l \langle \vec{v}_l \rangle^l \langle \vec{v}_l \rangle^l\right) = -g_l \nabla \tilde{p} + \nabla \cdot \left(g_l \nu_l \nabla \langle \vec{v}_l \rangle^l\right) - \frac{\nu_l g_l^2}{K} \langle \vec{v}_l \rangle^l + g_l \tilde{\rho}_l \vec{g}, \tag{2}$$

$$\text{where } \tilde{\rho}_l = (1 - \beta_C (C_l - C_0)) \text{ and } K = \frac{l_c^2 g_l^3}{180(1-g_l)^2}.$$

g_l is the liquid fraction, $\langle \vec{v}_l \rangle^l$ the liquid velocity, $\tilde{p}$ the kinematic pressure, ν_l the kinematic liquid viscosity, K the hydrodynamic permeability of the dendritic grains, l_c a characteristic length of the porous structure, and β_C the solutal volume expansion coefficient. C_0 is the initial solute concentration. Additionally, the solute conservation equation is also needed:

$$g_l \frac{\partial C_l}{\partial t} + g_l \langle \vec{v}_l \rangle^l \cdot \nabla C_l = D_l \nabla \cdot (g_l \nabla C_l) + C_l (k_p - 1)\frac{\partial g_l}{\partial t} \tag{3}$$

This equation (Eq. 3) is identical to the one in the diffusive model [9] but has an additional convection term. It gives C_l outside the envelope and g_l inside the envelope. C_l inside the envelope is equal to $C_l^* = (T - T_f)/m_l$ where T_f is the melting temperature of the solvent. All transport equations are solved with the finite-volume method with a backward Euler time stepping.

3. The cases of study

3.1. In situ experimental observations: the XRMON-SOL experiment of Murphy et al.

We have numerically simulated a small sub-domain of the experiment of isothermal equiaxed dendritic solidification in a thin sample performed by Murphy et al. [10]. In this experiment the growth of 15 equiaxed dendrites under the influence of solutal interactions was observed in situ by X-ray transmission projection imaging. Growth rates of individual grain envelopes and of dendrite tips were measured from the projected images. Murphy et al. concluded that flow

IOP Conf. Series: Materials Science and Engineering **529** (2019) 012040 doi:10.1088/1757-899X/529/1/012040

convection is negligible in this experiment, despite being performed in terrestrial conditions, since the experiment is perfomed in such a way that the $200\,\mu$m thin thickness direction, z, is parallel to gravity. Similar growth characteristics of terrestrial and microgravity conditions are shown in Ref. [16].

The experimental images do not provide information on the 3D structure of the sample, namely the position of the grains in the sample thickness and their 3D orientation. Modeling can help to identify these parameters. We therefore investigate their influence on the grain growth, using mesoscopic simulation.

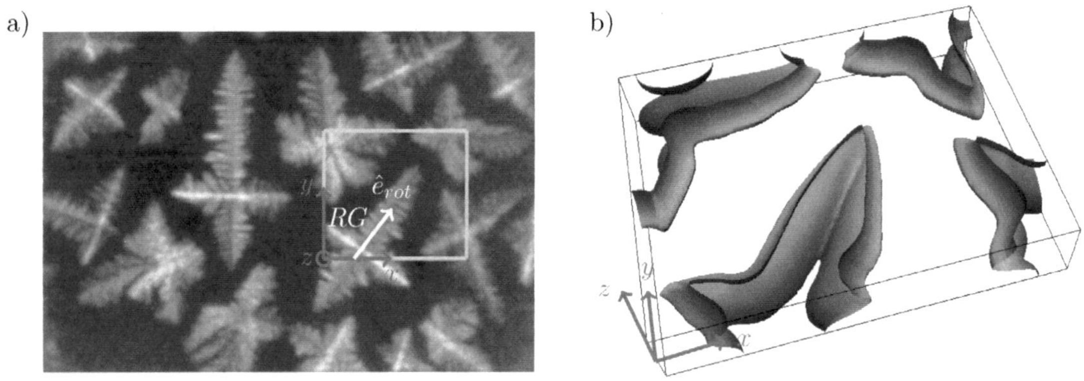

Figure 1. a) X-ray transmission projection images of growing grains from the experiment of Murphy et al. [10]. Images courtesy of D.J. Browne. b) 3D snapshot of the simulation domain and dendritic grains (RC) at the end of simulation.

3.2. Mesoscopic modeling

We establish a rectangular prismatic simulation domain of thickness equal to that of the experiment of Murphy et al. [10], $L_z = 200\,\mu$m, and length and width of $L_x = 1137.7\,\mu$m and $L_y = 883.2\,\mu$m, respectively (see Fig. 1). In the model, four grains are initialized as spherical seeds of diameter $D_{ini} = 90\,\mu$m. The nucleation time (onset of growth) is different for each of the four grains. We have used the nucleation time data available in [10]. The x and y coordinates of the seed centers are known from the experiments and are given in [10], whereas the position in the sample thickness, the coordinate z, is unknown. The grain orientation about the axes x and y is also unknown. The temperature field is assumed uniform with an initial temperature of $T_0 = 601.7°$C and a constant cooling rate of $\dot{T} = -0.05\,$K/s. A time step of 0.01 s and a uniform mesh of cubes of side $\Delta x = 10\,\mu$m are used.

The work in this paper is composed of four pairs of simulation cases. For each of these pairs, the reference grain, RG, (Fig. 1) has a different position along the sample thickness, z, and/or a different rotation about the rotation axis $\hat{e}_{rot}$ (Fig. 1). The other three grains in the simulation domain (Fig. 1b) have a vertical position of $z = 0$ (mid-thickness) and only rotations about the vertical axis, z. Each pair contains a simulation without gravity (NoGrav) and a simulation with gravity (Grav). The cases are collected in Table 1. The common thermophysical properties for the 8 different simulation cases are summarized in Table 2.

4. Simulation results and discussion

In this section we present and discuss the mesoscopic simulation results. We focus on the effect of the flow convection in each of the pairs of simulation cases (with gravity and without gravity).

Table 1. Position and rotation of the reference grain (RG) in the simulation cases.

Case Pairs	vertical position (z)	rotation about $\hat{e}_{rot}$
RC (reference case) | NoGrav / Grav	0 (mid-thickness)	0°
C30 | NoGrav / Grav	0 (mid-thickness)	30°
C100 | NoGrav / Grav	100 μm (top wall)	0°
C45+100 | NoGrav / Grav	100 μm (top wall)	45°

Table 2. Thermophysical properties used in the simulations.

Nominal concentration	C_0	20 wt%
Liquidus slope	m_l	-3.5 K/wt%
Solvent melting temperature	T_f	946.37 K
Partition coefficient	k_{Cu}	0.12
Liquid diffusion coefficient	D_l	$2.8 \times 10^{-9}\,\mathrm{m^2/s}$
Gibbs-Thomson coefficient	Γ_{sl}	$2.4 \times 10^{-7}\,\mathrm{Km}$
Kinematic viscosity	ν_l	$8.43 \times 10^{-7}\,\mathrm{m^2/s}$
Solutal expansion coefficient	β_C	$-1.65 \times 10^{-2}\,\mathrm{1/wt\%}$
Char. length of the porous structure	l_c	$1.0 \times 10^{-6}\,\mathrm{m}$

We first present the results of the 3D simulation cases with gravity at the end of the growth stage ($t = 130\,\mathrm{s}$) in form of simulated projection radiographs (where the transmission is obtained with the Beer-Lambert law) that can be compared to the experimental images (see Fig. 2, left column). The difference of the RG's projected shape in the radiograph between the four cases (a-d) is basically a consequence of the position and orientation of RG. For all the pairs, the difference between the case with gravity and its homologue without gravity remains small in the radiographs, this is why only the radiographs for the gravity cases are shown.

In Fig. 2, right column, we show, for each of the gravity cases (a-d), the RG's envelope and a slice containing the center and the primary tip P1 of RG, and the vertical axis, z, at time $t = 60\,\mathrm{s}$. The colormap of the RG's envelopes refers to the normal growth velocity of the envelope, V_n, and the colormap of the slice shows the melt velocity field, U_l. The counterclockwise direction of the melt convection flow in the images is due to the higher density of the solute-enriched melt nearby the RG's tip which descends in presence of gravity. A maximum velocity of melt of approximately $30\,\mu\mathrm{m/s}$ is obtained with tip growth velocities lower than $10\,\mu\mathrm{m/s}$ for all four cases. In those cases where the primary arm grows along the sample wall the flow direction is anti-parallel to the tip growth velocity, whereas for those cases where the grain is located at mid-thickness the flow direction is perpendicular to the tip growth velocity.

Afterwards, we show quantitative results for the growth of the primary arm P1. In Fig 3a we show the length evolution of P1 over time for the four pairs of simulation cases and compare to the experiment of Murphy et al. (XRMON). All simulation cases show a trend similar to the experimental one, although the initial transient is slower in the simulations. For the cases RC and C30, the effect of gravity (convection flow) on the growth kinetics is negligible. For the cases C100 and C45+100 a more important effect of gravity is found, obtaining approximately a 10 % longer primary arm when gravity is active. The cases C100 and C45+100 with gravity give the most similar trend to that of the experiment.

In Fig 3b, we show the relative difference between the projected final area of P1 in our simulations and in the experiment Murphy et al. As before, for the cases RC and C30, the effect of gravity (convection flow) on the final area of P1 is negligible. For the cases C100 and

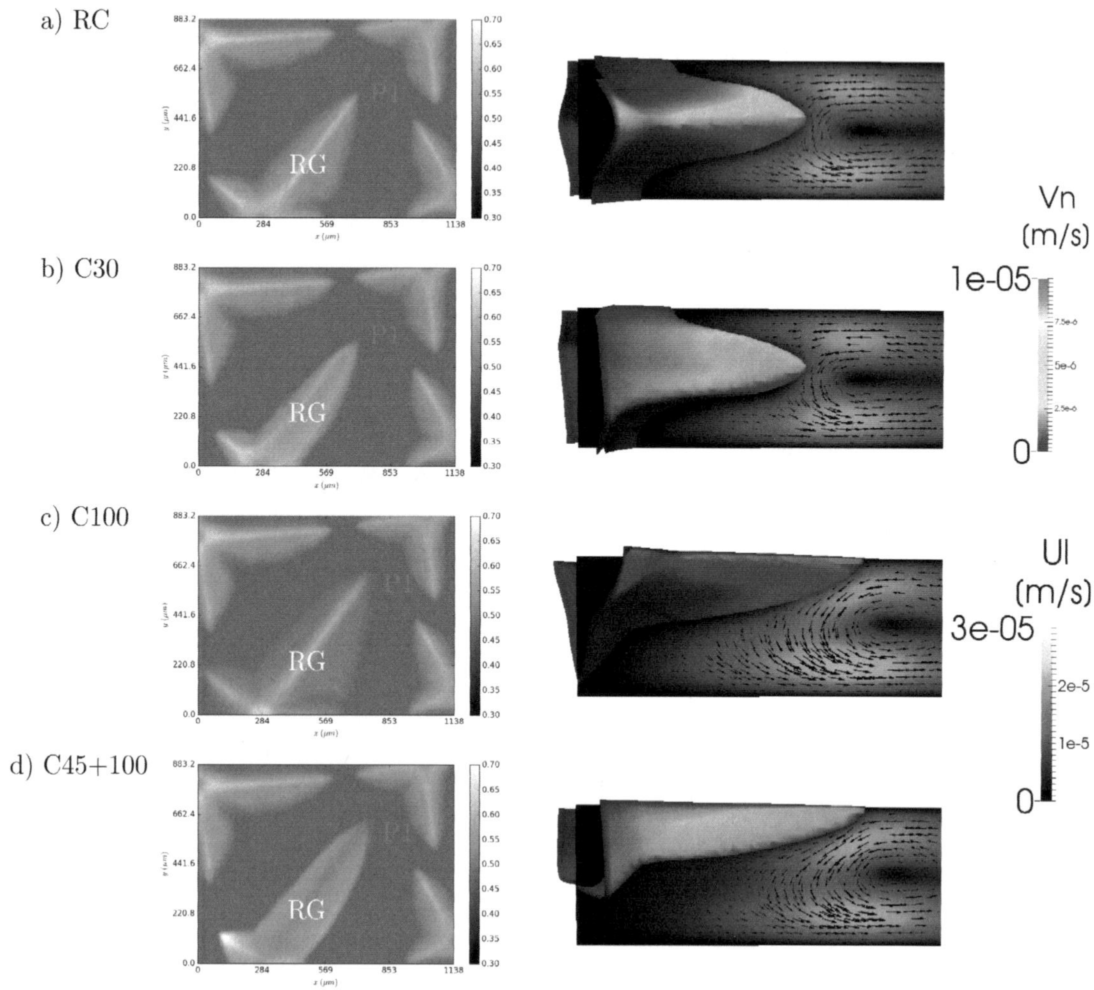

Figure 2. Left: simulation radiographs (where the grayscale represents the transmission obtained by the Beer-Lambert law) at the final instant, $t = 130\,\mathrm{s}$, for the cases: a) RC | Grav, b) C30 | Grav, c) C100 | Grav, and d) C45+100 | Grav. The reference grain is indicated in the images, RG, as well as the primary arm P1. The gray scale bar refers to the transmission coefficient. Right: RG's envelope and slice containing the center and P1 of RG and the vertical axis, z, for each of the cases a, b, c and d, at time $t = 60\,\mathrm{s}$. The normal growth velocity of the RG's envelope, V_n, and the melt velocity, U_l, are shown.

C45+100 a more important effect is found.

In this way, for cases where the vertical position of the RG's center is at the top wall (C100 and C45+100) a much important influence of the convection flow is found than for those cases where the RG's center is at the mid-thickness. This is a consequence of the higher flow of solute-depleted melt convected to the tip in cases C100 and C45+100 which increases the tip growth speed.

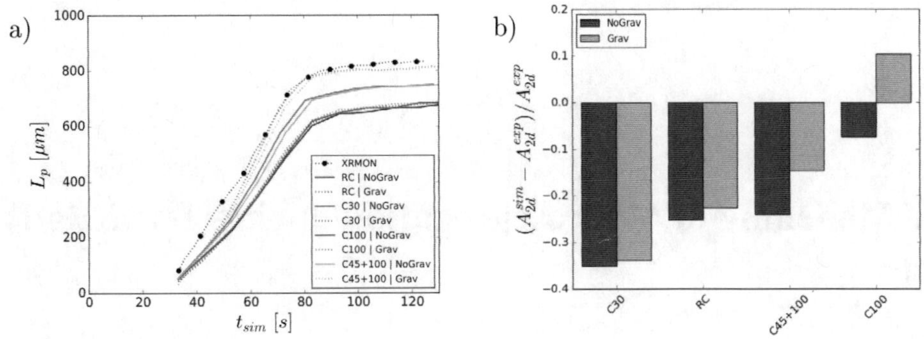

Figure 3. a) Length evolution of the primary arm P1 over time for the simulation cases RC, C30, C100, C45+100, with and without gravity, and the experiment (XRMON) of Murphy et al. (with gravity). b) Relative difference between the projected final area of P1 in the simulation cases and the experiment of Murphy et al. [10].

5. Conclusions

The mesoscopic envelope model is shown to be an useful tool to numerically simulate the equiaxed dendritic isothermal solidification including the convection flow. In case of isothermal solidification of a thin sample of Al-20 wt%Cu, we show that the effect of the thermo-solutal convection flow – consequence of the presence of gravity parallel to the sample thin thickness direction – on the growth kinetics of the reference grain depends on its initial grain position along the sample thickness.

Acknowledgments

This work was supported by the French State through the program "Investment in the future" operated by the National Research Agency (ANR) and referenced by ANR-11 LABX-0008-01 (LabEx DAMAS). We thank David J. Browne, Henri Nguyen Thi and Guillaume Reinhart for helpful discussions.

References

[1] Boukellal A K, Debierre J M, Reinhart G and Nguyen-Thi H 2018 *Materialia* **1** 62–69
[2] Jacot A and Rappaz M 2002 *Acta Materialia* **50** 1909 – 1926
[3] Voller V 2008 *International Journal of Heat and Mass Transfer* **51** 823 – 834
[4] Li D, Li R and Zhang P 2007 *Applied Mathematical Modelling* **31** 971 – 982
[5] Tourret D and Karma A 2016 *Acta Materialia* **120** 240 – 254
[6] Carozzani T, Digonnet H and Gandin C A 2012 *Modelling and Simulation in Materials Science and Engineering* **20**
[7] Steinbach I, Beckermann C, Kauerauf B, Li Q and Guo J 1999 *Acta Materialia* **47** 971 – 982
[8] Souhar Y, Felice V F D, Beckermann C, Combeau H and Založnik M 2016 *Computational Materials Science* **112** 304 – 317
[9] Viardin A, Založnik M, Souhar Y, Apel M and Combeau H 2017 *Acta Materialia* **122** 386 – 399
[10] Murphy A, Mathiesen R, Houltz Y, Li J, Lockowandt C, Henriksson K, Zimmermann G, Melville N and Browne D 2016 *Journal of Crystal Growth* **440** 38 – 46
[11] Ivantsov G 1947 *Dokl Akad Nauk SSSR* **58** 567 – 570
[12] Cantor B and Vogel A 1977 *Journal of Crystal Growth* **41** 109 – 123
[13] Langer J and Müller-Krumbhaar H 1978 *Acta Metallurgica* **26** 1681 – 1687
[14] Sun Y and Beckermann C 2007 *Journal of Computational Physics* **220** 626 – 653
[15] Založnik M and Combeau H 2010 *Computational Materials Science* **48** 1 – 10
[16] Murphy A, Mathiesen R, Houltz Y, Li J, Lockowandt C, Henriksson K, Melville N and Browne D 2016 *Journal of Crystal Growth* **454** 96 – 104

On the Modelling of Macrosegregation during Twin-Roll Casting

C M G Rodrigues[1], A Ludwig[1], A Kharicha[1,2], M Wu[1] and A Vakhrushev[2]

[1] Department Metallurgy, Montanuniversitaet Leoben, Leoben, Austria
[2] Christian Doppler Laboratory for Metallurgical Applications of Magnetohydrodynamics, Montanuniversitaet Leoben, Leoben, Austria

christian.gomes-rodrigues@unileoben.ac.at

Abstract. Solidification during twin-roll casting happens by cooling of the melt between two counter-rotating rolls, where the molten alloy is constantly fed in. For an inoculated Al-melt, grain growth leads to a gradual increase of solid fraction, so that a coherent solid network forms. Depending on the process condition, this solid network might be subjected to compression within the gap between the two rolls. By using a two-phase volume average model that accounts for (i) transport and growth of spherical grains within a flowing melt, (ii) formation of a coherent solid network above a specific solid fraction and (iii) viscoplastic flow of the solid network saturated with interstitial melt during casting and compression, the process is numerically analysed. It is found that an optimum process with minimum macrosegregation can be achieved for conditions where the kissing point of the two viscoplastic semi-solid shells nearly coincides with the roll nip. It is demonstrated how casting speed, cooling intensity and strand thickness must be related to hit the optimum process window.

1. Introduction

Twin-roll casting is an emerging commercial material processing technique that produces thin strips directly from molten metal. As the casting and hot rolling processes are performed at the same time in the production line, the technology requires less initial investment and ensures a more energy efficient process than conventional continuous casting techniques.

As a result, interest has grown among engineers and researchers, and many studies have been carried out over the past years. For instance, several experiments have been conducted to optimize the operating conditions, such as the pouring temperature, set-back distance, strip thickness or casting speed [1,2]. Other researchers have instead studied micro- and macro-defects, ranging from surface bleeds, internal defects, hot tearing and buckling [3,4]. However, there is obviously a limit to how much information one can obtain solely from measurement techniques applied to experimental trials.

Comprehension of the twin-roll casting process requires a thorough understanding of the complex flow dynamics that develop during the production of the strip. That is why numerical models have been employed to accomplish this task. Accordingly, due to the stiff nature of the problem, finite element simulations are widely used for that purpose, although they often rely on a single phase thermo-mechanical model originally designed for the analysis of pure rolling [5].

However, deformation of the partly solid shells is usually an important mechanism in twin-roll casting. It is known that the deformation of a solid-liquid mixture above a certain solid fraction can lead to macrosegregation [6,7]. In order to capture these flow dynamics, it is essential that the solid-liquid mixture is considered as a contiguous solid skeleton saturated with interstitial liquid when the solid fraction exceeds the specific threshold. In the present paper, this stage is referred to as the viscoplastic regime and the major distinction – compared to the lower solid fraction regime – is that the rheology of the flow is now dictated fundamentally by the solid phase. Therefore, in the case of an alloy in the semi-solid state being subjected to deformation, the corresponding motion of the coherent structure drives the liquid flow, which means that at this stage, the mechanical character of the process takes over the fluid flow dynamics, and consequently, the melt becomes the passive phase which reacts to the solid behavior.

In the present work, a two-phase volume-average model is used to replicate a twin-roll casting scenario. The model considers the entire spectrum of solid fractions, and so it is able to deal with an almost completely molten alloy and turn it into a fully solid material, and vice-versa. Furthermore, above a specific criterion, the equiaxed crystals are assumed to behave according to a dense coherent network which interacts with the interstitial liquid. This means that mush deformation is accounted for by mean of a viscoplastic model that is employed when the solid fraction is above 0.57.

2. Numerical implementation

The simulation of the twin-roll continuous casting presented in this paper relies on a two-phase Eulerian-Eulerian volume-average model. The general approach solves mass, momentum, species, and enthalpy conservation equations for each phase. Additionally, a transport equation is also taken into account for the calculation of the number density of grains across the domain. A detailed description of the corresponding governing equations has been made in a previous publication [8], so the interested reader can refer to those contributions for further details. For the sake of conciseness, only the most relevant equations will be highlighted here.

The important novelty of this model is the implementation of the viscoplastic regime, which is imposed when the solid fraction exceeds $g_s^t = 0.57$. The change in material behavior is transmitted through the viscous stress tensor, which is then defined differently depending on the solid fraction.

$$g_s \boldsymbol{\tau}_s^{eff} = \begin{cases} 2 g_s \mu_s \mathrm{dev}(\dot{\boldsymbol{\varepsilon}}_s) & \text{for } g_s \le g_s^t \\ 2\frac{\mu_s^{\mathrm{app}}}{A}\mathrm{dev}(\dot{\boldsymbol{\varepsilon}}_s) + \mu_s^{\mathrm{app}}\left(\frac{1}{9B}\right)\mathrm{tr}(\dot{\boldsymbol{\varepsilon}}_s)\mathbf{I} & \text{for } g_s > g_s^t \end{cases} \tag{1}$$

Below the viscoplastic threshold, the equation is defined by the general Newtonian viscous stress tensor (assuming that bulk viscosity is zero). This expression is analogous to the one adopted in the liquid phase with the strain rate tensor for either of the phases (i) being given by $\dot{\boldsymbol{\varepsilon}}_i = 1/2\left(\nabla \mathbf{v}_i + (\nabla \mathbf{v}_i)^T\right)$. In the viscoplastic regime, on the other hand, experimental studies on the behavior of metallic alloys at high temperatures show that the solid phase can be described by the equation presented above [6].

Comparison between the two equations defined for each regime shows that above the transition criterion a second term is considered in the formulation. This term is related to the bulk viscosity and corresponds to a normal compression force considered in this regime. Note that A and B are two rheological parameters proposed by Nguyen et al. [6], which can be written as follows:

$$A = \frac{3}{g_s^{6.47}} \quad \text{and} \quad B = \frac{1}{g_s^{6.94}} - 1 \tag{2}$$

In Eq. (1), the terms μ_s and μ_s^{app} represent the solid viscosities in each of the regimes. For low solid fractions, Ishii [9] postulated that the mixture viscosity of a set of particles can be expressed according to the power-law viscosity model. Assuming the general mixture rule as $\mu_{mix} = g_s\mu_s + g_l\mu_l$, the solid viscosity can be given by:

$$\mu_s = \frac{\mu_\ell}{g_s}\left(\left(1-\frac{g_s}{g_s^p}\right)^{-2.5g_s^p} - (1-g_s)\right) \tag{3}$$

where g_s^p is considered as a fitting constant that represents the random loose packing limit for spheres. Following the experimental results from Olmedilla et al. [10] g_s^p= 0.585 has been adopted here.

Regarding the viscoplastic regime, the apparent solid viscosity generally accepted in the literature takes the form of the Herschel–Bulkley model:

$$\mu_s^{\text{app}} = 3K_v\left(\sqrt{3}\dot{\varepsilon}_s^{\text{eq}}\right)^{m-1} \tag{4}$$

where K_v and m are the viscoplastic consistency and the strain-rate sensitivity, respectively. The latter parameter is usually assumed as a fixed value which depends on the material properties. In the present work, it is set as 0.213. On the other hand, the viscoplastic consistency can be calculated by combining the strain-rate tensor proposed by Nguyen et al. [6] and the Norton-Hoff stress-strain-rate relation. It takes the value of 6.31×10^6 Pa s.

As for the equivalent strain rate presented in Eq. (4), it is given by the following expression:

$$\dot{\varepsilon}_s^{\text{eq}} = \sqrt{\tfrac{2}{A}(\dot{\boldsymbol{\varepsilon}}_s : \dot{\boldsymbol{\varepsilon}}_s) - \left(\tfrac{2}{3A} - \tfrac{1}{9B}\right)\text{tr}(\dot{\boldsymbol{\varepsilon}}_s)^2} \tag{5}$$

which again depends on the rheological parameters defined in Eq. (2).

Details on the microscopic phenomena and the coupling between the conservation equations have been described in detail in companion papers [7,8].

3. Results and discussion

The schematic diagram of the geometry replicated in the simulations is illustrated in Fig. 1. The results have been obtained for an inoculated Al-4wt.%Cu alloy ($\cong 2.5$wt.%). Since the width of the sheet is in concept much larger than its thickness, the test case is assumed to be 2D. A mesh composed of 5700 cells was used, which corresponds to an average cell size of approximately 0.85 mm per 0.47 mm. The time step was variable, but stayed in the order of 10^{-4} s a few moments after the start of the simulation. Also, the effect of gravity is neglected at this point, so the interpretation of the results is not influenced by the sedimentation of the crystals. The densities of the solid and liquid phases are assumed to be equal (ρ=2743 kg/m^3), so solidification-induced feeding is also neglected at this point.

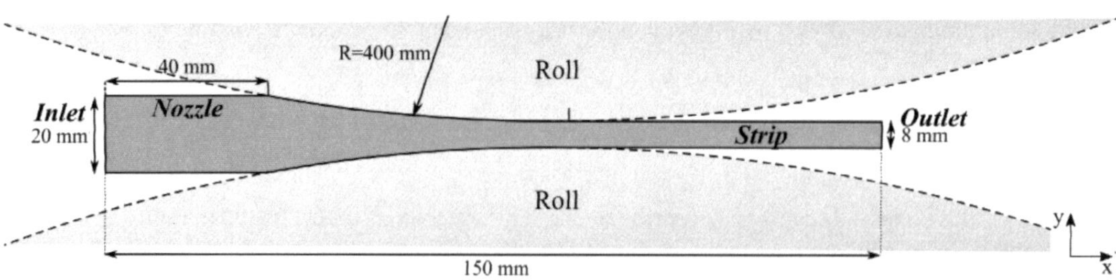

Figure 1. Schematic representation of the twin-roll continuous casting process.

The melt is injected into the domain through the inlet with a fixed pressure field (p=10^5 Pa) at 925 K. Dirichlet boundary conditions are imposed at the inlet to solid fraction (10^{-5}, i.e., almost negligible amount), initial crystal diameter (d=5 μm), and liquid and solid species mass fractions (c_l=0.025 and c_s=0.0036, with the values taken from the phase diagram). Regarding the temperature, it is set as 925 K at the nozzle surface, and zero gradient at the outlet. In the roll and strip sections, on the other hand, a heat flux is imposed between the surface and each of the two phases with a constant heat transfer coefficient of 5.5 kW/m^3/K. Even though the assumption of equal heat transfer coefficient in the strip and roll is generally not precise in practical environments, it has been adopted in the present work to simplify the simulation as much as possible for a proper understanding of the flow dynamics and macrosegregation patterns.

The velocity boundary conditions for the liquid phase are assumed as free-slip at the nozzle, roll and strip sections, whereas for the solid phase they are set as no-slip instead. At the outlet, two different scenarios are considered: Case A has a casting speed of 0.040 m/s, whereas Case B is set with a casting speed of 0.025 m/s.

The steady state results for the solid fraction in both Case A and Case B are illustrated in Fig. 2 a) and b), respectively. The black contour line in each of the snapshots represents the solid fraction at which the viscoplastic regime starts. In Fig. 2b), besides this, also the contour line from Case A is presented with white dots.

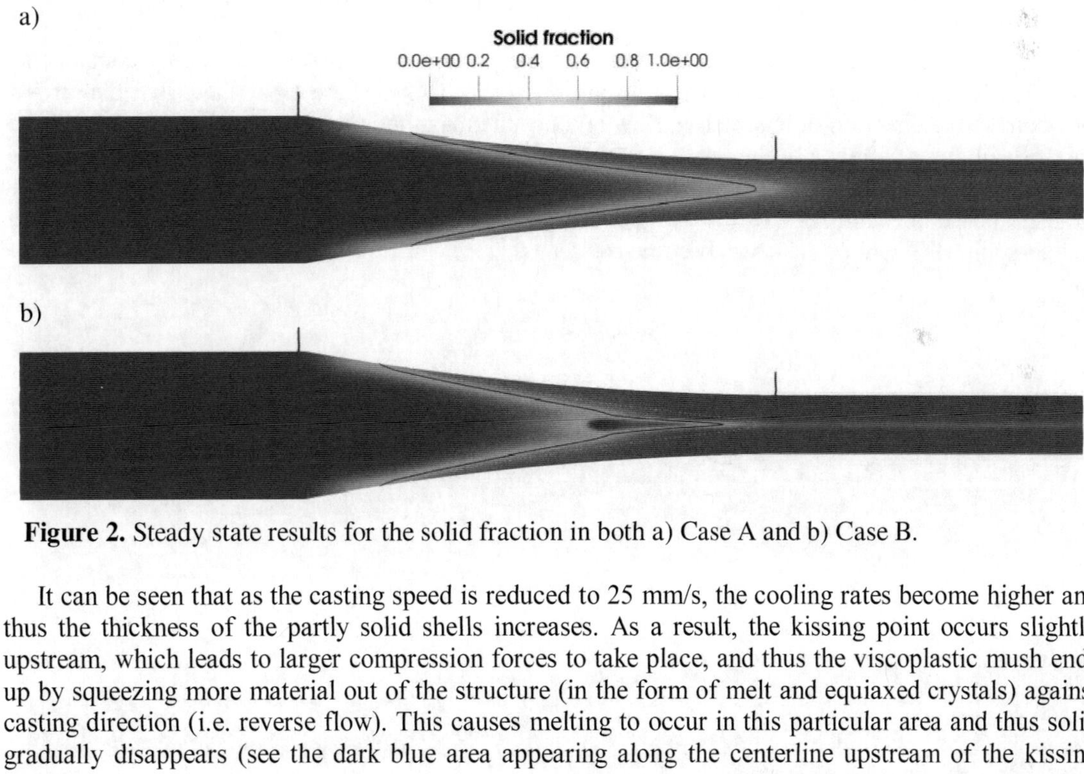

Figure 2. Steady state results for the solid fraction in both a) Case A and b) Case B.

It can be seen that as the casting speed is reduced to 25 mm/s, the cooling rates become higher and thus the thickness of the partly solid shells increases. As a result, the kissing point occurs slightly upstream, which leads to larger compression forces to take place, and thus the viscoplastic mush ends up by squeezing more material out of the structure (in the form of melt and equiaxed crystals) against casting direction (i.e. reverse flow). This causes melting to occur in this particular area and thus solid gradually disappears (see the dark blue area appearing along the centerline upstream of the kissing point).

Obviously, such flow dynamics have an important effect on macrosegregation. Deviations from nominal composition are shown in Fig. 3 for both cases examined here. It becomes clear that in Case B the significant positive segregated centreline appears as a result of the reverse flow discussed above. Furthermore, negatively segregated bands located near the surface are predicted next to positively segregated bands in both Case A and Case B. To understand these findings, one has to combine information regarding the general direction of the flow with the relative velocity between solid and liquid.

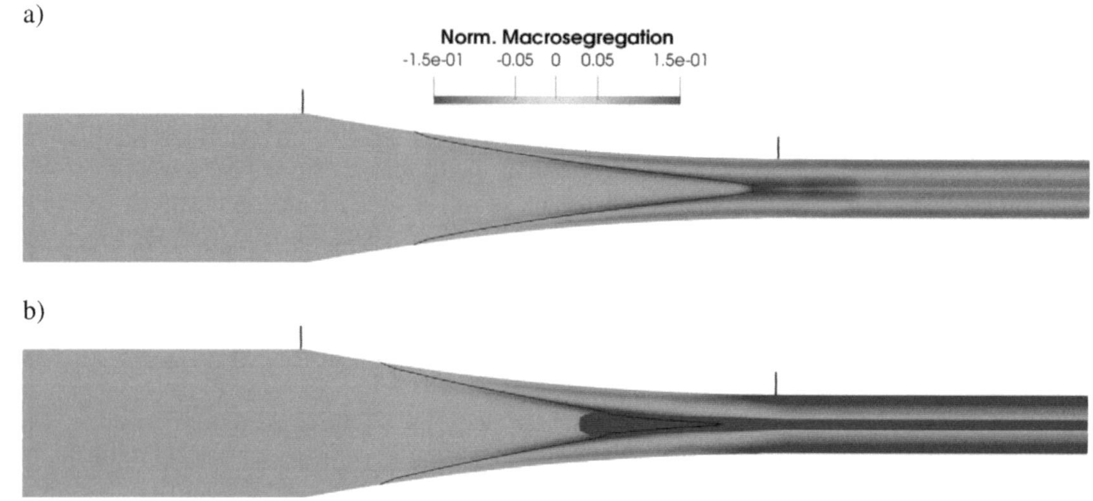

Figure 3. Steady state results for solid fraction in a) Case A and b) Case B. Data range has been limited for better understanding of the deviations from nominal composition.

Figure 4 presents the relative velocity between solid and liquid in x- (top half) and y-components (bottom half) with black arrows representing the relative velocity of the flow. Note that the arrows only express the direction of the relative flow and not the magnitude as the arrows have all the same size. Instead, the magnitude of the relative flow between phases can be identified through the relative velocity snapshots. The arrows, on the other hand, allow one to keep a clear perception of the dominant phase in each region of the domain. Also, note that the range of the color legend has been reduced so the different mechanisms become clearer.

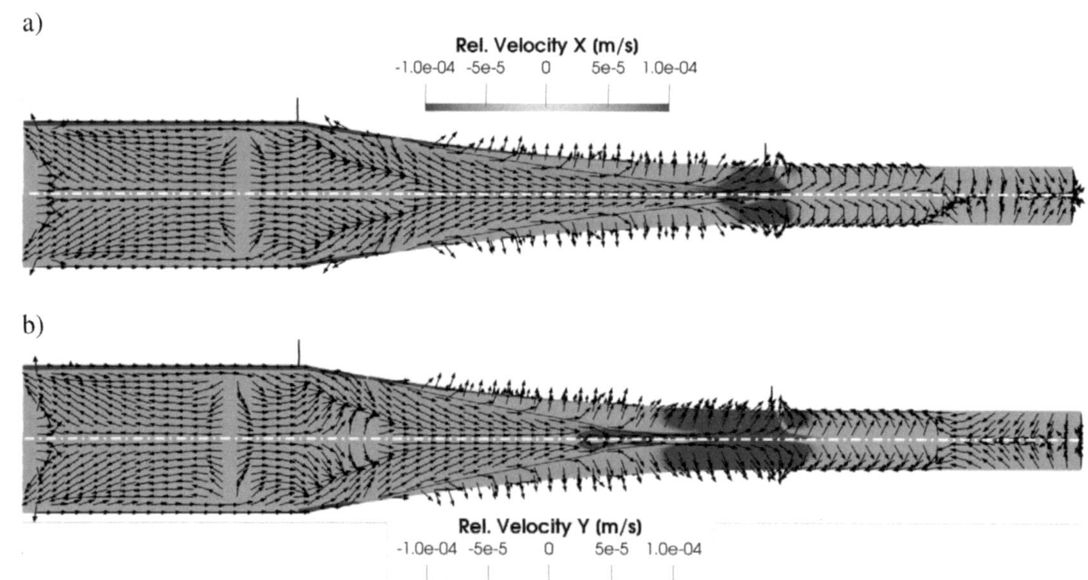

Figure 4. Steady state results of liquid velocity for both both x- (top half) and y-component (bottom half) for a) Case A and b) Case B. Arrows representing the relative velocity between solid and liquid phases are shown on top of the relative velocity snapshots.

The change in relative velocity at the walls is obviously due to the different boundary conditions imposed to both phases, as described in the previous chapter. Then, in the section of the domain where the rolls are being enforced (between the two vertical markers), one can notice in both cases a slight reddish area appearing in the non-viscoplastic regime that fades away as we move towards the centreline of the domain. This corresponds to an area where the solid phase is faster than the liquid both inwards and downstream. This is because the abrupt decrease in velocity of the melt (between casting speed near the surface to almost zero velocity along the centreline) is not accompanied immediately by the solid phase because of the much higher viscosity.

Notice that right after the contour line which establishes the transition to the viscoplastic regime, the arrows start pointing outward instead of inward in both Fig. 4 a) and b). This finding marks the start of the viscoplastic regime, and means that (contrary to the previous state) the liquid is now moving inwards faster than the solid. It also correlates well with the negative macrosegregation bands shown in Fig. 3, near the surface. It can be explained by a compression-induced expulsion mechanism occurring due to hot rolling of the solid-liquid mixture, which causes segregated melt to be squeezed out of the viscoplastic skeleton. As a result of these occurrences, the enriched melt travels to adjacent regions in the domain, causing positive macrosegregation to develop in those regions. This justifies the appearance of bands with positive deviation from nominal composition located next to bands with negative deviation in Fig. 3 a) and b).

At the kissing point, the two semi-solid shells merge together and form a single continuous casting strip. Comparison between Case A (Fig. 4 a) and Case B (Fig. 4 b) shows that in the latter, the relative velocities in both components are much bigger in this region because of the larger compression forces, as discussed above. In the region where the reverse flow develops in Case B, melt is moving upstream faster than the solid, which means that the compression-induced expulsion is such that interstitial liquid is squeezed out against casting direction. On the other hand, after the two solid shells have merged together, the blue area in the y-component and red area in the x-component of the velocity mean that the liquid is moving faster than the solid inwards (because of compression-induced expulsion) and upstream (because it is easy for the flow to travel through lower viscosity regions). Such flow dynamic leads to an increase in positive macrosegregation near the kissing point region, while inducing a slight negative macrosegregation along the centreline because the enriched melt there is moving upstream (arrows are horizontal along centreline).

4. Conclusion

Even though twin-roll casting is a promising processing technique, the material properties depend greatly on the operation parameters (e.g. casting speed, strip thickness and casting temperature), which in turn are difficult to control. In the present paper, two different casting speeds are presented, and the outcome of the simulations are discussed in terms of macrosegregation. The two cases illustrate significant casting speed differences, and the corresponding effect on the quality of the strip.

The present results show that, for a specific strip thickness, there is a corresponding casting speed that provides minimal deviation from nominal composition. Below this threshold, further cooling of the semi-solid is enforced which eventually increases the strength of the compression-induced expulsion mechanism and amplifies macrosegregation.

Acknowledgments

This work was financially supported by the FWF Austrian Science Fund (P28785-N34) which the authors gratefully acknowledge.

References

[1] Zhang X M, Jiang Z Y, Yang L M, Liu X H, Wang G D and Tieu A K 2007 Modelling of coupling flow and temperature fields in molten pool during twin-roll strip casting process *J. Mater. Process. Technol.* **187–188** 339–43

[2] Sahoo S, Kumar A, Dhindaw B K and Ghosh S 2012 Modeling and experimental validation of

rapid cooling and solidification during high-speed twin-roll strip casting of Al-33 wt pct Cu *Metall. Mater. Trans. B Process Metall. Mater. Process. Sci.* **43** 915–24
[3] Jin I, Morris L R and Hunt J D 1982 Centerline Segregation *J. Met.* 70–5
[4] Lockyer S A, Yun M, Hunt J D and Edmonds D V. 1996 Micro- and macrodefects in thin sheet twin-roll cast aluminum alloys *Mater. Charact.* **37** 301–10
[5] Lee Y S, Kim H W, Cho J H and Chun S H 2017 Coupled thermal-fluid-mechanics analysis of twin roll casting of A7075 aluminum alloy *Met. Mater. Int.* **23** 923–9
[6] Nguyen T G, Favier D and Suery M 1994 Theoretical and experimental study of the isothermal mechanical behaviour of alloys in the semi-solid state *Int. J. Plast.* **10** 663–93
[7] Rodrigues C M G, Ludwig A, Wu M, Kharicha A and Vakhrushev A 2019 A Comprehensive Analysis of Macrosegregation Formation during Twin-Roll Casting *Metall and Materi Trans B* 1–17 https://doi.org/10.1007/s11663-019-01527-x
[8] Ludwig A and Wu M 2002 Modeling of globular equiaxed solidification with a two-phase approach *Metall. Mater. Trans. A* **33** 3673–83
[9] Ishii M 1977 *One-Dimensional Drift Flux Model and Constitutive Equations for Relative Motion Between Phases in Various Two-Phase Flow Regimes*
[10] Olmedilla A, Založnik M and Combeau H 2017 DEM simulation of dendritic grain random packing: application to metal alloy solidification *EPJ Web Conf.* **140**

IOP Conf. Series: Materials Science and Engineering **529** (2019) 012042 doi:10.1088/1757-899X/529/1/012042

Atomistic simulation of crack propagation along γ-TiAl lamellar interface

W Li[1,2], H Nan[2], X W Qian[1]*, Y J Yin[1]*, Q Xu[1], X Shen[1], W Yu[1,2], X Feng[2], L P Zhu[2] and J X Zhou[1]

[1] State Key Laboratory of Materials Processing and Die & Mould Technology, Huazhong University of Science and Technology, Wuhan 430074, China.
[2] AECC Beijing Institute of Aeronautical Materials, Beijing 100095, China.

E-mail: intecastqxw@163.com, yingyajun436@hust.edu.cn

Abstract. Due to start-up and shut-down operations of engine, TiAl structural components usually undergo not only static but also cyclic mechanical loading. The crack propagation mechanisms of γ-TiAl under two types of loading are studied in this work to reveal the differences of the mechanisms under constant strain rate and cyclic loading. Since the crack prefers to nucleate at the interface, two types of loadings are applied to a γ-TiAl interface system with a pre-existing micro-crack at the interface by the means of classical molecular dynamics simulation, the loading direction is along [111] perpendicular to the interface. The evolution of crack tip and dislocation is observed in atomistic scale. The results show that, under both loading types, the crack propagates asymmetrically, Shockley dislocations emit on the (-1-11) slip plane from the right crack tip and slip along [-1-1-2] direction. The dislocations blunt the extension of crack while the left crack tip propagates in a brittle way. During the cyclic loading, different with constant strain rate condition, the crack advances and dislocations slip with increasing loads and retreat during unloading. In addition, the stress decreases and the crack length increases with the increase of the cyclic loading number.

1. Introduction

TiAl intermetallic based alloy has potential for replacing heavier materials in high temperature applications such as automotive and aerospace engine components, due to their remarkable thermomechanical properties and low density. Unfortunately, neither the ductility nor the fracture toughness of TiAl-alloys is particularly high. Thus, it's of significant importance to study the mechanism of crack propagation in TiAl components.

TiAl alloy formed entirely by lamellar colonies, which is accordingly named "fully lamellar (FL)", has the most promising mechanical properties over a wide range of temperature[1]. Previous experimental investigations of the influence of lamellar spacing on ductility, creep resistance and fracture toughness show that increasing the number of lamellar interfaces is advantageous[2,3], and enhances the ductility significantly[4–6] while also increasing the creep resistance[7,8]. Meanwhile, micro-cracks prefer to initiate at and propagate along the lamellar interfaces or lamellar colony boundaries[9]. Therefore, investigation of crack propagation in lamellar interface system is of significant importance.

Typical TiAl lamellar microstructures consist of many sharp γ/γ and few γ/α2 interfaces which are all perpendicular to the [111]γ or [0001]α2 directions. Different γ/γ variants can be created by rotating half of the initial supercell around the [111]γ direction by 60°, taking into three different variants, the pseudo-twin (PT), the rotational-boundary (RB) and the true-twin (TT) by 60°, 120° and 180° respectively[10].

Previous investigations on crack propagation of TiAl intermetallic based alloy mainly using experiment method. Study by experiments are, however, time consuming and costly due to sample fabrication and thus only a limited number of tests per condition can be generally performed. Classical molecular dynamics simulation provides a physical insight into understanding the deformation phenomena in the atomic scale and enables one to predict certain properties of materials. Numerous researches have been done via molecular dynamics to study the crack propagation of various materials, and their studies show that molecular dynamics simulation provides a physical insight into understanding the crack propagation phenomena in the atomic scale[11,12]. In addition, the influences of grain boundary or twin boundary on crack propagation have also been investigated in atomic scale.

In this work, molecular dynamics simulations of crack propagation in a γ/γ interface system under static and cyclic loading have been performed to investigate the different crack extension mechanisms under different loading types.

2. Simulation Details

A semi-empirical embedded atom method (EAM) potential by Zope and Mishin[13] is used to model the interatomic interactions of Ti-Al intermetallic system, using an open code Large-scale Atomic/Molecular Massively Parallel Simulator (LAMMPS)[14]. The simulation box contains about 522240 atoms with a size of 9.8*30*30 nm^3. The simulation box consists of two lamellae with a PT interface and a pre-exist crack, as shown in figure 1(a).

All of the simulations are performed at 0.01 K to avoid the complexity due to the thermal activity of atoms. At first the system is equilibrated through applying NVE ensemble. And then the top and bottom layers are fixed for applying Mode I loading under a uniaxial tensile loading process with a strain rate of $10^9 s^{-1}$. The NVE ensemble is applied during the tensile deformation, and Langevin thermostat is used to control the temperature. Static and cyclic loading is performed in this work respectively, as it is shown in figure 1(b). The periodic boundary condition is assigned in the x and y direction. For the sake of eliminating the stress oscillation resulting from the sudden loading, the loading is applied such that the velocity was linearly distributed along the z-direction. The velocities are maximum positive value and minimum negative value at the top and bottom fixed layers. The time step is chosen to be 0.001 ps.

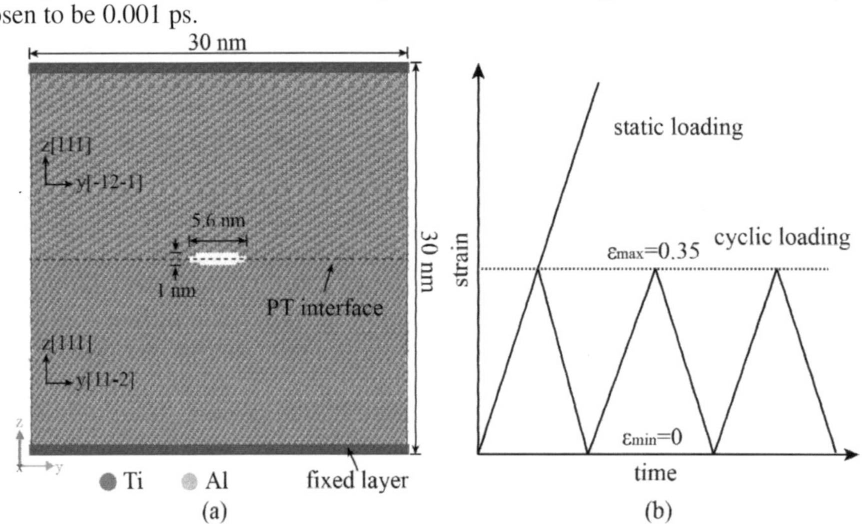

Figure 1. Schematic of the (a) simulation model and (b) two types of loading regimes

The visualization of the simulation results is performed by OVITO[15]. Dislocation extraction algorithm (DXA) and common neighbor analysis (CNA) are used to identify the structure and dislocations[16].

3. Result and Discussion

Figure 2 lists the tress-strain curve and snapshots of the microstructure at different loading stages under static loading. Imagines are colored according to CNA analysis. Green and red atoms represent the perfect FCC and HCP (interface or stacking faults here) structures respectively, and the dark blue atoms are of the other structure (crack boundary and dislocation core here).

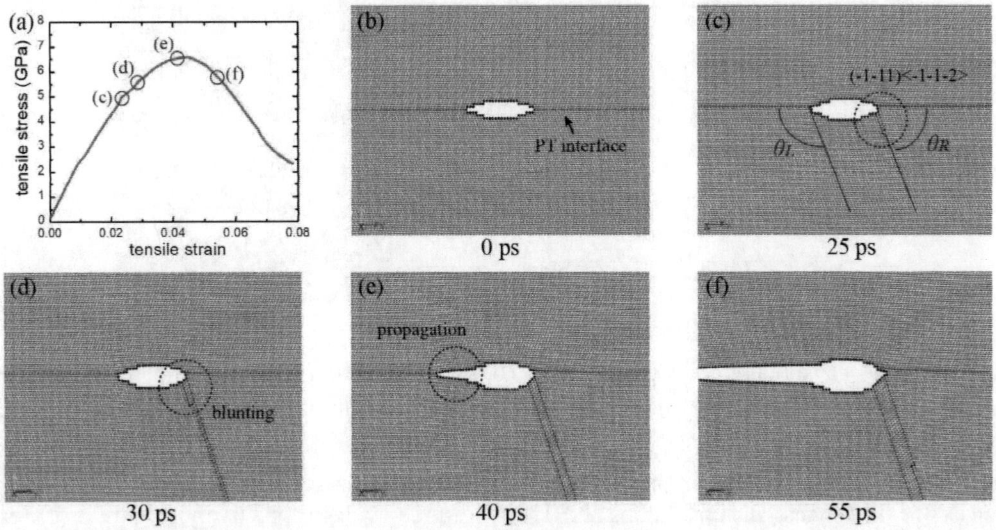

Figure 2. PT interface system under static tensile loading along z direction: (a) stress-strain curve and (b)-(f) snapshots of microstructure evolution around the micro-crack.

There are three major competing deformation modes in γ-TiAl with $L1_0$ structure: 1/2<110] ordinary dislocation, <101] superlattice dislocation and {111}<11-2] true twin which serves the chemical order of γ phase. And the dominant slip system is (-1-11)<-1-1-2> under loading along <111> direction according to the Schmid factor analysis. According to the DXA analysis which is used to compute the Burger vectors of dislocations, it can be found that the Shockley dislocation with Burger's vector=1/6[-1-1-2] emits from the right crack tip and slip on the (-1-11) plane at 25ps (figure 2 (c)). Thus, the simulation result agrees well with the Schmid factor analysis. And the propagation of the right crack tip is blunted by the slip plane (figure 2 (d)).

The $L1_0$ intermetallic TiAl alloy has an asymmetric structure, which makes only one type of slip system can be found under loading along <111> direction. In the upper TiAl lamella, the preferred slip system is not parallel to the crack tip due to the twist angle of 60°. Therefore, the upper lamella requires more energy for dislocation nucleating from the crack tip than the lower lamella. In the lower TiAl lamella, the relative angles between the preferred slip plane and the interface plane are θ_L and θ_R respectively (figure 2 (c)). The left crack tip with a larger angle requires a larger amount of energy to nucleate and propagate dislocation away from the crack tip[17]. Therefore, only one slip system can be found at the right crack tip, and the crack propagates in a brittle way along -y while in a ductile way along y.

Under cyclic loading along z direction, as it shown in Figure 1 (b), the crack propagation under loading before 35ps is similar to it under static loading. Figure 3 (a) lists the snapshots of the crack evolution in the first cyclic loading. At the beginning of the unloading process (40ps), the crack is still growing along -y due to the high stress. And then the left crack tip and the slip system at the right crack tip retreat (60ps). In addition, at the end of the unloading (70ps), the left crack tip returns to its initial state while the dislocation emitted from the right crack tip doesn't disappear.

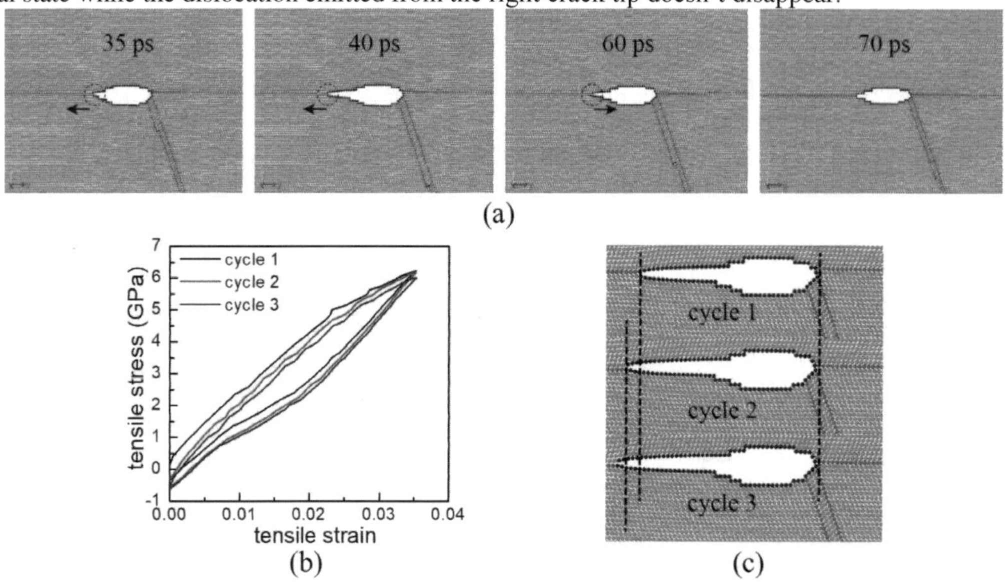

Figure 3. (a) the snapshots of crack propagation; (b) stress-strain curves of the three loading cycles; (c) cracks at the longest stages in the three loading cycles under cyclic loading

It can be observed from the stress-strain curves of the 3 cycles (figure 3 (b)) that the stresses decrease as the cyclic loading number increase. According to the simulation result, the crack length propagates to the longest at 50ps, 120ps and 190ps in three cycles respectively. Figure 3 (c) lists the longest cracks in the three cycles. It can be speculated that the maximum of the crack length is longer as the increase of the cyclic loading number.

4. Conclusion

In summary, crack propagation in a γ-TiAl intermetallic lamellar system with a PT interface is investigated by classical molecular dynamics under static and cyclic loading. The following conclusions can be drawn:

(1) Due to the different energy required to nucleate dislocation of the two crack tip, the crack propagates asymmetrically along the PT interface of γ-TiAl under high strain rate.

(2) Under cyclic loading, the stress decreases and the rack length increases with the increase of the cyclic loading number.

Acknowledgements

Thanks for the financial support by National Natural Science Foundation of China (Grant No. 51775205 and No. 51605174), Investment Casting Collaborating Laboratory on Advanced Aeronautical Lightweight Alloy Materials of Huazhong University of Science &Technology and AECC Beijing Institute of Aeronautical Materials, and thanks to SCTS/CGCL HPCC of HUST for providing computing resources support.

References

[1] S.C. Huang, Microstructures and property tradeoffs in wrought TiAl-base alloys, Metall. Trans. A. 23 (1992) 375–377.

[2] K.S. Chan, D.S. Shih, Fatigue and fracture behavior of a fine-grained lamellar TiAl alloy, Metall. Mater. Trans. A. 28 (1997) 79–90.

[3] H. Umeda, K. Kishida, H. Inui, M. Yamaguchi, Effects of Al-concentration and lamellar spacing on the room-temperature strength and ductility of PST crystals of TiAl, Mater. Sci. Eng. A. s 239–240 (1997) 336–343.

[4] G. Villard, D. Pelloquin, A. Maignan, A. Wahl, R.J. Dimelfi, Grain boundary sliding and its relation to ductility and fracture in fine-grained polycrystalline materials with a particular focus on γ-TiAl, Mater. Sci. Eng. A. 237 (1997) 141–149.

[5] W.T. Marketz, F.D. Fischer, H. Clemens, Deformation mechanisms in TiAl intermetallics—experiments and modeling, Int. J. Plast. 19 (2003) 281–321.

[6] M. Yamaguchi, H. Inui, TiAl-base Alloy Crystals in the Polysynthetically Twinned (PST) Form, Encycl. Mater. Sci. Technol. (2005) 1–6.

[7] C.E. Wen, K. Yasue, J.G. Lin, Y.G. Zhang, C.Q. Chen, The effect of lamellar spacing on the creep behavior of a fully lamellar TiAl alloy, Intermetallics. 8 (2000) 525–529.

[8] T.A. Parthasarathy, M. Keller, M.G. Mendiratta, The effect of lamellar lath spacing on the creep behavior of Ti-47at%Al, Scr. Mater. 38 (1998) 1025–1031.

[9] Y. Hao, J. Liu, S. Li, J. Li, X. Liu, X. Feng, Effects of nano-twinning on the deformation and mechanical behaviours of TiAl alloys with distinct microstructure at elevated loading temperatures, Mater. Sci. Eng. A. 705 (2017).

[10] M. Kanani, A. Hartmaier, R. Janisch, Stacking fault based analysis of shear mechanisms at interfaces in lamellar TiAl alloys, Acta Mater. 106 (2016) 208–218.

[11] L. Ma, S. Xiao, H. Deng, W. Hu, Molecular dynamics simulation of fatigue crack propagation in bcc iron under cyclic loading, Int. J. Fatigue. 68 (2014) 253–259.

[12] W. Fang, H. Xie, F. Yin, J. Li, D.F. Khan, Q. Fang, Molecular dynamics simulation of grain boundary geometry on crack propagation of bi-crystal aluminum, Mater. Sci. Eng. A. 666 (2016) 314–319.

[13] R.R. Zope, Y. Mishin, Interatomic potentials for atomistic simulations of the Ti-Al system, Phys. Rev. B. 68 (2003) 24102.

[14] S. Plimpton, Fast parallel algorithms for short-range molecular dynamics, J. Comput. Phys. 117 (1995) 1–19.

[15] A. Stukowski, Visualization and analysis of atomistic simulation data with OVITO-the Open Visualization Tool, Model. Simul. Mater. Sci. Eng. 18 (2010) 2154–2162.

[16] A. Stukowski, Structure identification methods for atomistic simulations of crystalline materials, Model. Simul. Mater. Sci. Eng. 20 (2012) 45021–45035(15).

[17] I. Adlakha, M.A. Tschopp, K.N. Solanki, The role of grain boundary structure and crystal orientation on crack growth asymmetry in aluminum, Mater. Sci. Eng. A. 618 (2014) 345–354.

Numerical simulation of wave-like nucleation events

Lang Yuan[1], Arvind Prasad[2], Peter D Lee[3] and David StJohn[2]

[1] Department of Mechanical Engineering, University of South Carolina, Columbia, SC, 29201, USA
[2] Centre for Advanced Materials Processing and Manufacturing, The University of Queensland, St Lucia, Queensland, 4072, Australia
[3]Department of Mechanical Engineering, Univesity College London, London, WC1E 6BT, UK

[1]langyuan@cec.sc.edu

Abstract. The Interdependence model [1] predicted that nucleation would occur in waves of events with regions of no nucleation in between each wave. The waves continue to form until nucleation covers the sample. The cause of this phenomenon was attributed to the formation of a nucleation-free zone which incorporates solute suppressed nucleation and inhibited nucleation zones. Recent real-time synchrotron x-ray studies by Prasad et al [2], Liotti et al [3] and Xu et al [4] have confirmed this hypothesis showing nucleation occurs in a step-wise fashion with a number of events occurring followed by little or no nucleation for a short period before another set of events occurs. A microscale solidification model that predicts diffusion-controlled dendritic growth has successfully shown the effect of the developing constitutional supercooling on the selection of nucleation events. In this study, we use this model to predict the solidification behaviour under the conditions experienced during these real-time synchrotron studies.

1. Introduction

Solidification processes that result in a microstructure of equiaxed grains are preferred with the associated better properties of the cast product and also ease of casting. Grain refinement using suitable inoculants is the standard practice in industry, but the exact mechanism of equiaxed grain formation leading to grain refinement still eludes us. An analytical model, the Interdependence Model [1], was developed which combines the particle and alloy chemistry characteristics, suggesting that while particles play a role in triggering a nucleation event, solute diffusion from an already growing neighboring grain provides constitutional supercooling (CS) that may also prevent nucleation events. The Interdependence model has been verified in different experimental scenarios as well as numerically [2,5,6].

Figure 1 shows a schematic that illustrates the Interdependence Model in the presence of a small thermal gradient (temperature increasing from left to right). The inoculant in the middle is triggered when there is adequate undercooling available. This CS is provided by the already growing grain on the left when it reaches a certain minimum size, rejecting solute ($k < 1$) as it continues to grow. The sum of the growth radius of the growing grain, x_{cs}, and the solute diffusion length, x_{dl}, creates a distance where nucleation is unlikely and is termed the Nucleation Free Zone (NFZ).

One of the natural extensions of the model is that wave-like nucleation can be predicted since the solute diffusion from the nucleated grain in the middle of the figure would provide the required CS for triggering the inoculant placed on the right. Such wave-like nucleation events have been observed in in-situ experiments on different alloy systems performed at different X-ray synchrotron facilities – Al-Si [2], and Al-Cu [3,4]. These results reveal the nucleation wave-front as predicted by the Interdependence Model. Figure 2 shows the results from Prasad et al's work with Al-15Cu[7].

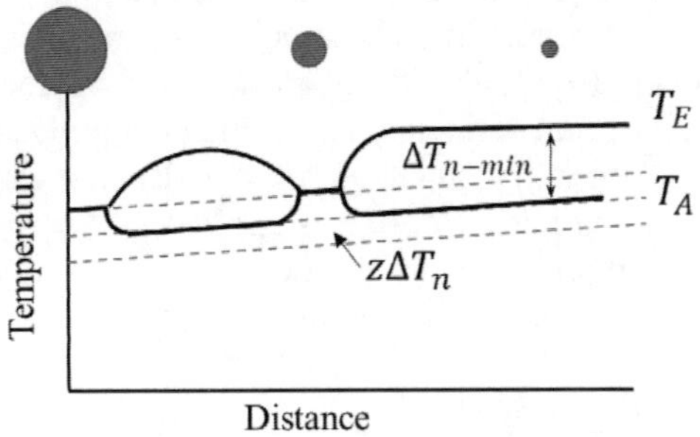

(a)

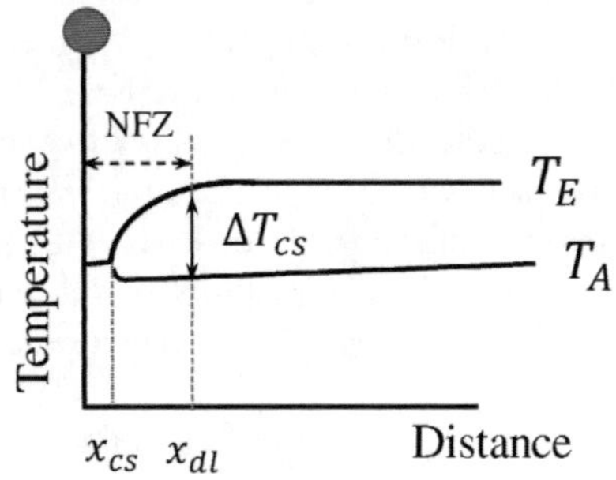

(b)

Figure 1. (a) The already growing grain on the left rejects solute and provides adequate constitutional supercooling (CS) to activate the inoculant in the middle triggering a nucleation event at a distance from the S-L interface. The now growing middle grain provides adequate CS for the rightmost inoculant to trigger another nucleation event. The process repeats and a cascade of nucleation events may be observed from left to right. T_A represents the melt temperature with a small thermal gradient from left to right. (b) In the Interdependence model the Nucleation Free Zone includes the radius of the growing grain and the diffusion length of the rejected solute as defined in [1].

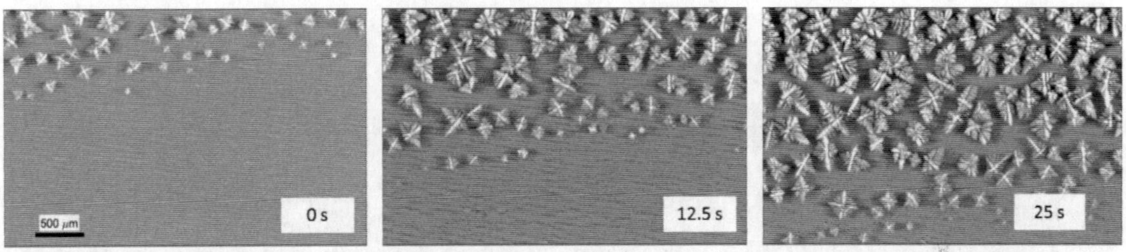

Figure 2. The cascading nucleation events observed during the in-situ X-ray synchrotron experiments with grain refined Al-15Cu. [adapted from [7]].

While the Interdependence Model is an analytical model, a numerical simulation technique is employed in this study to elucidate the effect of cooling rate and thermal gradient on the NFZ and thereby confirm the wave-like nucleation sequence. A cellular automaton based open source numerical model [8], which has been used to study the Interdependence model, was modified and applied to simulate the experimental conditions from the synchrotron experiments of [3] and [4].

2. Simulations of heterogeneous nucleation

The solidification model simulates nucleation events, dendritic growth with solute partitioning and diffusion. It has been validated against experiments in terms of the solute diffusion at the interface, the predictions of primary dendrite arm spacing, the columnar to equiaxed transition and freckle initiation during directional solidification processes [9–11]. The model implementations are not presented here, which were described in previous publications [12,13]. To evaluate the current model capability, a set of numerical simulations were calculated under the reported solidification conditions that have been examined through in-situ X-ray radiography by Xu et al.[4]. The experiments were performed with well-controlled cooling rates and thermal gradients. In addition, the observation direction is parallel to the gravity direction where convection and grain motion can be largely ignored. Thus, the experiments provide a reliable set of data for model comparison and further examination of nucleation behaviour.

Simulations are performed in a domain of 3×6 mm^2 ($x \times y$ to cover the field of view in the experiments (2.877×1.916 mm^2). Periodic boundary conditions in the x direction for both thermal and solutal fields are applied to accommodate the wider experimental setup (5 × 50 mm^2). The extended dimension (6 mm vs. 1.919 mm) in the y direction allows the initial development of grains to get close to steady state where the observation most likely was done during the experiments. Since thickness (z) is 0.2 mm in the experiments, only 10 cells across the z direction were used in the simulation for thin 3D simulation envelopment to allow extra freedom of grain orientations. The simulation parameters are listed in Table 1. For nucleation, the Free Growth model [14], one of the most accepted models for heterogeneous nucleation of crystals with the addition of grain refiners, is simplified to a Gaussian distribution to account for stochastic nucleation as a function of the solute adjusted undercooling. The mean nucleation undercooling of 0.5K with a standard deviation of 0.1K was applied to control the nucleation events during directional solidification. The emphasis of the model is on constitutional supercooling due to segregation. Convection effects and the eutectic reaction are ignored.

Table 1. Simulation Parameters for Al-20wt%Cu alloy

Parameter	Unit	Value
Initial composition, C_0	wt%	20
Partitioning coefficient, k		0.14
Liquidus slope, m	K/wt%	-3.4
Liquidus Temperature, T_{liq}	K	875
Diffusion coefficient in solid, D_s	m/s^2	1.0e-12
Diffusion coefficient in liquid, D_L	m/s^2	4.65e-09
Cell size, d_{xyz}	M	5.0e-6

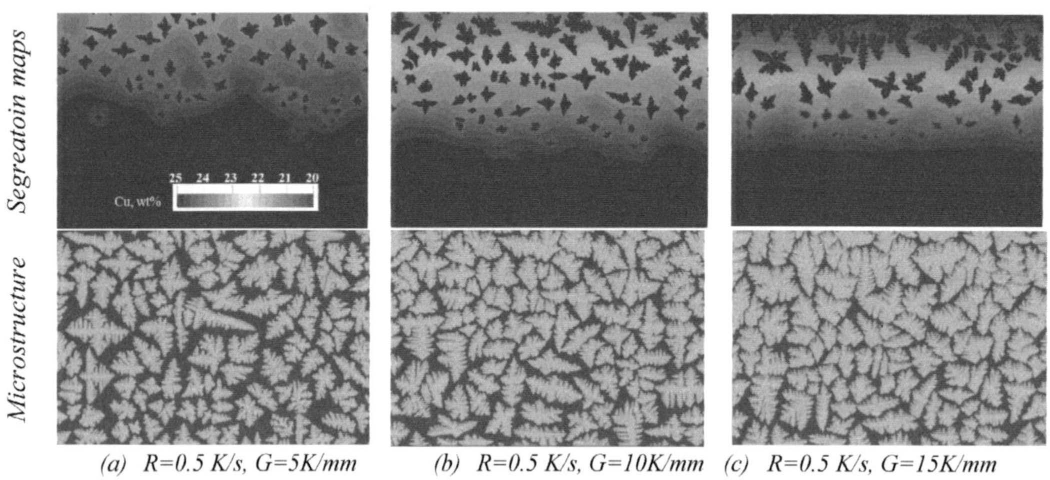

Figure 3. Predicted segregation and microstrcture for inoculated Al-20wt% Cu alloy at 0.5K/s cooling rate under three different thermal gradients: (a) 5 K/mm, (b) 10 K/mm and (c) 15K/mm

Figure 3 shows the predicted segregation at the solidification front and the dendritic structures after grains fill the space during the directional solidification process for the cooling rate (R) of 0.5 K/s and increasing thermal gradients. With increasing thermal gradient, the withdraw velocity was reduced to keep the cooling rate constant. Note that the simulated domain was cropped to the same size as the experimental observation window and the maximum nucleation density (N_{max}) was set to 4.0E+12/m^3. The number of grains was also extracted from the simulations and plotted against the experimental data, as shown in Figure 4(a). Both the grain size and grain morphologies agree well with the experimental observations for the current conditions. This also indicates that the final grain size predicted from the current model agrees with the experimental results of Xu et al. [4]. The morphologies also correspond well with Liotti et al's experiment (Figure 2). A nucleation front is seen in all cases. The diffusion length which is proportional to D_l/v (where v is the solid-liquidus interface velocity), increases with increasing thermal gradient. This increases the size of NFZ, resulting in a decrease in the number of grains, in other words, an increase of grain size.

Two other experimentally examined cooling rates (0.1 K/s and 0.025 K/s) were also simulated with the same model setup. However, to match the evolution of the number of grains, the maximum nucleation densities were reduced to 1.2E+12/m^3 and 0.8 E+12/m^3, respectively. With the same thermal gradient of 5K/mm, the number of grain reduces from ~110 to ~ 10 when the cooling rate decreases from 0.5K/s to 0.025K/s. Note that, within the same cooling rate, the model parameters remain unchanged.

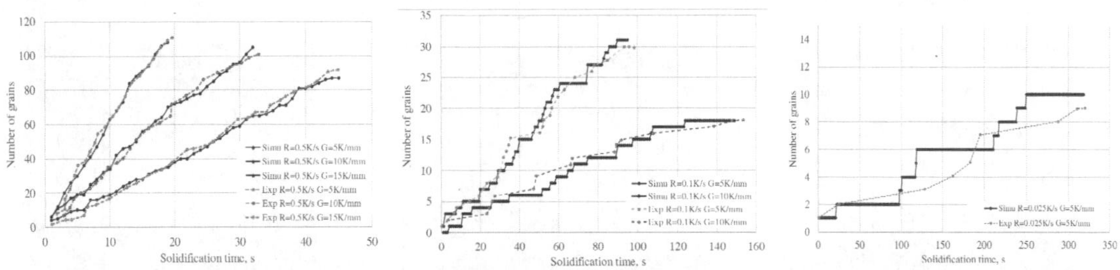

(a) $R=0.5K/s$, $N_m = 4.0E+12/m^3$ (b) $R=0.1K/s$, $N_m = 1.2E+12/m^3$ (c) $R=0.025K/s$, $N_m = 8E+11/m^3$

Figure 4. Evaluation of grain numbers for the same thermal gradeint (G = 5K/mm) at three different cooling rates, compared to the experimental data from the in-situ observations by Xu et al. [4]

3. Discussion

It has been proposed by the Free Growth model that dendrites are formed preferentially on larger nucleant particles first at relatively low undercooling and then progressively on smaller particles at a higher undercooling [14]. With inoculated Al alloys, a certain distribution of particle sizes is expected for the potential TiB_2 nucleants in the melt, where the population of small particles is much higher than that for the large particles. When the temperature starts to decrease, nucleation occurs first on large nucleants with low undercooling. The growing nuclei reject solute into the liquid for partitioning coefficients of $k < 1$. For a given thermal gradient, with a low cooling rate, solute has more time to diffuse compared to that with a high cooling rate giving the same thermal undercooling. This leads to a thicker solute segregation layer with higher solute concentration ahead of the solid-liquid interface and therefore lower CS. Based on the Interdependence Theory, nucleation events can be suppressed with lower CS, preventing nucleation on less-efficient nucleants. In contrast, with high cooling rate, a thinner solute segregation layer forms at the solidification front. The associated NFZ is reduced allowing nucleation on smaller nucleants which are more populous. Nucleation is further assisted by a faster heat extraction rate, which results in higher thermal undercooling. Based on this understanding, it implies that melts subjected to higher cooling rates have a higher potential to activate smaller nucleants with higher undercooling, compared to the situation when low cooling rates are applied. Moreover, Liotti et. al [3] observed that the mean and spread of undercoolings decrease with increasing cooling rate. This justifies the lower maximum nucleation density applied to the low cooling rate cases in the simulations which matched the experimental results. Since nucleation undercooling in the current simulations do not change, the increase of maximum nucleation density also reflects the increase of potential nucleants at higher cooling rates. This implies that a more sophisticated nucleation model that accounts for the actual size distribution of TiB_2 particles needs to be implemented to account for the variation of nucleation undercoolings.

In-situ X-ray studies have shown the wave-like nucleation under low cooling rates with low thermal gradients [2,3]. This is also observed in the simulations. From Figure 4(b) and (c), the incremental increase of nuclei number is in a step-wise manner which is enhanced at the lower cooling rate. The step-wise behaviour indicates that the formation of grains is paused for a certain period before a new set starts. As mentioned earlier, at lower cooling rate, the rate of CS development is reduced and the NFZ is extended. This requires additional heat extraction to increase the total undercooling and activate a

IOP Conf. Series: Materials Science and Engineering **529** (2019) 012043 doi:10.1088/1757-899X/529/1/012043

new set of particles, which leads to the pause in nucleation events. The new set of nuclei is well spaced from the previous sets due to the existence of NFZ, leading to the wave-like nucleation event as schematically shown in Figure 1(a). This phenomenon has been explained by the Interdependence model that includes the effect of the temperature gradient and the NFZ imposed by solute segregation resulting in the wave-like nucleation front in the refined alloys. The spatial distance of nuclei in the thermal gradient direction increases with the size of NFZ. Thus, the wave-like nucleation behaviour is observed more readily with large NFZ formed at low cooling rate. To better illustrate the wave-like nucleation event, a test case was run with R=0.01K/s and G=0.5K/mm, and shown in Figure 5. The nuclei are well spaced and grow in bands.

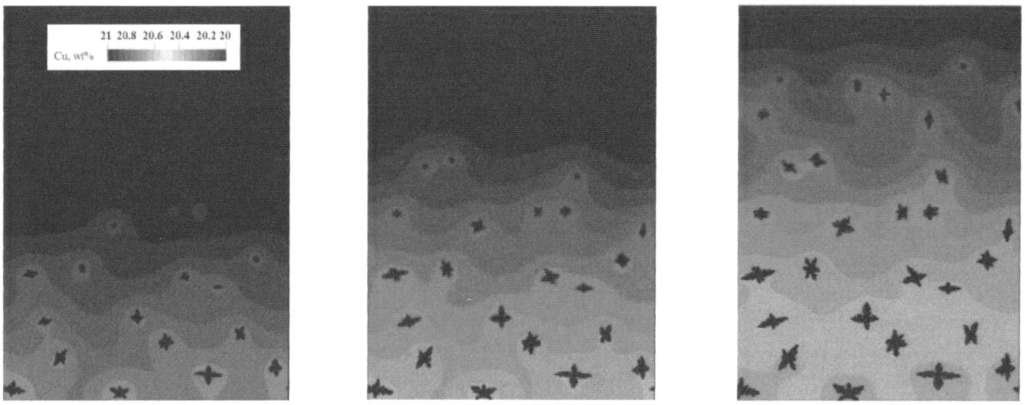

(a) t=150s *(b) t=200s* *(c) t=250s*

Figure 5. Wave-like nucleation for R=0.01K/s and G=0.5K/mm

For the same cooling rate (Figure 4(b)), the step-wise increments of grain number are more frequent at a high thermal gradient, indicating the promotion of the wave-like nucleation event. Figure 6(a) shows the solute profiles extracted at a steady state solidification front for G=10K/mm and G=15K/mm. A small difference in terms of solute distribution is observed along the length. (Note that these profiles can vary location by location, which are also affected by the surrounding grains. However, they can represent a typical scenario for each case). Neglecting the curvature undercooling and kinetic undercooling, the total undercooling for nucleation is plotted in Figure 6(b) by combining thermal and constitutional undercoolings.

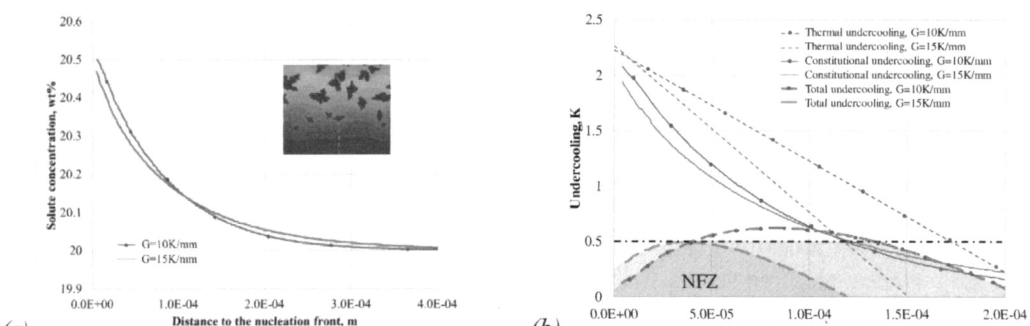

Figure 6. Illustration of undercooling at the nucleation front: (a) solute profile extracted from simulation for R=0.5K/s. and (b) the illustration of NFZ.

As the thermal gradient increases, the total undercooling in the local melt decreases and the volume of the potential nucleation zone reduces. Assuming the inoculant undercooling of 0.5K, the entire length ahead of the solidification front for G=15K/mm is located under the limit, where there is less potential for grain formation and NFZ is expected throughout the length. Therefore, for the directional

IOP Conf. Series: Materials Science and Engineering **529** (2019) 012043 doi:10.1088/1757-899X/529/1/012043

solidification conditions [4], high thermal gradient increases the size of NFZ, which delays the nucleation events or totally prohibits nucleation. At lower thermal gradient some part of the total undercooling is above the inoculant undercooling limit and therefore exhibits a reduced NFZ. Thus, with increasing thermal gradient, the nucleation zone can be completely suppressed depending upon the cooling rate.

4. Conclusions

Wave-like nucleation events as predicted by the Interdependence Model have been revealed by in-situ X-ray experiments and validated by a microscale solidification model. The model agrees well with published experimental results in terms of both dendrite morphology and evolution of grain numbers by adjusting the maximum nucleation density, where the melt with a high cooling rate includes a larger number of potential nucleants of smaller size but in greater numbers. This indicates that the nucleation model in the simulation needs to be improved to account for the actual particle size distribution in a commercial grain refiner. Step-wise nucleation behaviour was predicted for the melts with low cooling rate. A demonstration case shows that the nucleation events are well-spaced, and the nuclei grow in bands. This agrees well with experimental observations that these wave-like nucleation events occur at low cooling rate. In addition, the nucleation events are also affected by the thermal gradient.

References

[1] StJohn D H, Qian M, Easton M A and Cao P 2011 The Interdependence Theory: The relationship between grain formation and nucleant selection *Acta Mater.* **59** 4907–21
[2] Prasad A, McDonald S D, Yasuda H, Nogita K and StJohn D H 2015 A real-time synchrotron X-ray study of primary phase nucleation and formation in hypoeutectic Al–Si alloys *J. Cryst. Growth* **430** 122–37
[3] Liotti E, Arteta C, Zisserman A, Lui A, Lempitsky V and Grant P S 2018 Crystal nucleation in metallic alloys using x-ray radiography and machine learning *Sci. Adv.* **4**
[4] Xu Y, Casari D, Mathiesen R H and Li Y 2018 Revealing the heterogeneous nucleation behavior of equiaxed grains of inoculated Al alloys during directional solidification *Acta Mater.* **149** 312–25
[5] Prasad A, Yuan L, Lee P D and StJohn D H 2013 The Interdependence model of grain nucleation: A numerical analysis of the Nucleation-Free Zone *Acta Mater.* **61** 5914–27
[6] StJohn D H, Prasad A, Easton M A and Qian M 2015 The Contribution of Constitutional Supercooling to Nucleation and Grain Formation *Metall. Mater. Trans. A* **46** 4868–85
[7] Prasad A, Liotti E, McDonald S D, Nogita K, Yasuda H, Grant P S and Stjohn D H 2015 Real-time synchrotron x-ray observations of equiaxed solidification of aluminium alloys and implications for modelling *IOP Conf. Ser. Mater. Sci. Eng.* **84** 12014
[8] The microstructural modelling of materials, http://www.imperial.ac.uk/engineering-alloys/research/software/
[9] Dai H J, Dong H B, D'Souza N, Gebelin J-C and Reed R C 2011 Grain Selection in Spiral Selectors During Investment Casting of Single-Crystal Components: Part II. Numerical Modeling *Metall. Mater. Trans. A* **42** 3439–46
[10] Dong H B and Lee P D 2005 Simulation of the columnar-to-equiaxed transition in directionally solidified Al–Cu alloys *Acta Mater.* **53** 659–68
[11] Karagadde S, Yuan L, Shevchenko N, Eckert S and Lee P D 2014 3-D microstructural model of freckle formation validated using in situ experiments *Acta Mater.* **79** 168–80
[12] Wang W, Lee P D and McLean M 2003 A model of solidification microstructures in nickel-based superalloys: predicting primary dendrite spacing selection *Acta Mater.* **51** 2971–87
[13] Yuan L and Lee P D 2012 A new mechanism for freckle initiation based on microstructural level simulation *Acta Mater.* **60** 4917–26
[14] Greer A, Bunn A, Tronche A, Evans P and Bristow D.2000 Modeling of inoculation of metallic melts: application to grain refinement of aluminium by Al–Ti–B *Acta Mater.* **48** 2823–35

IOP Conf. Series: Materials Science and Engineering **529** (2019) 012044 doi:10.1088/1757-899X/529/1/012044

Increasing the Liquidus Temperature by Employing the Controlled Diffusion Solidification (CDS) Process: A Potential Route to Improved Castings

Mohammad Pourgharibshahi[1], Mehdi Divandari[1], Hassan Saghafian[1], Alberto Fabrizi[2], Giulio Timelli[2]

[1]School of Metallurgy and Materials Engineering, Iran University of Science and Technology, Narmak 16846-13114, Tehran, Iran

2 Department of Management and Engineering, University of Padova, Stradella S. Nicola, 3 36100, Vicenza, Italy

Divandari@iust.ac.ir

Abstract. Recent theories suggest the existence of an incubation time, over which a liquid alloy prepares for nucleation by decomposing into compositional fluctuations. Accordingly, in a recent work by the present authors, the solidification path of a Controlled Diffusion Solidification (CDS) mixture was calculated. The calculated CDS path begins at a higher liquidus temperature comparing to conventional solidification and the fraction solid values are achieved at a relatively higher temperature. To provide information on the CDS mechanism and physical structure of the CDS mixture in the mushy zone, Al-7.8Zn-2.6Mg-2Cu alloy was solidified, in this study, via conventional and CDS process in the presence and absence of recalescence. Typical grain structures obtained via the two solidification conditions is characterized using Electron Back Scattered Diffraction. Results showed that the nucleation continues to occur in the presence of recalescence, while it is suppressed in its absence. According to the two step nucleation theory, the increase in the nucleation temperature causes sufficient recalescence in the mixture, allowing the unnucleated liquid phase to decompose into chemical fluctuations and prepares for further nucleation. As a result, in the presence of recalescence, nucleation in a CDS mixture is not as readily halted as during the conventional solidification, which is in contradiction with the recent theories developed based on the classical theory of nucleation.

1. Introduction

Controlled Diffusion Solidification (CDS) process is a relatively novel casting method to be used for near net shape casting of wrought aluminum alloys [1,2]. Wrought aluminum alloys are single phase alloys, the low fraction of non-equilibrium liquid during the later solidification stages, leave them very brittle and prone to hot tearing, thereby non-castable [3]. By taking the advantages of mixing two precursor melts with different thermal mass at a definite mass ratio, which are held nearly above their respective liquidus temperatures, the CDS process results in a microstructure mostly composed of globular equiaxed grains [4]. The facilitated feeding associated with the globular structure is claimed as the main reason of reduction in the hot tearing susceptibility of the alloys by CDS processing which enables shape casting with wrought aluminum alloy [4–6].

The intrinsic difficulty associated with experimental and numerical study of the dynamic events occurring in the liquid state and during the course of mixing led researchers to follow some hypothetical mechanisms based on the phenomenology of the process, and to focus pragmatically on the improvement of properties in the cast parts produced via CDS process. Therefore, the main mechanism of reduction in hot tear susceptibility by CDS, which is probably dependent on some special nucleation and growth phenomena in the process, remains to be revealed.

A recent attempt by the authors showed that the early stage of solidification path follows the Scheil solidification curve of the high thermal-mass (HTM) alloy [7]. Using this approach and assuming the concept of continuity of the solidification path, the absolute value of fraction solids was determined. From this work it was possible to determine the onset temperature of solidification (zero fraction solid) of the CDS mixture which is the same as that of the HTM alloy. It is noteworthy that the major problem with assessing the solidification path (nucleation and growth) of a CDS mixture is the impossibility of detecting the onset of solidification. As such, the complete solidification path of a CDS mixture cannot be experimentally extracted but it can be calculated in some way [7]. Regarding the physical structure of the alloy within the semi-solid region, however, no information can be deduced from the solidification path obtained by the phenomenological method of Computer Aided Thermal Analysis (CATA). That is, whether solidification proceeds by non-stop nucleation or growth of some already nucleated ones remains to be understood more thoroughly.

To investigate more deeply and provide information concerning the physics of the alloy during the CDS processing, it seems necessary to make a distinction between the isothermal and non-isothermal solidification. This can help to integrate the CDS process with the conventional casting methods. The effect of recalescence becomes considerable with approaching from solidification at high cooling rates towards isothermal solidification [8]. Even though the isothermal solidification is not something hypothetical and occurs in some cases including the center of DC-cast ingots [9,10].

The recalescence has been reported as the main nucleation suppression mechanism [8]. When recalescence is absent, i.e. in real casting conditions of industrial cooling rates, however, nucleation is also stopped after a given solid fraction value. This has been related to formation of a "solute suppression nucleation" (SSN) zone [11] or "nucleation-free" zone (NFZ) [12] caused by the segregation of solute elements. As a result, more time is required for nucleation events to occur, over which the liquid must cool further to fulfill the undercooling required for nucleation. From another point of view, solute segregation reduces potency of nucleants deactivating them when surrounded by the segregated liquid [9].

These theories assume Classic Nucleation Theory (CNT) as the mechanism responsible for nucleation during every stages of solidification [8]. The CNT approach behaves alloy liquid like a pure one and considers simultaneous occurrence of the chemical and structural fluctuations in the liquid at some supercooling. This last is the heart of the CNT approach. A supercooled condition in the liquid is favorable to make structural fluctuations, which is an athermal transition. At the same time, however, it disfavors chemical fluctuations which is a diffusion controlled process dependent on the temperature. Moreover, due to the presence of heterogeneous nucleation sites or intentionally added grain refiners, there requires only negligible amount of supercooling. It should be noted here that only 1% of the total added inoculants take part in the nucleation process [13]. Hence, the suppression effect by inoculant deactivation caused by solute segregation which necessitates only a marginally higher constitutional supercooling may not be significant.

In contrast to the CNT, Two-Step Nucleation (TSN), or Homophase Fluctuation Mediated (HFM), theory [14,15] believes in the formation of a metastable phase (homophase), wherein the stable solid nuclei will be formed subsequently. More accurate understanding of the nucleation suppression phenomenon is required to be considered under the TSN concept to improve the prediction models for grain size.

The main purpose of this study is to provide information regarding the physical structure of the mush of an AA 7xxx alloy during the CDS process. The recently obtained solidification path for the CDS processed alloy [7] is used in this study to describe the nucleation suppression mechanism according to the TSN approach.

IOP Conf. Series: Materials Science and Engineering **529** (2019) 012044 doi:10.1088/1757-899X/529/1/012044

2. Experimental procedure

Precursor melts used in this investigation as the high thermal-mass (HTM) and the low thermal-mass (LTM) alloys were respectively Al-Mg3.5 and Al-Zn31.2-Cu8, which were mixed at the mass ratio (ratio between the mass of the HTM melt and that of the LTM melt) of 3 to result in the target alloy composition, Al-Zn7.8-Mg2.6-Cu2, during the CDS process.

The temperature of the melts before mixing were set to be marginally above the corresponding liquidus temperatures (5 °C). It should be noted that the successful response of this precursor couple to CDS process has been previously reported in Ref. [16].

The CDS mixture was immediately cast into a metallic mould which was used for cooling curve measurement and thermal analysis. This mould was a steel cup having a uniform thickness of 1 mm, coated by BN and preheated at 500 °C. The aim was to investigate isothermal solidification of CDS mixture. Another die with a cylindrical shape, and an inner diameter of 15 mm and a height of 100 mm was used to investigate the solidification behaviour of the CDS mixture under a low recalescence condition. The cooling rate of metal in the solidification range was 3 °C/s in the die and 0.17 °C/s in the steel cup.

The cross section of the castings was examined using a field emission scanning electron microscope (FE-SEM) equipped with an electron back-scattered diffraction (EBSD) probe to study the grain structure of the specimens. For EBSD characterization, the samples, preliminary prepared according to standard metallography procedure, were etched in 1 M NaOH solution at 60 °C for 90 s and rinsed in HNO_3 for 30 s in order to remove Beilby layer on the specimens [17].

3. Results and discussion

The nucleation suppression phenomenon in the alloys has been described so far based on the CNT approach, which cannot provide information concerning the pathway of atoms leading towards nucleation. A better realization of the phenomenon is needed keeping in mind the theory of Ostwald's rule of stages [18] or a TSN approach being able to provide information regarding pre-nucleation events occurring in the liquid state.

The calculated CDS path evidences that the first nuclei form at higher temperature, and the solidification proceeds to its end at a relatively higher temperature comparing to conventional solidification. At any given temperature, the value of fraction solid is higher in the CDS case comparing to conventional solidification [7]. Whether the increase in solid fraction is by nucleation or growth of the present nuclei can be realized through post-mortem examination of the grain structure and grain size distribution.

Fig. 1 shows the quaternion misorientation (QM) maps of a relatively large area of the samples solidified in the absence (die) and presence (steel cup) of recalescence, respectively shown in Figs. 1a and 1b. The different colours in the QM maps refer to different orientations, and therefore individual grains. The grain structure of both the specimens is equiaxed, even though it seems the number of large grains is greater in absence of recalescence. This can be better addressed through the statistical distributions of the grain size, which are shown in Fig. 2. As observed the log-normal distribution fits well with both the graphs, showing that nucleation suppression occurs to some extent in both the solidification conditions. The first inflection point can be interpreted as the onset of nucleation suppression or the size at which the growth begins to dominate the process of solidification [19]. Accordingly, nucleation is suppressed at a grain size of 38 μm in the absence of recalescence. The corresponding size in the recalescence condition is 63 μm. This comparatively larger value of the grain size at the onset of nucleation suppression can be directly related to the coarsening of grains.

According to the grain size distributions and the QM maps, it seems the grains are uniformly graded in the presence of recalescence as compared with those in the recalescence-free condition, where they are well-graded. That is, recalescence caused the grains to be all about the same size. Moreover, the coarse tail of the fit curve (Fig. 2b) approaches to a non-zero value showing that the coarse grains are formed as a result of a re-nucleation phenomenon leading to this higher-kinetics nucleation. It should be noted that the total number of grains in the constant area of the QM maps shown in Fig. 1, is higher in the recalescence condition.

Recalescence and solute segregation are well-known as the main nucleation suppression mechanisms during conventional solidification of alloys. It seems, however, they cannot suppress nucleation events in the isothermally solidifying CDS mixture. Moreover, athermal nucleation is believed as the mechanism by which nucleation keeps continuing during the non-isothermal solidification; i.e. higher cooling rates. As discussed, in the absence of recalescence, nucleation get ceased at a relatively smaller grain size or solid fraction. The athermal nucleation is therefore not effective to promote nucleation rate in a CDS mixture. This clearly demonstrates that CNT fails to predict the nucleation rate for CDS process.

This non-conventional behaviour can be explained by assuming the TSN theory which in contrast with the CNT requires some chemical fluctuations to precede the formation of stable solid nuclei. According to the TSN mechanism, the nucleation of stable crystalline nuclei occurs in two steps: formation of a metastable (intermediate) phase followed by appearance of the crystalline nuclei inside it. In contrast to the later step which is a diffusionless transformation, the former is diffusion-controlled and its kinetics depends upon the temperature.

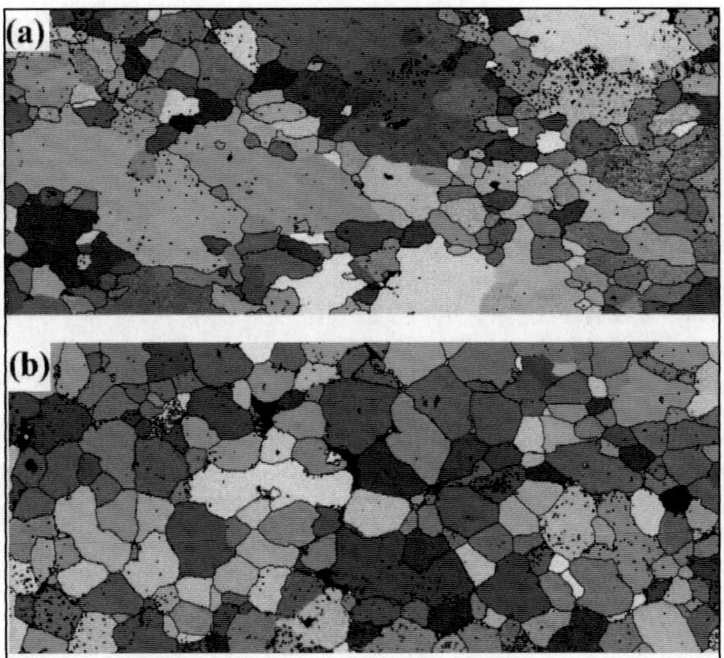

Figure 1. Quaternion maps of the typical grain structure of the specimens solidified in the (a) absence and (b) presence of recalescence.

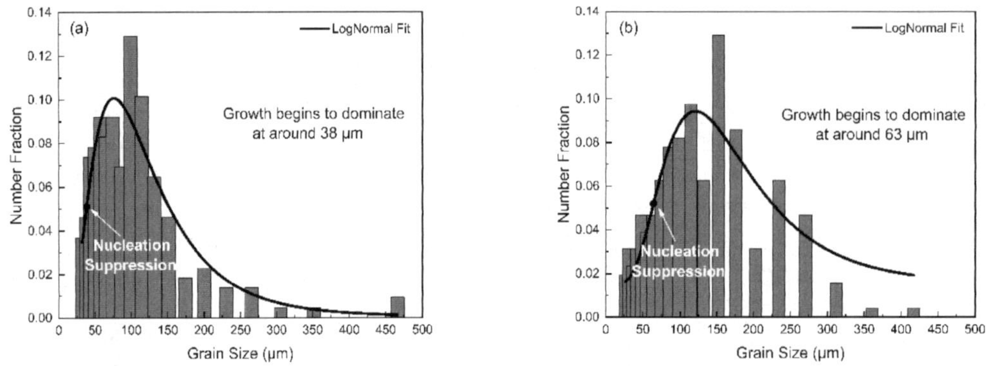

Figure 2. Distributions of the grain size referred to quaternion maps shown in Figure 1 at (a) absence and (b) presence of recalescence. The log-normal distribution fits on the data and the growth begins to dominate respectively at around size of 38 and 63 μm.

The SSN and NFZ theories are developed based on the assumption of the CNT approach. Accordingly, the retardance or suppression of nucleation is because of the insufficient constitutional supercooling in the liquid caused by decreasing either the liquidus temperature or nucleant's potency. Based on TSN, for non-stop nucleation in the liquid alloy, the formation of a metastable phase (chemical fluctuations) is mandatory. Suppression of nucleation in the CDS mixture in the absence of recalescence means that, it is this step which retards further nucleation. As the liquid is not prepared for nucleation, the atoms prefer to add to the already formed nuclei through atom-by-atom absorption [20] (a manner which is completely different from the concept of wetting which is required according to the CNT) and continue the solidification process in this way. Similarly, the latent heat of solidification leading to the recalescence in the case of very low cooling rate or isothermal solidification accelerate the first step by increasing the atomic mobility. As a result, nucleation in the CDS mixture is not ceased in the presence of recalescence. This, however, may raise a new question: why does recalescence cease nucleation in the case of conventional solidification? The answer would be the as-decomposed state of the liquid at the time when recalescence become significant in a CDS mixture. During conventional solidification, the recalescence and solute segregation are concurrent. In other words, the liquid is heating while it is enriching by segregating atoms. The un-nucleated liquid in the CDS mixture, is reached in the supercooled state where it is spontaneously decomposed to chemical fluctuations. This decomposition become accelerated by latent heat induced due to solidification of already formed grains. As a result, the unnucleated liquid prepares for formation of stable nuclei.

Conclusions

Two-step nucleation theories are demanded to be used instead of the somehow ill-defined classical nucleation theory to justify the nucleation suppression phenomenon in metallic alloys. The results of the current study show that, in contrast to the understanding based on the classical nucleation theory (CNT) approach, recalescence can help avoiding nucleation suppression in a controlled diffusion solidification (CDS) mixture.

The solute suppression nucleation (SSN) and nucleation free zone (NFZ) theories developed based on the CNT approach should be revised, considering Ostwald's rule of stages, in order to provide better prediction of nucleation rate and as-solidified grain size. Here, it was proposed that the recalescence induced by latent heat of solidification in the high thermal-mass melt, during the mechanical mixing stage, helps the un-nucleated liquid to decompose into chemical fluctuations and through this, further nucleation becomes possible in the newly formed solute-poor fluctuations.

Dealing with shape casting with wrought aluminum alloys, the CDS process can result in a higher nucleation rate in the relatively thicker sections, where solidification proceeds at the presence of recalescence, and a lower nucleation rate in the thinner sections. As this is the shrinkage of the hotspot which imposes stress on a hot tear-sensitive region of a casting, this higher nucleation rate in the hotspot could postpone coherency point and thereby onset of shrinkage-force buildup, which would be the main mechanism of reduction in hot tearing susceptibility by the CDS process.

References

[1] US 7201210B2: 2003.
[2] M. Pourgharibshahi, M. Divandari, H. Saghafian Larijani, and P. Ashtari: *J. Mater. Process. Technol.*, 2017, vol. **250**, pp. 203–19.
[3] S. Lin: PhD Thesis, Universitie Du Quebec Cfflcoutim, Canada, 1999.
[4] D. Saha: PhD Thesis,Worcester Polytechnic Institue, USA, 2005.
[5] K. Symeonidis: PhD Thesis,Worcester Polytechnic Institue, USA, 2009.
[6] A. Khalaf: PhD Thesis, McMaster University, Canada, 2010.
[7] M. Pourgharibshahi, H. Saghafian, M. Divandari, and G. Timelli: *Metall. Mater. Trans. A*, 2019, vol. **50**, pp. 326–35
[8] I. Maxwell and A. Hellawell: *Acta Metall.*, 1975, vol. **23**, pp. 229–37.
[9] H. Men and Z. Fan: *Acta Mater.*, 2011, vol. **59**, pp. 2704–12.
[10] D.H. StJohn, A. Prasad, M.A. Easton, and M. Qian: *Metall. Mater. Trans. A Phys. Metall. Mater. Sci.*, 2015, vol. **46**, pp. 4868–85.
[11] D. Shu, B. Sun, J. Mi, and P.S. Grant: *Acta Mater.*, 2011, vol. **59**, pp. 2135–44.
[12] D.H. StJohn, M. Qian, M.A. Easton, and P. Cao: *Acta Mater.*, 2011, vol. **59**, pp. 4907–21.
[13] A.L. Greer: *Philos. Trans. R. Soc. A Math. Phys. Eng. Sci.*, 2003, vol. **361**, pp. 479–95.
[14] B. Cini, E. Vinet, and P.J. Desre: *Philos. Mag. A*, 2000, vol. **80**, pp. 955–66.
[15] P.J. Desré, E. Cini, and B. Vinet: *J. Non. Cryst. Solids*, 2001, vol. **288**, pp. 210–7.
[16] R. Ghiaasiaan, X. Zeng, and S. Shankar: *Mater. Sci. Eng. A*, 2014, vol. **594**, pp. 260–77.
[17] D. Feenstra: MSc Thesis, McMaster University, Canada, 2016.
[18] J.W.P. Schmelzer and A.S. Abyzov: in *Thermal Physics and Thermal Analysis*, J. Šesták, P. Hubík, and J.J. Mareš, eds., Springer, Cham, 2017, pp. 195–211.
[19] I. Leubner: *Precision Crystallization: Theory and Practice of Controlling Crystal Size*, CRC Press/Taylor & Francis, 2009.

IOP Conf. Series: Materials Science and Engineering **529** (2019) 012045 doi:10.1088/1757-899X/529/1/012045

Influence of minor additions on Icosahedral Short-Range Order (ISRO) and its consequences on nucleation and solidification growth kinetics in fcc alloys

J. Zollinger[1], M. Rappaz[2]

[1] Institut Jean Lamour, Department of Metallurgy and Materials Science & Engineering, Université de Lorraine, Campus ARTEM, Allée André Guinier, F-54011 Nancy, France

[2] Institute of Materials, École Polytechnique Fédérale de Lausanne, Station 12, CH-1015 Lausanne, Switzerland

julien.zollinger@univ-lorraine.fr

Abstract. The present contribution reviews the recent progress related to the influence of Icosahedral Short-Range Order (ISRO) and icosahedral Quasicrystals (i-QC) formation on the solidification of fcc alloys through minor solute element additions. From intensive crystallographic analysis of multi-twinned regions in as-cast Al-based and Au-based fcc alloys, Kurtuldu *et al.* have shown recently that a so-called "iQC-mediated" nucleation mechanism occurs when a few hundred ppm of Cr and Ir, respectively, are added to the melt [1] [2]. Similarly, it appears that the growth directions of dendrites in Al-Zn:Cr is also influenced by ISRO in the liquid, thus showing an attachment kinetics effect [3]. In a recent contribution, we have shown that iQC-mediated nucleation also occurs in pink gold alloys with Ir-additions, but two additional phenomena at high solidification speed [4]: (i) a spinodal-type decomposition of the liquid, leading to the formation of twinned Cu precipitates in addition to multi-twinned Au-rich grains; (ii) a change of the microstructure of the Au-rich grains, from ⟨100⟩ dendrites to ⟨111⟩ textured cells in the columnar zone.

1. Introduction

Icosahedral Short Range Order (ISRO) in the liquid phase, conjectured by Frank in 1952 [5], was thought to be responsible for the large undercooling measured in metallic alloys. This preferred local arrangement of atoms in the liquid, due to local covalent bonding, was confirmed to occur by atomistic calculations and by small angle scattering [6]. ISRO is also key in understanding the formation of QuasiCrystals (QC) discovered by Shechtman *et al.* in Al-Mn alloys [7], and later in many other alloys [8]. Through several recent papers, Kurtuldu and Rappaz have shown that ISRO can also promote the formation of fine equiaxed fcc grains through a mechanism which they referred to as "*iQC-mediated nucleation*", the "*i*" standing for icosahedral [1][2][9]. This mechanism is summarized in Figure 1 for Al-Zn:Cr. In the liquid state (a), Al and Zn atoms tend to be localized in ISRO configuration around smaller Cr atoms. As the liquidus of the Al_7Cr phase is first crossed during cooling, even at very low Cr composition, iQC tend to form easily (b), since the interfacial energy between the liquid and iQC is at least an order of magnitude lower than that with the fcc phase [8]. As the temperature is further lowered,

IOP Conf. Series: Materials Science and Engineering **529** (2019) 012045 doi:10.1088/1757-899X/529/1/012045

the liquid surrounding the iQC is depleted in Cr atoms, thus creating a driving force to form the fcc phase (c), with the following heteroepitaxy relationships with the iQC template:

$\langle 111 \rangle_{fcc}$ *// 3-fold symmetry of iQC* and $\langle 110 \rangle_{fcc} \perp$ *2-fold symmetry axis of iQC*

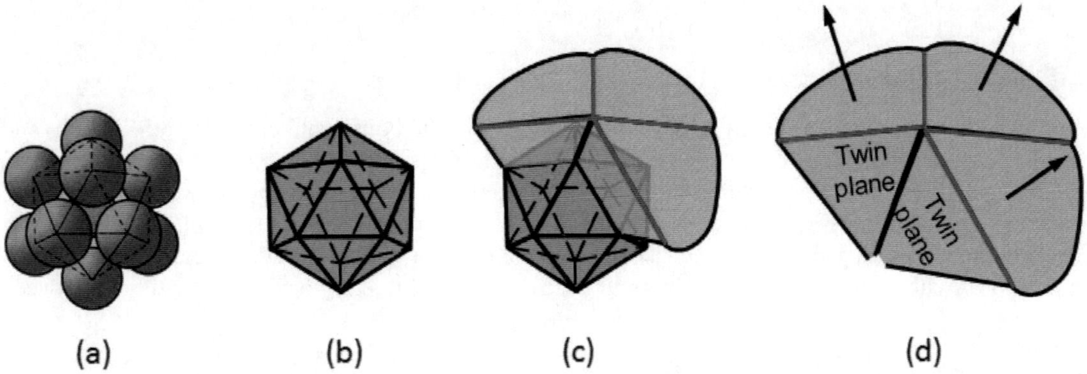

Figure 1: Summary of the iQC-mediated nucleation mechanism (from [2]).

As a result, grains forming on the facets of the iQC have Multiple Twin (MT) relationships among them, resulting in symmetry relationships similar to what was observed in the 1960's in small nanocrystals of gold [10]. During further solidification, the central Cr-rich iQC can even disappear since Al-Cr is peritectic (d) [11]. As a matter of fact, Kurtuldu *et al.* discovered this mechanism from the excessive number of twinned grain boundaries in Al-20wt%Zn when 0.1wt% Cr is added. They also found MT relationships among several nearest-neighbor (nn) fcc grains, respecting the icosahedral symmetry of the iQC template. In some cases, they found the nn fcc grains to have symmetry relationships compatible with the so-called interlocked icosahedron (two icosahedra linked via pentagonal edges).

The same mechanism was then shown to operate in yellow gold (Au-12.5wt%Ag-12.5wt%Cu) when minute amounts of Ir are added to the melt (from 0.001 to 0.02 wt%), a technique used since many years in jewelry and watch industries for grain refining [12].

When ISRO occurs in the liquid, Kurtuldu *et al.* also showed that solute diffusion is slowed down since the mobility of these clusters is reduced compared to the situation when localized covalent bonds are not present [13]. If such is the case, ISRO should also impact the growth kinetics of the fcc phase. Already in his PhD thesis, Kurtuldu reported that the growth directions of dendrite trunks and arms are changed in Al-Zn alloys when Cr is added, and become a function of the solidification speed [3].

Recently, we have shown that iQC-mediated nucleation occurs also in pink gold alloys (Au-20.5wt%Cu-4.5wt%Ag) when 0.01wt% Ir atoms are added to the melt [4]. While this mechanism could be expected at low solidification speed, since it is also used routinely in jewelry for grain refining, two additional phenomena were found at high solidification rate: (i) the formation of fine Cu precipitates, sometimes with twinning relationships, in addition to MT Au-rich grains; and (ii) a change of the internal morphology and texture of columnar grains. These last results are briefly summarized in the present contribution.

2. Experimental method and analysis

Pink gold alloys with and without 100 wt. ppm Ir additions, were supplied by PX Group. Low solidification rate (typically -0.25 K/s) was achieved in a DSC calorimeter, while rapid cooling experiments were performed by letting a small liquid drop (about 4g) fall onto a water-cooled Cu substrate. In this case, the cooling rate is typically -10^4 K/s near the copper chill and -10^2 K/s near the top of the solidified drop. After cutting and polishing, the grain and microstructure were observed using optical microscopy, SEM and EBSD. Further experimental details are given in [4] and we report below only the results concerning rapidly cooled specimens, with and without Ir addition.

3. Results and discussion

Figure 2 summarizes the BSE analysis results of rapidly solidified pink gold droplets without (top) and with (bottom) Ir. In this figure, the copper chill is located on the left (i.e., gravity is from right to left) and the left figures have been taken in the part of the droplets which was in contact with the chill. The two figures on the right correspond to the equiaxed zone, in the upper part of the droplets. Small inserts have been added in each figure to emphasize better the internal microstructure of the grains. Au-rich grains appear with various shades of light grey in BSE, while Cu-rich regions located either in between dendrite arms or as precipitates (see discussion below) are much darker.

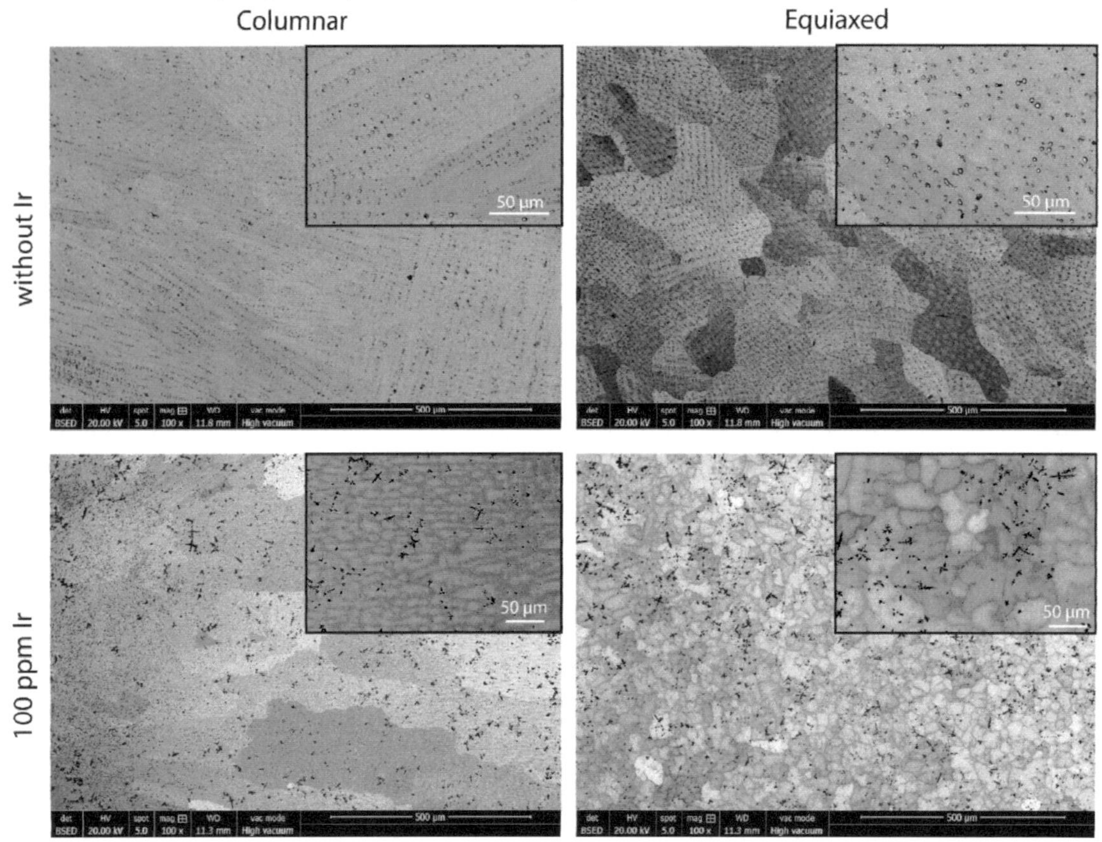

Figure 2: Backscattered electrons image showing the microstructure of Au-20.5wt.%Cu-4.5wt.%Ag specimens rapidly solidified without (top) and with (bottom) 0.01wt% Ir. The left and right figures correspond to the columnar and equiaxed zones, respectively, the copper chill being located on the left. For each figure, an enlarged view of the internal microstructure of the grains is shown in the small inserts.

The grain structure of the droplets was also analyzed by EBSD in order to determine the orientation of the grains and detect twin boundary relationships, i.e., rotation of 60 deg. around a common ⟨111⟩-axis, within an accuracy of 5 deg. as in [1,2]. Figure 3 (a) and (b) show false-color reconstruction of cross sections of rapidly solidified pink gold droplets. The part of the droplet in contact with the Cu plate is on the left and exhibits columnar grains, while the upper part of the droplet (slowest cooling rate) is on the right and exhibit equiaxed grains. Twinned grain boundaries are shown in white. Multiple twin relationships were also found between nn Au-rich equiaxed grains, as illustrated in Figure 3(c) for 4 grains with 0.01wt%Ir addition. The corresponding ⟨110⟩ pole figure shows their multiple twin relationships, with the common ⟨110⟩ direction corresponding to the 5-fold symmetry axis of an icosahedron, the common {111} planes being shown with arcs of circle.

(a)

(b)

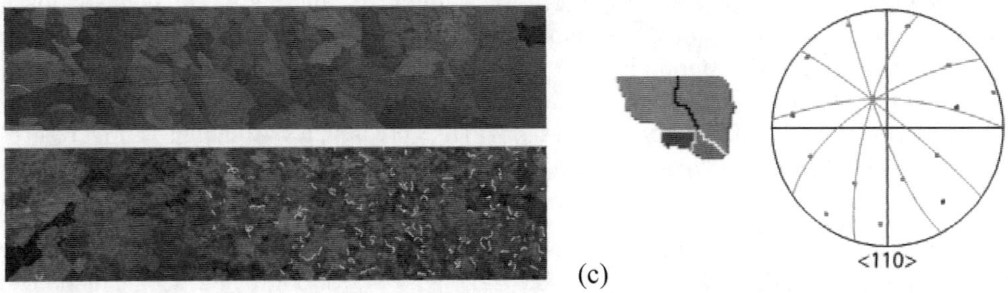

(c)

Figure 3: EBSD reconstructed false-color map of a section through a rapidly solidified pink gold droplet (a) without and (b) with 0.01wt%Ir. Twin grain boundaries are shown in white. (c) An example of 4 nn multiple twin pink gold grains present in (b) with corresponding ⟨110⟩ pole figures. The common {111} planes are shown with arc of circles.

These observations first show that minute additions of Ir induce similar i-QC and ISRO effects in pink gold alloys [2]:

- Iridium refines the grain structure of Au-rich equiaxed grains (compare Figure 2 right, top and bottom, or Figure 3(a) and (b)). This grain refinement is accompanied by a large percentage of twin grain boundaries (7.6%), much larger than the percentage obtained for a population of randomly oriented grains (0.2%). This percentage of grain boundaries is almost independent of the cooling rate, as observed in slowly cooled pink gold specimens with the same amount of iridium [4].
- Multiple twin or near-twin relationships between several nn Au-rich grains have been observed in the Ir-containing specimen. As illustrated in Figure 3(c) for 4 grains, they respect the 5-fold symmetry of the icosahedron with a common ⟨110⟩ crystallographic axis.
- It might be surprising that the columnar zone of the specimen with Ir does not exhibit any twin, and so does that of the Ir-free specimen. However, this is expected even if nucleation at or on the copper chill surface is mediated by the formation of iQC. Indeed, during growth, grain competition cannot favor two adjacent grains coming from the same iQC template and thus observed twins, if any, can only result from randomness. Surprisingly, while the equiaxed grains are refined with Ir addition, the columnar zone becomes more extended: it is about 1.2 mm long with Ir and only 0.5 mm without Ir. This seems to imply that the growth kinetics is also affected by Ir addition.

Indeed, with the high solidification rate associated with the falling droplet, several new phenomena have been observed showing that ISRO in the liquid also influences growth:

- First, the internal microstructure of the columnar grains with Ir is markedly different from that without Ir (compare Figure 2 left, top and bottom). Without Ir, the columnar grains are clearly dendritic and exhibit a ⟨100⟩ texture, as commonly observed in most metallic fcc alloys where competition of ⟨100⟩ dendrite trunks occurs [14]. With Ir, the dendritic structure is replaced by cells growing in a direction more or less parallel to the thermal gradient direction (i.e., from left to right in Figure 2). Furthermore, these columnar grains exhibit a ⟨111⟩ texture, thus showing an effect of

ISRO in the liquid on the attachment kinetics during growth. To some extent, equiaxed grains exhibit a similar change, i.e., dendritic without Ir and globular-dendritic with Ir, but it is less evident. First, the grains are very small and the microstructure becomes naturally more globular-dendritic. Second, the cooling rate is reduced in this part of the droplet which, associated with the reduced grain size, reduces the interface velocity.

- Finally, besides the MT Au-rich grains appearing in various light grey shades in Figure 2, one can observe that small black precipitates or even dendrites appear in BSE when Ir is added. EDX measurements have revealed that these precipitates are a Cu-rich phase, i.e., composition higher than 95wt% in Cu. It is to be noted that this phase does not appear in DSC when the solidification speed is low. The size and morphology of these Cu-rich particles/dendrites clearly indicate that they did not form by precipitation in the solid state and solidify even before the Au-rich MT grains, as shown in Figure 4(a). The EBSD analysis of these precipitates is shown in Figure 4(b-c). It appears that these equiaxed Cu-rich dendrites most often exhibit twins and probably result from a similar nucleation mechanism induced by ISRO.

This nearly simultaneous formation of Cu-rich precipitates/dendrites and Au-rich grains, which only occurs at high solidification rate when 0.01wt% Ir is added to the melt, seems to indicate that Ir induces a spinodal-type decomposition of the liquid via ISRO. The Cu-Ir phase diagram is peritectic, with a peritectic composition around 5at% [15], and thus the mechanism is similar to that prevailing in Al-Zn:Cr, i.e., depletion of Ir around iQC and driving force to form Cu. Unfortunately, the Au-Ir phase diagram is not clearly established: a reference mentions a eutectic with a very low eutectic composition (50 at ppm) [16], and another a peritectic with a much larger peritectic composition (0.5 at%) [17].

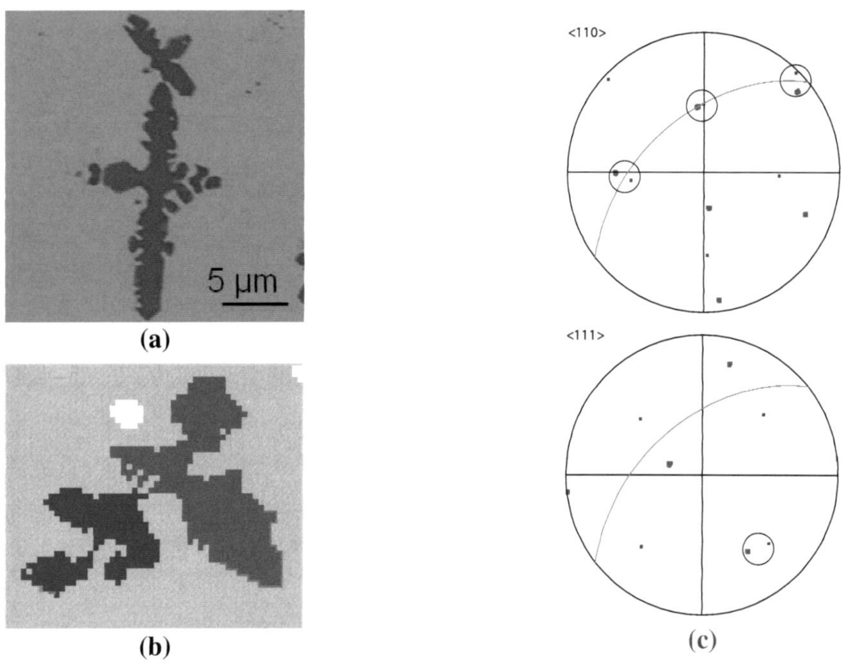

Figure 4: (a) Equiaxed copper dendrites observed 500µm from the chill in the sample with 0.01wt% Ir. The matrix is fcc-gold and corresponds to the columnar zone in Fig. 3(b). (b) False color EBSD map of similar copper dendrites, and (c) corresponding pole figures.

4. Conclusion

Addition of small amounts of Ir in pink gold induces the same iQC-mediated nucleation of the fcc phase, similarly to what has been reported in Al-Zn:Cr and in yellow gold with Ir. First, the Au-rich grains are dramatically refined with a large percentage of twinned grain boundaries. Second, multiple twins relationships between Au-rich grains have been found: they reproduce the symmetry of the icosahedron template. These observations are not surprising and while MT boundaries have been the only phenomenon to be seen at low solidification speed, two new phenomena have been observed at high solidification rate. First, the internal microstructure of the grains is clearly different, especially for columnar grains for which a change of the texture from $\langle 100 \rangle$ to $\langle 111 \rangle$ is observed when Ir is added. This can be explained if one considers the attachment kinetics of ISRO clusters in the liquid on the growing fcc phase. Similar observations were made by Kurtuldu in directional solidification of Al-Zn:Cr dendrites when the solidification speed is increased. Secondly, it is shown that this mechanism, in undercooled melt, can lead to the nucleation of metastable phase, here copper-rich dendrites/precipitates that exhibit similar twin boundary relationships.

References

[1] G. Kurtuldu, P. Jarry, and M. Rappaz, 'Influence of Cr on the nucleation of primary Al and formation of twinned dendrites in Al-Zn-Cr alloys: Can icosahedral solid clusters play a role?', Acta Mater., vol. 61, no. 19, pp. 7098–7108, Nov. 2013.
[2] G. Kurtuldu, A. Sicco, and M. Rappaz, 'Icosahedral quasicrystal-enhanced nucleation of the fcc phase in liquid gold alloys', Acta Mater., vol. 70, pp. 240–248, May 2014.
[3] G. Kurtuldu, 'Influence of trace elements on the nucleation and solidification morphologies of fcc alloys and relationship with icosahedral quasicrystal formation', PhD thesis, EPFL, Lausanne, 2014.
[4] J. Zollinger, B. Rouat, S. K. Pillai, and M. Rappaz, 'Influence of Ir additions and icosahedral short range order (ISRO) on nucleation and growth kinetics in Au-20.5wt%Cu-4.5wt%Ag alloy', Metal. Mater. Trans. A, accepted publication, 2019.
[5] F. Frank, 'Supercooling of liquids', Proc. R. Soc. Lond. Ser. -Math. Phys. Sci., vol. 215, no. 1120, pp. 43–46, 1952.
[6] H.-R. Trebin, Quasicrystals: Structure and physical properties. John Wiley & Sons, 2006.
[7] D. Shechtman, I. Blech, D. Gratias, and J. Cahn, 'Metallic phase with long-range orientational order and no translational symmetry', Phys. Rev. Lett., vol. 53, no. 20, pp. 1951–1953, 1984.
[8] K. F. Kelton, 'Quasicrystals: Structure and stability', Int. Mater. Rev., vol. 38, no. 3, pp. 105–137, Jan. 1993.
[9] M. Rappaz and G. Kurtuldu, 'Thermodynamic aspects of homogeneous nucleation enhanced by icosahedral short range order in liquid fcc-type alloys', JOM, vol. 67, no. 8, pp. 1812–1820, Aug. 2015.
[10] S. Ino, 'Stability of multiply-twinned particles', J. Phys. Soc. Jpn., vol. 27, no. 4, pp. 941–953, Oct. 1969.
[11] G. Kurtuldu, P. Jessner, and M. Rappaz, 'Peritectic reaction on the Al-rich side of Al-Cr system', J. Alloys Compd., vol. 621, pp. 283–286, Feb. 2015.
[12] D. Ott and C. J. Raub, 'Grain size of gold and gold alloys', Gold Bull., vol. 14, no. 2, pp. 69–74, Jun. 1981.
[13] G. Kurtuldu, P. Jarry, and M. Rappaz, 'Influence of icosahedral short range order on diffusion in liquids: A study on Al-Zn-Cr alloys', Acta Mater., vol. 115, pp. 423–433, Aug. 2016.
[14] M. Rappaz and J. A. Dantzig, Solidification. EPFL Press, 2nd Edition, 2016.
[15] E.Rauband, E.Roeschel.Diekupfer-iridium-legierungen.Z.Metallkd.,60:142–144,1969.
[16] Au-Ir binary phase diagram. http://ipandat.computherm.com/Diagram/Details/7556, 2017.
[17] J.P. Nielsenand, J.J. Tuccillo. Grain size in cast gold alloys. Journal of Dental Research, 45(3):964–969,1966.

IOP Conf. Series: Materials Science and Engineering **529** (2019) 012046 doi:10.1088/1757-899X/529/1/012046

Mechanisms of TiAl alloys isomorphic inoculation from cryo-milled Ti-Al-Nb powders

J. Zollinger[1,2], J.R. Kennedy[1,2,3], E. Brodu[2,3], B. Rouat[1,2], E. Bouzy[2,3]
1 Université de Lorraine, Institut Jean Lamour, Department of Metallurgy & Materials Science and Engineering, All_ee Andr_e Guinier - ARTEM Campus, F-54000 Nancy, France.

2 Laboratory of Excellence on Design of Alloy Metals for low-mAss Structures (DAMAS), Université de Lorraine, 57073 Metz, France.

3 Université de Lorraine, CNRS, Arts et Métiers ParisTech, LEM3, F-5700 Metz, France.

E-mail: julien.zollinger@univ-lorraine.fr

Abstract. Isomorphic inoculation has recently been introduced by the authors as a successful method to grain refine cast titanium aluminides [1]. Analyses of the cast grain size together with introduced particle size distributions revealed anomalously high grain refinement efficiency which was attributed to the particles breaking up during the holding stage prior to solidification [2]. In the present work, the microstructure of the inoculant powders is investigated in both the cryo-milled state as well as after simulated thermal cycles to reproduce their heating and holding in the melt. Results show that milling time does not impact the grain size in the particles, only their size distributions. Heat treatments between 1500 and 1600°C for short periods of time allowed the activation energy for grain growth and evaluation of the grain size evolution in the particles during the isomorphic inoculation process to be determined. Assuming that grain boundary melting is the predominant break up mechanism, a model to estimate dissolution of the powders is presented which includes diffusion and fluid flow. Despite its relative simplicity, the predicted number of particles remaining after heating and holding, which lead to grains in the as-cast structure, are in good agreement with the measured grain size. Finally, the paper summarizes the main features and mechanism making isomorphic inoculation a promising route for grain refining as-cast alloys.

1. Introduction

Inoculation is an efficient way for grain refining as-cast structures by adding exogenous particles on which heterogeneous nucleation will be favored, or even controlled. These particles must be stable in the melt and are typically borides, carbides or nitrides. For some alloys, such particles can be detrimental to the as-cast properties, such as γ-TiAl alloys. For these alloys, the authors developed the process of isomorphic inoculation consisting in inoculating the alloy with particles having the same phase as the primary solidification phase [1, 2]. Isomorphic inoculation has been successful in reducing the grain size in cast alloys. Analyses of the cast grain size together with introduced particles size distributions revealed anomalously high grain refinement efficiency, i.e. one introduced particle would lead to more than one new grain in the final ingot. This was attributed to the particles breaking up during the holding

IOP Conf. Series: Materials Science and Engineering **529** (2019) 012046 doi:10.1088/1757-899X/529/1/012046

stage prior to solidification by preferential melting at the grain boundaries. In this contribution, the mechanism leading to these anomalous efficiencies is highlighted. First, the heat treatment undergone by the particles prepared by cryo-milling is simulated in salt baths at 1500 and 1600°C. A model for the particle dissolution is then introduced and used to predict the evolution of the particle size distributionin the melt.

2. Experimental

All the alloys were prepared by induction melting from commercially pure metals (purity>99.7%) as described previously [3]. A Ti-46Al alloy was inoculated with a Ti-Al-Nb powder cryo-milled in a Retsch CryoMill apparatus at -196°C for 3, 6, 9 or 11h. The as-milled powders were characterized by on-axis transmission Kikuchi diffraction (TKD) [4]. High purity calcium fluorine (> 99.95 %) was heated and melted at 100 K/min in alumina crucible at 1500 and 1600°C and held for 30 minutes to reach thermal equilibrium under argon atmosphere. The powders were then introduced into the molten salt for different holding times ranging from 20 to 300 s before quenching the crucible into a water bath. The microstructure of the heat-treated powders was analysed from SEM-BSE images to determine the grain size after heat treatment.

3. Modelling

3.1. Grain growth

The grain growth kinetics, neglecting the heating stage of the particles, can be modelled using the approach of Malow and Koch [5]:

$$d^n - d_0^n = k.t = k_0 e^{\left(-\frac{Q}{RT}\right)}$$

where d is the grain size, d_0 is the initial grain size, n is the grain growth exponent, k_0 is a constant and Q is the activation energy for grain growth.

3.2. Dissolution

To model the dissolution of the particles, the model developed by Hsu and Lin [6] was used. The model is based on diffusion fluxes and accounts for fluid ow. The main feature of the model is that the diffusion boundary layer is controlled by the intensity and length of the flow boundary layer. The reader is invited to read ref. [6] for full details.

4. Results and discussion

4.1. Evolution of the powder microstructure

Figure 1(a) and (b) shows the microstructure of the powders after 3h and 9h of cryo-milling, respectively. As can be seen the grain size is relatively similar, in the order of a few hundred nanometers, but the grains after 3h of milling are more elongated. This means that after 3h of cryo-milling, the grain size inside a particle will not evolve drastically, thus further milling will only decrease the particle size (see ref. [1, 2] for full details about particle size distributions) but not the grain size within the particle.

In Figure 2 the microstructures of the heat-treated powders are shown. Depending on the efficiency of the quench in the calcium fluoride bath, some α-phase can form inside the particles from which the former β grains can still be easily identified, while the α phase present at the particle periphery is likely due to oxygen contamination. For the analysis, only the grain size at the center of the particles were considered, and lead to values of k_0= 10 m^2/s and Q=320 kJ assuming a parabolic growth (n=2) in eq. 1, which is consistent with values found in the literature [7].

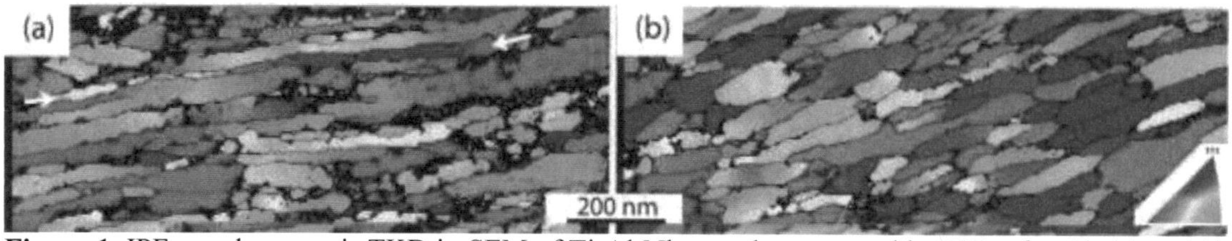

Figure 1. IPF map by on-axis TKD in SEM of Ti-Al-Nb samples prepared by FIB, after 3h (a) and 9h (b) cryomilling.

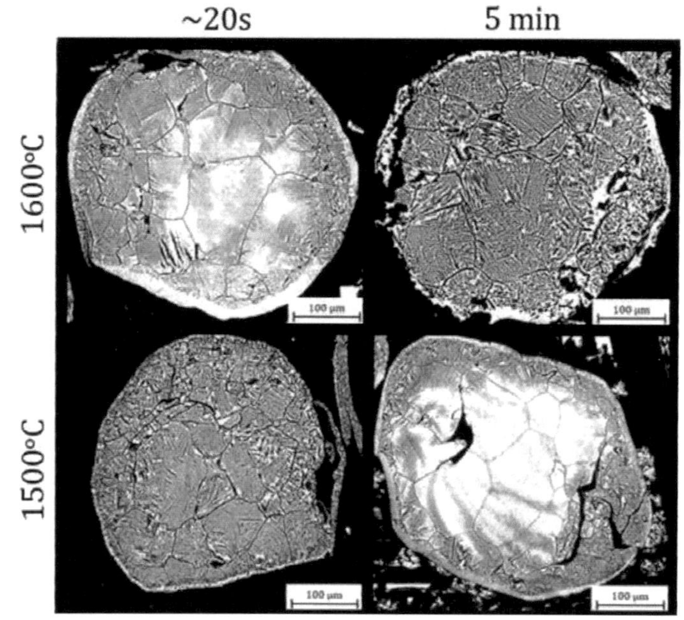

Figure 2. Grain structure in the heat-treated powders at different temperature and for different times for particles with 3h cryo-milling.

It must be mentioned that no difference between in grain size was found between the different milling times, meaning that whatever the particle size and/or the milling time, all the particles contain β grains with the same average grain size of 37 μm. The consequence is rather important, since the particles dissolve preferentially at grain boundaries [2]. This means that the cryo-milling time is affecting the particle size distribution, but after heating up to the melting temperature, all the particles will have the same grain size. The knowledge of the particle size distribution and considering the preferential dissolution at the grain boundary will allow the number of grains (rather than particles) that will act as inoculant and contribute to the grain refinement in cast TiAl alloys to be calculated. However, since the particles do not have diffusional stability in the melt, the complete dissolution of some particles must also be accounted for. The approach used here is presented and discussed in the next section.

4.2. Application to isomorphic inoculation

For each particle size class, the dissolution and the grain growth steps are treated independently. As illustrated in Figure 3, for each particle of size dp, the length dissolved during the holding in the melt is calculated. In parallel, the final grain size within a particle in calculated. For a given particle population, a particle size distribution after dissolution is obtained. Knowing the grain size within a particle and assuming a complete melting of the grain boundaries, a number of grains that will act as grain initiator in the cast alloys can be found. Figure 4 shows the grain density found in the as cast-ingots vs. the number of grains introduced (left) for the case were only the grain growth is considered (black circles) and when dissolution and grain growth are both considered (white circles). The concept of isomorphic inoculation is that the solidifying grain does not nucleate but grows by epitaxy from the inoculant particle, such that each inoculant should produce a single grain in the as-cast materials, which correspond to the 1:1 black line in Figure 4. As can be seen from the figure, accounting for dissolution and grain growth most of the points determined from the different milling time (and thus different particle size distributions) are quite close from the 1:1 ratio. The only point which is farther corresponds to the 3-hour milling time, in which particles are the largest and contain many cracks that might add a supplementary phenomenon that would have to be accounted for.

Figure 4 also shows the efficiency factor, defined as the number of inoculant particles introduced divided by the number of grains in the as-cast alloy vs. the number of particles introduced (right). The factor being higher than one indicates that a single particle produces more than one grain in the final casting. It can be seen again that considering dissolution and break up along grain boundaries leads to a reasonably good agreement with the experimental data. It indicates that these two phenomena are sufficiently well accounted for; while it could potentially be improved by coupling dissolution and grain boundary melting.

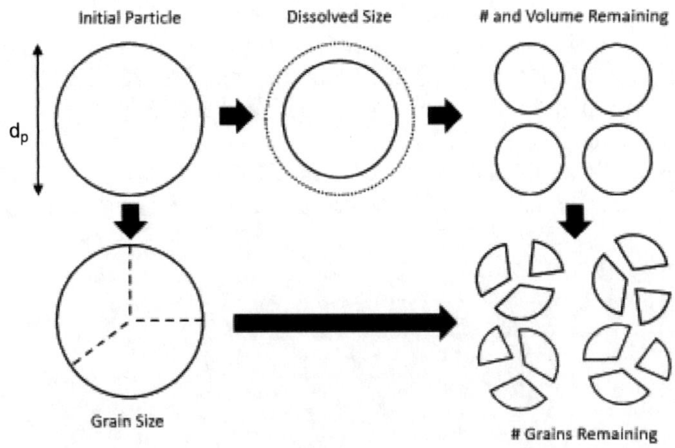

Figure 3. Method to determine the number of grains available for solidification after heating and holding in the melt.

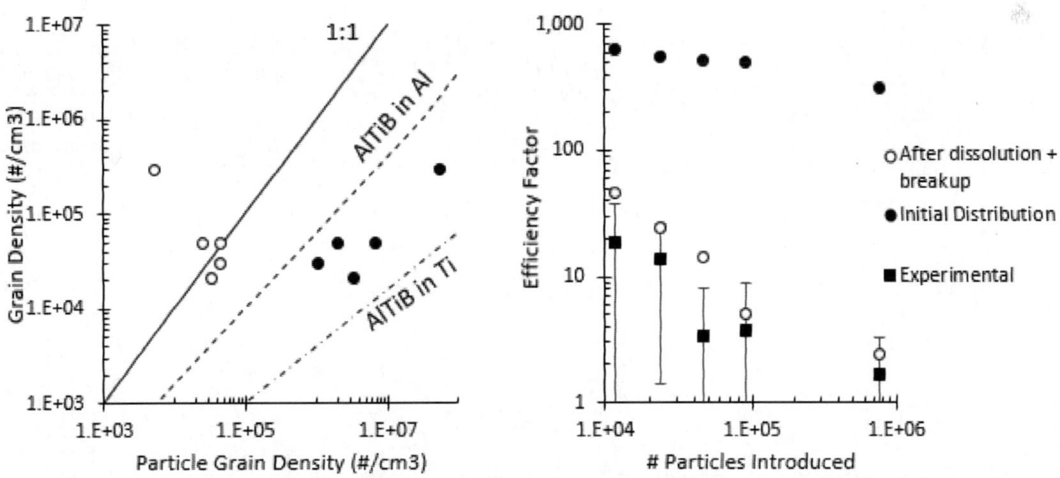

Figure 4. Grain density found in the as cast-ingots vs. the number of grains introduced (left) and efficiency factor vs. the number of particle introduced (right), both considering the initial particle size distribution and the distribution after dissolution and break-up of particles at the grain boundaries.

IOP Conf. Series: Materials Science and Engineering **529** (2019) 012046 doi:10.1088/1757-899X/529/1/012046

5. Conclusions

In this short contribution, the mechanism by which grain refinement is achieved through isomorphic inoculation by cryo-milled powder was demonstrated. Analysis of the microstructure evolution of the powder during the thermal cycle has shown that grain size in the particles does not change with milling time and remains around 200 nm and does not change after heating and high temperature holding remaining around 40 µm independent of the particle size. Accounting for the grains rather than the particles and using a simple model for dissolution of the particles, it has been shown that knowing the particle size distributions is necessary but not sufficient to predict the finale grain size in the ingot, as an important parameter is the grain size within the particles. The obtained results showed a good agreement with the experiments despite the simplicity of the model.

References

[1] Kennedy J, Daloz D, Rouat B, Bouzy E and Zollinger J 2018 Intermetallics **95** 89-93
[2] Kennedy J R, Rouat B, Daloz D, Bouzy E and Zollinger J 2018 Materials **11**
[3] Reilly N, Rouat B, Martin G, Daloz D and Zollinger J 2017 Intermetallics **86** 126-133
[4] Brodu E, Bouzy E and Fundenberger J 2017 Microscopy and Microanalysis **23** 530-531
[5] Malow T and Koch C 1997 Acta Materialia **45** 2177-2186
[6] Hsu J P and Lin M J 1991 Journal of Colloid and Interface Science **141** 60-66
[7] Wang T, Guo H, Tan L, Yao Z, Zhao Y and Liu P 2011 Materials Science and Engineering: A **528** 6375-6380

Molecular dynamics simulation of heterogeneous nucleation via grain refiner inoculated in aluminium melt

Takuya Fujinaga and Yasushi Shibuta
Department of Materials Engineering, The University of Tokyo, 7-3-1, Hongo, Bunkyo-ku, Tokyo 113-8656, Japan

E-mail; fujinaga@mse.mm.t.u-tokyo.ac.jp

Abstract. Since it is not straightforward to directly observe nucleation at the initial stage of solidification in experiments, investigation from computational approach is strongly desired. In this study, influence of grain refiner in heterogeneous nucleation of undercooled Al melt is investigated by molecular dynamics calculations. Particularly, we focus temperature, size of the refiner, and anisotropy in surface orientation of the grain refiner. It is confirmed that the growth rate of FCC Al at the (0001) plane is much larger than that at the other surfaces at all temperatures calculated in this study. Moreover, epitaxial growth of HCP Al appears on the surface of $(10\bar{1}0)$ and $(11\bar{2}0)$ planes at large undercooling temperature.

1. Introduction

Since most metallic materials are formed through solidification from undercooled melt, it is indispensable to control nucleation at the initial stage of solidification [1]. However, it is not straightforward to directly observe nucleation process in experiments and therefore computational researches have contributed the understanding a fundamental aspect of nucleation [2,3]. Especially, molecular dynamics (MD) simulations have shed light on atomistic details of nucleation [2-4]. At present, a billion-atom MD simulation is achieved to investigate homogeneous nucleation from undercooled melt of pure metals owing to recent advance in high-performance computing [4]. On the other hand, heterogeneous nucleation usually occurs in actual production process. Moreover, heterogeneous nucleation is often promoted by inoculating grain-refiners in the melt to produce fine microstructure [5,6]. Although most of the current MD studies on nucleation focus on homogeneous nucleation mainly due to the limitation of interatomic potentials, recent advance in development of interatomic potentials of multi-component systems [7] expands the availability of MD simulations on metallurgical processes such as phase transformation in alloy system and heterogeneous nucleation from hetero-walls or grain refiners. In this study, heterogeneous nucleation of Al melt from a surface of grain refiner is investigated by an MD simulation, which is not straightforward to be observed by experiments. Here, Ti and Ti_3Al are employed as grain refiners since Ti-based compounds are often used as grain refiners [6].

2. Methodology

In this study, MD simulations are performed using LAMMPS (Large-scale Atomic/Molecular Massive Parallel Simulator) [8]. An EAM (embedded atom method) potentials fitted by Zope and Mishin [9] is employed as an interatomic potential of Al-Ti interaction. Parameter files for the EAM potentials from the database of NIST-IPR project [7] are utilized for simulations by the LAMMPS. The velocity-

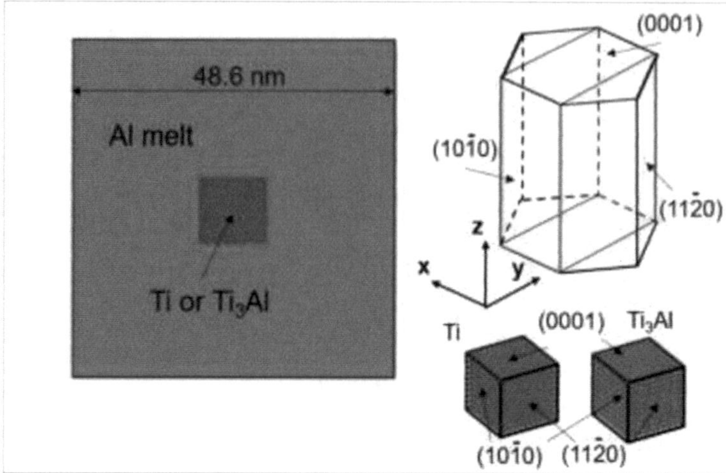

Figure 1. Initial configuration of the calculation system.

Verlet method is used to integrate the classical equation of motion with a time step of 1.0 fs. The Nose-Hoover thermostat and barostat [10,11] are employed to control temperature and pressure. OVITO (Open visualization tool) [12] is used for visualization of atomic configuration and post-analyses of calculation results such as adaptive common neighbor Analysis (a-CNA) [13].

Figure 1 shows an initial configuration of the calculation system, which is prepared as follows. Firstly, a liquid Al is obtained by annealing a face-centered cubic (FCC)-Al crystal consisting of 120 × 120 × 120 unit cells (6,912,000 atoms) at 2000 K for 10 ps with the number of atoms, volume and temperature (NVT)-constant ensemble. Separately, Ti or Ti_3Al cubic particles with various sizes (approximately 5, 8, 10 and 15 nm on a side) are cut out from a hexagonal column of hexagonal-close-packed (HCP)-Ti or $D0_{19}$-Ti_3Al and annealed at 500 K during 10 ps. Here, cubic particles are employed to investigate the effect of surface orientation on nucleation behavior. Then, one cubic particle is inserted in the center of Al melt, while omitting liquid atoms located within 2.5 Å from Ti atoms and energy minimization is performed for the combined structure to avoid unexpected proximity between liquid Al and solid Ti at the interface. Prepared structures are relaxed at 700 and 500 K for 300 ps with the number of atoms, pressure and temperature (NPT)-constant ensemble and growth behavior of solid Al on the surface of the cubic particle is investigated. Moreover, a system filled out with Al melt only (i.e. no grain refiner) is annealed at 700 and 500 K for comparison. Prior to main MD simulation, melting point of pure Al of this EAM potential is estimated by the convergence temperature technique [14, 15] and it is estimated to 870 K. Therefore, annealing temperatures, 700 and 500 K correspond to the undercooling temperature, 170 and 370 K, respectively. Note that the undercooling temperature in this study is larger than that of typical experimental conditions to accelerate nucleation event within the time scale of MD simulation.

3. Results and discussion

Figure 2 shows snapshots of atomic configuration during heterogeneous nucleation at surfaces of Ti and Ti_3Al particles of ca. 8 nm at 700 and 500 K. In general, FCC phase of Al is nucleated from the (0001) plane of both Ti and Ti_3Al particles and it grows hemispherically at both temperatures. In the case of 700 K, there is no nucleation at other planes but only a few layers of HCP Al appear on the surface of these planes. Then, area of FCC phase of Al nucleated on the (0001) plane comes around the other planes and finally, almost all surfaces of the particle are covered by FCC phase of Al. On the other hand, HCP phase of Al grows on $(10\bar{1}0)$ and $(11\bar{2}0)$ planes until it is trapped by FCC phase, which comes in front proactively in the case of 500 K. Then, the area of HCP phase gradually changes into FCC phase. It is considered that purple lines in FCC phase represent stacking faults rather than residual HCP phase. Regarding species of the grain refiner, there is no significant difference in the growth behavior of FCC

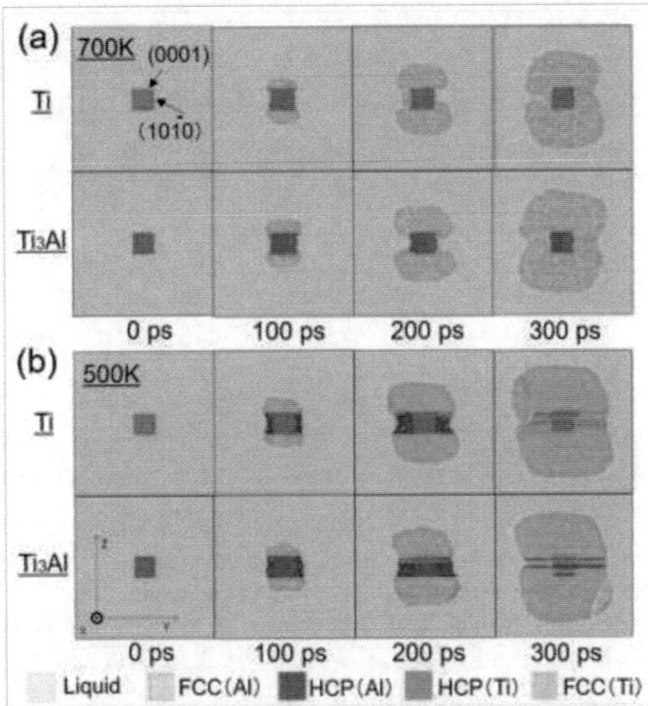

Figure 2. Snapshots of atomic configuration during heterogeneous nucleation at the surface of Ti and Ti_3Al particles of 8 nm at (a) 700 and (b) 500 K.

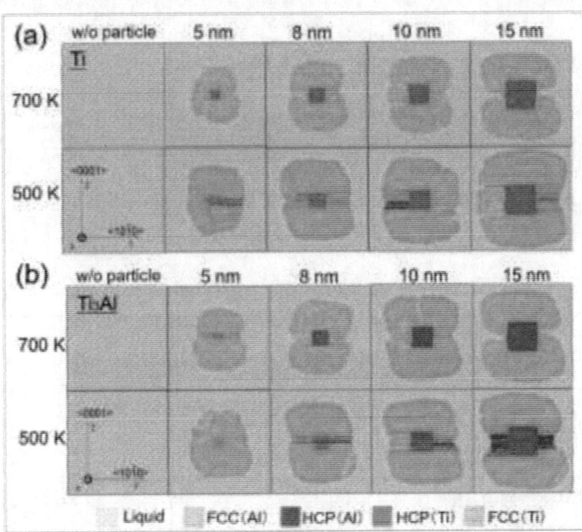

Figure 3. Snapshots of atomic configuration after 300 ps simulation with various sizes of particles: (a) Ti and (b) Ti_3Al particles

phase of Al between Ti and Ti_3Al particles within the range of simulation time. However, growth velocity of FCC phase of Al on Ti_3Al particle is slightly faster than that on Ti particle. Figure 3 shows snapshots of atomic configuration after 300 ps simulation with various sizes of particles. At a glance, there is no significant difference in growth behavior with respect to particle size within the range of calculation. As shown in Figure 2, FCC phase of Al is nucleated from the HCP(0001) plane, which is

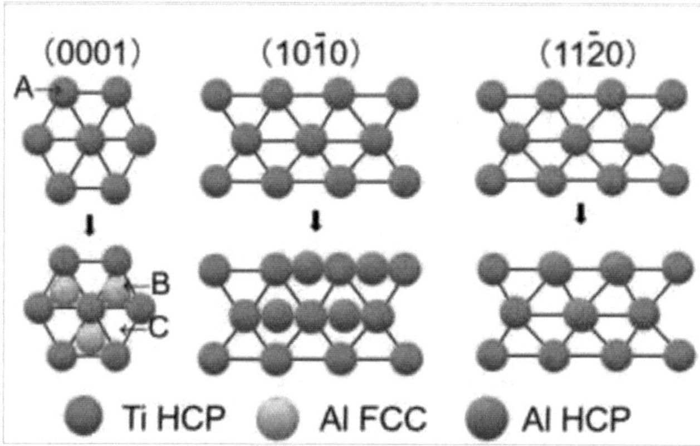

Figure 4. Schematic image of epitaxial growth of Al on various Ti surfaces.

equivalent to the FCC(111) plane. Furthermore, HCP phase grown from (10$\bar{1}$0) and (11$\bar{2}$0) planes changes into FCC phase when the size of the Ti particle is 8 nm or less, at 500 K. It is confirmed that HCP structure appears just after the nucleation but it could not be retained.

Figure 4 shows a schematic image of generation process of Al nuclei on the Ti surface. On the (0001) plane, Al atoms come at a B-layer position of FCC Al on the HCP Ti plane first. Once B-layer positions are filled, Al atoms come at a C-layer positon next, which forms a FCC structure. In other aspects, Al crystal is generated in accordance with the crystal structure of Ti or Ti_3Al particle, which is regarded as epitaxial growth. In the same manner, epitaxial growth occurs on (10$\bar{1}$0) and (11$\bar{2}$0) planes. That is, the HCP layer appears as illustrated in Figure 4. However, the growth of HCP layer stops after only a few layers appears at 700 K since HCP phase is not the most stable for Al. However, HCP phase keeps growing until it is trapped by the FCC phase at 500 K (i.e., a large undercooling temperature) as confirmed in Figures 2 and 3. To discuss it from energetic viewpoint, we calculate cohesive energies of FCC-Al, HCP-Al with a same bond length with FCC-Al (HCP1 in Table 1), and HCP-Al with a lattice constant of Ti (HCP2 in Table 1) using the present EAM potential. Since the cohesive energy of FCC-Al takes the minimum value (i.e. the most stable), it is reasonable that FCC-Al nucleates preferentially on the Ti and Ti_3Al surface. However, epitaxial growth of HCP-Al is partially observed on (10$\bar{1}$0) and (11$\bar{2}$0) surfaces since the energy difference between FCC-Al and HPC-Al is relatively small.

Table 1. Cohesive energy of Al crystal for FCC and HPC structures by present EAM potential.

Crystal structure	FCC	HCP1	HCP2
Lattice constant [Å]	4.05	2.83	2.95
Cohesive energy [eV/atoms]	-3.35998	-3.31983	-3.31558

In order to compare a growth rate of solidified Al structures on each plane of Ti and Ti_3Al particles, a kinetic coefficient of Al is estimated from temperature dependency of solid-liquid interfacial velocity of Al as follows. The propagation velocity of solid-liquid interface of Al is measured on the system connecting Ti or Ti_3Al solid with a target orientation and Al liquid. This biphasic system is relaxed at 400, 500, 600, 700, 800 and 900 K at 300 ps. Then, the slope of the interfacial velocity as a function of temperature is defined as a kinetic coefficient. Technical details of this method is shown in elsewhere [16]. Note that the interfacial velocity of HCP-liquid interface is measured for $(10\bar{1}0)$ and $(11\bar{2}0)$ planes whereas that of FCC-liquid interface is measured for (0001) plane. Figure 5 shows kinetic

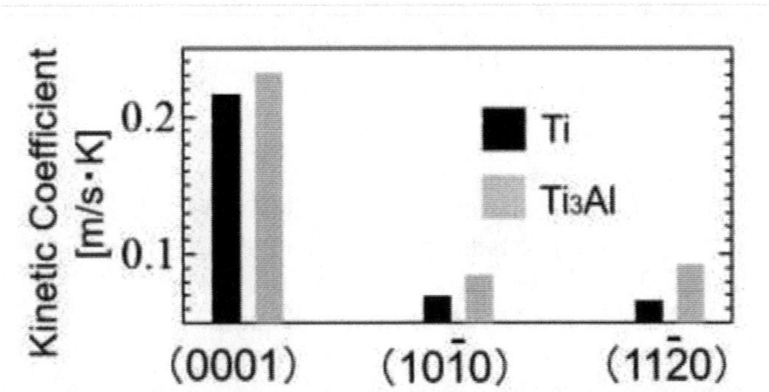

Figure 5. Kinetic coefficients of solid-liquid interface of Al solidified from Ti and Ti_3Al surfaces with various orientations.

coefficients of solid-liquid interface of Al solidified from Ti and Ti_3Al surfaces with various orientations. The kinetic coefficient for the (0001) plane is much larger than those at the other planes, which is consistent with our MD results as shown in Figures 2 and 3. Also, kinetic coefficients for Ti_3Al surfaces become larger than those for Ti surface. It is also consistent with our MD results that Al growth from Ti_3Al particles is faster than that from Ti particles.

4. Conclusions

In this study, MD simulations are performed to investigate the influence of grain refiner in heterogeneous nucleation of undercooled Al melt. It is confirmed from simulation results that the growth rate of FCC Al at the (0001) plane is much larger than that at the other surfaces at all temperatures calculated in this study. Although only a few layers of HCP Al appear on the surface of $(10\bar{1}0)$ and $(11\bar{2}0)$ planes at 700 K, epitaxial growth of HCP Al appears at 500 K. For all orientations, growth rate of Al from Ti_3Al particles is faster than that from Ti particles, which is also confirmed from the kinetic coefficient.

In summary, it is significant in this study to shed light on the heterogeneous nucleation at grain refiners inoculated in Al melt since it is not straightforward to observe it at atomistic scale from experimental measurement. The relationship between undercooling temperature (especially in the range of small undercooling temperature) and particle size, which is a key factor of a free growth model [6,17], will be investigated in the next step.

Acknowledgements

This work was supported by Grant-in-Aid for Scientific Research (B) [No.16H04490] from Japan Society for the Promotion of Science (JSPS), Japan.

IOP Conf. Series: Materials Science and Engineering **529** (2019) 012047 doi:10.1088/1757-899X/529/1/012047

References

[1] Dantzig J A and Rappaz M, Solidification 2009 (Lausanne: EPFL Press) pp. 249-285.
[2] Shibuta Y, Ohno M and T. Takaki 2015 *JOM* **67** 1793.
[3] Shibuta Y, Ohno M and T. Takaki 2018 *Adv. Theo. Sim.* **1** 1800065.
[4] Shibuta Y, Sakane S, Miyoshi E, Okita S, Takaki T and Ohno M 2017 *Nat. Commun.* **8** 10.
[5] Hamilton R W, Forster M F, Dashwood R J and Lee P D 2002 *Scr. Mater.* **46** 25.
[6] Greer A L, Cooper P S, Meredith M W, Schneider W, Schumacher P, Spittle J A and Tronche A 2003 *Adv. Eng. Mater.* **5** 81.
[7] Becker C A, Tavazza F, Trautt Z T and Buarque de Macedo R A 2013 *Curr. Opin. Solid State Mater. Sci.* **17** 277, http://www.ctcms.nist.gov/potentials.
[8] Plimpton S J 1995 *J. Comput. Phys.* **117** 1, http://lammps.sandia.gov.
[9] Zope R R and Mishin Y 2003 *Phys. Rev. B* **68** 024102.
[10] Nose S 1984 *J. Chem. Phys.* **81** 511.
[11] Hoover W G 1985 *Phys. Rev. A* **31** 1695.
[12] Stukowski A. 2010 Model. Simul. Mater. Sci. Eng. **18** 015012.
[13] Stukowski A. 2012 Model. Simul. Mater. Sci. Eng. **20** 045021.
[14] Shibuta Y and Suzuki T 2007 *Chem. Phys. Lett.* **445** 265.
[15] Shibuta Y, Takamoto S and Suzuki T 2008 *ISIJ Int.* **48** 1582.
[16] Watanabe Y, Shibuta Y and Suzuki T 2010 *ISIJ Int.* **50** 1158.
[17] Greer A L and Quested T E 2006 *Philo. Mag.* **86** 3665.

IOP Conf. Series: Materials Science and Engineering **529** (2019) 012048 doi:10.1088/1757-899X/529/1/012048

Segregation of Ca at the Mg/MgO interface and its effect on grain refinement of Mg alloys

S H Wang[1], F Wang[1], Y Wang[1], Q M Ramasse[2,3] and Z Fan[1]

1 BCAST, Brunel University London, Uxbridge, Middlesex UB8 3PH, UK
2 SuperSTEM Laboratory, SciTech Daresbury Campus, Daresbury WA4 4AD, UK
3 School of Chemical and Process Engineering and School of Physics, University of Leeds, Leeds LS29JT, UK

Email: shihao.wang@brunel.ac.uk

Abstract. It has been reported that native MgO particles in Mg alloy melts can act as heterogeneous nucleation substrates such that grain refinement of Mg alloys is achieved. A recent study showed the addition of Ca, combined with the native MgO particles, significantly improves grain refinement of Mg and its alloys. However, the mechanism underlying the grain refining phenomenon is not well understood due to the lack of direct experimental evidence. In this work, we investigated the segregation of Ca atoms at the Mg/MgO interface and its effect on grain refinement in Mg-0.5Ca alloys by utilizing advanced analytical electron microscopy. The experimental results focus on the chemical and structural information at the interface between MgO and the Ca solute. Adsorption layers rich in Al, N and Ca have been detected on *{1 1 1}* facets of MgO particles, with the lattice structure resembling the structure of MgO. It is suggested that the significant grain refinement improvement can be attributed not only to the growth restriction due to the presence of Ca addition but also to the specific chemistry and structure of the adsorption layers.

1. Introduction

As the lightest structural metal, magnesium and its alloys have found a wide variety of applications in the industry due to their high specific strength and excellent castability [1,2]. The grain refinement of Mg alloys has received much attention over the past few decades since fine grain size is one of the factors improving the alloys' mechanical properties [3,4]. To achieve grain refinement, chemical and physical methods have been extensively studied from both experimental and simulation perspectives [5-7]. More than the improvement of mechanical properties, the addition of solute elements such as Al, Zr, Si and Ca is also a common and simple way to achieve grain refinement [8-10]. Among them, Ca produces a strong grain refining effect in pure magnesium [9], whose mechanism has been interpreted by the interdependence theory where Ca has the second highest grow restriction factor (Q) [10]. In Al-bearing Mg alloys, the intermetallic compound Al_2Ca was believed to act as a potent substrate for Mg nucleation, supported by the strong geometric misfit calculated in the edge-to-edge matching model [11].

Recently, Fan et al [12] reported the successful application of the native MgO particles found in Mg alloys as a grain refiner, with an extremely high number density of the oxide particles [13] well dispersed by the high shear melt conditioning (HSMC) device. This was corroborated by the experimental observation of well-defined orientation relationships (OR) between the MgO particles and the Mg grain in both CP Mg and AZ91D alloys [12,14].

It is expected that the further addition of various alloying elements in Mg alloys would interact with the native MgO particles, as the local segregation of solutes has been shown to affect many properties such as wetting, catalysis, electronic structure, embrittlement of grain boundaries etc… Previous work on the manipulation of substrates by adsorption on the exposed surfaces showed that the structural or chemical variance brings about changes of nucleating potency [15-21]. Both the atomic structure and chemistry on the exposed surface play a key role in determining the nucleation potency. However, the interaction between MgO and alloying elements has been rarely studied; for example, Lee [8] found that the combined addition of MgO and strontium (Sr) in Mg-1Al alloys was more effective than the sole MgO or Sr addition. It was also demonstrated by Peng *et al*[22] that there was a significant decrease in the grain size following the marginal addition of Ca (0.01%). The authors suggested that a Mg_2Ca compound formed on the surface enhances the potency of MgO. However, the enhanced potency was not confirmed by atomic-level experimental results, *i.e.* the chemical and structural information at the interface between MgO/Mg.

So far, research on the interaction of MgO with different elements is still at an early stage. Few experiments and theoretical studies have been undertaken. In this study, the Mg-0.5Ca alloys were taken as an example to investigate the effect of the addition of Ca on the grain refining performance of MgO. Advanced electron microscopy is utilized to determine both the structure and chemistry of the Mg/MgO interface at atomic resolution, providing evidence to understand the mechanisms underlying the heterogeneous nucleation and grain refinement.

2. Experiments

Commercial purity magnesium (Mg-0.04Al-0.02Mn-0.013Si) (all compositions are in wt.% unless specified otherwise) and pure calcium (99.9%) were used in this work. Grain-refining tests were followed by a standard TP-1 test [23] proving a consistent cooling rate of 3.5K/s. Pure Mg was melted at 680°C and protected by N_2 plus 0.5 vol pct SF_6; the addition of 0.5% Ca was followed by holding for 20mins. The melt was cast into a TP1 mould pre-heated to 350 °C. For alloys treated by high shearing, the high shear device was applied at a rotation speed of 4000 rpm for 10mins before casting. Specimens for grain size measurements were quantified by the mean linear intercept technique. The samples treated by HSMC are suffixed with '-HS'.

All the MgO particles for characterization were collected by a pressurized melt filtration technique [12]. The melt was first transferred to the pre-heated filtration unit isolated from air, and argon gas was then to force the liquid metal flowing through a filter at the bottom. The remaining melt in the crucible solidified steadily afterwards and left the concentrated oxides above the filter.

The morphology and chemistry of MgO particles were analyzed by scanning electron microscopy (SEM) using Carl Zeiss Crossbeam 340 and Super 35V microscopes, equipped with energy dispersive x-ray spectrometers (EDS). Conventional selected area diffraction (SAD) and high-resolution transmission electron microscopy (HRTEM) were conducted on a JEOL 2100F instrument operated at an accelerating voltage of 200 kV. For advanced characterization, scanning transmission electron microscopy (STEM) was carried out in a Nion UltraSTEM100 (operated at 100 kV) equipped with a Gatan Enfina electron energy loss (EEL) spectrometer. All high angle annular dark field (HAADF) images were collected under image detector semiangles of 89-195 mrad. All elemental peaks were identified following the reference [24]; chemical maps were produced by integration over a suitable window of the recorded intensity above the characteristic edge threshold for each element, after noise reduction by Principal Component Analysis (PCA) [25] and subtraction of the decaying background using a conventional power law model.

3. Results and discussion

3.1. Enhanced grain refinement by high shear melt conditioning and additions of Ca

Figure 1 shows the as-cast microstructures of CP-Mg and Mg-0.5Ca alloys, demonstrating the effect of Ca addition and HS treatment. As we can see, CP-Mg has a columnar grain structure with an average grain size of 816.4±95.9μm (figure 1a). Moreover, either the addition of 0.5% Ca or the application of HS leads to the columnar-to-equiaxed transition (CET) and reduces the grain size to 296.6±34.1μm in Mg-0.5Ca (figure 1b) and 235±18.1μm in CP-Mg-HS (figure 1c), respectively. Furthermore, significant grain refinement was achieved by the joint effect of Ca addition and HS application, as the grain size of Mg-0.5Ca-HS is further decreased to 103.6±7.1μm (figure 1d).

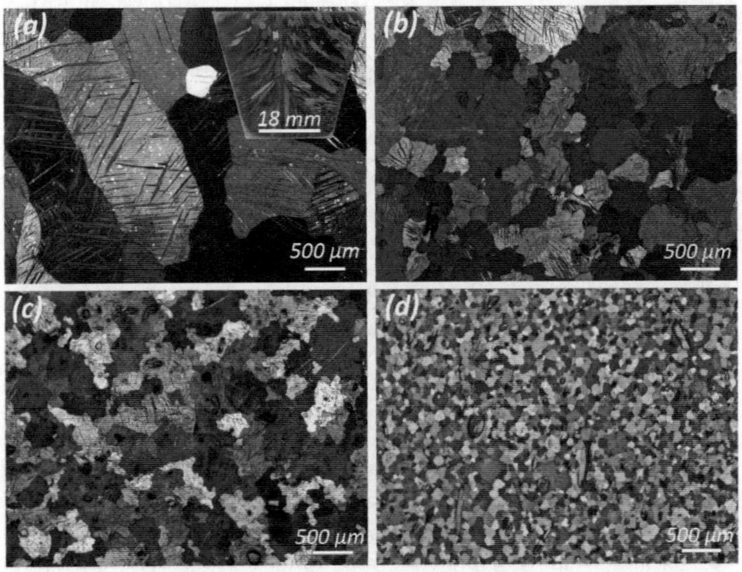

Figure. 1 Optical micrographs showing the microstructures of Mg-xCa alloys solidified in TP-1 condition. (a) CP-Mg; (b) Mg-0.5Ca; (c) CP-Mg-HS and (d) Mg-0.5Ca-HS.

3.2. Native MgO particles as endogenous grain refiners

The positive effect of HS on grain refinement had been investigated by Fan *et al.* [12] who proposed that the enhanced grain refinement comes from the dispersion of native MgO particles in CP-Mg and AZ91D [12,14]. As presented in figure 2a, MgO particles are indeed present with an irregular morphology, with some distinct facets being observed in particular particles (as circled). A previous study [14] had suggested that all such particles would have octahedral shapes bounded by *{1 1 1}* facets. Here, their observed morphology was only negligibly affected by the addition of 0.5Ca, as

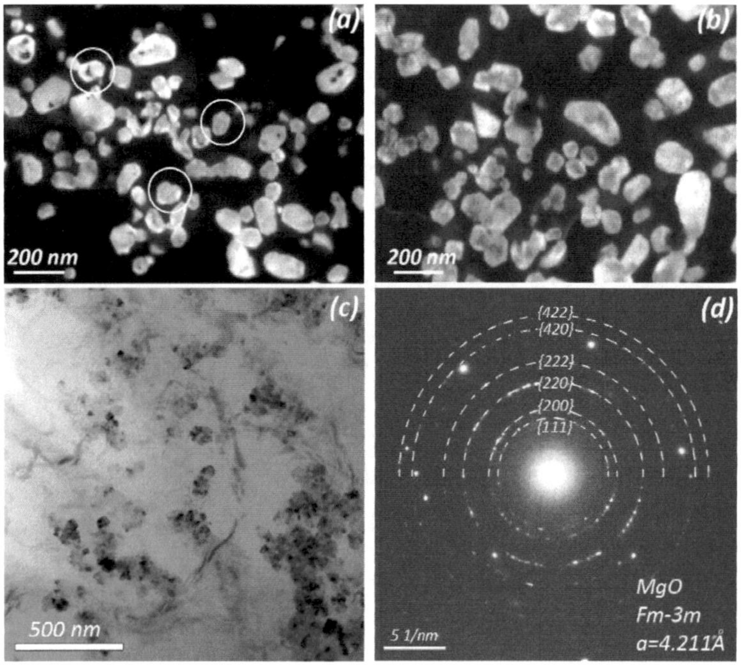

Figure. 2 (a, b) SEM images showing the native MgO particles in CP Mg and Mg-0.5Ca, respectively; (c) TEM bright field image of MgO particles involved in part of a large MgO film; (d) the related SAD pattern taken from the area covering the major part of (c).

consisting of discrete particles in Mg-0.5Ca alloy. The corresponding SAD pattern acquired on the oxide film (figure 2d) has the feature of diffraction rings which can be indexed as MgO (F m-3m, a=4.211Å [26]). It indicates that the oxide film is composed of discrete MgO particles in random orientations.

Further characterization on over a hundred of particles in this study never detected the presence of any possible CaO particles or other oxides. It is thus believed that the MgO particles are the only oxide type in the melt of Mg and Mg-0.5Ca. Moreover, considering the casting condition of both alloys in this work, MgO is expected to be the predominant inclusions outweighing other particles like nitrides or fluorides [27] due to the high affinity of Mg with oxygen. So it is reasonable to suggest that the inoculant in both melts should be native MgO particles during solidification, at least as the dominant species.

3.3. Nature of the adsorption layer

Recent studies have demonstrated that the alloying element can affect nucleation *via* the adsorption onto the surface of nucleation particle [16, 28]. It is therefore essential to investigate the possible interaction between Ca and MgO. This was examined by chemically sensitive HAADF imaging and EELS spectroscopy performed using a Cs-corrected Nion UltraSTEM100. The HAADF image (figure 3a) shows an adsorption layer with a thickness of around 1 nm being observed at the Mg/MgO interface, which appears in a relatively lower contrast compared with MgO. The EELS spectra acquired from different positions (figures 3b&c, averaged over the regions indicated by coloured boxes) suggest the composition of the adsorption layer differs from MgO and Mg matrix: it contains Ca, N and Al, and seems mostly devoid of O since the strengths of O K EELS edge from the interface and Mg matrix are nearly the same (figure 3b), likely only arising from the native oxide at the surface of the thin TEM sample.

IOP Conf. Series: Materials Science and Engineering **529** (2019) 012048 doi:10.1088/1757-899X/529/1/012048

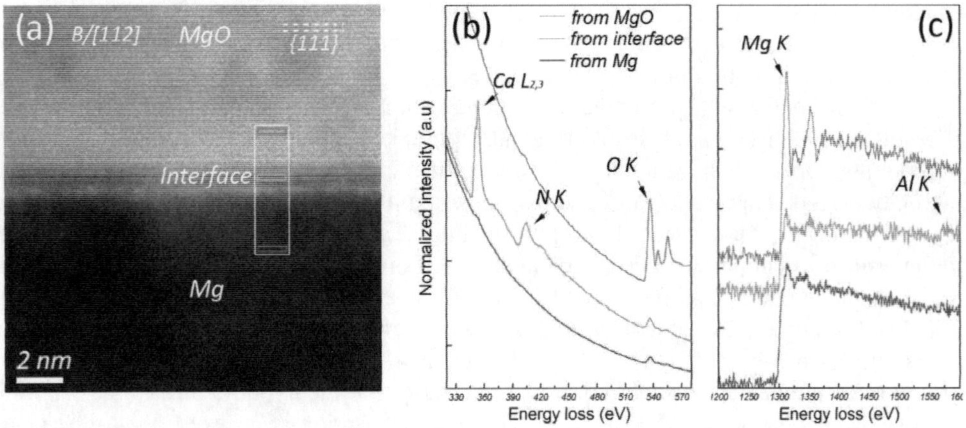

Figure. 3 (a) STEM-HAADF image presenting an adsorption layer at the interface; (b, c) the spectra extracted from different areas in (a). Note that background subtraction had been applied for the spectra in (c).

It has been shown in previous studies that crystallographic misfit plays an important role in determining the grain refining potency of a particle [29-32], which makes it important to determine the precise lattice structure of the adsorption layer at the interface. As shown in figure 4a. the faceted MgO particle shows the adsorption layers on both *{1 1 1}* facets. HRTEM images corresponding to both facets reveal that the adsorption layer has the same structure as the adjacent MgO (figures 4b and 4c). Therefore, the lattice misfit between MgO with adsorption layers and Mg remains, the same as that between normal MgO and Mg (8.19% [14]). This suggests that the nucleation potency should also remain unchanged from the crystallography point of view.

According to the adsorption and the epitaxial model [18,29], it is possible for the segregated atoms to form an ordered solid phase on the substrate surface with mono- or multi- layers (less than critical thickness) which resemble the structure of the substrate even at a temperature above the alloy liquidus.

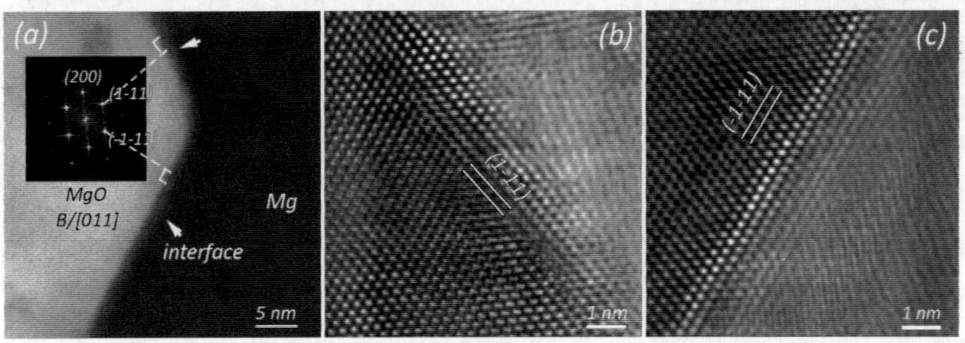

Figure. 4 (a) ADF image of a {1 1 1} faceted MgO (inset: the related FFT pattern) with adsorbed layers; (b, c) the corresponding HRTEM images taken from the facets in (a).

3.4. The enhanced grain refinement under the combined effect of Ca addition and HS application

As shown in figure 1, without application of HS, the addition of 0.5 Ca provides growth restriction and the constitutional undercooling at the solidification front, promoting MgO particles to nucleate and then decreases the grain size (figure 1b). Coupled with the dispersion of native MgO particles by HS, more particles could be activated for grain initiation and hence more significant grain refinement as shown in figure 1d. Comparing figures 1c and 1d shows that the presence of Ca in the Mg melt is important to provide sufficient constitutional undercooling to promote heterogeneous nucleation by activating more MgO particles, which have been well dispersed by HS. However, the previous study by Peng *et al.* [22] has demonstrated that further addition of Ca in Mg melt from 0.5 to 1% produces marginal further grain refinement effect only. This implies that, besides providing growth restriction and constitutional undercooling affecting the growth process of solidification, the presence of Ca in the Mg melt may also have effects on the solidification process.

From the experimental results presented in Section 3.3, it is revealed that the interaction between Ca and MgO results in the formation of isostructural adsorption layers on the surface of MgO particles. The atomically thin layer seems to have a relatively small effect on the grain initiation process based on the free growth model [33]. This model suggests that the grain initiation is a deterministic process and the undercooling for free growth is inversely proportional to the particle size. In the current study, the size of the MgO particles, including the adsorption layer (itself only around 1nm in thickness), did not vary significantly: the undercooling for free growth should thus be constant.

Nevertheless, it is now well recognized that the adsorption layers will affect the nucleation potency of substrate particles, as reported by Fan *et al* and Wang *et al* [16,28]. Although elucidating the precise mechanism at play in the current work requires further effort, a few hypotheses can be proposed. One possibility is that the potency of MgO for heterogeneous nucleation is changed due to the change of surface roughness [34], which might be induced by the presence of different elements in the adsorption layers. Wang *et al.* [28] showed that a mono-layer of Ti_2Zr forms on the surface of TiB_2, which induces surface roughness and then impedes the heterogeneous nucleation of Al [28, 34]. However, it is still unclear whether or not the surface roughness of the MgO increases or decreases with adsorption based on current results. Another possibility is that the surface energy of MgO particles could be decreased by the solute segregation in the form of adsorption layers. This decrease of surface energy could cause further dispersion of MgO particles during HS application, which may further increase the number density of MgO particles available for nucleation, leading to significant grain refinement.

In summary, it is proposed that the enhanced grain refinement produced by the combined effect of Ca addition and HS application can be attributed not only to the growth restriction generated by Ca partitioning but also to the formation of an adsorption layer on the surface of MgO particles.

4. Conclusions

1. Combining the application of HS and Ca addition can produce significant grain refinement in commercial pure Mg alloys.
2. Ca atoms tend to segregate at the Mg/MgO interface. The interaction between Ca and MgO results in the formation of distinct adsorption layers at the interface with a thickness of around 1nm, containing Ca, N and Al. The crystal structure of the adsorption layer resembles the lattice structure of the MgO particle.
3. It is suggested that the considerable grain refinement induced by HS application and Ca addition is the result from not only the growth restriction but also from the formation of the adsorption layers caused by the interaction between Ca and MgO particles.

5. Acknowledgment

EPSRC is gratefully acknowledged for financial support under grant number EP/N007638 /1. The SuperSTEM Laboratory is the U.K National Research Facility for Advanced Electron Microscopy, supported by EPSRC. SHW gratefully acknowledges the China Scholarship Council (CSC) for financial support and ETC in Brunel University for providing access to the facilities. The author is also grateful to Mr. Ewan Lordan for help on the manuscript preparation

References

[1] Joost W J 2014 *Magnesium Technology* (Wiley, New York) p 3–4.
[2] Rooy E L 1988 *Metals Handbook* vol 15 (Metals Park, ASM) p 750.
[3] Nelson C E 1948 *Trans AFS.* **56** p1e23.
[4] Emley E F 1966 *Principles of magnesium technology.*
[5] Murty BS, Kori SA, and Chakraborty M 2002 *International Materials Reviews.* **47** 3–29.

[6] Pearce BP and Kerr HW 1981 *Metal. Mater. Trans. B.* **12B** 479–86.
[7] Fan Z et al 2009 *Int. J. Cast Met. Res.* **22** 318–22.
[8] Lee YC, Ph.D. thesis, 2002 *Grain Refinement of Magnesium*, Queensland University.
[9] Lee YC, Dahle AK. and StJohn DH 2000 *Metal. Mater. Trans. A.* **31** 2895–2906.
[10] StJohn DH, Ma Q, Easton MA, Cao P and Hildebrand Z 2005 *Mater. Trans. B.* **36** 1669–79.
[11] Jiang B, Liu W, Qiu D, Zhang MX. and Pan F 2012 *Mater. Chem. Phys.* **133** 611–16.
[12] Fan Z, Wang Y, Xia M and Arumuganathar S, 2009 *Acta Materialia.* **57** 4891–01.
[13] Men H, Jiang B, and Fan Z 2010 *Acta Materialia.* **58** 6526–34.
[14] Wang Y, Peng G and Fan Z 2017 *Magnesium Technology.* p. 99–106.
[15] Schumacher P, and Greer AL 1994 *Materials Science and Engineering: A.* **181** 1335–39.
[16] Fan Z et al 2015 *Acta Materialia,* **84** 292–304.
[17] Wang Y, Zhou L, and Fan Z 2016 *Light Metals* p. 725–29.
[18] Cantor B, 2003 *Philosophical Transactions of the Royal Society of London A: Mathematical, Physical and Engineering Sciences*. **361** 409–17.
[19] Zhang DL and Cantor B 1990 *Journal of crystal growth.* **104** 583–92.
[20] Ho CR and Cantor B 1992 *Philosophical Magazine A*. **66** 141–149.
[21] Fang CM, Men H and Fan Z 2017 *solidification processing* 52–55.
[22] Peng GS, Song GS, Wang Y, Chen KH and Chen SY, 2018 *Int J Cast Metal Res* **31** 99–107.
[23] Standard test procedure for aluminium alloy grain refiners (TP-1). 1990 *Washington, DC: The Aluminium Association.*
[24] Ahn CC and Krivanek OL 1983 *EELS Atlas*, Warrendale Gatan Inc.
[25] Lucas G et al 2013 *Micron* **52** 49–56.
[26] Hazen RM 1976 *American Mineralogist* **61** 266–271.
[27] Sin SL 2013 *International Materials Reviews* **58:7** 419–436.
[28] Wang Y et al 2018 *Acta materialia* **164** 428-439.
[29] Fan Z 2013 *Metal. Mater. Trans. A.* **44** 1409–18.
[30] Bramfitt BL 1970 *Metallurgical Transactions* **1** 1987–95.
[31] Turnbull D, Vonnegut B 1952 *Ind. Eng. Chem.* **44 (6)** 1292–98.
[32] Zhang MX and Kelly PM 2005 *Acta materialia* **53** 1085–96.
[33] Greer AL 2000 *Acta materialia* **48** 2823-2835.
[34] Jiang B et al 2018 *Comput. Mater. Sci.* **153** 73–81.

IOP Conf. Series: Materials Science and Engineering **529** (2019) 012049 doi:10.1088/1757-899X/529/1/012049

Grain refinements of magnesium alloys inoculated by additions of external SiC particles

Yuanding Huang*, Jian Gu, Sihang You, Karl Ulrich Kainer, Norbert Hort

MagIC-Magnesium Innovation Centre, Helmholtz-Zentrum Geesthacht, Max-Planck-Str. 1, 21502 Geesthacht, Germany

*yuanding.huang@hzg.de

Abstract. A homogeneous microstructure of as-cast magnesium alloys is desired to improve the formability during their subsequent thermomechanical processing. Owing to its similar crystal structure to Mg, the part of Zr formed by peritectic reaction during solidification was considered to be the most effective nucleants for alpha-Mg. However, regarding the Al-containing magnesium alloys, up to now no suitable and effective external nucleants were found for them. Recently, it was demonstrated that the additions of SiC worked in refining both the Mg-Al and Mg-Zn alloys. The SiC particles acted as nucleants in magnesium alloys are attracting more attentions. The present work investigated and compared the effects of external SiC particle additions on the grain refinements of Mg-Al and Al-free Mg-Zn (Mn) alloys. Their microstructures were characterized using XRD, SEM and TEM. It was found that the additions of SiC particles could refine the grains of both Mg-Al and Mg-Zn alloys. The SiC particles cannot act as a direct heterogeneous nucleant for the nucleation of alpha-Mg in both Mg-Al and Mg-Zn (Mn) alloys. The responsible mechanisms for their grain refinements are different. Regarding for Mg-Al alloys, the grain refinement caused by the addition of SiC particles is attributed to the formation of a ternary intermetallics Al_2MgC_2, which has a very similar crystal structure to that of Mg. As for Mg-Zn (Mn) alloys, the grain refinement is attributed to the formation of a (Mn, Si)-enriched intermetallics by the interactions between SiC and impurity Mn in alloys.

1. Introduction

A homogeneous microstructure of as-cast magnesium alloys is necessary to improve the formability during their subsequent thermomechanical processing. In Al-containing magnesium alloys, the grain refinement by carbon inoculation is considered to be the best approach until now [1]. Besides carbon inoculation, the addition of alloying element Ca can also produce a strong grain refining effect for the Al-bearing Mg alloys. Carbon inoculation can be carried out by adding materials such as graphite, wax, C_2Cl_6 and SiC particles etc. Due to the fact that carbon inoculation only functions with the existence of aluminum, the enhanced nucleation of alpha-magnesium on the surface of Al_4C_3 particles was popularly treated as the reasons responsible for the grain refinement [2-4]. Another speculation is the modified version of Al_4C_3, which was suggested by Yano et al. [3]. They interpreted Al-C-O particles as Al_2CO, acting as the nucleants for the grain refinement by carbon inoculation. Moreover, it was also reported that the grain refinement in Mg-Al alloys inoculated by carbon is associated with (Al,Mn)-containing intermetallics [5].

Recently, additions of SiC in aluminum-free Mg-Zn alloys could also result in their grain refinement. Cao et al. successfully prepared Mg-Zn/SiC nano-composites by ultrasonic cavitation-based solidification processing [6, 7]. Their results showed that the grain size of Mg–Zn alloy was reduced considerably by the addition of SiC nanoparticles. They attributed the grain refinement to SiC nanoparticles served as nuclei for heterogeneous nucleation, but no direct evidence was obtained. Inem et al. investigated the nucleation and crystallographic orientation of both the Mg matrix and the eutectic $Mg(ZnCu)_2$ phase at the β-SiC particle surface in ZC63 and ZC71 Mg matrix composites [8, 9]. They found that the eutectic nucleates at the particle surface with an identical crystallographic orientation, but no distinct crystallographic orientation of α-Mg with SiC particle has been resolved. Jian Gu et al. added the external SiC particles to Mg-Zn alloys. They found that the as-cast microstructures of Mg-Zn alloys were also refined [10]. With microstructural observations and EDX analysis, they attributed the grain refinement of Mg-Zn alloys caused by SiC particles to the possible interactions between impurities Fe, Mn and SiC particles. Inside the grains the particles Fe-Mn-Si intermetallics were observed.

In summary, the explanations for the grain refinements in both Mg-Al and Mg-Zn alloys caused by the addition of external SiC particles are still controversial. Previous investigations show that their grain refinements are due to the physical or chemical effects of SiC particles. The present work is to investigate and clarify the mechanisms responsible for the grain refining effects of SiC additions in Mg-Al and Al-free Mg-Zn (Mn) alloys.

2. Experimental procedures

Mg alloys with different Al, Zn and Mn contents inoculated with different contents of SiC were prepared using pure Mg Al, Zn and Mn. SiC particles (supplied by Alfa Aesar GmbH & Co KG, Germany) with an average size of 2 μm were used as refiner. Mg and Zn ingots were melted at 700 °C in an electrical resistance furnace using a mild steel crucible under a protective gas mixture of high pure Ar + 0.2% SF6. The melt was manually stirred for 2 min and then its surface was skimmed. SiC particles preheated to 500 °C under Ar atmosphere were added into the melt directly. After that the melt was stirred vigorously at 100 rpm for 5 min to ensure a good dispersion of SiC particles. Before casting the melt was held at 700 °C for 15 min. Each ingot was cast by pouring the melt into a steel mold preheated to 200 °C with a diameter of 70 mm at the bottom and 80 mm at the top and a height of 250 mm.

Metallographic samples were transversally sectioned from the same position of 20 mm from the bottom of the castings. They were prepared according to a standard procedure. The samples for optical observations were etched with a solution of 8 g picric acid, 5 ml acetic acid, 10 ml distilled water, and 100 ml ethanol. The average grain size was measured by the linear intercept method from the micrographs taken using polarized light in an optical microscope. A Zeiss Ultra 55 (Carl Zeiss GmbH, Oberkochen, Germany) scanning electron microscope (SEM) equipped with energy dispersive spectroscopy (EDS) was also used to observe the microstructures of selected samples. X-ray diffraction (XRD) investigations were carried out using a Siemens diffractometer operating at 40 kV and 40 mA with Cu radiation. Measurements were obtained by step scanning from 20 to 90° with a step size of 0.02°. A count time of 3 seconds per step was used. PANalytical X'pert HighSocre software with the International Center for Diffraction Data (ICDD) PDF2-2004 database was used to analyze the diffraction peaks of identified phases based on crystal structures.

The differential scanning calorimetry (DSC) experiments were carried out to investigate the solidification process. DSC measurements were done using calorimeter Netzsch 404C with crucibles made of chromium steel under the protection of Ar gas. A heating and cooling rate of 2 °C/min was used. The total weight of each sample was about 80 mg.

3. Results and discussion

3.1. Mg-Al alloys with SiC particles

Previous microstructural investigations demonstrated that after the addition of SiC particles the grains of Mg-Al-SiC alloys were refined from milometers order to about 200 μm [11-13]. Microstructural observations and XRD diffractions confirmed that SiC particles reacted with Mg-Al melt because Mg_2Si intermetallics was detected in Mg-Al-SiC. Inside the grains the possible nucleants were identified as Al_2MgC_2 by FIB cutting technique and TEM, which was also supported by XRD results. This ternary intermetallics Al_2MgC_2 has a very similar crystal structure to Mg. Thus, its formation was considered to enhance the nucleation of alpha-Mg and to result in grain refinement. Base on DSC experiment results and previous results, the following will explain the grain refinement mechanism for Mg-Al alloys inoculated with SiC particles schematically.

DSC cooling curves are presented in Figure 1. No peaks are observed on the cooling curve for carbon in the temperature range from 560°C to 720°C. The interaction between carbon and chromium steel crucible can be excluded. For the binary alloys Mg-10C, Al-10C and Mg-3Al, only one peak appears on their cooling curves. These peaks correspond to the solidification of magnesium, aluminum and Mg-Al solid solution, respectively. The starting temperatures of solidification are 655°C and 645°C for Al-10C and Mg-10C, respectively. As the liquidus temperature difference between these alloys (10°C) is the same as the difference in melting temperatures between pure Al (660°C) and pure Mg (650°C), it seems that no interactions have occurred between Al and C or between Mg and C in the binary Mg-C and Al-C alloys. In contrast, after the Mg-3Al alloy is inoculated by C or SiC particles, a small peak close to the solidification peaks of alpha-Mg is observed (Figure 1(b)). The temperatures corresponding to these two small peaks are 636.5°C and 639.8°C for Mg-3Al-0.3SiC and Mg-3Al-0.3C alloys, respectively. Considering that these small peaks are not observed in the binary alloys, their appearance in the case of ternary alloy systems must be attributed to the interactions among Mg, Al and C to form a ternary carbide. Since the DSC experiments were carried out under Ar gas, the possibility of the formation of the ternary oxide such as Al_2CO is ruled out. Moreover, based on the previous XRD and TEM results [11], it can be concluded that these two small peaks correspond to the formation of the Al_2MgC_2 particles.

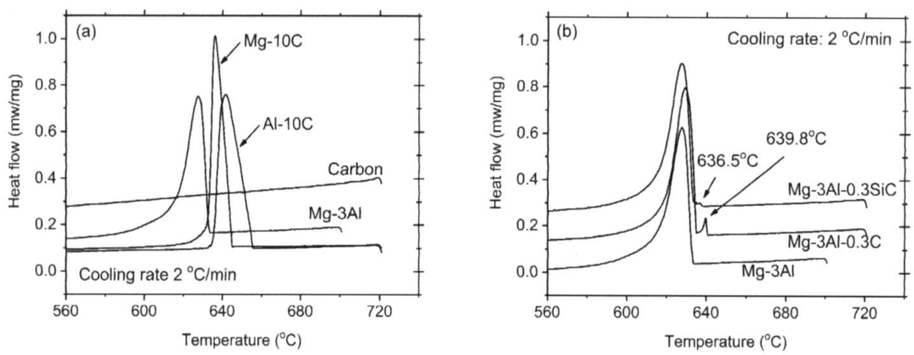

Figure 1. (a) DSC cooling curves for carbon, Mg-10C, Al-10C and Mg-3Al alloys and (b) DSC cooling curves for magnesium alloys. DSC cooling curves showing a small peak at a temperature little higher than that of solidification of alpha-Mg for the alloys inoculated by carbon or SiC particles. Please notice that the DSC curves have been shifted in the vertical direction in order for clear reading.

In this paper a schematic model is proposed based on the experimental evidences to explain the grain refinement of Mg-Al alloys inoculated by SiC particles (Figure 2). In difference to the Al_4C_3 hypothesis, the present mechanism is based on the nucleation of alpha-magnesium on the surface of ternary carbide Al_2MgC_2 which has a crystal structure more closely to that of magnesium. Its planar disregistry with magnesium is 5.32%. In contrast to Al_2MgC_2, the lattice parameter of Al_4C_3 with a value of 2.4996 nm is much larger than that of Mg. Hence from the crystal structure point of view, Al_2MgC_2 is a more suitable core for the nucleation of alpha-magnesium than Al_4C_3.

DSC results illustrate that the ternary carbide Al_2MgC_2 starts precipitation just before the solidification of alpha-magnesium (Figure 1), which makes it possible to act as the nucleant for the solidification of alpha-Mg. The grain refinement process is illustrated schematically in Figure 2. First, during that time when the temperature decreases from the melting temperature 700°C to about 640°C, the aluminum segregates around the SiC particles. The aluminum enrichment has been observed in the area near the "hillock" when Mg-Al alloys were inoculated by SiC particles. Second, SiC particles react with liquid magnesium and aluminum to form the ternary Al_2MgC_2 and the silicon is released in this process. The segregation of silicon around the remaining SiC particles has also been observed (Figure 3). The released silicon diffuses into the molten magnesium alloys and at the eutectic temperature it reacts with magnesium to form the Mg_2Si particles. Finally, with the solidification temperature decreasing, the alpha-magnesium starts nucleation on the surface of Al_2MgC_2. Due to their similar crystal structure between Al_2MgC_2 and magnesium, the nucleation of alpha-Mg on the surface of Al_2MgC_2 is enhanced, and then the grains are refined.

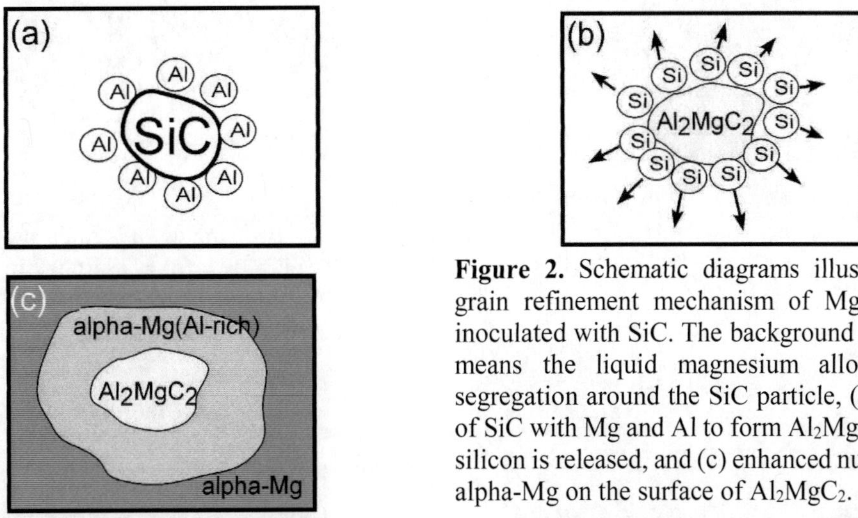

Figure 2. Schematic diagrams illustrating the grain refinement mechanism of Mg-Al alloys inoculated with SiC. The background with white means the liquid magnesium alloy: (a) Al segregation around the SiC particle, (b) reaction of SiC with Mg and Al to form Al_2MgC_2, and the silicon is released, and (c) enhanced nucleation of alpha-Mg on the surface of Al_2MgC_2.

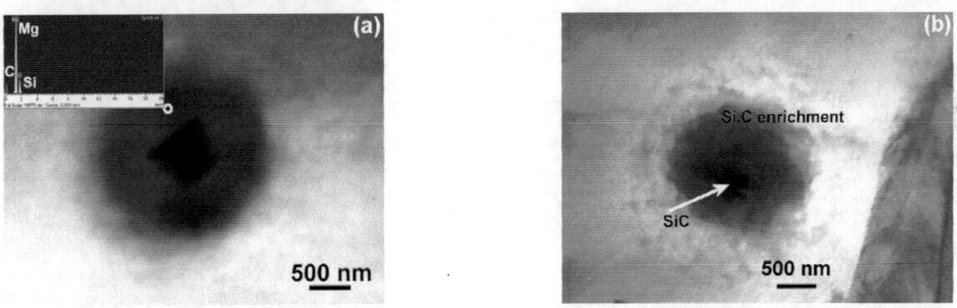

Figure 3. TEM micrograph showing the interactions between the SiC particle with magnesium matrix: (a) Mg-1.2Al-0.8SiC with stirring and holding for 5 min and (b) Mg-3Al-10SiC.

3.2. Mg-Zn alloys with SiC particles

Previous microstructural observations demonstrated a slight decrease of average grain size from 330 ± 12 to 285 ± 10 μm with increasing the SiC content from 0 to 0.2% in Mg-3Zn alloy [10]. A sharp decrease of average grain size from 285 ± 10 to 180 ± 9 μm was obtained with increasing SiC content from 0.2% to 0.3%. After that, the average grain size remains relatively stable with the increase of SiC content, even up to 10% SiC.

SEM observations frequently found that inside the grains some particles with several micrometers exist. They seem to be the nucleants for the formation of alpha-Mg during solidification. **Figure *4*** presents such a typical particle inside the grain in Mg-3Zn-0.5SiC alloy and its EDS line analysis. The grain boundaries were indicated by the white arrows. Under the back-scattered mode, the black part enriched with Si and C can be confirmed as a SiC particle. Firstly, this particle has even darker contrast than that of Mg matrix suggesting that its average atomic number is less than that of Mg matrix. In addition, the length of black part particle is about 3 μm, which is close to the original size 2 μm of the SiC particle. The inside white part is enriched with Fe and Mn, indicating that SiC is likely to associate/react with Fe and Mn to serve as nuclei for alpha-Mg grains. In addition, Zn enriched area was also detected by EDS line-scan analysis at the bottom of the right corner, which is closely attached to the SiC particle.

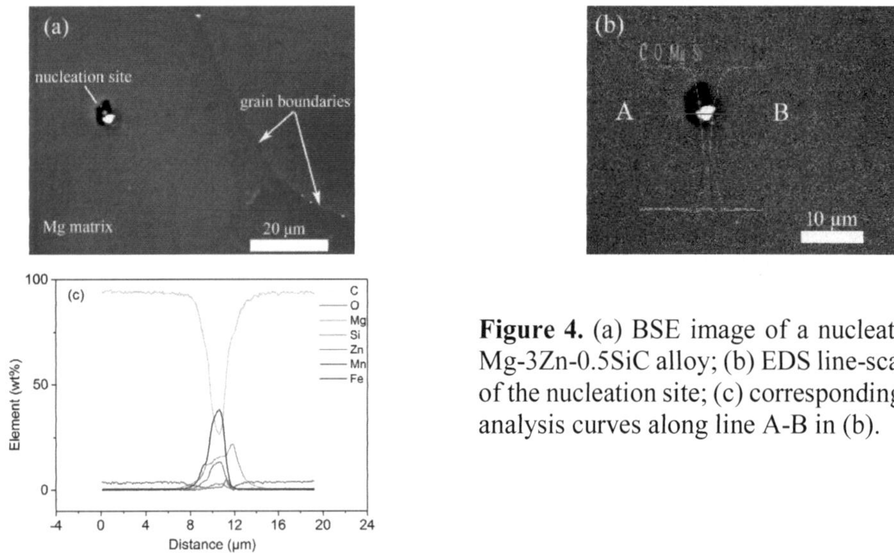

Figure 4. (a) BSE image of a nucleation site in Mg-3Zn-0.5SiC alloy; (b) EDS line-scan analysis of the nucleation site; (c) corresponding line-scan analysis curves along line A-B in (b).

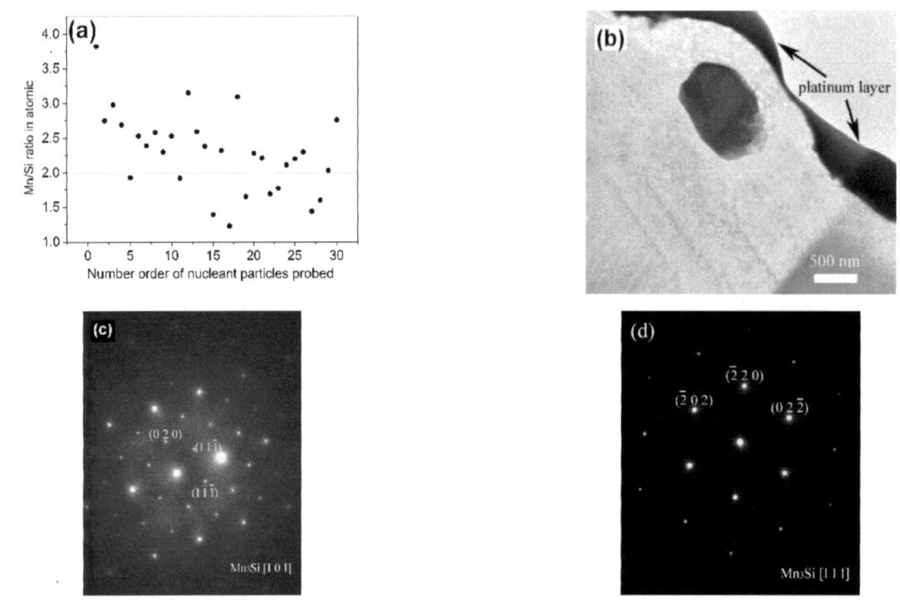

Figure 5. (a) Mn/Si ratio in atomic percent from the 30 probed nucleant particles by presenting in number order, (b) bright-field TEM image of the nucleant particle, (c) and (d) are the diffraction patters collected from the nucleant particle.

In order to characterize and analyze the nucleant particles in detail, 30 nucleant particles were randomly probed. Those nucleant particles have almost the same morphologies but with different diameters. The corresponding EDS analysis results of Mn/Si ratio from the 30 probed nucleant particles are shown in Figure 5(a). It was found that the majority of the nucleant particles detected falls in the size range of 1-2 μm, and its frequency reaches to 57%. The detected Mn/Si ratios were not uniform, which could be affected by the surrounding environment such as the Mg matrix and also the size of the probed nucleant particle. However, it clearly shows that most measurements of the Mn/Si ratios were found to fall in the range between 2 and 3. Those new formed nucleant particles could be Mn_3Si. Figure 5(b) is the morphology of the nucleant particle observed by TEM characterization. Diffraction patters of Figure 5(c) and (d) confirm that its structure is Mn_3Si. The lattice parameters of Mn_3Si phase are a = b = c = 0.5724 nm with a space group of $Fm\bar{3}m$. The lattice parameters of Mg are a = 0.3209 nm and c = 0.5210 nm with a space group of P63/mmc. Apparently, their lattice parameter "a" values are close, which means that the formation of Mn_3Si could enhance the nucleation of alpha-Mg and refine the microstructure.

After inoculation in Al-free Mg-Zn(Mn) alloys, part of SiC particles can react with Mn and/or Fe atoms to form in-situ potent Mn_3Si or $(Mn,Fe)_3Si$ nuclei, which not only promotes heterogeneous nucleation at the beginning of the solidification, but also restricts the grain growth through continuing nucleation within the CS (Constitutional supercooling) zone at front of the S-L interface. However, as the formation of CS zone depends on the solute segregation at the front of the S-L interface, the solute element Mn with relatively low Q value (Growth restriction factor) has little contribution to the formation of the CS zone. Thus, the grain refining efficiency is moderate in Mg-Mn-SiC alloy system compared with that of Mg-Zn-SiC alloy system, because Zn with high Q value can provide additional undercooling to activate and accelerate the heterogeneous nucleation process.

4. Conclusions

The additions of external SiC particles in Mg-Al and Al-free Mg-Zn(Mn) alloys can refine their as-cast microstructures. In Mg-Al alloys, SiC reacts with Mg-Al melts to form a ternary intermetallics Al_2MgC_2 which has a same crystal structure and very close lattice parameters to that of Mg. Its formation can enhance the nucleation of alpha-Mg on its surface and then the microstructure is refined. Different from the refinement mechanism of Mg-Al alloys, in Al-free Mg-Zn(Mn) alloys the alloying element Mn and/or impurities Mn, Fe plays very important role in interacting with the external particles SiC. Their chemical interaction can result in the formation of Mn_3Si-type intermetallics which also has a close lattice parameter to that of magnesium. Consequently, the nucleation of alpha-Mg is enhanced and grains are refined.

References

[1]. StJohn DH, Qian M, Easton MA, Cao P 2005 *Metall Mater Trans* **36A**1669.
[2]. Qian M, Cao P 2005 *Scripta Materialia* **52** 415.
[3]. Yano E, Tamura Y, Motegi T, Sato E 2001 *J. Jpn. Inst. Light Metals* **51** 599.
[4]. Emley EF 1966 *Principles of Magnesium Technology. Oxford: Pergamon.*
[5]. Kim YM, Yim CD, You BS 2007 *Scripta Materialia* **57** 691.
[6]. Cao G, Choi H, Konishi H, Kou S, Lakes R, Li X 2008 *J Mater Sci* **43** 5521.
[7]. Cao G, Kobliska J, Konishi H, Li X 2008 *Metallurgical and Materials Transactions A* **39** 880.
[8]. Inem B 1995 *J Mater Sci* **30** 5763.
[9]. Inem B, Pollard G. 1993 *J Mater Sci* **28** 4427.
[10]. Gu J, Huang Y, Kainer KU, Hort N 2016 *Magnesium Technology 2016*, ed. Singh A, Solanki K, Manuel MV, Neelameggham NR (Cham: Springer International Publishing) p.177.
[11]. Huang Y, Kainer KU, Hort N 2011 *Scripta Materialia* **64** 793.
[12]. Huang YD, Zheng XH, Liu B, Anopuo O, Hort N, Kainer KU, Kim GS 2010 *Magnesium Technology 2010*, ed. Agnew SR, Neelameggham NR, Nyberg EA, Sillekens WH (Warrendale, Pa.: Minerals, Metals and Materials Society) p.419.
[13]. Huang YD, Zheng XH, Okechukwu A, Kim GS, Kainer KU, Hort N 2010 *Materials Science Forum* **vol. 654-656** p.675.

Optimization of ultrasonic cavitation processing in the liquid melt flow

T Subroto[1], D G Eskin[1,2,*], I Tzanakis[3,4], G S B. Lebon[1], A Miranda[1], K. Pericleous[5]

[1]Brunel Centre for Advanced Solidification Technology, Brunel University London, Kingston Lane, Uxbridge UB8 3PH, UK
[2]Tomsk State University, Tomsk, 634050, Russia
[3]Faculty of Technology, Design and Environment, Oxford Brookes University, Oxford OX33 1HX, UK
[4] Department of Materials, University of Oxford, Oxford OX1 3PH, United Kingdom
[5]Computational Science and Engineering Group (CSEG), Department of Mathematics, University of Greenwich, London SE10 9LS, United Kingdom

Email: dmitry.eskin@brunel.ac.uk

Abstract. Ultrasonic processing (USP) during direct-chill (DC) casting of light metal alloys is typically applied in the sump of a billet. This approach, though successful for structure refinement and modification, has two main drawbacks: (a) mixture of mechanisms that rely heavily on dendrite fragmentation and (b) a limited volume that can be processed by a single ultrasonic source. We suggest moving the location of USP from the sump to the launder and applying it to the melt flow for continuous treatment. The apparent benefits include: (a) degassing of the melt volume, (b) grain refinement through activation of non-metallic inclusions, fragmentation of primary crystals, and deagglomeration of grain refining substrates, and (c) a possibility to use a single ultrasonic source for processing large melt volumes. To optimize this process with regard to the acoustic intensity and melt residence time in the active cavitation zone, flow modification with baffles as well as informed location of the ultrasonic source are required. In this paper, we demonstrate the results of experimental trials where the degassing degree and grain refinement have been the indicators of the USP efficiency for two aluminium alloys, i.e. LM25 and AA7050. The results are supported by acoustic measurements and computer simulations.

1. Introduction

Treating liquid metal with ultrasound, i.e. ultrasonic melt processing (USP), produces beneficial results that include melt degassing and structure refinement of the as-cast alloy. It is an efficient, economical, and clean melt treatment technology. For example, instead of injecting an inert or chemically inactive gas such as argon to extract hydrogen from the melt, cavitation bubbles are generated during USP and leave the melt through buoyancy when they grow due to rectified diffusion of hydrogen, i.e., a process where hydrogen diffuses into the bubble during the rarefaction acoustic cycle [1]. In the case of structure refinement, the cost of common grain refinement agents such as TiB_2 for aluminium alloys [2] or zirconium for some Mg alloys [3] could be reduced or avoided by using USP. These benefits are highly attractive for industrial applications.

Due to its current limitation in treating large melt volumes, USP is mostly confined to laboratory or small-scale processing despite its potential advantages [4]. To circumvent this problem, multiple ultrasonic sources are utilized [1]. However, this solution reduces its economic feasibility thereby hampering its widespread industrial adoption. The imminent challenge for USP technology is to improve melt treatment efficiency at industrial scale using a single ultrasonic source.

Recent developments in USP modelling [5,6] enabled us to design a flow modification system in the launder (i.e. using various partition configurations and different sonotrode locations) that maximized the melt residence time in the cavitation active zone. In addition, such models also predicted that resonance conditions in a launder can increase the bubble concentration [7] and thus cavitation activity. Recent technological developments in cavitation intensity measurements [8] enabled us to quantify acoustic pressures and map the pressure field in the melt. These measurements allowed further numerical model development and supported the design of a smart flow modification system in the launder for optimized USP.

In this work, we verify the resonance phenomenon by varying the distance between partitions in a launder. Resonance is an attractive trait for improving USP efficiency on a single sonotrode configuration. The effect of sonotrode placement with respect to partitions is also studied. Acoustic pressures were measured in a water model. Metallurgical indicators such as degassing efficiency and grain refinement analysis were used to assess USP efficiency in the liquid metal experiments.

2. Materials and methods

2.1. Cavitation activity measurement

Acoustic pressures were measured in water with a calibrated cavitometer. A detailed description of cavitometer, data acquisition process and pressure conversion procedure has been previously described in [5, 9,10]. A 1-kW Hielscher piezoelectric ultrasonic transducer with a titanium cylindrical sonotrode (diameter 40 mm) was used in water experiments at a power of 70% (corresponding to 12 μm peak-to-peak ultrasonic tip amplitude displacement). The sonotrode tip was immersed 1 cm below the free surface.

A launder with partitions made from transparent plastic material was used to study the acoustic behaviour of water under USP. The length and width of such a launder were 100 cm and 9.4 cm, respectively. More details of this launder has been previously described in [7]. The gap between the partitions and the launder where water can flow through were 2 cm and 1 cm for upstream and downstream partition, respectively. The configuration of the partitions (flow passed below upstream partition and above downstream partition) was intended so there would be a recirculation of the liquid between the partitions. This ultimately maximizes pressures and cavitation activity to attain more inclusion activation and particle fragmentation by higher pressures and more efficient redistribution of the acoustic energy.

To observe acoustic resonance effect in water, acoustic measurements at three partition distances were carried out. The measurements were performed under the sonotrode. Two partition separation distances represent *n*-fold water wavelengths at ultrasonic frequency of 20 kHz—15 cm (2 λ_{water}) and 30 cm (4 λ_{water}), and one width (20 cm) represents an off-wavelength partition width. The effect of sonotrode placement with respect to the partitions was also studied. Three sonotrode locations were used: next to the upstream partition, next to the downstream partition, and equidistant between both partitions. Experiments were repeated with water flow (of 2 l/min) and without flow. The case corresponding to a partition separation of 15 cm is simulated using a numerical model that was validated with Particle Image Velocimetry (PIV) measurements in water [11].

2.2. Metallurgical indicator

The experiments with liquid aluminium alloys were performed in a launder made from refractory material. The length and width of such a launder were 90 cm and 10 cm, respectively. A liquid alloy was poured to the launder at a flow rate of 1 l/min. A thermocouple was placed in the vicinity of the launder exit, where samples were collected. USP was performed with a Reltec water-cooled magnetostrictive ultrasonic transducer (5 kW) operating at 17.5 kHz. A conical Nb sonotrode tip

IOP Conf. Series: Materials Science and Engineering **529** (2019) 012050 doi:10.1088/1757-899X/529/1/012050

(diameter 20 mm) was immersed 2 cm below the free surface of the melt. USP power of 3 kW corresponding to 33 μm peak-to-peak USP tip amplitude was used. The partitions were placed at 1 wavelength distance for liquid aluminium at 17.5 kHz, which is 26 cm. The experimental setup for liquid metal experiments is shown in Figure 1.

The degassing efficiency of USP process was assessed with an LM25 alloy (Al–7.5 wt% Si). This is because Al–Si alloys can have relatively high content of hydrogen in melt [12], thus may provide a clear indication of degassing efficiency. Samples were collected at different process conditions: without USP, with sonotrode at different positions while one reference sample was taken when the melt was still in the furnace. Density index (DI%) (the difference between the densities of the sample solidified under atmospheric pressure and the sample solidified under vacuum related to the density of the sample solidified under atmospheric pressure) that can be regarded as a measure of the end-cast product cleanliness was calculated from reduced pressure tests (RPT) which was performed using RPT (3VTMKGmbh) machine. At least three samples from independent experiments were obtained from each position and process conditions.

AA7050 alloy was used for assessing grain refinement efficiency of USP as it contains the necessary elements (Zr and Ti) that can trigger effective grain refinement under cavitation condition [1,13]. Three samples were collected as the melt left the launder. The samples were solidified in a TP1 mould. For this analysis, samples were obtained at different sonotrode positions and one sample obtained without USP as a reference. Standard metallographic sample preparation procedure and optical microscopy technique was utilized to measured grain size.

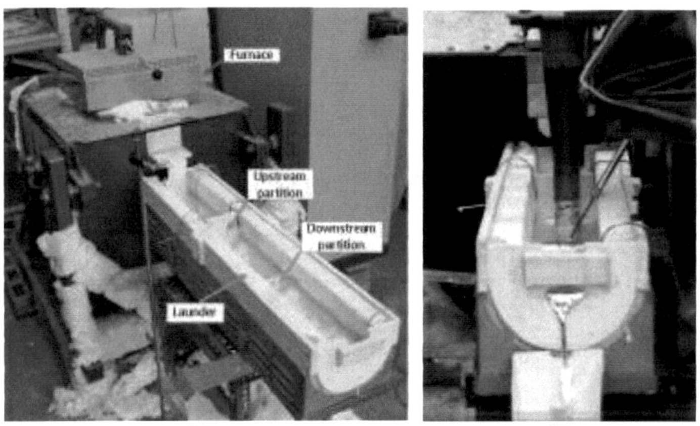

(a) (b)

Figure 1. Photos of experimental setup used for liquid metal experiments. (a) Position of furnace, launder and partitions (b) with ultrasonic sonotrode and liquid metal experiment running.

3. Results and discussion

3.1. Acoustic measurements

Figure 2a and 2b show the measured maximum and root mean square (RMS) pressures in water, respectively. The highest maximum pressure with flow was recorded when the sonotrode was positioned next to the downstream partition (1.47–1.49 MPa). The pressure at the centre position is 1.36–1.42 MPa and the lowest pressure was recorded with the sonotrode next to the upstream partition (1.24–1.38 MPa). Figure 2 also shows that the highest pressure was measured when the partition distance was $2\lambda_{water}$, and followed by $4\lambda_{water}$. The lowest pressure was obtained when the partition separation was at off-wavelength distance (20 cm). These results support the prediction that a resonance length promotes an increase in pressures [10]. This result also reveals that this resonance effect does not only occur at one liquid wavelength, but also at n-fold liquid wavelengths (where n is an integer number). According to Figure 2, the resonance effect decreases with increasing n, presumably due to attenuating pressures as the sound waves have to travel over a longer distance in a bubbly medium.

Figure 2 also shows that the presence of the flow leads to an increase in the measured acoustic pressure regardless of the position of sonotrode and the distances between partitions. In terms of maximum pressure (Figure 2a), the upstream position has the most significant pressure increase, followed by downstream position and the centre position has the least pressure increase. The trend between maximum pressure and RMS is similar on different sonotrode positions and partition distances. RMS value represents the average of maximum RMS acoustic pressure measured from 60 individual waveforms. Figure 2b shows that without flow, the RMS value is approximately less than one-fifth of the maximum pressure value, while measurements with flow produced RMS value close to 20% of the maximum pressure value. This is probably due to the decrease in the shielding effect under the sonotrode as bubbles are advected downstream. With less bubbles to surround the sonotrode surface, pressure waves can propagate more freely and over longer distances in the liquid medium with less attenuation.

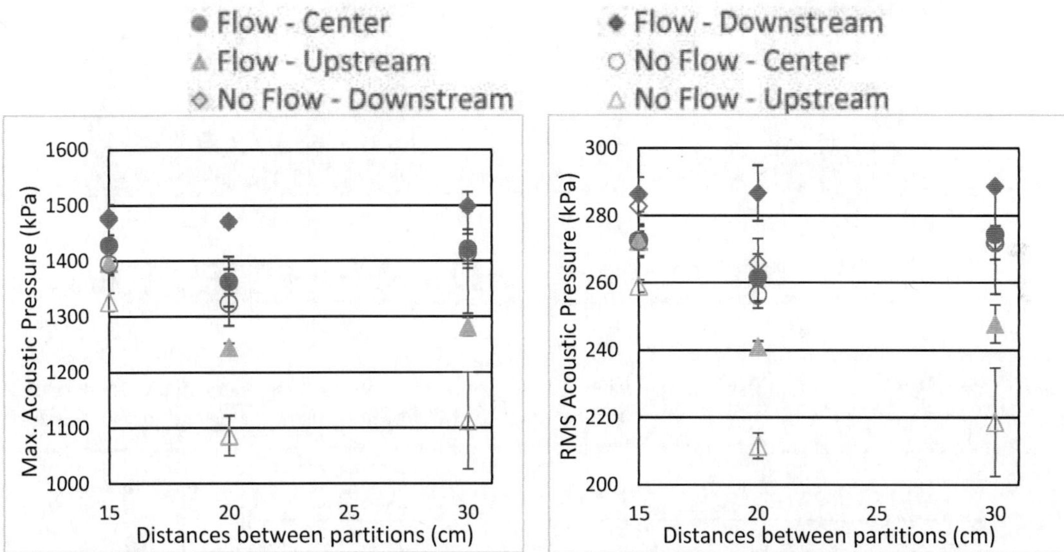

(a) (b)

Figure 2. Effect of flow in water experiments, sonotrode positions with respect to partitions (upstream, centre and downstream), and distances between partitions – comparison between n-fold λ_{water} and non-λ_{water} partition distances. (a) Maximum pressure measurements and (b) RMS measurements.

3.2. Numerical results
Figure 3 shows the predicted acoustic pressure in a 3D representation of the water launder when the sonotrode is placed centrally between the two partitions. The results show that the highest pressures are recorded downstream, in qualitative agreement with the results from the previous section. Note that these results are highly sensitive to the assumed bubble density in water, which is an unknown parameter [11]. Numerical discrepancies can be explained by the lack of knowledge of this crucial factor: an empirical validation of this crucial parameter is currently being studied.

3.3. Metallurgical indicator
Figure 4 shows the effectiveness of degassing using USP assessed using density index (DI%) parameter. Higher value represents less degassing efficiency and lower value otherwise. The positive effect of USP for degassing can be clearly seen as samples taken without USP (launder) has highest average DI index. Among three different sonotrode positions, the best degassing result is obtained when the sonotrode was positioned next to the downstream partition that is in agreement with the experimental measurements of acoustic pressure in Figure 2. The DI is higher when the sample obtained in the launder without UST compared with the one in the furnace, which is logical as there is less water vapour in the furnace, thus less hydrogen pickup by the liquid metal.

Figure 5 shows the grain refinement efficiency of USP. USP clearly promoted grain refinement with samples after USP having smaller grain size. Figure 5a suggests that the best grain refinement was achieved when the sonotrode was positioned either in the centre between partitions or nearby downstream partition, which an average grain size reduction of up to 50% as compared to upstream position or without USP. The effect of melt temperature is demonstrated in Figure 5 on two sets of data taken at different temperatures. It is clear that USP treatment temperature has a critical influence on the structure refinement outcome, as reported by previous researches [1,14]. If intermetallic fragmentation is the main mechanism for grain refinement in this type of USP, treatment at lower temperature would yield better structure refinement due to the larger presence of intermetallics and due to a higher cavitation activity.

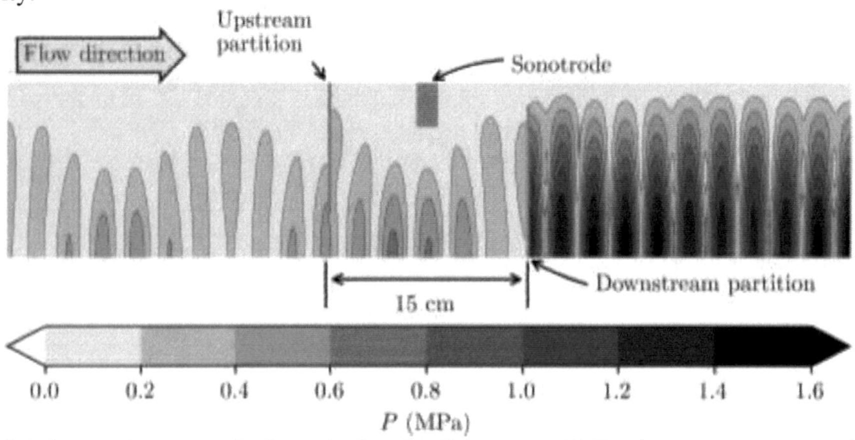

Figure 3. Predicted acoustic pressure in the water launder for a case with flowing water, representative of the experiment with distance between the partitions equal to two wavelengths of water and the sonotrode placed at the center. The aspect ratio is altered by a factor of 5.0 for clarity.

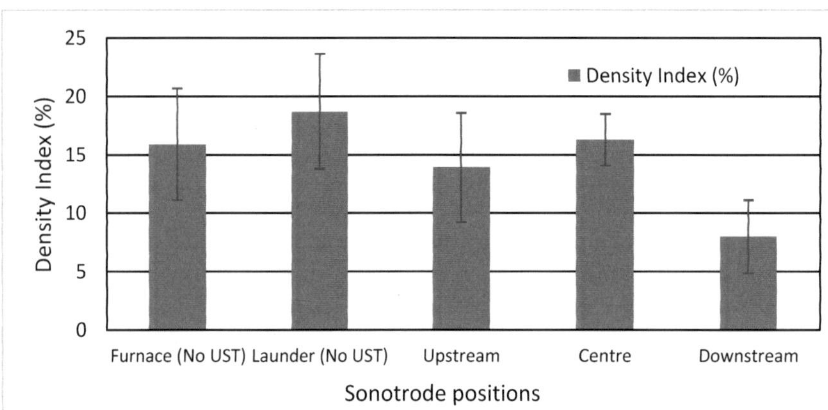

Figure 4. Density index (DI %) obtained through reduced pressure test for samples taken at different locations and processed with USP at different sonotrode positions.

From both acoustic measurements in water and liquid metals experiments, we found a coherent response. The better USP efficiency was achieved by placing the sonotrode nearby downstream partition. This corresponds to the highest acoustic pressure built up under resonance, and along with the flow pattern can increase re-circulation and residence time [14]. The second effect is the propagation of acoustic waves and cavitation activity beyond the downstream partition, which prolongs USP of the incoming melt to the downstream part of the launder. These mechanisms can ultimately improve the USP efficiency.

4. Summary

The resonance effect was observed in water experiments. Larger acoustic pressures were measured when the partition distance corresponds to integer wavelengths and lower pressures were measured when the distance is not a wavelength integer. The resonance effect does not only occur at one liquid wavelength but also at n-fold liquid wavelengths (higher-order resonance), though with attenuation.

The optimum location of the sonotrode is next to the downstream partition. This is well supported by various indicators, both from water and liquid metal experiments. The findings from this work clearly demonstrate that flow modification through variations of partition distances and placements of the sonotrode could provide improvement in USP efficiency, which could be worth to explore further for bringing the USP technology to industry.

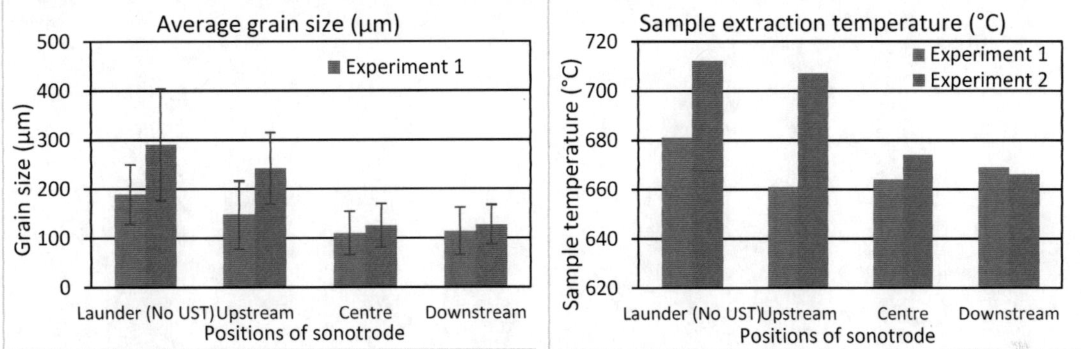

Figure 5. Structure refinement efficiency analysis from two independent experiments for samples taken at different process conditions. (a) Grain size analysis and (b) temperature where samples were obtained.

5. Acknowledgments
Financial supports from EPSRC UK through grants UltraMelt 2 (EP/R011001/1, EP/R011044/1, EP/R011095/1) and LiME Hub (EP/N007638/1) are gratefully acknowledged.

References
[1] Eskin G I and Eskin D G 2017 *Ultrasonic treatment of light alloy melts, second edition.* (CRC Press)

[2] Wang X and Han Q 2016 Grain Refinement Mechanism of Aluminum by Al-Ti-B Master Alloys *Light Metals 2016* ed E Williams (Cham: Springer International Publishing) pp 189–93

[3] Qian M and Das A 2006 Grain refinement of magnesium alloys by zirconium: Formation of equiaxed grains *Scr. Mater.* **54** 881–6

[4] Eskin D G 2017 Ultrasonic processing of molten and solidifying aluminium alloys: overview and outlook *Mater. Sci. Technol.* **33** 636–45

[5] Lebon G S B, Tzanakis I, Pericleous K and Eskin D 2018 Experimental and numerical investigation of acoustic pressures in different liquids *Ultrason. Sonochem.* **42** 411–21

[6] Lebon G S B, Tzanakis I, Djambazov G, Pericleous K and Eskin D G 2017 Numerical modelling of ultrasonic waves in a bubbly Newtonian liquid using a high-order acoustic cavitation model *Ultrason. Sonochem.* **37** 660–8

[7] Lebon G S B, Pericleous K, Tzanakis I and Eskin D 2016 A model of cavitation for the treatment of a moving liquid metal volume *Int. J. Cast Met. Res.* **29** 324–30

[8] Tzanakis I, Hodnett M, Lebon G S B, Dezhkunov N and Eskin D G 2016 Calibration and performance assessment of an innovative high-temperature cavitometer *Sens. Actuators Phys.* **240** 57–69

[9] Hurrell A M and Rajagopal S 2017 The Practicalities of Obtaining and Using Hydrophone Calibration Data to Derive Pressure Waveforms *IEEE Trans. Ultrason. Ferroelectr. Freq. Control* **64** 126–40

[10] Tzanakis I, Lebon G S B, Subroto T, Eskin D and Pericleous K 2019 Acoustic cavitation measurements and modeling in liquid aluminum *accepted for Light Metals2019*

[11] Lebon G S B, Tzanakis I, Pericleous K, Eskin D and Grant P S 2019 Ultrasonic liquid metal processing: The essential role of cavitation bubbles in controlling acoustic streaming *Ultrason. Sonochem.* doi.org/10.1016/j.ultsonch.2019.01.021

[12] Alba-Baena N and Eskin D 2016 Kinetics of Ultrasonic Degassing of Aluminum Alloys *Light Metals 2013* ed B A Sadler (Cham: Springer International Publishing) pp 957–62

[13] Atamanenko T V, Eskin D G, Sluiter M and Katgerman L 2011 On the mechanism of grain refinement in Al–Zr–Ti alloys *J. Alloys Compd.* **509** 57–60

[14] Zhang L 2013 *Ultrasonic processing of aluminium alloys* PhD Thesis (Delft University of Technology)

IOP Conf. Series: Materials Science and Engineering **529** (2019) 012051 doi:10.1088/1757-899X/529/1/012051

Online Modelling of Heat Transfer, Solidification and Microstructure in Continuous Casting of Steel

S Louhenkilpi[1,2], J Miettinen[1], J Laine[†], R Vesanen[3], I Rentola[4], J Moilanen[5], V-V Visuri[1], E-P Heikkinen[1] and A Jokilaakso[2]

[1] Process Metallurgy Research Unit, University of Oulu, P.O. Box 4300, 90014 University of Oulu, Finland
[2] Department of Chemical and Metallurgical Engineering, Aalto University, P.O. Box 16200, 00076 Aalto, Finland
[3] Casim Consulting Oy, Sepetlahdentie 12 F 54, 02230 Espoo, Finland
[4] SSAB Europe Oy, Rautaruukintie 155, 92101 Raahe, Finland
[5] Outokumpu Stainless Oy, 95490 Tornio, Finland
[†] In memoriam Jukka Laine

Email: seppo.louhenkilpi@oulu.fi

Abstract. Advanced numerical simulation models for continuous casting of steel were developed in Finland by Casim Consulting and Aalto University in cooperation with the University of Oulu and the steel industry. The aim was to develop models that are scientifically rigorous, but also computationally fast enough to be used in online applications. The models developed are a transient three-dimensional heat transfer model, CastManager, and a solidification and microstructure model, IDS. The computing time of these models are short, and they are integrated together in one online concept. This concept is installed in the automation systems of four slab casters in Finland. Testing and validation work is in progress. The system simulates the important heat transfer, solidification and microstructural phenomena in continuous casting online. The future aim is that this information will be used for online quality control and for optimizing the process conditions to avoid formation of defects. Many quality indices have already been developed. A steady state version of the CastManager tool has also been developed, called Tempsimu.

1. Introduction

Computational simulation and modelling of different phenomena in casting have greatly helped to solve practical problems in industrial casters and to improve process practices and control. Altogether, a deeper understanding of the complex solidification phenomena and transformations of the microstructure is still needed to respond to increasing requirements. For this purpose, advanced numerical simulation models for continuous steel casting were developed in Finland by Casim Consulting and Aalto University in cooperation with the University of Oulu and the steel industry. The aim was also to develop models that are scientifically ambitious but also to make these models computationally fast enough to be used in online applications of process control and quality prediction. In online applications, fast computing is of special importance. During the years, many kinds of models for continuous casting have been introduced in international journals, such as models for heat transfer and microstructural phenomena. The published microstructural models are often based on concepts, which are so time consuming that they cannot be applied in online applications. Typical concepts of this kind are cellular automata and phase field methods. Fast computing is also important for online heat transfer models because it allows a more detailed determination of the boundary conditions, leading, thus, to more accurate and realistic results.

The models developed and presented in this paper are a transient three-dimensional heat transfer model, CastManager [1], and a solidification and microstructure model, IDS [1–3]. The computing time of these models are short, and they are now integrated together in one online concept. The system simulates the important phenomena in continuous casting online. The future aim is to use this information for online quality control and for optimizing the process conditions to avoid defect formations. Many quality indices have already been developed. A steady state version of the CastManager tool has also been developed, called Tempsimu. This paper presents the developed models and the integrated on-line concept with some case examples.

2. CastManager and Tempsimu heat transfer models

CastManager and Tempsimu are heat transfer modelling software for continuous steel casting. They calculate strand temperatures, mold temperatures and temperature-related data (such as the shell thickness and total mold heat flux) as a function of the following main data:

- secondary cooling components: rolls, nozzles and radiation zones,
- steel grade (thermophysical material properties),
- strand geometry, mold geometry, mold material properties including the possibility to have different material layers (e.g. coating layers),
- casting variables: casting speed, superheat, spray cooling water flow rates.

CastManager is a software for transient online applications, while Tempsimu is a software for steady-state modelling. Tempsimu can be applied for instance to analyze the existing secondary cooling system and to study new alternatives (spray and roll configurations, cooling water flow rates). The programs consist of two separate modules: a three-dimensional mold model and a three-dimensional strand model. They are run iteratively, and the so-called gap heat transfer coefficient, which is a function of the strand temperature, is used to link them. In the calculations, the Leidenfrost effect and the effect of melt convection (fluid flow) are taken into account. Material data is calculated with the InterDendritic Solidification (IDS) package, which directly gives the data file needed. In many heat transfer models, a fixed grid is used. Our tools also use the fixed grid approach. To take the shrinkage and the mass balance into account, CastManager and Tempsimu use constant density. It is important to take care that the mass inlet is equal to the mass outlet in the casting model. To do this, we define the grid horizontal dimensions to be the dimensions of the mold at the meniscus level and, for the density, we give its value at the solidus temperature. With the latter treatment, we take into account the solidification shrinkage taking place during solidification.

The fluid flow, heat transfer, and solidification processes in continuous casting are interlinked and therefore, for a complete heat transfer model, a fully linked approach is needed. These kinds of linked models have been published in the literature and typically they have been made using a commercial CFD tool. Their use is very time consuming and needs a professional user. In addition, the calculation times are very long, and the models cannot be used online. A simplified approach uses the so called effective thermal conductive concept, in which the effective thermal conductivity takes into account the effects of fluid turbulence and convection on heat transfer. It is widely agreed that this approach gives reasonable results. An additional benefit is that the models based on this concept are very fast. CastManager and Tempsimu use the effective thermal conductive concept.

In CastManager, heat transfer and solidification in the strand are defined by equation (1), which can be described according to equation (2) when the enthalpy and Kirchhoff's transformations are applied; they are described by equations (3) and (4), respectively.

$$\rho c \frac{\partial T}{\partial t} + \frac{\partial v \rho c T}{\partial z} = \frac{\partial \left(k_{\text{eff}} \frac{\partial T}{\partial x}\right)}{\partial x} + \frac{\partial \left(k_{\text{eff}} \frac{\partial T}{\partial y}\right)}{\partial y} + \frac{\partial \left(k_{\text{eff}} \frac{\partial T}{\partial z}\right)}{\partial z} + Q, \tag{1}$$

$$\frac{\partial H}{\partial t} + v \frac{\partial H}{\partial z} = \Delta K(T), \tag{2}$$

$$H(T) = \rho \int_0^T c(\xi)\, \mathrm{d}\xi + \rho L\big(1 - f_{\mathrm{s}}(T)\big), \tag{3}$$

$$K(T) = \int_0^T k_{\text{eff}}(\xi)\, \mathrm{d}\xi, \tag{4}$$

where $H(T)$ is the enthalpy function, $K(T)$ is the Kirchhoff's function, t is time, k_{eff} is the effective thermal conductivity, c is the specific heat capacity, L is latent heat, f_s is the solid fraction, T (and ξ) is the temperature, ρ is the density and v is the actual casting speed in the casting direction. The enthalpy function H includes the effect of all phase transformations. Equation (5) describes the mold model where all variables are the same as in the strand model except that there are no phase transformations, so $L = 0$. In a numerical solution, the strand and mold models are iterated together so far that convergence is reached. In Tempsimu, the transient terms are not needed.

$$\frac{\partial H(T)}{\partial t} = \Delta K(T). \tag{5}$$

The boundary condition between the strand and the mold in the strand model is defined as:

$$-\nabla K(T) \cdot \vec{n} = h_{\text{gap}}(T - T_{\text{ext}}), \tag{6}$$

where $\vec{n}$ is the unit vector of outward normal to boundary of the strand, h_{gap} is the gap heat transfer coefficient, T is the strand surface temperature and T_{ext} is the external temperature, in this case the temperature of the mold surface. The gap heat transfer coefficient is a function of the strand surface temperature, but it also depends on the steel grade and the type of the mold powder. Equation (6) is also used as the boundary condition for the hotter side of the mold model, where h_{gap} is the same gap heat transfer coefficient as for the strand model, but T_{ext} is now the temperature of the strand surface, which is calculated by the strand model. For the colder side, a similar equation is used, but T_{ext} is the temperature of the cooling water and h is a constant. In the secondary cooling zone, the strand model has been divided into calculation domains between a pair of support rolls. This domain has been divided into four different cooling areas that are: the roller contact area, pre-nozzle area, spraying area, after spray and pool water area (post-nozzle area). The boundary equation for these areas is defined as:

$$-\Delta K(T) = h(T - T_{\text{ext}}) + \varepsilon\sigma\big(T^4 - T_{\text{air}}^4\big), \tag{7}$$

where h is heat transfer coefficient, T_{ext} is external temperature, T_{air} is the external temperature for the radiation term, ε is the emissivity value of strand surface and σ is the Stefan–Bolzmann constant. Heat transfer coefficients are defined with specific data and equations and these depend on the type of the roll or spray nozzle, amount of spray water, surface temperature. Accurate boundary conditions form the basis of good predictive capability. Laboratory trials, plant measurements and published data can be used for definitions. The model equations are discretized using fully implicit three-dimensional finite difference (FDM) and upwind schemes. An implicit Crank–Nicholson method is used for time discretization. The resulting algebraic equations are solved using a parallel Gauss–Seidel–Newton–Raphson method.

3. IDS tool

IDS is a thermodynamic-kinetic-empirical software package, which simulates solidification and cooling related microstructural phenomena, such as solute micro segregation, phase transformations, inclusion and precipitate formation, grain growth and austenite decomposition. IDS can be applied from the liquid state to the as-cast room temperature state and during the reheating treatments of a reheating furnace. The cooling rate can vary during cooling and reheating. The calculation times are very short making the tool suitable for online applications.

The present elements available in IDS are C, Si, Mn, P, S, Cr, Ni, Mo, Al, Cu, N, Nb, Ti, V, Ca, B, O, Ce, Mg and H. IDS consists of several modules and not all elements are included in all modules. As input, IDS requires the nominal steel composition and cooling-heating rates. IDS includes default equations for calculating the dendrite arm spacing and the austenite grain size, as a function of the composition and cooling rate, but the user can use own values too. The simulation of austenite decomposition is based on empirical Continuous Cooling Transformation (CCT) diagrams. The phases considered by IDS are δ-ferrite, α-ferrite, eutectic ferrite, austenite, cementite, pearlite, bainite, α-martensite (bct structure) and ε-martensite (hcp structure).

IDS applies thermodynamic and kinetic theories assuming thermodynamic equilibrium at the phase interfaces. During the austenite to proeutectoid ferrite or cementite transformation, the thermodynamic para-equilibrium condition is assumed, indicating diffusion only for the interstitial elements (e.g. B, C and N). In the austenite decomposition module, phase transformations are mainly based on regression equations optimized from empirical CCT diagrams. The considered elements of this module are C, Mn, Si, Cr, Ni, Mo and B, though Al and Cu will soon be added to the simulation. The user can define whether the nominal composition or the grain boundary dissolved composition is used in the calculations.

The formation and growth of various inclusions from the liquid is based on thermodynamic equilibrium calculations. For precipitates, forming below the solidus temperature, the calculation concept is more complicated, because kinetic phenomena and additional energies, like that of misfit must also be taken into account. For the austenite grain growth, an empirical formula is used. In this formula, the grain growth is reduced by high cooling or heating rates (lack of time), low temperature, and formation of ferrite and precipitates.

IDS also includes a module for the calculation of material properties, such as enthalpy, thermal conductivity and density, as a function of steel composition and cooling rate. More information about the IDS tool, including its databases and additional modules, are available in Refs. [1-3].

4. Integrated system IDS and CastManager/Tempsimu

The IDS tool has been coupled with CastManager for online applications and with Tempsimu to study offline steady state situations. The IDS and CastManager integrated system has been implemented for four slab casters in Finland. Testing and validation work is now in progress. The start and the end of casting, as well as the ladle change, the steel grade change and the width change are also included in the system. The system simulates heat transfer, solidification and microstructural phenomena in continuous casting and the aim is to use it for quality prediction and for optimizing the casting conditions. First, IDS calculates the material data of the steel for CastManager, using the online measured steel composition of the ladle. Then, during the casting process, CastManager gives temperatures and cooling rates to IDS, which calculates the microstructural phenomena and the quality (Figure 1). For describing the strand quality, several quality indices have already been developed. Their testing and validation is in progress. One way to speed up the system calculations-is to use a graphics processing unit (GPU) instead of a central processing unit (CPU). This work is also in progress. Similar work has been carried out for the developed reheating system. Correspondingly, this system is called the IDS and FurnaceManager integrated system. It has been installed for one reheating furnace in Finland. Using a GPU for that system was found to accelerate simulations by a factor of 5–13 compared to simulations using the CPU.

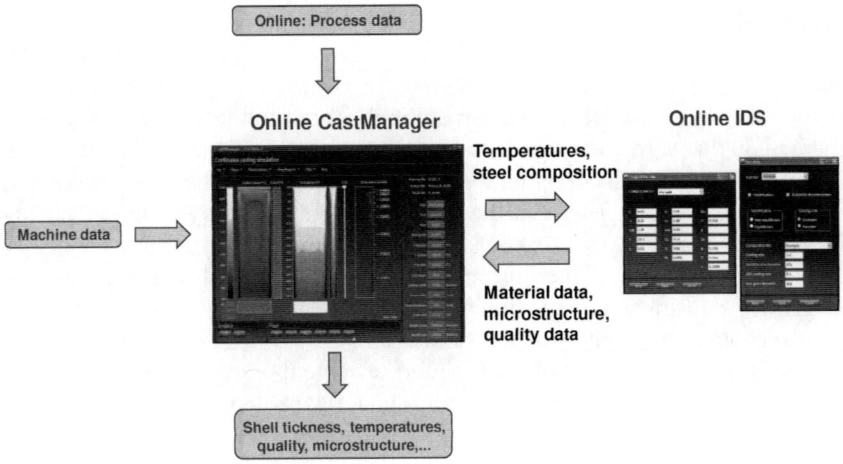

Figure 1. IDS and CastManager integrated online system for continuous casting.

5. Case examples: Quality criteria and grain growth

IDS simulates many microstructural phenomena as a function of the steel composition and cooling/heating rates. From this information, combined with the results of CastManager, several quality indices can be derived. Figure 2 shows two quality criteria (QI_{she} and QI_{str}) as function of carbon content.

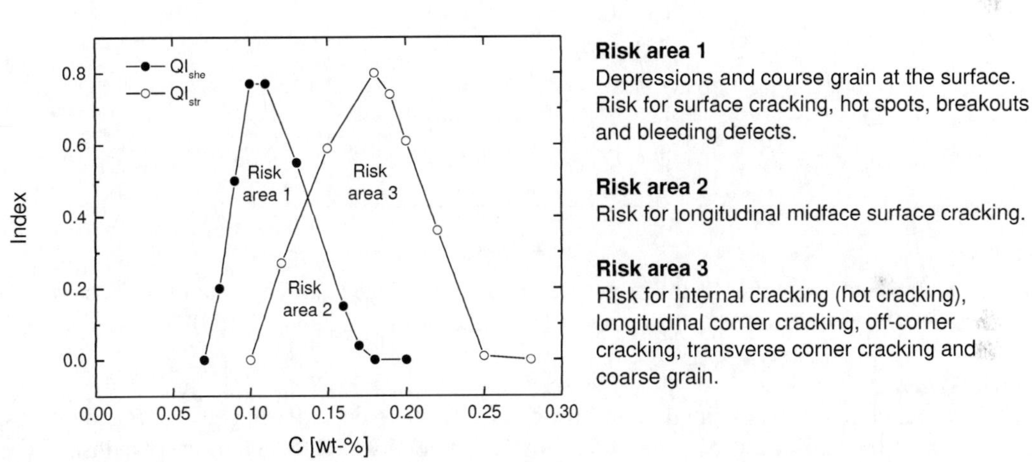

Figure 2. IDS quality criterion values of QI_{she} and QI_{str} for a certain steel composition with increasing carbon content. For stainless steels, these criteria overlap more. High index means high risk for defects.

The QI_{she} is the fraction of delta ferrite at the solidus, minus the fraction at T_{sol}–30 °C. It is known that ferrite to austenite transformation starting just below the solidus causes many problems, because of the shrinkage in this phase transformation. High values of QI_{she} indicate high risks of several types of surface defects, such as hot spots, surface cracking, bleeding, breakout, etc. Typically, problems occur between 0.08 and 0.15 wt-% C in carbon steels, but the alloying and tramp elements tend to change the carbon range. For stainless steels, the highest risk is when the Cr/Ni-equivalent ratio is about two. QI_{str} is the fraction of delta ferrite at T_{ZST} (zero strength temperature, related to the solid fraction of 0.8) minus that fraction at the solidus. If austenite starts to form (from liquid and ferrite) between solid fractions of about 0.8 and 1, the shrinkage of this phase transformation can cause stresses above the solidus, increasing the risk for surface and internal cracks.

The QI_{sol} criterion describes the reduced hot ductility (around the solidus) because of the strong enrichment of S, P and B in the interdendritic liquid, increasing the risk for internal hot cracking. The

QI_{duc} criterion describes the effects of precipitates in reducing the ductility in the low ductility areas of II and III due to the grain boundary sliding (area 2) or the deformation induced proeutectoid ferrite/cementite present at the grain boundaries (area 3). These all increase the risk of transverse corner cracks. Particularly, if QI_{str} and QI_{duc} both have a high value, the risk of transverse corner cracking becomes very high. In that case, the secondary cooling needs to be controlled so that during the bending and unbending processes, the slab corner temperatures remain outside the temperature region of the low ductility area. It is also possible to fine-tune the steel composition, already in the ladle, to pass the highest criteria peaks and thus, to reduce the risk of defects. Many other quality criteria are also under work.

Figure 3 shows an example of the austenite grain growth in carbon steel with different carbon contents. The growth is restrained by the grain boundary ferrite and the Nb(C,N) precipitation-Carbon has clear effects on both of these phenomena. Note that in the 0.05 wt-% C steel, the grain growth starts over 50 °C below the solidus, because of the high ferrite content present. Its final grain size however, is highest, due to the low amount of Nb(C,N), which is not capable of restraining the grain growth so effectively.

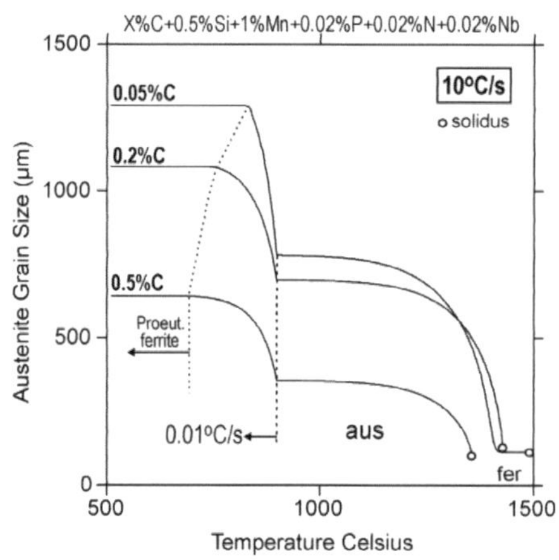

Figure 3. Calculated austenite grain growth in three low-alloyed steels containing 0.05 wt-% C, 0.2 wt-% C and 0.5 wt-% C, cooled with a rate 10 °C/s above 900 °C and with 0.01 °C/s below 900 °C.

6. Conclusions

Numerical simulation models for continuous casting of steel have been developed in Finland by Casim Consulting and Aalto University in cooperation with the University of Oulu and the steel industry. These models are intended to provide a detailed description of the main phenomena, while being sufficiently fast for online use. The main tools developed are the CastManager and Tempsimu heat transfer models and the IDS package, which can be coupled for a given task or online concept. The potential of the model was illustrated with predictions for quality criteria and grain growth.

Acknowledgements

This study was executed within the framework of the Genome of Steel profiling project and under the SIMP and FLEX research programs coordinated by DIMECC. The Academy of Finland (project 311934), Business Finland and the Finnish Cultural Foundation are acknowledged for funding this work.

References

[1] Louhenkilpi S, Laine J, Miettinen J and Vesanen R 2013 *Mater. Sci. Forum* **762** 691
[2] Miettinen J, Louhenkilpi S, Kytönen H and Laine J 2010 *Math. Comput. Simul.* **80** 1536
[3] Miettinen J, Louhenkilpi S, Visuri V-V and Fabritius T 2019 *Proc. 5th Int. Conf. on Advances in Solidification Processes and 5th Int. Symp. on Cutting Edge of Computer Simulation of Solidification, Casting and Refining*

IOP Conf. Series: Materials Science and Engineering **529** (2019) 012052 doi:10.1088/1757-899X/529/1/012052

Characterization of heat transfer and its effect on solidification in water cooled LPDC of wheels

AU Saadah, C Reilly, DM Maijer, and SL Cockcroft

Department of Materials Engineering, The University of British Columbia, 6350 Stores Road, Vancouver, BC, Canada V6T 1Z4

Abstract. Computational process modelling has become an important engineering tool in the casting industry to predict the solidification sequence in complex castings. Used properly, this tool can help reduce manufacturing costs. One of the challenging issues in developing casting simulations of the low pressure die casting (LPDC) process for automotive wheels is to quantify the heat transfer coefficients (HTC) within the cooling channels in a die. When water is used as the cooling media, the HTCs exhibit a complex, non-linear behaviour due to the boiling phenomena that occur making it possible to extract a significant amount of heat from the die in a short period of time and influence the solidification of a wheel. Primarily, constant heat transfer coefficients have been used to describe this heat transfer in casting models up until now, but an opportunity exists to improve the transient description of heat transfer in channels cooled with water. In this paper, HTC's in a lab-scale physical analogue model of die cooling will be characterized as a function of initial die temperatures.

1. Introduction

Wheels are one automotive component where weight reduction may be achieved by selecting lightweight materials over the traditionally used steel. In addition to their structural performance, styling and aesthetic appearance are important characteristics of automotive wheels. The most common process used to manufacture aluminium alloy wheels is the low pressure die casting (LPDC) process. LPDC is a permanent-mould, die casting technique used industrially to produce rotationally symmetric parts such as wheels where cosmetic considerations are important [1]. However, castings produced by this technique are often affected by defects that can compromise structural performance under cyclic loading conditions. These include microporosity, hot tearing, entrained oxide films [2] and cold shuts [3]. The formation of these types of casting defects is closely linked to the casting process. Careful control of solidification can be used to eliminate porosity and to achieve a finer cast microstructure which translates to improved performance for cast aluminium wheels. Controlled solidification may be achieved through superior die design and tuned cooling (cooling channel geometry, cooling media (water, air mist), cooling duration, and cooling media flow rate). Accurate mathematical models of the LPDC process are necessary to allow engineers to develop effective strategies to achieve controlled solidifications and to enhance aluminium wheel quality.

Casting simulations of components produced via the low pressure die casting (LPDC) process are typically based on the Finite Element or Finite Volume Methods. The information required to develop these models includes geometry, material properties, initial conditions and boundary conditions. One of the challenging issues of developing casting simulations is to accurately quantify the boundary conditions involved which are often dependent on time, temperature, location, surface treatment and cooling media [4]. In particular, the HTCs within the cooling channels in the die are one of the most critical and sensitive model parameters. In the LPDC process, temperatures can reach ~500°C at some locations in the die during steady-state casting conditions. When a fluid, such as water, contacts a surface with a wall temperature, T_w, above the liquid saturation temperature, T_{sat}, at a specified pressure, a liquid-to-vapor phase change (e.g. boiling) will occur. In a die cooled with water flowing through enclosed channels, internal flow boiling conditions are present where the combined effects of water flow, channel orientation and boiling influence heat transfer. Once the channel surface has cooled down below the saturation temperature forced convection conditions will prevail and will only occur if the cooling cycle is long enough. Forced convective heat transfer is a complex transport process involving the combined effects of heat conduction through a fluid and heat transport due to movement of the fluid. This work aims to characterize heat transfer occurring in a physical analogue of a cooling channel and thereby improve the description of HTCs in cooling channels. The expectation is that this work will enable significant improvement in casting process model accuracy.

2. Experimental Methodology

An experimental apparatus has been designed, fabricated and commissioned to mimic the heat transfer conditions experienced in an LPDC process. The cooling block experimental apparatus consists of an H13-tool steel block (83 x 83 x 305 mm) with a round cooling channel (25 mm dia.) machined into the center of the block. The block is heated by 8 channel heater strips (50 kW/m^2 each) that are bolted to the outside of the block (2 on each side). The block is instrumented with 16 type-K thermocouples and the incoming water flow rate is measured. A schematic of the cooling block setup is shown in Figure 1. A portable data acquisition system using National Instruments CompactRIO hardware and the LabView DAQ software is used to monitor and record temperatures during each experiment. This experimental setup allows the effects of parameters such as fluid type (air and water), flow rate, and initial block temperature to be assessed.

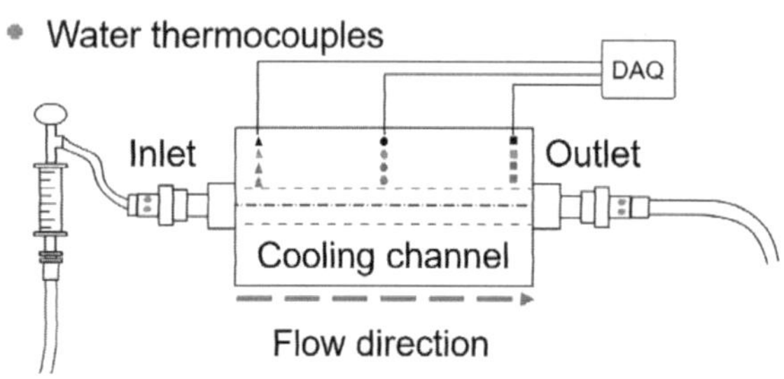

Depth (mm)	Axial Location: Inlet	Mid-length	Outlet
3.2	▲	●	■
9.5	▲	●	■
15.9	▲	●	■
22.2	▲	●	■

Figure 1. Schematic of Cooling Block Experimental Setup. Symbols represent location of thermocouples.

During each experiment, the block is heated to a desired starting temperature and held until a steady state temperature condition is reached. The data acquisition system, which records data from the thermocouples at 20 Hz, is used to verify that the steady-state condition is reached before introducing coolant to the block. The criterion used to assess whether steady-state has been reached is that the block temperature does not change more than ±2°C / min. There are 12 thermocouples embedded in the block at three different axial positions (near the Inlet end, Mid-length and Outlet end of the block). At each axial position, thermocouples were located 3.2, 9.5, 15.9 and 22.2 mm away from the cooling channel interface. The locations of the thermocouples in the cooling block are shown in Figure 1. The temperature of the coolant at the inlet and outlet to the block was recorded by 4 thermocouples (in the fluid near the wall and at the centre of the fitting at each location).

3. Mathematical model

A transient heat conduction model of the cooling block was developed based on the cooling block experiments using the general-purpose finite element modelling software, ABAQUS. The domain of the model was reduced to a 1/8th section of the block by assuming symmetric heat flow conditions exist during the experiment. To approximate the heat transfer occurring at the boundaries of the cooling block, a common form of the heat transfer coefficient boundary condition was used. The resulting heat flux applied to each surface can be calculated as:

$$q'' = h\,(T_w - T_\infty) \qquad (1)$$

where q'' is the heat flux(W/m^2), h is the heat transfer coefficient (W / (m^2·K)), T_w is the surface temperature of the block (°C), and T_∞ is the temperature of the environment adjacent to the block surface (°C). The h and T_∞ were manipulated to represent the different conditions on the block surfaces. User-written subroutines were used to describe the effects of the heaters and cooling (temperature dependent) from the central channel. Uniform HTCs and a constant ambient temperature were used to describe heat loss to the surrounding environment. A trial and error approach was used to determine appropriate values for the boundary conditions and the end results were verified by comparing the predicted and measured temperatures.

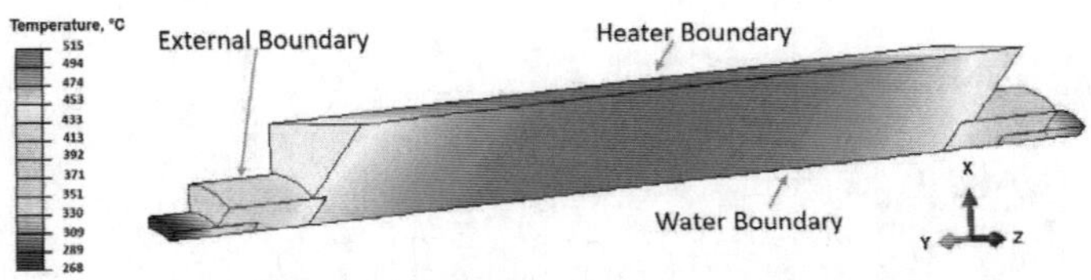

Figure 2. Cooling Block Model 1/8th domain

The initial condition (e.g. t = 0 s) for the cooling block model, shown in Figure 2 as the block temperature distribution, is based on data obtained from the cooling block experiments. Prior to a test, the cooling block is heated until a steady state temperature is reached. At the time when the coolant first reaches the block, the temperature at 3 points within the block closest to heater boundary layer (22.2 mm from the cooling interface at the Inlet, Mid-length, Outlet locations) and one point (3.2 mm from the cooling interface at the Mid-length location) furthest from the heated surface are used to generate the initial condition.

4. Result and Discussion

4.1. Cooling Block Experimental Result

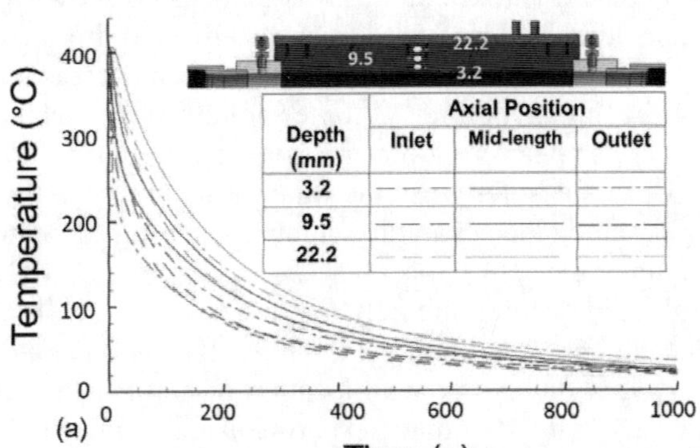

Figure 3. Measured temperatures in the cooling block for an initial block temperature of 400°C and a flow rate of 20 l/min.

Experiments were performed with initial block target temperatures of 300°C, 400°C and 500°C. A constant water flow rate of 20 l/min was used for these experiments. The experimental results consist of measured temperatures from the H13 block at different depths from the block/water interface and measured water temperatures at the inlet and outlet of the block. The measured temperatures in the block for the 400°C initial block temperature condition is shown in Figure 3.

During a cooling experiment, the locations nearest to the cooling channel (3.2 mm depth at the Inlet, Mid-length and Outlet) exhibit the first and most rapid temperature decrease followed sequentially by the other thermocouple locations increasing in distance from the H13/ water interface (9.5 and 22.3 mm). At each time during the experiment, the Mid-length position thermocouples at all depths have the highest temperatures while the Inlet positions have the lowest temperatures. This has been attributed to end effects caused by exposure of the ends of the block to the surrounding environment and the fittings connecting the coolant supply to the block (contact cooling), resulting in lower temperatures relative to the Mid-length block temperature. The Outlet positions of the block exhibits higher temperatures than the Inlet positions because the water at the outlet has been heated as it flows through the block.

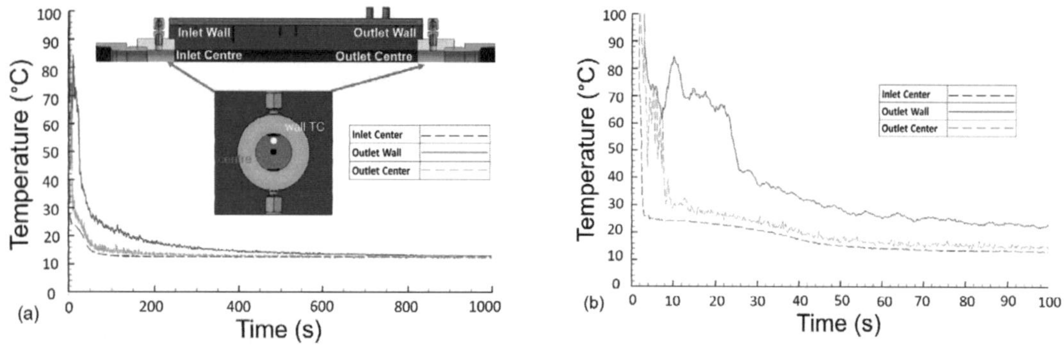

Figure 4 – Measured temperatures in the (a) water for 1000 s and (b) the magnification of (a) for 100s for an initial block temperature of 400°C and a flow rate of 20 l/min.

The measured water temperatures, shown in Figure 4(a), are from the locations within fittings attached at the inlet and outlet sides of the block. The temperatures are measured near the wall (~3 mm) and at the centre of the inlet and outlet fittings. Due to abnormalities in the measured temperature data from the Inlet Wall location, the data from this thermocouple was omitted from Figure 4(a). Previous experimental data has shown that the temperatures at the Inlet Wall and Inlet Centre locations are almost identical. A magnified view of the measured water temperature data is shown in Figure 4(b). At the Inlet Centre, the initial water temperature was measured to be ~100°C and this rapidly decreased to 26°C in the first 3 s during the experiment. The inlet water temperature then decreases to ~12°C. These changes are most likely caused by the initial boiling of water followed by decrease in the source water temperature. The initial water temperature from the source is around 26°C due to heat from the environment being transferred to the water when the water is stationary in the buildings plumbing and a small amount of heat transfer along the apparatus's inlet pipe during preheating.

The outlet water temperatures for both the centre (Outlet Centre) and wall (Outlet Wall) locations generally exhibit a period of rapid decrease initially (for times < 50 s), followed by a reduction in the cooling rate. Significant fluctuations (very short pulses, high fluctuations in temperatures) in the outlet water temperature are observed at the Outlet Wall location during the first 10 s. This may be due to transient phenomena taking place at the start of the experiment and / or due to boiling phenomena. After 10 s, fluctuations in the measured water temperatures at the Outlet location are present but at a reduced level as boiling phenomena are fully established at this time. An additional cause of these fluctuations may be the turbulent flow conditions (e.g. based on Reynold's number) that exist in the cooling channel. The Outlet Wall thermocouple was placed at a location within the channel where high temperature gradients were expected due to the boundary layer formed at the wall. During boiling and / or turbulent flow, significant mixing will occur resulting in fluctuations in the measured temperatures.

4.2. Cooling Block Model Result

The model was used to estimate the HTCs for each initial block temperature by adjusting the HTCs at the H13/heater interface and H13/water interface, in the form of data tables, until a reasonable fit was achieved. The predicted temperatures (model) are compared with the measured temperatures (experiment) for each initial block temperature at the Mid-length axial position in Figure 5. The model curves (solid lines) and the experimental curves (dashed lines) are observed to correlate closely. The biggest difference between the model and experiment at each thermocouple location ranges from approximately 0.6°C to 6.95°C.

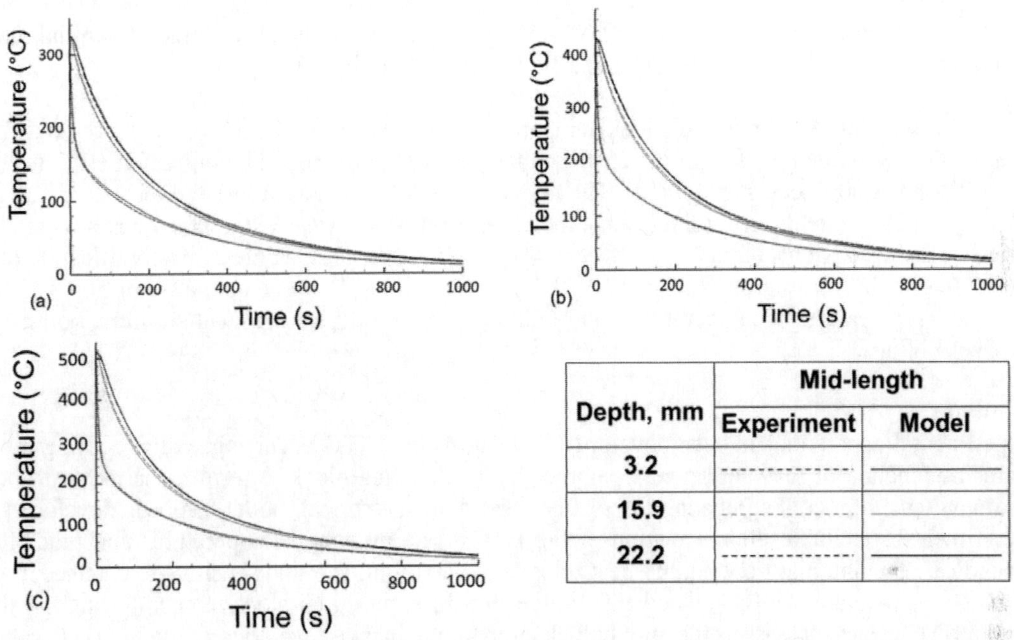

Figure 5 - Comparison of the predicted and measured temperatures at different depths from the water channel at the Mid-length position in the block for an initial block temperature of (a) 300, (b) 400 and (c) 500°C, for a flow rate of 20 l/min.

The heat flux at the cooling channel interface at the Mid-length position has been plotted in Figure 6(a) for the first 100 s of the cooling process for initial block temperatures of 300, 400 and 500°C. The heat flux increases rapidly to a high value within the first ~5 seconds before decreasing approximately 80 – 90% (t < 15s) and then slowly decreasing linearly after that. The heat fluxes are observed to be higher for higher initial block temperatures because of the increased driving force for heat transfer (e.g. ΔT) and larger amount of enthalpy in the cooling block at the start of the experiment which continues to diffuse to the cooling channel interface.

The HTC values as a function of wall temperature used to obtain the fits shown in Figure 5 are shown in Figure 6(b) and range from 2300 - 7509 W / (m^2·K), 2000 - 9074 W / (m^2·K),

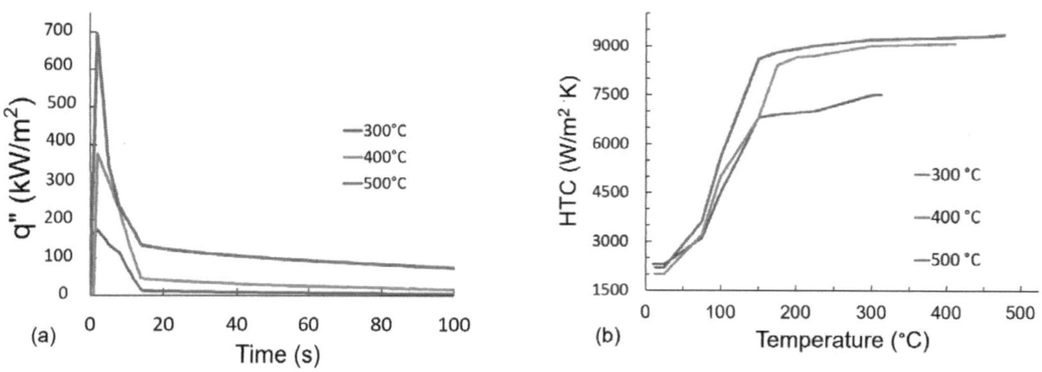

Figure 6 - (a) Corresponding heat flux history and (b) comparison of the HTC values for initial block temperatures of 300°C, 400°C and 500°C and a flow rate of 20 l/min

2200 - 9355 W / (m^2·K) for initial block temperatures of 300, 400 and 500°C, respectively. The maximum HTC value increases with increasing initial block temperature. The minimum HTC for the 300°C,400°C and 500°C cases are similar. The high heat extraction rates at higher block temperatures is consistent with the transient boiling phenomena, and the lower heat extraction rates suggest the occurrence of convective heat transfer. Between 74°C – 175°C, the HTC values drop significantly (e.g. from ~8600 W / (m^2·K) to around 3800 W / (m^2·K) for an initial block temperature of 500°C). The decrease in HTC values at temperatures within this range result from transition from boiling to convective cooling.

5. Summary

An experimental apparatus has been developed to quantify the HTC occurring within cooling block apparatus as function of relevant process parameters. The cooling block experiment is meant to be a physical analogue of a cooling channel in a LPDC die. A mathematical model has been developed of the experiment to facilitate the calculation of the HTCs. The measured temperatures and modelling results indicate that internal flow boiling heat transfer is very nonlinear and sensitive to changes in the initial block temperature. The calculated HTC values for different initial block temperatures show that maximum HTC values increase as the initial block temperature increase however minimum HTC values show minimal change and remain within the lower range of HTC. Further investigation is needed to quantify the boiling phenomena in greater detail. Additional experiments and modelling are planned to examine the effects of other process parameters. It is expected that the knowledge developed from the experimental characterization of heat transfer can be adapted for use in the industrial LPDC applications and that this improved description of the heat transfer occurring in the LPDC process will enable improvements to the manufacturing process through reduced casting defects.

References

[1] Miller A E and Maijer D M 2006 Investigation of erosive-corrosive wear in the low pressure die casting of aluminum A356 *Materials Science and Engineering: A* **435–436** 100–11

[2] Zhang B, Cockcroft S L, Maijer D M, Zhu J D and Phillion A B 2005 Casting defects in low-pressure die-cast aluminum alloy wheels *JOM* **57** 36–43

[3] Çetinel M 2001 *Investigation and development of the quality control of Al-Wheel in rim production process* Doctoral Dissertation (Izmir Institute of Technology)

[4] Zhang B, Maijer D and Cockcroft S 2007 Development of a 3-D thermal model of the low-pressure die-cast (LPDC) process of A356 aluminum alloy wheels *Materials Science and Engineering: A* **464** 295–305

IOP Conf. Series: Materials Science and Engineering **529** (2019) 012053 doi:10.1088/1757-899X/529/1/012053

Effects of strain rate on hot tear formation in Al-Si-Cu alloys

S Bhagavath[1], B Cai[2], R Atwood[3], PD Lee[4,5], S Karagadde[1*]

[1]Department of Mechanical Engineering, Indian Institute of Technology Bombay, Mumbai 400076, India

[2]University of Birmingham, Edgbaston, Birmingham, B15 2TT UK,

[3]Diamond Light Source, Harwell Campus, OX11 0DE, UK,

[4]Research complex at Harwell, Harwell Campus, OX11 0FA UK, [5]University college of London, London WC1E 6BT, UK

*Corresponding author: **s.karagadde@iitb.ac.in**

Abstract. The alloy casting process is one of the major manufacturing processes to produce near net shape components. The casing process is prone to a wide variety of defects, with hot tear being one of the most detrimental. The two main factors generally recognized as the primary cause for formation of hot tears are the mechanical response of the mush (which effects its permeability), and the solidification range (solidification time). The response of the mushy zone under deformation is mainly affected by the solid fraction, strain rate and grain morphology. Even though the science behind the formation of hot tear is understood, there is no general criterion to quantify the hot tear formation under varying casting conditions. The development of ultra-fast X-ray imaging has facilitated the means to quantify the effects of the critical parameters in-situ and develop better correlations for hot tear prediction. The in situ experiments will also provide insights into mush rheology, which has significant influence on hot tear formation. In this study, isothermal semi solid compression studies of Al-Si-Cu alloys were carried out using specially built thermo-mechanical rig. We studied the effects of the strain rate in the range of 2×10^{-4}– 0.02/s and solid fraction (~0.6-0.9) on the mechanical response of the mushy zone. The sample were characterized before and after deformation using X-ray micro tomography. The data was subjected to an image processing routine and the amount of porosity and hot tear was quantified. The stress-strain curve of the semisolid alloys showed a characteristic strain softening behaviour for semi solid samples with ~0.6-0.7 solid fraction, irrespective of loading rates, whereas the behaviour at higher fractions were that of constant flow stress. Additionally, in situ compression experiments were carried out, wherein the liquid channel thickness at various strain values were measured. Isolated liquid channels were formed under loading, from where the hot tears were found to nucleate. Hot tear susceptibility was found to increase with increasing strain rate and rheology of the mush, which is dependent on solid fraction.

1. Introduction

To achieve the continuing stride for improving the fuel economy and reducing the carbon footprint, development of lightweight high strength alloys for automobiles have become a necessity[1]. The mechanical properties of these alloys is controlled by the microstructure and the solidification defects like hot tear [2][3], segregation[4], porosity[5] etc. The root cause of the defects can be traced to the semi-solid state during solidification, at relatively high solid fractions. At these solid fractions, a solid network is formed and as a consequence, the permeability of the mushy zone will decrease resulting in difficulty in feeding. The response of this network under mechanical deformation has a role in the formation of afore mentioned defects

Both laboratory and *in situ* semi-solid deformation tests have been extensively reported in literature which describes the behaviour of several aluminium alloys under various form of deformation loads, deformation rates, alloy composition and solid fraction. While strength and ductility of the network was determined by conducting tensile tests [6], the rheology of the semi-solid mush is studied by conducting shear [7] and compression. The rheological behaviour of various aluminium alloys at different solid fractions and strain rate were studied by Kim et al [8], Kang et al [9], Kapranos et al[10] and reported liquid segregation and cracks at the edge of the specimens. Parametric study of semi-solid compression test of Al-4 wt.% Cu alloys to understand the effect of solid fraction, strain rate, grain morphology is reported by Tsimas and identified different factors affecting the flow resistance[11].Based on these experimental observations, several theories of micro-mechanisms of defect formation have been proposed and criteria for cracking have been developed [12][13].

In this work, semi-solid compression experiments of Al-Si-Cu die-cast alloys at different strain rates has been performed. The deformed sample was characterised using 3d tomography to observe the crack morphology. The interplay between the deformation rate and dilatancy is studied from the stress strain diagram.

2. Materials and Methods

Die cast aluminium alloy ADC12, and a modified ADC12 (MADC12) alloy with copper content increased to 10 wt. % was used in the study. The modified alloy was used as to achieve an enhanced contrast between the primary phase and the inter-dendritic liquid. A secondary objective of modifying the alloy was to investigate the influence of the change in freezing range on hydrogen pore growth and the susceptibility of hot tear. The chemical composition of the alloy was assessed using inductively coupled plasma atomic emission spectroscopy (ICP-AES) technique and the major elements are reported in Table 1.

Table 1: Chemical assay of the alloys under study

Sample	Cu (%)	Fe (%)	Si (%)
ADC12	1.82	0.74	11.9
Modified ADC12	10.426	0.794	10.5

The sample alloys were machined into 3 mm diameter cylinders using wire electro discharge machining (EDM). The length of the sample was 5.4±0.3 mm. The Scheil solidification model in Thermocalc® was used to obtain the exact freezing range for determining experimental conditions, and is shown in figure 1(a).

2.1 Test setup

The semisolid compression tests were performed in the bespoke P2R mechanical compression test rig fitted with proportional integral derivative (PID) controlled resistance furnace. The P2R is custom made testing machine that had been used for conducting a variety of semisolid deformation experiments as reported in [14]. The setup has the capability of applying a maximum load of 500N, at a minimum strain rate of 0.1μ/s. The resistive furnace is powered by a high current DC power supply and the temperature is controlled by a Reconfigurable Input Output (RIO) controller (National Instruments, Model –NI cRIO-9068) The furnace can reach a maximum temperature of 1100°C and the temperature can be raised at a maximum rate of 1°C /s. K type thermocouple is used to record and control the furnace temperature. During the semisolid compression, the sample is heated to pre-determined isothermal temperature. A preload of approximately 7 N was applied to keep the sample in place during compression experiments. The loading ram was adjusted during the experiment to account for thermal expansion during heating and preload is maintained. The thermal cycle followed during the experimentation along with the setup and sample arrangement is represented in figure 1(b).

IOP Conf. Series: Materials Science and Engineering **529** (2019) 012053 doi:10.1088/1757-899X/529/1/012053

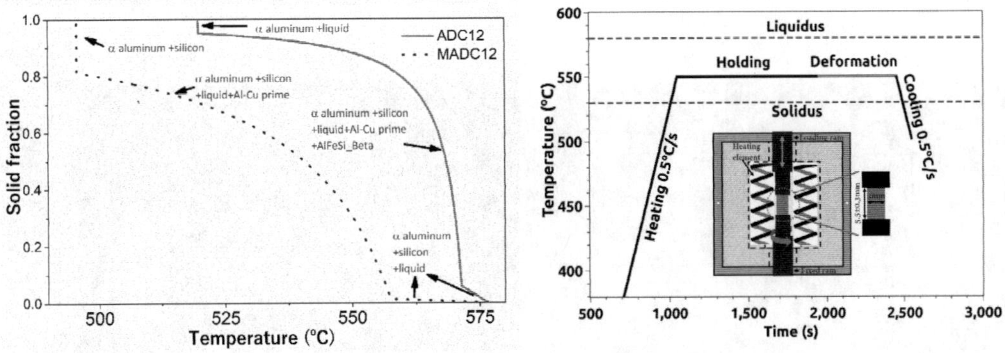

Figure 1: (a) Solid fraction change with temperature for (1) Modified ADC12 (2) ADC12 alloy calculated using Thermocalc® Scheil solidification module (b) Thermal cycle followed during Compression test. Inset- furnace used for the semis solid compression experiment with sample dimensions.

2.2 Post deformation X-Ray tomography characterization

The deformed sample were investigated using 3D using Nikon XT H 225 ST, X-ray tomography machine to know the morphology and distribution of pores. The voltage of the X-Ray source for the ADC-12 samples was set between 70-85kV, whereas for the modified alloy it was between 70-92kV. The sample scan was taken at 40.3 magnification with voxel size of 3.72μm. 3143 projections were taken within 360° with a projection at every 0.116° angle . The exposure time for each scan was 500ms. Flats and darks were taken to minimize the artefacts. The reconstruction of the scan was done using CT agent software, which uses filtered back projection algorithm. The image stacks from the reconstruction was subjected to 3D anisotropic diffusion filter followed by median filter in Fiji Imagej [15]. The initial volume was cropped in to a suitable representative volume. The cropped volume was segmented using Trainable weka segmentation tool [16], a Fiji-Imagej plugin to distinguish porosity and Si particles.

3. Results and Discussion

The radiographs of the deformed sample at various strain rates is shown in figure 2. During the compression, the barrelling effect is seen in the sample, which results in increase area at the centre of the sample. The extent of the damage, as seen in the radiograph is significantly larger at higher strain rates. The corresponding 3D rendering for the sample without deformation and 2×10^{-3}/s case is shown in figure 3.

The stress-strain behaviour for ADC12 and MADC12 at 60±5% solid fraction for MADC12 alloy is shown in figure 4. The stress strain curve can be divided into two zones. In zone one, there is a sharp increase in load followed by a slight decrease. This zone follows what is called as compaction mode, and in which dilatancy is observed. When the sample is subjected to loading, the grains pin up against each other (strain localization). This is indicated by the initial peak in the stress strain curve. The strain localization has been observed by Cai et.al [17], who had used Digital Image Correlation to quantify it. This is followed by a movement of grains to relieve stress, and inter dendritic liquid filling the voids formed by the movement of these grains, as indicated by the drop in the peak stress. There is a large change in stress for small change in strain. In the second zone, called the flow mode zone the solid starts flowing under a constant stress as indicated in the curve.

It was observed that the strain value at which peak stress occurs shifts to a higher value with increased strain rate (0.06048 for 2×10^{-4}/s, 0.22167, for 0.02/s). During 2×10^{-4}/s strain, the liquid flow due to a local pressure differential is comparable to that of the loading rate, and thus the peak stress value is significantly smaller at a lower strain value. With increasing the loading rate to 2×10^{-2}/s, the liquid flow lags behind, and the strain value for the peak stress increases. Thus the interplay between deformation rate and the local strain values which determine the response of the interdendrtic liquid is vital to understand the crack formation during casting.

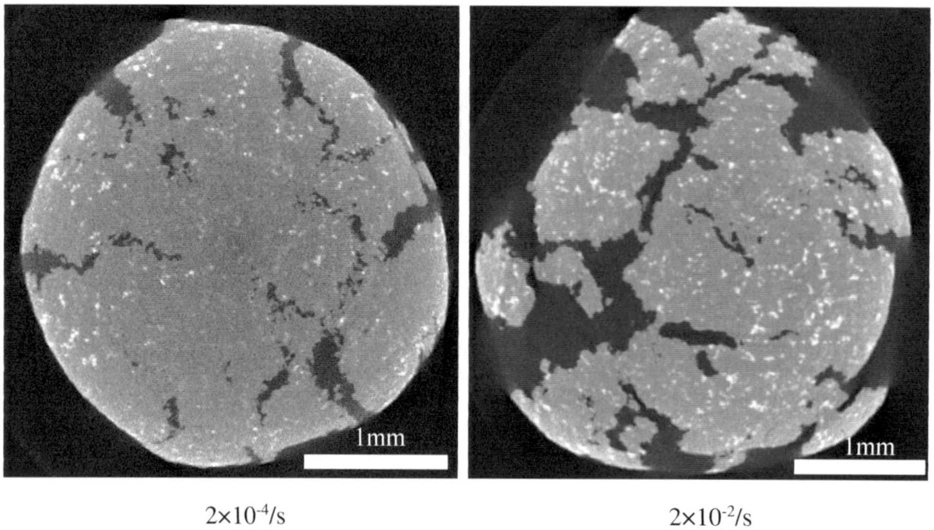

Figure 2: Radiographs of the deformed sample at (a) 2×10^{-4}/s (b) 2×10^{-3}/s and (c) 2×10^{-2}/s strain rate

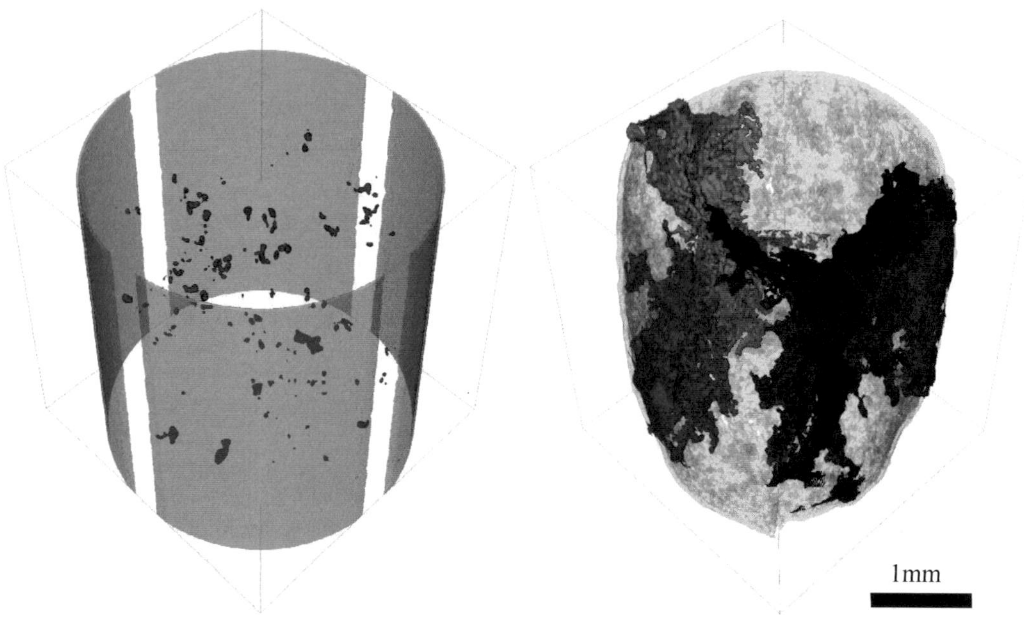

Figure 3: 3D rendering of the sample before and after deformation at 2×10^{-3}/s

The effect of temperature on ultimate compression load at 2×10^{-3}/s strain rate for ADC12 is shown in the figure 5. The behaviour of stress strain behaviour at 60% is explained in the previous section. With the increase in the solid fraction to 75%, the stress initially increases, but remains largely constant after that. However for the samples with 88% and there on the load increased continuously. This can be attributed to the formation of coherent dendritic network and on the application of load the dendrites move and entangle thus resisting the applied force.

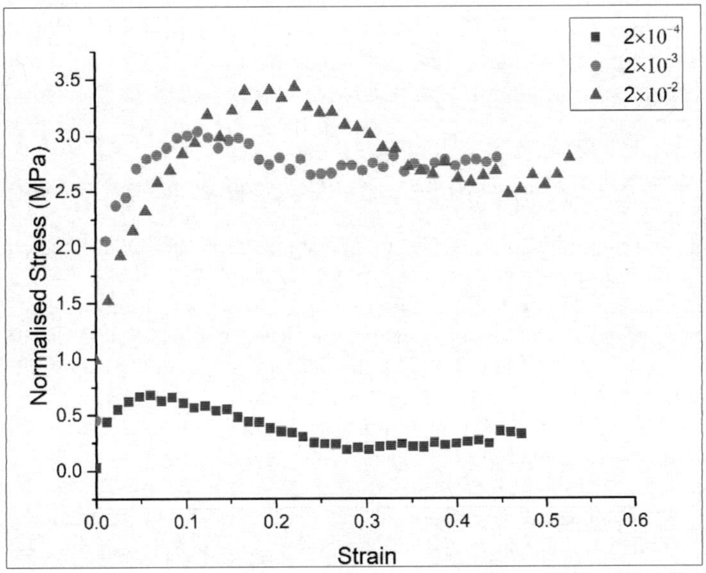

Figure 4: Influence of strain rates on the stress-strain behaviour at 60±5% solid fraction for the modified alloy

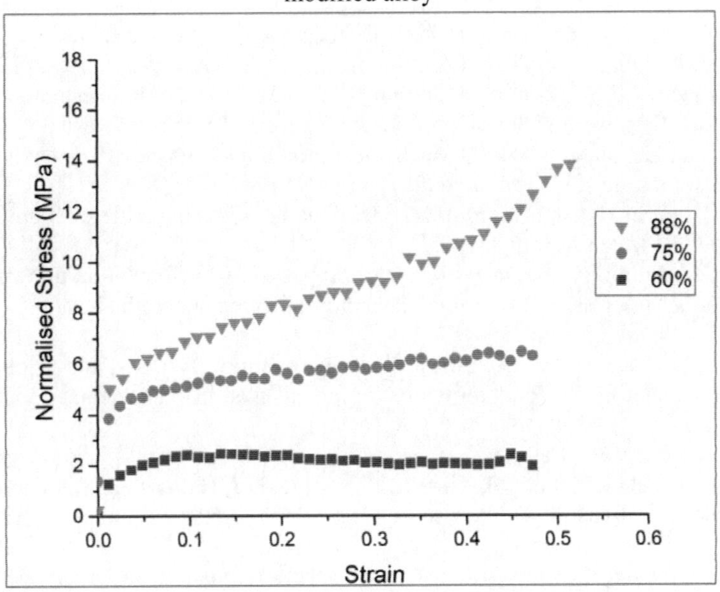

Figure 5: Influence of solid fraction on the stress-strain behaviour at 2×10^{-3}/s strain rate for ADC-12

4. Conclusion

The ADC12 alloy and the modified ADC12 alloy were subjected to the semisolid compression load to study the role of strain rates. The deformed sample was characterized in 3D using X-Ray tomography technique. This was followed by *in situ* experimentation by utilizing ultrafast synchrotron X-ray imaging to observe the hot crack formation 2 different Al-Si-Cu based die cast alloy specimens *in situ.* From, parametric studies the stress strain curve for ADC12 and MADC12 sample showed a characteristic hump followed by a constant stress region. The strain value at which peak stress occurs was found to shift to a higher value with increased strain rate. This was attributed to the competing deformation rate and liquid flow due to dilatancy.

Acknowledgement
SB is grateful to Dr. Sara Nonni, and Sebastian Marucci for their help during the experiments. The support from ferrous metallurgy lab and Machine tools lab, IIT Bombay in preparing samples is appreciated. Authors thank Ford USA for their partial financial support.

Reference

[1] Jason, Rowe P K M 2012 *Advanced materials in automotive engineering* ed R Jason (Woodhead Publishing Limited)
[2] Eskin D G, Suyitno and Katgerman L 2004 Mechanical properties in the semi-solid state and hot tearing of aluminium alloys *Prog. Mater. Sci.* **49** 629–711
[3] Puncreobutr C, Lee P D, Kareh K M, Connolley T, Fife J L and Phillion A B 2014 Influence of Fe-rich intermetallics on solidification defects in Al-Si-Cu alloys *Acta Mater.* **68** 42–51
[4] Gourlay C M, Dahle A K and Laukli H I 2004 Segregation band formation in Al-Si die castings *Metall. Mater. Trans. A* **35** 2881–91
[5] Lee P D and Hunt J D 2001 Hydrogen porosity in directionally solidified aluminium-copper alloys: A mathematical model *Acta Mater.* **49** 1383–98
[6] Terzi S, Salvo L, Suéry M, Limodin N, Adrien J, Maire E, Pannier Y, Bornert M, Bernard D, Felberbaum M, Rappaz M and Boller E 2009 In situ X-ray tomography observation of inhomogeneous deformation in semi-solid aluminium alloys *Scr. Mater.* **61** 449–52
[7] Gourlay C M and Dahle a K 2007 Dilatant shear bands in solidifying metals. *Nature* **445** 70–3
[8] Kim W Y, Kang C G and Kim B M 2007 Deformation behavior of wrought aluminum alloys in incremental compression experiment with a closed die *J. Mater. Process. Technol.* **191** 372–6
[9] Kang C G, Choi J S and Kim K H 1999 Effect of strain rate on macroscopic behavior in the compression forming of semi-solid aluminum alloy *J. Mater. Process. Technol.* **88** 159–68
[10] Kapranos P, Liu T Y, Atkinson H V. and Kirkwood D H 2001 Investigation into the rapid compression of semi-solid alloy slugs *J. Mater. Process. Technol.* **111** 31–6
[11] Tzimas E and Zavaliangos A 1999 Mechanical behavior of alloys with equiaxed microstructure in the semisolid state at high solid content *Acta Mater.* **47** 517–28
[12] Rappaz M, Drezet J M and Gremaud M 1999 A new hot-tearing criterion *Metall. Mater. Trans. A Phys. Metall. Mater. Sci.* **30** 449–55
[13] Sistaninia M, Phillion A B, Drezet J M and Rappaz M 2011 Simulation of semi-solid material mechanical behavior using a combined discrete/finite element method *Metall. Mater. Trans. A Phys. Metall. Mater. Sci.* **42** 239–48
[14] Cai B, Lee P D, Karagadde S, Marrow T J and Connolley T 2016 Time-resolved synchrotron tomographic quantification of deformation during indentation of an equiaxed semi-solid granular alloy *Acta Mater.* **105** 338–46
[15] Schindelin J, Arganda-Carreras I, Frise E, Kaynig V, Longair M, Pietzsch T, Preibisch S, Rueden C, Saalfeld S, Schmid B, Tinevez J Y, White D J, Hartenstein V, Eliceiri K, Tomancak P and Cardona A 2012 Fiji: An open-source platform for biological-image analysis *Nat. Methods* **9** 676–82
[16] Arganda-Carreras I, Kaynig V, Rueden C, Eliceiri K W, Schindelin J, Cardona A and Seung H S 2017 Trainable Weka Segmentation: A machine learning tool for microscopy pixel classification *Bioinformatics* **33** 2424–6
[17] Cai B, Karagadde S, Marrow T J, Connolley T and Lee P D 2015 Synchrotron X-ray tomographic quantification of deformation induced strain localisation in semi-solid Al-15wt.%Cu *IOP Conf. Ser. Mater. Sci. Eng.* **84**

Interfaces at internal chills in solidifying steel sections

Charles Monroe[1], Robin Foley[1], John Griffin[1], Sean Gibbons[2], Alec Saville[3]

[1]University of Alabama at Birmingham, Birmingham, AL 35294-4461, USA
[2]Air Force Research Laboratories, Munitions Directorate, Eglin AFB
[3]Colorado School of Mines

camonroe@uab.edu

Abstract. Chills are used in the production of metal castings as a thermal aid to promote directional solidification in casted sections. This study will review microstructure and thermal circumstances of the internal chill interface where the chill is intended to be incorporated into the structure of the cast section. The conditions for interface coherency of cast steel sections from 25 to 50mm will be shown. Solidification proceeds away from the interface chill growing to a maximum thickness and then melting back to the original chill geometry. Furthermore, it is shown that promoting section solidification prior to the complete melting of the internal chill can lead to the formation of interfaces will be coherent across the chill and cast sections for compatible steel alloys. A computer model of the heat transfer and interface evolution show the possibility of using the coherent interface of internal chills in the design of other cast components.

1. Background

In the heating of materials, there are many methods and sources to add energy. During cooling, the number of methods and sources to accelerate heat extraction are more limited. Chills represent one of these few methods to enhance the cooling process. The enhancement in cooling using a chill is relative to the increase in conductivity and heat capacity of the chill material. Chills are standard manufacturing processing aids utilized within the metal casting industry [1]. The intent of the chill is to promote directional solidification which improves material soundness and properties. In some mining and wear applications, internal chills have been used in various methods of creating bimetallic, cast bonded, or hard-faced components [2-6]. This process takes the existing solid section (i.e., the internal chill) and through thermal bonding with the solidifying cast material creates one coherent monolithic structure, with components capable of exhibiting properties of both materials. In all of the cases, cooling rate of liquid metals during solidification and the respective configuration play a key role in the resulting interfacial reaction, but to this point only limited work has been completed to computational represent the reaction and to predict the coherency of the interface when using a chill in a cast solidification process.

Process modeling of metal casting is common in industry to understand solidification sequence and design the gating system for cast sections. Using computational fluid dynamics (CFD) to calculate the fluid flow and heat transfer is essential to the final results studied using MAGMAsoft [7]. However, assumptions typical to this process are limited to standard process conditions. A typical process models does not allow the melting of the chill and therefore conditions for a coherent interface are not predicted. To better understand the evolution of solidification at the interface of internal chills, two models for the interface of an internal chill are shown. First an analytical solution for a chill interface is proposed and

IOP Conf. Series: Materials Science and Engineering **529** (2019) 012054 doi:10.1088/1757-899X/529/1/012054

used to show the governing parameters and second an axisymmetric one-dimensional enthalpy method for solving the energy balance is used and implemented in MATLAB [8].

2. Method

To motivate the stability of the interface at an internal chill interface, an analytical solution of interface velocity including material properties and initial temperatures is shown. This method was developed in the book by Dantzig and Tucker and is reproduced here for completeness [9]. The internal chill for this example is calculated as one-dimensional solution in an infinite medium. The material used in this example is assumed to be a pure material with a single melting temperature with no diffusion. The first region of the internal chill, second region of solid growth, and third region of superheated liquid have equations for energy and initial/boundary conditions that are shown as,

$$\frac{\partial u}{\partial t} = \alpha_k \frac{\partial^2 u}{\partial x^2}, \tag{1a}$$

$$u = u_0(x \to -\infty),\ u = u_s(x = 0),\ u = u_m(x = \delta(t)),\ u = u_\infty(x \to \infty), \tag{1b}$$

where u, x, and t are the temperature, distance, and time respectively and α is the thermal diffusivity with subscript *k* substituting for the internal chill with *i*, solid growing away from chill with *s*, and superheated liquid with *l*. The initial/boundary conditions are shown for the internal chill as u_0 with increasing temperature the surface temperature u_s, the melt temperature is u_m, and the superheat temperature is u_∞. The solution for the three regions, chill, solid, and liquid are

$$u = u_s + (u_0 - u_s) erf\left(\frac{-x}{2\sqrt{\alpha_i t}}\right),\ -\infty < x \le 0. \tag{2a}$$

$$u = u_s + \frac{(u_m - u_s)}{erf(\phi)} erf\left(\frac{x}{2\sqrt{\alpha_s t}}\right),\ 0 \le x \le \delta(t) \tag{2b}$$

$$u = u_\infty + \frac{(u_m - u_\infty)}{erf\left(\phi\sqrt{\frac{\alpha_s}{\alpha_l}}\right) - 1}\left[erf\left(\frac{-x}{2\sqrt{\alpha_l t}}\right) - 1\right],\ \delta(t) \le x < \infty \tag{2c}$$

where $\delta(t)$ is the distance of the growth of this solid/liquid interface which changes as a function of time. The region of solid growth grows assuming a constant surface temperature determining both position of the interface and temperature of the surface equated the fluxes at *x* equals zero shown as

$$\delta(t) = 2\phi\sqrt{\alpha_s t},\ \text{and}\ u_s = \frac{u_0\sqrt{k_i \rho_i c_{pi}} erf(\phi) + u_m\sqrt{k_s \rho_s c_{ps}}}{\sqrt{k_i \rho_i c_{pi}} erf(\phi) + \sqrt{k_s \rho_s c_{ps}}}, \tag{3}$$

where ϕ is a constant that will be found using the Stefan condition, k, ρ, and c_p are the thermal conductivity, density, and specific heat for materials. The combination of these properties has been termed the heat diffusivity in contrast to the thermal diffusivity. For the purpose of this discussion, these are considered different because of the average temperatures of these regions however from the assumptions of the internal chill and solid growth regions these materials are substantively the same. Using the Stefan condition at the location of the solid/liquid interface, the final transcendental equation which completes the analysis is found as

$$\frac{c_{ps}(u_m - u_0)}{L_f\sqrt{\pi}} = \left(\phi \exp(\phi^2) + \frac{c_{ps}(u_m - u_\infty)}{L_f\sqrt{\pi}} \frac{\exp\left[\left(1 - \frac{\alpha_s}{\alpha_l}\right)\phi^2\right]}{\left[erf\left(\phi\sqrt{\frac{\alpha_s}{\alpha_l}}\right) - 1\right]} \frac{\sqrt{k_l \rho_l c_{pl}}}{\sqrt{k_s \rho_s c_{ps}}}\right)\left(erf(\phi) + \frac{\sqrt{k_s \rho_s c_{ps}}}{\sqrt{k_i \rho_i c_{pi}}}\right). \tag{4}$$

To determine the stable interface condition, the solid growth constant must be zero or $\phi = 0$. Therefore, equation (4) reduces to the relationship

$$(u_m - u_0)\sqrt{k_i \rho_i c_{pi}} = (u_\infty - u_m)\sqrt{k_l \rho_l c_{pl}} \tag{5}$$

IOP Conf. Series: Materials Science and Engineering **529** (2019) 012054 doi:10.1088/1757-899X/529/1/012054

A quick survey of liquid and solid material properties will demonstrate that the ratio of the heat diffusivity of liquid and solid is less than one but still on the order of one. This results in a stationary interface condition when the initial temperature of the chill is closer to the melt temperature than when the superheat temperature is to the melt temperature. This condition is opposite of metal casting practice where the initial temperature of the chill can be two orders of magnitude less than the superheat temperature. This result also shows that the solid growth velocity is slowed by increasing superheat temperatures. Therefore, it is expected that the solid/liquid interface will grow with a rate between 0 and a maximum when the superheat is zero. The conclusion is that only in the case of significant chill preheat temperatures will an internal chill melt rather than grow solid from the chill interface. This fundamental result also gives some insight into a parameter for a coherent interface, surface undercooling or $u_m - u_s$. At a stationary interface the solution of the surface undercooling is zero and the surface temperature is the same as the melt temperature. At all positive interface velocities, the undercooling increases and can be significant as

$$u_m - u_s = (u_m - u_0)\left(1 + \frac{\sqrt{k_s \rho_s c_{ps}}}{\sqrt{k_i \rho_i c_{pi}} erf(\phi)}\right)^{-1}. \quad (6)$$

The surface undercooling will vary from zero when the interface is stable to a maximum approximately half of the temperature difference between the melt temperature and the initial chill temperature. From this result, the use of an internal chill will initially form solid until/unless the chill soaks with heat raising the far field temperature resulting in a reversal in the solidification. After this point the progression of solidification will reverse and melting toward the chill will begin. It is this sequence of events that is of interest in this work to develop conditions for nucleation and grow away from chill as well as melting and coherence across this interface.

Given the analysis above, it was decided to analyze a similar condition using the one-dimensional axisymmetric model implemented within MATLAB is evaluated using the PDEPE command. The material properties in this domain can be calculated for the chill and matrix material using calculation of phase diagram or CALPHAD approaches with respect to the measured composition [10]. A mesh with uniform 100 elements across the chill dimension, uniform 100 elements across the cast dimension, and logarithmic spaced 40 across the mold dimension and result is shown in Figure 1. A typical result of temperature as a function of time and space from this model is shown with the heating of the chill near the axis of symmetry, solidification in the cast region, and heating in the mold according to the initial temperature. This result can be expressed in the chill and cast materials as a function of solid fraction shown with a similar angle as the temperature field as well as a contour plot with regions identifying the chill and cast materials. This result shows that the solid fraction initially grows from the chill interface but eventually the solidification front arrests reversing direction and melting toward the chill. This melting front continues until the outside solidification front freezes the entire structure.

This approach is a continuum solution which assumes a fixed thermodynamic path for the heating, cooling, and solidification. The materials for the matrix and for the internal chill material are summarized in Table 1 below. The chill material, 4330V, alloy targets include significant nickel, chromium, and molybdenum content [11]. These elements are absent from the matrix materials of WCB [12] with a maximum carbon target of 0.3 weight percent which will result in a material with similar or lower carbon than the 4330V chill and thus a higher melting temperature.

Table 1. Compositions selected for material in hybrid experiment.

ID	C wt%	Si wt%	Mn wt%	Cr wt%	Mo wt%	Ni wt%	Cu wt%	V wt%	Pour °C
4330V	*0.32*	*0.25*	*0.9*	*0.8*	*0.45*	*1.8*	*0*	*0.08*	*30*
WCB	*0.228*	*0.481*	*0.955*	*0.042*	*0.01*	*0.0299*	*0.0552*	*0.0295*	*1620*

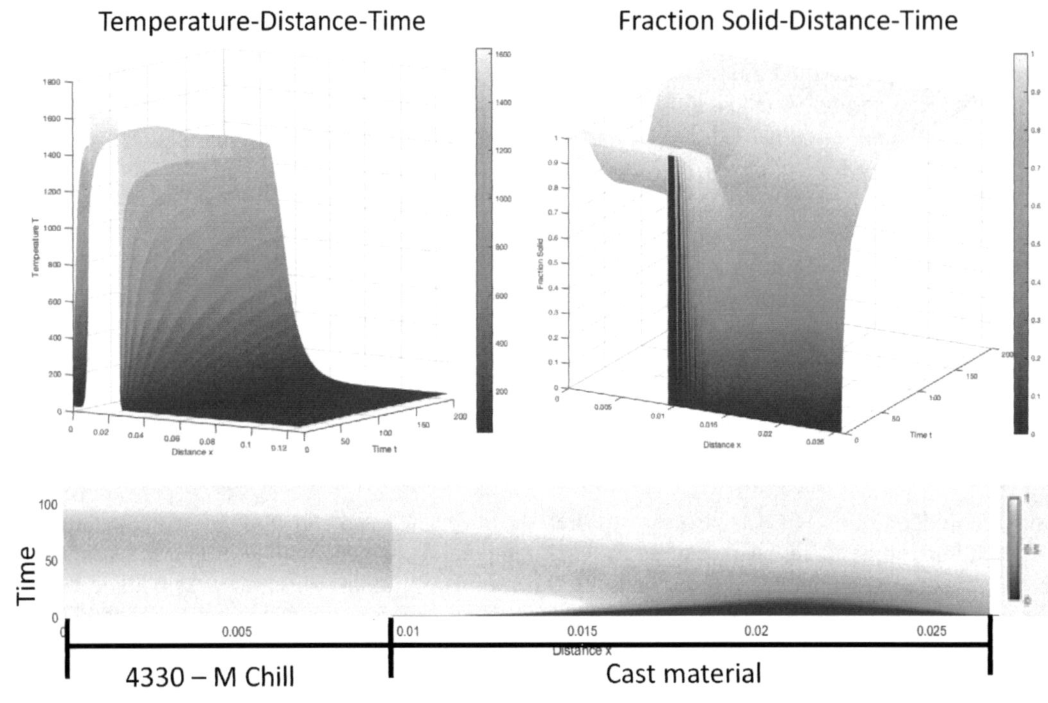

Figure 1. Example solution of energy equation for a chill, matrix, and mold domain is shown.

The geometry shown in Figure 2 illustrates the configuration of the chill and matrix material. The casting was manufactured using a sand mold bonded with a phenolic urethane binder system assembled with the internal chill rods. The rods were prepared prior to pouring using sand blasting. The melting was accomplished using an induction melting system with a liquid argon drip protective atmosphere. Alloys were prepared using low carbon steel feedstock material, ferromanganese, and ferrosilicon which was poured directly from the furnace. Target pouring temperature was ~100 degrees above the liquidus of the matrix material. The pour time was approximately 9 seconds. After shakeout, the gating system was removed, and the castings were sectioned at three heights as shown on Figure 2 labeled. To estimate the preheat temperature distribution in the MATLAB code the surface of the chill and the mold was held at the pour temperature for 0.1, 2, and 4 seconds respectively in the "A", "B", and "C" sections.

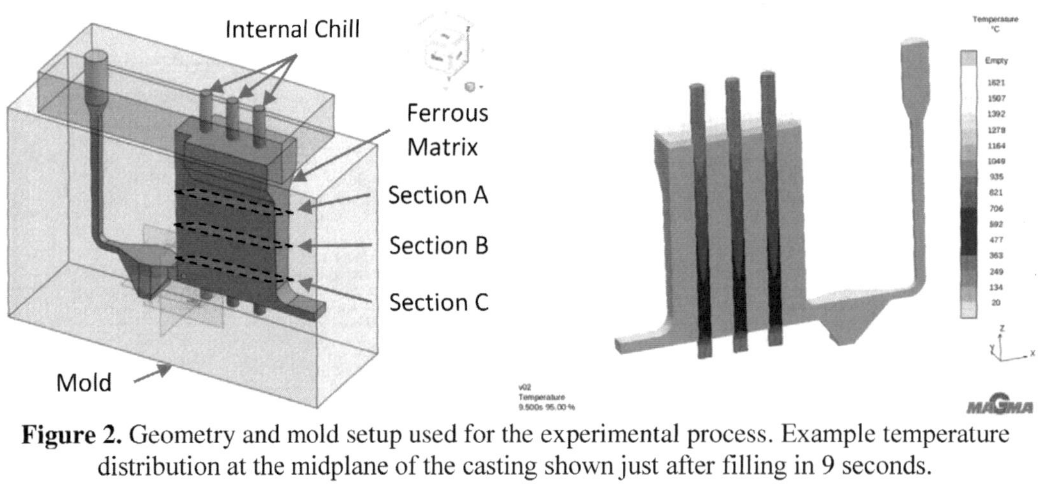

Figure 2. Geometry and mold setup used for the experimental process. Example temperature distribution at the midplane of the casting shown just after filling in 9 seconds.

3. Results

The commercial process simulation of the fluid flow shows significant heating during the filling process even after a short filling time of 9 seconds. Figure 2 shows an example of the temperature along the midplane of the casting just after filling with an estimate of temperature in the internal chill which exceeds 500 degrees Celsius. These results from the various combinations of size and temperature are compared to the experimental samples in Figure 3. All sections formed a coherent interface across the chill and matrix with a martensitic structure in the 4330V chill material and a pearlite structure in the WCB matrix. The microstructure is revealed using a nital etch. Figure 3 shows arrows the progression of solidification ending at the last to solidify area of the matrix. These solidification paths all follow the same progression of solid fraction shown and discussed for Figure 1. In the modeling result shown corresponds to the section size (A, B, or C) changed with the height. The initial temperature in the model was taken from the experimental value specified in Table 1 with one modification. The surface of the chill and mold was held at the pour temperature for 0.1, 2, and 4 seconds at sections A, B, or C respectively. In each case, the pattern of solidification away from the chill, arrest, and melting back was observed with resulting microstructures solidified from the outside to the interior. Due to the rapid heating at the thinnest section "C" during filling the microstructure had the slowest cooling rate and largest solidification features. In contrast, the shortest solidification time was near the top of the wedge in section "A", which showed evidence of fastest cooling rate and finer microstructural features. The center section "B" had mixed results with formation of a crack that extends through both the matrix and chill material. The orientation of the crack through the chill may confirm the coherence across the interface.

4. Conclusions

It appears that by controlling the flow, temperature, geometry, and composition of ferrous alloy that conditions for a coherent interface can be obtained. The key feature of the expected coherence of the interface appears to be the sequence of solidification determined by the surface undercooling. In the WCB and 4330 case, bonding was found at all locations and a significant chill layer and melting back of the chill layer was seen in the modeling. This mechanism looks like an important criterion to the prediction of bonding in ferrous/ferrous hybrid composites. Future work to couple the compositional effect as well as measurements of the microstructure demonstrating the cooling rate and other verification is ongoing.

References

[1] Ou, S., Carlson, K. D., Hardin, R., & Beckermann, C. (2002). Development of new feeding-distance rules using casting simulation: Part II. The new rules. *Metallurgical and Materials Transactions B*, *33*(5), 741–755.

[2] Lucey, T., Wuhrer, R., Moran, K., Reid, M., Huggett, P., & Cortie, M. (2012). Interfacial reactions in white iron/steel composites. *Journal of Materials Processing Technology*, *212*(11), 2349–2357.

[3] Qian, M., Harada, S., Yanxiang, L., & Dongjun, M. (1996). On the fabrication of steel-wire-reinforced white cast irons. *Materials Science and Engineering A*, *206*(1), 104–109.

[4] Wróbel, T. (2014). Characterization of Bimetallic Castings with an Austenitic Working Surface Layer and an Unalloyed Cast Steel Base. *Journal of Materials Engineering and Performance*, *23*(5), 1711–1717

[5] Cingi, C. (2010). Cast bonding of cast irons to ferritic stainless steels, *Materials Science Forum*

[6] Lucey, T., Wuhrer, R., Moran, K., Reid, M., Huggett, P., and Cortie, M., (2012). Interfacial reactions in white iron/steel composites, *Journal of Materials Processing Technology*, vol. 212(11), 2349–2357.

[7] MAGMA GmbH. *MAGMAsoft v5*, 52072 Aachen, Germany

[8] MATLAB Release 2017b, The MathWorks, Inc., Natick, Massachusetts, United States.

[9] Dantzig, J. A., & Tucker, C. L. (2001). Modeling in Materials Processing. Cambridge University

Press.
[10] Sente Software Ltd. *Jmatpro 9.0*, Surrey Research Park, United Kingdom
[11] ASTM A646 / A646M-17, Standard Specification for Premium Quality Alloy Steel Blooms and Billets for Aircraft and Aerospace Forgings, ASTM International, West Conshohocken, PA, 2017, www.astm.org
[12] ASTM A216 / A216M-18, Standard Specification for Steel Castings, Carbon, Suitable for Fusion Welding, for High-Temperature Service, ASTM International, West Conshohocken, PA, 2018, www.astm.org

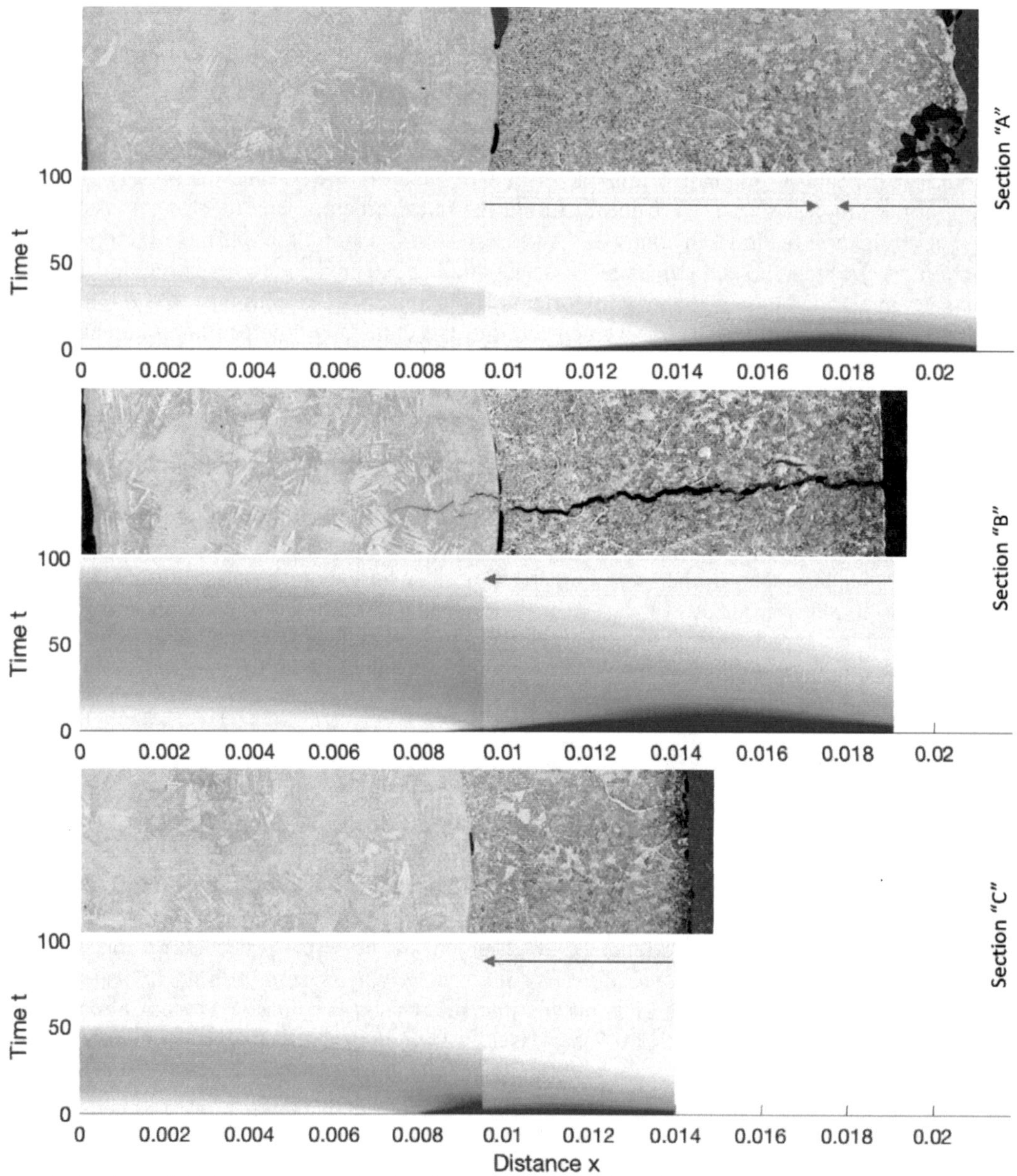

Figure 3. WCB and 4330 sections etched with nital compared to simulations, distance x in meters and time in seconds.

IOP Conf. Series: Materials Science and Engineering **529** (2019) 012055 doi:10.1088/1757-899X/529/1/012055

On Analytical Concepts of Novel Multi-Resolution Casting Simulations

Amir M Horr

Senior Scientist, Center for Low Emission Transport, Light Metals Technologies Ranshofen, Austrian Institute of Technology, Giefinggasse 2, 1210, Vienna, Austria

Amir.Horr@ait.ac.at

Abstract. One of the crucial ingredients of today's numerical simulation technologies is their cross-scale and cross-platform capabilities for material processes applications. Handling of simultaneous evolution paths at various scales\times (i.e., multi-scaling) during complex material processes where materials microstructure & microchemistry interact with meso and macro events, is one of awkward challenges of computational material engineering. The material phase change and also its thermal energy evolutions are also drastically increasing the complexity of the numerical simulations. The introduction of multi-resolution and multi-scale numerical schemes in recent years and their ground-breaking potentials for computational material science applications have vividly raised the expectations for more resourceful future virtual tools. A crucial point in implementing these novel numerical techniques for simulation of material processes is their flexibility towards the modelling approach (i.e., discrete, continuous...) and also their compatibility with solver-independent platforms. As these multi-resolution\physical techniques should provide some answers to the best ways of designing future high-performance materials along with optimisation of new & existing material processes and also improvement of their life-time performance, a broad & well-structured research work is required. Hence, the proposed multi-resolution framework herein, has been developed based on analytical & numerical techniques built on sound physical and mathematical foundations developed during the last few decades. The combination of recently developed concepts of dynamic\evolving domains along with cross-scale grid overlapping\interfacing and also sound parallel computing routines have been employed to address the multi-scale challenges of material processes simulations. In the research work herein, analytical and computational aspects of multi-resolution simulation framework for dynamic casting processes (i.e., continuous and semi-continuous casting) are presented and physical\mathematical basis of the analytical-computational solidification and cooling modules are elaborated. Industrial applications of the techniques are also envisaged using parallel-processing and fast computing facilities for full-scale casting applications.

1. Introduction

An efficient numerical simulation tools designed in a way to help industries to solve their problems and to find & optimise the most agile solutions for their material processing technologies is foreseen for today's fast and competitive industries. With a carefully-designed simulation system, it is possible to automate many aspects of industrial process optimisations in an integrated computerized environment where sophisticated physical & mathematical rules, boundaries, data processes\handling and goal seeking analyses are performed. The research work herein is a part of a general framework which will deliver a numerical simulation tool for implementation of novel and innovative material and process modelling into industrial mainstream procedures for boosting efficiency and competitiveness of market-

oriented products. The multi-resolution & scale-bridging computational framework and its new overlapping and interfacing concepts [1, 2] combine advanced simulation techniques, material models at different length\time scales and modelling methods from diverse scientific disciplines to handle the wide range of material engineering applications.
A series of interactive sub modules including solidification, cooling and thermal-mechanical routines have formed an integrated computer-based platform which helps engineers to solve challenges in casting technologies. The goal of research work herein is to contribute\respond to some fundamental industrial needs for:

- Optimising and tailoring material processes for increase part quality\performance;
- Improving the confidence\reliance of industrial processes on advanced material modelling and;
- Amplifying the use of computer-based virtual modelling and simulation systems for industries

For the case studies herein, using the analytical-numerical simulation system, the optimisation of industrial casting process under real-world conditions have been conducted. In these case studies, problems of phase changes and thermal energy evolution due to cooling (i.e., air & water cooling) effects have been considered. Moreover, special attention has been devoted to the development of more practical and general solidification and cooling modules. Different sets of in-house experimental casting trials have been used for the verification\calibration of the simulation system. The characteristics of the simulation modules are briefly described herein, where the efficiency of the simulation technique is highlighted. The purpose of this paper is to present a new horizon, which improves the modelling of thermal energy evolution and material solidification during industrial casting processes.

2. Conventional macro-scale casting simulations

The conventional use of fixed Lagrangian, updated Lagrangian and stationay\moving Eulerian techniques for simulation of industrial casting processes have traditionally been popular among engineers and scientists. In the conventional pure Lagrangian and also updated Lagrangian approaches (with adaptive re-meshing), the deformation process of a material body is resembled as a virtual flow where each material element\particle carries its own properties (such as microstructure, temperature, density…). For the simulation of dynamic casting systems where solidified billets are cast in time, as the flow front advances, its properties may change in time. However, the inability of these approaches to follow large deformation\distortions of the computational grid without comprehensive re-meshing [3] is its main deficiency. Alternatively, in the stationary\moving Eulerian approaches, rather than tracking each material element, the evolution of the material flow properties at every point in the numerical grid (i.e., vortices) can be calculated (at sequential time steps). This means that the melt flow properties at a specified location within the grid (i.e., mould) depend on its location and time (advection). In this approach, the conservation of material is satisfied at spatial and time points and the computational mesh is mostly fixed where material moves with respect to the grid. The large deformation\distortions (and melt flow) can be handled with relative ease using convective feature of the method where relative motion of the deforming front in respect to the discretized grid is calculated [3].
The basic concepts of discretization in space and time to define simpler subspaces (i.e., cells) with exact\approximate solutions have also been employed in almost all of the existing numerical solvers. Different types of the meshing\gridding methods which discretise the virtual domain into small individual cells\nodes have been developed over past seventy years [3]. The technology which is able to discretize a wide variety of complex geometries, adapt into large deformation of the numerical domain (i.e., adaptive re-meshing) and cope with non-conformal aspects of multi-domain grids has already been mobilised. There have also been various attempts to introduce innovative discretization techniques for dynamic systems with evolution in time [4] and interfacing technology for moving grids (i.e., continuous casting). Most of these techniques have been developed for numerical simulation of complex geometrical systems based on the solution technique (i.e., discrete, continuous…), accuracy, CPU time and complexities. Although, the fixed Lagrangian grid has extensively been used for domain discretization in solidified part of the billet [4], its application for melt part of the domain with its internal microstructure evolutions is limited.

3. Novel evolving casting simulations

To improve the simulation technique for continuous and semi-continuous casting process a novel dynamic mesh technique has already been developed [4]. In the dynamic mesh approach, the generation of the billet is modelled using new subdomains\zones which are generated & joined at specific time points to resemble the extension of the billet. These newly generated zones are overlapping\attached to the main domain at discrete time points where comprehensive mapping procedures (including grid overlapping, nodal condensation, propagation interface…) are applied to re-assemble the domain to handle the new material (i.e., melt material) and energy input. The added dynamic subdomains\zones can be stationary or moving depending on requirement of the numerical model for the industrial process. As the numerical solution is performed on a full parallel-processing machine, multiple instances of the same solver (or alternatively different solvers) can independently be employed for the main domain and its attached\overlapping zones. The generated subdomains (which can be at different length\time scales) would gradually integrated into the main domain (and become a part of main domain) as the phases within the subdomains\zones come to the end (i.e., microstructure, solidification…). Figure 1 shows a casting simulation case study (3D quarter model) using dynamic mesh technique. More comprehensive discussion about the dynamic meshing and finite zoning techniques can be found in [1, 2, 4].

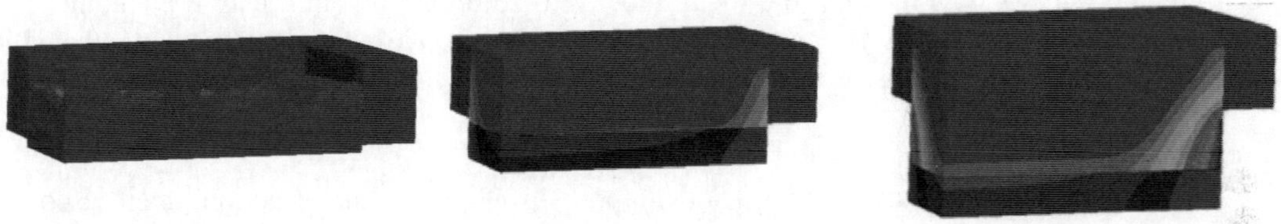

Figure 1. Dynamic meshing technique for continuous casting process (symmetric quarter model)

One of the challenges for the simulation of continuous casting processes is that the location and the shape of the solidification interface are dynamic (time-dependent). During the simulation time these boundaries evolve and may have translational and rotational movements. These moving boundaries can be taken into account using multiple mesh resolutions (overlapping grids) within the simulation domain using some hybrid\mixed techniques [2].

4. Solidification: enhanced analytical-numerical approach

The concept of numerical simulation of material processes and in particular casting processes has already attracted many interests among scientists, material engineers and industries. One of the challenging aspects of numerical simulations for casting applications is the change of phases and solidification phenomena during the process. The solidification phenomena during casting process can be modelled at different levels of scales and sophistication. For macro scale simulations, there are different approaches for defining the phase transition. This section highlights a brief description of the analytical-numerical technique which has been developed herein to enhance the accuracy and efficiency of solidification numerical modelling.

Most popular scheme for the industrial macro-scale solidification modelling is the application of the Scheil-Gulliver equation [5, 6] where a global non-equilibrium approach is formulated for solute redistribution during solidification (no solute would diffuse back into the solid in the standard form). The solid fraction is defined by melt flow, solidus and liquidus temperatures as well as the partition coefficient. The Scheil-Gulliver equation in its simple and standard form can be written as [5, 6];

$$\begin{cases} C_L = C_0\, f_l^{k-1} \\ C_S = kC_0(1 - f_S)^{k-1} \end{cases} \tag{1}$$

Where C_L , C_S and C_0 are liquid, solid and initial compositions and f_l , f_S and k are solid & liquid mass fractions and coefficient $k = \frac{C_S}{C_L}$ (which can be determined from phase diagram), respectively. The formulation is valid for local equilibrium state between the solid and liquid at interface, and moderate to low cooling rates. Different numerical methodologies have been developed [4] to handle the phase change phenomena within numerical domains, namely; conduction-based, conduction-convection based and enthalpy-based methods. The enthalpy-based scheme is to use the enthalpy concept where enthalpy (which is a thermodynamic quantity defining the total heat energy of a system) can be combined with other numerical features to account for change of phase during solidification. In the mainstream industrial software tools, the enthalpy method is generally combined with Volume of Fluid (VOF) and domain porosity features (to damp the fluid momentum when it turns into solid). Although VOF has been developed and used extensively to treat problems in fluid dynamics with liquid–gas interfaces but it has been extended to solidification problems in material science [7]. For the VOF technique within the mold, the surface tension forces are implemented in the momentum equation (as a source term) using technique introduced in [8]. The domain porosity would also enable the smooth change of phase from liquid to solid by damping the momentum equation using momentum sink terms. The introduction of momentum sink terms in the energy equation might introduce more numerical challenges since the convergence of the system become more difficult. Consider the mass continuity and momentum equation for the melt flow inside the mold as [3];

$$\nabla . U = 0\ , \qquad \frac{\partial(\rho U)}{\partial t} + \nabla . (\rho U \otimes U) = -\nabla \mathrm{p} + \nabla . (\mu \nabla \mathrm{U}) \tag{2}$$

A momentum sink term is traditionally introduced into melt momentum equation (for temperatures between liquidus and solidus) to damp the melt velocity to zero due to solidification phenomena. A practical approach for momentum damping can be defined using the Kozeny, Carman approach and its modification by Blake [9]. Let's assume the equation for fluid flow in resistive media as;

$$\frac{\Delta p}{L} = \alpha \left(\frac{(1-k)^{\beta}}{k^3}\right)\left(\frac{\mu v}{d^2}\right) \tag{3}$$

Where $\frac{\Delta p}{L}$, k, μ, d and v are pressure gradient, domain porosity (liquid volume fraction), melt viscosity, primary channel width and melt superficial velocity. Variables α, β can be defined for material processes during solidification based on the alloy and process properties. Two different types of momentum sink terms can be defined for the melt solidification, namely; viscous and quadratic terms. The basic viscous sink term for fluid flow in a porous media can be defined as [10];

$$L_V = \frac{\mu}{K}\, U \tag{4}$$

Where K is the permeability of the resistive domain. Voller et al. has modified [11] the fluid viscous loss model and the permeability of the domain as;

$$L_q = -\frac{\mu}{K} . (u - u_s) \tag{5}$$

Where u_s is the solidified slab speed (can be assumed as casting speed) and K is permeability which can be calculated based on the resistance against the melt flow inside the dendrite channels during solidification. Ignoring the re-melting and the solute concentration effects (temperature exchange at interface…), Gu et al. [12] has proposed the following equation for the permeability:

$$K = C \frac{{f_l}^3}{{f_s}^2}\, \lambda_1^2 \tag{6}$$

Where f_l, f_s and λ_1 are volume fractions of liquid and solid phases (at the solidification front) and primary dendrite arm spacing. The constant C can be assumed as $C=6\times10^{-4}$ [12]. The secondary arm spacing and its resistance effect have not been explicitly defined in the equation (5). The broader equation for estimation of permeability have been derived by the author using the combination of physical and analytical approach as [13];

$$C_{LV} = C * \lambda^2 \left(\frac{(1-f(\mathrm{T}))^{\alpha}}{f(T)^{\beta}}\right) \qquad f(T) = C_1 * \left(1 - \left(\frac{(\mathrm{T}-\mathrm{T}_s)}{\Delta T}\right)^{\gamma}\right) \qquad \Delta T = \mathrm{T}_L - \mathrm{T}_s \tag{7}$$

Where C, C_1 are constant coefficients, λ is the primary dendrite arm spacing and α, β, γ are power coefficients which can be calibrated for the alloy processes (for Aluminum alloys α= 3, β=2 and γ =1). Coefficients C and C_1 can be assumed as C= 0.0006 (similar to Gu et. al) and C_1 can be assumed between 1 & 1.2. A more detailed version of the simplified loss term can also be developed [13] using primary & secondary dendrite arm spacing and melt isolation phenomena. Similarly, based on physical concepts of energy damping and a mathematical formulation of melt flow kinetic energy, a quadratic sink term can also be derived as [13],

$$C_{LQ} = \begin{cases} C.\beta.\left(1-\left(\frac{T-T_E}{T_E-T_S}\right)^{\alpha}\right) & T_E < T < T_L \quad \alpha = 1 \\ \beta + C.(1-\beta).\left(1-\left(\frac{T-T_S}{T_E-T_S}\right)^{\alpha}\right) & T_S < T < T_E \quad \alpha = 2 \end{cases} \tag{8}$$

Where T_E is a temperature which denotes the starting point for dendrites bridging phase during solidification. For aluminum alloys, the simple assumption can be that $T_E = \frac{T_L - T_S}{2}$ and $\beta = \gamma.\left(\frac{T_E - T_S}{T_L - T_S}\right)$ where γ is a coefficient which depends on the material microstructure evolution during solidification (when growth and progression of dendrites starts to create significant resistance against the melt flow near the interface). For Aluminum alloys where equiaxed grain growth are perceived, it can be assumed that $\gamma = 0.6$. The coefficient C can be calculated\measured and calibrated from the total momentum of the melt within the mold prior to the solidification phenomena. More comprehensive discussions about these mathematical derivations of momentum sink terms can be found in [13].

5. Enhanced analytical HTC calculation

Numerical modelling of cooling processes (i.e., water spray, air cooling…) is one of the most challenging part of the casting simulations. Many thermal evolution studies have been carried out in last two decades to develop an estimation technique for the dynamic Heat Transfer Coefficient (HTC) on the cast billet surface. These works are considering different water boiling regimes (nucleate, unstable and stable filming regimes) and also Leidenfrost effects. The variation of HTC against the billet surface temperature can be divided into four distinctive zones, namely; below water saturation temperature, between water saturation and Critical Heat Flux (CHF) temperature (nucleate boiling regime), from CHF temperature to Leidenfrost temperature (unstable boiling regime) and finally above Leidenfrost temperature (film boiling regime). In the work herein, the analytical HTC calculation developed by the author [14] has been implemented using in-house coding. Figure 2 shows the variation of the HTC with respect to cast billet surface temperature (i.e., at the impinging point) using analytical formulation for different water boiling regimes. More comprehensive discussions about the mathematical derivation of HTC functions can be found in [14].

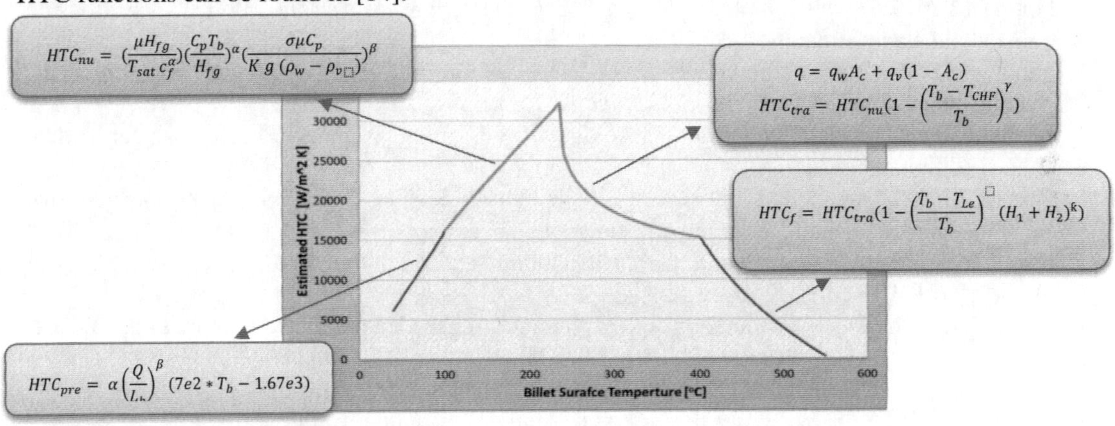

Figure 2. Variation of HTC with billet surface temperature using analytical calculations (T_{Le}, T_b, μ_w, H_{fg}, C_p, σ, ρ_w, K_w ρ_v, Q , L_b , are water Leidenfrost temperature, billet surface temperatures, dynamic viscosity, latent heat of evaporation, specific heat, surface tension, density, thermal conductivity and water flow rate and billet perimeter, respectively. ρ_v and T_{CHF} are the vapor density & critical heat flux temperature and $\alpha, \beta, \gamma, \kappa$ and k are film cooling system parameters)

6. Industrial-scale casting simulation

The development of new numerical framework for industrial casting simulation has motivated the comprehensive sets of parallel simulation & experimental studies herein. The thermal-fluid CFD simulation for mould filling\solidification and evolving thermal-mechanical simulation [13] for stress\strain states have been carried out for simulation case studies. The most challenging parts of thermal-fluid CFD applications for the industrial casting processes are the numerical solidification and cooling simulations. The numerical-analytical aspects of solidification and cooling has briefly been described in previous sections where the change of phase from fluid to mushy and solid state under the thermal gradient caused by cooling process have been modelled. In the research work herein, new porosity functions derived for the momentum sink terms have been implemented and coded in-house. The cooling effects of water spray cooling during casting process and its solidification phenomena has also been implemented using an in-house code which includes the newly developed HTC functions for different water boiling regimes.

Two sets of casting trails which have been carried out for verification and measurement purposes, namely, horizontal and vertical direct chill casting procedures. These continuous and semi-continuous casting trials are performed in-house at semi-industrial scale where round and rectangular aluminium billets were cast using air & water spray cooling. Figure 3 shows the horizontal direct chill casting trails in which a round cross section billet were cast. The parallel numerical simulations trials have also been carried out to examine and calibrate the parameters for the derived mathematical models for solidification and HTC calculations.

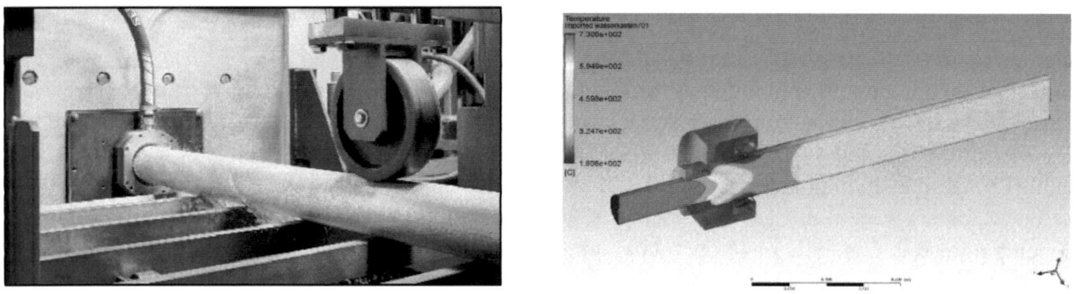

Figure 3. Horizontal semi-industrial direct chill casting process and its numerical 3D half model

7. References

[1] Horr A M 2015 *Computational Methods and Experimental Measurements*, WIT Transactions on Modelling and Simulation, **59** 201

[2] Jäger S, Horr A M, Scheiblhofer S 2017 *6th Decennial International Conference on Solidification Processing*, Old Windsor, in: "*Solidification Processing 2017*" p 675

[3] Ferziger JH, Peric M 1999 *Computational Methods for Fluid Dynamics*, (Berlin, Springer Verlag, 2nd edition)

[4] Horr A M, Kronsteiner J, Mühlstätter C, Scheiblhofer S 2017 *Proceeding of Aluminium Two Thousand World Congress*, Verona, Italy

[5] Scheil E 1942 Bemerkungen zur Schichtkristallbildung, *Z. metallk*, **34** 70

[6] Gulliver G H 1913 J. Inst. Met. **9** 120

[7] Hirt C and Nichols B 1981 Method for the Dynamics of Free Boundaries. J. Comp. Phys, **39** 201.

[8] Brackbill J, Kothe D & Zemach C. A 1992 J. Comput. Phys, **100**, 335.

[9] Blake F C 1922 *Transactions of the Institution of Chemical Engineers*, **14** 415

[10] Carman P C 1956 *Flow of gases through porous media*, (London, Butterworths,)

[11] Voller V R, Brent A D & Prakash C 1989 *Int. J of Heat & Mass Transfer*, **32** 1719

[12] Gu J P, Beckermann C 1999 *Metall. Mater. Trans. A*, **30A** 1357

[13] Horr A M 2018 Analytical and Numerical Analyses of Horizontal Direct Chill Casting, *PSHeRo Project Report II*, LKR, (Austrian Institute of Technology, Vienna).

[14] Horr A M 2016 Calculation of Thermal Evolution and Heat Transfer Coefficient during EMC Process, *Amoree Project Report IV*, LKR, (Austrian Institute of Technology, Vienna).

Acknowledgment

The author would like to thank the Austrian Federal Ministry for Transport Innovation and Technology (BMVIT), Land of Federal State of Upper Austria (within PSHeRo project OÖ Fin-010104/187) and Austrian Institute of Technology (AIT) for the technical/financial support. The author would also like to thank Dr Stephan Ucsnik & Dr Christian Chimani of AIT, for their contributions and support.

Quantification of microstructure to reveal the solidification path of an alloy

A-A. Bogno[1], W.Hearn[1], J.E.Spinelli[1, 2], J. Valloton[1], H.Henein[1]
[1] Department of Chemical and Materials Engineering
University of Alberta, Edmonton, AB T6G 2G6 Canada

[2] Department of Materials Engineering, Federal University of São Carlos, São Carlos, SP 13565-905 Brazil

Email: bogno@ualberta.ca / hhenein@ualberta.ca

Abstract. This paper reports on the development of Solidification Continuous Cooling Transformation diagrams that relate the solidification paths to the inherent microstructures of binary and ternary alloys. The methodology is based on the quantification of a solidified microstructure for its various phase fractions. This measured data is combined with well-established solidification models and phase diagrams to yield undercooling temperatures of individual phases. The thermal history and undercooling of different phases in the solidified alloy are estimated for a wide range of cooling rates (from 10^{-2} °Cs^{-1} to 10^{5} °Cs^{-1}). To describe the methodology, dedicated samples of Al-Cu, Al-Cu-Sc and Al-Si alloys were studied. With said alloys being solidified in a controlled manner, over a wide range of cooling rates and undercoolings, via Impulse Atomization, Electro-Magnetic Levitation, and Differential Scanning Calorimetry

1. Introduction

Cooling rate and undercooling are the most critical process parameters during a solidification process. For some alloys, solidification by cooling too slowly results in the precipitation of undesirable phases or the formation of undesirable phase morphologies. Reactions leading to such outcomes must be avoided in order to achieve optimum properties from the resultant microstructure. For most alloys, when significant undercooling is achieved during solidification, the solid can form using numerous paths from the liquid, yielding a number of possible phases, morphologies, structures and/or supersaturation of solute. Thus, cooling rate must be done as fast as needed coupled with undercooling in order to suppress or promote precipitation of phases and morphologies that are of interest.

The influence of cooling rate and undercooling on the phase transformation behaviour (during solidification) can be described by a solidification continuous cooling transformation (SCCT) diagram. For most industrially relevant metallic alloys, the generated information from such a diagram could be used to optimize the solidification process and thus the properties of the resulting components.

So, as we continue to seek a general way of relating the resultant microstructure to macro-solidification conditions SCCT diagrams could prove a useful asset in alloy and process design. This approach has been achieved in steels with the development of Continuous Cooling Transformation (CCT) diagrams [1,2]. These CCT diagrams have also been used for aging aluminum alloys[3,4] and for representing the crystallization conditions for bulk metallic glasses [5].

In this work, we propose a methodology to formulate and quantify the solidification paths of binary and ternary alloys with an SCCT diagram. The methodology is based on the quantification of a solidified microstructure for its various phase fractions. This measured data is combined with well-established phase diagrams and solidification models to yield undercooling temperatures of individual phases [6] . The thermal history and undercooling of different phases in the solidified alloy are measured either directly or estimated for a wide range of cooling rates (from 10-2 °Cs-1 to 105 °Cs-1). In this paper, the methodology will be described using examples of Al-Cu, Al-Cu-Sc and Al-Si alloys. For the Al-Cu /Al-Cu-Sc alloys, the effect of Sc is investigated and the precipitation of an undesirable phase is analyzed. For the Al-Si alloy, the variation in the eutectic Si morphology is investigated.

2. Methodology

Samples were solidified in a controlled manner, under a wide range of cooling rates and undercoolings, using three very well-known techniques: Impulse Atomization (IA) [7,8], Electro-Magnetic Levitation (EML) [9–11], and Differential Scanning Calorimetry (DSC) [12,13]. In this work, the term cooling rate refers to the cooling rate of the liquid whether in droplet form or in a crucible. DSC is a thermal analysis technique used to determine the amount of energy absorbed or released by a sample as it is heated or cooled inside a crucible. As a non-containerless solidification technique, it yields low nucleation undercooling and its cooling rate (which is directly measureable) is limited to a narrow range (up to 0.8 °C s-1). IA and EML are both containerless solidification techniques that have the ability to induce high nucleation undercooling during solidification. EML generates a single 6 mm diameter droplet per run, which solidifies at relatively low but measurable (via a pyrometer) cooling rates (101 °C s-1), triggered by jets of inert He gas. In comparison, IA generates numerous droplets of various sizes during a single run. The cooling rate in IA is defined by the droplet size and the surrounding cooling gas (He, Ar or N2), and can reach cooling rates up to 105 °C s-1. However, in-situ measurement of the cooling rate in IA is complicated, as the spray of droplets fall within an inert atmosphere with an initial velocity of 0.5ms-1 [7,14]. Consequently, a solidification model for atomization developed by Wiskel et al. [7,14] was used to estimate the cooling rate of each investigated IA droplets.

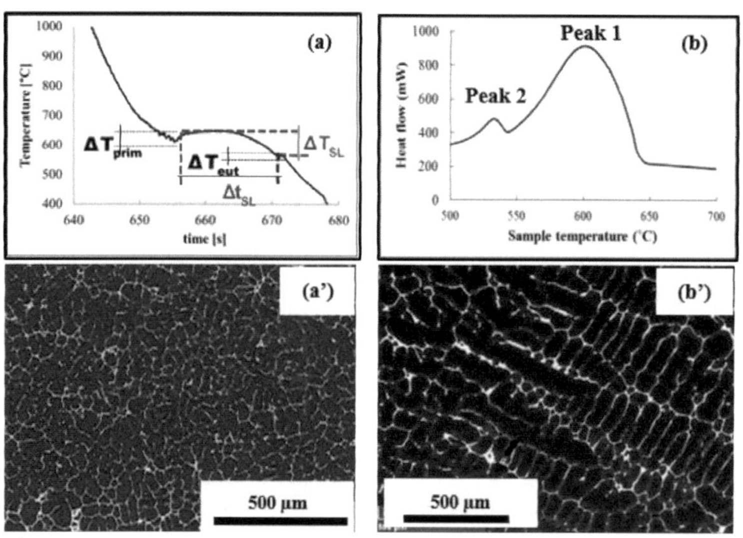

Figure 1: Solidification curve and microstructures of Al-4.5 wt% Cu samples obtained by (a-a') EML with a measured cooling rate of 9.2 °Cs^{-1} and primary and eutectic nucleation undercooling ΔT_{prim} =*34°C*

and ΔT_{eut}= 13°C respectively; (b-b') DSC cooling at the rate of 0.8°Cs^{-1}, peak 1 and peak 2 correspond respectively to the precipitation of primary α-Al and the eutectic structure (α-Al +θ-Al_2Cu).

Figure 1 shows cooling curves of the Al-4.5wt% Cu samples solidified using EML and DSC. The corresponding sample micrographs were obtained by Scanning Electron Microscopy (SEM), and the application of backscattered electron (BSE) imaging. Thus, atomic number (Z) contrast was obtained so that α-Al is identified by the dark color while the Cu rich θ-Al_2Cu is white.

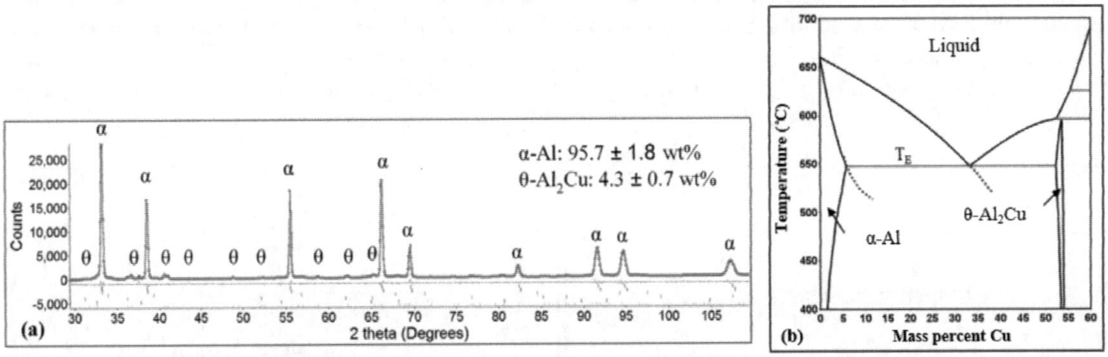

Figure 2: (a) Neutron diffraction diagram of Impulse Atomized droplets of Al-4.5wt% Cu. The phase fractions are obtained by rietveld refinement analysis of the diffraction pattern [15] (b) Al-richer corners of Al-Cu binary phase diagrams. The dashed lines represent the extensions of the solidus and liquidus lines obtained by Thermo-Calc, TTAL7 version 1.1, 2008 [16].

Primary and secondary phase nucleation undercoolings are determined using our new methodology described in [6]. The methodology is based on experimental determination of phase fraction, preferably using a 3D technique such as neutron diffraction (Figure 2). In addition, this methodology includes, secondary dendrite arm spacing (SDAS) measurements, metastable extension of the solidus and liquidus lines of the alloy and a semi-empirical coarsening model that uses the SDAS [17].

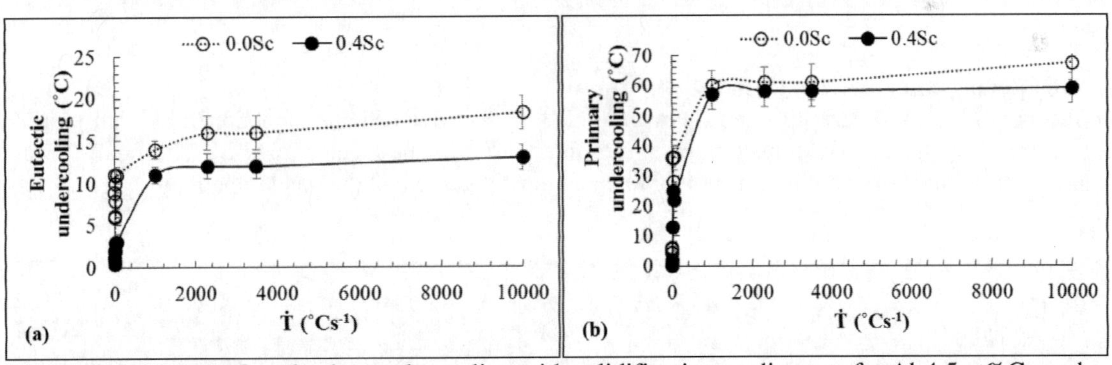

Figure 3: Variation of nucleation undercooling with solidification cooling rate for Al-4.5wt%Cu and Al-4.5wt%-0.4wt%Sc alloys.

Figure 3 shows the variation of primary phase and eutectic nucleation undercooling with cooling rate for the Al-Cu and Al-Cu-Sc alloys. It is worth noting that Sc addition causes a decrease in the nucleation undercooling that is independent of the cooling rate.

3. SCCT diagrams of Al-4.5wt%Cu and Al-4.5wt%Cu-0.4wt%Sc

Temperature, during DSC experiments, is directly measured by thermocouples that are placed underneath the sample and reference crucibles. During EML, the temperature of the sample is recorded via pyrometry, including the rise of temperature due to the release of solidification latent heat. However, as stated earlier, during IA, direct measurement of the thermal history is difficult to achieve in-situ due to the experimental setup. A numerical model is therefore, used to determine the thermal history of an atomized droplet. The model, based on the quantification of heat exchange between an IA droplet and gas in an inert environment, has been developed from Wiskel's heat transfer model formulation [7] [14] and validated in [8].

Figure 4a and 4b show the solidification continuous cooling transformation (SCCT) curves of the investigated Al-4.5wt%Cu and Al-4.5wt%Cu-0.4wt%Sc which are in preparation for publication [18], respectively. From these results, it can be concluded that both primary phase and eutectic nucleation undercoolings increase as the cooling rate is increased. Also, to avoid the formation of the ternary W-phase, which is reported to be detrimental for Al-Cu-Sc alloys[19]–[25], cooling rates of the order of 101 °Cs-1 or higher are necessary.

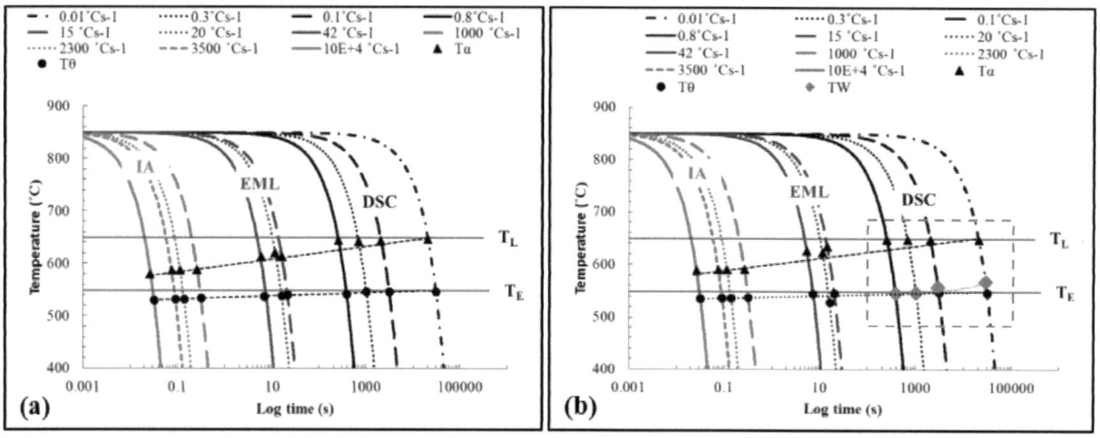

Figure 4: Solidification Continuous Cooling Transformation curves of (a) Al-4.5wt%Cu (b) Al-4.5wt%Cu-0.4wt%Sc. The nucleation temperatures zone of the ternary W-phase is indicated by the square.

4. SCCT diagram of Al-10wt%Si

In recent work by W.Hearn [26] and Hearn et al [27], an Al-10wt%Si alloy, generated by both IA and DSC, was investigated. The transition of Si morphology in this near eutectic Al-Si solidification was analyzed as a function of the liquid cooling rate and the eutectic Si nucleation temperature.

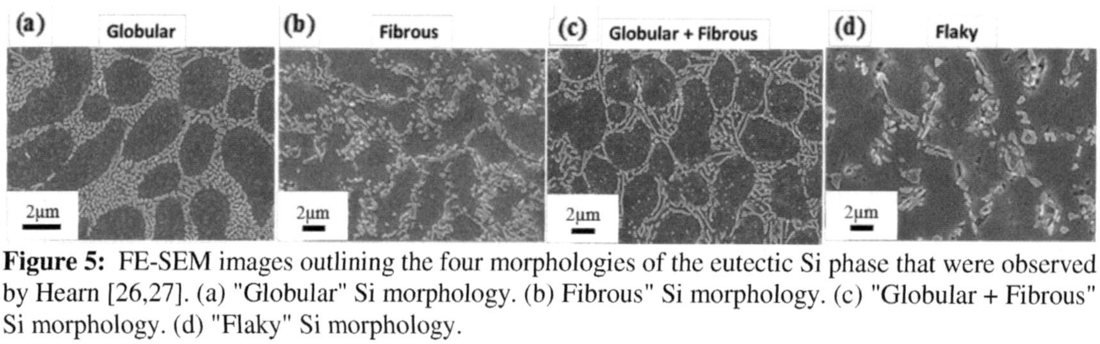

Figure 5: FE-SEM images outlining the four morphologies of the eutectic Si phase that were observed by Hearn [26,27]. (a) "Globular" Si morphology. (b) Fibrous" Si morphology. (c) "Globular + Fibrous" Si morphology. (d) "Flaky" Si morphology.

Figure 5(a through d) shows the four types of Si morphologies observed in W. Hearn's work. Typically, as the cooling rate decreases with the droplet size, the Si morphology will transition from globular → fibrous → flaky. Consequently, an SCCT diagram, shown in Figure 6, was generated in order to correlate the solidification parameters and the inherent Si morphologies. Thus, the primary and eutectic nucleation temperatures, along with the Si morphology, are shown as a function of liquid cooling rate. From the SCCT diagram it can be seen that, the Si morphology transitions from globular → fibrous → flaky corresponding to the decrease in undercooling with decreasing cooling rate. Also, it is worth mentioning that a more refined Si is obtained as the morphology transitions from flaky to globular.

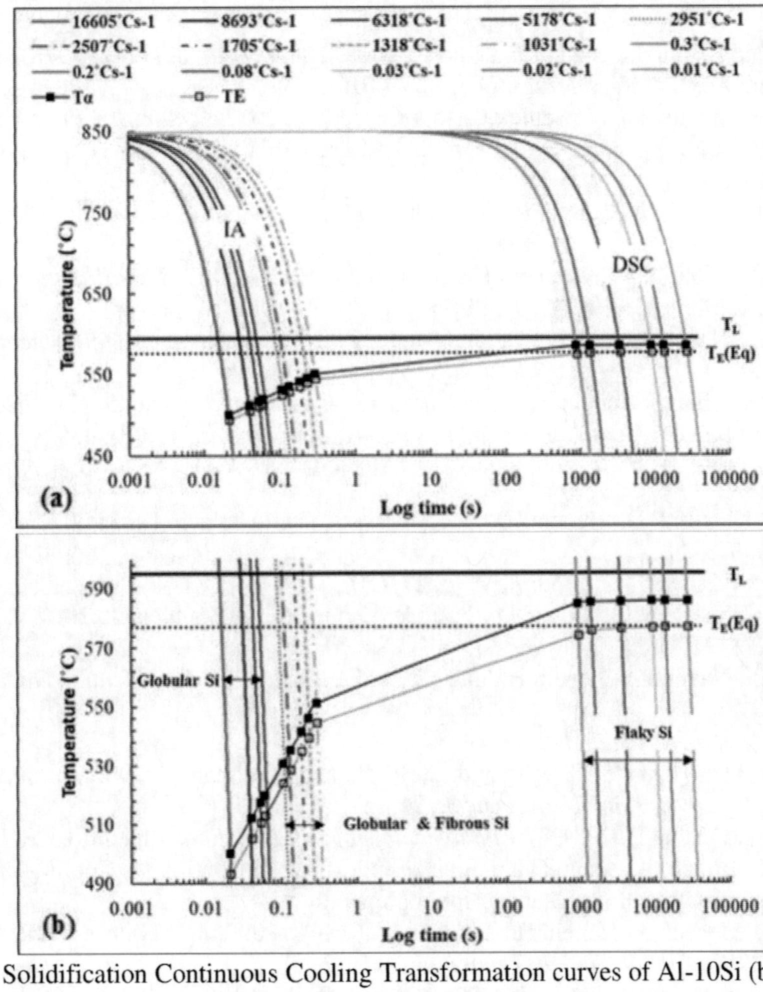

Figure 6: (a) Solidification Continuous Cooling Transformation curves of Al-10Si (b) A zoom on the variation of primary and eutectic nucleation temperature with cooling rate and the corresponding Si morphologies [27].

5. Summary

A new methodology to determine the solidification path of an alloy over a wide range of cooling rates has been presented. The methodology is based on quantitative analysis of the solidification microstructures of samples generated by DSC, EML and IA. Phase fraction measurements were combined with solidification models to determine the nucleation undercooling of individual phases. In this study, Al-Cu, Al-Cu-Sc and Al-Si of hypo-eutectic compositions were investigated. Sc is found to lower the nucleation undercooling and induce the precipitation of the detrimental W-phase at lower cooling rates (<10°Cs-1) and undercooling (<10°C). For the Al-Si alloy, it was found that the Si morphology would transition from flaky →fibrous → globular, where a globular Si morphology was favoured at high cooling rates and undercoolings.

Acknowledgements
The authors express their gratitude to Dimitry Sediako and Ahmed Nawabi for their valuable help during neutron diffraction experiments carried out at the Canadian National Laboratory in Chalk River, Canada. The authors are also grateful to the German Aerospace Center (DLR) and staff for the electromagnetic levitation experiments. The Natural Sciences and Engineering Research Council (NSERC) of Canada and the European Space Agency (ESA) are gratefully acknowledged for their financial support.

References

[1] Xu, G., Wan, L., Yu,S., Liu,L., and Luo,F., 2008, *Mater. Lett.*, **62**, (24), 3978.
[2] Kong,J, and Xie,C., 2006, *Mater. Des.*, **27**, (10), 1169.
[3] Milkereit,B., Wanderka, N., Schick,C.,and Kessler, O., 2012, *Mater. Sci. Eng. A*, **550**, 87.
[4] Kessler, O., von Bargen, R., Hoffmann, F., and Zoch,H.W., 2006, *Mater. Sci. Forum*, **519–521**, 1467.
[5] Schroers, J., Masuhr,A., Busch, R., and Johnson,W.L., 1999, *BULK METALLIC GLASSES*, **554**, 263.
[6] Bogno,A.-A., Khatibi,P.D., Henein,H., and Gandin, Ch-A., 2016, *Metall. Mater. Trans. A Phys. Metall. Mater. Sci.*, **47A,** (9), 4606
[7] Wiskel, J.B., Navel, K., Henein,H., and Maire, E., 2002, *Canadian Metallurgical Quarterly*, **41**, (2), 193.
[8] Prasad, A. Mosbah,S. Henein, H. and Gandin, Ch.-A. 2009, *ISIJ Int.*, **49**, (7), 992.
[9] Liu, R.P., Volkmann, T., and Herlach,D.M., 2001, *Acta Mater.*, **49**, (3), 439.
[10] Tourret,D. Gandin, Ch - A.,Volkmann, T., Herlach, D. M., 2011, *Acta Mater.*, **59**, (17), 6658.
[11] Ilbagi, A., Khatibi,P.D., Henein,H., Gandin, Ch.-A., and Herlach, D.M., 2012, *Int. Conf. 3D Mater. Sci.*, 67.
[12] Hayoune, A., 2013, *Int. J. Mod. Phys. B*, **27**, (20), 1350084.
[13] Gao, Y.L., Zhuravlev,E., Zou,C.D., Yang,B., Zhai,Q.J., and Schick,C., 2009, *Thermochim. Acta*, **482,** (1–2), 1.
[14] Wiskel, J.B., Navel, K., Henein,H., and Maire, E., 2002, *Canadian Metallurgical Quarterly*, **41**, (2), 97.
[15] Young, R.A., *The Rietveld Method.* 1993.
[16] Andersson, J.O., Helander,T., Höglund,L., Shi,P., and Sundman,B., 2002, *Calphad Comput. Coupling Phase Diagrams Thermochem.*, **26**, (2), 273.
[17] Kurz, W., and Fisher,D.J., 1998, *Fundamentals of Solidification*, 4th editio. CRC Press.
[18] Valloton, J., Bogno, A.-A., and Henein, H., Solidification Continuous Cooling Transformation Diagrams of metallic alloys (Manuscript in preparation).
[19] Prasad, A., Mosbah,S., Henein,H., and Gandin,Ch.-A., 2009, *ISIJ International*, **49**, (7), 992.
[20] Røyset,J., 2007., *Metall. Sci. andTechnology*, **25**, (2),11.
[21] Røyset, J., and Ryum,N., 2005, *Int. Mater. Rev.*, **50**, (1), 19.
[22] Norman, A., Prangnell,P., and McEwen,R., 1998, *Acta Mater.*, **46**, (16), 5715.
[23] Bo, H., Liu,L.B., and Jin,Z.P., 2010, *J. Alloys Compd.*, **490**, (1), 318.
[24] Toropova, L.S., Eskin, D.G., Kharakterova,M.L., and Dobatkina, T.V., 1998, *Advances in Aluminum Alloys Containing Scandium.* Amsterdam: Gordon and Breach Science publishers.
[25] Kharakterova, M.L., 1991, *Izv. Akad. Nauk SSSR, Met.*, **4**, 191.
[26] Hearn, W., 2018, “The Microstructure, Morphology and Mechanical Properties of Rapidly Solidified Al-10wt%Si Alloy,” University of Alberta.
[27] Hearn, W., Bogno,A.-A., Spinelli, J., Valloton,J., and Henein, H., 2018, *Metall. Mater. Trans. A.,(in press).*

Effect of cooling rate on the porosity defect in the thick aluminum casting by 3D computed tomography analysis

I S Cho[1*], S Y Kwak[1], Y H Kim[2], H K Lee[2], S M Yoo[1]

[1]*Korea Institute of Industrial Technology, 156 Gaetbeol-ro, Yeonsu-gu, Incheon, 21999, South Korea*
[2]*Daeshin Metals. Co., Ltd. 24-11, Pallyong-dong, Changwon-si, Gyeongsangnam-do, South Korea*

*Corresponding author's e-mail : ischo@kitech.re.kr

Abstract. In the present study, effect of cooling rate on the formation of the porosity in the thick aluminum sand casting was investigated. Nowadays large scale thick aluminum casting replaces steel frame for vacuum chamber for semiconductor production, with the consideration of weight and cost reduction. Several thick aluminum castings were manufactured using chill with temperature measurements. The castings were inspected by using 3D computed tomography in order to quantify the porosity defect density in the castings. Effect of the thickness of the chill on the porosity defect density were discussed.

1. Introduction

Vacuum chambers for manufacturing solar cells, LCD display and semiconductors are fabricated by cutting and machining of steel or aluminum plate and welding with them. This costs too much time and money, so it will be very important to reduce the cost for vacuum chamber manufacturing. Nowadays machining and welding process are replaced with sand casting process. When the sand casting is applied to manufacturing large vacuum chamber rather than machining and welding, over 40% cost reduction is possible [1]. However in the cast products, gas defects inside the casting can be leaked during operating the vacuum chamber, the casting technique minimizing the gas defect and inclusions inside the casting and its inspection technique is necessary.
Generally large cast product has thick walls, therefore cooling is crucial for taking sound castings. Chills are usually used for accelerating cooling rates, and have several advantages as follows

- Smooth melt feeding to shrinkage cavities
- Increase of temperature gradient
- Fine grain structure
- Decrease of dendrite spacing
- Decrease of large shrinkage cavities

When the molten aluminum solidifies, solubility of hydrogen in the melt is decreased and gas porosities are formed. When chills are applied, fast cooling rates can be obtained and gas in the melt is supersaturated. This avoids nucleation of gas porosity, so amount of gas porosity can be minimized.

Traditionally, there have been many studies on the shrinkage cavity of products according to cooling rate in aluminum casting process [1]. Internal defects of casting product were analyzed using industrial CT (Computed Tomography) [2] and a series of studies have been carried out by Kwak [3, 4] to evaluate the mechanical strength of fatigue and impact according to defects which were detected through CT. However, since the cooling rate depends on the location in the product, its characteristics also vary in the distribution of the cooling rate proportionally. Therefore, it is very important to obtain temperature distributions during cooling in the product. In this paper, an investigation was carried out the temperature history of the A356 cast aluminum alloy specimen of 40 mm thickness used for the vacuum chamber by varying the thickness of the chill and the simulation for the temperature history.

2. Experimental Method

2.1 Preparation of specimens by casting experiment

In the present study, size of 200x150mm with 40mm thickness specimens were cast. As shown in Fig.1. With computer simulation, gates and risers is found to be properly designed to obtain sound casting with no shrinkage defects.

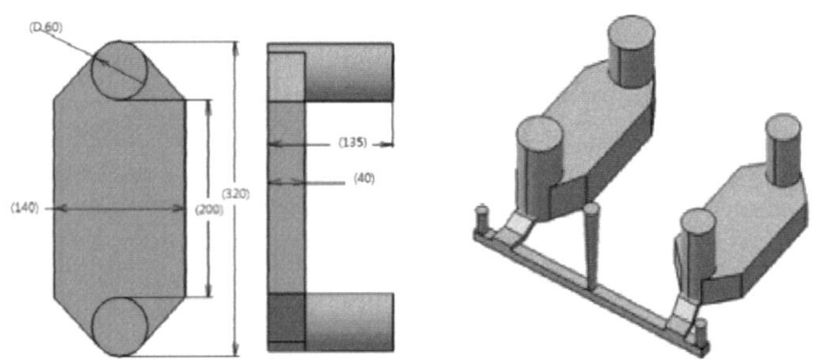

Figure 1. Schematic diagram of specimen and its casting design.

The chill material was SS400 steel and the thickness was 40mm, 60 mm, and 80 mm (hereafter 40T, 60T, 80T, respectively). The surface of the chill is machined with knurled shape, in order to provide more contact area than that of flat surface chill between casting and sand mold, as shown in Fig. 2. The position of the thermocouple was 10mm from the chill, same position with that of porosity defect analysis. In order to obtain the temperature distribution of the casting, 6 thermocouples are installed inside the mold cavity, and the temperature is measured during casting.

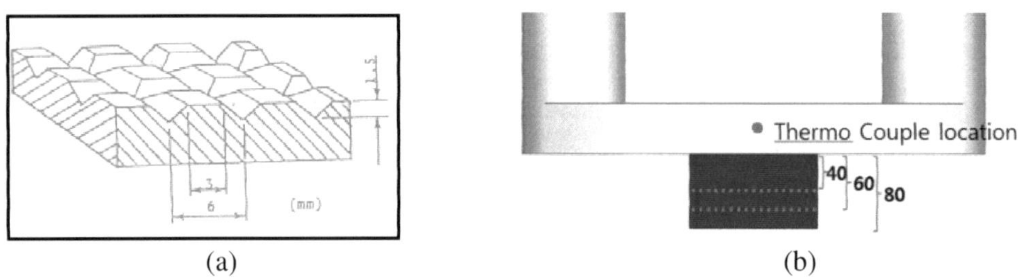

Figure 2. (a) Schematic diagram of knurl shaped chills and (b) thickness of the chills and the position of the thermocouple and chill.

A356 alloy was used for casting materials and compositions were shown in Table 1.

Table 1 Chemical composition of A356 proposed in the present study

	Al	Si	Fe	Cu	Mg	Ti
wt%	92.5	7.0	0.09	0.002	0.35	0.12

Fig. 3 shows the photograph of castings in the present study. Knurl shape in the bottom of the casting was designed in order to achieve good contact of the casting and mold.

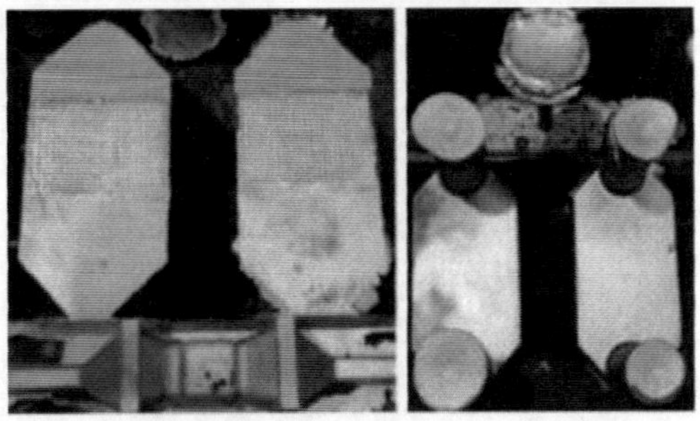

Figure 3. Photographs of the casting used in the present study.

2.2 Analysis by simulation and computed tomography (CT)
In order to investigate the relationship between cooling rates and porosity in the specimen, computer simulation analysis and 3D industrial CT was used. For the simulation of the casting process, in-house developed Z-Cast software is used, and Rayscan 250E CT system is used for CT scan and in-house developed VX3D software is used for porosity analysis. Fig. 4 shows the photograph of Rayscan 250E and VX3D used in the present study.

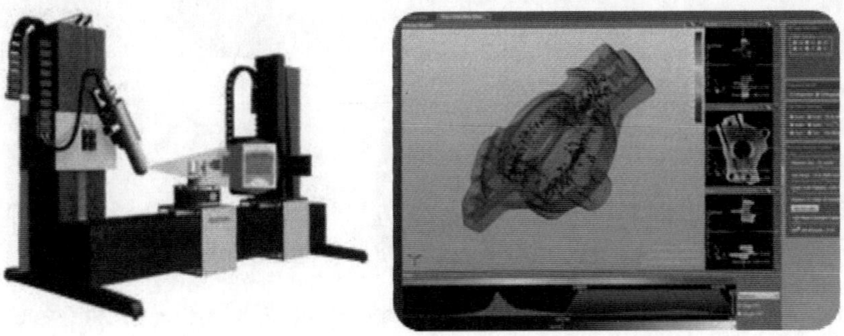

Figure 4. Photographs of Rayscan 250E CT system and VX3D analysis software.

3. Result and Discussions

3.1 Cooling Rate by Experiment & Numerical Data
Fig. 5 shows experimental cooling curves for chill thickness 40T, 60T, and 80T. Fast cooling and short solidification time was found when the chill thickness increases.
Since it is difficult to install thermocouples in all areas of the specimen, the numerical conditions for heat transfer analysis those can simulate temperature history correspond to the thermocouple experimental data was set. Then the cooling rates for each area inside the product were obtained through the heat transfer analysis. Fig. 6 shows the experimental and numerical values at the thermocouple location 1 in Fig. 7. The vertical positions of cooling rate analysis in the specimen are 10 mm, 20 mm,

IOP Conf. Series: Materials Science and Engineering **529** (2019) 012057 doi:10.1088/1757-899X/529/1/012057

and 30 mm from the chill, and each positions are located horizontally from 12.5mm (position 2, 4, 6) and 37.5mm (position 1, 3, 5) from the right edge of the chill, as shown in Fig. 7. The simulation and experiment cooling curves is shown to be a relatively good agreement.

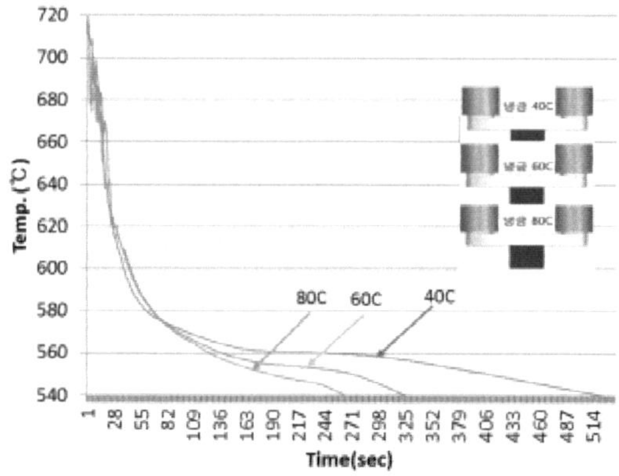

Figure 5. Cooling curves after pouring with chill thickness variance.

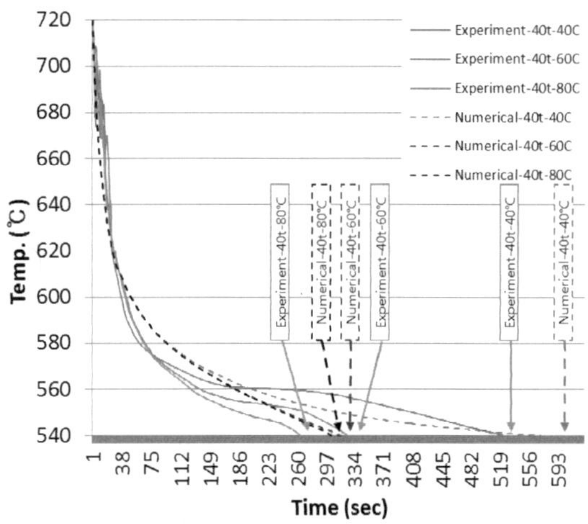

Figure 6. Cooling curves after pouring at thermocouple location 1

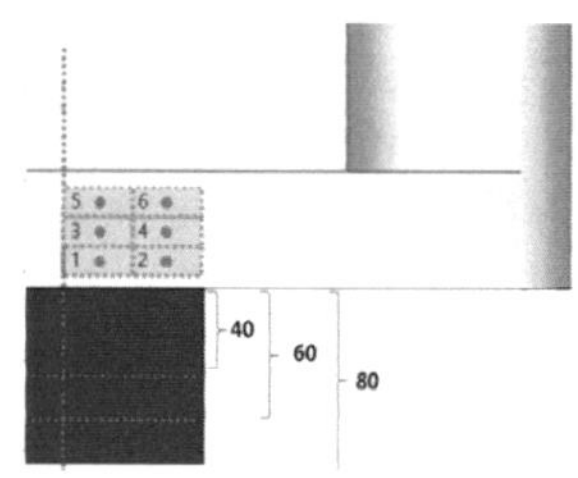

Figure 7. The locations of thermocouples for simulated cooling rate analysis.

The cooling rate can be evaluated as the temperature interval divided by the cooling time to 542°C from 613°C which are the solidus and liquidus temperatures of A356.

Fig. 8 shows cooling curves for every 6 positions in the casting in the case of using 40T chill, by simulation analysis. The fastest cooling rates is found in the thermocouple position 1, and then 3 → 5 → 2 → 4 → 6. The slowest cooling rates is shown in the thermocouple position 6.

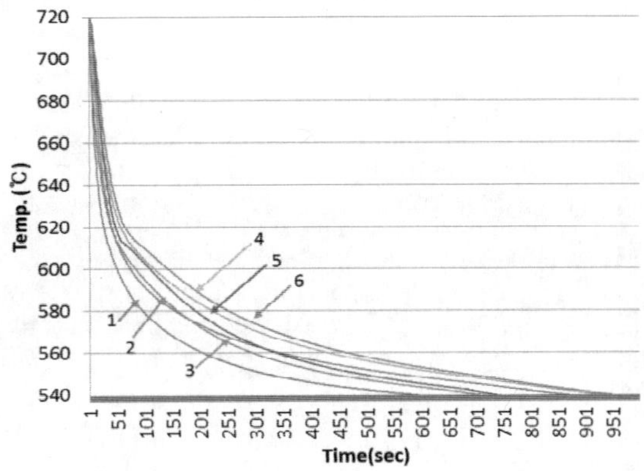

Figure 8. Cooling curve of 40mm chill

The cooling rates of all specimens are listed in Table 1. The position 1 of 80T has the fastest cooling rate of 0.266°C/sec, and the position 6 of 40T was the slowest at 0.088°C/sec. As shown in Table 2, when comparing the cooling rates according to the thickness, 60T and 80T shows relatively similar cooling rates, but they were different from 40T. It is considered that maximum chill thickness for maximum cooling rate will be 60T.

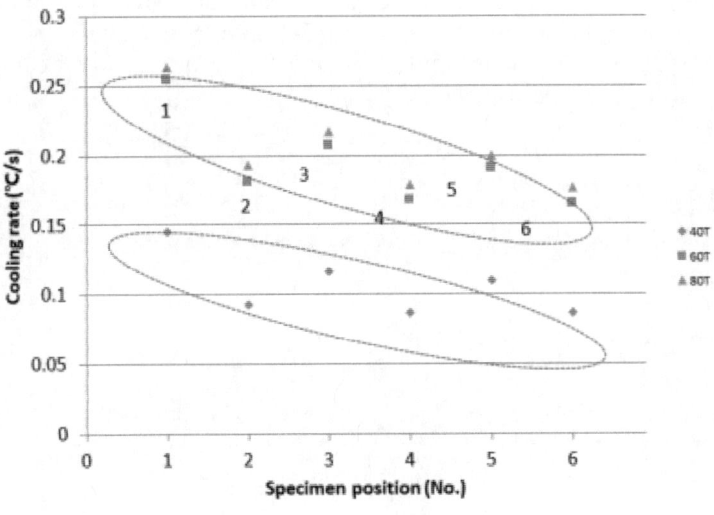

Figure 9. Effect of chill thickness on cooling rates.

Table 2. Cooling rate according to thickness of chill

No	Cooling Rate [°C/sec]		
	40T	60T	80T
1	0.158	0.257	0.266
2	0.098	0.184	0.196
3	0.125	0.210	0.220
4	0.094	0.171	0.182
5	0.118	0.193	0.202
6	0.088	0.168	0.179

3.2 Comparison of 3D CT results with chill thickness and cooling rate

Figure 10 shows an example of porosity analysis. Table 3 shows the amount of porosity and average diameter in the specimens. With the chill thickness of 40T, it is found the amount of porosity is averagely 0.23%, whereas 60T for 0.039% and 80T for 0.017%. Average diameter of the porosity is 0.42mm for 40T, 0.36mm for 60T and 0.35 for 80T. It is considered that the thin chill thickness would have larger amount of porosity.

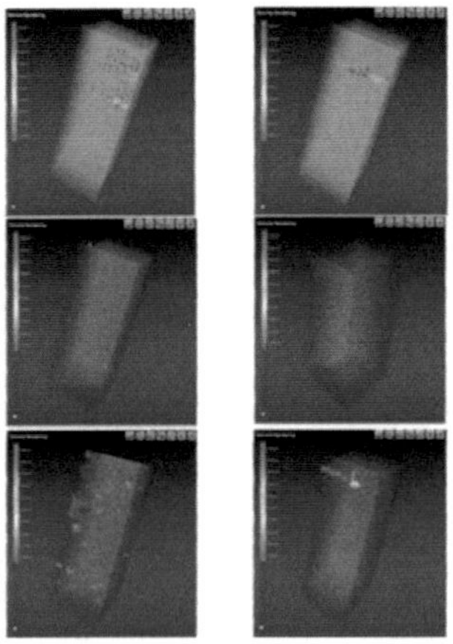

Figure 10. Example of posority analysis of the specimen.

Table 3 Amount and average diameter of the porosity in the specimens

TC Position	40T		60T		80T	
	porosity [%]	Avg. Diameter [mm]	porosity [%]	Avg. Diameter [mm]	porosity [%]	Avg. Diameter [mm]
1	0.1484	0.39	0.0045	0.36	0.0016	0.32
2	0.1793	0.39	0.1163	0.36	0.0094	0.33
3	0.1346	0.41	0.0065	0.33	0.0148	0.38
4	0.1703	0.41	0.0263	0.36	0.0115	0.36
5	0.5218	0.48	0.0793	0.40	0.0524	0.37
6	0.2564	0.42	0.0054	0.35	0.0126	0.35
Avg.	**0.2351**	**0.42**	**0.0397**	**0.36**	**0.0170**	**0.35**

Fig. 11 shows relationship of porosity with coolingrates. In the case of chill thickness 60T and 80T, porosity percentage is found to be below 0.1%, otherwise more than 0.1% porosity is found in the case of 40T. Chill thickness of 60T and 80T shows the cooling rate to be above 0.16oC/sec, so faster cooling rate than 0.16oC/sec is desirable to acquire less porosity below 0.1%.

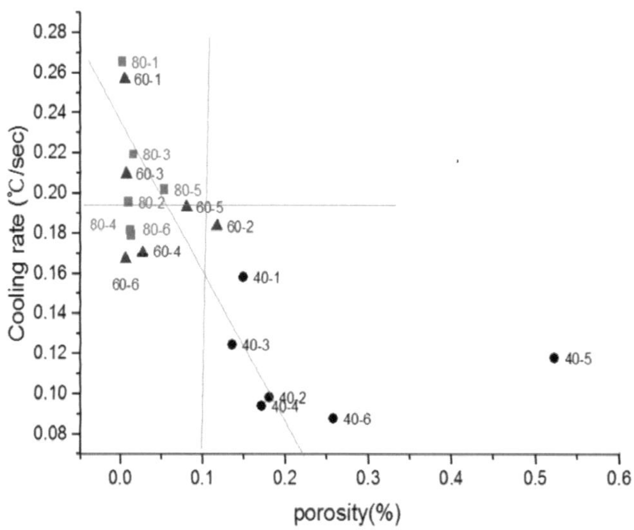

Figure 11. Porosity with cooling rates of the specimens.

4. Summary

It was not realistic to measure the temperature histories by installing thermocouples in all areas of the specimen. Therefore, the heat transfer analysis combined with the thermocouple temperature history data was performed to obtain the cooling rate distributions for the whole specimens. In case of 40mm aluminum block, 40 mm chill and 60mm chill were different at cooling rate, but the difference between 60mm and 80 mm was small. The cooling rate distribution inside the specimens could be obtained by using the commercial casting process analysis program Z-Cast. And it can be used to analyze the correlation between the metallographic or defect distribution results and the cooling rate.

5. Acknowledgments

This work was supported by the Ministry of Trade, Industry and Energy (MOTIE, South Korea) [Project Name: Development of large-scale high-quality aluminum sand casting technology].

6. References

[1] M.J. Oh, S.M. Yoo, I.S. Cho and Y.H. Kim, J. Korea Foundry Society, "A study on the Effect of Insulating sleeve on solidification characteristics of A356 aluminum alloy", 31-4 (2011) 205-211.
[2] I.S. Cho, S.Y. Kwak, M.J. Oh, S.C. Jang, C.H. Lim, 2nd Korea Japan Industrial CT Workshop,"Evaluation of effectiveness of casting defect on physical properties for diecast by using 3D computed tomography"17-23
[3] S.W. Lee H.K. Kim, H.Y. Hwang and S. Y. Kwak, J. Korea Foundry Society, "Fatigue Life Analysis for Al Alloy Casting Containing Internal Defects", 30-5 (2010) 25-28.
[4] S.C. Hwang and S.Y. Kwak, J. Korea Foundry Society, "Effect of Shrinkage Defect on Fracture Impact Energy of A356 Cast Aluminum Alloy", 34-1 (2014) 248-252.

IOP Conf. Series: Materials Science and Engineering **529** (2019) 012058 doi:10.1088/1757-899X/529/1/012058

Numerical simulation of fluid flow, solidification and defects in high pressure die casting (HPDC) process

K Dou[1], E Lordan[1], Y J Zhang[1], A Jacot[1] and Z Y Fan[1]

[1] Brunel Centre for Advanced Solidification Technology (BCAST), Brunel University London, Kingston Lane, Uxbridge, UB8 3PH, United Kingdom

E-mail: Kun.Dou@brunel.ac.uk

Abstract. The high pressure die casting process is extensively used to manufacture light metal parts with high productivity. A major drawback of the process is the relatively high variability in mechanical properties and poor repeatability between casting cycles, limiting the achievement of weight reduction through lighter design. Although it has been established that mechanical properties are adversely affected by casting defects, the origin of the relatively high randomness in the HPDC process is not well understood. Numerical simulation is a powerful and cost-effective tool to address this question, as it gives access to quantities that are difficult to obtain experimentally. A numerical simulation approach based on the finite element casting software ProCAST has been developed. The model was applied to the casting of aluminium tensile test samples, which were used to measure the tensile properties of the alloy. Simulation permitted the study of fluid flow, solidification and defect formation during each stage of the HPDC process: pouring, injection and cooling. Air entrapment and porosity distribution in the cast part were predicted. The results were compared with temperature measurements, porosity observations and solid distribution in the sleeve prior to injection. Although the results are still very preliminary, some trends could be established between the level of turbulence of the melt during injection and reduced elongation.

1. Introduction

The high pressure die casting process (HPDC) is widely used within the automotive industry to produce parts requiring high dimensional accuracy and low surface roughness, exhibiting great productivity and competitive cost for mass production [1,2]. The application of light metal castings manufactured by HPDC is an effective way to reduce vehicle weight and limit carbon emissions. HPDC products exhibit good mechanical properties due to fine grain formation under rapid cooling. However, some variability in mechanical properties among different positions in the die and batches is often observed[3-5]. Although it has been established that mechanical properties are adversely affected by casting defects[6], the origin of the relatively high randomness in the HPDC process is not well understood. Optimizing the mechanical properties and controlling their variability through trial-and-error experiments can prove to be time-consuming and it is likely that the procedure will have to be repeated for new designs. More fundamental understanding is thus desired to be better guided in this procedure. A combined approach based on both experiments and simulation was developed to address the topic. The present contribution is focused on the modelling effort to bring insight on fluid flow, solidification and defect formation

IOP Conf. Series: Materials Science and Engineering **529** (2019) 012058 doi:10.1088/1757-899X/529/1/012058

during each stage of the HPDC process, including the role of the sample position in the die. The results are compared with experimental data and the possible origin of the variability of the mechanical properties is analysed.

2. Experimental

Silafont-36 (Al-9Si-0.3Mg-0.5Mn) was used throughout this work. A 40 Kg crucible of Silafont-36 was melted in an electric resistance furnaces and held at 750°C for 30 min to maintain a uniform composition distribution. Then, the melt was degassed using the HSMC device[7]. This process involved two phases (i) degassing for 10 min at a rotor speed of 1500 rpm and an argon flow rate of 0.2 L/min, followed by (ii) conditioning via intensive melt shearing for an additional 20 min without argon flow. Following melt treatment, molten metal was poured into the shot sleeve of a Bühler 4500 kN locking force cold chamber HPDC machine using a transfer ladle. The temperature of the melt, shot sleeve and die cavity were maintained at 680°C, 180°C and 150°C respectively. The molten metal was then injected into the die cavity at a slow shot speed of 0.3 ms-1 and a filling speed of 3.6 ms-1 to produce eight round tensile samples with a nominal gauge diameter of ø6.35 mm in accordance with ASTM standards. The geometry of the samples and gating system is illustrated in Figure 1.

3. Modelling

As HPDC is a complex process involving high thermal gradients, high fluid velocities, and high probability for the melt free surface to fold and fragment, it is important to address all the stages of the process, starting with shot sleeve filling, followed by melt injection (using the actual piston kinematics), and finally solidification inside the die. The commercial casting simulation suite ProCAST was used to simulate the whole process. The CAD models for the casting volume, the shot sleeve, the fixed/mobile dies and the shot piston were built and assembled in VisualMesh[11]. A finite element (FE) mesh was then generated accordingly with a minimum mesh size of 0.5 mm to ensure an accurate description of fluid flow through the thin ingate (2 mm×5 mm). Details about the geometry and meshes are shown in Figure 1. Fluid flow, heat transfer and solidification of the melt during the different steps of the process were modelled with the ProCAST solver, which calculates the evolution of the melt free surface with a Volume of Fluid (VOF) method and solves coupled conservation equations of momentum, mass, and heat [8-10].

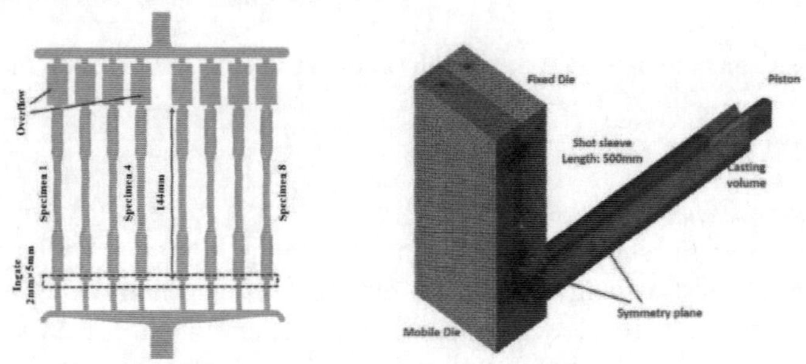

Figure 1 Geometry of the casting and gating system (left) and finite element mesh used for the simulation (right).

(left: tensile sample casting in full symmetry, right: HPDC configuration in half symmetry with FE meshes)

The Al melt was treated as an incompressible fluid. The amount of entrained air in the melt and final casting was predicted qualitatively using the *GAS* model[11,12], which is based on a prediction of the turbulence/fragmentation of the free surface. In addition, the *Advanced Porosity Module* (APM)[13-15] was also used to solve the Darcy equation and the segregation of gas in the mushy zone, in order to predict the distribution of microporosity in samples.

3.1 Simulation conditions

The pouring of the melt through the pouring hole in the shot sleeve was simulated with an inlet boundary condition, which specifies the amount of metal flowing through a selected surface section. In the HPDC process, the inner wall of the shot sleeve is typically coated with oil for protection and lubrication, which can influence the flow and heat transfer at the interface. In this work, a temperature dependent interfacial heat transfer coefficient (iHTC) and a dedicated function to account for the effect of the surface roughness on the boundary layer at the wall (*WALLF function*[11]) were used to account for these effects. The piston movement during the shot sleeve injection process was defined based on the displacement curves obtained from the HPDC machine control panel. To better simulate the intensification stage of the HPDC process, a critical solid fraction of 0.95 was assumed as a threshold under which the biscuit region can still be fed with melt continuously.

3.2 Model validation

Dedicated experiments were carried out to validate the heat transfer analysis and the solid distribution predicted by the model. The melt temperature at different locations in the shot sleeve was measured with thermocouples and the distribution of the solid was analysed by emptying the content of the shot sleeve instead of proceeding with injection. As can be seen in Figure 2 (left), the predicted and measured temperatures are in good agreement. The main discrepancies are observed when the thermocouples are reached by the melt, which is explained by their relatively high response time (~1 s). In addition, it is observed that the shape of the metal after partial solidification in the sleeve closely resemble those predicted (Figure 2, right).

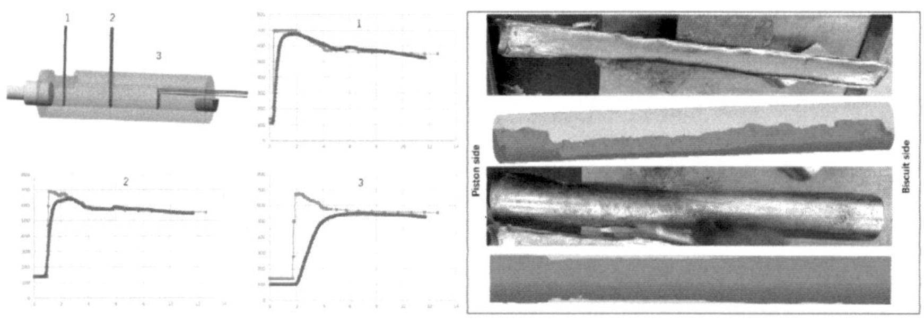

Figure 2 Comparison of temperature prediction (left) and solid shape (right) in the shot sleeve.

4. Results and Discussions

4.1 Pre-solidification in the shot sleeve

When the melt is poured into the shot sleeve, the contact with the relatively cold piston and shot sleeve leads to the formation of solid, which is usually referred to as externally solidified crystals (ESCs). Although the solid crystals were not directly modelled in this approach, solid formation in the sleeve could be predicted. Figure 3 shows the evolution of the solid fraction prior to melt injection. It could be seen that solid forms near the piston and at the bottom of the sleeve due to rapid cooling. As the melt front progresses towards the biscuit, more solid forms at the bottom of the shot sleeve. During the slow injection stage, the solid in contact with the piston is pushed with the potential for some crystals to re-melt. During the fast injection stage, the mixture is transported at high velocity into the die cavity and subsequently solidifies.

IOP Conf. Series: Materials Science and Engineering **529** (2019) 012058 doi:10.1088/1757-899X/529/1/012058

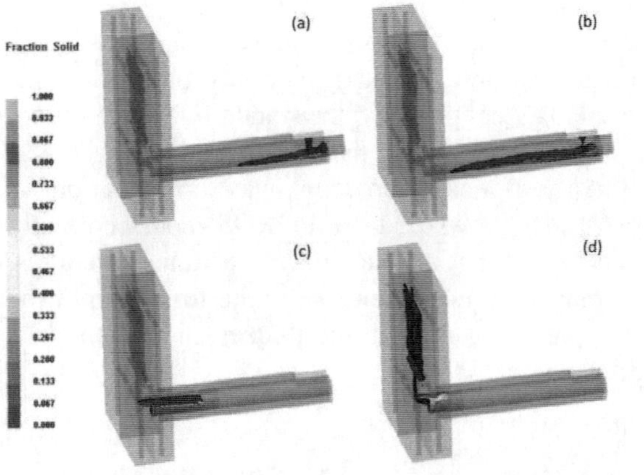

Figure 3 Evolution of the solid fraction in the shot sleeve, gating system and die during injection.

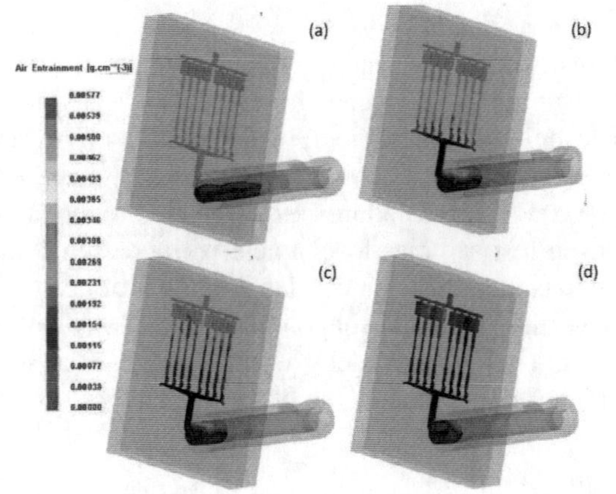

Figure 4 Predicted air entrainment during filling and injection.

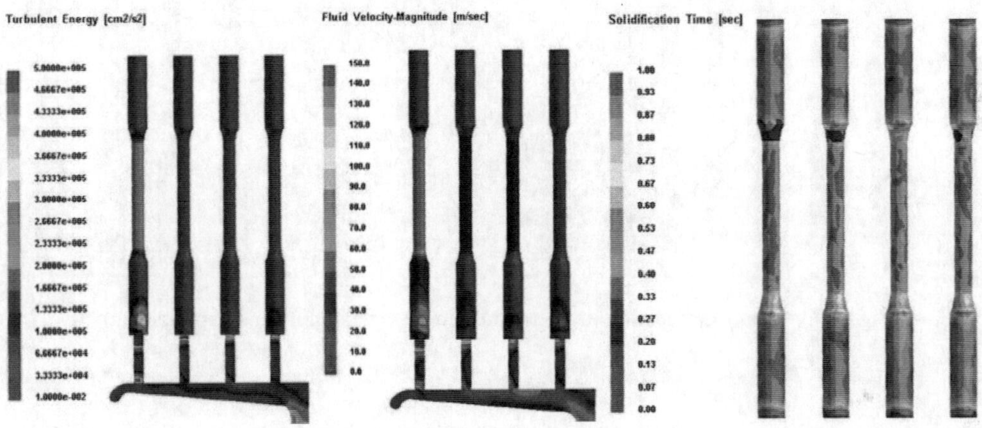

Figure 5 Calculated turbulence energy, melt velocity and solidification time in the tensile bars.

4.2 Air entrainment

Figure 4 illustrates the predicted evolution of entrained air during the process. As the plunger accelerates during the first stage, the melt is pushed towards the biscuit. Most of the air in the shot sleeve is driven out, but some can be entrapped at the melt front, the amount being largely determined by the shape of the wave. During the second stage, the melt is injected into the die at high speed. A certain amount of air accumulates and is trapped inside the final casting. Optimizing the design of venting systems, the motion of the plunger and the application of vacuum devices can minimize the amount of entrapped air, which is beyond the scope of the present study.

4.3 Fluid flow and solidification in die cavity

As shown in Figure 5, during the die filling process, vigorous turbulences are predicted at the ingate sections, which leading to substantial differences in turbulent energy and maximum velocity inside the tensile sample cavities. It can be seen that the maximum velocity magnitude exceeds 100 m/s at the ingate area. The fluid velocity magnitude reaches about 80 ms-1 in the left cavity however does not exceed 20 ms-1 in the right one. Such differences in fluid velocity magnitude can potentially influence the variability of properties between cast parts for a given shot.

4.4 Porosity

Porosity defects in casting can be categorized into macro- and micro-porosities based on their sizes[17-19]. Microporosities can themselves be categorized into gas and shrinkage porosity based on their morphology. In this paper, two distinct models were used to assess porosity: the standard POROS model of ProCAST[11, 12], which focuses on shrinkage and macro porosity, and the advanced porosity model (APM) [14] , which addresses different categories of porosity. Optical microscopy was carried out on the tensile test samples to characterize porosity. Figure 6 (left) shows that the POROS model correctly predicts the formation of a large amount of porosity at the bottom of the bars, due to feeding difficulties. The rather spherical shape indicates that gas certainly played an important role in the formation of the pores. The APM model predicts the presence of microporosities only in the thicker ends of the bar, which is confirmed by the observation (Figure 6, right).

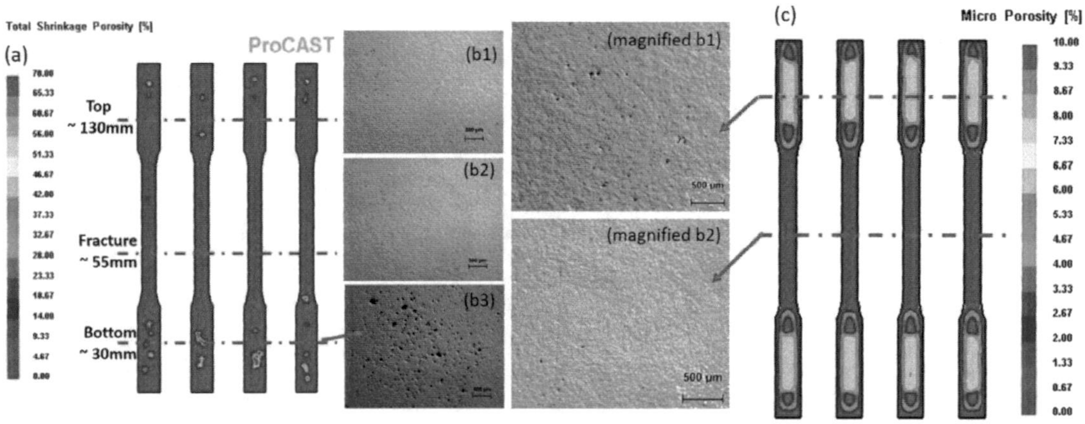

Figure 6 porosity distribution in the tensile samples (modelling and experiments)

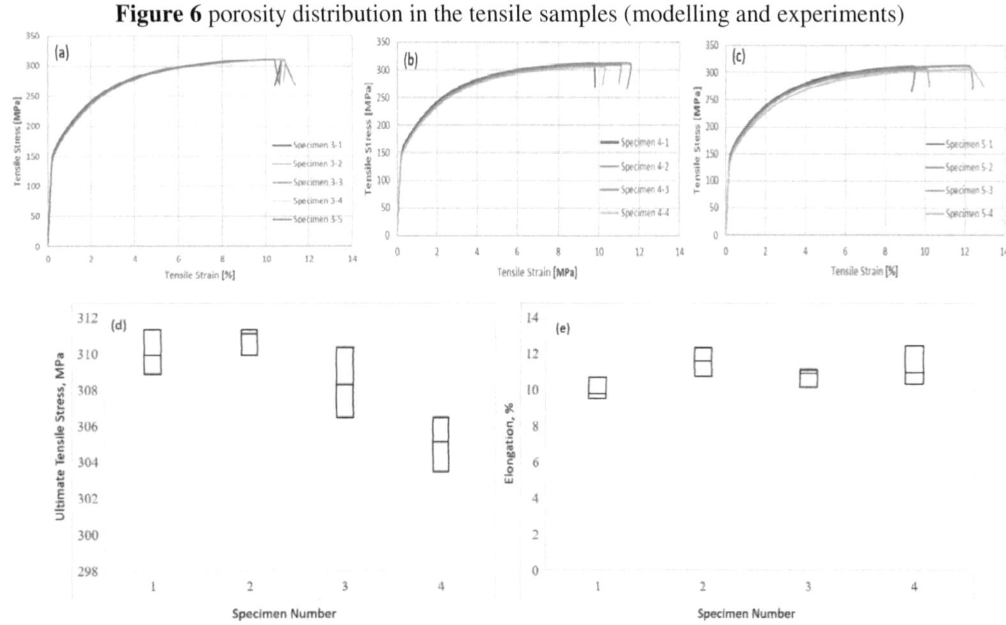

Figure 7 Test results for tensile stress and elongation of samples in different locations

4.5 Mechanical properties

The tensile properties of samples from three consecutive shots are shown in Figure 7a, 7b and 7c for

IOP Conf. Series: Materials Science and Engineering **529** (2019) 012058 doi:10.1088/1757-899X/529/1/012058

different positions in the die (See Figure 1 for the position labels). Scatter plots of the ultimate tensile stress and elongation as a function of position are presented in Figure 7c and 7d (Refer to Figure 1 for the position labels). It can be observed that elongation exhibits higher relative variations than UTS and that, globally, the position in the die does not seem to play an obvious influence on the properties considering the data scatter. However, a tendency for reduced elongation at position #1 can be pointed out. According to the simulation, the highest level of turbulence and maximum melt velocities are obtained at position #1, which might be at the origin of this tendency. However more data is required to confirm this correlation.

5. Conclusion

A simulation approach has been developed to gain insight into melt flow, solidification and defect formation during the high pressure die casting of aluminium alloys. The evolution of the solid formed in the shot sleeve was analysed and discussed. Predictions of porosity showed to be in good agreement with empirical measurements. The UTS and elongation of the 12 tensile samples were measured. A trend for reduced elongation where the level of turbulence and melt velocities are the highest could be detected. More tensile data and further investigation is however required before the origin of variability in mechanical properties can be evidenced.

References

[1] Bonollo F, Gramegna N and Timelli G 2015 *JOM* **67** 901.
[2] Mordike B L and Tü Ebert 2001 *Mater. Sci. Eng. A* **302** 37.
[3] Lee S G, Patel G R, Gokhale A M, Sreeranganathan A and Horstemeyer M F 2006 *Mater. Sci. Eng. A* **427** 255.
[4] Gunasegaram D R, Finnin B R and Polivka F B 2007 *Mater. Sci. Tech.* **23** 847.
[5] Chen Z W 2003 *Mater. Sci. Eng. A* **348** 145.
[6] Ji S, Yang W, Jiang B, Patel JB, Fan Z 2013 *Mat. Sci. and Eng. A* **566** 119.
[7] Zhang Y J, Patel J B, Lazaro-Nebreda J and Fan Z Y 2018 *JOM* **70** 2726.
[8] Vijayaram T R, Sulaiman S, Hamouda A M and Ahmad M H 2006 *J. Mater. Proc. Tech.* **178** 29.
[9] Cleary P W, Ha J, Prakash M and Nguyen T 2010 *Appl. Math. Model.* **34** 2018.
[10] Guo J Z and Samonds M 2011 *JOM* **63** 19.
[11] ProCAST version 2018.1 User Manual 2018 ESI Group France.
[12] Guo J Z, Makinde A and Bewlay B 2009 *Modeling of Casting Welding and Advanced Solidification Processes* (TMS: Warrendale) pp 337–344.
[13] Pequet C, Rappaz M and Gremaud M 2002 *Metall. Mater. Trans. A* **33** 2095.
[14] Guo J Z et al. 2015 *IOP Conference Series: Materials Science and Engineering* **84** (IOP Publishing).
[15] Carlson K D and Beckermann C 2009 *Metall. Mater. Trans. A* **40** 163.
[16] Gaël C and Rappaz M 2006 *TMS Annual Meeting*. No. LSMX-ARTICLE-2008-003.
[17] Lu Y, Taheri F, Gharghouri M A and Han H P 2009 *J. Alloy. Compd.* **470** 202.
[18] Lee S G and Gokhale A M 2006 *Scripta Mater.* **55** 387.
[19] Fan Z Y et al. 2009 *Int. J. Cast Metal Res.* **22** 103.

Acknowledgements

This project is financially supported by EPSRC UK in the EPSRC Centre for Innovative Manufacturing in Liquid Metal Engineering (The EPSRC Centre—LiME).

IOP Conf. Series: Materials Science and Engineering **529** (2019) 012059 doi:10.1088/1757-899X/529/1/012059

Solidification processing of scrap Al-alloys containing high levels of Fe

J Lazaro-Nebreda, J B Patel, I T H Chang, I C Stone and Z Fan

BCAST, Brunel University London, Uxbridge, UB8 3PH, UK

E-mail: jaime.lazaronebreda@brunel.ac.uk

Abstract. The accumulation of iron in molten aluminium is one of the main concerns for the recycling and casting industries because it leads to the formation of undesirable Fe-rich intermetallic compounds which are detrimental to mechanical properties. Many methods have been developed in the past to reduce the iron accumulated in molten aluminium scrap, but they all suffer from poor efficiency. Hence, a more efficient method is urgently needed to mitigate the negative effect of high iron levels in the melt, thereby avoiding downgrading secondary aluminium to low quality products or the dilution with expensive primary aluminium. This contribution provides a study of the Fe-rich intermetallic compounds developed in aluminium casting alloys with high levels of Fe as a function of melt processing conditions. Results show that the formation of the Fe-compounds is not only dependent on the cooling rate and holding time before solidification, but more on the initial melt treatment as it enhances the nucleation and growth of the Fe-phases. Elemental addition of Mn leads to the formation of large and compact intermetallic particles, but at slow rate. Physical melt treatment by intensive high shearing produces a much faster nucleation and results in a fine dispersion of smaller iron containing intermetallic particles. The latter could be used either to increase the tolerance to iron contamination or to facilitate the iron removal process, providing huge benefits for the recyclability of scrap aluminium alloys as it would allow the transformation of low-grade feedstock into a low cost and small carbon footprint material for high quality castings.

1. Introduction

Aluminium alloys are being widely used in various industrial sectors, especially in aerospace and automotive applications, for the purpose of weight saving and high performance [1]. Manufacturing components using primary aluminium requires 20 times more energy and produces 4 times more CO_2 emissions than directly using secondary aluminium scrap [2]. However, one of the main problems of repeated aluminium scarp recycling is the gradual accumulation of impurities in the molten scrap, mainly iron, due to the presence of high iron containing components (rivets, screws, nuts…) which cannot be fully separated prior melting or from the steel tools used during the casting processes [3].

Because Fe has low solubility in solid aluminium, it can lead to the formation of intermetallic compounds (IMCs) during solidification, which are detrimental to the mechanical properties [4]. First approach used in industry is to downgrade the contaminated material to low quality casting applications or to dilute with expensive primary aluminum, but this is non-sustainable short term solution. Molten refining processes include electrolysis, electroslag fluxing, fractional solidification and the precipitation of primary Fe-rich phases form when the melt is partially solidified [5].

Regarding the latter, many methods have been developed in the past (gravity, filtration, centrifuging and electromagnetic separation), all of them being very effective but also suffering from poor efficiency mainly because the long processing times required to first form and then separate the intermetallic compounds [5]. This limits their use to laboratory research and makes them unpractical for large scale processing in industry at an affordable cost and highlights that a better understanding of the intermetallic formation process is needed.

Among the primary Fe-rich compounds that can be formed in Al-Si based alloys widely used in automotive components, the most common are the needle-shape β-Al_5FeSi and the star-like α-$Al_{15}(Fe,Mn)_3Si_2$ in the presence of Mn [4-7]. Previous research has shown that these intermetallics tend to nucleate and grow on the wetted side of the natural occurring oxide bi-films [6, 8-11]. However, these oxides are normally agglomerated and not well distributed within the melt, something that hinders the nucleation and delays the growth of the Fe-rich intermetallics, making it difficult their separation from the low-Fe melt.

This work aims to present a more detailed study on the formation of primary Fe-rich intermetallic phases in an aluminium alloy (A380) containing increasing levels of Fe and Mn, by using both thermodynamic calculations and experimental evaluation. The influence of the solidification conditions and the use of melt treatment on the nucleation, growth and final type, size and morphology of the Fe-rich intermetallic phases has been also analyzed and discussed.

2. Experimental procedure

The alloy used in this study was a commercial Al-Si die casting alloy (A380) with a composition (wt.%) of Al-8.3Si-3.5Cu-0.15Mg-1.9Zn-1Fe-0.2Mn. Different additions were made to the base alloy using Al-80Fe and Al-60Mn master alloys, simulating typical compositions that can be found in recycled aluminum scrap [5] by adding Fe and to examine the formation of α-$Al_{15}(Fe,Mn)_3Si_2$ when adding Mn.

The alloys were prepared in SiC crucibles inside an electric resistance furnace and held at 700 °C for 1h before taking sample for analysis by optical emission spectroscopy (OES), using for casting the samples a steel mould cup (d30mmx40mm) preheated at 250 °C.

First set of samples (fast cooling) involved the ones taken for composition analysis. A second set of samples (slow cooling) was collected by pouring the melt into a SiC crucible (d50mmx60mm) preheated at 700 °C and placed on an insulated base and allowed to naturally cool down. In both cases the temperature during cooling was recorded by placing a K-type thermocouple in the middle of the sample and data was acquired at 100Hz using a data logger connected to a PC.

To evaluate the effect of time on the formation of the intermetallics, on a second set of experiments the melt was cooled controlled by pouring into launder that ended into a cylindrical steel mould (d50mmx250mm) externally heated controlled at 600 °C (above the temperature of α-Al formation) were it was held for different times before allowing the melt to fully solidify.

Physical melt treatment was performed on selected alloys by means of high shear melt conditioning technology (HSMC) developed at BCAST [12]. The system comprises a rotor-stator unit, attached to an electrical motor with speed control that was immersed into the melt and set at 3000rpm for 2min, being the melt later processed in the same way as described above.

Solidified samples were cut, mechanically polished and then examined using a combination of optical microscopy and EDX analysis to identify type and size of the intermetallic compounds.

Pandat software was used to calculate the phase diagrams of the alloys under consideration in order to know the expected primary Fe-phases and the concentration of Fe in the liquid phase as the precipitation of the IMCs occurs during the solidification [13, 14]. This was useful information about the thermodynamic limit for iron removal and a reference to compared with the experimental evaluation.

3. Results and discussion

3.1. Effect of Fe addition

The presence of iron in the aluminium alloy leads to the formation of a variety of Fe-rich intermetallic compounds as it is shown by the calculated phase diagram in figure 1a. With the increase of iron content some primary intermetallic phases quickly appear above the temperature of formation of α-Al (595 °C). Figure 1b shows the typical microstructure of the alloy with 1% Fe when directly solidified in the compositon mould. The material exhibits a small amount of compact or star like primary intermetallics, identified as α-$Al_{15}(FeMn)_3Si$.

With higher Fe content (figure 1c), the amount of α-$Al_{15}(FeMn)_3Si_2$ particles increases and become more compact, but their size remains the same, while the most significant effect is the appearance of high amount of large and needle shape intermetallics, identified as β-Al_5FeSi. Elongated porosity (cracks) is also appreciated in the microstructure, something consistent with the idea of nucleation of the intermetallics on the oxide bi-films present in the melt [8-11]. It is important to observe that in both cases the amount and size of α-$Al_{15}(FeMn)_3Si_2$ particles is small, suggesting that the formation of this primary phase is difficult [15].

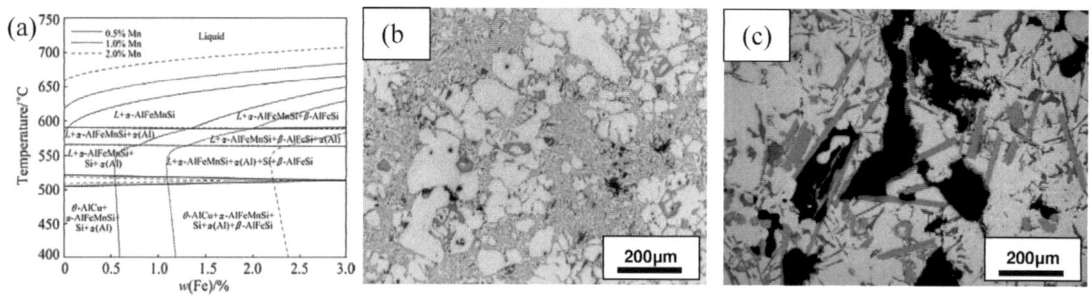

Figure 1. (a) Phase diagram for A380 alloy with different Fe and Mn contents [14]. (b) Fe-rich intermetallics formed in A380-0.2Mn with 1%Fe and with (c) 2%Fe.

3.2. Effect of Mn addition

Increasing the Mn content in the alloy promotes the formation of α-$Al_{15}(FeMn)_3Si_2$, in detriment of the β-Al_5FeSi phase [14]. The microstructure for the A380 alloy with 0.5 and 1% Mn content (figure 2), reveals the presence of only one type of primary compound, with compact shape and increased size. The average particle size is 83 μm for 0.5 %Mn (figure 2a) and 137 μm for 1%Mn (figure 2b), in comparison with the base alloy with only 0.2%Mn (28 μm, figure1b). Despite the increase in size is in agreement with previous findings, the area fraction of particles in the whole sample is below expected for these Fe and Mn contents [16,17], once again highlighting that the formation of the particles is a slow process and might not be fully completed under high cooling rates.

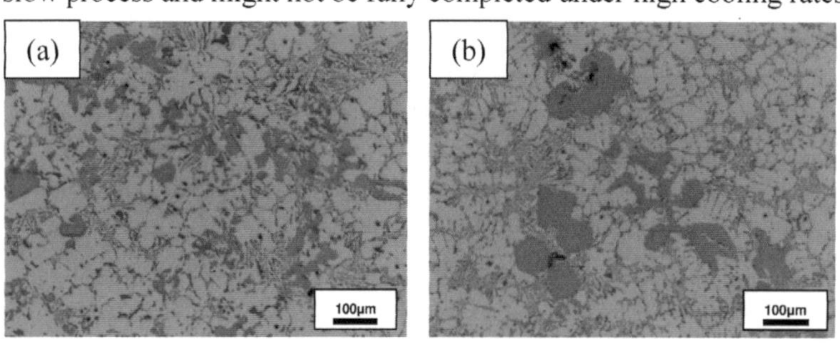

Figure 2. Fe-rich particles formed in A380-1Fe with increasing Mn, (a) 0.5Mn, (b) 1Mn.

3.3. Effect of physical melt treatment by HSMC
The application of intensive melt shearing to the molten alloy breaks up the oxide films present in the liquid and disperses them into very fine particles, thus increasing significantly the amount of nuclei for the intermetallic compounds. This not only increases the possibilities of forming more primary Fe-rich particles, but at the same time they are expected to be much finer [12]. Figure 3 shows the microstructure of the alloy with 0.2%Mn (figure 3a) and 1%Mn (figure 3b), similarly to figure 1b and figure 2b, but in this case when intensive melt shearing has been applied before solidification. In both cases the number of particles increases significantly and they become more rounded and compact.

Regarding the average size, it varies from 28 to 51 μm for low Mn content alloy (figure 1b vs. figure 3a). This can be explained considering that an early nucleation gives the particles more time to grow during the solidification. On the other hand, with increasing Mn content, by itself already promoting the formation of the particles, the average size decreases from 137 to 102 μm (figure 2b vs. figure 3b) when the melt has been intensively sheared.

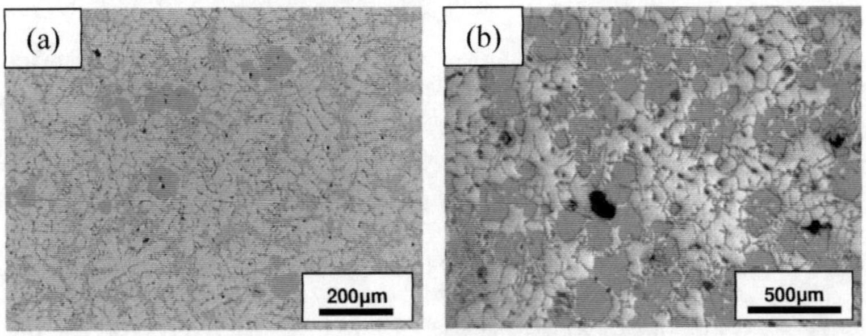

Figure 3. Fe-rich particles formed after intensive melt shearing on A380-1Fe (a) 0.2Mn, (b) 1Mn.

3.4. Effect of cooling rate on solidification process
In order to understand better the results from previous sections, it is important to evaluate the temperature evolution during the solidification process. Figure 4 presents the cooling curves for the A380 alloy with 1%Fe and 1%Mn (with and without HSMC) as function of the cooling rate, estimated as -3.2 °C/s and -0.3 °C/s for the fast and slow cooling respectively. It can be seen that all curves present an inflexion point between 600 and 650 °C, different in each case, which corresponds to the precipitation of the primary α-$Al_{15}(FeMn)_3Si_2$ [17]. Under natural conditions, the undercooling increases with the cooling rate from -30 to -50 °C, supporting the idea that nucleation of primary phases requires long time and that can be easily suppressed if the cooling rate is high enough [15]. On the other hand, the application of high shear to the melt reduces the undercooling to -10°C for slow cooling and to -28°C for fast cooling, much smaller in both cases than the reference material.

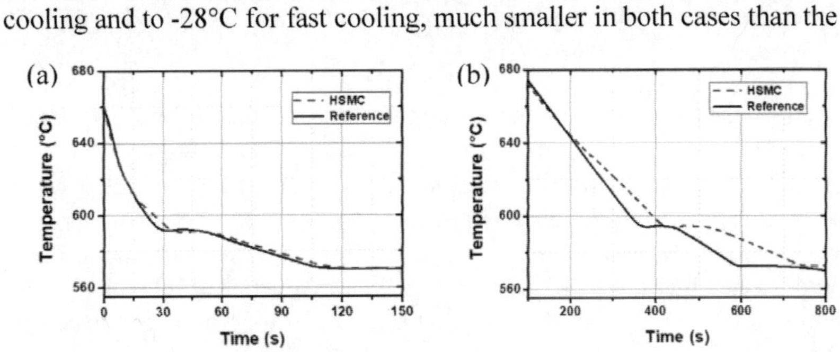

Figure 4. Cooling curves for A380-1Fe-1Mn under (a) fast and (b) slow cooling.

3.5. Effect of holding time
The density of α-$Al_{15}(Fe,Mn)_3Si_2$ particles is about 3.2 g/cm^3 [14], higher than the density of molten aluminium (2.3 g/cm^3). As a consequence, they naturally tend to settle in the melt, leaving the top part with a lower Fe level. Without any treatment (figure 5a) the direct solidification produces a very small gradient from top (0.95%) to bottom (1.1%), because the intermetallic do not have enough time to

grow, in case they have nucleated on the agglomerated oxides, and then settle. It is only when the melt is held at 600 °C that the particles can grow and a increasing with time gradient is observed. Complete sedimentation of the intermetallic phase requires in this case between 10 to 15 min.

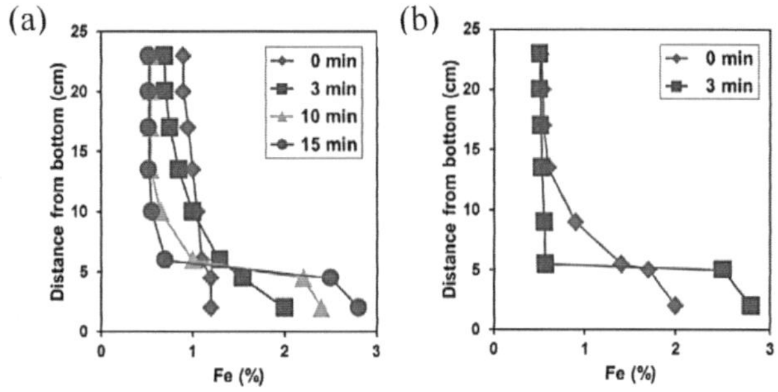

Figure 5. Fe concentration along the vertical section of samples after keeping the melt at 600 °C for different times before solidification. (a) without and (b) with physical melt treatment by HSMC.

On the other hand, when high shear melt treatment is applied, the sedimentation is much faster (figure 5b). Even for a direct cooling the sedimentation has taken place in most of the casting, leaving the top half with Fe levels close to the thermodynamic limit for this alloy (0.5% [14]). Full sedimentation takes place in less than 3 minutes in this case, therefore 3 to 5 times faster, evidencing the earlier formation and growth of the primary Fe-rich compounds during the cooling stage.

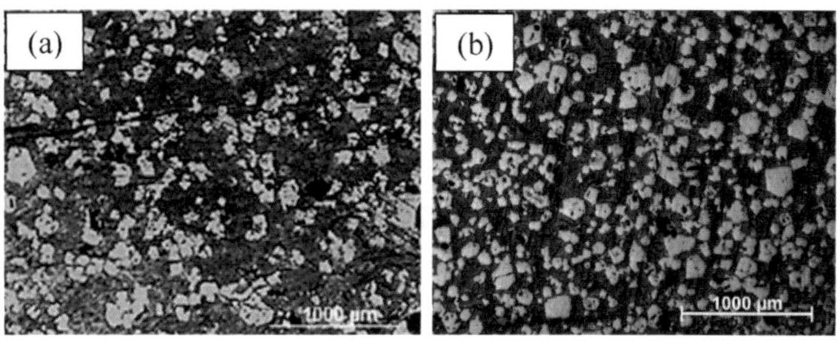

Figure 6. Microstructure of the bottom of the samples after complete sedimentation of the intermetallic particles by holding at 600°C (a) without and (b) with physical melt treatment by HSMC.

The sludge in the reference material consists of irregular particles of 180 µm with poor distribution as it is shown in figure 6a. For the treated material (figure 6b) it consists of smaller intermetallics of 119 µm, with more rounded shape and higher density. This result supports the idea that a low nucleation rate (progressive) promotes higher growth and branched shape, while a high nucleation rate (explosive) results on low growth but compact particles [18].

4. Conclusions

Due to the natural agglomeration of the oxides in molten aluminium, the formation and growth of primary Fe-rich intermetallic compounds on top of them is a slow process and tends to produce more needle shape β-Al_5FeSi than compact α-$Al_{15}(Fe,Mn)_3Si_2$ particles at high levels of Fe. The addition of Mn promotes the formation of α-$Al_{15}(Fe,Mn)_3Si_2$ in detriment of β-Al_5FeSi, for Mn/Fe ratio over 0.5, resulting in compact and bigger particles. However under natural cooling conditions the fraction of these intermetallics is below expected, suggesting the process is still slow even with chemical melt treatment. The nucleation and growth is favored by a slow cooling rate and/or long holding times, but this solution is far from efficient.

Intensive melt shearing efficiently disperses the oxide bi-films, resulting in the explosive nucleation of primary α-$Al_{15}(Fe,Mn)_3Si_2$, something that has two important consequences to improve the recyclability of the high-Fe alloys. First, increased iron tolerance, as the final particles tend to be smaller and better dispersed, improving the mechanical properties of castings. Second, faster iron removal from the melt, as the earlier formation speeds up the sedimentation of Fe-rich intermetallics.

Acknowledgments
The financial support from EPSRC Future Lime Hub (UK) under grant number EP/N007638/1 and from the European commission Seventh Framework (FP7) under grant number 603577 is appreciated.

References
[1] Lumley R N 2018 *Fundamentals of Aluminium Metallurgy: Recent Advances* Woodhead Publishing Series in Metals and Surface Engineering, Volume 2
[2] Polmear I J 1995 *Light Alloys, Metallurgy of light alloys 3rd edition*
[3] Green J 2007 *Aluminium Recycling and Processing for Energy Conservation and Sustainability* **274** 53
[4] Mondolfo L F 1976 *Aluminium Alloys: Structure and Properties* Butterworths, London
[5] Zhang L, Gao J, Damoah L N W and Robertson D G 2012 Removal of iron from aluminum: A review. *Mineral Processing and Extractive Metallurgy Review: An International Journal* **33** 99–157
[6] Cao X and Campbell J 2000 Precipitation of primary intermetallic compounds in liquid Al-11.5Si-0.4Mg alloy *Int. J. Cast Metals Res.* **13** 175–184
[7] Shabestari S G 2004 The effect of iron and manganese on the formation of intermetallic compounds in aluminium-silicon alloys *Mater. Sci. Eng. A* **383(2)** 289–98
[8] Cao X and Campbell J 2003 The nucleation of Fe-rich phases on oxide films in Al-11.5Si-0.4Mg cast alloys *Metallurgical and Materials Transactions A* **34(7)** 1409-1420
[9] Cao X and Campbell J 2004 The solidification characteristics of Fe-rich intermetallics in Al-11.5Si-0.4Mg cast alloys *Metallurgical and Materials Transactions A* **35(5)** 1425–1435
[10] Khalifa W, Samuel F H, Gruzleski J E, Doty H W and Valtierra S 2005 Nucleation of Fe-Intermetallic Phases in the Al-Si-Fe alloys *Metallurgical and Materials Transactions A* **36(4)** 1017–1032
[11] Orozco-González P *et al.* 2015 Nucleation of Fe-Rich Intermetallic Phases on α-Al_2O_3 Oxide Films in Al-Si Alloys *Journal of Minerals and Materials Characterization and Engineering* **3** 15–25
[12] Fan Z, Zuo Y and Jiang B 2012 Apparatus and method for liquid metals treatment WO 2012035357 A1
[13] Chen S L, Daniel S, Zhang F, Chang Y A, Yan X Y, Xie F Y, Schmid-Fetzer R and Oasted W A 2002 The PANDAT software package and its applications *Calphad* **26** 175–188
[14] Yang W, Gao F and Ji S 2015 Formation and sedimentation of Fe-rich intermetallics in Al-Si-Cu-Fe alloy *Trans Nonferrous Met Soc China* **25** 1704–1714
[15] Que Z, Wang Y and Fan Z 2018 Formation of the Fe-containing intermetallic compounds during solidification of Al-5Mg-2Si-0.7Mn-1.1Fe alloy *Materials and Metallurgical Transactions A* **49** 2173-2181
[16] Ferraro S, Fabrizi A and Timelli G 2015 Evolution of sludge particles in secondary die-cast aluminium alloys as function of Fe, Mn and Cr contents *Materials Chemistry and Physics* **153** 168–179
[17] Liu Y, Huang G, Sun Y, Zhang L, Huang Z, Wang J and Liu C 2016 Effect of Mn and Fe on the formation of Fe-rich intermetallics in Al-5Mg-Mn alloys solidified under rapid cooling *Materials* **9** 88
[18] Bjunrenstedt A, Casari D, Seifeddine S, Mathiesen R H and Dahle A K 2017 In-situ study of morphology and growth of primary a-Al(FeMnCr)Si intermetallics in an Al-Si alloy *Acta Materialia* **130** 1–9

Microstructure evolution in an Al-Si piston alloy under ultrasonic melt processing

Suwaree Chankitmunkong[1, 2*], Dmitry G. Eskin[2, 3], and Chaowalit Limmaneevichitr[1]

[1]Department of Production Engineering, Faculty of Engineering, King Mongkut's University of Technology Thonburi, 126 Pracha-Utid Rd., Bangmod, Tungkhru, Bangkok, 10140 Thailand
[2]Brunel University London, BCAST, Uxbridge, Middlesex UB8 3PH, United Kingdom
[3]Tomsk State University, Tomsk 634050, Russian Federation

*suwaree.03@mail.kmutt.ac.th

Abstract. Piston Al-Si eutectic alloys are used to produce direct-chill cast billets for subsequent forging. Because of a very complex composition and multi-phase heterogeneous structure, it is necessary to control the formation of primary and eutectic compounds either through alloying or casting conditions (or both). In this study we used ultrasonic melt processing above or across the liquidus line to affect the occurrence and size distribution of primary Si as well as morphology of primary Al dendrites and high-temperature eutectic phases. The refinement of these particles has potential benefit for mechanical properties and formability during forging.

1. Introduction

Al-Si piston alloys have been widely used in automotive industry due to their high strength, excellent castability, good wear resistance and low thermal expansion. These alloys are also used to produce hot forged engine parts. Generally, the mechanical properties of an Al-Si piston alloy are dominated by the complex microstructure with primary and eutectic phases such as α-Al, primary and eutectic Si, Mg_2Si, Al_2Cu, $Al_5Cu_2Mg_8Si_6$, Al_3Ni, Al_3CuNi and Al_7Cu_4Ni phases. Moreover, the addition of Cu and Mg is also used to improve the mechanical properties of Al-Si piston alloys through heat treatment [1, 2]. The amount, size and morphology of primary silicon and primary Al dendrites play role in the deformation process and affect the mechanical properties of final products. Furthermore, the β-phase ($Al_9Fe_2Si_2$) intermetallic is considered as a harmful phase in Al-Si alloys, particularly the large size and elongated morphology of this phase play detrimental role in mechanical properties [3]. Thus, it is important to refine or modify the hard particles such as primary Si and β-phase intermetallics, as well as achieve a uniformly equiaxed grain structure of primary aluminium in a billet. Generally, the addition of alloying elements to aluminium is a conventional method used to modify the microstructure. Ultrasonic melt treatment (UST) is shown to be efficient in refining grains, primary particles and small intermetallics [4]. However, the mechanisms and effects of UST relate to the temperature ranges in which it is applied to the melt [5, 6].

The purpose of this work is to study the microstructural evolution of an Al-Si piston alloy with high Cu content upon ultrasonic treatment during solidification, and relate the observed effects to the mechanisms of ultrasonic treatment, for improving the microstructure and hardness.

2. Material and methods

The experimental alloys were prepared using high-purity Al (99.9 wt%), commercial 99.5 wt% pure Mg and Al–45 wt% Si, Al–45 wt% Cu, Al–20 wt% Ni and Al–45 wt% Fe master alloys. The composition of the alloy was analysed by emission spectrometry and given in Table 1. This is essentially an AA4032 alloy with increased Cu concentration. The alloy was melted in a clay graphite crucible in an electric resistance furnace. The molten metal was maintained at 750 ± 5 °C. UST was performed with a 5-kW ultrasonic generator (Reltec) and a 5-kW water-cooled magnetostrictive transducer (Reltec) with a niobium sonotrode 20 mm in diameter at a working frequency of 17.5 kHz (amplitude ±20 μm). The experimental set up is designed to observe the effect of UST at different stages during solidification.

IOP Conf. Series: Materials Science and Engineering **529** (2019) 012060 doi:10.1088/1757-899X/529/1/012060

The steel mould (Figure 1) was covered by insulation material from outside to minimize heat losses from the mould wall, while being placed on a massive steel base to assure the progressive solidification from the bottom to the top. The sonotrode preheated to 600 °C was located at the top part of the mould, so that it will be submerged into the liquid metal by approximately 5 mm when the mould is filled with the melt. The molten metal was poured at 720 °C. When the mould was completely filled, the ultrasonic transducer was turned on. Therefore, the solidifying metal at the bottom part of the sample was not affected by UST (No UST). As the solidification progressed from the bottom towards the sonotrode, the solidifying alloy experienced action of ultrasound at different stages of solidification: first in the liquid phase only, then (in the middle part of the mould) across the liquidus, and finally in the mushy zone in the upper part of a billet as can be seen in Figure 1.

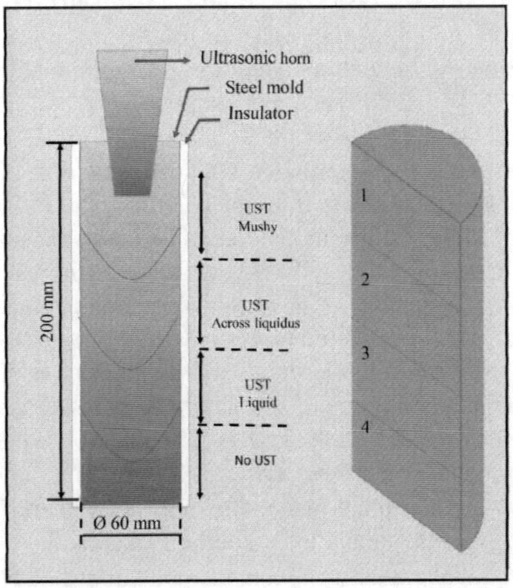

Figure 1. Schematic diagram of experimental set up of casting at different stages of ultrasonic treatment

Table 1. Chemical composition of Al-Si piston alloy with high Cu content as determined by emission spectrometry

Alloy	Chemical composition (wt.%)					
	Si	Cu	Mg	Ni	Fe	Al
3.5Cu	12.7	3.56	0.92	1.07	1.29	Balance

The microstructures were examined in vertical sections reflecting each stage of UST, from the mushy zone, across the liquidus, in liquid and without UST by standard metallographic procedures. The microstructure and intermetallics were observed by optical microscope (OM) and scanning electron microscope (SEM). Prior to observation, the samples were etched by Keller's reagent (5 ml HNO_3, 3 ml HCl and 2 ml HF in 190 ml distilled water) for 1 min. Polished samples were also anodized in Barker's solution (5% HBF4 water solution) for about 2 min at 20 VDC and were then examined under polarized light. The grain sizes were measured using the linear intercept method according to ASTM E112-10. The quantitative measurement of primary Si particles size and grain size were carried out using image analysis software. The iron-containing intermetallics were also analysed for the chemical composition in a ZEISS Supra 35 SEM with an energy dispersive spectroscopy (EDS). Hardness tests were conducted by a Brinell hardness testing machine with 187.5 N load, 2.5 mm ball indenter, and 20 s dwell time.

3. Results and discussion

3.1. Primary Si particles and α-Al dendrite transformation

To investigate the effect of UST treatment on the microstructure evolution at different stages of solidification, the macrostructure of the billet cross-sections was observed, and can be seen in Figure 2 (a-b). The results show that the grain macrostructure started to change when UST had been applied in the liquid stage, as compared to the bottom section of the billet where no UST had been felt and where the grain morphology was coarse. After the ultrasonic transducer was turned on and the solidification front moved upwards to the ultrasonic source, the macrostructure started to refine with a clear transition seen as the dashed line region in Figure 2c (the transition between UST and no UST treatment). This indicates that the UST was effective even when applied in the liquid state. As the solidification front was moving upwards, the grain size of Al dendrites progressively decreased, while the amount of primary Si increased, see Figure 2 (a-c, e-g).

The quantitative analysis of Al grains and primary Si is given in Figure 3. The smallest size of primary Si particles is about 35μm and is achieved when the alloy was treated with ultrasound in the liquid state, which is smaller than in the alloy treated by UST across the liquidus (42 μm), and in the mushy stage resulted in the size of primary Si particles become slightly increased. In all cases of UST the primary Si was significantly refined. However, the amount of primary Si particles is highest when UST was applied across the liquidus and in the mush, which also corresponded to a more homogenous distribution of these particles in the structure (Figure 2e, f). This is a desired effect as it makes the stress concentration on brittle particles less probably in the microstructure under load, and may improve the toughness of the alloy. The increased amount and simultaneous refinement of primary Si particles can be attributed to the effect of acoustic cavitation upon UST on the nucleation process of primary Si (when applied above and across the liquids) and fragmentation of these particles (when applied across and below the liquids). Similar effects have been observed upon UST for primary Si [5], Al_3Zr and Al_3Ti primary particles [7, 8]. Moreover, the UST treatment might also increase the melt temperature (by a few degrees) that later limit the growth of primary particles.

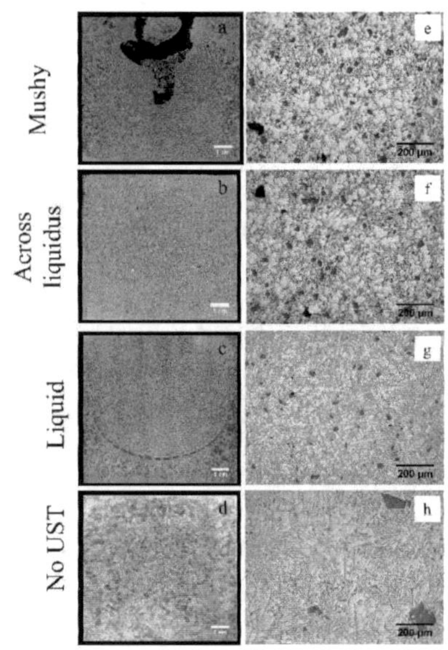

Figure 2. Macrostructure and microstructure of the Al-Si piston alloy with applied UST at different stages of solidification. (The red-dashed line depicts the beginning of UST effect)

Figure 3b also confirms that the grain size was progressively and significantly decreased as the solidification front was moving towards the sonotrode. Figure 4 (a-d) shows the effects of UST treatment on the refinement and the growth morphology of Al grains. It is clear that the morphology of Al grains changed from coarse dendritic to tiny globular when UST was applied in the mushy stage as compared to the alloy without UST treatment (Figure 4a, d), which correlated to refinement of the grain from about 250 to 50 μm. The Al dendrites also became finer when the UST was applied in the liquid state and across the liquidus. In these cases one can observe a mixed structure with dendritic and globular grains. It is well known that the grain refinement is directly related to the number of active nuclei in the melt. Hence, when UST was applied at high temperature (liquid stage), the main mechanism should be related to the cavitation-induced heterogeneous nucleation, e.g. on oxides [4]. While when UST was applied at lower temperature (across and below the liquidus), in addition to the already effected heterogeneous nucleation, dendrite fragmentation caused by cavitation and acoustic flows resulted in grain multiplication and refinement [9]. This led to the most significant grain refinement achieved upon UST in the mushy stage.

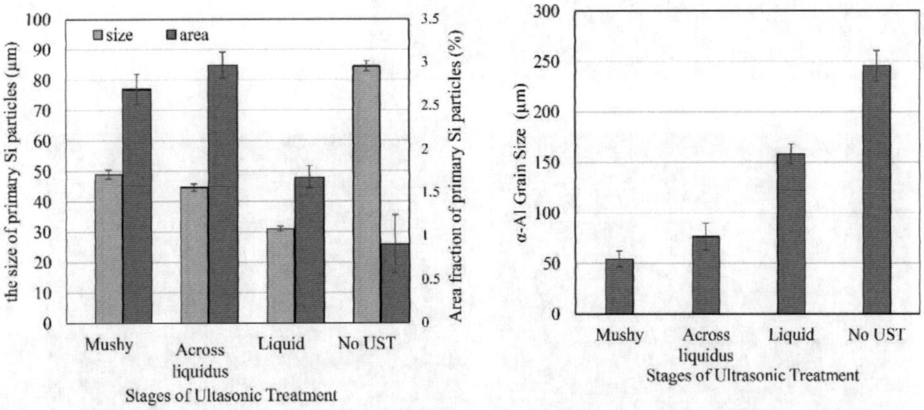

Figure 3. The quantitative analysis of the Al-Si piston alloy; (a) the size and amount of primary Si particles, (b) size of α-Al grains without and with UST applied at different stages of solidification.

3.2. *Effect of UST on the morphology of Fe-containing intermetallic*

Figure 5 shows SEM images of the microstructure with Fe-containing intermetallics without UST and with UST across the liquidus. The iron-containing intermetallic phases were quite large in the Al-Si piston alloy without UST, and can be clearly seen as elongated needle-like crystals in eutectic regions (Figure 5a, shown by red arrow). After UST the morphology of these phases changed to short rods as can be seen in Figure 5b (shown by red arrows). We assume that this change was due to the ultrasonic fragmentation mechanism similar to that observed in Ref.[10]. The chemical composition of Fe-containing intermetallics were measured by EDS spot analysis with averaging over seven measured particles, and the results are listed in Table 2. Although the chemical composition of the rod-like phase is slightly different from the needles-shaped particles, the compositional range reflect the β-$Al_9Si_2Fe_2$ phase [11]. For refinement and the morphology change to occur the UST had to be applied in the temperature range of peritectic reaction (585-597 °C [12]) L+δ-Al_4FeSi_2 → Si+β-phase +Si [13, 14].

Table 2. Chemical composition analysis of iron intermetallic without and with UST applied at across liquidus stage

Iron intermetallics phases	Elements (at.%)				Possible phase
	Al	Si	Fe	Fe/Si	
Needle plate-like (No UST)	63.8	18.9	17.3	0.92	β-$Al_9Si_2Fe_2$
Rod-like (UST)	72.6	14.1	13.3	0.94	β-$Al_9Si_2Fe_2$

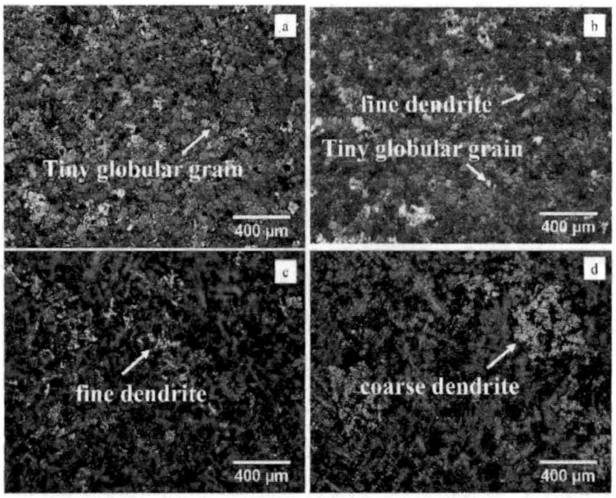

Figure 4. Optical micrographs illustrating the morphology and refinement of Al grains in the Al-Si piston alloy when UST was applied to: (a) mushy, (b) across liquidus, (c) liquid; and (d) No UST

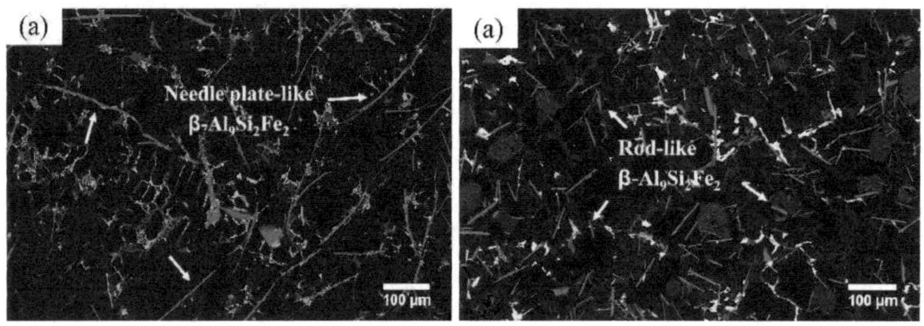

Figure 5. Microstructure showing the morphology of iron intermetallics in the Al-Si piston alloy: (a) without UST and (b) with UST at across the liquidus.

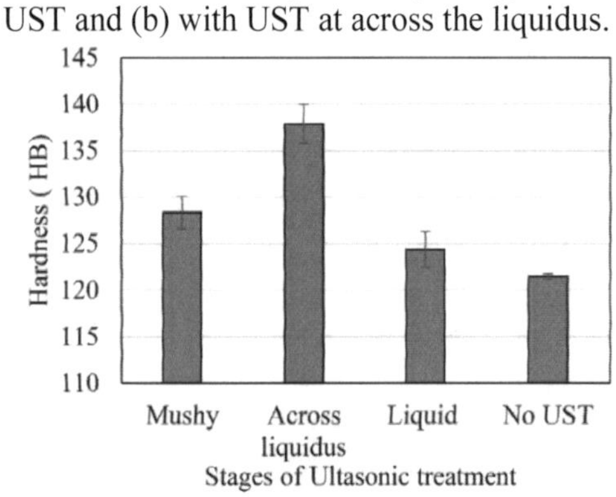

Figure 6. Hardness of the Al-Si piston alloy with high Cu content after UST applied at different stages of solidification.

3.3. Hardness measurements

In general, the hardness depends on the amount of hard particles and the size of Al grains. Figure 6 shows the effect of UST applied in different stages of solidification on the hardness of the high-Cu Al-Si piston alloy. In all cases the hardness of the alloys treated with ultrasound was higher than in the alloy solidified without UST. The largest gain in the hardness was achieved when UST was applied across the liquids (138 vs 122 HB). This correlates with the refinement of Al grains and increased amount of refined primary Si particles as well as finer intermetallics as can be seen in Figure 2f and Figure 4b. The alloy solidified without UST had coarser eutectic Si phases, large unevenly distributed primary Si and also large needle-like β particles, yet its hardness was the lowest, which gives indication of the significance of the size and distribution of hard particles for the overall alloy hardness.

4. Conclusions

- The refinement of primary Si particles can be achieved when the Al-Si piston alloy is treated by ultrasound in the liquid state (above liquidus).
- The application of UST during solidification (below liquidus and during eutectic reaction) refined Al grains through enhanced heterogeneous nucleation and dendrite fragmentation.
- The fragmentation mechanism is responsible for the modification of β-phase morphology from needle-like to rod-like structures, with simultaneous refinement of the particles.
- UST improves the hardness, particularly when UST is applied across the liquidus upon solidification.

Acknowledgements

This work was financially supported by The Research and Researchers for Industry (RRi) under the Thailand Research Fund (TRF) offers a scholarship for Ph.D. student (PHD57I0060) with Thai Metal Aluminium Co., Ltd and King Mongkut's University of Technology Thonburi through the "KMUTT 55th Anniversary Commemorative Fund". S.C. thanks BCAST (UK) for hosting a research visit.

References

[1] Y. Yang, K. Yu, Y. Li, D. Zhao, X. Liu, Evolution of nickel-rich phases in Al–Si–Cu–Ni–Mg piston alloys with different Cu additions, Materials & Design 33(1) (2012) 220-225.

[2] N. Belov, D. Eskin, N. Avxentieva, Constituent phase diagrams of the Al–Cu–Fe–Mg–Ni–Si system and their application to the analysis of aluminium piston alloys, Acta materialia 53(17) (2005) 4709-4722.

[3] S. Gencalp Irizalp, N. Saklakoglu, Effect of Fe-rich intermetallics on the microstructure and mechanical properties of thixoformed A380 aluminum alloy, Engineering Science and Technology, an International Journal 17(2) (2014) 58-62.

[4] G.I. Eskin, D.G. Eskin, Ultrasonic treatment of light alloy melts, CRC Press,2015.

[5] J.Y. Wang, B.J. Wang, L.F. Huang, Structural evolution of Al–8%Si hypoeutectic alloy by ultrasonic processing, Journal of Materials Science & Technology 33(11) (2017) 1235-1239.

[6] L. Zhang, D. Eskin, A. Miroux, L. Katgerman, Formation of microstructure in Al-Si alloys under ultrasonic melt treatment, Light Metals (2012) 999-1004.

[7] F. Wang, D. Eskin, T. Connolley, J.-w. Mi, Influence of ultrasonic treatment on the formation of primary Al3Zr in an Al-0.4 Zr alloy, (2017).

[8] F. Wang, D. Eskin, J. Mi, T. Connolley, I. Tzanakis, Refinement of Primary Al3Ti Intermetallic Particles in an Al-0.4 wt% Ti alloy by Ultrasonic Melt Processing.

[9] D.G. Eskin, Ultrasonic processing of molten and solidifying aluminium alloys: overview and outlook, Materials Science and Technology (2016) 1-10.

[10] F. Wang, D. Eskin, J. Mi, C. Wang, B. Koe, A. King, C. Reinhard, T. Connolley, A synchrotron X-radiography study of the fragmentation and refinement of primary intermetallic particles in an Al-35 Cu alloy induced by ultrasonic melt processing, Acta Materialia 141 (2017) 142-153.

[11] C.M. Dinnis, J.A. Taylor, A.K. Dahle, As-cast morphology of iron-intermetallics in Al–Si foundry alloys, Scripta Materialia 53(8) (2005) 955-958.

[12] L.-g. Hou, Y.-h. Cai, H. Cui, J.-s. Zhang, Microstructure evolution and phase transformation of traditional cast and spray-formed hypereutectic aluminium-silicon alloys induced by heat treatment, Int J Miner Metall Mater 17(3) (2010) 297-306.

[13] T. Maitra, S.P. Gupta, Intermetallic compound formation in Fe–Al–Si ternary system: Part II, Materials Characterization 49(4) (2002) 293-311.

[14] Y. Zhang, J. Jie, Y. Gao, Y. Lu, T. Li, Effects of ultrasonic treatment on the formation of iron-containing intermetallic compounds in Al-12%Si-2%Fe alloys, Intermetallics 42 (2013) 120-125.

Simulation of casting filling process using the lattice Boltzmann method

Y J Zhang[1], X W Qian[1]*, J X Zhou[1]*, Y J Yin[1], X Shen[1], X Y Ji[1] and J L Huang[2]

[1] State Key Laboratory of Material Processing and Die & Mould Technology, Huazhong University of Science and Technology, Wuhan, 430074, China
[2] Advanced Forming Research Centre, University of Strathclyde, 85 Inchinnan Drive, Renfrew PA4 9LJ, United Kingdom

E-mail: zhoujianxin@hust.edu.cn, intecastqxw@163.com

Abstract. Numerical simulation of casting filling process with complex shape is time-consuming. Compared with the traditional SOLA-VOF method, the lattice Boltzmann method (LBM) calculates the pressure field by particle distribution functions instead of the correction of the velocity and pressure fields, which greatly simplifies the calculation process. In addition, the LBM provides a flexible approach which can be easily parallelized. In this study, the LBM is employed to simulate casting filling process. An implementation of a volume-of-fluid (VOF) method within the lattice Boltzmann framework is proposed to capture the free surface of the casting filling process. A Smagorinsky large eddy simulation (LES) model is adopted to improve the numerical stability of the LBM. An adaptive time stepping technique is implemented to ensure an efficient and stable simulation. The model is validated by the experimental and simulation results of Campbell box filling process. The filling process of complex casting is simulated, and the result is compared with the filling process obtained by the SOLA-VOF method. The prediction accuracy and reliability of free surface profile is analysed.

1. Introduction

Casting process is an efficient manufacturing process, which can produce parts with complex shape. The filling process is crucial for the casting quality. An unstable filling process is accompanied with turbulences, splashes, and bubbles, which leads to casting defects. Therefore, the motion of liquid metal during filling process remains to be investigated. As the filling process is hard to capture, the numerical simulation becomes an alternative approach to understand the filling process. However, most simulation methods, such as SOLA-VOF method, the correction of velocity and pressure field is a time-consuming process. Compared with these conventional methods, the lattice Boltzmann method (LBM) calculates the pressure field by particle distribution functions, which simplifies the calculation process. Another advantage of the LBM is that it can be easily parallelized, which shows the potential for massive parallel simulation. But an evitable drawback of the LBM method is the numerical instability at low viscosity.

In the past years, several methods have been proposed to improve the numerical stability. Ginzburg et al. [1] proposed a generalized LBM model for free surface flow. Koerner et al. [2] presented a LBM model for the simulation of metal foaming process, in which the free surface flow including gas diffusion is captured. Thuerey et al. [3] applied the adaptively coarsened grids to the LBM model, which can simulate high Reynolds number flows with high efficiency. Janssen et al. [4] implemented a volume-of-fluid-based LBM model. The breaking dam case and flow past a weir case are validated. Zhao [5] performed the simulation of the continuous casting process by the LBM.

However, the above studies mainly focus on the simulation of LBM based free surface flow with a simple geometric shape. Numerical stability of the LBM is also influenced by the geometric shape [6]. Whether the LBM is capable of simulating the free surface flow with complex shape accurately remains to be investigated. Therefore, in this study, the LBM coupled with volume-of-fluid method is employed to simulate the casting filling process. The Smagorinsky LES model [7] is adopted to improve the numerical stability and the adaptive time stepping technique [3] is used to decrease simulation time. In addition, the explicit upwind scheme proposed by Ginzburg [1] is adopted due to its robustness for free surface simulations with high Reynolds number. The model is validated by the filling process of Campbell box [8]. The results are compared with the simulation results by SOLA-VOF method. Then the filling process with complex shape is simulated to analyze the accuracy and stability of the LBM.

2. Model description

2.1. Lattice Boltzmann model

In this study, the Bhatnagar-Gross-Krook (BGK) model is used and is defined by

$$f_i(\mathbf{x}+\mathbf{e}_i\Delta t, t+\Delta t)-f_i(\mathbf{x},t)=-\frac{1}{\tau}\left[f_i(\mathbf{x},t)-f_i^{eq}(\mathbf{x},t)\right]+F_i \tag{1}$$

where $f_i(\mathbf{x},t)$ is the distribution function at position $\mathbf{x}$ and time t in one particular lattice direction, $f_i^{eq}(\mathbf{x},t)$ is the equilibrium distribution function, $\mathbf{e}_i$ is the discrete lattice velocity, Δt is the time step, τ is the dimensionless relaxation time and F_i is the external force (e.g., gravity).

The non-dimensional relaxation time τ depends on the fluid kinematic viscosity ν

$$\tau = 3\frac{\nu\Delta t}{c^2}+\frac{1}{2} \tag{2}$$

where $c = \Delta x / \Delta t$ is the particle velocity.

The D3Q19 lattice model, which has better numerical stability than the D3Q15 lattice model [6], is used in the LBM model. The macroscopic quantities for density and velocity are calculated by the first two moments of the distribution functions

$$\rho=\sum_{i=0}^{18} f_i \tag{3}$$

$$\mathbf{u}=\frac{1}{\rho}\sum_{i=0}^{18}\mathbf{e}_i f_i \tag{4}$$

As the model is a weakly-compressible scheme for impressible flows, the fluid pressure is calculated by the equation of state

$$p=c_s^2\rho \tag{5}$$

where $c_s = c/\sqrt{3}$ is the sound speed in D3Q19 lattice model.

2.2. Smagorinsky large eddy simulation

Casting filling process usually occurs at very high Reynolds numbers in the turbulent state. The low kinematic viscosity of liquid metal and high velocity will cause the relaxation time τ close to 0.5, which decreases the numerical stability. While the Smagorinsky LES model introduces an additional turbulent viscosity ν_T. In a Smagorinsky model, ν_T is defined by

$$\nu_T=(C_s\Delta x)^2\|\mathbf{S}\| \tag{6}$$

where C_s is Smagorinsky constant, its value is set to be 0.12 and $\mathbf{S}$ is strain rate tensor, defined by

IOP Conf. Series: Materials Science and Engineering **529** (2019) 012061 doi:10.1088/1757-899X/529/1/012061

$$S_{\alpha\beta} = \frac{Q_{\alpha\beta}}{2\rho c_s^2 \tau} \tag{7}$$

where Q is the moment flux tensor [7]. The relaxation time in the Smagorinsky model is given by

$$\tau = 3\frac{(\nu_0 + \nu_T)\Delta t}{c^2} + \frac{1}{2} \tag{8}$$

where ν_0 is the relaxation time related to the kinematic viscosity. From Eq.(6), (7) and (8), the total relaxation time is obtained, which is given by

$$\tau = \frac{1}{2}\left(\sqrt{\tau_0^2 + 18C_s^2\Delta x^2 Q} + \tau_0\right) \tag{9}$$

After using the LES model, the simulation becomes robust for different problems due to the increase of the total relaxation time [1] and requires less calculation time than the multi-relaxation-time model.

2.3. Free surface model

An important aim of casting filling process simulation is the interface capturing. The filling process is accompanied by complex interaction between gas and liquid metal. As the large density difference between liquid phase and gas phase during filling process, the dynamics of the gas phase is neglected. Therefore, the filling process can be described as a single phase flow within an acceptable accuracy. Various kinds of VOF methods have been proposed for the LBM, however, these methods adopt a simplified advection scheme in which no interface velocity is included. Therefore the mass flux terms are not calculated exactly and the sharp interface is captured approximately. In this study, the implement of VOF-based method within the lattice Boltzmann framework is proposed to improve the accuracy for interface capturing. The VOF indicator function is given by

$$\frac{\partial \varepsilon}{\partial t} + \nabla \cdot (\mathbf{v}\varepsilon) = 0 \tag{10}$$

where ε is the fluid fraction and $\mathbf{v}$ is the fluid velocity in a control volume. Eq.(10) can be discretized with a control volume method and the advection scheme is obtained

$$\varepsilon^{t+\Delta t} = \varepsilon^t - \Delta t \sum_{k=1}^{6} \varepsilon_k \mathbf{v}_k \cdot \mathbf{n}_k \tag{11}$$

where $\mathbf{v}_k$ is the fluid velocity at the kth face of the control volume, $\mathbf{n}_k$ is the outer normal vector of the kth face and ε_k is the fill level at the kth face, which can be calculated using the upwind scheme. The time step and space step are set to be 1.0 with non-dimensional form.

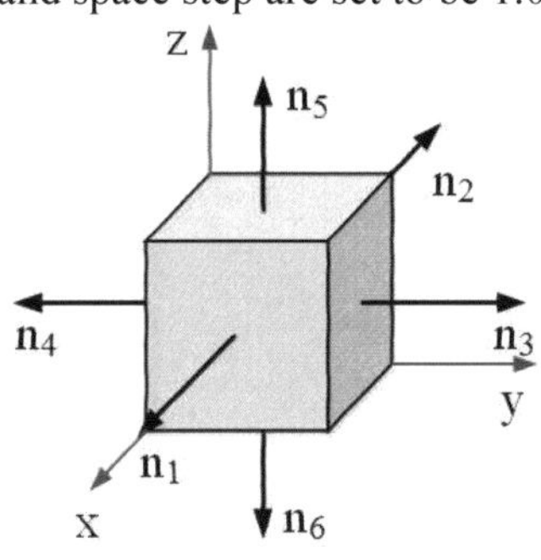

Figure 1. Control volume

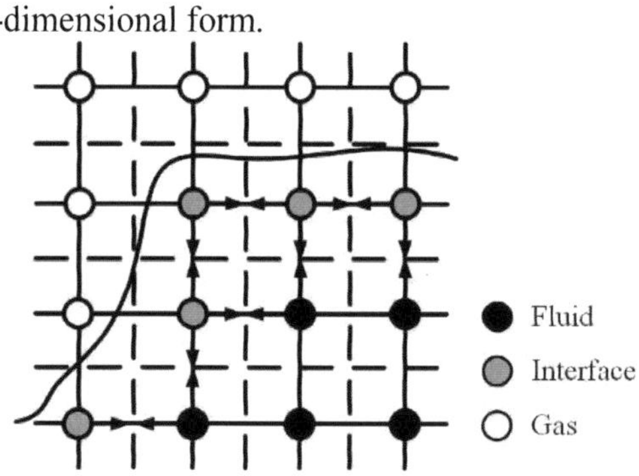

Figure 2. Fluid, interface and gas cell

The last term of Eq.(11), which represents the mass flux between the control volumes, shown in figure 1, with six neighboring control volumes, can be expressed as

$$\sum_{k=1}^{6}\sum_{i=0}^{18} \varepsilon_k \mathbf{n}_k \cdot \left[\mathbf{e}_i f_i(\mathbf{x}) + \mathbf{e}_i f_i(\mathbf{x}+\mathbf{e}_i)\right]/2 \tag{12}$$

IOP Conf. Series: Materials Science and Engineering **529** (2019) 012061 doi:10.1088/1757-899X/529/1/012061

The mass flux is calculated between fluid cell and interface cell. The mass exchange shown in figure 2, doesn't occur between interface cell and gas cell as the fluid fraction of gas cell is zero, which is similar to the treatment in reference [2, 3]. In addition, the adaptive time stepping technique [3] is used to speed up the simulation without reducing the numerical stability.

3. Numerical simulation

3.1. Model validation

A benchmark test of Campbell box [8] is simulated to validate its ability to capture the free surface. Instead of adjusting its value to satisfy the stability condition, the kinematic viscosity is set to be 5.5e-7 m^2/s, which is the physical parameter of the liquid aluminum. The turbulent model and upwind scheme are used to avoid numerical instability, which is different from the treatment of the benchmark test in the simulations of Ginzburg [1]. The inlet velocity is 0.7 m/s. The space step is set to be 1 mm, which results in the total grid number of $246 \times 19 \times 411$ (including the mold cavity). The time step is set to be 1×10^{-5} s. The Smagorinsky constant is set to be 0.12 within a suitable range. The free-slip boundary conditions are applied to the solid wall of the mold cavity.

(a) 0.52 s (b) 0.76 s (c) 1.00 s (d) 1.52 s

Figure 3. The filling sequences of Campbell box by the experiment

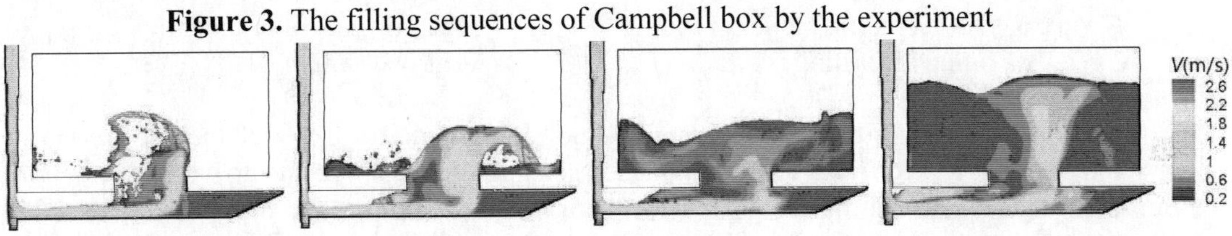

(a) 0.52 s (b) 0.76 s (c) 1.00 s (d) 1.52 s

Figure 4. The predicted filling sequences of Campbell box by the LBM simulation

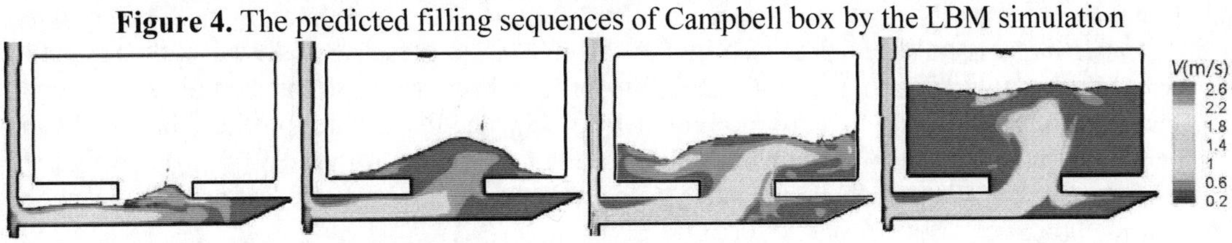

(a) 0.52 s (b) 0.76 s (c) 1.00 s (d) 1.52 s

Figure 5. The predicted filling sequences of Campbell box by the SOLA-VOF simulation

The experimental results of the filling sequence are shown in figure 3. The filling sequences shown in figure 4 are simulated by the LBM model proposed in this study. Figure 5 shows the simulation results obtained by the casting simulation software InteCAST, in which SOLA-VOF method is integrated. The sprue fountains to the bottom of the mold firstly. When the liquid metal begins to fill the casting zone, splashes can be clearly observed at the time t=0.52 s, both in the LBM simulation and the experiment results. The splash shape simulated by the LBM agrees well with the X-ray radiograph of the filling process. When the liquid metal begins to enter the bottom casting zone, the hole is formed due to the falling of the flow frontier, which is well predicted by the proposed LBM model. The irregular interface is caused by the turbulent flow as the Reynold number reaches 25000 approximately. In figure 5 the filling process is consistent with the laminar flow predictions [8], where no splash occurs during the whole filling process. Therefore, the LBM model coupled with the Smagorinsky LES model is capable of predicting the turbulent filling process. When the filling process becomes stable and the flow slows down after t=1.0 s, the interface shape predicted by these two methods is consistent. Compared with the laminar prediction, the suitable turbulent model, such as the Smagorinsky LES model, contributes to capturing interface with bubbles and splashes. The LES-coupled LBM model performs acceptable accuracy and numerical stability.

3.2. Practical application

The filling process of a practical thin gray iron casting shown in figure 6 is studied. The wall thickness of the casting is 10 mm. The gating system consists of four ingates and the total weight is 60 kilogram. The inlet velocity is 1.0 m/s. The grid size is 3 mm and the total grid number is $151\times217\times179$ (including the mold cavity). The grid model used for calculation is shown in figure 7. The filling time is 5.3 s.

Figure 6. The geometric model of the thin gray casting

Figure 7. The grid model of the thin gray casting

The filling sequences at different time simulated by the LBM and the SOLA-VOF method are shown in figure 8 and figure 9, respectively. The liquid metal runs through the sprue and arrives at the bottom of the runner. In figure 8 (a) and figure 9 (a), the two ingates at the bottom are fully filled with liquid metal and the two upper ingates are partially filled due to the small flow rate. The low velocity at the upper ingates causes the discontinuous flow at t=0.76 s. At t=1.4 s, fours ingates are fully filled and the liquid fountains to the cavity quickly which can be clearly observed in figure 8 (b). In the casting cavity, the asymmetric liquid metal frontier caused by the casting structure can be observed in the results of these two methods. At t=2.87 s and t=3.94 s, the filling process becomes steady as the fluid slows down, which can be observed in figure 8 (c), (d) and figure 9 (c), (d). The free surface profiles obtained by the LBM model are in agreement with the SOLA-VOF method, as shown in figure 8 (c), (d) and figure 9 (c), (d). Therefore, the LBM model has an acceptable prediction accuracy of the free surface during the casting filling process.

The LBM simulation is time-consuming due to its high requirement of computation resource. To obtain better robustness, the discrete velocity model D3Q19 is used with the cost of high memory requirement. A casting system model consists of one million nodes and requires approximately 300 Mb memory with D3Q19 model. The calculation time of this simulation is 8 h with the adaptive time stepping technique. Thus, the LBM model is promising for practical casting filling process with an acceptable calculation speed.

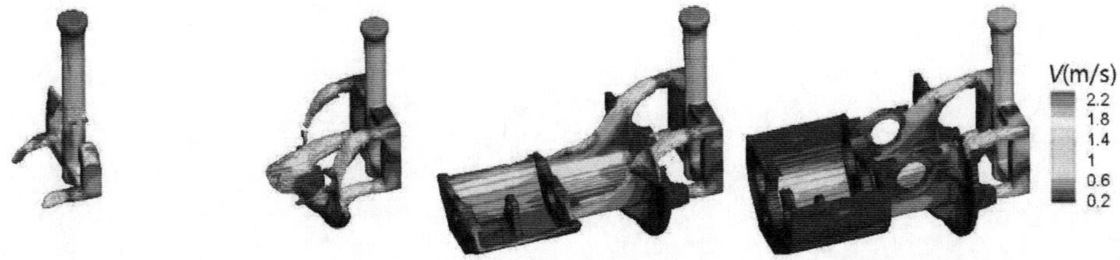

(a) 0.76 s (b) 1.4 s (c) 2.87 s (d) 3.94 s

Figure 8. The predicted filling sequences of the thin gray iron casting simulated by the LBM

(a) 0.76 s (b) 1.4 s (c) 2.87 s (d) 3.94 s

Figure 9. The predicted filling sequences of the thin gray iron casting simulated by the SOLA-VOF method

4. Conclusions

The VOF-based LBM model is implemented and applied to the simulation of the casting filling process. The comparison with the available experiments and SOLA-VOF simulation results proves the proposed model is reliable to capture the free surface during casting filling process. Also, the turbulent flow at high Reynold number is well described with Smagorinsky LES model implemented in the framework of the LBM. Thus, the model is prospective for high accuracy simulation of the casting filling process.

Acknowledgements

This research was financially supported by National Natural Science Foundation of China (No. 51775205).

References

[1] Ginzburg I and Steiner K 2003 Lattice Boltzmann model for free-surface flow and its application to filling process in casting *J COMPUT PHYS* **185** 61-99

[2] Körner C, Thies M, Hofmann T, Thürey N and Rüde U 2005 Lattice Boltzmann Model for Free Surface Flow for Modeling Foaming *J STAT PHYS* **121** 179-96

[3] Thürey N and Rüde U 2009 Stable free surface flows with the lattice Boltzmann method on adaptively coarsened grids *Computing & Visualization in Science* **12** 247-63

[4] Janßen C and Krafczyk M 2011 Free surface flow simulations on GPGPUs using the LBM *COMPUT MATH APPL* **61** 3549-63

[5] Zhao P, Li Q, Kuang S B and Zou Z 2017 LBM-LES Simulation of the Transient Asymmetric Flow and Free Surface Fluctuations under Steady Operating Conditions of Slab Continuous Casting Process *Metallurgical & Materials Transactions B* **48** 456-70

[6] Krüger T, Kusumaatmaja H, Kuzmin A, Shardt O, Silva G and Viggen E M 2017 The Lattice Boltzmann Method

[7] Hou S, Sterling J, Chen S and Doolen G D 1994 A Lattice Boltzmann Subgrid Model for High Reynolds Number Flows *Fields Institute Communications* **6** 151

[8] Sirrell, Holliday and Campbell 1995 Benchmark testing the flow and solidification modeling of Al castings *Journal of the Minerals, Metals and Materials Society* **48** 20-3

Using morphology-equivalent method to simulate the evolution of shrinkage in Ti6Al4V alloy castings during HIP

Q Xu[1], J X Zhou[1*], W Li[1,2], X Feng[2], H Nan[2], Y J Yin[1]

[1] State Key Laboratory of Materials Processing and Die & Mould Technology, Huazhong University of Science and Technology, Wuhan 430074, China
[2] Beijing Institute of Aeronautical Materials, Aviation Industry of China, Beijing 100095, China

Email: zhoujianxin@hust.edu.cn

Abstract. Hot isostatic pressing(HIP) is an effective method to eliminate the shrinkage in castings. The morphology of shrinkage is complex, and there are structures such as sharp corners and small passages which would lead to a large number of elements and easily divergent calculation results. Therefore, the application of numerical simulation in HIP is limited. To solve the non-convergence problem, the real shrinkage is often simplified as a sphere. However, this simplification ignores the characteristics of the shrinkage and makes the simulation results unreliable. In this paper, the technique of morphology-equivalent ellipsoid is applied to the numerical simulation of shrinkage evolution during HIP. Firstly, the 3D morphology of shrinkage in Ti6Al4V alloy castings is obtained by micro computed tomography. The radius of sphere and the geometric size and orientation of morphology-equivalent ellipsoid are calculated by corresponding equivalent techniques. Secondly, the numerical simulations of HIP for the Ti6Al4V castings before and after the equivalent method are carried out. The volume evolution of three kinds of shrinkages are recorded and compared. The results show that the volume evolution of the morphology-equivalent ellipsoid is closer to that of the real shrinkage, the feasibility of the morphology-equivalent ellipsoid and the limitations of sphere are verified.

1. Introduction

Ti6Al4V alloy is widely used in the aerospace industry due to their low density, high specific and corrosion resistance. During the casting of Ti6Al4V alloy, the metal liquid is difficult to feed as a result of poor fluidity, and the defect such as shrinkage is easily generated. Hot isostatic pressing (HIP) is a well-known method to eliminate the shrinkage in castings and improve material properties[1].

HIP is a process that the castings or powder is densified under the action of high temperature and high pressure. The current research on HIP of castings is mainly based on experimental methods. Anton du Plessis et al. [1] firstly applied x-ray microcomputed tomography (Micro-CT) to quantify the shrinkage changes before and after HIP of as-cast Ti6Al4V rods. Chung-Hung Tam et al.[2] have studied the effects of the HIP temperature on Cr-Si targets and found suitable HIP temperatures. X.G. Zheng et al.[3] explored the healing process of shrinkage in a Ni-based superalloy by HIP. Numerical simulation is another effective way to study the process of shrinkage elimination during HIP due to their low cost and predictive ability. However, there are few researches on numerical simulation of HIP. Alexander Epishin et al.[4] gave the kinetics of pore annihilation during HIP by experiment and modelling. Yu Zhou et al.[5] proposed a healing model for creep cavities under HIP. In our previous research[6], the effects of temperature on the shrinkage densification and microstructure of Ti6Al4V alloy have been studied by experimental and numerical simulation methods. In all existing numerical simulations of HIP, it is assumed that the shrinkages in the castings are spherical, ignoring the morphology of the real shrinkage. That is because the real shrinkage morphology often has complex geometric features such as sharp corners and small passages which will lead to the diverging simulation results. Therefore, it is meaningful to solve the non-convergence problem during HIP.

In previous studies, an equivalent method using ellipse or ellipsoid to replace the real shrinkage was employed in forging peocess. Kieran F. Mulchrone et al.[7] fitted an ellipse to an arbitrarily shaped region for strain analysis. M. Saby et al.[8] used the morphology-equivalent ellipsoid to analyze the void closure during forge. This paper refers to the equivalent method of shrinkage in forging, and applies this method to the numerical simulation of HIP for Ti6Al4V alloy. The volume evolution of real shrinkage, volume-equivalent sphere and morphology-equivalent ellipsoid are compared and discussed.

2. Equivalent method

The morphology of the castings and shrinkage was obtained from the Micro-CT, as shown in figure 1.

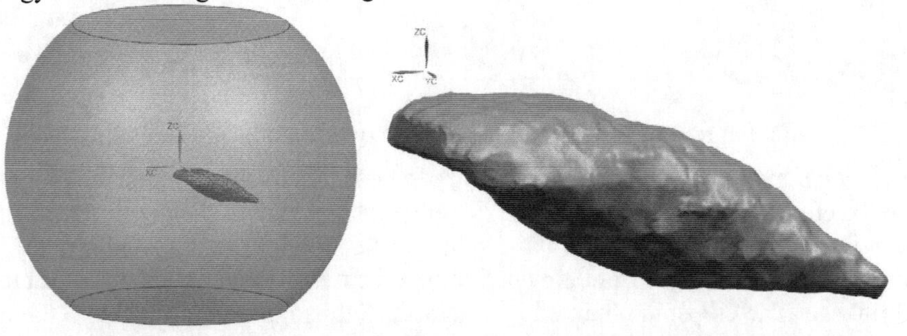

Figure 1. The morphology of the casting and the shrinkage.

2.1. The volume of the shrinkage

The binary STL files were obtained from the Micro-CT, which contain the information of triangles. The volume of the shrinkage can be expressed as Eq. (1):

$$\frac{1}{6}\sum_{\tau\in\Gamma}\begin{vmatrix} x_{\tau,1}-x_0 & y_{\tau,1}-y_0 & z_{\tau,1}-z_0 \\ x_{\tau,2}-x_0 & y_{\tau,2}-y_0 & z_{\tau,2}-z_0 \\ x_{\tau,3}-x_0 & y_{\tau,3}-y_0 & z_{\tau,3}-z_0 \end{vmatrix} \tag{1}$$

where τ is a triangle of the surface mesh T of the shrinkage, $x_{\tau,i}$, $y_{\tau,i}$ and $z_{\tau,i}$ (i = 1,2,3) are the coordinates of the three vertices of triangle τ, x_0, y_0 and z_0 are the coordinates of origin.

Encode Eq. (1) with Python, the volume of the shrinkage is 8.4802mm^3. The equivalent spherical shrinkage has the same volume with the real shrinkage. Therefore, the radius of the equivalent spherical shrinkage is 1.2650mm.

2.2. Generation of the morphology-equivalent ellipsoid

The morphology-equivalent ellipsoid is generated in order to represent shape factors and principal orientations of real shrinkage accurately[8]. The schematic diagram of the construction process of the morphology-equivalent ellipsoid is shown in the figure 2. First, the shape factors and principal orientations of inertia ellipsoid are obtained by computing the inertia matrix of the real shrinkage. Then, the morphology-equivalent ellipsoid is obtained by volume correction of the inertia ellipsoid. And the inertia matrix can be represented as:

$$\begin{bmatrix} I_{xx} & -I_{xy} & -I_{xz} \\ -I_{yz} & I_{yy} & -I_{yz} \\ -I_{zx} & -I_{zy} & I_{zz} \end{bmatrix} \tag{2}$$

where I_{xx}, I_{yy} and I_{zz} are the moment of inertia of the rigid body relative to the coordinate axis, I_{xy}, I_{xz}, I_{yz}, I_{yx}, I_{zx} and I_{zy} are the inertia product of the rigid body relative to the coordinate axis, and they are defined as:

$$\begin{cases} I_{xx} = \iiint_G \left(y^2 + z^2\right) dxdydz \\ I_{yy} = \iiint_G \left(x^2 + z^2\right) dxdydz \\ I_{zz} = \iiint_G \left(x^2 + y^2\right) dxdydz \end{cases} \text{ and } \begin{cases} I_{xy} = I_{yx} = \iiint_G xydxdydz \\ I_{yz} = I_{zy} = \iiint_G yzdxdydz \\ I_{zx} = I_{xz} = \iiint_G zxdxdydz \end{cases} \tag{3}$$

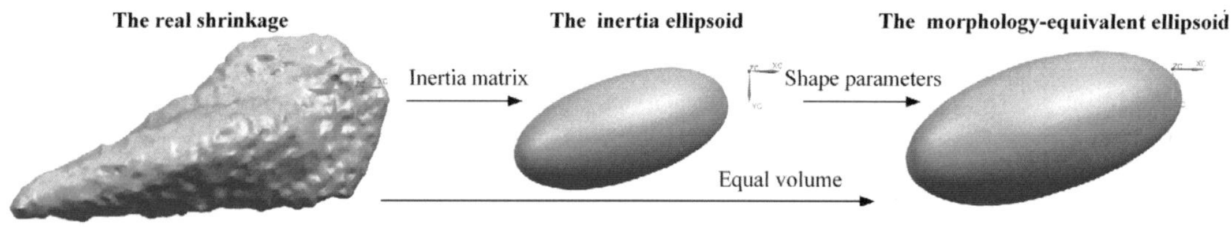

Figure 2. Schematic diagram of the process for calculating the morphology-equivalent ellipsoid.

The matrix shown by Eq. (2) is a symmetric matrix whose eigenvalues are the lengths of three semi-axes of the inertia ellipsoid, and the eigenvectors are the principal axes of the inertia ellipsoid. The length ratio and the principal orientations of the three semi-axes of the inertia ellipsoid are kept unchanged, and the semi-axes of the inertia ellipsoid is scaled so that the transformed volume is equal to the volume of the real shrinkage. The ellipsoid obtained by the inertia ellipsoid transformation is the morphology-equivalent ellipsoid. The morphology-equivalent ellipsoid is shown in figure 3 with red color superimposed with the real shrinkage in green color and volume-equivalent sphere in blue color.

Figure 3. The real shrinkage, morphology-equivalent ellipsoid and volume-equivalent sphere.

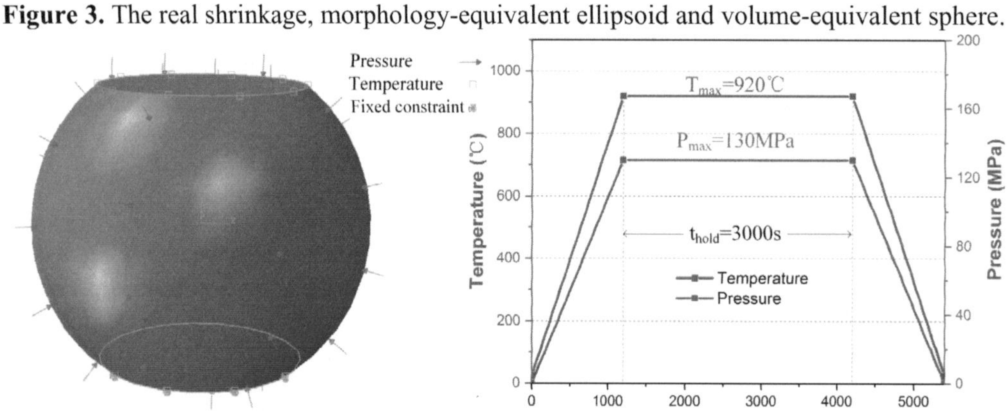

(a) (b)

Figure 4. Boundary conditions and variation of pressure and temperature with time in a HIP cycle.

3. FEM simulation

The numerical simulations of HIP process for the Ti6Al4V alloy castings with three kinds of shrinkage were carried out using the finite element software ABAQUS. The fixed boundary was applied to the bottom surface of the geometry, a temperature boundary and a pressure load were applied to each face of the geometry, as shown in figure 4(a). The curves of temperature boundary and pressure load versus time are shown in figure 4(b). During the HIP process, the temperature and pressure simultaneously increased to 930 °C and 120 MPa, and maintained for 50min, and then decreased to 25 °C and $1E^{-5}$ MPa, respectively.

4. Simulation results and discussion

The morphology evolution of real shrinkage, morphology-equivalent ellipsoid and volume-equivalent sphere are shown in figure 5(a). The closure process of the casting with real shrinkage cannot reach the end of simulation due to the mesh distortion. While it still can be found that the morphology-equivalent ellipsoid has the same closure tendency as the real shrinkage hole, that is, the change of shrinkage is small in the long axis direction, and the shrinkage closure is mainly dependent on the collapse in the short axis direction. However, the volume-equivalent sphere exhibits equal shrinkage in almost all directions.

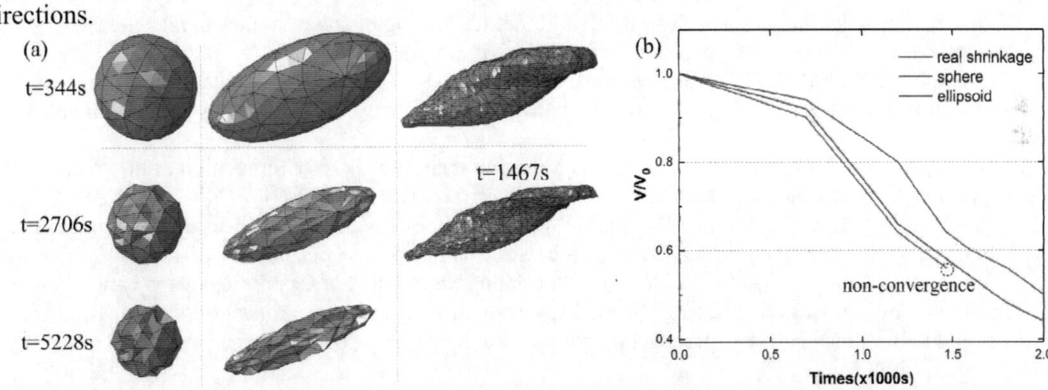

Figure 5. The FEM analysis of the morphology change and the volume evolution of real shrinkage, the morphology-equivalent ellipsoid and volume-equivalent sphere under HIP.

The volume evolution of real shrinkage, the morphology-equivalent ellipsoid and volume-equivalent sphere are plotted in figure 5(b). It can be seen that the volume-equivalent sphere underestimates the real closure degree during HIP which may be ascribed to the tortuosity of the surface of real shrinkage. Some of the initial morphology must be favorable to void closure[8]. The morphology-equivalent ellipsoid underestimates slightly the real closure degree and it agrees well with the real shrinkage. Therefore, using the morphology-equivalent ellipsoid to replace the real shrinkage will have better prediction results. The result also reveals that the flatten-shaped shrinkage[9] can be healed easily.

5. Conclusion

In this study, the morphology-equivalent method was used to simulate the elimination of shrinkage in Ti6Al4V alloy castings during HIP. Important conclusions are presented below.

(1) The morphology of real shrinkage is so complicated that the numerical simulation of HIP for castings is difficult to be carried out. Replacing the real shrinkage with the morphology-equivalent ellipsoid can effectively solve the problem in convergence.

(2) The morphology-equivalent ellipsoid is generated by considering the volume, shape factors and principal orientations of the real shrinkage. It can be found from the simulation results that using the morphology-equivalent ellipsoid is proved to have better prediction results than the volume-equivalent sphere.

(3) The volume-equivalent sphere underestimates the real closure degree during HIP which may be ascribed to the tortuosity of the surface of real shrinkage. The morphology-equivalent ellipsoid

underestimates slightly the real closure degree. It is inferred that some of the initial morphology must be favorable to void closure.

Acknowledgement

This work was financially supported by the National Natural Science Foundation of China (No. 51475181) and AECC Beijing Institute of Aeronautical Materials.

References

[1] Plessis A and Rossouw P 2015 Investigation of Porosity Changes in Cast Ti6Al4V Rods After Hot Isostatic Pressing *J. Mater. Eng. Perform.* **24** 3137–41

[2] Tam C H, Lee S C, Chang S H, Tang T P, Ho H H and Bor H Y 2009 Effects of the temperature of hot isostatic pressing treatment on Cr-Si targets *Ceram. Int.* **35** 565–70

[3] Zheng X G, Shi Y N and Lou L H 2015 Healing Process of Casting Pores in a Ni-based Superalloy by Hot Isostatic Pressing *J. Mater. Sci. Technol.* **31** 1151–7

[4] Epishin A, Fedelich B, Link T, Feldmann T and Svetlov I L 2013 Pore annihilation in a single-crystal nickel-base superalloy during hot isostatic pressing: Experiment and modelling *Mater. Sci. Eng. A* **586** 342–9

[5] Zhou Y, Zhang Z, Zhong Q and Zhao Z 2012 Model for healing of creep cavities in nickel-based superalloys under hot isostatic pressing *Comput. Mater. Sci.* **65** 320–3

[6] Xu Q, Zhou J, Nan H, Yin Y, Wang M, Shen X and Ji X 2017 Effects of hot isostatic pressing temperature on casting shrinkage densification and microstructure of Ti6Al4V alloy *China Foundry* **14** 429–34

[7] Vancheeswaram R, Queheillalt D T, Elzey D M and Wadley H N G 2001 Simulation of the creep expansion of porous sandwich structures *Metall. Mater. Trans. A* **32** 1813–21

[8] Saby M, Bernacki M, Roux E and Bouchard P-O 2013 Three-dimensional analysis of real void closure at the meso-scale during hot metal forming processes *Comput. Mater. Sci.* **77** 194–201

[9] Yuan C, Li J, Dong K-X and Guo J-T 2017 On Healing Mechanism of Cast Porosities in Cast Ni-Based Superalloy by Hot Isostatic Pressing BT - Energy Materials 2017 *Energy Materials 2017* ed X Liu, Z Liu and K Brinkman (Cham: Springer International Publishing) pp 265–75

IOP Conf. Series: Materials Science and Engineering **529** (2019) 012063 doi:10.1088/1757-899X/529/1/012063

Advances in Modeling of Steel Solidification with IDS

J Miettinen[1], S Louhenkilpi[1,2], V-V Visuri[1] and T Fabritius[1]

[1] Process Metallurgy Research Unit, University of Oulu, P.O. Box 4300, 90014 University of Oulu, Finland
[2] Department of Chemical and Metallurgical Engineering, Aalto University, P.O. Box 16200, 00076 Aalto, Finland

E-mail: ville-valtteri.visuri@oulu.fi

Abstract. IDS (Inter-Dendritic Solidification) is a thermodynamic-kinetic software package that simulates phase changes, compound formation/dissolution, and solute distribution during solidification of steels as well as during their cooling/heating process after solidification. The software package also simulates solid-state phase transformations related to the austenite decomposition process at temperatures below 900/600 °C, and calculates thermophysical material properties from the liquid state down to room temperature. These data are needed in other models, such as heat transfer and thermal stress models, whose reliability heavily depends on the input data. The software package also features a database for thermodynamic, kinetic and microstructure data, as well as for several material properties. Owing to the short calculation times, the IDS tool is suitable for online applications. This paper presents IDS and its modules with the latest developments and validations, along with examples of modeling results.

1. Introduction

IDS is a thermodynamic-kinetic solidification package for steels [1] developed since 1984. It has been validated with extensive data from solidification experiments, with good results [2]. The calculations of the IDS package are made in one volume element set on the side of a dendrite arm [1] based on two assumptions: (1) The dendritic structure formed throughout the strand is regular. (2) There are no strong liquid flows causing macrosegregation. Due to the simple-format thermodynamic sub-models of the IDS package and the simultaneous solution of thermodynamic, kinetic and material balance equations during the simulation, the calculation times of IDS are relatively short.

2. Description of the IDS model

The present IDS package includes six calculation modules: (1) SOL (solidification), (2) ADC (austenite decomposition), (3) MAT (material properties), (4) GAS (hydrogen and nitrogen behavior), (5) PRF (precipitate formation at the ferrite/austenite phase interface), and (6) SCA (oxide scale formation). It also includes an extensive database containing the thermodynamic, kinetic, microstructure and material data needed in calculations. In addition, an additional module for QIS (steel quality indices) is under development to provide information about the sensitivity for various casting problems, such as cracking. Figure 1 shows the main input and output data of the offline version of IDS, as well as its modular structure. Also included is the coupling of IDS with some other model packages, discussed later in the text.

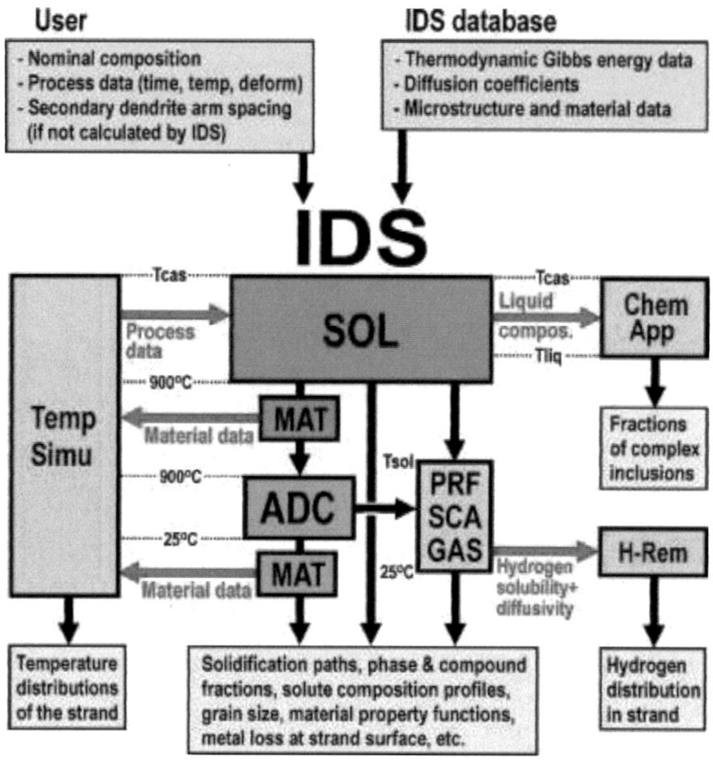

Figure 1. Main input and output data of the offline IDS model package and its modules and their coupling to other model packages.

2.1. *SOL module*

The SOL module is the main module of the IDS package. It is applied in the temperature range 1,600–900 °C for low-alloy steels and 1,600–600 °C for high-alloy steels. The steel is regarded as high-alloy steel whenever at least one of the following conditions for the nominal composition is true: wt%Cr $\geq$ 5, wt%Ni $\geq$ 5, wt%Mn $\geq$ 5, wt%Mo $\geq$ 3 or wt%Si $\geq$ 3. The module applies thermodynamic chemical potential equality equations (all components), interfacial material balance equations including Fick's first law of diffusion (solutes), and a finite difference application of Fick's second law of diffusion (solutes). The computational routine has been described in [1].

Using the IDS input data (see Figure 1), the SOL module simulates or calculates the following phenomena: (1) solidification of steel (ferritic, austenitic, and peritectic solidification); (2) enrichment or impoverishment of solutes in liquid; (3) formation of various compounds (inclusions) from liquid; (4) homogenization of solute microsegregation below the solidus; (5) ferrite-austenite transformation below the solidus; (6) formation and dissolution of eutectic ferrite (in high-alloy steels only); (7) formation and dissolution of various compounds (precipitates) below the solidus; and (8) growth of austenite grains.

The possible solute or impurity components and their recommended maximum compositions (wt%) are given for low-alloyed (L) and high-alloyed (H) steels as: C = 3.5(L), 1.5(H); Mn = 5(L), 25(H); Cr = 5(L), 30(H); Mo = 3(L), 6(H); Ni = 5(L), 25(H); Al, Cu, Si = 3; Nb, Ti, V, N = 0.5; P, S = 0.05; Ca, Ce, Mg = 0.02, and B, O, H = 0.005. For the present, the possible compounds (precipitates) of the simulation are: AlN, BN, CrN, Cr_2N, Si_3N_4, TiB_2, CaS, MgS, H_2(g), N_2(g), $FeMo_2B_2$, Fe_3Mo_3C, FeNbB, Ti_2CS, (Mn,Fe)S, (C,N)Nb, (C,N)Ti, (C,N)V, $(Cr,Fe)_2B$, $(Fe,X)_2B$ (X=Cr,Mn, Ni, V), $(Cr,Fe)_{23}C_6$, $(Cr,Fe)_7C_3$, $(BC)_6Fe_{23}$, $(Fe,X)_3C$ (X=Cr,Mn,Mo), $(Cr,Ni)_8Fe_9$ (~sigma), together with several simple oxides.

2.2. *ADC module*
The ADC module is used for low-alloy steels in the temperature range 900–25 °C to simulate the decomposition of austenite to proeutectoid ferrite or cementite, pearlite, bainite and BCT martensite. Also simulated is the formation or dissolution of different precipitates (see the compounds of the SOL module). The ADC module applies thermodynamic chemical potential equality equations of paraequilibrium condition, material balance equations of carbon, boron, and nitrogen for the ferrite/austenite interface advancing in a spherical austenite grain, and regression formulas optimized from the German and British CCT measurements and some additional measurements on boron steels. In these formulas, the influencing solutes are C, Si, Mn, Cr, Mo, Ni, and B. Using the IDS input data (see Figure 1), the ADC module simulates or calculates the following phenomena: (1) fractions of austenite and its decomposition phases, (2) formation and dissolution of various precipitates, (3) Ae3 and AeC temperatures, and (4) austenite grain growth.

Due to the limited solute contents in the original CCT data, the recommend composition ranges for the ADC calculations are relatively low. A tentative suggestion is to apply compositions of $0.001 < wt\%C < 3.5$, $wt\%Si < 1$, $wt\%Mn < 2$, $wt\%Cr < 3$, $wt\%Mo < 1$, and $wt\%Ni < 3$, while the compositions of the other solutes (taking no action in the CCT formulas) should be as low as possible. In further work, the present CCT formulas will be re-optimized to include a much larger set of CCT measurements for alloys with higher solute contents, and also including two new solutes: Al and Cu.

For high-alloy steels, the ADC module is used in the temperature range 600–25 °C, and the only decomposition phases are BCT and HCP martensites. In this case, the recommended composition ranges are the same as those of the SOL module, and no simulation is made for the compound formation.

2.3. *MAT module*
The MAT module calculates temperature and time functions for important material properties using the material data of the IDS database and the phase fractions and compositions calculated by the SOL and ADC modules. Here, the most important material properties are enthalpy, thermal conductivity, and density, which are needed as input data in other model packages in order to simulate the macroscopic heat transfer in a strand. In the case of enthalpy, the thermodynamic Gibbs energy data of the IDS database is used directly, and in the case of other material properties, the regression parameters optimized for the database are used. The advantage of the method is that discontinuities caused by different phase changes are automatically included in the material property functions. The present IDS package calculates material data for the Tempsimu package [3], but the data can be "tailored" to other packages, as well. The new MAT module also calculates the strain rates at different stress values (0.1–400 MPa) and the elastic modulus for the considered steel, as a function of temperature. Calculations are made also to give the surface tension and the stacking fault energy (in high-alloyed steels).

2.4. *PRF module*
The PRF module is applied to simulate the precipitate formation at the ferrite/austenite phase interface, if present in the steel. Only the final amounts of the precipitates are reported (at 25 °C) together with the temperature and time intervals of their appearance and disappearance. This output data gives an opportunity to study the tendency of Cr (enriching strongly on the ferrite side of the boundary in stainless steels) to form Cr carbides or even Cr nitrides, and their possible effect on cracking.

2.5. *GAS module*
The GAS module is applied to simulate the behavior of gaseous elements (H and N) in solid steels below the solidus. Temperature (and time) functions are calculated for (1) solubility of H and N in solid (ferrite and austenite), (2) excess (insoluble) composition of H and N in solid, and (3) diffusion coefficients of H and N in solid. Above the solidus, part of H and N may take a gaseous form, leading to porosity formation in the mushy zone. This is handled by the SOL module, allowing gaseous $H_2(g)$ and $N_2(g)$ "compounds" to form from the liquid phase. The gas formation depends on the ferrostatic pressure in

the liquid phase and its decrease from the mushy zone shrinking. The resulting H and N contents (their solubility at the solidus) thus represent the start composition of the GAS simulation.

2.6. *SCA module*
The SCA module is applied to simulate oxide scale formation on the surface of the continuous casting strand, depending on the oxygen pressure. This module is a typical furnace treatment application of continuous casting. Temperature and time functions are calculated for (1) metal loss, (2) oxide phase weight gain, and (3) approximate fractions of different oxide types. In addition, the oxide type, whether it is detrimental or not, is tentatively characterized.

2.7. *Database*
During this decade, a great amount of new thermodynamic, kinetic and microstructure data has been stored in the IDS database to make IDS simulations for different steel grades. Also included are new material data parameters permitting the calculation of thermal conductivity, density, viscosity, surface tension, strain rate, and elastic modulus for these steels as a function of temperature. All these data have been verified with experimental measurements. Most of the data is the Gibbs energy data of solution phases and compounds taken from thermodynamic assessments of iron-based alloys.

2.8. *Coupling of IDS to other model packages*
The IDS package can be coupled with some other model packages: ChemApp, H-Rem and Tempsimu, as illustrated in Figure 1. ChemApp [4] is a thermodynamic programming library that can be used to simulate the formation of multiphase inclusions in the liquid steel. H-Rem is a package for calculating the hydrogen distribution in continuous casting slabs. The model is used to find optimal holding temperatures and times for different strand geometries, which can decrease the hydrogen content of the strand to an acceptable level. Tempsimu [3] is a package that simulates the heat transfer in continuous casting strands. In this case, IDS calculates the temperature functions of the material property data (including liquidus and solidus temperatures) for Tempsimu, while Tempsimu calculates the nodal process data (time, temperature, deformation) of the strand for IDS. The deformation effect comes from the bending and unbending processes. In online processes of continuous casting, the output of IDS can be used in casting and strand simulators visualizing the state of the strand for the user. So far, IDS and online version of Tempsimu, called CastManager, have been installed in four slab casters for online use.

2.9. *Latest developments of IDS.*
During the two last years, several steps have been made to improve the IDS package. Some essential improvements are summarized below.

2.9.1. Database. The thermodynamic and material property data of the IDS database has been extended considerably in comparison to earlier work [1]. The focus in the thermodynamic data has been on systems containing nitrogen, and high-alloyed systems. In the material properties, the focus has been on the calculation of new properties (such as strain rates (as a function of stress), elastic modulus and stacking fault energy), and on new optimizations for thermal conductivity, density, surface tension and secondary dendrite arm spacing (SDAS).

2.9.2. Extension of IDS to cast irons. The recommended maximum carbon composition of low-alloyed steels has been increased from 1.5 to 3.5 wt% C, to extend the applicability of IDS simulations to cast irons. Also, the material property algorithms have been updated in this respect. The calculations have been validated with numerous temperature measurements on cast irons [2]. Note that IDS cannot simulate the morphology of cast iron, but only the formation of graphite and cementite phases from liquid and austenite. The effect of nodularity on thermal conductivity, however, can be taken into account.

2.9.3. Cooling/heating processes and precipitation. The IDS procedure linking a complex cooling/heating curve to the simulation of precipitation, phase transformation, and scale formation in steels has been improved. Particularly, the problems related to the sensitivity of the precipitation calculations to diverge (as a result of sudden changes from cooling to heating and vice versa) have been solved. Also, the treatment of precipitate formation itself has been improved, to get a clear "nose" for their formation. This makes it possible to seek a critical cooling rate above which we get no formation for a precipitate. The results have been validated with experimental measurements.

2.9.4. Solidus of high-alloyed steels. In heavily alloyed steels beyond the recommended composition ranges of the SOL module, the solidus may sometimes get far too low and even crash the simulation. That trend has now been reduced by specific thermodynamic modifications made close to the solidus temperature, when its value gets exceptionally low.

*2.9.5. Eutectic ferrite***.** The treatment of the eutectic formation has been simplified and allowed for any high-alloyed steels, including their equilibrium solidification. In the latter case, the extension concerns only the purely austenitic solidification, since there is no entrance from the austenitic solidification back to the ferritic solidification in IDS. In other words, the simulation of eutectic ferrite formation is then the only way to get ferrite back in the structure.

3. Validation and results

The calculations of the IDS model have been validated with numerous experimental measurements of solidification [2]. Table 2 summarizes the average deviation between calculated and measured temperatures of liquidus (T_{LIQ}), austenite formation (T_{AUS}), zero-strength (T_{ZST}), and solidus (T_{SOL}) of low-alloyed and stainless steels, and for the residual ferrite contents of stainless steels (ferrite content). The agreement is reasonably good, and much better than can be obtained with analytic literature equations, e.g. for the liquidus and solidus temperatures.

Table 1. Average deviation between IDS calculations and measurements for steels [2].

		Average deviation				
Alloy type	Number of measurements	T_{LIQ}	T_{AUS}	T_{ZST}	T_{SOL}	Ferrite content
Ternary Fe–X–C alloys	283	5.6 °C				
Ternary Fe–Cr/Ni–X alloys	274	3.9 °C				
Steels	387	3.2 °C				
Steels	47		7.4 °C			
Low-alloyed steels	25			5.6 °C		
Steels	170				10.5 °C	
Stainless steels	93					1.11%

Figure 2 shows examples of calculations mimicking phenomena in continuous casting and reheating furnace. The figure shows how (a) some precipitates form in an AISI 410-type stainless steel (containing exceptionally 0.1%C and 0.1%Nb), and how (b) the metal loss proceeds in the slab surface of the 0.15% C, AISI 410 and AISI 316 steels, as the applied cooling/heating histories are those shown in the upper figures. Figure 2a reveals the importance of fixing the cooling/heating history and the straightening (or bending) position so that fewer precipitates would form in the latter region. Otherwise, the probability for cracks may clearly increase. Figure 2b shows that in the case of low-alloyed carbon steels, a high holding temperature in combination with a high oxygen pressure leads to higher metal loss. In ferritic stainless steel (AISI 410), the metal loss is much smaller, and it is smaller still in AISI 316 austenitic stainless steel. The weaker oxidation in the latter steels is due to the formation of protective layers of Cr_2O_3 and MoO_2 oxides. The cases are illustrative and not from any real industrial process.

IOP Conf. Series: Materials Science and Engineering **529** (2019) 012063 doi:10.1088/1757-899X/529/1/012063

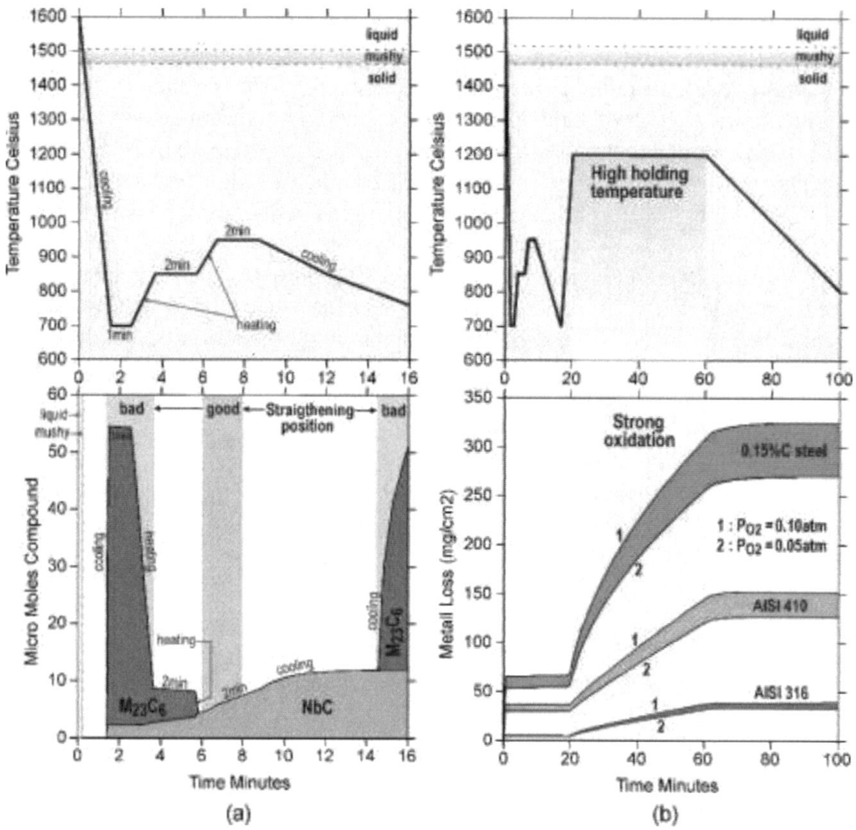

Figure 2. Effect of temperature histories (upper figures) on the formation of precipitates in (a) AISI 410-type stainless steel containing 0.1%C and 0.1%Nb, and (b) on the metal loss in a plain carbon steel (0.15%C) and in AISI 410- and 316-type stainless steels.

4. Conclusions

The aim of this paper was to present IDS and its modules with the latest developments. Validation of the model suggests a good match with experimental data for different steel grades. The latest developments have been directed at extending the thermodynamic and material property database as well as improving the descriptions for cooling/heating processes and precipitation, solidus of high-alloyed steels and eutectic ferrite formation. Owing to its short calculation times, IDS is suitable for online applications of continuous casting.

Acknowledgements

This study was executed within the framework of the Genome of Steel profilation project. The Academy of Finland (project 311934) and Finnish Cultural Foundation are acknowledged for funding this work.

References

[1] Miettinen J, Louhenkilpi S, Kytönen H and Laine J 2010 *Math. Comput. Simul.* **80** 1536.

[2] Miettinen J 2017 *Validation of IDS calculations with experimental data* (Espoo: Casim Consulting Oy).

[3] Louhenkilpi S, Laine J, Miettinen J and Vesanen R 2013 *Mater. Sci. Forum* **762** 691.

[4] Erikson G and Königsberger E 2008 *Pure Appl. Chem.* **80** 1293.

IOP Conf. Series: Materials Science and Engineering **529** (2019) 012064 doi:10.1088/1757-899X/529/1/012064

Investigation on Band Segregate Formation during the Electroslag Remelting of H13 Die Steel

Q Wang[1, 2]**, X J Wang**[1]**, G Q Li**[1]**, Y Liu**[1]

1.Wuhan University of Science and Technology, Wuhan 430081, Wuhan, China;
2. Delft University of Technology, Delft 2628 CD, Delft, Netherlands.

Abstract. Band segregation has been found in the H13 die steel produced by the electroslag remelting (ESR) technology. Chemical and metallographic studies have been carried out on a one ton ESR ingot of H13 die steel, so as to understand the formation mechanism of the band segregation. The results indicate that the T.O content and S content decreased because of cleanliness improvement of ESR process. Transverse macrosegregation of S content decreased after ESR. The overall removal ratio of the inclusion is around 65.8%. The original complex inclusions would be modified to the $CaO{\cdot}Al_2O_3$ inclusions. Al_2O_3 and MnS inclusions can be found after ESR. Both of Al_2O_3 and MnS inclusions were found to be the core of primary carbides. The net like structure in ESR ingot and banded structure in the forged steel were observed. V, Mo, Cr and S are rich in the segregation areas of ESR ingot. Besides, black and white segregation bands can be observed on the forged steel samples after etching. Uneven distribution of carbides rich in V, Mo and Cr was observed in banded structure.

1. Introduction

H13 steel is a common hot-working die material, for its good mechanical properties. In order to avoid abnormal failure, it is important to decrease the defects of the H13 steel such as inclusions and segregation[1-4]. Segregation is unavoidable during the solidification, severe segregation which cannot be completely eliminated by heat treatment would generate adverse effect on the final products. During solidification of H13 steel, elements like C, S and V would diffuse from solidified steel to molten steel causing enrichment of these elements in inter-dendritic, that is what we call dendritic segregation, a kind of microsegregation, large size precipitates can also be found in the segregation areas. After forging or rolling, banded structures always exist in H13 steel. Those precipitates and banded structures mentioned above will lower the impact toughness of H13 steel, the isotropy also can be decreased strongly.

Electroslag remelting (ESR) is usually adapted to produce H13 steel for the excellent ability in removing inclusions and improving solidification structure[5-7]. However, ESR still can't solve the microsegregation entirely.

Lots of researches have been done to decrease segregation. Li *et al.* [8] have studied the effect of trace magnesium addition on the inclusions modification and carbide improvement, they found that the disregistry between $MgO{\cdot}Al_2O_3$ inclusions and γ-Fe is smaller than that between Al_2O_3 inclusions and γ-Fe, which means $MgO{\cdot}Al_2O_3$ inclusions are more effective in promoting nucleation of γ-Fe. Comparing with Al_2O_3 inclusions, the modified inclusions $MgO{\cdot}Al_2O_3$ are less possibly to cluster, and the finer and better distributing $MgO{\cdot}Al_2O_3$ inclusions would contribute to the improvement of solidification structure. Besides, the primary carbides could be also refined. Liu *et al.*[9] found that MnS inclusions can be the nucleation agent of primary carbides, the segregation decrease with desulphurization. MA *et al.*[10] investigated the thermal homogenization treatment. After being kept in 1200℃ for 8 hours, the isotropy of H13 was improved.

Those works mentioned above did show obvious improvement on the segregation in as-cast steel, but there is still lacking of the researches about segregation in forged steel. In the present work, the inclusions in the H13 steel were studied. The segregation in both the as-cast steel and forged steel were investigated.

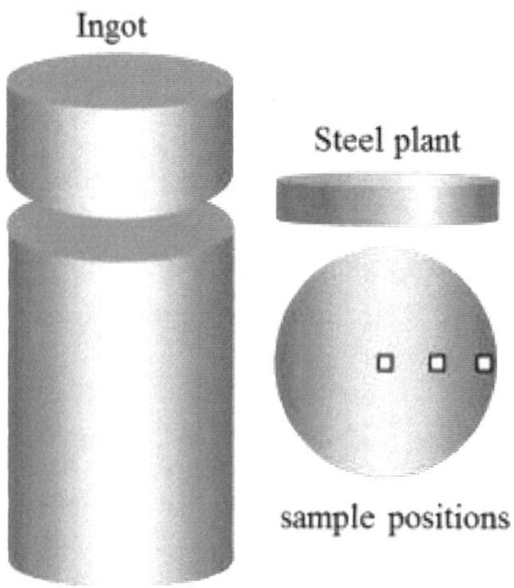

Figure 1. Schematic diagram of the ingot and sample positions

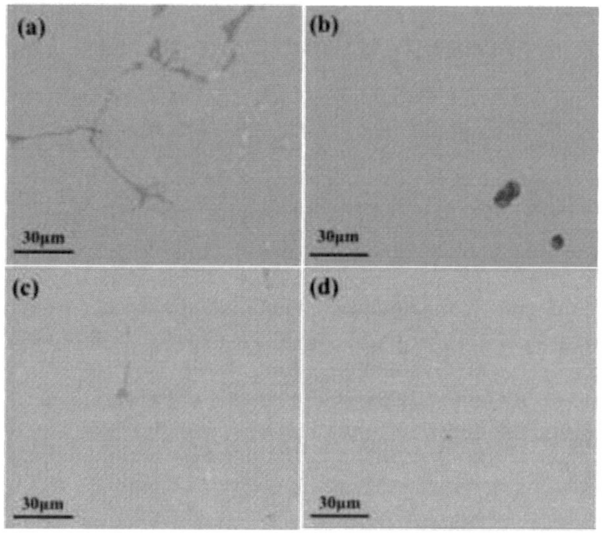

Figure 2. images of the inclusions and precipitates in the steel samples: (a) and (b) electrode, (c) and (d) ESR ingot.

2. Experiment

H13 steel was manufactured with EAF→ LF→VD→Ingot Casting→ ESR (an one ton electroslag furnace under air atmosphere)→Forging procedure. The slag used in ESR process consists of 70% CaF_2+30%Al_2O_3 (38 Kg). The current of the remelting process is about 9000A, and the remelting process lasted for 4 hours. Steel plants were taken from head of the electrode (200mm in diameter), ESR ingot (350mm in diameter) and forged steel (150mm in diameter). The samples were taken from the center, half of the radius and surface of electrode and ESR ingot with a size of 4×4×4mm^3, which were used to analyze the chemical compositions. One sample of 10×10×10mm^3 in size was taken from the center of each plant to study the inclusions, and segregation. Figure 1 is the schematic diagram of the ingots and positions where the samples were taken from. The C, S, O and N content and other elements were determined by C-S analyzer, O-N analyzer and ICP-AES. The size, morphology and composition of the inclusions were investigated by SEM (Nova 400 Nano) equipped with EDS (Le350PentaFETx-3). Fifty photos taken randomly by SEM were employed to count the number and size of inclusions by Image Pro-Plus at 2000 magnification. The morphology of segregation areas in ESR ingot and Forged steel was investigated by OM and SEM.

3. Results and Discussion

3.1 Chemical composition

Table 1 shows the chemical composition of electrode and ESR ingot. Comparing the S content in electrode plant with that in ESR ingot plant, there is a significant decrease, which should be attributed to gasifying desulfurization. In air atmosphere, desulfurization process of ESR would happen as Equations (1) and (2) show, and this is gasifying desulfurization mentioned above[6]. It can be also seen that there are no obviously changes among center, 1/2 radius and surface of the electrode except for the

S content. However, the concentration of sulphur along the transverse of the ESR plant is more homogeneous macroscopically. Thus, the macrosegregation of Sulphur along the transverse of electrode could be solved by electroslag remelting.

$$(O^{2-}) + [S] = (S^{2-}) + [O] \quad (1)$$

$$(S^{2-}) + 3/2O_2 = SO_2 + (O^{2-}) \quad (2)$$

N content increased after electroslag remelting which means that the steel was polluted by air. This can also explain the decrease of silicon. It has to be known that the oxidation is detrimental to the desulfurization reaction[6]. As to the total O content, its average value varied from 46ppm(electrode) to 35ppm(ESR ingot), that should owe to the removing of inclusions by electroslag remelting.

Table.1 Chemical composition of different positions in steel plant of ingot (wt. %).

Electrode	C	Si	Mn	Cr	Mo	V	S	T.O	N
Center	0.39	0.96	0.36	5.24	1.30	1.04	0.0110	0.0044	0.0287
1/2 Radius	0.40	0.97	0.36	5.27	1.33	1.06	0.0093	0.0047	0.0286
Surface	0.40	0.98	0.37	5.31	1.34	1.06	0.0088	0.0047	0.0274
ESR ingot	C	Si	Mn	Cr	Mo	V	S	T.O	N
Center	0.41	0.86	0.36	5.21	1.28	1.06	0.0044	0.0034	0.0315
1/2 Radius	0.40	0.85	0.35	5.16	1.32	1.01	0.0045	0.0035	0.0293
Surface	0.41	0.87	0.37	5.45	1.35	1.08	0.0042	0.0036	0.0295

3.2 Inclusions characteristics and distribution

Figure 2 shows the inclusions in the samples observed by SEM， it can be seen that the size of inclusions are extremely large (Figures 2a, 2b). Inclusions became finer after ESR (Figures 2c, 2d). The number and size were counted, and 50 images with a total area of 0.845mm^2 were counted for each sample. The average diameter of inclusions measured by Image Pro-Plus is average equal-area-circle diameter. The results are listed in Table 2, about 63.9 percent in number of inclusions were removed, the area fraction (total area of inclusions/total area observed×100%) of inclusions reduced from 0.313% to 0.107%. The effect of electroslag remelting on cleanliness improvement is remarkable.

Table.2 Statistical results of inclusions

Number of observed inclusions					Area fraction of inclusions to total area
Average diameter	1~2μm	2~3μm	3~5μm	5μm~	
Electrode	189	57	30	29	0. 313%
ESR ingot	52	29	12	17	0. 107%

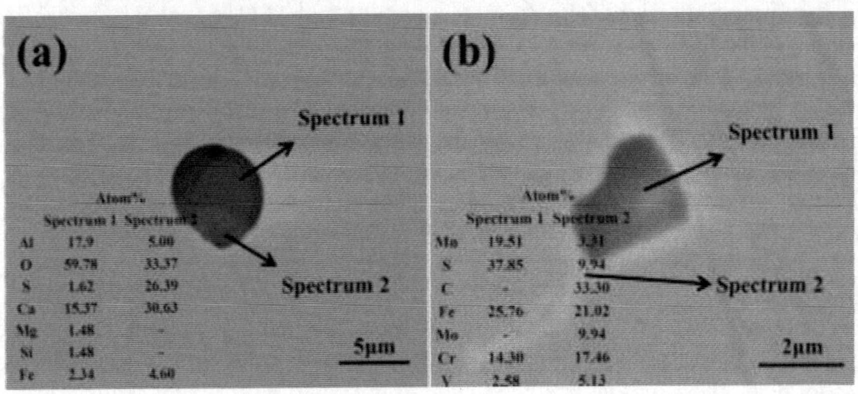

Figure 3. Typical inclusions in electrode sample

IOP Conf. Series: Materials Science and Engineering **529** (2019) 012064 doi:10.1088/1757-899X/529/1/012064

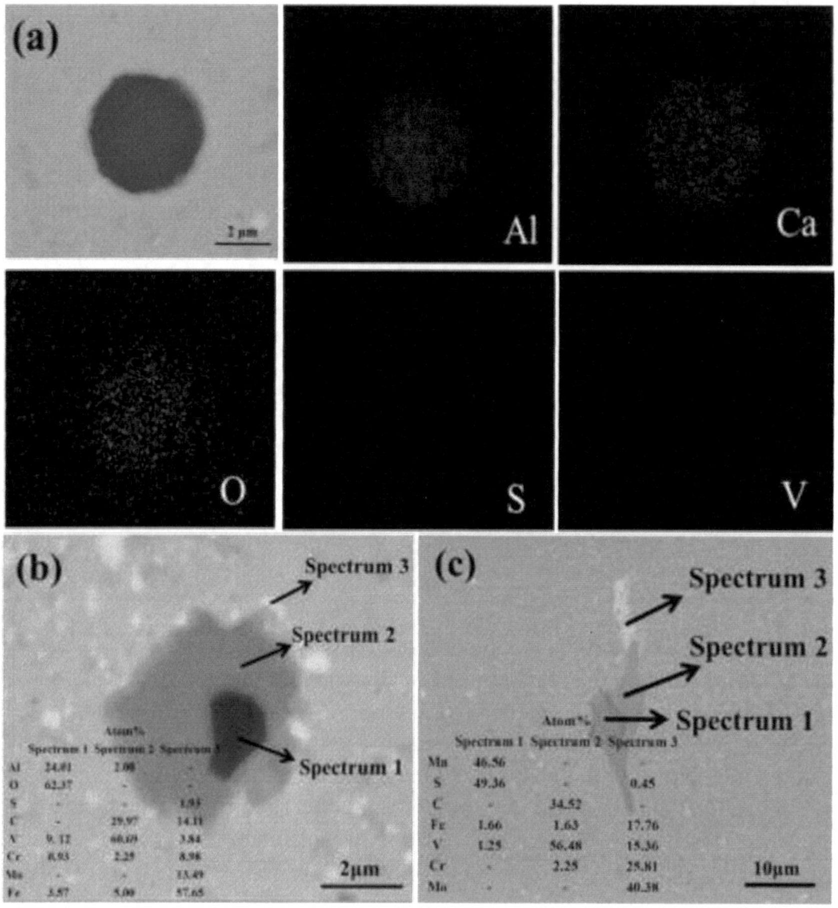

Figure 4. Typical inclusions in ESR ingot sample

Figure 3 shows the typical inclusions in electrode. Figure 3(a) is a complex inclusion with a (Ca-Mg-Al-Si) oxide matrix attached by sulfide. (Ca-Mg-Al-Si) oxide is common inclusion in traditional refining process[2]. Figure 3(b) shows a sulfide in electrode, the brighter area around sulfide is the primary carbide of Mo and Cr, and this kind of long stripe carbide usually forms at the end of the solidification because of the high content of elements in molten steel. After remelting, complex inclusions like Figure 3(a) disappeared, parts of these were modified into CaO-Al_2O_3 (Figure 4a), and Al_2O_3 can also be seen after ESR.

MnS inclusions would form before primary carbides during solidification[6]. A solid MnS inclusion is possible to be the nucleation agent of primary carbide, and there are lots of primary carbides with MnS core observed before and after ESR. The disregistry between Al_2O_3 and M(C,N) has been calculated by others[8]. A larger value of the disregistry mean more supercooling is needed during heterogeneous nucleation, The study shows that Al_2O_3 is effective in promoting nucleation of M(C, N). Thus it can be inferred that both Al_2O_3 and MnS could act the nucleation agent of carbides.

3.3 Microsegregation and banded structure

One of the most important functions of ESR is enhancing the solidification structure. However microsegregation still cannot be solved entirely. Figures 5(a) and 5(b) is the metallographic image of the center part in ESR ingot plant, the sample was etched by 4% nital. Net like structure can be seen in Figure 5(a), in the dark area, much more granular carbides can be observed, and large primary carbides formed in the dark area (Figure 5b). Element map scanning of the segregation area show clearly that the main segregation elements is V, Cr, S and Mo.

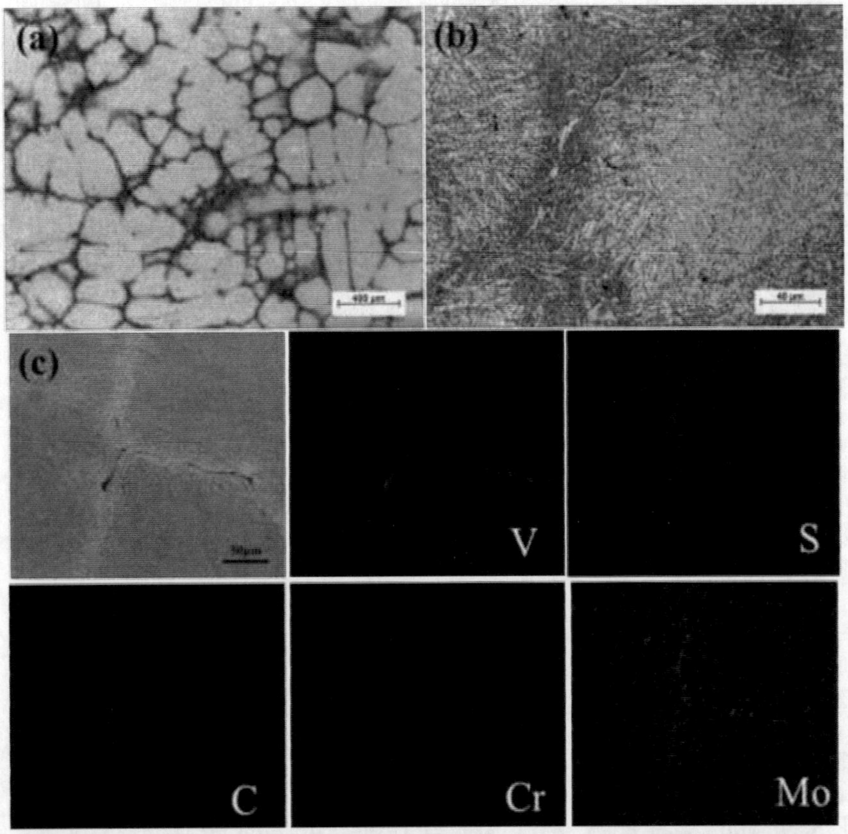

Figure 5. Microsegregation in the ESR ingot

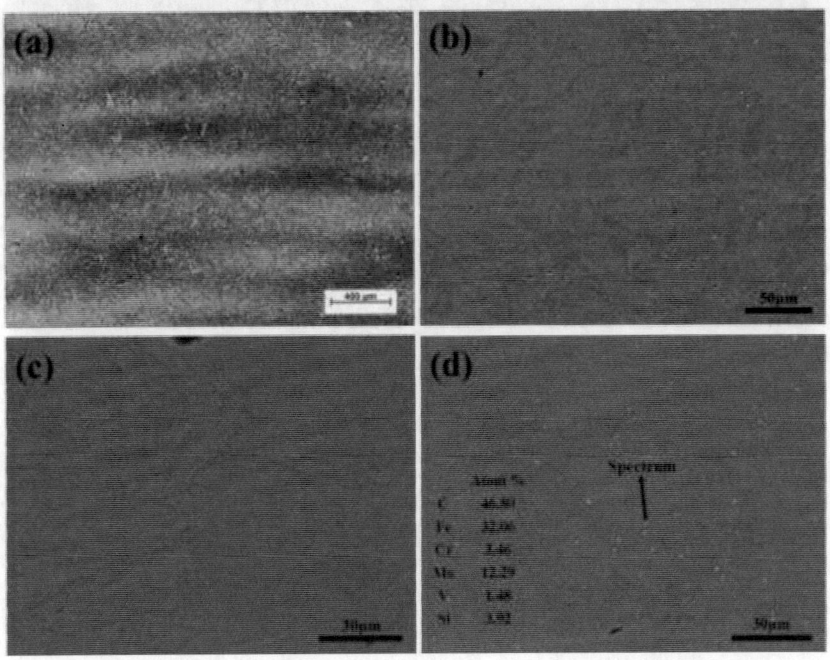

Figure. 6 Banded structure and carbides

The forged ingot sample etched by 4% nital was observed. Banded structure in the sample is showed in Figure 6(a) taken by OM at 50 magnification. Figure 6(b) was taken by SEM under back-scatter mode, lots of white spots could be observed on the right part of Figure 6(b). The left part (fewer white spots) and right part (more white spots) of Figure 6(b) are shown in Figure 6(c) and Figure 6(d) respectively. The number of the white spots in Figure 6(c) is 15 and in Figure 6(d) is 64. Actually, those white spots are carbides with higher atomic number than the matrix, it can be seen that that kind of carbides are rich in Mo, V, Cr and Si (Figure 6d). The uneven distribution of carbides is the behavior of segregation in forged H13 steel. Mo, V and Cr are rich in both the net like segregation area (as cast ingot) and the white spot (forged ingot), which indicates that the banded structure in forged ingot is related to the microsegregation in as cast ingot.

4. Conclusions

(1) The S and T.O content, the number, size and area fraction of inclusions decrease owing to the cleanliness improvement of ESR process, N content increase and Si content of ESR ingot decrease because of the influence of air atmosphere. Complex inclusions were modified into $CaO\text{-}Al_2O_3$. Al_2O_3 and MnS can be found in ESR ingot, both of which could promote the nucleation of large carbides.
(2) The macrosegregation of S content was improved by ESR process, microsegregation and banded structure still can be found. V, Mo, Cr and S are rich in the segregation area in as cast H13 steel, besides, long stripe primary carbide can be also found. Uneven distribution of carbides rich in Mo, V, Cr and Si can be observed in the banded structure.

References

[1] Jin T Y， Liu Z Y and Cheng Y F 2010 *Int. J. Hydrogen Energy* **35** 8014.
[2] Dong Y W, Hua Z H, Cao Y L, Yu A and Hou D 2014 *Metall. Mater. Trans. B* **45** 1315
[3] Li J Y, Chen Y L and Huo J H 2015 *Mater. Sci. En. A* **640** 16.
[4] Fezi K, Yanke J and Krane M J M 2015 *Metall. Mater. Trans. B* **46** 766
[5] Qi Y F, Li J, Shi C B, Geng R M and J. Zhang, 2018 *ISIJ Int.* **58** 1275.
[6] Liu Y, Zhang Z, Li G Q, Wang Q, Wang L and Li B K 2017 *Steel Res. Int.* **88** 1700058.
[7] Liu Y, Zhang Z, Li G Q, Wang Q, Wang L and Li B K 2017 *Metals* **7** 185.
[8] Li J, Shi C B, Wang L L, Wu Z and Wang H 2016 *Can. Metal. Q.* **55** 321
[9] Liu Y. Wang X J, Li G Q, Wang Q, Zhang Z and Li B K 2018 *Vacuum* **154** 351
[10] Ma D S, Zhou J, Chen Z Z, Zhang Z K, Chen Q A and Li D H 2009 *J. Iron Steel. Res. Int.* **16** 56

IOP Conf. Series: Materials Science and Engineering **529** (2019) 012065 doi:10.1088/1757-899X/529/1/012065

Dendrite growth direction measurements: understanding the solute advancement in continuous casting of steel

A. SenGupta[1], B. Santillana[2], S. Sridhar[3] and M. Auinger[1]
[1]WMG, University of Warwick, Coventry CV4 7AL, UK
[2]Tata Steel Research and Development, 1970CA IJmuiden, The Netherlands
[3]Metallurical and Materials Engineering, Colorado School of Mines, Golden, Colorado 80401, USA

E-mail: A.Sengupta@warwick.ac.uk, begona.santillana@tatasteeleurope.com

Abstract. Maintaining competitiveness in steel manufacturing requires improving process efficiency and production volume whilst enhancing product quality and performance. This is particularly challenging for producing value-added advanced steel grades such as advanced high strength steels and electrical steels. These grades due to higher weight percentage of alloying elements cause difficulties in various stages of upstream and downstream processing, and this includes continuous casting, wherein high solute levels are critical towards macro-segregation. Interface growth direction in systems with more than one component is dictated by the solute profile ahead of the moving solidification front. Understanding the profile of growth direction with casting process parameters during the progress of casting will provide an important perspective towards reducing the macro-segregation in the cast product. In the present study, two steel slab samples from conventional slab caster under the influence of electromagnetic brake (EMBR) at Tata Steel in IJmuiden (The Netherlands) have been investigated for dendrite deflection measurements. The samples showed a transition zone where a change in the deflection behavior occurs. Also, the magnitude of the deflection angle decreases away from the slab surface. Correlating these experimental data with modeled fluid flow profile will help in improving the understanding of the dynamic nature of the solute advancement so that the casting parameters can be optimized to improve product quality.

1. Introduction

It is well known that the change of state of a system is governed by the systems thermodynamics[1] which is based on the fundamental principle of Gibbs's free energy minimization and isolation criterion. The growth direction of a solid phase during the process of solidification indicates the direction of the interface movement between solid (S) and liquid (L) phases. The growth kinetics of the solid phase is governed by thermodynamic equilibrium and diffusion processes occurring at the interface, primarily in the liquid phase. Bulk fluid flow can enhance the diffusion processes in the liquid phase. For solidification in single component metallic systems, the driving force is entirely of thermal nature [2] while in multi-component systems, the driving force for the S/L interface motion is also influenced by the solute transport. Thus the morphology and the distribution of the solid phase typically called the "microstructure" in the cast-product of the industrial alloys (mostly multi-component) will depend on the evolution of the solute profile in front of the moving interface. Forced convection due to bulk fluid flow[3] (also called macro-scale flow) within the liquid melt will further enhance the kinetics of the solute transport process which might have an additional influence on the interface growth direction.

Dendritic solidification[1] structure is quite common among the majority of metallic alloys. Due to the nature of their preferred crystallographic growth direction, dendrites are observed to grow in the form of branched tree-like morphology. In the continuous casting process, liquid steel transforms into solid, due to cooling, in a process wherein a liquid feedstock is continuously fed to an oscillating mold and a semi-solid product is extracted at the outlet. It is a quite complicated process where the phenomena of heat transfer and mass transfer take place simultaneously in presence of the incoming turbulent liquid jet, which after entering the mold creates a high degree of bulk convection. This incoming liquid stream on its way from the submerged entry nozzle (SEN) towards the mold exit takes away the rejected solute (rejected by the growing solid phase) thereby changing the local chemical driving forces and creating a washing effect. The liquid flow sweeps away the solute ahead of the growing interface from the upstream side to the downstream side. Thus it lowers the degree of undercooling on the downstream side (i.e in the direction of the outgoing fluid flow from SEN to the mold exit in the casting direction in **Fig. 1**) and promoting the driving force for solid growth on the upstream side (i.e towards the incoming fluid flow from SEN in **Fig. 1**). This has a direct effect on the evolving solidification microstructure. As the amount of solid fraction increases, the magnitude of the convection effect of the bulk flow decreases, thereby decreasing the effectiveness of the washing effect. During the process, if after a certain extent of time there exists a region with excessive solute accumulation, the undercooling will reduce drastically. This excess solute, if not further washed away, will lead to macro-segregation in the cast product. This macro-segregation is the inhomogeneity in the chemical composition when allowed to extend over large distances will lead to undesirable mechanical properties of the end product. Chemically driven precipitation can occur in this region between reactive metallic alloying additions and interstitial solutes. Thus the interface growth direction depends on the solute profile in front of the moving interface - which in turn depends on the casting process parameters. Hence to maintain the quality of the cast steel slab, it would be beneficial for the casting operators to have a good estimate about the interface growth direction and its dependency on the combination of heat transfer, mass transfer, and fluid flow. This will help to dynamically re-adjust the casting process parameters in order to minimize the degree of macro-segregation.

Due to stringent quality requirements, continuous casting of commercial advanced steel grades can be at the risk of having macro-segregation due to addition of a wide variety of alloying elements in higher proportions, and some of these very reactive elements are Al, Si or Ti. Thus it is important to find out the linkages between the fluid flow and the transient nature of the interface growth evolution from the fundamental principle of thermodynamics. The solidification microstructure consists of micron-scale dendrites spread over relatively large distances (of the order of several mm's). Under typical casting conditions, one can observe an almost fully columnar dendritic structure extending from the slab surface to the slab center. Experimentally[2,3,4] it has been observed that the dendrites in presence of melt convection, have a biased growth direction towards the upstream side. This phenomenon is called dendrite deflection. The variation in the angle of deflection from the slab surface towards the centre represents how the dendrite growth direction dynamically changes during casting. It is of scientific relevance how the solute is rejected ahead of the moving solidification front as Takahashi[3] showed that the distribution coefficient depends on solidification rate and bulk flow velocity. This mostly depends on the flow velocity, growth rate / undercooling and the solute content of the alloy. Very limited quantified experimental evidence[2-4] of the phenomenon of dendrite deflection is available in literature and the information on steels cast under industrial conditions is even more scarce. From his experimental data on steel ingots, Takahashi et al[3] proposed an empirical correlation, showing the dependency of the interface growth direction on the solidification rate and the bulk flow velocity. With steels of different carbon contents, Esaka et al[2] modified Takahashi's relation by adding a carbon concentration based pre-factor. The present work involves the measurement of dendrite deflection in industrially cast steel slab samples. The objective of this paper is to establish whether differences in casting conditions will result in different dendrite deflection angles, with the underlying hypothesis that this is caused by differences in convective flow causing differences in solute distribution. Higher casting speed increases the degree of convective flow while higher mold width would tend to provide more time

IOP Conf. Series: Materials Science and Engineering **529** (2019) 012065 doi:10.1088/1757-899X/529/1/012065

for liquid steel to flow laterally towards the narrow face walls after it exits from the SEN nozzle. Convective flow influences the superheat dissipation, which would affect the mold temperature distribution. However, the thermal diffusion coefficient being orders of magnitude higher than the solute diffusion coefficient, flow driven mass transport will be more critical than flow driven temperature towards macro-segregation.

2. Experimental Methodology

Two steel slabs namely – sample A (LCAK : low carbon aluminium killed) and sample B (LAP: low alloyed peritectic steel) from the conventional slab caster were used in the present study. The chemical composition of the slabs are given in **Table 1** and the casting process parameters are given in **Table 2** respectively.

Table 1. Chemical composition of the steel grades.

Steel	C, wt%	Mn, wt%
Sample A (LCAK)	0.06 – 0.08	0.2 – 0.4
Sample B (LAP)	0.11 – 0.14	0.5 – 0.8

Table 2. Casting process parameters.

Steel	Casting Speed, m/min	Superheat, °C	Slab thickness, mm	Width, mm
Sample - A	1.7	23	225	1300
Sample - B	0.76 – 1.55	16	225	1600

It can be seen that sample A had a stable casting speed of 1.7 m/min[6] while sample B had a wide variation in casting speed. This is because sample B was collected during the ramp-up period, i.e with continuous increase in casting speed. Both the slabs were cast under the influence of EMBR. **Fig. 1** shows the portion of the slab from where the samples were taken with the sample cutting plane shown as a green dotted line.

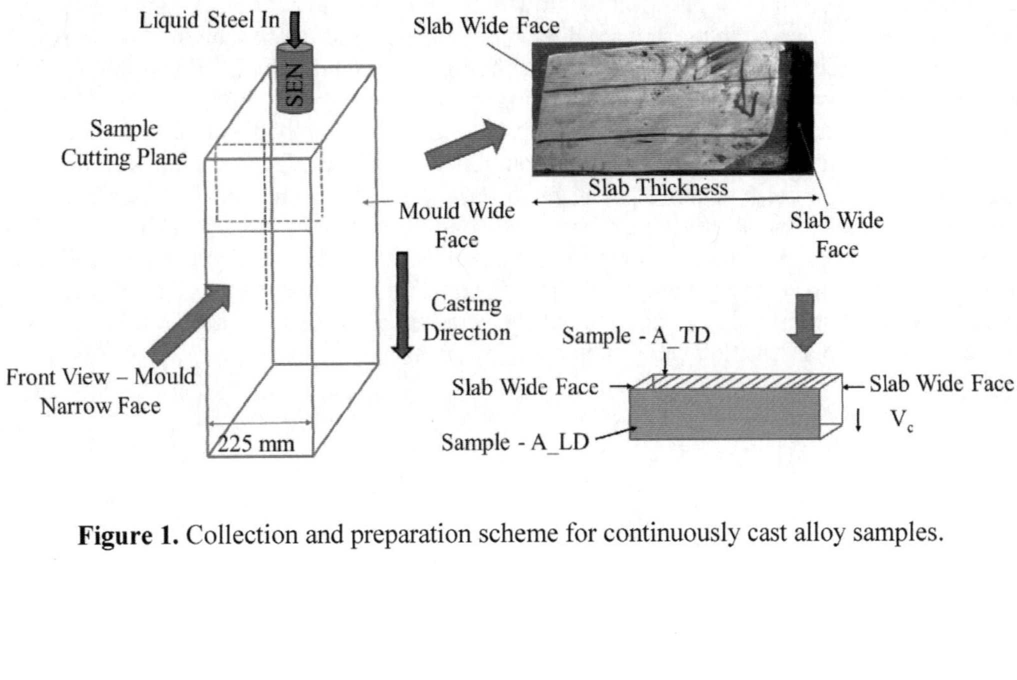

Figure 1. Collection and preparation scheme for continuously cast alloy samples.

The sample was cut approximately at a quarter distance inwards from one of the narrow face. Thus the sample cutting plane was parallel to the narrow face and the red dotted line in **Fig. 1** shows the slab centreline on the narrow face along longitudinal direction. Samples after cutting from the slab were cut further into small pieces of 35x35x10 mm. The area of the sample closest to the slab surface was selected for analysis and thus mounted as per the standard metallographic procedure so that it could be handled conveniently to be used for subsequent steps of metallographic sample preparation. Final samples for solidification structure analysis were prepared after coarse grinding and subsequent fine polishing with diamond abrasive particles followed by room temperature etching with Bechet–Beauchard etchant[7] to reveal the dendritic structure. The etched sample was observed by optical microscopy for visualization of the solidification structure. The dendrite deflection angle was measured by individually selecting the dendrites and measuring the height and width of the rectangle with dendrite as the diagonal. **Fig. 2** shows the convention of measuring the bending angle of a dendrite originating from the mold wall in presence of fluid flow (bending angle denoted as θ). Starting from the slab surface, the bending angle was measured at perpendicular distances. At each distance, the bending angle was measured at various points.

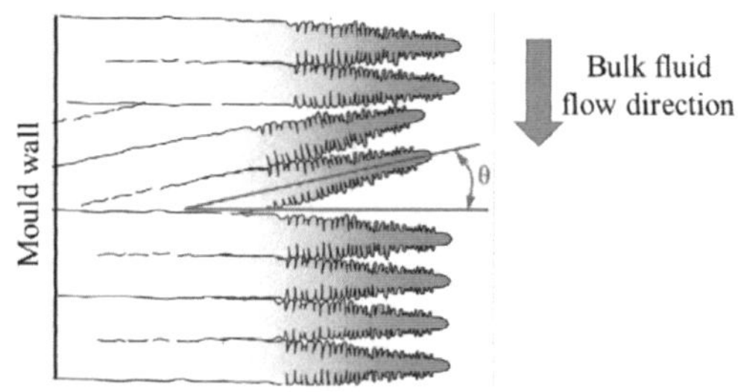

Figure 2. Schematic representation of the definition for the bending angle θ[7], used in this study.

3. Results and Discussion

Fig. 3 shows the variation of the dendrite growth direction from slab surface towards the slab center for sample - A solidified under a constant casting speed. It can be seen that near the surface, all primary dendrites (shown as white dotted lines) are oriented in the same downward direction. A small distance away from the slab surface, the primary dendrites undergo a change in growth direction from downwards to upwards. Some primary dendrites are seen to follow their original downward growth direction, while other dendrites are seen to have an upwards growth direction. Further away, it can be seen that all primary dendrites are oriented in the upward direction, which is exactly opposite in close proximity to the surface. Thus, it seems that the observed change in the dendrite growth direction might be due to a change in the fluid flow direction ahead of the moving front. Generally, a portion of the liquid steel after exiting from the submerged entry nozzle with relatively high flow velocity directly hits the narrow face mold walls[8]. After striking the mold walls, a portion of the liquid steel may travel upwards towards the meniscus and creating a recirculation loop at the top of the mold. The rest of the liquid steel travels down towards the mold exit after striking the moving solidification front. Thus at a particular distance away from mold walls, a change in the flow direction may occur, which is responsible for the experimentally observed change in the dendrite growth direction mentioned before. The transition in the preferred growth direction can be compared with the transition in the fluid flow profile and is thus helpful to study how this transition varies with casting process parameters.

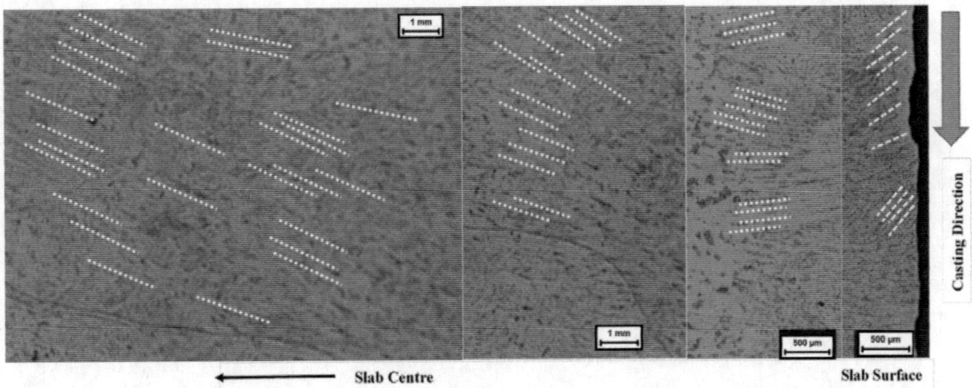

Figure 3. Variation in dendrite growth direction from the slab surface (right) towards the slab centre (left) for a low carbon aluminium killed steel (sample A).

Fig. 4a shows the variation in the bending angle from slab surface towards the center for both samples. Sample - A_LD represents the bending angle for sample A on the plane parallel to one of the narrow face walls (as shown in **fig. 1**). Sample - A_TD represents the bending angle on a plane perpendicular to the narrow face side (i.e along transverse direction) (as shown in **fig. 1**). Liquid steel, after exiting from the SEN port, travels towards the narrow face walls along width direction. It can be seen that for all the three samples, the magnitude of the mean bending angle near the surface is similar but within a wide variation of values. For the sample - B_LD, the mean deflection angle decreases up to 3 mm, after which the sign of the bending angle changes from positive to negative, indicating the change in the growth direction. This is likely caused by the change in the flow direction. For sample - A_LD, at about 1 mm from the surface, the bending angle is both positive and negative which again indicates a transition region where the bending angle is both positive and negative. Similarly for sample - A_TD, at about 9 mm from the surface, the bending angle is both positive and negative indicating a similar transition region of preferred dendrite growth. For sample - A_LD from 6.5 mm onwards, the bending angle is negative, which is opposite compared to near the surface. For all samples, the absolute value of the mean value and its standard deviation decrease from the surface towards the centre of the slab. This may be due to decrease in the magnitude of the flow velocity from surface towards the centre. Okano[9] also observed this decrease in the bending angle across the slab width. Though the sample - B_LD had a wide variation in casting speed, the variation in the bending angles are similar to the case of sample A. This may be due to EMBR[8], which tries to stabilize the turbulent flow within the mold region. Also, the slab thicknesses are same for both slabs and there is not much difference in the chemistry as well. For the sample - A with stable casting speed, the solidification rate within the mold region was estimated using the CON1D model[10]. The obtained solidification rate and the experimental bending angle was put into the Takahashi[3] correlation to estimate the fluid velocity. Only the maximum and minimum values of the bending angles at different positions for sample - A were used to get an idea of the range of the predicted flow magnitudes. **Fig. 4b** shows the range of the flow velocity magnitude at different distances from the slab surface for sample - A. The minimum predicted flow velocity was in the range 0.10 - 0.12 m/s. Many dendrites at the surface were found to have bending angles in the range of 40° – 45°. To get such high values, the corresponding estimated fluid velocity was found to be about 10 m/s even at a slower growth rate of 0.01 cm/s. Also at about 1 and 7 mm from the surface, the estimated maximum flow velocity was found to be in the range of 2 – 2.75 m/s. This extremely high flow velocity seems not realistic in a slab caster mold operating at a casting speed of 1.7 m/min with EMBR compared to what is reported in literature[8]. Also, the fluid flow simulations performed at Tata Steel Research and Development, IJmuiden, The Netherlands with EMBR at a casting speed of 1.7 m/min revealed much lower values of maximum flow velocity. Thus it seems that the empirical correlations by Takahashi[3] and Zhang[11] are not applicable for the entire slab geometry and needs to be modified in order to predict the high bending values obtained in the slab samples.

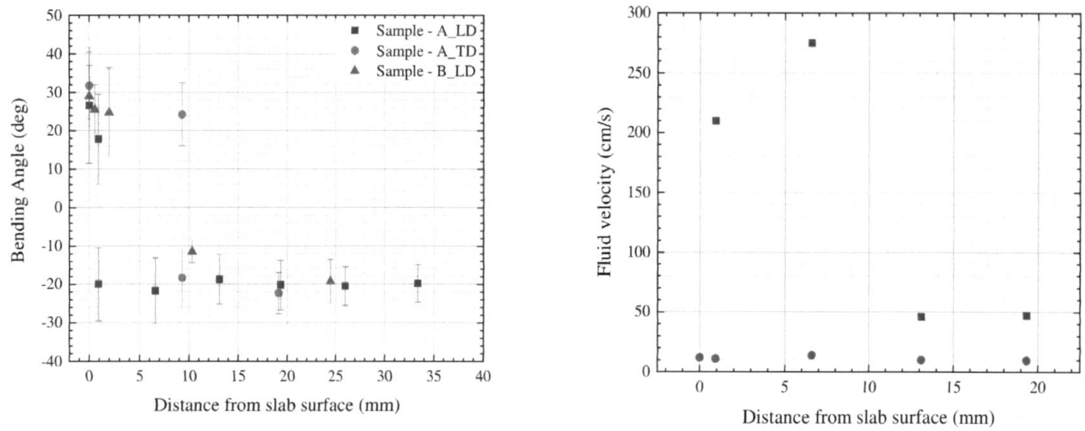

(a) **(b)**

Figure 4. (a) Variation in bending angle from the slab surface towards the slab centre; (b) Minimum (●) and Maximum (■) Fluid velocity for sample - A as calculated from Takahashi equation[3].

4. Conclusions

Two steel slab samples from an industrial thick slab caster were investigated for the variation in dendrite deflection from the slab surface towards the slab centre. Both alloys have a similar composition and casts were made under the influence of electromagnetic brake (EMBR). A clear change in the dendrite growth direction from surface towards the centre occurs, albeit over a wider zone, which can be correlated to changes in the fluid flow profile. Decrease in bending angle from surface towards the centre was observed. Takahashi relation needs to be revised to predict the high bending angle values. Correlating these deflection measurements with compositional analysis has been planned as a part of future activities, which would provide insight on the relation between the macro-segregation phenomenon and dendrite deflection. The transition area between downwards and upwards bending seems to be larger in the case of sample B, which fits qualitatively with the gradual increase of casting speed and the moving solid/liquid interface region.

5. References

[1] Glicksman M E 2011 *Principles of Solidification* (New York: Springer).
[2] Esaka H, Suter F and Ogibayashi S 1996 *ISIJ International* **36** 1264–1272.
[3] Takahashi T, Ichikawa K and Kuduo M 1975 *Tetsu-to-Hagane* **61** 2198–2213.
[4] Gandin Ch-A, Guillemot G, Appolaire B and Niane N T 2003 *Materials Science and Engineering A* **342** 44-50.
[5] Wang X, Wang S, Zhang L and Seetharaman S 2016 *Metallurgical and Materials Transactions A* **47A** 5496–5509.
[6] Park J, Ko E, Choi J and Sohn I 2014 *Met. Mater. Int.* **(20)6** 1103–1114.
[7] Santillana M B 2013 *Thermo-mechanical properties and cracking during solidification of thin slab cast steel* PhD thesis Copyright: Tata Steel Nederland Technology B.V.
[8] Thomas BG, Singh R, Chaudhary R and Vanka P 2013 *Fifth Baosteel Biennial Academic Conference, Shanghai, PRC, June 4-6,* 1–10.
[9] Okano S, Nishimura T, Ooi H, and Chino T 1975 *Tetsu-to-Hagané* **61**, 2982–2990.
[10] Meng Y. A. and Thomas B. G. 2003 *Metall. Mater. Trans. A* **34B** 685–705.
[11] Zhang H, Nakajima K, Xing W, Wang A and He J 2009 *ISIJ International* **49(7)**, 1010–1018.

IOP Conf. Series: Materials Science and Engineering **529** (2019) 012066 doi:10.1088/1757-899X/529/1/012066

Utilizing inclusion data in characterization of oxide-sulfide stringers in hot-rolled plates

T Alatarvas[1], T Antola[2], T Fabritius[1]

[1] Process Metallurgy Research Unit, P.O. Box 4300, FI-90014 University of Oulu, Finland

[2] SSAB Europe Oy, P.O. Box 93, FI-92101 Raahe, Finland

tuomas.alatarvas@oulu.fi

The targets of this study were to determine the effect of vertical location and the composition of inclusions to the occurrence of oxide–sulfide stringers, and to determine the calcium aluminate phases most prone to form stringers during hot rolling of aluminium killed, calcium treated steel. The phases present in the inclusions have a significant effect on the deformation of inclusions during hot rolling, and consequently, on mechanical properties of the steel. A MATLAB script is utilized to identify and locate detrimental stringers from the hot rolled plates. Inclusion analysis data gathered with a scanning electron microscope and exported from IncaFeature software is analyzed. The following properties are presented for each stringer: the number of inclusions and length of stringer, phase fractions and compositions, and the composition of the unfragmented inclusion before hot rolling. According to the results, the longest stringers have total lengths over 200 μm, with almost 20 inclusions. The overall composition of the longest stringers is between C12A7 and C3A calcium aluminates with minor MgO contents. The diameters of the unfragmented inclusions in the slabs, forming stringers during hot rolling, were estimated to be around 20 μm for the longest stringers. From the dataset, plenty of CaO–CaS stringers were also characterized, obviously a result of excess calcium treatment.

1. Introduction

Oxide-sulfide stringers are considered harmful in hot rolled steel plates, similarly to elongated manganese sulfides [1]. In aluminum-killed and calcium treated steels, the stringers are often a mixture of oxides in Al_2O_3-CaO-MgO system and calcium sulfides (CaS), occasionally accompanied with manganese sulfides (MnS) and titanium nitrides (TiN). The formation and characteristics of oxide-sulfide stringers is reported in recent studies [1, 2]. The phases present in the inclusions have a marked effect on the deformation of inclusions during hot rolling, and consequently, on mechanical properties of the steel. Still, no consensus exists on the deformation tendency of various calcium aluminate phases, for example.

The developed method is used to locate and characterize oxide–sulfide stringers by using Excel spreadsheets exported from Oxford Inca Feature software. A schematic view of the formation of oxide–sulfide stringers is illustrated in figure 1. The targets of this study were to determine the effect of vertical location and the composition of inclusions to the occurrence of stringers, and to determine the calcium aluminate phases most prone to form oxide–sulfide stringers during hot rolling.

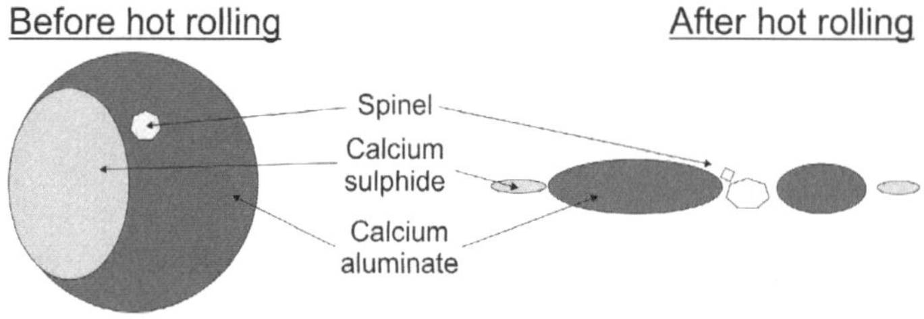

Figure 1. Schematic view of the formation of oxide–sulfide stringers during hot rolling.

2. Materials and Inclusion Characterization Methods

In this study, 25 heats of aluminum killed, calcium treated steels produced in the SSAB Europe Raahe steel plant were investigated. Two samples were prepared from the hot rolled product of each heat, making up a total of 50 samples.

The inclusion characterization was conducted in the SSAB Europe Raahe steelworks. Inclusions were analyzed with a Jeol JSM-7000F FESEM equipped with an EDS detector. Inclusion analyses were acquired with Oxford IncaFeature software using an acceleration voltage of 15 kV. In the INCA Feature runs, inclusions larger than 1.0 µm were analyzed with a livetime of 1.0 second with EDS. The total sample area comprised of product thickness, approximately 12 mm, width being 4 mm, resulting in approximately 48 mm^2 of scanned area per sample.

3. MATLAB script

3.1. Stringer detection

The oxide–sulfide stringers are located by a MATLAB script that handles Excel spreadsheets exported from Inca Feature software. The sample area is analyzed by going through the whole sample area with horizontal strips, parallel to the rolling direction. The strips cover the sample width, and the vertical thickness is defined as parameter. If multiple inclusions on a strip are close enough to each other, the inclusions are identified as a stringer. In figure 2, it can be seen that the consecutively analyzed strips are overlapping in order to ensure the stringer detection. In this case, the stringer is completely discovered in the strip illustrated in figure 2c, and taken into account.

To locate stringers in the samples, various parameters are used in the MATLAB script. In Table 1, the parameters used in this case, including the ones presented in figure 2, are presented. The parameters are chosen in a way that they eliminate other than oxide-based stringers, such as manganese sulfides. In addition, the stringers must be horizontal, i.e., assumedly formed during the hot rolling of the steel.

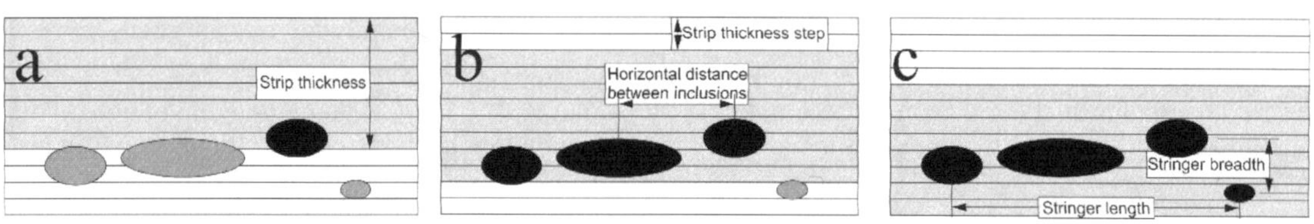

Figure 2. Procedure to locate stringer in the sample area, consecutive strips a, b, and c.

Table 1. Applied parameters in the stringer detection script.

Parameter	Threshold value	Note
Number of inclusions	≥ 3	After deducting inclusions which contain TiN + MnS in total over (90 mol.%)
Horizontal distance between inclusions	< 30 μm	
Average horizontal distance between inclusions	< 20 μm	
Strip thickness	8 μm	
Strip thickness step	2 μm	
At least one inclusion must contain oxides	> 10 mol.%	
Calculated TiN + MnS phase fractions in undeformed inclusion in total	< 50 mol.%	
Stringer aspect ratio OR	> 8	Eliminates clusters and inclined stringers
R^2 value of linear fit of inclusion coordinates AND	> 0.9	Stringer aspect ratio =
Stringer aspect ratio	2 < AR < 8	Stringer length / Stringer breadth

3.2. Original inclusion volume estimation

In this context, "original inclusion" refers to the assumed single inclusion in the slab, which has fragmented into several smaller inclusions, i.e., a stringer during hot rolling. The size of the original inclusion corresponds to the estimated total volume of stringered inclusions. For convenience, the size of the original inclusion is presented as a diameter of a sphere with the same volume.

The inclusions are assumed ellipsoids with the major axis parallel to the rolling direction, as presented by Fuhr et al. [3]. The observed inclusion area is considered the area of a projection of an ellipsoid. Further, the aspect ratios for the observed inclusion and the approximation are assumed equal.

3.3. Original inclusion composition estimation

The inclusions are taken as a mixture of CaS, MnS, TiN and oxide (Al_2O_3-CaO-MgO), which are the most common phases in inclusions in aluminum killed, calcium treated steels [4]. The phase fractions and oxide phases composition are calculated from the elemental analysis for each inclusion. The linear set of equations to calculate the compound fractions in each inclusion is presented in equation 1, where M_i is the molar mass of element or component i, n_i is the amount of compound i in inclusion, m_i is the mass of element i in inclusion, and $n_i \geq 0$.

$$\begin{bmatrix} 2M_{Al} & 0 & 0 & 0 & 0 & 0 \\ 0 & M_{Ca} & M_{Ca} & 0 & 0 & 0 \\ 0 & 0 & 0 & M_{Mg} & 0 & 0 \\ 0 & 0 & 0 & 0 & M_{Mn} & 0 \\ 0 & 0 & M_S & 0 & M_S & 0 \\ 0 & 0 & 0 & 0 & 0 & M_{Ti} \end{bmatrix} \begin{bmatrix} n_{Al2O3} \\ n_{CaO} \\ n_{CaS} \\ n_{MgO} \\ n_{MnS} \\ n_{TiN} \end{bmatrix} = \begin{bmatrix} m_{Al} \\ m_{Ca} \\ m_{Mg} \\ m_{Mn} \\ m_S \\ m_{Ti} \end{bmatrix} \quad (1)$$

The composition of the original inclusion is the average phase composition of the stringered inclusions, weighted with the volumes of the stringered inclusions. The oxide phases in each inclusion is calculated from the total composition, according to the lever rule. In this study, the focus is on various calcium aluminates, with compositions of CA, C12A7, and C3A.

4. Results and Discussion

The following properties are stored for each stringer: the number of inclusions and length of stringer, phase fractions and compositions, and the composition of the unfragmented inclusion before hot rolling.

IOP Conf. Series: Materials Science and Engineering **529** (2019) 012066 doi:10.1088/1757-899X/529/1/012066

In figure 3, an example of observed stringered inclusion and its geometrical visualization is illustrated. The calculated phase fractions and oxide compositions for each inclusion are also presented.

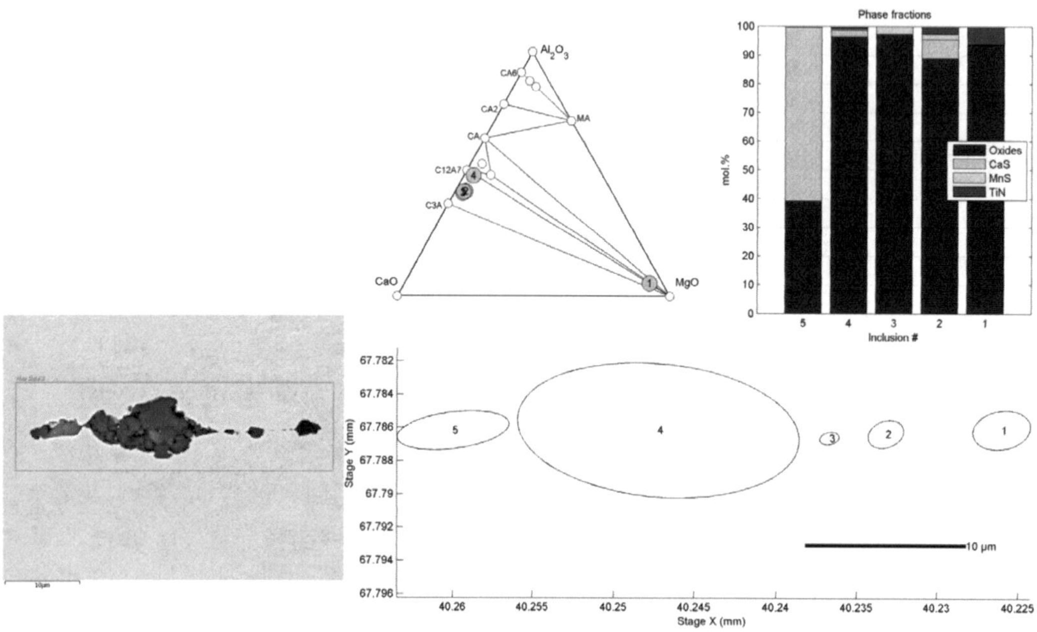

Figure 3. Stringered inclusion observed in hot-rolled product sample (left), phase fractions and oxide compositions (top) and geometrical visualization of inclusions (bottom right).

Most of the long stringers, with total lengths over 100 μm, locate on the upper half of the rolled product, as seen in figure 4 (left). On the other hand, shorter stringers are more evenly dispersed throughout the thickness of the product. The longest stringers have total lengths of over 200 μm in the studied samples. The longest stringers comprised of 19 inclusions (figure 4, right). On the lower side of the hot rolled product, only one stringer including more than ten inclusions is observed.

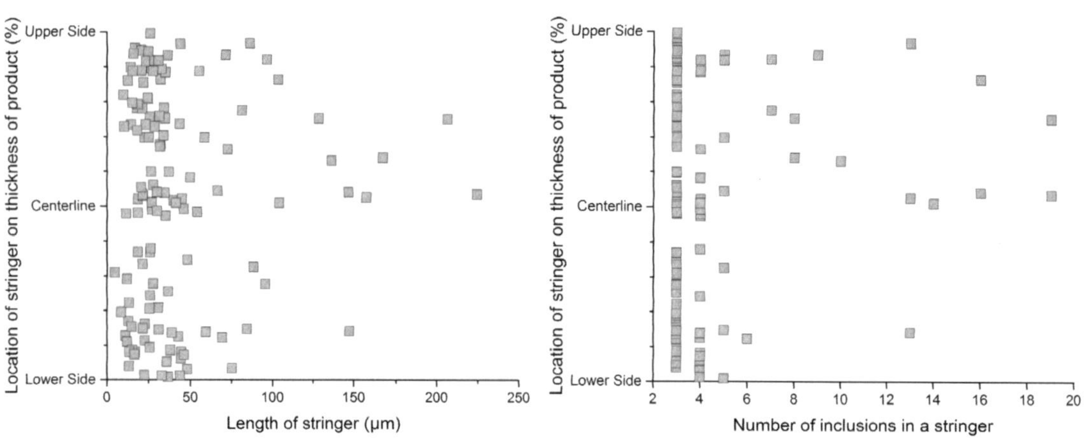

Figure 4. Length of stringers as a function of stringer location.

The size of original inclusions was estimated and shown in figure 5. The longest stringers have formed from inclusions with sizes of 10–17 μm, assuming they were spherical inclusions in the slab before hot rolling. The original inclusion sizes calculated here are in accord with the values reported by other authors [1, 3].

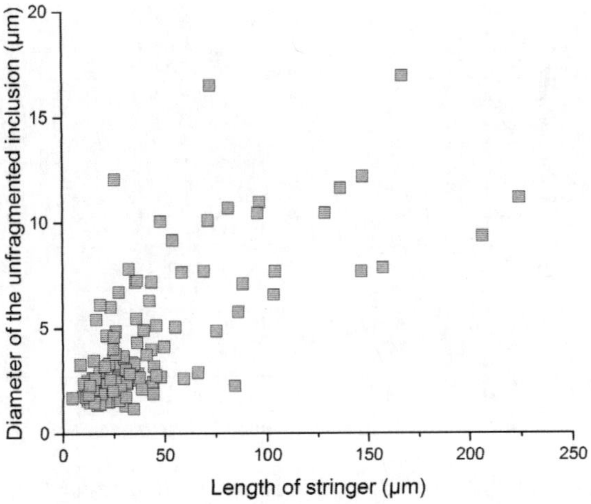

Figure 5. Estimated diameter of unfragmented inclusions.

Here, the correlation between fractions of calcium aluminates and calcium oxide in the original inclusions and the thickness location of stringers on the hot rolled product is assessed. Stringers containing CA calcium aluminate phases are often under 100 μm, illustrated in figure 6 (left). However, stringers with lengths over 150 μm contain marked amounts C12A7 (figure 6, right) or C3A phases (figure 7, left). Typically, the longest stringers contain also minor amounts of MgO, but negligible amounts of CaS phases. Plenty of CaO–CaS stringers were observed, obviously formed due to excess calcium treatment. The CaO-containing stringers are recurrently shorter than 50 μm, as seen in figure 7 (right).

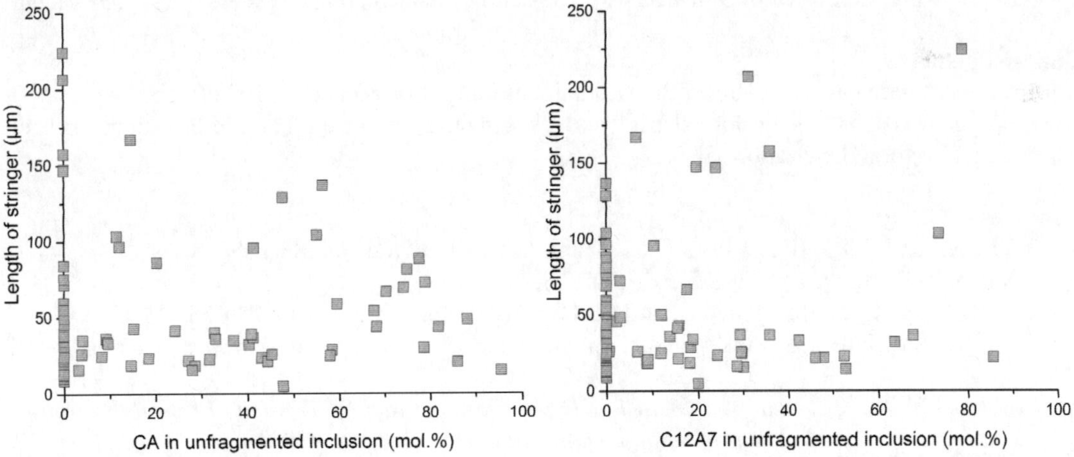

Figure 6. Calculated fractions of CA and C12A7 of original inclusion as a function of stringer length.

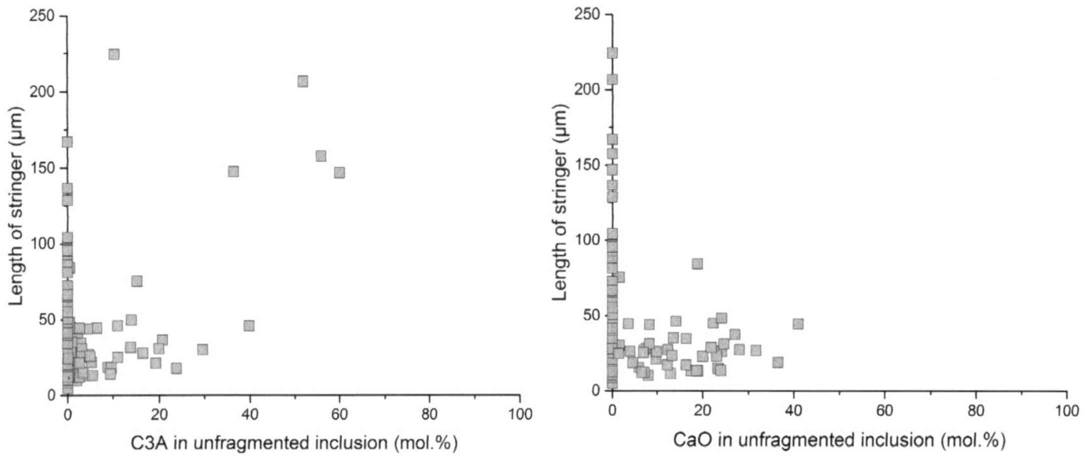

Figure 7. Calculated fractions of C3A and CaO of original inclusion as a function of stringer length.

5. Summary

Utilizing the MATLAB script, the number and properties of oxide–sulfide stringers are found. Excel spreadsheets exported from Inca Feature software act as input files without further formatting. The stringers are rapidly located from the dataset, providing additional information on the inclusion data in hot-rolled product samples. With a desktop computer with Intel i7 processor running at 3.40 GHz, a typical time to analyze the stringers in on sample takes less than two minutes.

According to the results, the longest stringers have total lengths over 200 μm, with almost 20 inclusions. The typical composition of long stringers is C12A7–C3A calcium aluminates with minor MgO contents, but negligible amount of CaS. The diameters of the unfragmented inclusions in the slabs, forming stringers during hot rolling, were estimated to be around 20 μm for the longest stringers. A large number of relatively short CaO–CaS stringers were observed, an indication of excess calcium treatment.

Acknowledgements

This research has been carried out under the Academy of Finland project no. 311934 (Genome of Steel). SSAB Europe Oy is acknowledged for sharing and allowing the use of data. Parts of this study has been published in a doctoral dissertation [5].

References

[1] Wang X, Li X, Li Q, Huang F, Li H and Yang J 2014 *Steel Res. Int.* **85** 155
[2] Zhao D, Li H, Bao C and Yang J 2015 *ISIJ Int.* **55** 2115
[3] Fuhr F, Torga G, Medina F and Cicutti C 2007 *Ironmaking Steelmaking* **34** 463
[4] Tervo H, Kaijalainen A, Pikkarainen T, Mehtonen S and Porter D 2017 *Mater. Sci. Eng. A* **697** 184
[5] Alatarvas T 2018 *Evolution of Inclusion Population in Calcium Treated Ultra-High Strength Steels. Novel Applications of Sample Data Treatment.* University of Oulu.

IOP Conf. Series: Materials Science and Engineering **529** (2019) 012067 doi:10.1088/1757-899X/529/1/012067

A Novel Powder Metallurgy Technique for Introducing Synthetic Inclusions into Liquid Steel

A Smith[1], T Tran[1], F Ji[1], W Tiekink[2], G Abbel[2] and S Verdier[2]

[1] Materials Processing Institute, Eston Road, Middlesbrough, UK
[2] Tata Steel Europe, 1970 CA IJmuiden, The Netherlands

E-mail: andy.smith@mpiuk.com

Abstract. Previous studies have shown that it is extremely problematic to add synthetic inclusions successfully into liquid steel for clean steel experiments. Small micro-particles encapsulated in a metallic parcel are difficult to pass through the melt-gas interface and inclusions tend to agglomerate then float up to the liquid surface. In this study, powder metallurgy is used to distribute cerium oxide particles of a known size from 1μm to 14μm within a small-scale steel ingot. Carbonyl iron powder has been mixed with 0.1 wt.% cerium oxide (CeO_2) then sintered in an electrical furnace to produce sintered ingots of 500 grams. A series of induction furnace melting trials using a total of 400 grams of electrolytic iron and cerium oxide sintered ingot have been undertaken. The work has included extensive FE-SEM analysis using the INCA Feature® to characterise synthetic cerium oxide inclusions from both the sintered ingots and trial ingots. The INCA Feature® results showed that the size distribution and number of cerium oxide inclusions agreed well between the sintered ingot and trial ingot. The synthetic cerium oxide inclusions are homogeneously dispersed through the bottom, centre and top of the trial ingot with approximately 16 number counts/mm^2. Analysis suggests that more than 95% of the cerium oxide from the sintered ingot has been distributed throughout the trial ingot. The work has also been upscaled from 400 grams up to 1.5 kg and has been successful. The method developed will be useful for further studies into steel cleanness of high value alloyed steels.

1. Introduction

It is well established that non-metallic inclusions are detrimental to the castability of clean steels and also to the mechanical properties of the downstream rolled and finished product. For example, tyre wire steel is drawn down to very fine wire (~150 μm diameter) and incorporated into the tread of automobile tyres to provide strength. An excessive level of hard non-deformable inclusions, such as alumina (Al_2O_3), can increase the risk of wire breakage during the drawing process. In bearing steels, non-metallic inclusions can act as stress raisers which can initiate micro-cracks and spalling leading to total bearing failure. Therefore, it is necessary to control and improve the operating practices during the steel manufacturing process to improve the overall cleanness of the steel.

Introducing synthetic inclusions into liquid steel is an important step in the study of inclusion removal at a laboratory level. There are only a few published papers for introducing synthetic inclusions into liquid steel. Dogan et al. [1] published some work in 2012, where they sealed cerium oxide (CeO_2) powder in-between two pieces of steel plate. The plates were than pressed together using 50 tonnes of pressure to seal the oxide particles into the compressed steel sample and remove any entrapped gas. This steel sandwich was then added to the steel liquid. Their results showed that 9% of total inclusions contained the cerium element. A similar pressing method was described in a Japanese patent

(JP2005091017) in July 2005 [2]. Alumina particles were spread on the surface between a steel plate and a metallic foil, then pressed at high pressure.

Powder metallurgy techniques [3, 4] have showed promising results where oxide particles and powder metal can be compacted by hot isostatic pressing [5,2] or hydraulic pressing [6]. A major advantage of these methods is that oxide inclusions of known chemical composition, inclusion size and density can be chosen [6].

The objectives of this study were to find a technique, which can introduce the synthetic inclusions into a steel melt for cleanness studies. The new method should allow the addition of inclusions into the steel melt of a known quantity and inclusion size and that these inclusions are homogeneously dispersed through the bottom, centre and top of the sample.

2. Experimental materials and methods

2.1. Manufacturing sintered cerium oxide ingots

The sintered cerium oxide ingots were manufactured by Tata Steel's Ceramics Research Centre (CRC), IJmuiden. In this study, powder metallurgy was used to distribute cerium oxide particles homogeneously in an iron matrix.

Carbonyl iron powder with a grain size of approximately 10 μm was mixed together with 0.1 wt.% cerium oxide powder with a grain size of from 1 μm to 14 μm and 1.0 wt.% of the binder bisstearoylethylene diamide in a container called a Turbula Mixing System. The amount of iron powder, cerium oxide and binder were mixed for approximate 1 hour.

Approximately 200 gram of the powder mixture was then pressed in a cylindrical mould (Ø 50 mm) with 60 MPa of pressure using a hydraulic press. The pressed sample was then heated to 600 °C at a heating rate of 3 °C/min in atmospheric conditions to remove the binder. The sample stayed at this temperature for 1 hour before it was allowed to cool down following a natural cooling curve.

After debinding, the sample was put in a hydrogen furnace (see Figure 1), to sinter the sample. A temperature program was applied with a heating rate of 5 °C/min, and the sample held at 600 °C for one hour, before further heating until a top temperature of 1300 °C was reached. The holding time at the top temperature was 30 minutes, before the sample was cooled down following a natural cooling curve. The density of the ingot is approximately 7.2 g/cm^3 after sintering.

Figure 1. Hydrogen furnace at CRC, Tata Steel, IJmuiden, used for sintering the cerium oxide ingots.

2.2. Introducing synthetic inclusions into liquid steel

IOP Conf. Series: Materials Science and Engineering **529** (2019) 012067 doi:10.1088/1757-899X/529/1/012067

The sintered cerium oxide ingot was added together with electrolytic iron at a 400 gram scale. A 45 kW induction furnace based at the Materials Processing Institute (the Institute) was used for melting the samples. The induction furnace was purged with argon to remove any traces of oxygen. Alumina crucibles were used in these experiments. A graphite sleeve manufactured in the workshop of the Institute was placed around the alumina crucible to improve melting performance.

Several techniques of placing the sintered cerium oxide ingot within the electrolytic iron have been tested. The electrolytic iron and sintered cerium oxide ingot were melted out in the induction furnace and the temperature raised to 1610 °C and held for 5 minutes. The induction furnace was then shut down and the melt temperature quickly cooled down to room temperature. The melt temperature was measured using a type-S thermocouple submerged in the liquid steel.

Figure 2 shows a schematic diagram of the experimental induction furnace set-up used at the Institute.

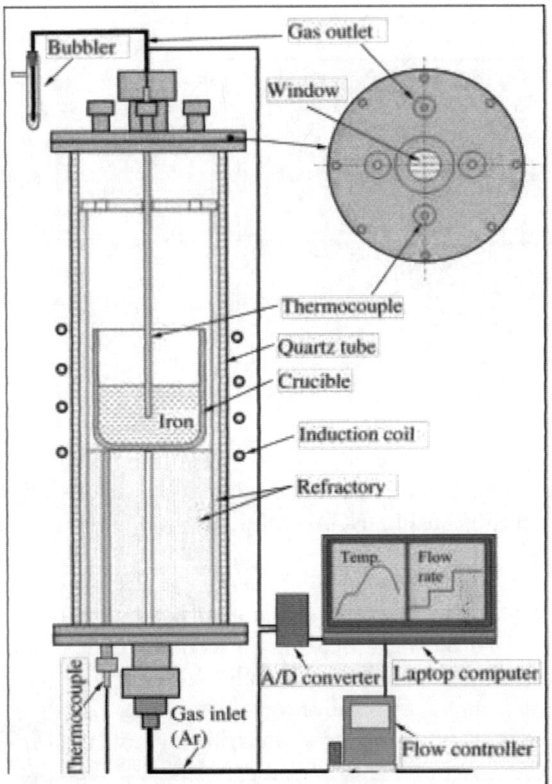

Figure 2. Schematic diagram of induction furnace experimental set-up.

2.3. *SEM mapping technique*

The work has included extensive Field Emission Gun Scanning Electron Microscopy (FEG-SEM) using the Oxford Instruments INCA Feature® software to characterise synthetic cerium oxide inclusions from both the cerium oxide sintered ingots and trial induction furnace ingots. INCA Feature® is an automated procedure that analyses particle features detected within a designated sample area. The data can then be processed to give several parameters including (a) inclusion counts/mm², (b) inclusion area %, (c) inclusion size and (d) particle shape for each detected phase analysed. The analyser was able to measure very small sized particles < 1µm upwards within the designated analysis areas.

3. Results and discussion

3.1. *Metallography of the sintered ingots*

Samples of the sintered cerium oxide ingots, manufactured at CRC, have been cut, ground and polished to a 1µm finish, suitable for SEM investigation. The samples have been examined using the Institute's FEG-SEM, equipped with an Oxford Instruments Energy-Dispersive X-ray Spectroscopy (EDX) analyser.

Each sample was mapped to detect every cerium oxide particle, so that a distribution of cerium oxide inclusions within the ingot could be determined. Figure 3 shows the location of cerium oxide inclusions detected within a sample area of 17 mm². It was seen that the cerium oxide inclusions were dispersed homogenously in the sintered ingot, from a small inclusion size of 1 µm through to 50 µm. The increase in inclusion size was caused by agglomeration during pressing. The INCA Feature® analysing software was able to detect the size of each cerium oxide inclusion, so that a size distribution could be made for the ingot, which is shown in Figure 4. The major inclusion sizes were from 1 µm to 7 µm. No contamination of the inclusions in the sintered ingot was found.

The INCA Feature® results showed that the number of cerium oxide inclusions was around 45 number counts per millimeter square.

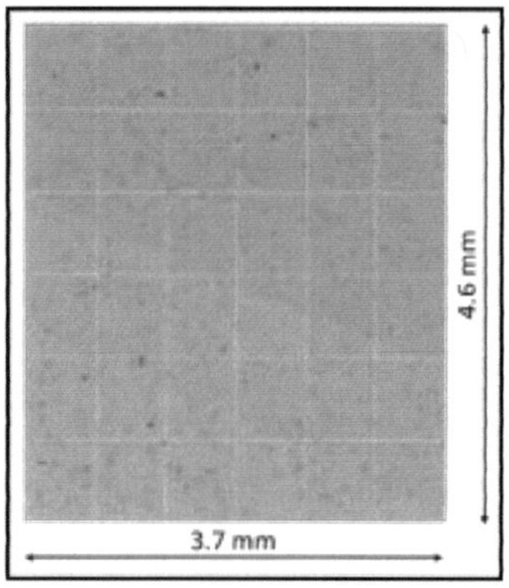

Figure 3. SEM map of cerium oxide inclusions using INCA Feature®.

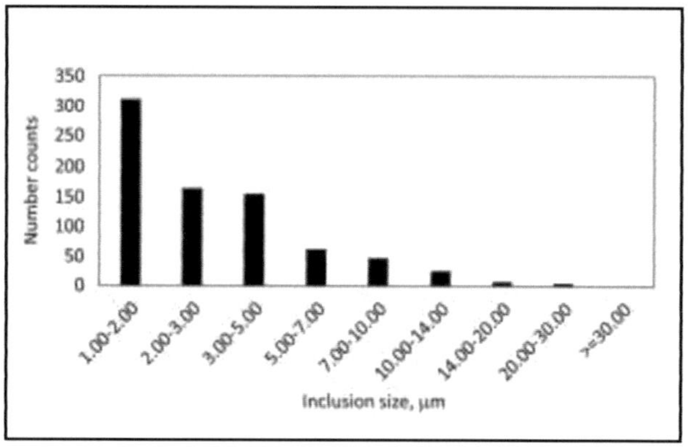

Figure 4. Size distribution of cerium oxide inclusions in the sintered ingot.

3.2. Metallography of the induction furnace samples

To find the best yield of the cerium oxide inclusions in the steel samples, three trials using 280 gram of electrolytic iron and 120 gram of the sintered cerium oxide ingot were undertaken.

Each trial used a different technique by placing the sintered cerium oxide ingot in different positions, as seen in Figure 5 and as follows:

(1) CeO_2-iron ingot at the bottom of the electrolytic iron in the crucible.

(2) CeO_2-iron ingot in the middle of the electrolytic iron in the crucible.

(3) CeO_2-iron ingot closer to the bottom but surrounded by electrolytic iron in the crucible.

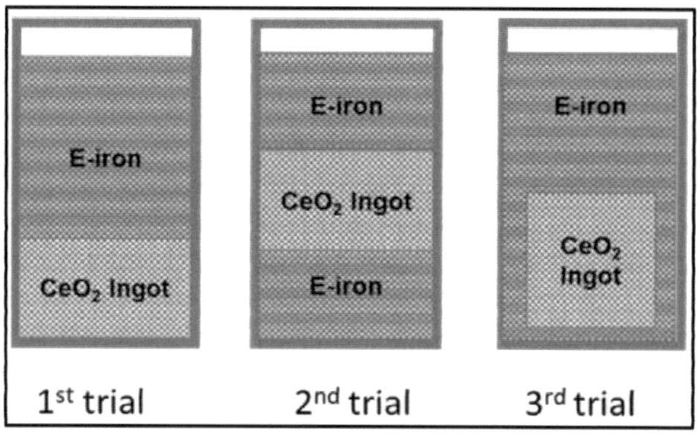

Figure 5. Position of material for the three induction furnace trials.

The third control trial showed that the maximum recovery of cerium oxide from sintered cerium ingot was achieved in the melted-out samples, as detailed in Table 1.

Table 1. Inclusion counts for different cerium oxide ingot positions.

Trial	Ingot Top (number counts/mm^2)	Ingot Centre (number counts/mm^2)	Ingot Bottom (number counts/mm^2)	Ingot Average (number counts/mm^2)
1st	10.4	12.4	9	10.6
2nd	15.9	15.5	12.1	14.5
3rd	20.3	16.7	15.2	17.4

A sample area of 17 mm^2 was scanned using INCA Feature® for the three trials. Figure 6 shows the distribution of the cerium oxide inclusions at the centre of each sample. The SEM maps showed similar results to the sintered cerium oxide ingot, in that the cerium oxide inclusions were homogenously distributed within the samples of the three trials.

The second and third trials showed that the number of cerium oxide inclusions were around 16 number counts per mm^2, which had dropped from 45 number counts per mm^2 for the sintered cerium oxide ingot, which was initially added to the melt. This was consistent with 30% of the sintered cerium oxide ingot having been added to the induction furnace trial samples. It was therefore concluded that 95% of the cerium oxide inclusions were recovered from the sintered cerium oxide ingot into the induction furnace trial samples.

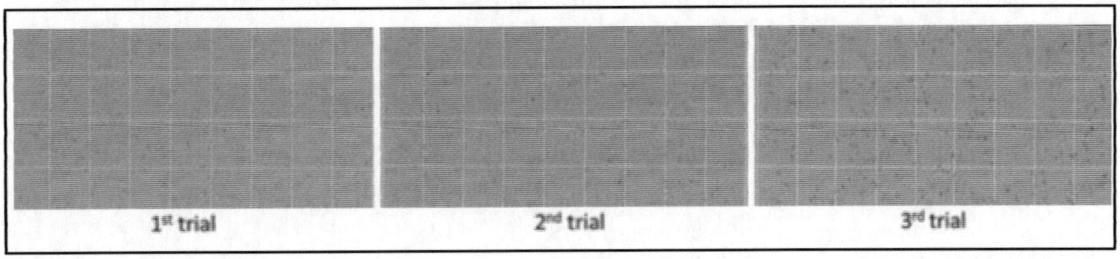

Figure 6. INCA Feature® map of cerium oxide inclusions at the centre position of the induction furnace samples.

Figure 7 shows the size distribution of cerium oxide inclusions for the three trials at the centre location of the melted-out ingot. Analysis showed that the major cerium oxide inclusion sizes were from 1 µm to 7 µm. No larger inclusion sizes from 14 µm to over 30 µm were recorded from the melted-out ingot. This can be explained if the larger size inclusions detected from the original sintered ingot were due to the agglomeration of particles during the sample preparation process. Once these large oxide inclusions get into the liquid melt, the agglomerated inclusions are separated into their original particle size.

More in-depth calculations of the stability of the CeO_2 inclusions in the liquid melt was conducted using the thermodynamic software FactSage 7.2 utilising databases FactPS, FToxid and FTmisc. The oxygen activity was calculated with an input O_2 content of 0.01 ppm, coming from the argon used in the induction furnace. The temperature ranged used in the calculations was 1200 °C to 1650 °C. Results of the FactSage calculations, indicated that Ce_2O_3 is stable in the liquid iron and would be duplexed with FeO after solidification. SEM elemental results confirm that FeO-Ce_2O_3 inclusions were found, which is shown in Figure 8.

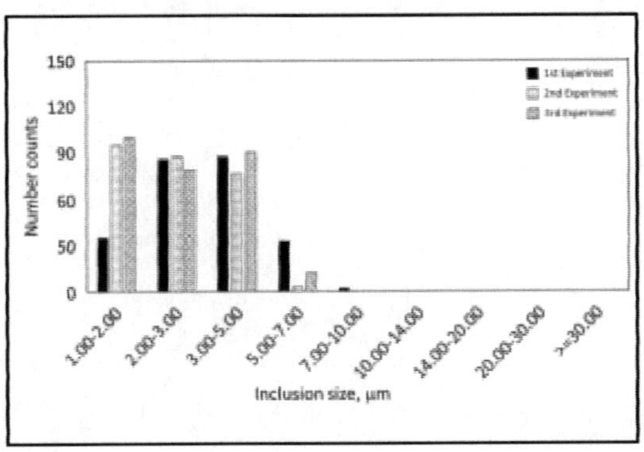

Figure 7. Size distribution of cerium oxide inclusions at the centre of the induction furnace samples.

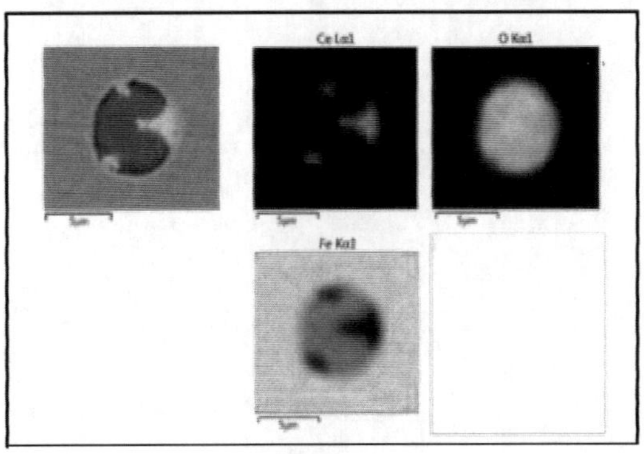

Figure 8. SEM elemental maps confirming inclusions to be FeO-Ce_2O_3

IOP Conf. Series: Materials Science and Engineering **529** (2019) 012067 doi:10.1088/1757-899X/529/1/012067

4. Conclusions

A powder metallurgy technique has been developed to enable the addition of synthetic inclusions of known size, composition and density into liquid steel. The sintered cerium oxide ingot was added with electrolytic iron and melted-out in an induction furnace. Metallographic assessment of the samples showed that the cerium oxide inclusions were homogenously dispersed through the sample from top, centre and bottom. Analysis suggested that more than 95% of the cerium oxide from the sintered ingot had been distributed throughout the trial ingots produced during the experiments

This technique is therefore useful for future steel cleanness studies for high value alloyed steels.

References

[1] Dogan N, Monaghan B, Longbottom R, Reid M and Tsekouras X 2012 Scanmet IV pp 1-10
[2] Pruvcheva K and Tsolov T 1978 Metallurgiya (Sofia) 33(9) pp 21-3
[3] Flórián K, Fischer W and Nickel H 1994 Fresenius J. Anal. Chem 349 pp 174
[4] Flórián K, Fischer W and Nickel H 1994 J. Anal. At. Spectrom 9 pp 257
[5] Mittelsädt H, Nazikkol C and Muller G 2002 Patent EP1227312
[6] Boué-Bigne F, Hoehne M, Sturm V, Műller G, Menut D, Ruby-Meyer F and Forrest R 2010 European Commission Research Fund for Coal and Steel ISBN 978-92-79-14228-4

Acknowledgements

The authors would like to thank Tata Steel Europe for funding this research work and for kind permission to present this paper. The authors also would like to thank Tata Steel Europe, Ceramics Research Centre (CRC), IJmuiden for their support in manufacturing the sintered cerium oxide ingots.

Numerical Simulations of Electromagnetic Counteractions to Mold Fluid Flow Asymmetry Deviations

Martin Sedén[1] and Bengt Rydholm[1]

[1] ABB AB, Industrial Automation, Process Industry, Metallurgy Products, Terminalvagen 24, 721 59 Vasteras, Sweden

martin.seden@se.abb.com

Abstract. Asymmetric flow of the molten steel in a continuous slab casting mold may be detrimental to the quality of the solidified end product. Depending on the severity of the biased flow, local mold powder entrainment, non-homogeneous solidification around the perimeter of the initial shell or non-optimal inclusion seclusion may occur. In particular, nozzle clogging in the last slabs before an SEN exchange may cause strong and asymmetries in the upper regions of the mold. To avoid costly downgrades of steel quality in these slabs, it is vital to maintain stable and symmetric fluid flow conditions in the mold. The EMBR and the FC Mold are two flexible electromagnetic devices able to produce asymmetric braking/stirring along the width of the mold in slab casting, and in this way allow counteraction against e.g. biased mold flow or local excessive flow speeds. Numerical computations of mold fluid flow and magnetic flux have been carried out to quantify the required fields to symmetrize biased flow scenarios caused by e.g. SEN clogging. In conjunction with steel plant trial feedback, the simulation results have been used to setup control algorithms for the EMBR and FC Mold.

1. Introduction

Process stability during continuous casting is crucial in order to obtain homogeneous and constant conditions for the fluid steel to solidify evenly in the mold. In highly productive slab casters, the utilization of the submerged entry nozzle (SEN) is often maximized. In the last slabs before SEN exchange in non-calcium treated steels, clogging in the nozzle may be severe and inhomogeneous, leading to strongly asymmetric and biased molten steel flows in the upper portion of the mold. Not only SEN clogging, but also asymmetric stopper or slide-gate positioning, anisotropic argon injection, mold eigenfrequency excitations or turbulence in the flow through the SEN may trigger asymmetric flow patterns. As a consequence of the asymmetric mold flow pattern, local mold powder entrainment or non-optimal inclusion seclusion may occur with deteriorated steel quality as a result [4]. To avoid downgrades of steel quality, it is vital to maintain stable and symmetric conditions in the mold.

Building on the confirmed results of the electromagnetic actuators EMBR and FC Mold to stabilize and optimize the mold flow [2] in the continuous slab casting process, the next step is an increasingly flexible EM device able to catch and react on local casting deviations such as meniscus fluctuations, undesired fluid flow velocities or patterns, crack indications or flow asymmetries. Recent developments in mold sensor technology [7]-[9] now enable detection capabilities of much more localized phenomena in mold fluid flow. A highly flexible EM device is the other cornerstone in creating an automatic mold flow control system that can handle both local and transient deviations from ideal conditions.

Applying spatially asymmetric mold flow control, the casting process becomes more stable and symmetric. E.g., excessive meniscus fluctuations and flow speeds on one side of the mold can be mitigated by extra stabilization and braking in this area, or uneven speeds in the SEN jets due to clogging can be homogenized by applying more braking or stirring on one side of the mold.

Counter measures of biased flow deficiencies allow quality upgrades with productivity and financial benefits for the steelmaker. Generally, a homogeneous and stable casting process yields a homogeneous solidified end product, with improvements in quality and product performance for the end-consumer.

2. Electromagnetic Flow Control in Slab Casting Mold with EMBR and FC Mold

The EMBR is in continuous casting a well-established electromagnetic actuator [2] that exposes a static magnetic field $\boldsymbol{B}$ underneath the SEN, perpendicular to the flow in the mold, and as such applies a braking Lorentz force $\boldsymbol{F}$ in the opposite direction of the flow velocity $\mathbf{v}$,

$$\boldsymbol{F} = \sigma(\boldsymbol{E} + \mathbf{v} \times \mathbf{B}) \times \boldsymbol{B}. \quad (1)$$

Recent ABB developments in EMBR technology offer the possibility to apply a strong magnetic braking field along the full width of the thin slab mold, but with a varying field magnitude on left on right sides as the coil currents are controlled by independent L/R drives. The ferromagnetic cores of the left and right sides can also be arranged flexibly in a bowl-shaped fashion to better accommodate for different incident flow jet directions from the SEN [12].

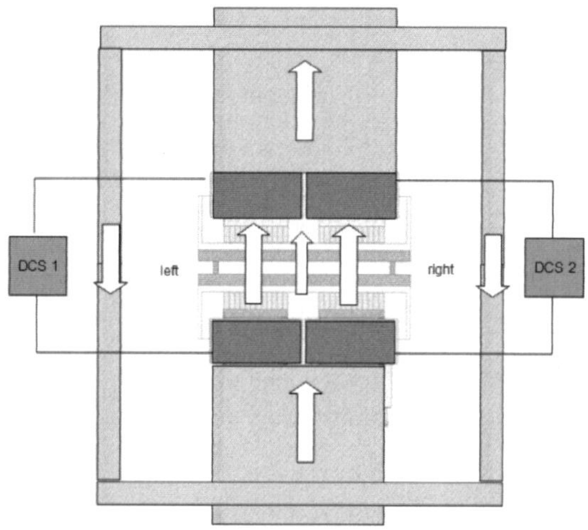

Figure 1. Configuration of EMBR with L/R independent field control.

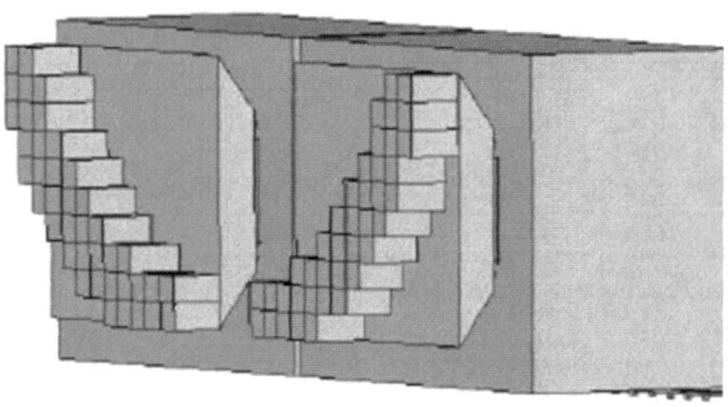

Figure 2. Bowl-type EMBR front cores.

The FC Mold is a double-level EM device used for conventional slabs, producing a horizontal upper magnetic field distribution close to the meniscus, above the SEN outlet ports, and a bottom horizontally located field below the SEN ports in the lower part of the mold. In the lower level, a static DC magnetic field is created across the mold thickness. In the upper level, another DC field is applied together with a superimposed AC travelling magnetic field to drive rotational stirring [8].

Electromagnets allow gradual field changes, and the EMBR and FC Mold provide seamless transitions between function modes in order to customize the applied magnetic forces according to the casting conditions at hand. For the FC Mold, DC and AC fields are available near the meniscus flow to enable braking if flow speeds are excessive, or acceleration by stirring if the flow speeds in this region are stagnant. Both DC and AC fields are applied simultaneously and the meniscus fluctuations can be stabilized at the same time as rotative stirring homogenizes the perimeter [8]. In modern EMBR and FC Mold generations, the

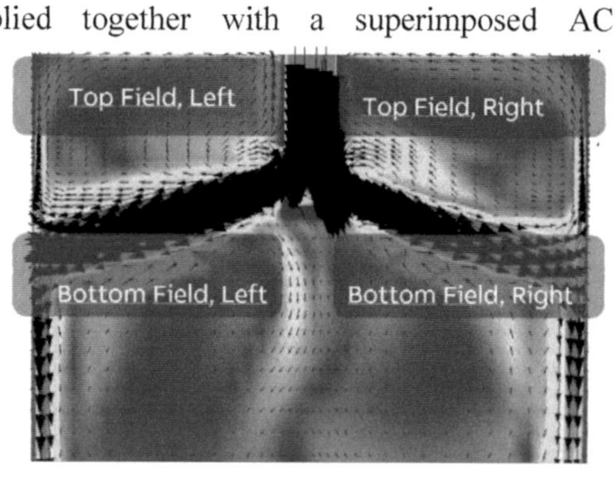

Figure 3. Positioning of independent FC Mold top and bottom, left and right fields.

magnets and their electrical drives allow independent control of the fields in both the upper and lower levels, as well as on left and right sides.

3. Numerical Simulations of Molten Steel Flow in Casting Mold

A platform for numerical modelling of the fluid steel flow in the strand, incorporating a magnetohydrodynamics coupling to the CFD equations [6], has over the past decades been continuously developed by ABB and refined with improvements in computational hardware, software as well as feedback from real steel plant measurements of magnetic flux, fluid flow, solidification characteristics and steel quality [2], [8], [11]. Mold flow is in the computational fluid dynamics model [5] governed by SEN jet momentum, argon bubble buoyancy and the forces from applying static braking and alternating accelerating electromagnetic fields of different magnitudes.

The magnetic fields and induced eddy-currents from the EMBR and FC Mold are obtained by solving the Maxwell field equations for the magnetic vector potential $\boldsymbol{A}$ with a FEM model incorporating various non-linear dependencies in material properties in and around the casting mold:

$$\boldsymbol{\nabla} \times \frac{1}{\mu} \boldsymbol{\nabla} \times \boldsymbol{A} = -\sigma \left(\frac{\partial \boldsymbol{A}}{\partial t} + \boldsymbol{\nabla} V \right). \quad (2)$$

A long history of manufacturing of ABB EM devices for metals production has enabled an accurate numerical prediction model based on a large database of measured data for validation and tuning [3].

The fluid dynamics model utilizes a transient, unsteady RANS model with an Eulerian representation of the two-phase flow. A Reynolds stress model has been used for turbulence approximations with an additional transport equation for the anisotropy of the turbulence caused by the magnetic field [1]. The magnetohydrodynamic coupling for static magnetic fields in the molten steel uses the electric potential method and a velocity dependent Lorentz force source term in the transport equations for the travelling alternating fields. The boundary conditions are given by a predescribed solidified shell of and a flat free surface (steel/air) at the meniscus level. Isothermal conditions are assumed in the evaluation of the flow.

4. Flow Control Under Asymmetric Casting Conditions

A set of numerical CFD simulations has been conducted on the platform described above to illustrate the effect of asymmetric magnetic field distributions acting on an asymmetric flow field inside the mold. Nozzle clogging is difficult to predict as the flow of steel and argon through the SEN is highly turbulent and the buildup of Aluminum oxides on the inner walls of the nozzle is more or less stochastic. Both the spatial location of the clog and the time when it occurs are hard to foresee. The result is however many times a strongly biased flow out of the nozzle leading to an asymmetric flow pattern in the mold and inhomogeneous solidification and particle seclusion conditions.

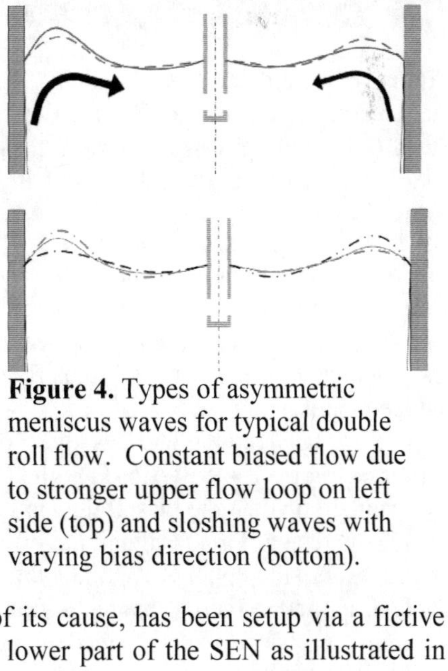

Figure 4. Types of asymmetric meniscus waves for typical double roll flow. Constant biased flow due to stronger upper flow loop on left side (top) and sloshing waves with varying bias direction (bottom).

Even casting for a perfectly new SEN with a smooth inner bore can lead to low frequency sloshing of the steel in the mold from one narrow face to another. The inherently turbulent nature of the flow through the SEN excites asymmetries that may trigger the eigenfrequencies of the bulk steel confined by the mold walls.

A simple model of an asymmetric mold flow, regardless of its cause, has been setup via a fictive linearly skew inlet flow condition over a cross section in the lower part of the SEN as illustrated in Figure 5. This creates a greater relative mass flow through the right nozzle port leading to a stronger SEN jet on the right side. This simple inlet condition has been applied to a set of different casting conditions to show the effect on the mold flow and to enable evaluation of relevant counter measures created by varying the magnetic field distributions.

4.1. Asymmetry Counteraction in Conventional Slab Caster

In the simulation model, a 237x1100 mm slab format cast at 1.9 m/min in a rectangular mold with 8 NL/min of injected argon, has been evaluated for the asymmetric inlet flow conditions with a +/-50% variation.

A snap-shot of the simulated velocity distribution in the center slice of the mold during these conditions is shown in Figure 6 without the impact of an FC Mold on the left and for an asymmetric FC Mold field distribution on the right [13]. In Figure 7 the transiently simulated meniscus flow speeds at the center thickness of the mold, y= ±0.385 m for the left and right sides, are shown for a time period of 210 s. In the first 60 s of the simulation, the FC Mold is turned off.

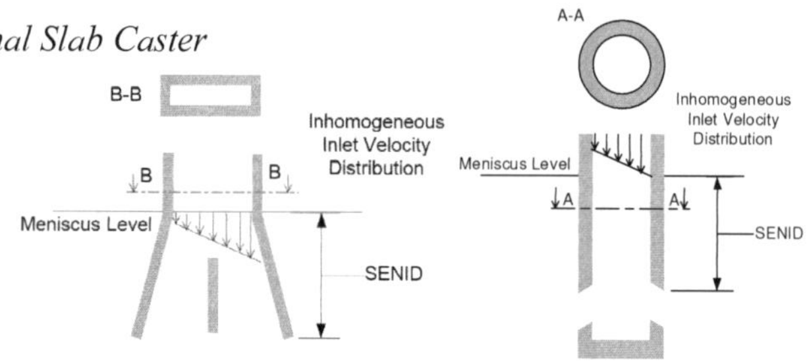

Figure 5. Models of skew inlet flow in the lower part of 2-port SENs for thin (left) and conventional (right) slab casting.

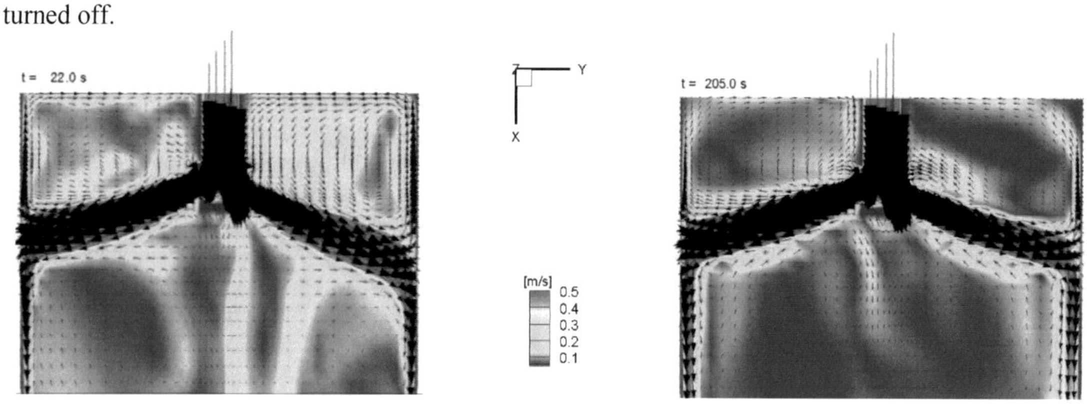

Figure 6. Simulated flow velocity distributions in mold center slice for skew inlet flow. Without FC Mold (left) and with asymmetric FC Mold DC field distribution with stronger DC field in the top right corner (right).

From here until t=75 s, horizontally homogeneous and symmetric DC magnetic field distributions in both levels of the FC Mold are ramped up, with the result that in 75-135 s, the meniscus flow speeds are reduced and fluctuations stabilized.

The gap between the left and right flow speed curves of Figure 7 is only closed in the last 60 s of the simulation when the biased flow is symmetrized. This is the result of a horizontally asymmetric field distribution in the upper FC Mold level where a 25% boost is applied to the DC magnetic flux density on the top right. The stronger flow speed on the right side of the meniscus is in this way subject to extra braking to obtain the same flow speed as on the left side. The

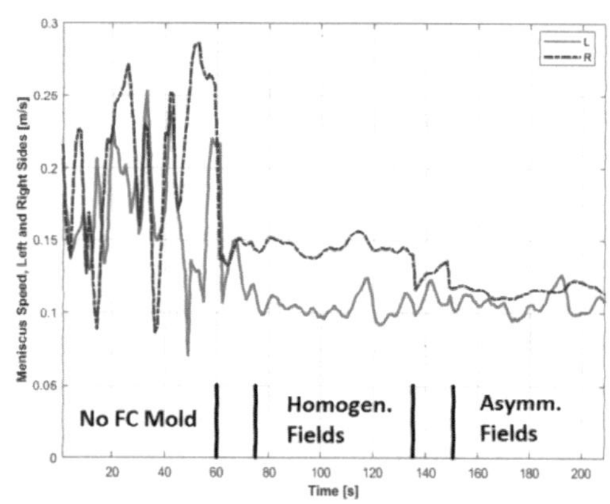

Figure 7. CFD simulated meniscus flow speeds on left (L) and right (R) sides in modelled skew inlet flow case for varying EM field configurations;
a. 0-60 s: No field,
b. 75-135 s: Homogeneous DC fields,
c. 150-210 s: Asymmetric DC fields.

flow speed pattern in the mold during asymmetric braking is illustrated on the right in Figure 6.

4.2. Asymmetry Counteraction in Thin Slab Caster

For a thin slab casting format of 100x1400 mm and casting speed 5.5 m/min, numerical simulations have been carried out with the above described linearly varying inlet speed method in a 2-port fish-tail nozzle to generate an initial mold flow asymmetry. This casting scenario is in the higher throughput range, calling for stabilization of the process. The DC field of the EMBR is applied below the SEN in a bowl-shape to stabilize the flow and to help guide momentum toward the meniscus in a double roll flow pattern, but at the same time minimize meniscus fluctuations and regulate meniscus flow speeds.

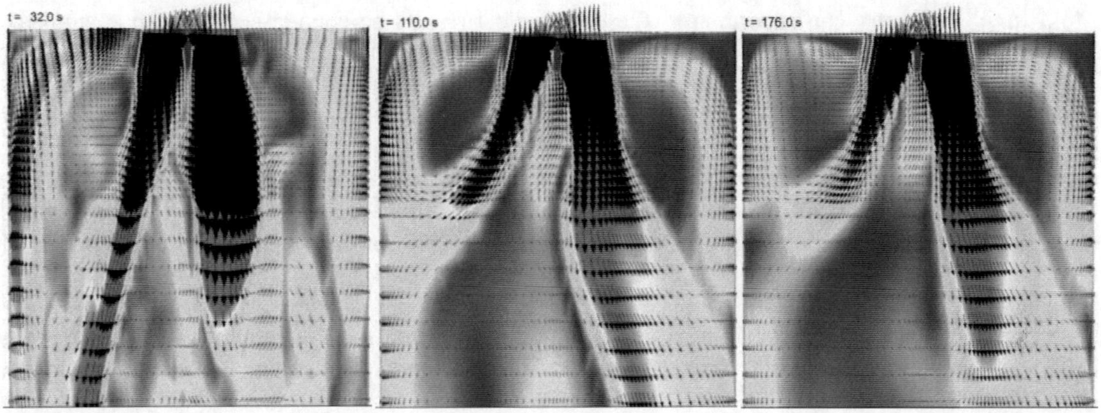

Figure 8. Simulated 100x1400 mm center strand flow velocity snap-shots for skew inlet flow. Without EMBR (left), with symmetric EMBR bowl-shaped fields (center) and with asymmetric EMBR DC field distribution with boosted DC field in the left side (right).

The application of the bowl-shaped EMBR fields generally leads to a decreased and stabilized meniscus flow speed, but when a skew velocity inlet condition with a ±50% variation is applied, the flow in the mold, even with a symmetric EM field, becomes asymmetric, see Figure 8. Figure 9 shows the transiently simulated sub-meniscus flow speeds at the center thickness of the mold, y= ±0.440 m for the left and right sides.

Within a 230 s period, three different constant field exposure modes are simulated where the described asymmetric flow situation is: a. unaffected by magnetic fields, b. under the influence of symmetric EMBR fields, c. counter-acted by an asymmetric EMBR field. The DC field in c. is 23% stronger on the left side to level out the meniscus flow asymmetry and dampen the flow speed peaks on the left side of the meniscus from time period b.

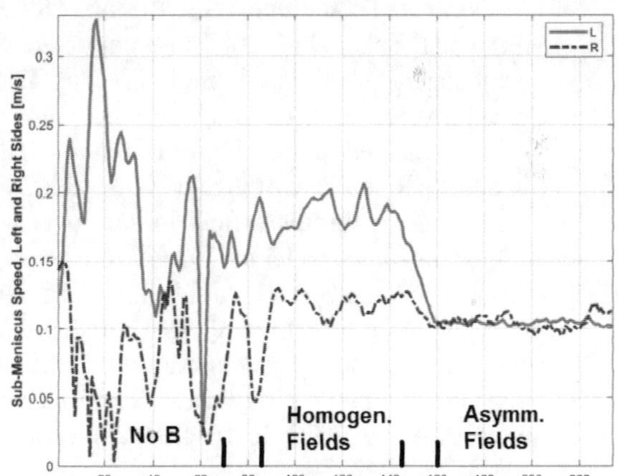

Figure 9. CFD simulation results of meniscus flow speeds on left (L) and right (R) sides in modelled skew thin slab inlet flow distribution for varying electromagnetic field configurations;
a. 0-70 s: No field,
b. 85-145 s: Symmetric EMBR DC fields,
c. 160-230 s: Asymmetric EMBR DC fields.

4.3. Summary

A flexible control of the EMBR/FC Mold would not apply the rugged method of constant field strengths over time as in this investigation, but rather a proportionate field strength variation depending on the response from an online mold flow sensor. However; to shed some light on the underlying control mechanisms, these two examples show that asymmetric meniscus flow can be counter-acted by appropriate magnitudes of the left and right DC magnetic fields.

In combination with an online flow measurement sensor such as mold level sensors or the OptiMold Monitor® which detects flow parameters such as the meniscus speed on the left and right sides of the mold [10], an automatic asymmetry control based on these principles can be set up. In addition to flow symmetrization, such a control set-up can also regulate the meniscus flow speed to a desired level.

5. Conclusions

Mold flow asymmetry can according to these numerical results be counter-acted by an asymmetric magnetic field distribution to achieve a homogeneous flow speed distribution over the entire meniscus level. This is valid both for conventional slab casting cases with the FC Mold and for thin slab casting with the EMBR. With modern ABB EM equipment, the magnetic fields are configurable independently, not only in the upper and lower levels, but also on the left and right side regions. This, in combination with a flow speed detection sensor such as the OptiMold Monitor®, makes it possible to control the magnetic fields of the EM actuator autonomously against flow speed and mold flow asymmetry unlocking potential for improved process stability, slab cleanliness and surface quality of cast steel.

References

[1] Widlund O et al 2000 Modeling of anisotropic turbulent transport in simulations of liquid metal flows in magnetic fields *3rd Int. Symp. on Electromagnetic Processing of Materials (Nagoya)* pp 97-102

[2] Yamamura H Toh T et al 2001 Optimum magnetic flux density in quality control of casts with level DC magnetic field in continuous casting mold *ISIJ Int.* **41** 10 pp 1229-1235

[3] Löfgren P and Bel Fdhila R 2003 Modeling of the continuous slab casting process. Part3: Comparison between algebraic slip model and 2-phase model & water case validation (Västeras) Tech. Report SECRC/AT/TR-03/023

[4] Miki Y and Takeuchi S 2003 Internal defects of continuous casting slabs caused by asymmetric unbalanced steel flow in mold *ISIJ Int.* **43** 10 pp 1548-1555

[5] Chaudhary R 2011 *Studies of Turbulent Flows in Continuous Casting of Steel With and Without Magnetic Field* Ph.D. Thesis (Urbana-Champaign: Univ. of Illinois at Urbana-Champaign)

[6] Chaudhary R, Thomas B G and Vanka S P 2012 Effect of electromagnetic ruler braking (EMBr) on transient turbulent flow in continuous slab casting using large eddy simulations *Metall. Mater. Trans. B* **43** 3 pp 532–553

[7] Lieftucht D et al 2013 HD Mold – a new fiber-optical-based mold monitoring system *AIST Trans. Iron and Steel Technology* **10** 12 pp 87-95

[8] Sedén M, Jacobson N, Lehman A and Eriksson J 2014 Control of flow behavior by FC Mold G3 in slab casting process *Proc. 8th European Continuous Casting Conf. (Graz) pp* 558-569

[9] Spierings A et al 2017 Development and application of fiber bragg gratings for slab casting *Proc. AISTech2017 (Nashville)*

[10] Sedén M and Jacobson N 2017 Securing dynamic mold flow control with FC Mold and OptiMold Monitor *Proc. 9th European Continuous Casting Conf. (Vienna)* pp 67-75

[11] Jacobson N, Sedén M, Go Y, Lim S, Kim K and Kim Y 2017 Flow control by FC Mold G3, operation and results from slab casting process *Proc. 9th ECCC (Vienna)* pp 199-208

[12] Hwang J Y, Cho M J, Thomas B G and Cho S M 2018 Numerical simulation of turbulent steel CEM mold under high mass flow condition *9th Int. Symp. on Electromagnetic Processing of Materials (Hyogo) IOP Conf. Ser.: Mater. Sci. Eng.* **424** 012033

[13] Sedén M and Jacobson N 2018 Online flow control with mold flow measurements and simultaneous EM braking and stirring *9th Int. Symp. on Electromagnetic Processing of Materials (Hyogo) IOP Conf. Ser.: Mater. Sci. Eng.* **424** 012015

IOP Conf. Series: Materials Science and Engineering **529** (2019) 012069 doi:10.1088/1757-899X/529/1/012069

Continuous Casting of High Carbon Steel: How Does Hard Cooling Influence Solidification, Micro- and Macro Segregation?

A. Nicholas Grundy[1], Steve Münch[2], Stephan Feldhaus[2], Johan Bratberg[1]

[1] Thermo-Calc Software AB, Solna, Sweden
[2] SMS Group, Zurich, Switzerland

nicholas@thermocalc.com

Abstract. One technology that is often employed in continuous casting of high-carbon steel billets to minimize centre- (or macro) segregation is hard secondary cooling. Investigations unanimously show, that hard cooling significantly reduces macro-segregation, but a mechanism for the reduced segregation is rarely given. In this paper the solidification of high carbon tire cord grade C80D cast as a 150x150 mm billet is calculated using the proprietary SMS Group solidification simulation package CHILL using steel properties calculated with the Thermo-Calc Software package and TCFE steels database. The obtained cooling rates in the billet for hard and soft secondary cooling are used to run solidification simulations considering solute redistribution using the diffusion module DICTRA. It is shown that for cooling rates achieved in continuous casting the steel solidifies far away from equilibrium. The solidification profile and solidus temperature lie in between the Scheil solidification model and the para-equilibrium Scheil model with carbon defined as a fast diffusing element. The calculated cooling rates and temperature gradients are used to simulate the solidification microstructure 20 mm from the billet surface using the phase field approach and the MICRESS® software package linked to Thermo-Calc through the TQ interface. This model clearly shows, that the most probable mechanism by which hard cooling reduces segregation is trapping of solutes between the intricately branched dendrite microstructure that results from the steep temperature gradients achieved when applying hard secondary cooling.

1. Introduction

The chemistry of tire cord steel is simple, the main alloying element being carbon only. It is however challenging to produce as macro- or centre segregation during continuous casting must be minimized and very stringent requirements on steel cleanness must be met. These factors are important, because the steel is drawn to very thin gauges of 0.15 mm and even below. A single 150x150 mm square billet thus yields ~15’000 km of wire and the thinner the gauge the higher the likelihood becomes of wire breakages, initiated either by non-deformable inclusions or by martensite spots that result from high centre segregation.

A technology to reduce segregation in continuous casting of billets is “hard-cooling” ([1] and references therein). In this process the billet surface is cooled to 300-500°C in the spray chamber with high intensity water spray cooling. The mechanism by which this reduces segregation has still not been fully clarified. This paper by no means has the ambition to present a detailed and exhaustive investigation, it merely aims to introduce various fundamental concepts that are at play during the solidification of continuously cast billets at different cooling rates. Simulations are performed use thing the Thermo-Calc Software package [2], Thermo-Calc’s diffusion module DICTRA [3][4], the phase-field based microstructure evolution simulation package MICRESS® [5] and the SMS Goup’s proprietary solidification simulation program for continuous casting CHILL [6]. For all calculations the high carbon steel grade C80D (EN 1.0622, SAE 1080, DIN C80D) is used with a highly simplified composition of 0.8 wt% C and 0.6 wt% Mn so that all equilibrium and diffusion calculations can be reproduced by the interested reader using the freely available Thermo-Calc DEMO version and the free databases FEDEMO and MFEDEMO that can be downloaded from the Thermo-Calc web page [7]. For calculations with more realistic alloy compositions including other elements the commercial TCFE9 and MOBFE4 are required.

2. Equilibrium Solidification of C80D

Phase diagrams are the basis for understanding any metallurgical process. ***Figure 1*** shows the calculated Fe-C isopleth with constant Mn content of 0.6 wt% (left) and the corresponding equilibrium phase fractions for C80D with 0.8 wt% C and 0.6 wt% Mn (right).

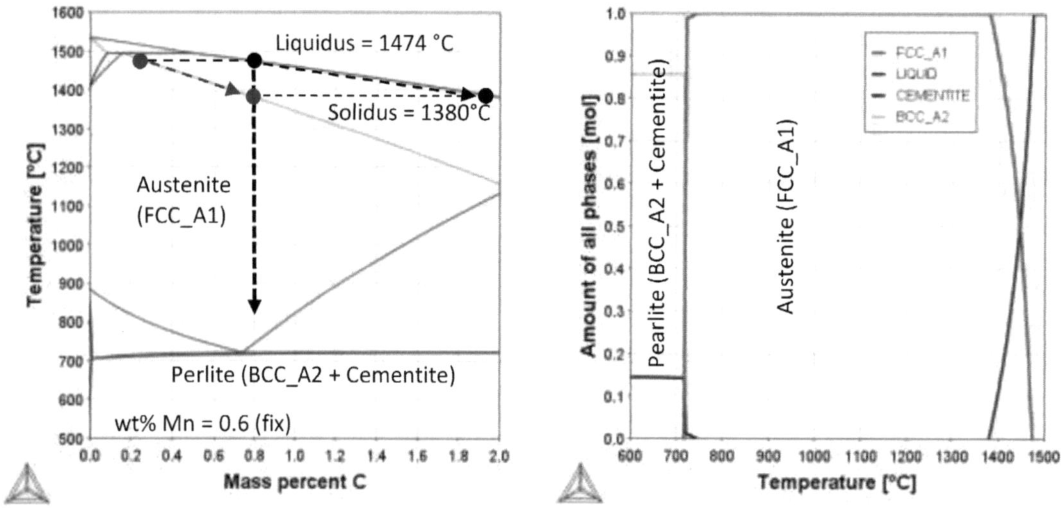

Figure 1. Isopleth of the Fe-C-Mn system at 0.6 wt% Mn (left) and equilibrium phase fractions in function of temperature of the composition Fe - 0.8 wt% C - 0.6 wt% Mn.

Under equilibrium conditions (full solute redistribution, infinitely slow cooling rate) pure Austenite (FCC_A1) forms on solidification with a liquidus- and solidus temperature of ~1474°C and 1380°C respectively. The Austenite decomposes into BCC and Cementite (Perlite) close to the eutectoid composition at ~720°C. The points and arrows in the phase diagram indicate the change in composition required in the solid and liquid phase under equilibrium conditions. In reality, the steel chemistry of grade C80D will be much more complex with tramp and trace elements resulting in slightly different solidus and liquidus temperatures and many other phases including oxides, carbides, nitrides or sulphides might precipitate. All these reactions can be accurately calculated using Thermo-Calc and the commercial TCFE9 database.

3. Scheil-Gulliver Solidification Model and Solidification Considering Back Diffusion

Real solidification processes are too fast for complete homogenization of the solid phase to occur. This means, the equilibrium solidification path (lever-rule) as shown in ***Figure 1*** is not applicable. An alternative solidification calculation is the Scheil-Gulliver (or simply Scheil) model [8]. It considers local equilibrium at the solid-liquid interface but assumes that no diffusion takes place in the solid phases. These thus become progressively richer in the solute elements from centre to shell, like an onion, as solidification progresses (***Figure 2***). In the liquid uniform solute distribution is assumed. The liquid becomes richer and richer in the solute as solidification progresses resulting in a marked depression of the solidus temperature.

Carbon dissolves interstitially in steels and diffuses very quickly. So Scheil's assumption of no diffusion in the solid is invalid. The assumption of equilibrium solidification is also invalid as the substitutional elements (Mn in our case) diffuse very slowly. This dilemma can be resolved by applying a partial equilibrium- or para equilibrium Scheil model, where one or more elements are defined as "fast diffusing". This is done by setting the chemical potential of the fast diffusing element(s) in the whole phase equal to their chemical potential at the phase boundary. Mathematically this corresponds to infinitely fast diffusion of these elements. In ***Figure 3***

Scheil solidification of the simplified C80D steel is compared assuming no diffusion of carbon (left) and fast diffusion of carbon (right).

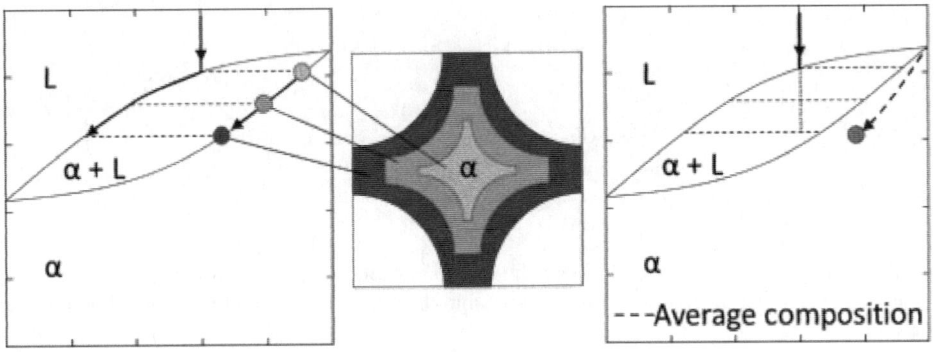

Figure 2. Visualisation of the Scheil solidification model.

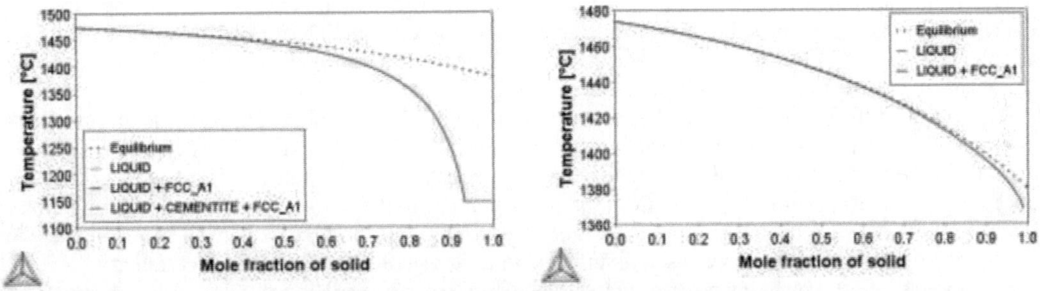

Figure 3. Scheil solidification simulation assuming no diffusion in the solid phase (left) showing large depression of the solidus temperature and precipitation of primary cementite (left) compared to Scheil solidification simulation with C defined as fast diffusing element. Please note the different temperature scales!

The former predicts a solidus temperature of ~1150°C and the precipitation of primary cementite. In normal casting conditions this is not observed. Setting C as a fast diffusing element results in a more realistic depression of the solidus temperature by about 20°C to 1360°C compared to equilibrium solidification. Which of the solidification models outlined above is more appropriate depends on the cooling rate. Equilibrium solidification corresponds to infinitively slow cooling, Scheil solidification without fast diffusion of C corresponds to such a fast cooling rate that even C does not have enough time to diffuse. Scheil solidification with fast C diffusion corresponds to some intermediate cooling rates that allows enough time for C to diffuse, but not enough time for all other element (in our case Mn) to diffuse. If we want to understand which model is most appropriate, we must first know the cooling rates obtained in continuous casting. Then quantitative diffusion calculations using Thermo-Calc's diffusion module DICTRA can be performed.

4. Solidification simulation of continuously cast 150x150 mm billet

The state-of-the-art solidification simulation model CHILL developed by the SMS Group for the on-line and real-time simulation of the continuous casting process of long products was used to calculate the solidification of a 150x150 mm square billet. Casting speed is 2.2 m/min and the specific secondary cooling volume is 0.8 l/kg for soft cooling and 2.2 l/kg for hard cooling. A standard 9 m billet caster design is used with 1 m CONVEX® copper mould tube and standard water-only secondary spray cooling design. The steel properties used for the simulation (enthalpy, thermal conductivity and volume) are all calculated by equilibrium calculations using Thermo-Calc and the TCFE9 steels database.

Figure 4 shows the calculated billet solidification profiles (fraction solid 0 and fraction solid 1 isotherms in the billet, top inserts) and the surface temperature and the temperature 20 mm below the billet surface for the two secondary cooling intensities.

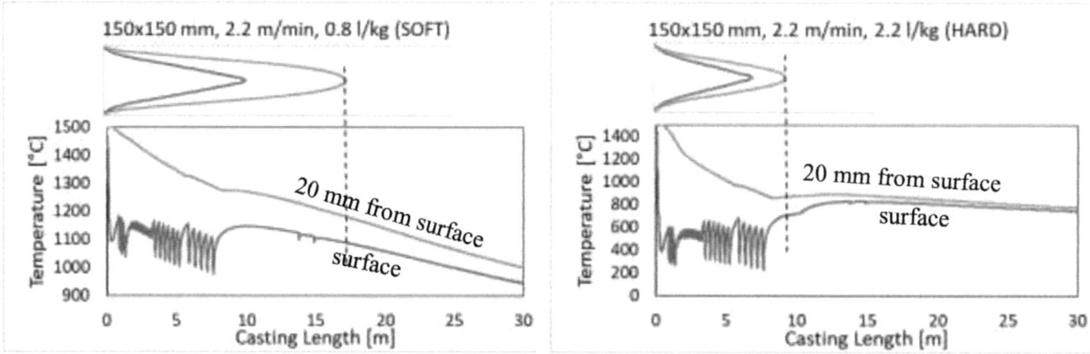

Figure 4. Solidification profiles for 150x150 billet. The calculated solidus and liquidus profiles in the billet are shown at top, at the bottom the surface temperature (blue) and the temperature 20 mm from the billet surface is shown. Please note the different scale!

5. Solidification Simulation Considering Full Diffusion

The time-temperature relation 20 mm from the surface of the 150x150 mm billet as calculated in the preceding section is used for the following solidification calculated under full quantitative consideration of the back diffusion in both the liquid and solid phase. The calculation domain is a small pool of liquid with a width of 0.5 mm that is trapped between secondary dendrite arms 20 mm below the billet surface. The diffusion equation is solved using the diffusion module DICTRA with the free mobility database MFEDEMO. The driving force for diffusion (gradients in chemical potentials) are calculated using Thermo-Calc and the free database FEDEMO. ***Figure 5*** (left) shows how the solid-liquid interface moves for the two cooling rates. This could be interpreted as the thickening of the dendrites in function of time and temperature. In the case of hard cooling, solidification of the small 0.5 mm volume is completed in ~40 sec. For soft cooling solidification takes ~160 sec. Note that the movement of the solid-liquid interface has nothing to do with the "shell growth" of the billet, rather it corresponds to the rate at which tiny pools of liquid that are trapped between the secondary dendrite arms solidify and disappear. Or in other words, it corresponds to the time required for our small volume of liquid to cool from the liquidus temperature down to the solidus temperature.

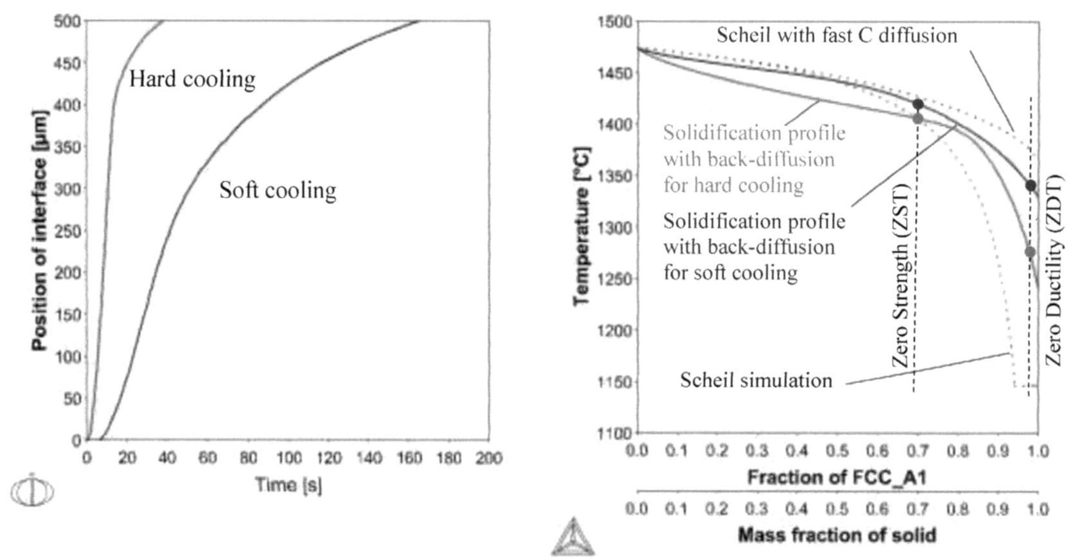

Figure 5. Position of solid-liquid interface in function of time for hard and soft cooling (left) and fraction solid curves (right) considering back-diffusion compared to the results from the Scheil calculations

As the solidification front moves it pushes highly segregated liquid ahead of itself that depresses the solidus temperature as shown in ***Figure 5*** (right). This thin but pronounced layer of C and Mn rich liquid is more pronounced for hard cooling as the interface moves faster (***Figure 6***) leaving less time for the back-diffusion of the elements. This layer is also the source of interface instability that results in the formation of dendrites due to constitutional undercooling.

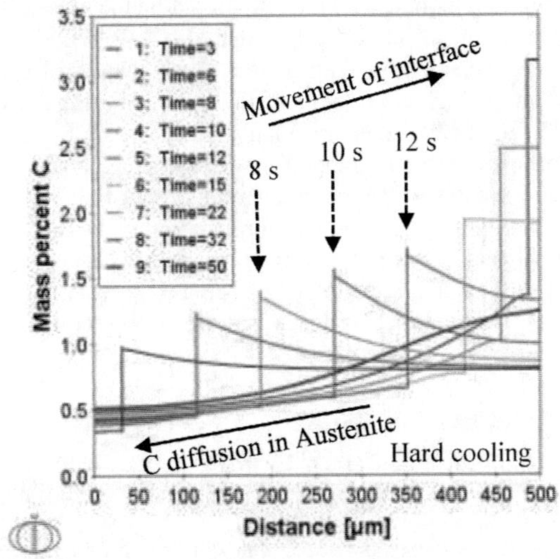

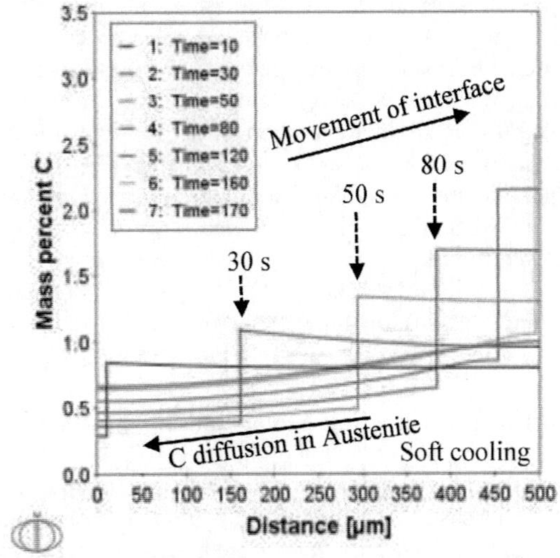

Figure 6. Carbon concentration in function of position and time. For hard cooling the pronounced concentration peak ahead of the solidification front can be clearly seen.

Internal cracks or hot tears (***Figure 7***) are a serious problem in continuous casting. The susceptibility to crack formation is closely related to the solidification process. When the solid fraction is low the steel behaves like a liquid. The steel flows and consequently has no mechanical strength and cracks cannot form. As the solid fraction gets higher, the space between dendrites decreases until at a certain temperature bridges start forming between the dendrites. This temperature is known as the "Zero Strength Temperature" (ZST) and is usually assumed to correspond to a fraction solid of ~0.6 – 0.8. Below this temperature brittle fracturing of the steel can occur: the weak bridges between dendrites are broken and the crack is usually filled by highly segregated liquid from between the dendrite arms. On lowering the temperature further, the amount of inter-dendritic liquid is reduced and the bridges between the dendrites become stronger until at one point the ductility is no longer zero. This temperature is referred to as the "Zero Ductility Temperature" (ZDT) and is generally assumed to correspond to a fraction solid of 0.99 or 1 [9] [10].

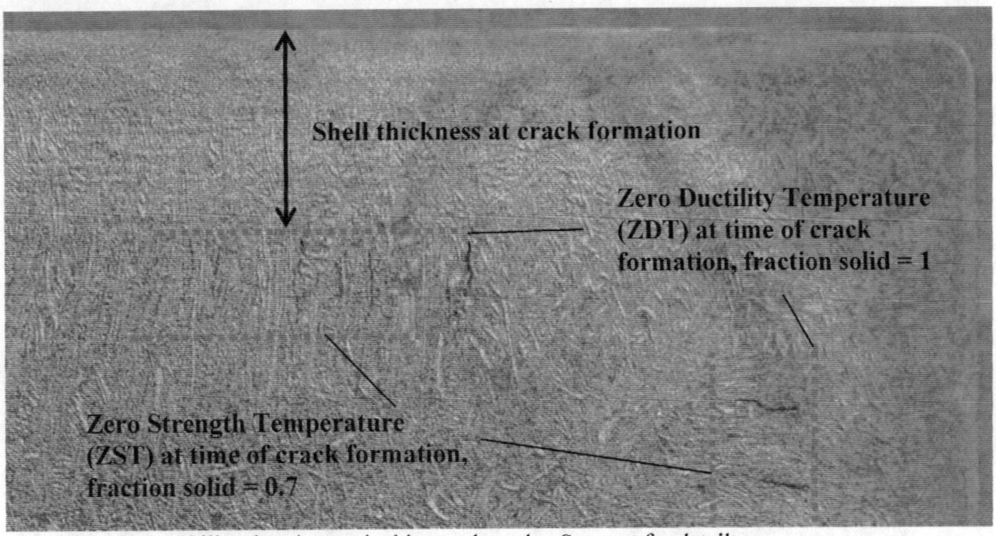

Figure 7. 150x150 mm billet showing typical internal cracks. See text for details.

Basically, this means that internal cracks occur when the fraction solid is between 1 and ~0.7. ***Figure 6*** shows that the difference between ZST and ZDT is greatly increased for hard cooling, because the solidus temperature is depressed, and highly segregated liquid remains between the dendrites down to very low temperatures (of course only under the assumption of the same liquid pool size, which might not be the case). If the complete steel chemistry would be considered in the calculations if would be seen that also S, P, and all other elements strongly segregate between dendrite arms making the cracking tendency far more severe for hard cooling. What makes the problem of cracking even worse is that the high cooling intensities applied during hard cooling also result in higher thermal stresses and strains, further aggravating the cracking problem.
The calculated fraction solid curve shown in ***Figure 5*** can thus directly be used together with the calculated temperature profiles of the cast billet shown in ***Figure 4*** to interpret the position in the continuous casting machine where the cracks are formed. This information is of great practical value for eliminating the internal cracking problem by optimizing casting parameters (balance between secondary cooling intensities of zones, adjust specific water volume, etc...), by improving maintenance (nozzle alignment, nozzle clogging, etc...) or improving the design of the continuous caster (longer secondary cooling zone, reduced gaps between secondary cooling zones, etc...).

6. Billet Solidification Microstructure for Different Solidification Rates

So, if hard cooling results in higher micro-segregation, makes the steel more susceptible to hot cracking and also results in higher thermal stresses and strains putting very high requirements on the design and maintenance of the continuous casting machine, why should anyone be interested in applying this technology?
Up to now we have only investigated micro-segregation that occurs between the secondary dendrite arms. This micro-segregation is not really a problem for downstream processing of the billet. The big problem is macro-segregation, that is found close to the centre. It is widely recognized that hard-cooling does in fact reduce this macro-segregation, however the mechanism by which this happens seems poorly understood. It is sometimes suggested that the higher cooling rate and steeper thermal gradients result in a finer branching dendrite microstructure at the solidification front that traps solutes and prevents them from being pushed ahead of the growing shell towards the centre of the billet.

To investigate this assumption, a multiple phase field simulation for the simplified C80D steel-grade investigated in this paper was set-up using the MICRESS simulation package that has already been applied to other continuous casting problems [11] [12]. The cooling rate and the temperature gradients in the billet are taken from the solidification simulation shown in ***Figure 4*** for a point 20 mm below the billet surface. All thermodynamic materials properties and the diffusion kinetics required for the phase field simulation are calculated using Thermo-Calc and are passed to MICRESS through the TQ programmers application interface [13] [14]. ***Figure 8*** compares the dendrite morphology for the thermal gradients obtained for hard cooling and soft cooling.

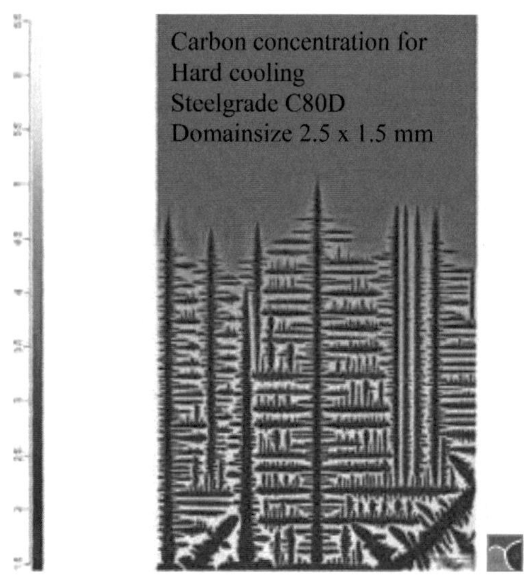

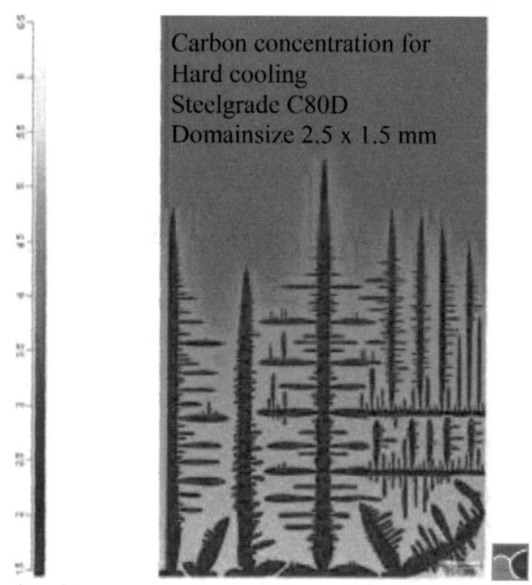

Figure 8. Solidification microstructure of C80D steel-grade with temperature gradient and cooling rate corresponding to hard cooling (left) and soft cooling (right).

IOP Conf. Series: Materials Science and Engineering **529** (2019) 012069 doi:10.1088/1757-899X/529/1/012069

Macro-segregation cannot be fully understood without considering the flow of liquid steel, in particular the suction effects induced by the volume shrinkage on final solidification resulting in the well-known V-segregation. Diffusion only takes place in the micro-meter range and is irrelevant. The phase field simulations clearly demonstrate the well-known finer dendrite microstructure that evolves when the thermal gradient and cooling rates are higher. This finer microstructure has a lower permeability for the liquid steel. Therefore the flow of highly segregated interdendritic melt towards the billet centre is obstructed and macrosegregation is reduced.

7. Conclusions and Outlook

This paper by no means has the ambition for an in-depth description of the billet solidification. The aim is to outline the linking together of readily available materials simulation tools, namely thermodynamic calculations, diffusion calculations, finite element solidification simulations and phase field modelling to elucidate a real-life industrial example from several angles in what could be considered an integrated computational materials engineering approach (ICME) [15]. All calculations are performed with state-of-the-art software packages and materials properties databases, but with many simplifications such as 2D calculations instead of 3D, strongly simplified steel chemistry, linearized thermal gradients, neglected fluid flow and gravity effects, no coupling between the models, etc… The simplifications have however been carefully chosen, so that all calculations still correctly capture the essence of the hard-cooling process. The next step is to expand and adapt the outlined models and apply them to the far more complex real-world industrial process.

8. References

[1] V. Ludlow *et al.*, "Strategy to minimise central segregation in high carbon steel grades during billet casting," *Ironmak. Steelmak.*, vol. 32, no. 1, pp. 68, 2005.

[2] B. Sundman, B. Jansson, and J.-O. Andersson, "The Thermo-Calc databank system," *Calphad*, vol. 9, no. 2, pp. 153, 1985.

[3] A. Borgenstam, L. Höglund, J. Ågren, and A. Engström, "DICTRA, a tool for simulation of diffusional transformations in alloys," *J. Phase Equilibria*, vol. 21, no. 3, pp. 269, 2000.

[4] J.-O. Andersson, T. Helander, L. Höglund, P. Shi, and B. Sundman, "Thermo-Calc & DICTRA, computational tools for materials science," *Calphad*, vol. 26, no. 2, pp. 273, 2002.

[5] www.micress.de (accessed Nov 15th 2018)

[6] L. Cestari, C. Dratva, A. N. Grundy and C. Tercelli, "Simulation Techniques for the Design and Process Control of Continuous Casting Machines," *Mat. and Man. Proc.*, Vol. 25, no. 1-3, pp 142, 2010.

[7] www.thermocalc.com (acceessed Nov 15th 2018)

[8] Q. Chen and B. Sundman, "Computation of Partial Equilibrium Solidification with Complete Interstitial and Negligible Substitutional Solute Back Diffusion," *Mater. Trans.*, vol. 43, no. 3, pp. 551, 2002.

[9] C. Cicutti and R. Boeri, "Analysis of solute distribution during the solidification of low alloyed steels," *Steel Res. Int.,* vol. 77, no. 3, pp. 194, 2006.

[10] T. W. Clyne, M. Wolf, and W. Kurz, "The effect of melt composition on solidification cracking of steel, with particular reference to continuous casting," *Metall. Trans. B*, vol. 13, no. 2, pp. 259, 1982.

[11] B.Böttger, M.Apel, B.Santillana and D.G.Eskin, "Phase-field modelling of microstructure formation during the solidification of continuously cast low carbon and HSLA steels", *Materials Science and Engineering* vol. 33, pp. 3765, 2010

[12] B.Böttger, S.Stratemeier, E.Subasic, K.Göhler, I.Steinbach and D.Senk, "Modeling of Hot Ductility During Solidification of Steel Grades in Continuous Casting – Part II", *Advanced Engineering Materials* vol. 12 no. 4, pp. 101, 2010

[13] J. Eiken, B. Böttger, I. Steinbach, "Multiphase-field approach for multicomponent alloys with extrapolation scheme for numerical application", *Phys Rev E,* vol. 73, no 6, pp 066122, 2006

[14] B. Böttger, J. Eiken, M. Apel, "Multi-ternary extrapolation scheme for efficient coupling of thermodynamic data to a multiphase-field model", *Comput Mater Sci.,* vol 108, pp 283–292, 2015

[15] G.J.Schmitz and U.Prahl (eds): Handbook of Software Solutions for ICME; Wiley VCH, Weinheim (2016), ISBN 978-3-527-33902-0

IOP Conf. Series: Materials Science and Engineering **529** (2019) 012070 doi:10.1088/1757-899X/529/1/012070

Numerical simulations of liquid steel alloying in the three strand continuous casting bloom tundish

A Cwudziński

Associate professor, Department of Metals Extraction and Recirculation, Faculty of Production Engineering and Materials Technology, Czestochowa University of Technology, Armii Krajowej 19, Czestochowa, PL

E-mail: cwudzinski@wip.pcz.pl

Abstract. This paper aims to obtain fundamental information on chemical homogenisation process of liquid steel with alloy additions in the bloom tundish. For alloy feeding to liquid steel pulse step alloying method was applied. Author checked the effect of hydrodynamic conditions occurs in the internal working space of tundish on the process of alloy mixing with liquid steel. Within the work basic and proposed tundish equipments were considered. Numerical modelling technique was employed to demonstrate the process of alloy addition mixing during continuous casting process. Ansys-Fluent computer program was used for numerical simulation. For particular continuous casting strands time mixing was calculated.

1. Introduction

The continuous steel casting process is used for the manufacture of semi-finished products constituting feedstock for rolling mills. The intensive development of advanced steel grades, in which a decisive role is played by alloy additions [1], encourages searching for new technological solutions in the area of steel metallurgy. As standard, alloy additions are fed at the stand of the ladle furnace which enables the chemical and thermal correction of the liquid steel [2-7]. However, in order to optimise the time and costs of semi-finished product manufacturing, which consists of the smelting, treatment and casting of the steel grade, it is justified to aid the process of correcting chemically the steel or non-metallic inclusion immediately prior to the liquid steel pouring into the mould. Blooms, billets or slabs cast by the continuous method are characterised by a varying shape and dimensions [8-10]. Therefore, often the same continuous steel casting (CSC) machines are used in industrial practice to enable different blooms, billets or slabs types to be cast. To this end, the mould module and the tundish are exchanged and the secondary cooling zone is corrected. Because of the specificity of the process of liquid steel flow in the ladle–tundish–mould system, the tundish seems to be a device the most appropriate to being used in the process of alloying. Within the additional treatment of alloying it seems feasible to make a correction to the chemical composition of the liquid steel or non-metallic inclusions. This process will obviously be restricted by the time that is limited by the tonnage of the tundish and the speed of casting. Previous studies carried out for an one strand slab tundish have shown that the duration of mixing alloy additions with steel will be influenced not only by the mode of feeding the addition, but also by the addition feed position and equipping the tundish with flow control devices (FCDs) [11-13]. A newly developed pulse–step method (PSM) of feeding alloy additions to the liquid steel as it flows through the three strand bloom tundish is employed in the present study.

IOP Conf. Series: Materials Science and Engineering **529** (2019) 012070 doi:10.1088/1757-899X/529/1/012070

2. Continuous casting tundish

Detailed description of tundish was presented in the previous authors work [14]. Previous studies have shown that with the appropriate tundish equipment similar hydrodynamic conditions can be obtained for individual tundish nozzles [14]. The examined tundish was equipped with FCDs in the form of a subflux turbulence controller (STC) and a dam (figure 1). Through its action on the liquid steel motion, the employed STC generated optimal conditions for the process of chemical homogenisation in the tundish designed for casting slabs [11-12]. In two tundish equipment variants, the height of the dam was changed from 0.12 (low dam LD) to 0.25 m (medium dam MD). The modified dam height was also taken from the previous studies [12]. The basic tundish variant with locations for introducing an alloy addition to liquid steel is shown in figure 1a. Four alloy addition feeding positions (AAFP) situated in the immediate vicinity of the tundish pouring zone were considered.

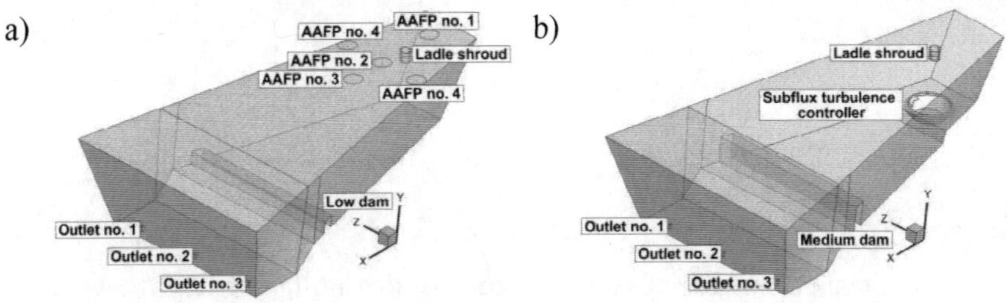

Figure 1. Three strand bloom tundish and considered alloy addition positions: a) low dam tundish, b) medium dam and STC tundish.

3. Numerical simulation methodology

Numerical simulation of the liquid steel flow and alloy addition behaviour in turbulent motion conditions was done using the Ansys-Fluent® 12.1 computer program. The Navier-Stokes equation was solved for liquid steel. While for the alloy, the species element for prediction a local mass fraction was added to continuity equation. All numerical simulations were done by employing a double-precision solver (3ddp) using discretisation of the second order for momentum, turbulent motion, species, energy and transient formulation. Numerical simulations were performed for the sequence of casting 0.28 m × 0.28 m blooms at a speed of 0.75 m/min and with the initial steel temperature of 1823 K. The aim of alloying was to achieve the 0.096 wt% correction to the chemical composition with metallic nickel and aluminium. To illustrate the chemical homogenisation process, mixing curves were recorded at individual tundish outlets. The level of chemical homogenisation was determined using the following relationship:

$$C_{PSM} = \frac{(C_t - C_0)}{(C_f - C_0)} \cdot 100\% \tag{1}$$

where: C_f - final concentration of alloy at tundish outlet (wt%), C_{PSM} - dimensionless concentration of alloy for PSM, C_t - temporary concentration of alloy (wt%), C_0 - initial concentration of alloy (wt%).

Based on the mixing curves, the mixing time was calculated. The mixing time is defined as the time, after which the minimum required liquid steel chemical homogenisation level is maintained, which should amount to at least 95%. The time interval was expressed by dimensionless time (DT) defined by the ratio of the actual time to the average time. The average time for the tundish under examination was 1270 seconds. The previous studies have confirmed the usefulness of the numerical model for simulating the chemical homogenisation process during steel flow through the tundish and the possibility of simulating the alloying process and recording the mixing process on a macro scale, with omitting the stage of the alloy addition melting in the steel [11-12]. Table 1 shows computer simulation variants (cases) discussed in the article.

Table 1. Considered variants of tundish configuration and liquid steel alloying

Case No.	Tundish flow control devices variants				Alloying variants					
					Alloy addition feeding position				Type of alloy	
	LD	LD + STC	MD	MD + STC	1	2	3	4	Ni	Al
1	×				×				×	×
2	×					×			×	×
3	×						×		×	×
4	×							×	×	×
5		×			×				×	×
6		×						×	×	×
7			×		×				×	×
8			×					×	×	×
9				×	×				×	×
10				×				×	×	×

Work [15] has shown that the chemical homogenisation process is also influenced by the type of alloy addition, chiefly in terms of its density as related to the steel density. Therefore, two metals, Ni and Al, were tested within the investigation; the former of which is slightly heavier while the latter is much lighter than steel. The physicochemical quantities that describe the steel and the metals introduced to it are given in table 2.

Table 2. Properties of considered alloy additions and steel

Materials	Density (kg m^{-3})	Viscosity (Pa s)	Heat capacity (J kg^{-1} K^{-1})	Thermal conductivity (W m^{-1} K^{-1})	Solute diffusivity (m^2 s^{-1})
Aluminum	2100	0.00052	1180	91	$8.6 \cdot 10^{-09}$
Nickel	7790	0.00159	556	50	$5.3 \cdot 10^{-09}$
Steel	7010	0.007	750	41	-

4. Computational results

4.1. Hydrodynamic conditions

From the performed computer simulations, a picture of liquid steel flow in the tundish working volume was obtained. Figure 2 shows the pathlines of liquid steel flow in the volume of the tundish. In the tundish with no STC, steel motion in the central part shows a descending behaviour (figure 2a). Only in the bottom lowering zone the steel streams flow horizontally towards the tundish front wall. Whereas, increasing the dam height results in the formation of an additional steel circulation under the influence of reverse streams flowing in from the tundish nozzles side (cases 5-6). A clearly horizontal arrangement of the streams circulating within the working space is noticeable when the tundish supply zone is equipped with STC, regardless of the dam height (figure 2b). The effect of the tundish equipment and the change in hydrodynamic conditions is also visible in locations, where the alloying process is initiated.

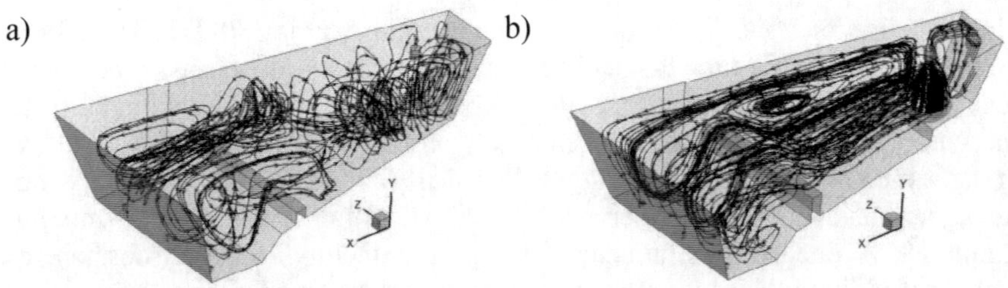

Figure 2. Path lines of liquid steel flowed through tundish: a) cases 1-4, b) cases 9-10

In the case of a tundish equipped with a dam only, an alloy addition is transferred to the central part of the working space, whereas, introducing the alloy addition simultaneously at both tundish side walls should provide a reasonably uniform distribution relative to the longitudinal axis of the tundish (cases 1-4). However, as can be seen, in the case of introducing the addition in the location denoted with the number 2 there is the risk of its asymmetric distribution within the liquid steel volume (case No. 2). This fact may clearly interfere with obtaining consistent chemical homogenisation conditions at all tundish nozzles. By contrast, in the case of retrofitting the tundish with the STC, the steel streams located in the plane transverse to the tundish axis, transform into ascending streams and a clear asymmetry in steel flow shows up. For multi-nozzle tundishes, the symmetry of hydrodynamic patterns forming in the working space gains added importance, because not only the mixing time span, but also the uniformity of the obtained time at individual tundish nozzles becomes important. Since a reasonably consistent picture of steel flow relative to individual tundish nozzles should provide similar conditions of chemical homogenisation. Therefore, the steel flow pattern obtained immediately above the nozzles will be an additional factor affecting the chemical homogenisation process. The most uniform distribution of liquid steel flow in the mould supply zone was obtained for tundish with STC and LD. By contrast, the most differential patterns for outlets zone were observed in tundish variants with only MD and MD with STC. In the case of the examined tundish, the hydrodynamic pattern is determined both by the streams supplying, as well as by the FCDs. In view of the above, the proposed tundish equipment modification will certainly influence the process of chemical steel homogenisation.

4.2. Influence of alloy additions locations on time mixing

The idea behind the proposed alloy addition feed method is to obtain the mixing curve characteristic of stationary reactors in optimally shortest possible time considering the continuous casting blooms forming in the mould. For the facility under discussion, it becomes additionally important also to uniform the chemical composition with respect to all nozzles and to supply the moulds with steel of identical chemical composition. This section provides results concerning the effect of the alloy addition feed location on the mixing process in the tundish with low dam (Figure 3).

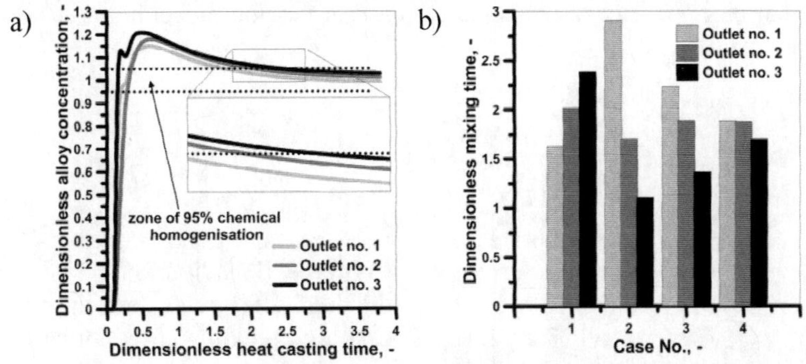

Figure 3. Behaviour of nickel in the bulk of liquid steel during casting: a) mixing curve for case No. 1, b) mixing time for different AAFP in the tundish with low dam

The process of casting square blooms is characterized by a lower casting speed, compared to casting slabs, which considerably extends the average time of steel residence in the tundish and thus the period, in which the alloying process can be carried out. Figure 3a shows mixing curves with the indicated 95% chemical homogenisation zone. The time of casting a single heat is 3.78 DT. Therefore, exceeding this duration in the case of mixing time disqualifies a given tundish variant as possible to be used in the alloying process. While it can be seen that the alloy addition shows up at individual nozzles at the same time, attaining the maximum concentration or a similar distribution of alloy addition concentration in the same period of time is difficult. Especially for the steel streams flowing into nozzle no. 3 a clearly higher nickel content is observed. Figure 3b illustrates results for the mixing time necessary for obtaining 95% chemical homogenisation. In the facility under discussion, a considerable asymmetry of alloy addition distribution is observed for nozzles situated on either side of the longitudinal tundish axis. This is especially true for the alloy addition feed location denoted with the number 2. The difference in mixing time between nozzles situated on the opposite sides relative to the longitudinal tundish axis amounted to over 1.8 DT. The most advantageous mixing time, fairly uniform for all nozzles, was obtained during feeding nickel at both side tundish walls simultaneously.

4.3. Influence of flow control devices on time mixing

Based on the results obtained for the steel alloying process during slab casting, the effect of the FCD on the behaviour of the alloy addition in the bloom tundish was verified. For the facility under consideration, the STC has been proposed, whose task is to attenuate turbulences in the pouring zone. Therefore, for TFCD variants 2 and 4, the average steel flow velocity was, respectively, 0.0159 and 0.0153 m/s. These values are by 0.006 and 0.0051 m/s smaller than those obtained for the tundish equipped only with a low or medium dam. Figure 4a shows mixing curves for the tundish equipped with the STC. For this tundish variant, the attenuation of the turbulence in the pouring zone has brought about considerable changes in the hydrodynamic pattern of liquid steel flow and has been reflected in the distribution of nickel in the bulk of the melt being cast. Much faster equalisation of the nickel content level in the bulk of steel is visible already at a time below 1 DT. However, for the extreme nozzles, too large an asymmetry of nickel distribution within the steel has occurred. The too high concentration of nickel in the liquid steel on the side of nozzle 3 prevents the 95% chemical homogenisation level from being attained in the period of time, during which the steel grade under consideration is being cast. In this connection, an additional factor in the form of individual streams supplying particular nozzles occurs in multi-nozzle tundishes, which modifies the hydrodynamic patterns influencing the chemical homogenisation process. The additional reduction of steel flow velocity by using the STC also seems unfavourable (lesser mixing energy apart from pouring zone). This is especially so because non-isothermal conditions predominate in the CSC process, and the forming temperature gradients together with low metal flow velocities cause localised modifications to the hydrodynamic patterns due to the action of natural convection forces. Figure 4b shows mixing time for different tundish equipment variants. It can be seen that the STC prevents the 95% chemical homogenisation level from being attained during melt casting. Whereas, increasing the dam height equalizes the mixing time recorded at individual nozzles also in the case of feeding nickel in the zone between the ladle shroud and the rear wall.

a)
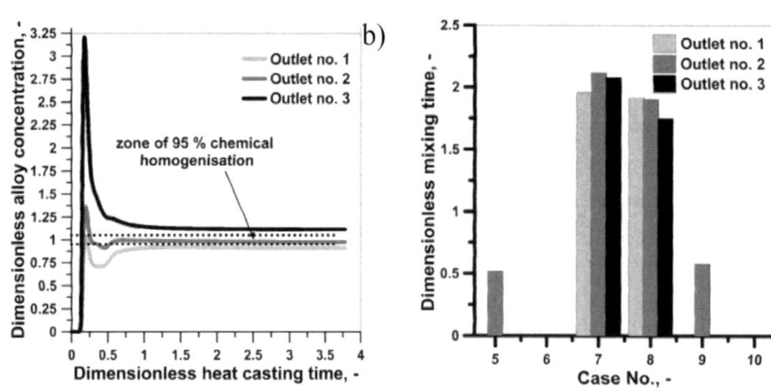

Figure 4. Behaviour of nickel in the bulk of liquid steel during casting: a) mixing curve for case No. 5, b)) mixing time for considered tundishes and AAFPs

IOP Conf. Series: Materials Science and Engineering **529** (2019) 012070 doi:10.1088/1757-899X/529/1/012070

4.4. Behaviour of lighter alloy in the bulk of liquid steel

This section provides results for the behaviour of aluminium in the liquid steel. Due to its density being much lower than that of steel, aluminium is an extremely attractive constructional material. However, as against the ladle furnace, in the tundish, only the supply stream energy can stimulate the process of alloy addition spread in the volume of steel. Therefore, the success of the chemical homogenisation process will be determined by the hydrodynamic pattern created by the supply stream and the FCD. Figure 5a shows the curves of mixing aluminium with steel. The hydrodynamic patterns existing in the tundish working space do not allow the efficient spread of the aluminium being lighter than steel. It can be seen that a batch of aluminium poorly disperses in the steel volume, and the aluminium concentrations recorded at individual nozzles are by almost 100% higher than the nickel contents. For this reason, the process of homogenisation of aluminium concentration was considerably extended. Like for nickel, also for aluminium the most favourable equipment turned out to be TFCD variant 3, and the alloy addition locations, 1 and 4 (Figure 5b). That being said, it should be noted that the mixing time for aluminium is near to the time of casting the entire heat.

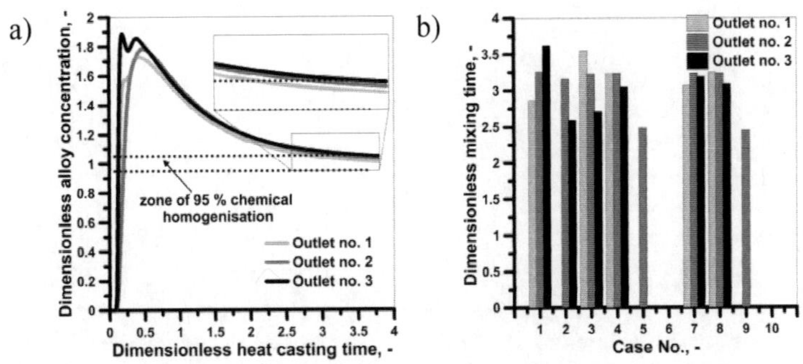

Figure 5. Behaviour of aluminium in the bulk of liquid steel during casting: a) mixing curve for case No. 1, b) mixing time for considered tundishes and AAFPs

5. Conclusions

Based on the computer simulations carried out, it has been found that:

- Reducing the intensity of liquid steel flow through the use of the STC causes localized modifications to the hydrodynamic pattern, whereby obtaining the 95% chemical homogenisation level at individual tundish nozzles becomes impossible.
- For the three-nozzle tundish, the optimal hydrodynamic conditions for the steel alloying process were achieved by increasing the dam height to 0.25 m.
- The most advantageous mode of feeding the alloy addition to the steel when casting square blooms is introducing it simultaneously in two places localized at the side walls of the tundish. This variant allows similar mixing times to be attained at individual tundish nozzles.
- In the case of an alloy addition in the form of a metal much lighter than steel, the mixing time reaches values close to the heat casting time, which will surely contribute to the chemical non-uniform of square blooms being cast.

References

[1] Campbell F C 2008 *Elements of Metallurgy and Engineering Alloys* (Ohio: ASM International)
[2] Mazumdar D, Dhandapani P, Sarvanakumar R 2017 *ISIJ Int.* **57** 286
[3] Van Ende M A and Jung I H 2017 *Metal. Mater. Trans. B* **48** 28
[4] Zhang L and Oeters F 1999 *Melting and mixing of alloying agents in steel melts* (Düsseldorf: Verlag Stahleisen GmbH)
[5] Derda W, Lawenda P, Pardela P 2015 *Arch. Metall. Mater.* **60** 281
[6] Zimmer A, Niedersberg Correia Lima A, Trommer R M, Braganca S R, Bergmann C P 2008 *J. Iron Steel Res. Int.* **15** 11
[7] Rodriguez-Avila J, Morales R D, Najera-Bastida A 2012 *ISIJ Int.* **52** 814
[8] Dong Q, Zhang J, Yin Y, Wang B 2017 *Metals* **7** 209

IOP Conf. Series: Materials Science and Engineering **529** (2019) 012070 doi:10.1088/1757-899X/529/1/012070

[9] Kong Y, Chen D, Liu Q, Chen H, Long M 2018 *Steel Res. Int.* **89** 1800091
[10] Ji C, Luo S, Zhu M 2014 *ISIJ Int.* **54** 504
[11] Cwudziński A 2015 *Steel Res. Int.* **86** 972
[12] Cwudziński A 2015 *Ironmak. Steelmak.* **42** 373
[13] Cwudziński A 2015 *Metall. Res. Technol.* **112** 308
[14] Cwudziński A 2013 *Arch. Metall. Mater.* **58** 1077
[15] Cwudziński A and Jowsa J 2019 *Metall. Ital.* **111** 20

Acknowledgments
This scientific work has been financed from the resources of National Science Centre, Poland in the years 2017-2019 as the Research Project No. 2016/23/B/ST8/01135

Evaluation of AHSS concepts with a focus on the product properties and appropriate casting characteristics of *Arvedi ESP thin slab casters*

C. Bernhard[1], **B. Linzer**[2], **P. Presoly**[1], **I. Watzinger**[2], **J. Watzinger**[2]

[1] Montanuniversitaet Leoben, Franz-Josef-Strasse 18, 8700 Leoben, Austria
[2] Primetals Technologies Austria GmbH, Turmstraße 44, 4031 Linz, Austria

Abstract. The Arvedi ESP process and a variety of produced materials have been continuously developed since the opening of the Arvedi ESP plant in Cremona in 2009 to meet market demands for more sophisticated steel grades. The development of grades for more advanced applications such as advanced high strength steels and multiphase grades is of interest. Dual phase grades such as DP600 are already produced through an ESP line on an industrial scale; additional multi-phase grades such as TRIP are under development. High-strength steels for the automotive industry have especially high demands on material properties. In addition to the mechanical material properties, an excellent surface quality is required. The fundamental basis for such material properties on rolled coils needs to be provided from continuous casting. This paper deals with the classification of different – either Si- or Al-based – alloying concepts for TRIP steels with respect to their prospective behaviour in a thin slab caster.

1. Introduction

Advanced high-strength steels (AHSS) are the most cost-effective lightweight alternative for mass produced vehicles. The global demand for AHSS is expected to increase by nearly 100 % between 2017 and 2025 according to estimates [1]. In cars, TRIP steels are mainly used for structural and reinforcement parts (e.g., B-pillar uppers, roof rails, and front rails), typically in strip gauges of 1.2 – 2 mm and are mostly produced via cold rolling and subsequent hot-dip galvanizing or continuous annealing.

Providing satisfying shape tolerances and microstructure homogeneity in hot rolled strips with thickness of 1.5 mm offers a cost-efficient alternative to cold-rolled TRIP steel on the market. Among the commercialized thin slab casting and direct rolling (TSDR) concepts, the Arvedi endless strip production (ESP) process has the highest potential to stably produce AHSS steel grades.

Steel grades such as DP 600 and DP 800 are commonly used and are frequently provided by TSDR processes. Manufacturing such products on an ESP line confirmed that the unique stable production conditions offered by the endless process are superior to those of batch processes [25].

Supporting the advantages of the Arvedi-ESP-process, **Figure 1** shows an excerpt of a DP 600 production sequence. The temperature control on the run-out table can be adjusted in a very accurate manner to obtain the correct microstructure as defined by phase selection and grain sizes and can be controlled afterwards for several hours with very minor deviations. The separation between the coils is achieved by using shear directly in front of the down coilers, which has no influence on the entire upstream process. With such production conditions, the pass schedules for the rolling mill part can be selected in a more optimized way to allow for products with width, thickness, shape and flatness comparable to conventionally cold rolled material. Threading and tailing out is eliminated, and the rolling parameters show similar stability characteristics to the given examples regarding temperature, which is in accordance with the typically higher requirements for the process conditions and narrowed production parameter fields of the steel grades of interest.

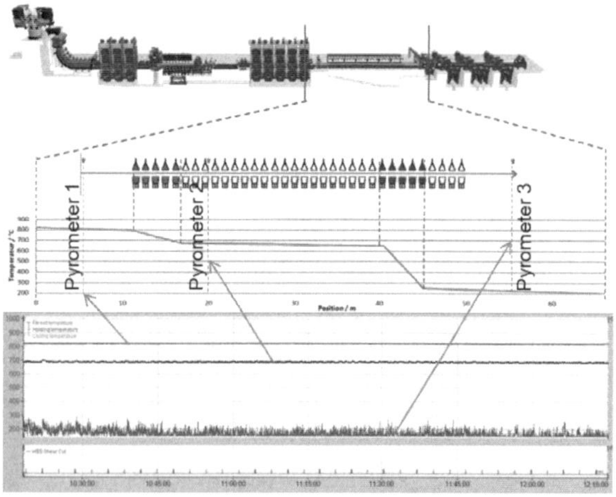

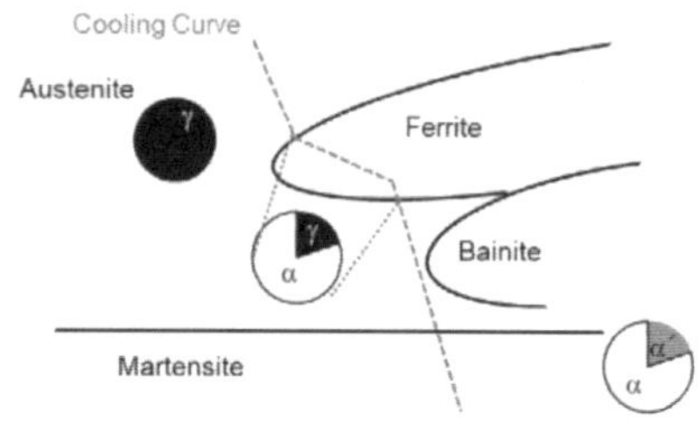

Stability of Temperature Profile:
(Stand. dev. during 2 hours of production)

End Rolling Temp	~ 1.4 K
Holding Temp	~ 2.4 K
Coiling Temp	~ 22 K (within Leidenfrost range)

Figure 1: Cooling pattern and temperature readings of pyrometers at the run-out table area during a two-hour production sequence (approx. 700 t) of DP 600 steel from an ESP line.

The present work focusses on the evaluation of common TRIP steel concepts with regard to the casting characteristics in an Arvedi ESP thin slab caster.

2. TRIP steels: Common alloying concepts

TRIP steels typically consist of a multiphase microstructure with a ferritic matrix, bainite, retained austenite and some martensite. The TRIP effect arises during plastic deformation and results in an outstanding combination of strength and total elongation.

The TRIP microstructure can either be achieved after hot rolling or for cold rolled strip, after continuous annealing or hot dip galvanizing. In hot rolling, the multiphase microstructure develops after the austenitic rolling by multistage cooling at the run-out table and coiler, which is rather difficult to control [e.g., 6]. Continuous annealing and hot dip galvanizing provide a multiphase microstructure through a two-stage heat treatment, including intercritical annealing above the Ar1 temperature to adjust a certain volume fraction of ferrite and, after accelerated cooling to the isothermal holding temperature, the partial transformation of austenite to bainite, which provides the partitioning of carbon and thus leads to the stabilization of the retained austenite at room temperature.

The classic alloying concept of TRIP steels contains 0.2 wt.% C, 1.0-2.0 wt.% Si and 1.5-2.0 wt.% Mn,. These steels are henceforth referred to as "*high-Si concept*" [2-5]. The main role of Si in this concept is to suppress the precipitation of carbides from the austenite. In addition, Si contributes to the strength of TRIP steels by solid-solution hardening. In contrast to the positive effects of Si, Si is also known to cause inner oxidation at elevated temperatures and causes hot-dip galvanizing problems [13].

The replacement of Si by other elements is thus an important issue in TRIP steel development, and Al is widely used today [8-14]. Al is a strong ferrite forming element but cannot fully replace Si, as Si more effectively suppresses cementite precipitation. In addition, Al does not significantly contribute to solution strengthening. Hence, low Si - "*high-Al concepts*" with typically 1.2 – 1.7 wt.% Al often contain microalloying elements such as Nb and V to increase the strength from precipitation hardening [e.g., 8, 9]. The main problem in the production of a steel with such a high Al content via the continuous casting

process is the interaction between the steel and mould flux [19-22]. The reduction of chemically less stable oxides at the steel/flux interface by the dissolved Al and the separation of Al_2O_3 from the steel into in the flux result in a significant increase in the Al_2O_3 content in the mould flux with increasing casting time. These effects cause steadily increasing viscosity and poor lubrication between the strand and the mould and may ultimately result in break-outs and deterioration of the surface quality.
An alternative ferrite stabilizer is phosphorus. P is highly effective at suppressing carbide precipitation and contributes to strength via solute solution strengthening. Thus far, TRIP steel concepts with up to 0.1 wt.% P have been investigated [10, 11, 13, 14]. In casting, phosphorus is a rather critical element; P is one of the strongest segregating elements and may cause problems with internal cracking [15]. Additionally, grain boundary segregation may cause embrittlement at lower temperatures and during annealing.
In summary, all available ferrite stabilizers have associated problems in the production route: high Si content is critical for galvanizing and heat treatment, whereas high Al and P contents are critical to the casting process. Therefore, a common strategy is to combine reduced Si and Al contents to achieve the TRIP microstructure. Si contents of approximately 0.5 wt.% and Al contents of 0.8 – 1.0 wt.% are typical for the "*Si-Al concepts*". The addition of microalloying elements such as Nb has also been proposed [8, 9].
With respect to the behaviour of TRIP concepts in a thin slab caster, three concepts were selected for detailed investigations, the chemical compositions of which are given in **Table 1**:

- a high-Si concept similar to [6] but without Nb and Ti,
- a Si-Al concept slightly modified from [8],
- and a high-Al concept according to [12].

Samples with a weight of 300 g were melted in a Linn High Therm Lifumat high-frequency re-melting and fusion unit The melts were prepared under argon atmosphere from high-purity raw-materials in an alumina crucible. The samples were characterized by excellent homogeneity due to rapid solidification in a copper mould and thus provided perfect testing materials for thermal analysis and all other laboratory experiments.
Several aspects of the prospective behaviour of the three alloying concepts for thin slab casting were investigated in experiments and theoretical considerations including:

- the phase transformation sequence during solidification to determine whether the alloying concept is hypo-peritectic or hyper-peritectic,
- the surface scale formation and tendency towards inner oxidation and the resulting risk for the formation of surface defects, and
- the sensitivity to the formation of internal defects and the homogeneity of the semi-finished product.

The results are discussed in detail in the following sections.

Table 1. TRIP steel concepts from literature used for further investigation.

No.	Wt.%C	Wt.%Si	Wt.%Mn	Wt.%Al	Others all in wt.%
High-Si concept	0.18	1.47	1.60	0.033	each element< 0.01 %
Si-Al concept	0.20	0.50	1.46	0.68	
High-Al concept	0.15	0.02	1.52	1.40	

3. Phase transformations during solidification in equilibrium

It is well-known that hypo-peritectic steels are extremely prone to the formation of defects in continuous casting [e.g., 18]: The uneven initial solidification due to the shrinkage induced by the phase transformation from ferrite to austenite immediately next to the solidus results in unsteady heat transfer in the meniscus region and mould level fluctuations. In addition, hypo-peritectics show a tendency to form deep oscillation marks. In consequence, peritectic steels are highly sensitive to the formation of internal defects and surface defects in the casting process. The number of countermeasures is limited and among them, the use of mould fluxes with low viscosity and high break temperature to reduce the heat transfer close to the meniscus is the most promising [18]. Due to these numerous casting problems, peritectic steels are generally considered unfavourable for high-speed casting in TSDR plants.

Thermal analysis by differential thermal analysis (DTA) and differential scanning calorimetry (DSC) were the most effective methods to accurately characterize the phase transformation sequence in the high temperature range. The DTA/DSC signal-based classification of Fe-alloys as hypo-peritectic, hyper-peritectic or low carbon steel correlates well with plant data, although the results apply only to equilibrium [16]. In the plant, hypo-peritectics are easy to identify by means of the measured temperature in the mould and the determined temperature variation coefficient [18]. A further finding from extensive DTA/DSC measurements was the apparent inaccuracy of commercial databases for the Fe-C-Si-Mn system [17] and for elevated Al content in the Fe-C-Mn-Al system, both of the highest importance for TRIP steel grades.

The available DTA/DSC-data was used to locally assess the SGTE alloy database for the system Fe-C-Mn-Si-Al [17]. This database excerpt was used to predict the influences of Si and Al on the hypo-peritectic range depending on carbon and manganese contents, as shown in **Figures 2a and 2b**. The classic high-Si concepts with 0.2 wt.% C are clearly located in the hyper-peritectic range, which also applies to the Si-Al concept. In general, the role of Si as a ferrite former in solidification of Fe-alloys is widely overestimated. Al is by far more effective, and for 0.2 wt.% C, an excess of approximately 1.3 wt.% Al results in hypo-peritectic solidification. Lowering the carbon content is, with respect to the castability, also a critical measure; the high-Al alloy is explicitly a hypo-peritectic steel and is unfavourable for thin slab casting. The role of P was not explicitly addressed in these calculations.

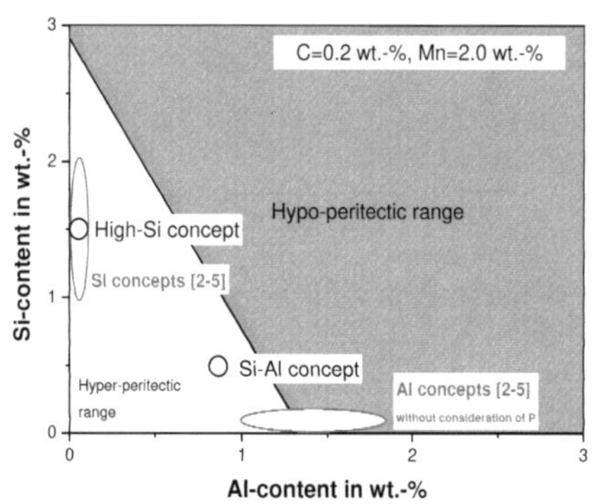

Figure 2a. Hypo-peritectic range for 0.2 wt.% C and 2.0 wt.% Mn depending on Al and Si contents.

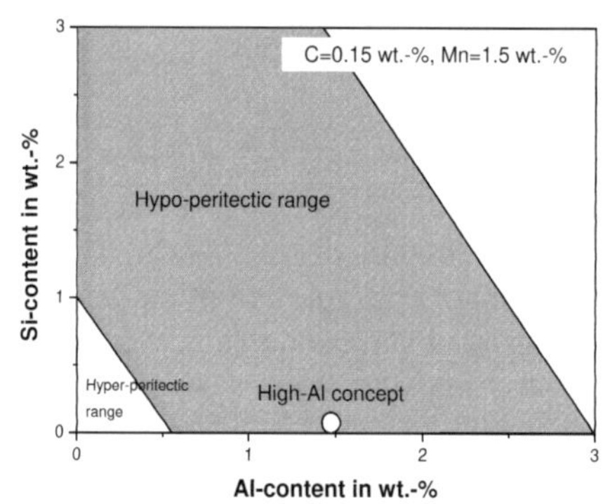

Figure 2b. Hypo-peritectic range for 0.15 wt.% C and 1.5 wt.% Mn depending on Al and Si contents.

Another key aspect for the casting of TRIP steels in a thin slab caster that is not thoroughly described here is the mould flux. In the casting of high-Al steel grades, SiO_2 and other less stable oxides are reduced by the dissolved Al in the steel, and the Al_2O_3 content in the flux accumulates [e.g., 19]. This effect results in increasing viscosity, decreasing mould flux consumption and a change in the primary crystallizing phase. The final consequences are unstable heat extraction, stronger slag rim, lower slag consumption, uneven shell growth resulting in poor surface quality and the occurrence of break-outs.

Plant observations report an increase in Al_2O_3 in the flux to between 15 and 35 wt.% depending on the Al content of the steel and the mould flux concept [20 - 22]. The Al_2O_3 accumulation rate depends on the dissolved Al content as long as the Al content is below 2 wt.%. The accumulation rate decreases significantly as soon as the Al content decreases to 0.5 wt.%.

IOP Conf. Series: Materials Science and Engineering **529** (2019) 012071 doi:10.1088/1757-899X/529/1/012071

With respect to sufficient lubrication in the mould, maximum Al contents between 0.5 and 0.8 % should not be exceeded to prevent an uncontrollable increase of Al_2O_3 in the mould flux.

4. Surface oxidation and surface defect formation

High-Si AHSS steels are well known to form highly adherent surface scales during casting and further processing. The main reason for these scales is the formation of a low melting FeO-2FeO·SiO_2 eutectic with a melting point below 1200 °C [24]. In Si-Al-AHSS concepts, the liquid phases close to the interface consist of Al_2O_3-SiO_2-FeO_n [25]. Even high-pressure hydraulic descaling has difficulty removing this oxide layer from the surface of the cast product. An additional phenomenon is selective oxidation along the grain boundaries and inner oxidation, i.e., the formation of oxides at the steel/scale interface. The high-Si TRIP concepts must therefore be critically evaluated with respect to the achievable surface quality of the hot rolled strip.

The oxidation of the three alloying concepts with the composition given in **Table 1** was simulated in a Netzsch STA 409 PG under simplified conditions, as shown in **Figure 3a**. The samples were heated under Ar-6.0 flushing, and after reaching a temperature of 1200 °C and temperature equalization, air was supplied to the furnace for 2 or 10 minutes before switching back to Ar 6.0 protection during cooling. Two thermal cycles were applied: cooling from 1200 to 1100 °C in either 2 minutes or 10 minutes, simulating the cooling in a thin slab caster ("ESP-cycle") or a conventional caster ("CC cycle"). The results give only a qualitative indication of the behaviour of the three alloying concepts with respect to scale formation and inner oxidation.

Figure 3b shows the final mass change of the samples due to the oxidation. The oxidation seems to mainly depend on the Si content of the sample, and the results confirm the role of Al as an inhibitor for oxidation [24]. The oxidation after 2 minutes is by far lower than that for 10 minutes of oxidation.

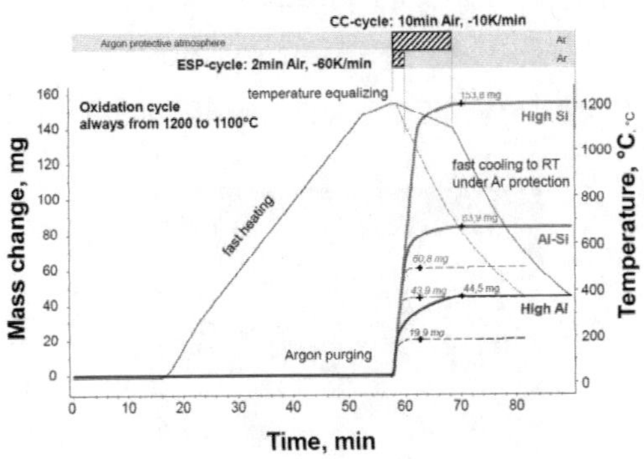

Figure 3a. Thermal cycle and mass change vs. time for the performed measurements

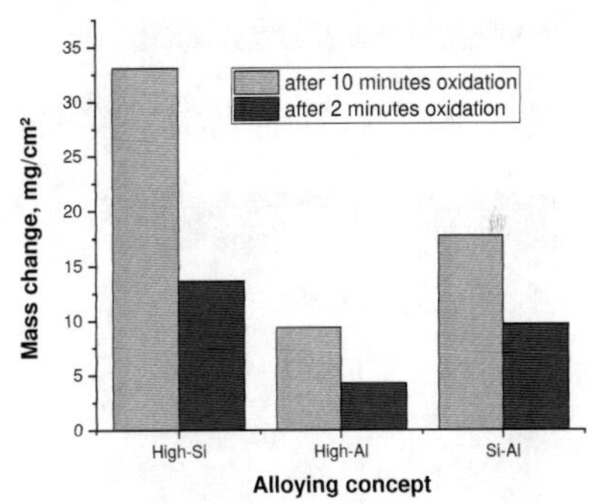

Figure 3b. Mass change of the sample after the oxidation experiments

Figures 4a-4f summarize the SEM results for the six measurements; the left side shows results after two minutes of oxidation, and the right side shows results after 10 minutes of oxidation. The magnification is identical for each of the micrographs. For the high Si concept, deep penetration of a mixture of SiO_2 and FeO/Fe_2O_3 along the former austenite grain boundaries becomes clearly visible. In the scale/bulk-interface, SiO_2/MnO particles nucleate. The scale layer closest to the interface consists of a FeO-2FeO·SiO_2 mixture with characteristics of a formerly liquefied layer. The oxide layer adheres to the interface after the experiment.

The oxidation of the high-Al samples results in the formation of an Al_2O_3/MnO/FeO_n layer surrounding the former austenite grains. The oxide particles in the bulk have a similar composition. The former austenite grains are very small in diameter, indicating the strong pinning effect of the AlN precipitations. The thickness of the interface layer increases over time.

For the Si-Al concept, the oxide particles in the interface layer consist mainly of Al_2O_3 and MnO. It is surprising that no significant enrichment of Al or Si along the former grain boundaries is detectable and that the whole interface is nearly free from any significant enrichment of Si. Only the first oxide scale layers contain a significant amount of SiO_2.

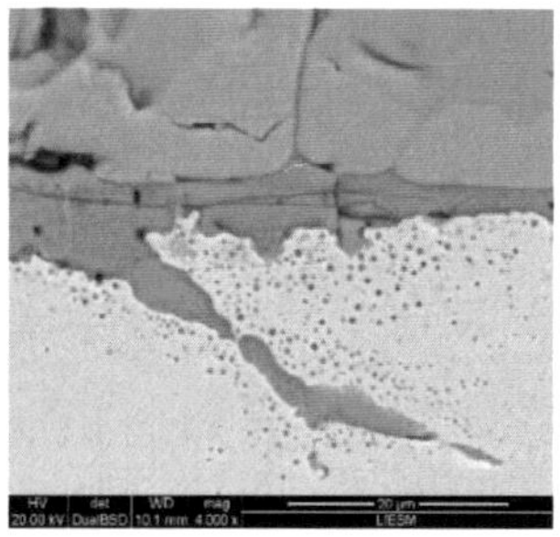

Figure 4a. High-Si concept, 2 minutes

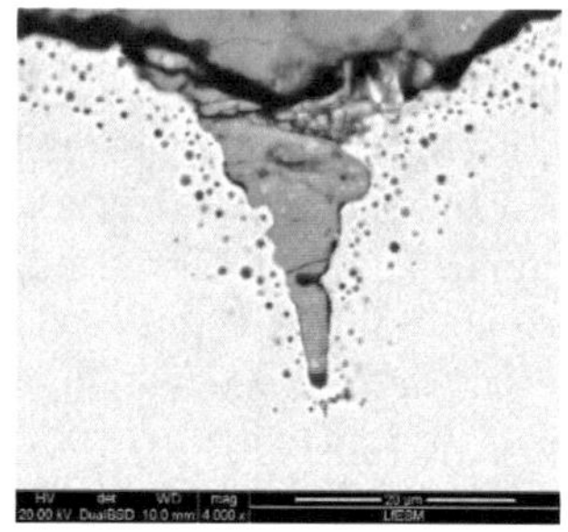

Figure 4b. High-Si concept, 10 minutes

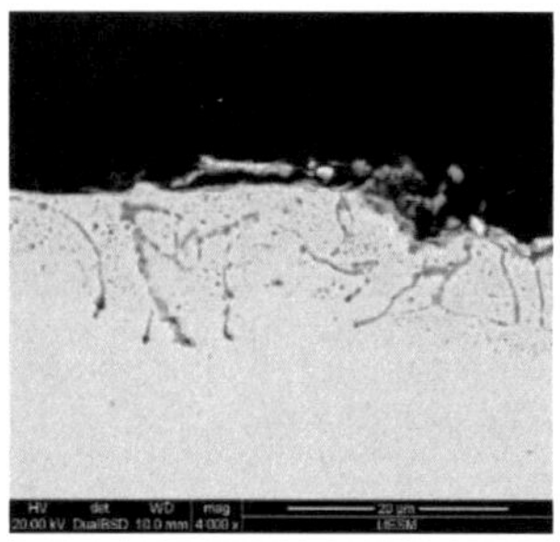

Figure 4c. High-Al concept, 2 minutes

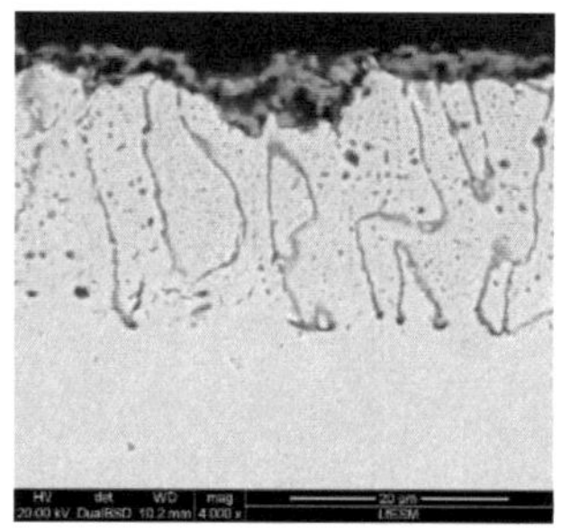

Figure 4d. High-Al concept, 10 minutes

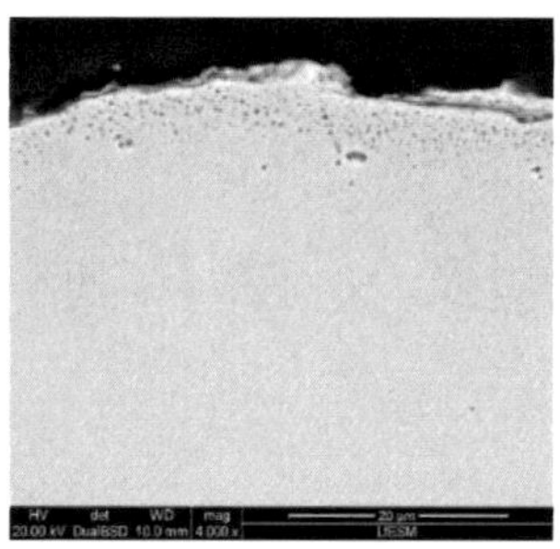

Figure 4e. Si-Al concept, 2 minutes

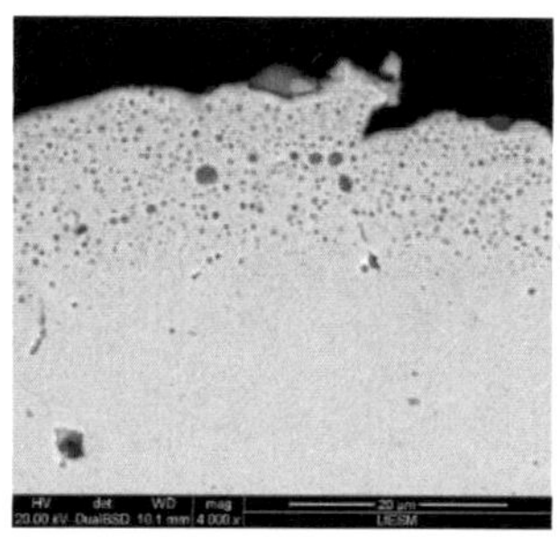

Figure 4f. Si-Al concept, 10 minutes

The adherence of the oxide layer after the experiment is very low for the high-Al and Si-Al-alloys; the oxide layer dropped off the samples from only handling after removing the samples from the TG.
Even if the results provide only qualitative indications, the results confirm that a high-Si concept is critical for TSDR plants due to the formation of a highly adherent scale and a deep penetration of the interface by oxide layers. The lower time at higher temperature for the "ESP-cycle" (2 minutes of oxidation) greatly improves the oxidation. The Si-Al concept seems to receive the most benefit, and the risk of the formation of surface defects due to the rolling in of adhered scale or oxide-penetrated grain boundaries seems to be rather low.

5. Internal defects and homogeneity

Regarding the formation of internal defects such as hot tear segregations, microsegregation is a key factor. The most common approach for predicting these kinds of defects in continuous casting is a strain criterion. The deformation of the strand by bending, straightening and bulging between the supporting rolls causes tensile strain in the two-phase liquid/solid region. Under the assumption of columnar growth of dendrites, which is a given for high-speed thin slab casting of steel, the straining by every pair of rolls is accumulated in a so-called *critical temperature range* (CTR). For the CTR, the temperature ranges between the defined solid fractions and solid fractions of 0.8 (or 0.9) and 1.0 (or 0.99) are frequently used as the upper and lower limits [e.g., 27]. For the following hot tearing analysis, the range of preferred strain accumulation (ΔT_A) is defined as fs = 0.96 – 1.0 [26]. The corresponding solidus temperatures are calculated with a numerical solution of Ohnaka's microsegregation model in ChemApp under the assumption of local (temperature dependant) equilibrium partition coefficients from the optimized FactSAGE Fe-C-Mn-Si-Al database [28]. The resultant temperatures and ΔT_A are listed in **Table 2**. Together with the three selected samples, the Al-P concept according to [13] was analysed. In addition to carbon and manganese, Si has a noticeable decreasing effect on the solidus temperature, at least in the high-Si concept. The influence of Al on the solidus is negligible, whereas P has a strong decreasing influence on the solidus as expected.

Table 2. Solidus temperatures for TRIP steel concepts, cooling rate 1 K/s.

No.	$T(f_s = 0.96)$ °C	$T(f_s = 1.0)$ °C	ΔT_A
High-Si concept	1432.3	1418.9	13.4
Si-Al concept	1453.1	1443.1	10.0
High-Al concept	1478.4	1470.1	8.3
Al-P concept [13]	1448.9	1427.7	21.2

The analysis of the hot tearing bases in the numerical simulation of solidification in a thin slab caster (casting speed 5 m/min, cooling for medium carbon steel, thickness in run-out area is 95 mm). The strain rate caused by bulging, liquid core reduction and straightening is predicted by a frequently used empirical formula [e.g., 27] and accumulated in the temperature range of preferred strain accumulation ΔT_A. The solidification is considered by the shell growth velocity of the solidus isotherm R_S and the temperature gradient between the liquidus and solidus G_{LS}. **Formula 1** summarizes this simplified approach for a strain rate-based hot tearing criterion. The critical strain ε_{crit} is determined from SSCT (submerged split-chill tensile) experiments according to [26]. For the presented analysis, the critical strain was assumed as 1.6 % for the Si and Al concept and as 1.0 % for the Al-P concept.

$$\Delta T_A \cdot \dot{\varepsilon} \cdot (R_S G_{LS})^{-1} < \varepsilon_{crit} \tag{1}$$

Figure 5a shows the calculated temperature of the surface and the temperatures at defined distances from the surface together with the calculated shell growth. **Figure 5b** summarizes the results of the hot tearing analysis: if the ratio between the maximum accumulated strain ε_{accum} and the critical strain ε_{crit} exceeds 1, the formation of hot tears under the assumed conditions is likely, as is the case for the Al-P concept due to the strong microsegregation of P. Among all the considered concepts, the Al concept is the least sensitive to hot tearing, but in general, the investigated steel grades should not be susceptible to hot tearing unless the phosphorus content is increased.

The analysis described above is only one-dimensional, but the results are significant for the general classification of steel grades regarding internal defect formation for a given caster design and defined operation conditions. Tensile strain may also be localized by the deformation of the shell in two dimensions, such as in liquid core reduction (LCR). LCR is a necessity for TSDR technologies when the thickness of the cast thin slab has to be reduced before the first rolling passes. The Arvedi ESP concept is based on the casting of 95 mm strand thickness and LCR of only a few millimetres, as the high hot forming potential of the high reduction mill makes higher reductions within the caster obsolete. This result is another way that Arvedi ESP presents excellent prerequisites for the casting of AHSS grades.

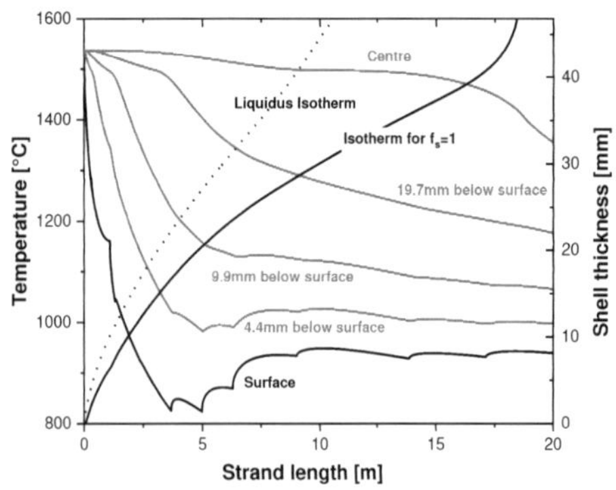

Figure 5a. Shell growth for thin slab caster, 5 m/min, 95 mm slab thickness.

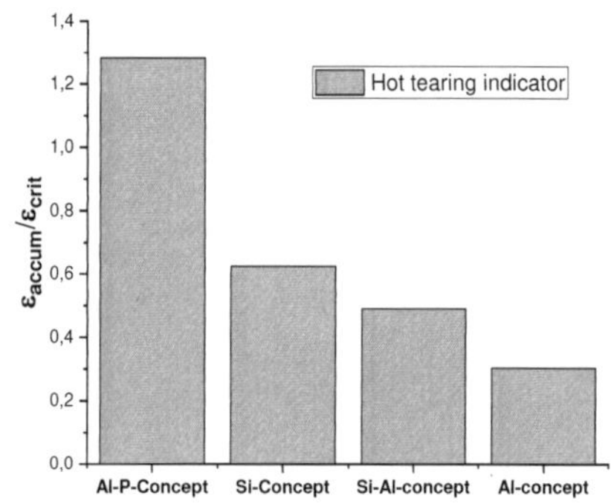

Figure 5b. Hot tearing indicator $\varepsilon_{accum}/\varepsilon_{crit}$ for the four selected steel grades.

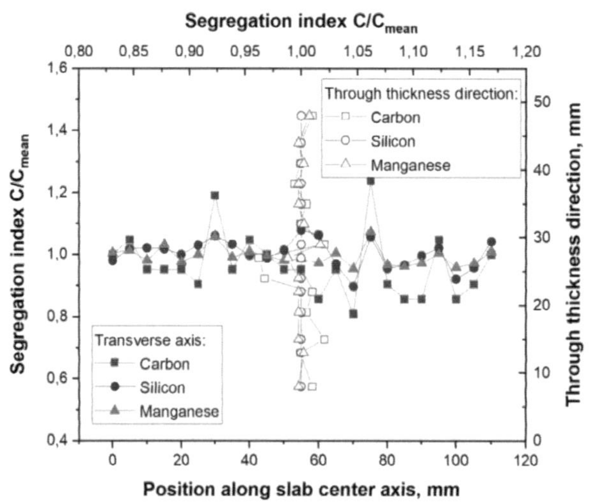

Figure 6. Local segregation index C/C_{mean} along centre axis and across the thickness of a thin slab.

The last aspect that is addressed by this study is the enrichment of alloying elements at the macroscopic scale. The demands on the homogeneity of the strip after hot rolling are high, and the prevention of pronounced centre segregation and a resulting banded structure in the hot rolled strip must be considered crucial. **Figure 6** presents the OES-measured concentrations of carbon, silicon and manganese for a thin slab from a plant trial (0.2 wt.% carbon and 2.1 wt.% Mn steel; 5 m/min). The measurement was performed in the axial direction and the through thickness direction, and the absolute values are given as the ratio between the local concentration and mean concentration (segregation index). The results show excellent Mn homogeneity with a minimum segregation index of 0.95 and a maximum of 1.07. The low diffusivity of Mn is a key factor for the homogeneity of the final product; neither holding at high temperature nor hot forming with the highest deformation significantly contributes to the homogenization of Mn, whereas the carbon concentration peaks are compensated during hot rolling due to the high diffusivity of carbon [29]. The ESP Arvedi thin slab therefore is an excellent material for the production of a homogeneous strip in the subsequent rolling process.

6. Conclusions

In the present work, three alloying concepts for TRIP steels were tested regarding their prospective behaviour in high-speed thin slab casting on an Arvedi ESP caster: a high-Al-grade, a high-Si-grade and a mixed Si-Al concept. The results showed that

- the *high-Al concept* is unsuitable due to the tendency towards hypo-peritectic solidification and the intensive chemical interaction with the mould fluxes. The oxidation of the high-Al-samples at temperatures between 1100 and 1200 °C results in grain boundary oxidation, which is a presumable reason for the formation of surface defects in the rolling process. The formation of internal cracks and inhomogeneity should be controllable due to the almost negligible enrichment of Al during solidification.
- The *high-Si concept* is hyper-peritectic and hence rather insensitive regarding the solidification in the mould. The oxidation experiments demonstrate the formation of a partly liquefied, adherent oxide layer, consisting of fayalite and associated with internal oxidation and selective oxidation along grain boundaries. This result could be an exclusion criterion for high-Si concepts. Si is a moderately segregating element and slightly increases the risk of the formation of internal cracks.
- The homogeneity of samples from plant trials is excellent, and the segregation indexes of Mn and Si close to the centre of the thin slab are almost equal to 1.
- *Phosphorus* is a rather unsuitable element to replace Si or Al as a ferrite former; the strong segregation results in a high risk for the formation of hot tears or overcritical centre segregations.
- The mixed *Si-Al concepts* show the highest potential among the investigated steel grades; hyper-peritectic solidification and a controllable interaction with the mould flux, the lowest tendencies towards inner oxidation and selective oxidation along grain boundaries and finally a low tendency towards the formation of internal defects provide an excellent basis for the further development of TRIP concepts for casting in thin slab casters.

7. Acknowledgements

Financial support by the Austrian Federal Government (in particular from Bundesministerium für Verkehr, Innovation und Technologie and Bundesministerium für Wirtschaft, Familie und Jugend) represented by Österreichische Forschungsförderungsgesellschaft mbH and the Styrian and the Tyrolean Provincial Government, represented by Steirische Wirtschaftsförderungsgesellschaft mbH and Standortagentur Tirol, within the framework of the COMET Funding Programme is gratefully acknowledged. This work is part of the ongoing K2-MPPE project P3.36 "Advanced ESP".

8. References

[1] Harris, T 2017 *SSAB Capital Markets Day* (www.SSAB.com)
[2] Ryu H B, Speer J G and J P Wise 2002 *Met. Mat. Trans. A* **33A** 2811
[3] Wu D, Zhuang L and H Lü 2008 *J. of Iron and Steel Research Int.* **15**(2) 65
[4] Zhang Z, Zhu F and Y Li 2010 *J. of Iron and Steel Research Int.* **17**(7) 44
[5] Zhang Y, Ma Y, Kang Y and H Yu 2006 *J. of Univ. Sci. Tech. Beijing* **13** 476

[6] Liu J, Zhang Z, Zhu F, Li Y and K Manabe 2012 *J. of Iron and Steel Research Int.* **19**(1) 43
[7] Timokhina I B, Pereloma E V and P D Hodgson 2001 Mat. Sci. Tech. **17** 135
[8] Fu B, Yang, W Y, Lu M Y, Feng Q, Li L F and Z Q Sun 2012 *Mat. Sci. and Eng. A* **536** 265
[9] Feng Q, Li L, Yang W and Z Sun 2014 *Mat. Sci. and Eng. A* **605** 14
[10] Hou X, Xu Y, Zhao Y and D Wu 2011 *J. of Iron and Steel Research Int.* **18**(11) 40
[11] Barbe L, Verbeken K and E Wettinck 2006 ISIJ Int. **48** 1251
[12] Chen Y, Wang L, Dong L, Li F and Y Zhang 2007 *Proc. Sino-Swedish Structural Materials Symp.* 308
[13] Ding W, Hedström P and Y Li 2016 *Mat. Sci. Eng.* A **674** 151
[14] Jing C, Suh D W,Oh C S, Wang Z and S J Kim 2007 *Met. and Mat. Int.* **13**(1) 13
[15] Pierer R and C Bernhard 2010 *Proc. AISTech* 193
[16] Presoly P, Pierer R and C Bernhard 2012 *Proc. TMS* 299
[17] Presoly P, Bernhard C and R Pierer 2012 *Met. Mat. Trans. A* **44** 5377
[18] Xia G, Bernhard C, Ilie S and C Fürst 2011 *Steel research int.* **82** 230
[19] Presoly P and C Bernhard 2017 *Iron & Steel Technology* **14** 100
[19] Kim M S and Y B Kang 2018 *Calphad* **61** 115
[20] Cho J W, Blazek K, Frazee M, Yin H B, Park J H and S W Moon 2013 ISIJ Int. **53** 62
[21] Harmuth H and G Xia 2014 *Prod. 8th European Cont. Casting Conf.* Graz/Austria
[22] Rudnitzki J, Shepherd R, Balichev E, Karrasch S and F Krueger 2014 *Prod. 8th European Cont. Casting Conf.* Graz/Austria
[23] Sauerhammer B, Senk D, Schmidt E, Safi M, Spiegel M and S Sridhar 2006 *Met. Mat. Trans.* **36B** 503
[24] Takeda M, Onishi T, Nakakubo S and S Fujimoto 2009 *Mat. Trans.* **9** 2242
[25] Bragin S, Großeiber S, Linzer B and A Rimnac, 2016 *Proc. 10th Intern. Rolling Conf. and 7th Europ. Rolling Conf.* Graz/Austria
[26] Bernhard C and R Pierer Proc. 2007 *Proc. 5th Decennial International Conference on Solidification Processing* Sheffield, UK 525
[27] Won Y M, Yeo T J, Seol D J and K H Oh 2000 *Met. Mat. Trans.* **31B** 779
[28] You D, Bernhard C, Wieser G and S Michelic 2016 *Steel research int.* **87** 840
[29] Presslinger H, Ilie S, Schiefermueller A, Pissenberger A, Parteder E and C Bernhard 2006 *ISIJ Int.* **46** 1845

IOP Conf. Series: Materials Science and Engineering **529** (2019) 012072 doi:10.1088/1757-899X/529/1/012072

Effect of different process parameters on non-metallic inclusions during electro-slag remelting of a tempering steel

R Schneider[1], S Schroft[1,2], M Molnar[1,2], G Klösch[3] and C Schüller[3]

[1] Univ. of Appl. Sciences Upper Austria, Stelzhamerstrasse 23, 4600 Wels, Austria
[2] K1-Met GmbH, Stahlstrasse 14, 4020 Linz, Austria
[3] voestalpine Stahl Donawitz GmbH, Kerpelystrasse 199, 8700 Leoben, Austria

stefan.schroft@fh-wels.at

Abstract. The remelting behavior of the tempering steel 50CrMo4, was investigated with several experimental melts on a lab-scale ESR-plant. The investigated parameters included a variation of the slag compositions and the use of a protective nitrogen atmosphere. Variations of the slag composition comprised slags with different contents of CaF_2, CaO and Al_2O_3 as well as a variation of the SiO_2-content. The remelted ingots were forged and analyzed regarding their chemical composition as well as their distribution and composition of the non-metallic inclusions (NMI) by automated SEM-EDX method. The chemical composition of the slag after remelting was analyzed as well. The results clearly show a relationship mainly of Si and Al in the steel with the process parameters. NMI changed in their total amount, type and size distribution. The protective atmosphere reduced the Si-losses during remelting. The majority of the NMI were of the Al_2O_3- & MnS-type. In general, remelting lead to an almost complete removal of sulfides, a reduction of oxisulfides and a shift towards more oxides. The total amount of NMI was most strongly reduced by the high CaF_2-containing remelting slag.

1. Introduction

The electro-slag remelting process was and today is still used to minimize inhomogeneities like segregations or shrink holes and to improve the cleanliness level, in particular reducing the amount of larger non-metallic inclusions (NMI) [1-5]. Thereby the remelting slag has a direct effect on the cleanliness level as well as on the type of inclusions formed after remelting.[1,2,6-10] Furthermore, the slag composition and its basicity has a significant effect on silicon losses.[1,2,9,11]

With rising demand on steel quality, the effect of various remelting parameters on NMI has found renewed interest in recent years. The effect of a slag deoxidation treatment and of a protective atmosphere on a corrosion resistant die steel is described in [5], showing that both procedures can provide improvements when using electrodes with a high oxygen contents. Investigations on a similar additionally Ni-alloyed grade resulted in more, but finer NMI and in a corresponding increase of the oxygen content at higher melt rates.[12] This behavior was explained by differences in the precipitation and growth kinetics. Results from industrial trials on hot work tool steels and a martensitic Cr-Ni-steel in [13,14] demonstrate the strong refining capabilities of especially protective gas ESR. The effect of higher SiO_2 contents in the remelting slag on the formation of NMI is documented in [15] for a hot work tool steel. Thereby larger mixed-type inclusions with higher SiO_2 contents are formed, but, similar to results in [13], MA-spinels form the majority of the inclusions. This confirms also earlier results in [1,9].

Laboratory scale experiments on a die steels in [16] show advantages of low filling ratios, a protective gas (Ar) and a multi-component slag compared to a plain CaF_2-Al_2O_3-slag. Remelting a high-temperature steel, with slags containing up to 10 % MgO lead to a stronger reduction of larger NMI, but the number of small inclusions (< 4 μm) increased when the amount of magnesia was raised [10]. The different effects of slag compositions and a protective N_2-atmosphere on the remelting of a hot-work tool steel was comprehensively demonstrated in [17]. The main effect of protective gas was the elimination of the scaling behavior of the electrode and corresponding Si-losses. The strongest effect of the slag composition was found with a CaO-free almost plain CaF_2-Al_2O_3-slag.

With this work the effects of two different slag compositions and of a protective N_2-atmosphere were investigated with a tempering steel, focusing on changes in the chemical composition, especially Al and Si and on corresponding changes in content, size distribution and chemical composition of the NMI.

2. Experimental

All experiments were conducted on a laboratory scale ESR plant. A more detailed description on the plant configuration and typical operation modes can be found in [17,18].

The continuous cast and rolled tempering steel 50CrMo4 (Mat. No. 1.7228) was used. All electrodes had a diameter of 101 mm and a machined surface. Two different slags with the nominal compositions according to Table 1 were applied. The first slag (3C3A), was applied twice, once in direct contact with air, and once under protective gas (N_2). The other slag (2C2A) was slightly enriched with SiO_2 and had an increased CaF_2-content.

Table 1. Chemical composition of the slags after remelting in wt.%.

Trial, Slag	CaF_2	CaO	Al_2O_3	SiO_2	MgO
3C3A	32	30	33	2	3
3C3A+N_2	32	30	33	2	3
2C2A	54	18	21	5	2

All experiments were performed with a frequency of 4.5Hz. The parameters were selected to realize comparable melt rates, which required the adjustment of the melting current. The voltage is the result of the electrical conductivity of the slag as well as of an immersion depth of about 1 cm (see **Table 2**). No deoxidants were added to the slag during remelting.

Table 2. Process parameters of the remelting experiments.

Trial, Slag	Atmosphere	Slag [kg]	Current [kA]	Voltage [V]	Melt rate [kg/h]
3C3A	air	5	3.1	62 - 64	39
3C3A+N_2	N_2	5	3.3	58 - 62	39
2C2A	air	5	4.0	52 - 54	41

After remelting, the ingots were forged to a diameter of 40mm. Samples for chemical analysis and investigations on NMI were taken from the upper third of the forged ingot. Additionally, the slag was analyzed after remelting. The automated SEM-EDS NMI scans were acquired in standardized fields of approximately 150mm². All detected NMI with an oxygen content of ≥5 wt% and an S:O-ratio of < 0.15 were considered oxides. Inclusions with oxygen and sulfur contents of ≥2 wt% each and an S:O-ratio between 0.15 and 6.67 were categorized as oxisulfides (conglomerates of oxides and sulfides). NMI with a sulfur content of ≥2 wt% and an O:S-ratio of < 0.15 were taken as sulfides.

3. Results

3.1. Changes in the chemical composition during remelting

There were only minor changes of the alloying elements C, Mn, Cr and Mo. More significant changes could be observed at the deoxidants Si and Al as well as in the S-content, which was reduced almost independent from the process parameters from 105 ppm to 20-27 ppm at all remelting trials. The correlation between the Si- and the Al-content is visualized in Figure 1. While remelting under a protective atmosphere showed a stoichiometric exchange reaction between Si, Al and its corresponding oxides, there is a clear loss of Si for open remelting conditions, and in the case of the slag 2C2A also a significant loss in Al.

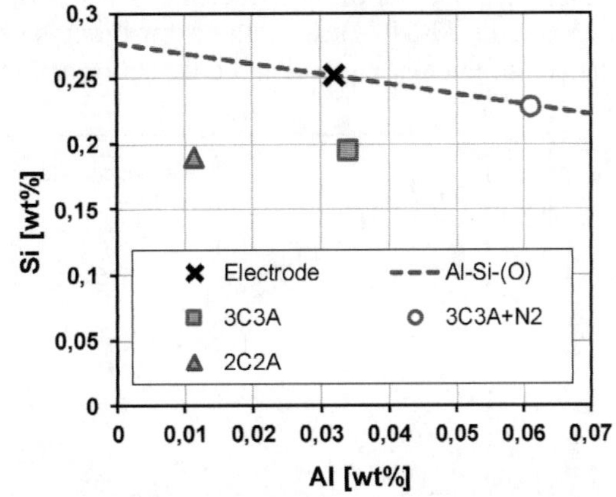

Figure 1. Silicon and aluminium contents prior and after remelting

The Chemical composition of the top slags after remelting are summarized in Table 3. The observed increase in SiO_2 at the open remelting processes correlates well with the Si-losses in the steel. There is also a clear correlation of the protective gas on the FeO content in the slag.

Table 3. Chemical composition of the slags after remelting in wt.%.

Trial, Slag	CaF_2	CaO	Al_2O_3	SiO_2	MgO	FeO
3C3A	32.8	29.0	31.4	2.9	3.9	0.13
3C3A+N_2	32.2	30.3	33.8	1.8	3.7	0.05
2C2A	50.7	19.5	20.4	7.6	1.7	0.09

3.2. Non-metallic inclusions

The specific amount of NMI per mm² and the corresponding area-% of the three main types of *NMI* are shown in Figure 2. Two general trend*s* can be observed. The first is a strong reduction of sulfides and a corresponding increase in oxides. The second is a general trend towards less NMI in total after remelting. Furthermore, it is evident that the best results could be achieved with the CaF_2-rich slag 2C2A.

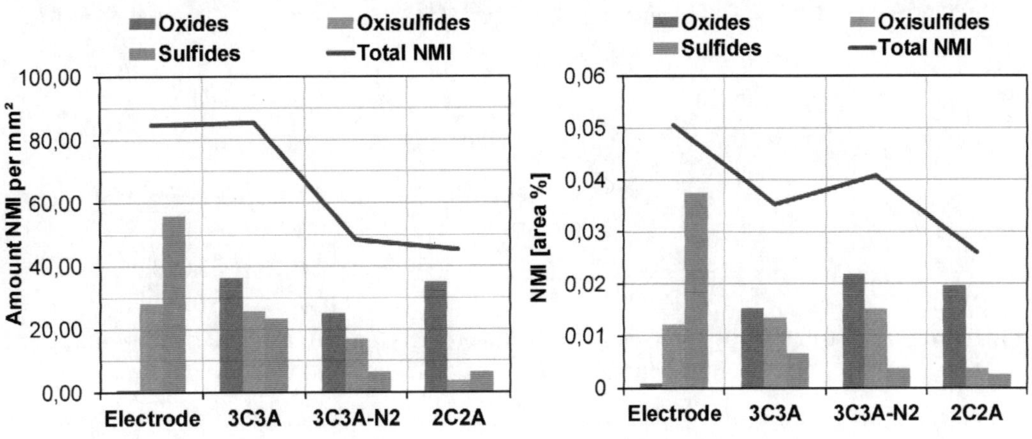

Figure 2. Content of non-metallic inclusion (NMI) left: amount/mm²; right: area %.

There was a strong increase in oxides after remelting for all inclusion sizes. Oxisulfides showed a reduction in small inclusions and an increase of larger inclusions. This effect is more pronounced for the slag C2A2. The sulfides were strongly reduced at all sizes. Combining all three types in Figure 3, there is a clear reduction in almost all inclusions sizes during remelting. This effect is better visible for larger inclusions (> 5 µm), despite the number of these inclusions is anyway quite low. Large inclusions (> 15µm) were completely removed by ESR. Despite the logarithmic sale, small differences between the results are within the accuracy of the test method and only major differences are of significance.

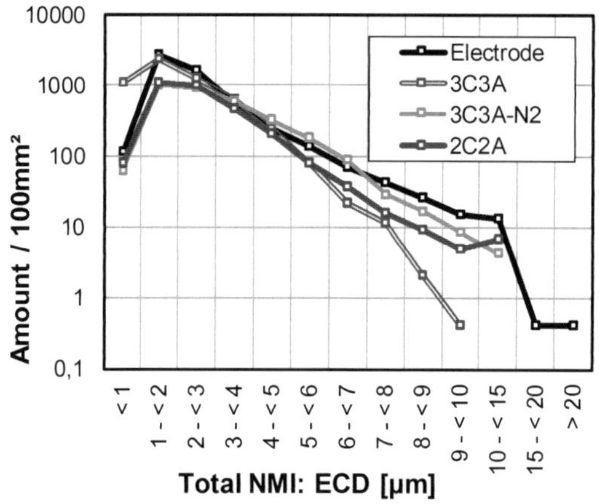

Figure 3. Size distribution of non-metallic inclusion (NMI)

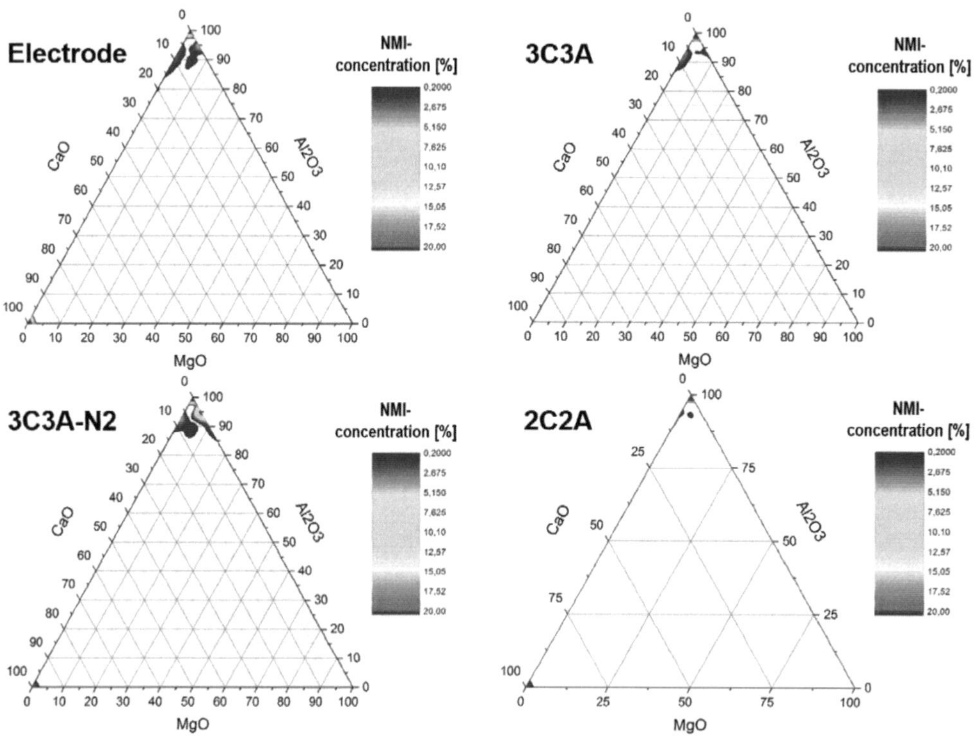

Figure 4. Size distribution of non-metallic inclusion (NMI) top-left: electrode; top-right: 3C3A; bottom-left: 3C3A-N2; bottom-right: 2C2A.

The chemical composition of the NMI is visualized by means of the ternary system Al_2O_3-MgO-CaO in Figure 4. Most of the oxide inclusions, including the oxidic share of the oxisulfides, were allocated to the diagram. The compositions were projected to 100% by keeping their specific shares. White areas represent compositions without NMI. Blue areas correspond to chemical compositions were only low concentrations of inclusions could be detected. On the other hand, red areas designate chemical compositions with a high share of the total inclusions. The chemical composition of the sulfides prior and after remelting was always in the range of pure MnS.

The NMI of the electrode show two different populations, one of pure CaO and the other in the range of high Al_2O_3-contents. After remelting the CaO-type inclusions have largely or completely disappeared. The detailed composition of the high-alumina-type inclusions changes slightly during remelting.

4. Discussion

According to [1,2,8-10,17,19] the loss of Si and the gain of Al in the steel can be partially explained by equation (1).

$$3[Si] + 2(Al_2O_3) \leftrightarrow 3(SiO_2) + 4[Al] \quad (1)$$

This equilibrium condition is the primary reason for the reduced Si- and the increased Al-contents after remelting under the protective atmosphere. This relationship is indicated as a dashed line in Figure 1. In case of the slag 2C2A the SiO_2-content and activity of the slag was increased while the Al_2O_3-content and activity was reduced, shifting the equilibrium to lower Al contents at almost similar Si-values. Both trials operated without protective atmosphere show a loss in total Al+Si and are therefore situated below the equilibrium line due to additional oxidation effects.

As described in [5,16,17] scaling at the electrode surface just above the slag from iron to iron oxide, which is the main reason for an oxygen uptake into the slag when remelting is performed in contact with air. Such uptake of FeO into the slag, results in an additional loss of Si (and/or Al). These oxidation reactions also contribute to the observed increase on SiO_2 and FeO in the slag for the open remelting conditions which can only be prevented by the use of a protective atmosphere. [8,17]

The strong desulfurization in all remelting trials is in agreement with earlier findings in e.g. [1,16,17,20]. A significant difference between open remelting and protective gas operation as described in [1,16,17] could not be observed. This indicates that the desulfurization reaction according to equation (2) did not lead to a saturation of the slag with sulfur and the regeneration of the slag in contact with oxygen from the atmosphere according to equation (3) is not yet dominating the result.

$$2[S] + 2(O^{2-}) \leftrightarrow 2[O] + 2(S^{2-}) \quad (2)$$

$$2(S^{2-}) + 3\{O_2\} \leftrightarrow 2\{SO_2\} + 2(O^{2-}) \quad (3)$$

The reason for this behavior might be found in the relatively small ingot size compared to the slag volume of this laboratory size experiments.

All remelting trials led to strong reductions in sulfides, partially also of oxisulfides and a significant increase in oxides. The reduction in sulfides (and oxisulfides) is directly related to the reduction of sulfur as described before. To some extent, and according to findings in [5,8,12,17], this effect can also be held responsible for the increase in oxides, as oxides often act as nucleus for the sulfide formation leading to mixed NMI. Therefore the determined threshold values for the S:O-ratio between oxides, oxisulfides and sulfides may lead to a general shift in the categorization from oxisulfides to oxides and from sulfides to oxisulfides when the S-content in such mixed NMI is reduced.

The chemical compositions of NMI, both prior and after remelting with and without protective atmosphere, differ from results of other steels as described in [9,13,14,16,17] where MA-spinel-type inclusions are dominating. A positive effect of higher CaF_2-contents on inclusion removal as described in [21] for a Ni-base Superalloy could be confirmed.

In this study the main content of inclusions is < 10μm and, similar to results in [17], can be regarded as newly formed during solidification. Therefore their composition is mainly based on equilibrium condition between aluminum and oxygen in the steel, which is similar to the high-Al_2O_3-type inclusions of the electrode but not a question of survival from the electrode. In contrast CaO-type inclusions were almost completely removed. Such small inclusions cannot be removed into the slag by floatation in the liquid melt pool and become trapped in the advancing solidification front [5,7,8].

5. Summary and Conclusions

Based on these results the following conclusions can be drawn.

Chemical reactions:

- The reactions between the melt (Al, Si) and the slag (Al_2O_3 and SiO_2) follow the corresponding equilibrium reaction.
- Except for remelting under protective atmosphere, all remelting trials showed a loss of Si+Al due to a FeO uptake into the slag by the scaling of the electrode.
- Open ESR leads to higher FeO- and SiO_2-contents in the slag.
- Sulfur was reduced at all remelting conditions.

Non-metallic inclusions:

- The total content of non-metallic inclusions was reduced by all remelting experiments.
- Higher CaF_2-containing slag lead to less NMI and therefore to a better level of cleanliness.
- While the content of oxides rose significantly, oxisulfides were partially reduced and sulfides strongly removed, this may be related to a shift of the S:O-ratio of mixed-NMI.
- Better cleanliness is based on the reduction of sulfides and on the elimination of larger NMI.
- Similar to one of the main populations in NMI-composition of the electrode, the highest content of NMI after remelting was of the high-Al_2O_3-type, with small concentrations of CaO and MgO.

Acknowledgments

The authors wish to thank voestalpine Stahl Donawitz GmbH for the financial, personnel and intellectual support. Likewise, the financial support of the Austrian Research Promotion Agency (FFG) for the "Competence Center for Excellent Technologies in Advanced Metallurgical and Environmental Process Development" (K1-Met) within the frame of the COMET-program is gratefully acknowledged.

References

[1] Holzgruber W and Plöckinger E 1968 *Stahl Eisen* **88** 638
[2] Holzgruber W and Plöckinger E 1968 *Berg- Huettenmaenn. Monatsh.* **113** 83
[3] Plöckinger E 1973 *J. Iron Steel Inst.* **211** 533
[4] Mitchell A 2005 *J. Mater. Sci. Eng.* A **10** 413
[5] Shi C B, Chen X C, Guo H J, Zhu Z J and Ren H 2012 *Steel Res. Int.* **83** 472-86
[6] Mills C K and Keene B J 1981 *Int. Met. Rev.* **1** 21
[7] Mitchell A 1974 *Ironmaking Steelmaking* **1** 172
[8] Mitchell A, Carmona F R and Wei C H 1982 *Iron Steelmaker* **9** 37
[9] Allibert M, Wadier J F and Mitchell A 1978 *Ironmaking Steelmaking* **5** 211
[10] Radwitz S, Scholz H, Friedrich B and Franz H 2015 *Proc. Int. Symp. Liq. Met. Process. Cast.* (Leoben) N71
[11] Miska H and Wahlster M 1973 *Arch. Eisenhuettenwes.* **44** 81
[12] Korp J 2012 *Berg- Huettenmaenn. Monatsh.* **157** 174
[13] Reiter G, Schuetzenhoefer W, Tazreiter A, Martinez C, Wuerzinger P, Loecker C 2013 *LMPC 2013, Proc. Int. Symp. Liq. Met. Process. Cast.* (Austin) p 213
[14] Liu J H, Wang G X, Bao Y P, Yang Y, Yao W and Cui X N 2012 *J. Iron Steel Res. Int.* **19** 1
[15] Schneider R, Schüller C, Würzinger P, Reiter G and Martinez C 2015 *Berg- Huettenmaenn. Monatsh.* **160** 117
[16] Dong Y W, Jiang Z H, Cao Y L, Yu A and Hou D 2014 *Metall. Mater. Trans. B* **45** 1315-24
[17] Schneider R, Molnar M, Gelder S, Reiter G and Martinez C 2018 *Steel Res. Int.* **89**
[18] Schneider R, Paar A, Zeller P, Reiter G, Schützenhöfer W and Würzinger P 2011 *Berg- Huettenmaenn. Monatsh.* **156** 112
[19] Mitchell A 1981 *Can. Metall. Q.* **20** 101
[20] Miska H and Wahlster M 1973 *Arch. Eisenhuettenwes.* **44** 19
[21] Manjili M H and Halali M 2017 *Metall. Mater. Trans. B* **49** 61

IOP Conf. Series: Materials Science and Engineering **529** (2019) 012073 doi:10.1088/1757-899X/529/1/012073

Solidifying shell waviness during continuous casting of AHSS slabs

Hongbin Yin

ArcelorMittal Global R&D East Chicago Center
3001 E. Columbus Dr., East Chicago, IN 46312, USA

Hongbin.yin@arcelormittal.com

Abstract. Transverse cracking of continuously cast products has been encountered at almost every caster operation. Enormous efforts have been carried out in the past aiming at identifying the cause and reducing the problem, especially on steel grades with peritectic chemistries. So far, however, there is still no cost-effective solution with good trade-off for internal quality and productivity. In this study, a new cracking formation mechanism is proposed based on observations of equally spaced crack pattern and the undulation (shell thinning pattern) observed on the inside and/or outside (surface depression) of the breakout shell with similar spacing. This wavy solidification shell forms at initial solidification stage and induces the local shell thinning and reheating which in turn causes the local "Blown grains" inside the solidification shell. As observed on slab surfaces, these blown grains are closely related to transverse cracking problem.

1. Introduction

Enormous efforts have been made to understand the cause of transverse surface cracking in continuously cast products[1]. For steel grades with peritectic chemistries, the problem is normally worse than other grade groups at a typical conventional caster. The basic guidance for resolving the transverse cracking problem is to avoid the hot ductility trough caused by the combination of weak ferrite film and large amount of tiny precipitates formed on austenite grain boundaries (AGB) during cooling to around bending and/or unbending temperature of cast product surface[2]. Slab surface cooling strategies and steel chemistry designs can be applied to avoid and/or to elevate the ductility trough. The secondary guidance is to reduce the depth of oscillation marks which behave like notch for crack initiation. Many studies have shown that transverse cracks initiate at the base of oscillation marks[3]. The third guidance relates to the reducing stress on slab surface in the casting machine. Mold lubrication and caster machine alignment become important[4]. With these guidelines in hands, casting operations around the world are able to reduce the occurrence of the transverse cracking in cast products.

Recently, a new group of steel grades called Advanced High Strength Steels (AHSS) are developed and heavily commercialized due to demands for weight reduction in cars and machineries[5]. Because of special chemistry design (2-3%Mn, 2-3%Si and 1 to 2%Al - unfortunately most of them are peritectic in nature during solidification) and heavy alloying to achieve certain mechanical properties, transverse cracking issue has re-appeared at casters making AHSS grades. Although these cracks in AHSS slabs have similar characteristics to the cracks in traditional HSLA and peritectic steel grades, there are unique features associated with the cracks in AHSS grades. This paper descripts the similarity and uniqueness

IOP Conf. Series: Materials Science and Engineering **529** (2019) 012073 doi:10.1088/1757-899X/529/1/012073

of the transverse cracks in AHSS slabs and tries to propose a new formation mechanism of transverse cracking based on observations of these cracks.

2. Equal spacing of transverse cracking

One of the unique features of transverse cracking in AHSS at one of the ArcelorMittal USA curved mold conventional slab casters is the equal spacing between these cracks. Due to the difficulties of viewing these cracks on as-cast slabs with width of 1.3 m and thickness of 0.24 m, the slab top surface has to be lightly scarfed by hand scarfing, grinding or machine milling to reveal them. One example of a hand scarfed slab surface with multiple transverse cracks is shown on the left in Fig.1 after sand blasting treatment in the area of slab quarter width. The scarfing can normally remove a layer of steel in about 2-3 mm thickness from the as-cast surface, which means these cracks are more than 3 mm deep. The interesting observation here is the roughly equal space at about 32 mm between those cracks. When this slab was made, the casting speed was at 1050 mm/min. This gives the timing interval between cracks at 1.8 s. It should be pointed that the mold oscillation frequency was 120/min, i.e., the timing between cracks is roughly 4 times of the mold oscillation cycle.

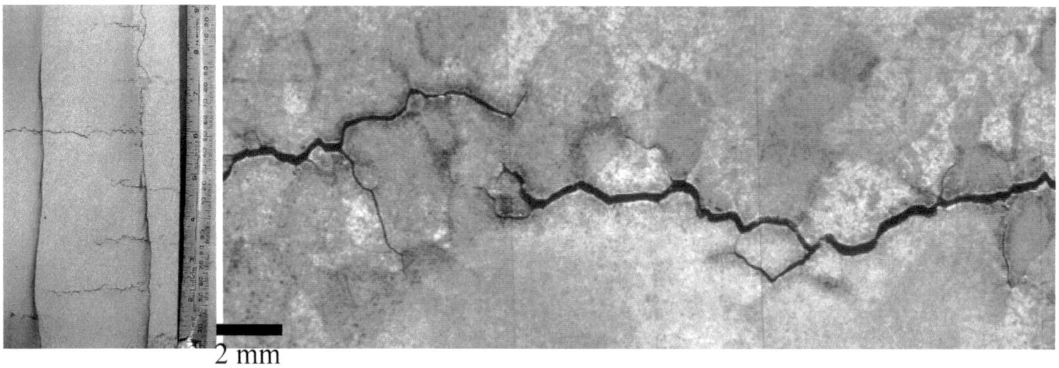

Fig.1 Equally spaced transverse crack exposed after lightly hand scarfed (left) and micro analysis of the crack showing "Blown grain" structure alone ferrite film on prior AGBs (right).

The micro-structure after polishing and etching one of these cracks by removing 2 mm from scarfed surface (to avoid torch heat affected zone) is given on the right in Fig.1. There are two significant features about these cracks. The first one is that the cracks follow ferrite film formed on the prior AGB. The second is the large size of these prior austenite grains in the range from 1 to 5 mm. This is in agreement with the "Blown grain" concept[6] which was proposed to explain severe transverse cracking due to reduced cracking path resistance.

Another example on the left in Fig.2 demonstrates transverse cracks at an equal spacing of about 48 mm in average in the area 150 to 250 mm from the slab corner. These cracks were discovered after surface milling of 2 mm depth applied on an as-cast slab top surface and they disappeared after the slab was milled down to 6 mm deep. At 1100 mm/min casting speed, the time interval between two adjacent cracks is 2.6 s, which is longer than the case in Fig.1. The magnified picture of one of these cracks on the right also showed "Blown grain" characteristics.

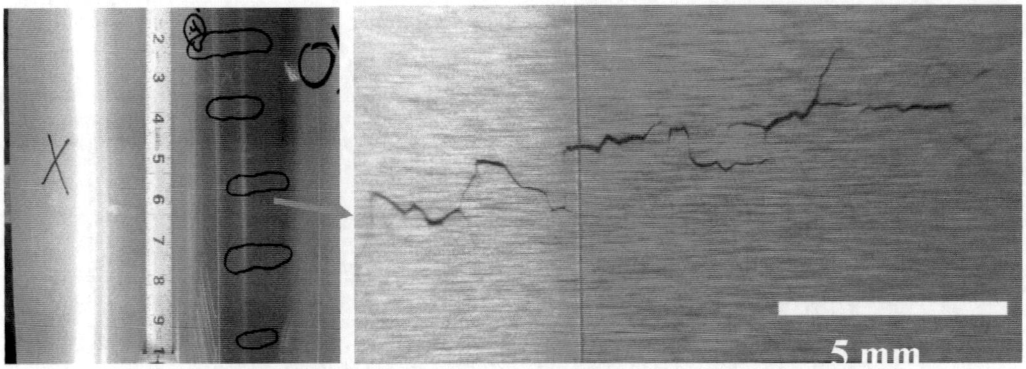

Fig.2 Transverse crack exposed after surface milled as-cast slab surface showing "Blown grain" feature.

One more example was found at the slab off-corner area as shown in Fig.3 after 2 mm surface milling and applied dye penetration to expose the cracks for better visualization. The bigger and darker the red dye shadow gets, the deeper and severer the crack is. These so-called off-corner transverse cracks were common for normal peritectic grades. The different in this AHSS slab case is that the most severe ones pointed by the black arrows demonstrate roughly equal spacing feature. The spacing is between 25 to 50 mm and lighter cracks are seen in between. Although the spacing is smaller compared with previous 2 cases, it is still significantly longer than the molds oscillation stroke.

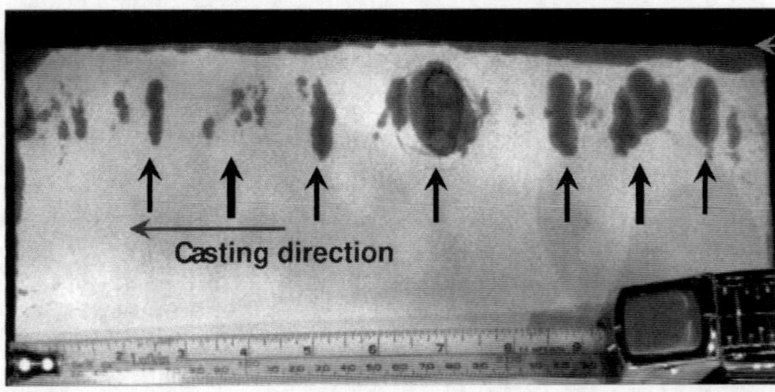

Fig.3 Transverse off-corner crack exposed after dye penetration at 2 mm from as-cast slab surface.

The detailed micro-etching of the off-corner area with cracks is given in Fig.4. This figure demonstrates the close correlation between blown grain to the transverse cracking problem. The larger prior austenite grains (blown grain) appears in a 30 mm zone which is 18 mm away from the slab corner and the severe cracks are only found in this zone. The neighboring areas on both sides of this blown grain zone have much smaller grain size and the cracks stop right at the transition lines from blown grains to fine grains.

It is common to find off-corner gutter formation in slabs close to the blown grain zone in Fig.4. The gutter is believed to form in the upper mold area when the slab corners rotate due to two-dimension heat and stress development near the corner area. Because of lack of heat removal in the gutter area, the temperature rebound will occur, which is potentially the reason for quick grain growth.

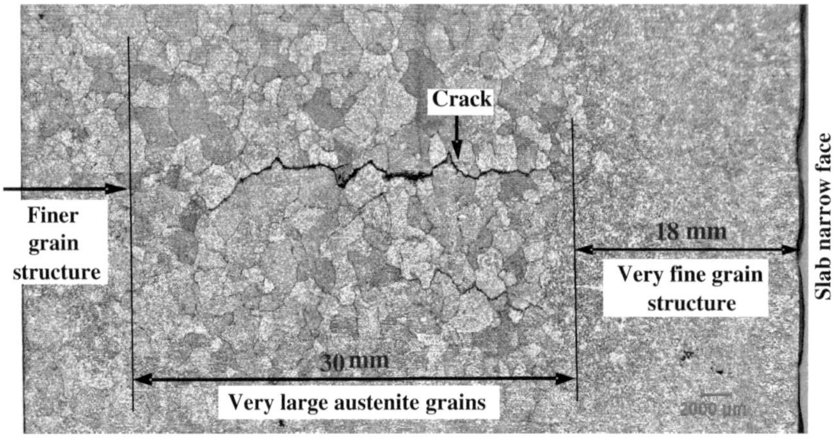

Fig.4 Transverse off-corner crack exposed after dye penetration at 2 mm from as-cast slab surface.

3. Equal spacing of transverse depressions

Transverse depression with equal spacing has been observed frequently on peritectic AHSS steel grades during continuous casting. An example is given in Fig.5 for an AHSS slab broad surface (BF) on the left from caster A and slab narrow face (NF) on the right from caster B. The spacing in-between these transverse depressions are roughly at 75 mm in average on the BF case and 70 mm on the NF case. It becomes clear that the equal spacing depression phenomenon is not caster dependent because when the similar AHSS grades are cast on different casters, the slab surfaces have very similar topographic featured depressions with very similar spacing distance when the casting speed is similar. With the use of laser line scan surface profile along the casting direction applied at 8 locations across the thickness to the narrow face of an AHSS slab from Caster C, the spacing in-between the measured big valleys (transverse depressions) can be determined at about 75 mm in average (See Fig.6).

Fig.5 Transverse depression on as-cast AHSS slab surface at an equal spacing (Left – BF and Right – NF).

The worst depression can be as deep as 4 mm while the oscillation mark depth (small peak-valley cycles along each colored line) is about 0.5 mm in average. After surface milling and dye penetration of the slab as-cast surface, severe transverse cracks were typically found almost exclusively aligned with the deepest oscillation mark inside the big valleys (transverse depressions). Micro-examination of these severe cracks shows the same blown grain features as mentioned above.

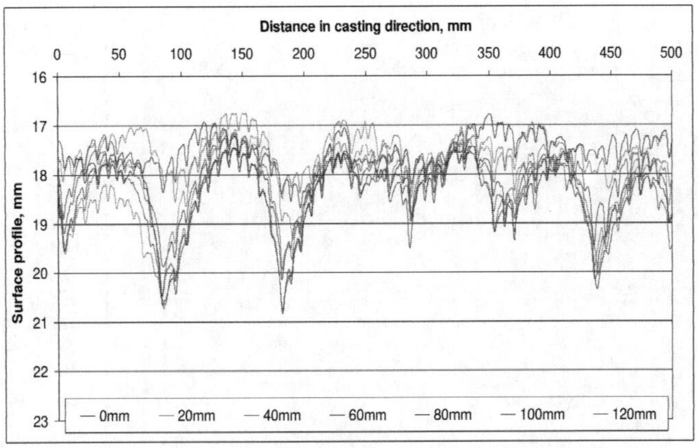

Fig.6 Transverse depression on as-cast AHSS slab narrow face at an equal spacing.

4. Equal spacing of shell thinning spots

Local thinning of the solidification shell during continuous casting has been often reported in the past. The direct result of shell thinning can be breakout of liquid steel. Strong liquid steel jet concentrating at a local area of the shell is often blamed for local thin shell in a larger area. For instance, when one of the SEN ports is clogged while most of the steel flow comes out from the other port. Inadequate mold flux feeding into the mold-shell gap at a local area or shell buckling due to inadequate taper setting can cause longitudinal local shell thinning, resulting in longitudinal cracks.
However, by carefully examine the breakout shells of an AHSS steel grade with peritectic chemistry, it was often found that a transverse shell thinning phenomenon exists at a roughly equal spacing along casting direction of the shell.

An example is shown in Fig.7 with 3D laser scan to obtain the breakout shell thickness profile. From this color map, it becomes obvious that the shell thinning in transverse direction (green horizontal lines in yellow area and yellow horizontal lines in red area) is well developed through the entire length of the shell inside the mold and the distance in-between the thinning lines is about the same, ~ 100 mm. Visual examination of the shell outside surface reveals transverse depressions very similar to the BF case in Fig.5 while the shell inside surface also shows smoother valleys mirroring the outside depressions. This can be seen on the cross-section image of the shell on the right in Fig.7.

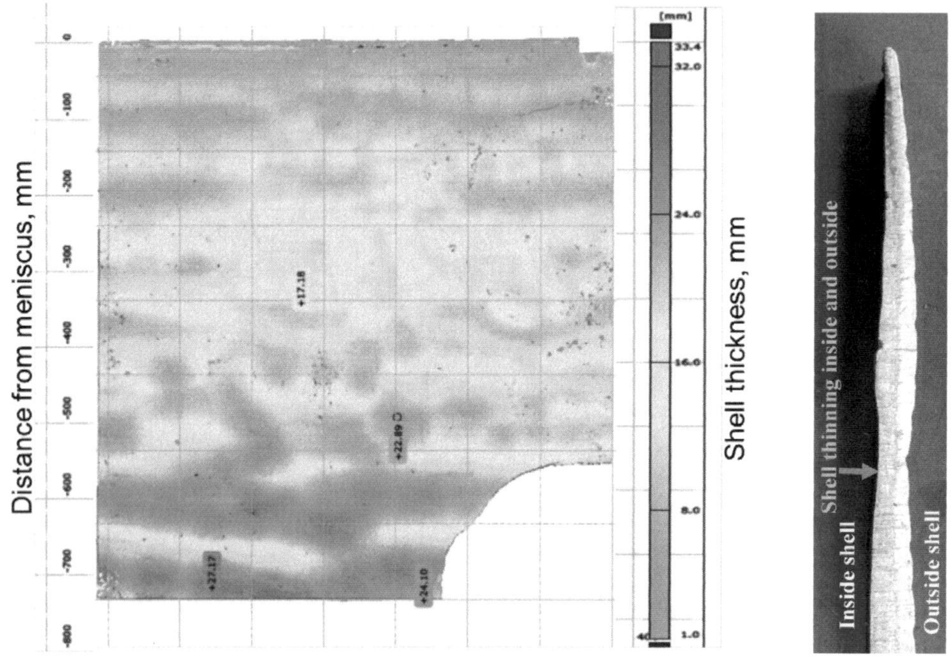

Fig.7 Shell thinning phenomenon on a breakout AHSS slab.

Actual measurements on shell thickness along the shell length direction were done on another breakout shell of an AHSS grade. The deviation of the measured data from the calculated theoretical shell thickness values at solidification coefficient of 28 is shown in Fig.8 for example. The severity of the shell thinning increases as the shell travels deeper in the mold and the maximum shell thinning can be as high as 12 mm from the expected value of a normal solidifying shell at 700 mm depth.

The important information in Fig.8 is that the shell thinning spots are at a roughly equal distance. The average spacing is about 63.5 mm. Under the casting condition when the breakout occurred, the time interval between the thinning spots is calculated at 4 seconds.

5. New concept for transverse crack formation in AHSS slab surface

Based on the above observations on the phenomena of transverse surface cracking at equal spacing, transverse depressions at equal spacing and the transverse shell thinning at equal spacing, all three are closely related for AHSS grades especially with peritectic chemistry from conventional slab casting process. In summary, the observed spacing distance between these defects is from 30 to 100 mm depending on steel grades and the casting time interval between these defects is in the range of 2.5 to 5 seconds depending on casting speed. After examining all other possibilities in casting process, none can be found that relates to a cycling pattern at a frequency of 0.2 to 0.4 Hz. It should be mentioned that the spacing distance or time interval only differs slightly at different casters as long as the casting conditions are similar. For conventional caster, the mold oscillation frequency is typically around 2 Hz and the caster roller pitch and roller periphery translate to roughly 0.06 Hz. The observed shell thinning cycling frequency of 0.2 to 0.4 Hz lies in between the two intrinsic mechanical cycling frequency of the casting machine. Therefore, this mid-frequency cycling phenomenon in 0.2 to 0.4 Hz is caster independent whereas reflects the nature of the initial solidification of the AHSS chemistry.

It is most likely that the solidification of steel gets into a rhythm to allow synchronized solidification shrinkage to occur and the cycling volume change of the liquid steel then results in a cyclic mold level disturbance. Hence, this mid-frequency mold level disturbance causes a transverse depression cycling pattern in the shell at initial solidification moment. At the depressed area, the shell temperature rebounds due to low local heat extraction from the mold, causing the shell to remelt partially from inside. This creates a wavy pattern in the shell at a roughly equal spacing.

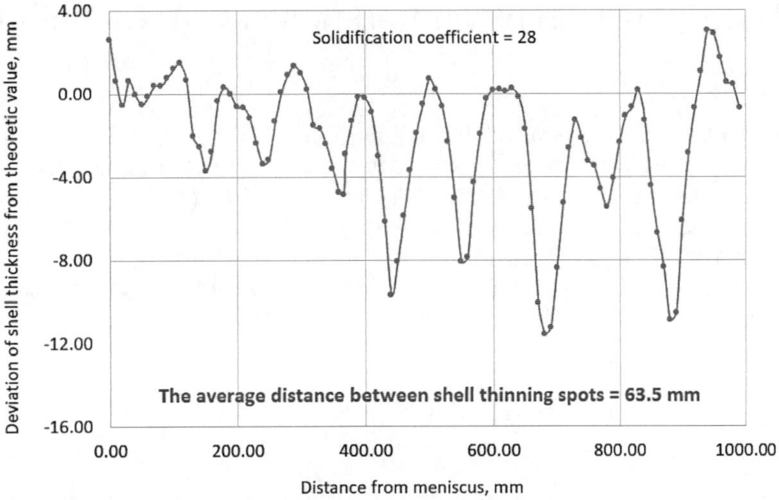

Fig.8 Shell thickness deviation from theoretical calculation along shell length.

For lean alloyed AHSS grades with lower high temperature strength, the shell at the local depressed area with thinner thickness and higher temperature can be pushed out against mold by static pressure at lower part of the mold. As a result, although the wave pattern exists at the inside of the shell, the as-cast slab surface depression on the outside of the shell can be barely observed. For rich alloyed AHSS grades which has much higher strength at high temperatures, the local thin shell cannot be easily bended out, and hence the surface depressions stay on the as-cast slab surface. In both cases, the local shell thinning effect exists which becomes much hotter than surrounding areas. This will cause local delta ferrite and/or austenite grains to grow much larger, so called "blown grains" (larger delta ferrite grains will induce larger austenite grains). This blown grains in turn will cause transverse cracks at bending and unbending process during casting.

According to this new concept, the reason for the mid frequency disturbance in the casting process needs to be fully understood. This will extend our knowledge on transverse cracking formation and develop countermeasures accordingly.

References

[1] N. A. McPherson, Continuous Casting Vol. VIII: Transverse cracking in continuously cast products, Iron & Steel Society, (USA), 1997
[2] B. Mintz, S. Yue and J. Jonas, International Materials Reviews, 1991, 36(5), 188-217
[3] H. Tsai, H. Yin, M. Lowry and S. Morales, Proc. AISTech 05, May, 2005, Charlotte, NC
[4] H. Yasunaka et al., Tetsu to Hagane, 81(9), 1995, 26-31
[5] H. Yin, Iron and Steel Technology, 3(6), 2006, AIST, 64-73
[6] E. S. Szekeres, Proc. 6th Intnl. Conf. Clean Steel, June 10-12, 2002, Balatonfured, Hungary

IOP Conf. Series: Materials Science and Engineering **529** (2019) 012074 doi:10.1088/1757-899X/529/1/012074

Study of elasto-plastic deformation in a cast AlCu7 alloy

M Schöbel[1,2], R Fernández[3] and HP Degischer[4]
[1]Forschungs-Neutronenquelle Heinz Maier-Leibnitz (FRM II), Technische Universität München, Lichtenbergstraße 1, 85747 Garching, Germany
[2]Linde Gas GmbH, Erdbergstraße 197-199, 1030 Vienna, Austria
[3]Nemak Linz GmbH, Zeppelinstraße 24, 4030 Linz, Austria
[4]Institute of Materials Science and Technology, Vienna University of Technology, Karlsplatz 13, 1040 Vienna, Austria

E-mail: michael.schoebel@frm2.tum.de

Abstract: The need for efficient and clean solutions, due to the increasing current environmental regulations puts extra pressure on new combustion engine development, to compete in a market with alternative driving concepts. Downsizing and weight reduction can reduce the engine emission and efficiency, but require light alloys with superior thermo-mechanical properties for high temperature exposure to maintain the same engine performance. Cast Al-Cu could be alternative to standard Al-Si alloys for new engine generations due to their higher temperature strength, creep-resistance and long term stability of engine components. In Al-Si and Al-Cu cast alloys with heterogeneous microstructures a composite-like deformation behavior is responsible for superior high temperature properties. Stiff Si or Al_2Cu particles, respectively reinforce a ductile α-Al matrix to a composite with improved thermo-mechanical strength. However, different Young's moduli and coefficients of thermal expansion are responsible for micro stress gradients and unpredictable micro crack formation under operation. These micro-mechanical deformation mechanisms in Al-Si and Al-Cu systems, responsible for crack initiation and growth, have been scarcely investigated so far.

This manuscript describes an example of elasto-plastic deformation mechanisms in an AlCu7 alloy. Tensile testing shows anomalous macroscopic deformation behavior indicating unknown internal micro-mechanical processes. External loading until yield strength and beyond are applied under laboratory conditions and during in-situ neutron diffraction. The results of macroscopic deformation and micro strain evolution are compared and correlated with the heterogeneous micro structure. High resolution synchrotron computed tomography reveals conclusions on the micro-mechanic deformation mechanisms and their effects on the macroscopic damage initiation and material's service performance.

1. Introduction

Modern combustion engines put high demands on materials thermo-mechanical properties and long term stability under service conditions [1]. Reduction of fuel consumption and emission going with higher power densities realized in downsized combustion engines is required to fulfill environmental regulations and to stay competitive with alternative drive concepts. Therefore Al-Cu cast alloys could be an alternative to the standard Al-Si alloys for cylinder head production due to their superior properties at elevated temperatures [2]. Al-Si and Al-Cu cast alloys consist of a heterogeneous microstructure which is responsible for their sophisticated thermo-mechanical properties [3]. This cast alloys show a composite-like deformation behavior with stiff and brittle reinforcing primary phase (Si particles in Al-Si alloys and Al_2Cu particles in Al-Cu alloys) embedded in a ductile α-Al matrix [4, 5]. High temperature strength and creep resistance are resulting from stress distribution in between the phases and dislocation accumulations at their interfaces [6]. Problems can rise with the formation of high micro stress gradients in between the phases under cycling loads, which act as potential sources for fatigue damage [7]. These micro-mechanical deformation mechanisms leading to micro cracks, crack propagation and failure of engine parts are strongly dependent on the type of the reinforcing phases, their interfaces with the matrix and their 3D architecture (i.e. interconnectivity) [8]. The investigation of such material properties require highly sophisticated non-destructive methods to study the materials fundamental micro-mechanical behavior under simulated service conditions.

A commercially cast AlCu7 alloy was studied with focus on micro structure, mechanical properties, micro stress evolution and crack initiation. AlCu7 contains primary Cu-Mn-Fe aluminides and eutectic regions of θ-Al_2Cu / α-Al in the as cast condition (T1) by non-equilibrium cooling: ~ 6, 5 vol.% of θ-Al_2Cu aluminides within the eutectic regions [9]. The mismatch in the elastic moduli ($E_{\alpha\text{-Al}}$ ~ 70 GPa, E_{Al2Cu} ~ 100 GPa) and the thermal expansion ($CTE_{\alpha\text{-Al}}$ ~ 25 ppm/K, CTE_{Al2Cu} ~ 17 ppm/K) leads to a composite structure which is also partially responsible for the superior thermo-mechanical properties of AlCu7 [10].

This work describes micro-mechanical deformation behavior in an AlCu7 alloy by application of complementary non-destructive testing methods using photons and neutrons as probe particles. Micro strains where measured phase-sensitively by neutron diffraction (ND), in-situ during tensile testing. Crack initiation and damage propagation was visualized in 3D by high resolution synchrotron computed tomography (SCT). Characterization of the microstructure was performed by light optical micrographs (LOM).

2. Experiment

2.1. Material

A cylinder head was cast with an AlCu7 alloy (NemAlloy HT200) at Nemak Linz GmbH by rotatory gravity die casting (Rotacast®) [11]. Test specimens were prepared from regions close to the heat exposed zones at the fire faced areas on top [1]. The average alloy composition is listed in Tab. 1.

Type	Cu [wt.%]	Mn [wt.%]	Ti [wt.%]	Zr [wt.%]	Al [%]
AlCu7	6.5	0.45	0.11	0.22	Bal.

Table 1.: Composition of investigated alloy.

Tensile test specimens were manufactured accordingly to DIN 50125: 2009-07 in cylindrical shape parallel to the surface. Light optical microscopy (LOM) shows AlCu7 in as cast (T1) condition in Fig. 1.

Dark primary θ-Al_2Cu particles appear embedded in in a bright α-Al matrix phase as a heterogeneous composite-like micro structure. Eutectic interconnectd intermetallics between the α-Al dendrites are formed during non-equilibrium solidification of the casting.

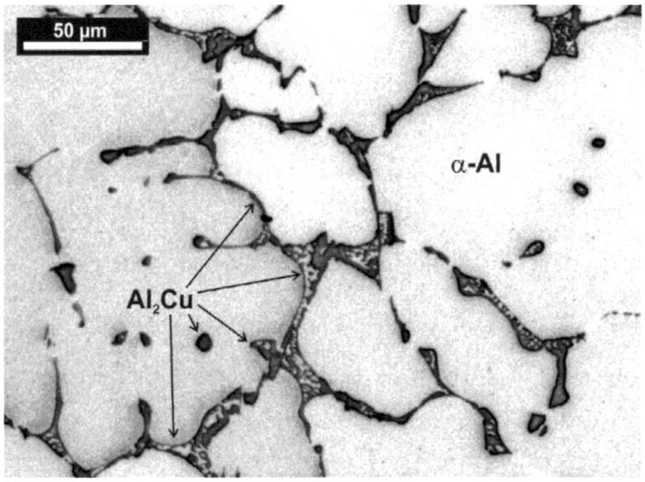

Figure 1.: Light optical micro graph (LOM) of AlCu7 alloy in T1 condition.

2.2. Tensile test

Tensile testing was performed on a Zwick Z050 [12] with loading capacity of 0.1 kN to 20 kN. Cylindrical tensile test specimens were tested in B 6 x 30 shape accordingly to DIN 50125 (6 mm diameter, 30 mm initial gage length, threaded heads M10). The tests included loading-unloading cycles from defined load steps. The upper load in the first cycle was 70 MPa, followed by unloading to 10 MPa. For each following loading the load was increased by 20 MPa (to 90 MPa, 110 MPa, etc.), whereas the lower point of the resulting hysteresis loops was kept at 10 MPa. Load was measured by a load cell with 50 kN nominal load which meets the demands for accuracy class 0.5 (EN ISO 7500-1) in the range from 100 N to 50 kN. Elongation was measured by a contacting extensometer with accuracy class 0.5 accordingly to EN ISO 9513 in the measurement range from 0.02 to 50 mm.

2.3. Neutron diffraction

Neutron diffraction (ND) was performed with the angle dispersive strain scanner STRESS-SPEC [13] at FRM II research reactor in Garching, Germany [14]. A monochromatic particle beam with a wavelength of 1.67 Å was chosen accordingly to the lattice spacing (bent Si(400) double focusing monochromator unit). The high neutron flux of 2 – 3 x 10^7 $ncm^{-2}s^{-1}$ diffracted from the samples allowed short acquisition times ~ 7 min. The gage volume was set to 5 x 5 x 10 mm and a ω-tilt of ± 5° (vertical rotation axis) to improve grain statistics sufficiently. The peaks were acquired with a position sensitive ^3He detector unit (PSD) covering an angle interval of 2θ = [82°, 94°] for simultaneous detection of Al(311), Al(222), Al_2Cu(332) and Al_2Cu(422) Debye-Scherrer ring segments in 1.035 m distance and one acquisition (i.e. 7 min).

The in-situ tensile test rig [15] was mounted on the x, y, z - table of the diffractometer and aligned in the beam accordingly. Force was applied and strains were measured in load direction, in transmission through the tensile test specimens. Reference was taken in unloaded initial condition, neglecting any residual micro stresses from casting.

2.4. Synchrotron tomography

Synchrotron computed tomography (SCT) was carried out on the ID19 high energy imaging beam line at ESRF Grenoble [16]. A pink beam setup was used in parallel geometry with an energy maximum at ~ 19 keV. A primary beam aperture of 2 x 2 mm² flooded the Al-Cu samples completely. The samples were manufactured in cylindrical shape from the center-top regions of the tensile test specimens as used for neutron diffraction. A region of interest scan was made from the sample at a diameter of ~ 0.6 mm and length ~ 10 mm showing a representative volume of 0.7 x 0.7 x 0.7 µm³. Image acquisition was made in transmission behind the sample by a PCO edge camera (CMOS) with a sample-to-detector distance of 13 mm. Up to 6000 projections were acquired during one 365° scan within 7 min acquisition time [17]. Reconstruction was made by radon transformation, single distance phase retrieval and combined with motion correction filtering of ring artefacts resulting in a 3D tomography with a voxel size of (0.32 µm)³. Further data processing for visualization was made with the ImageJ software toolkit [18].

3. Results

A tensile test was performed on AlCu7 alloy in T1 condition to investigate the macroscopic elastic properties after cycling loads (Fig. 2). External load was applied until 70 MPa are reached then unloaded, again loaded to 90 MPa and unloaded increased by 20 MPa each and repeated until fracture at ~ 210 MPa.

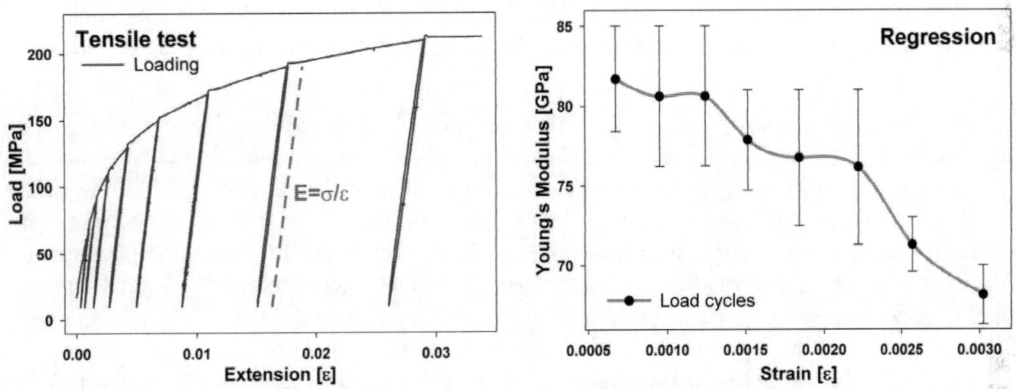

Figure 2.: Tensile test cycles of AlCu7 in T1 condition (left). Strain dependent Young's moduli evaluated by linear regression of the loading curve (right).

The Young's modulus was interpolated by linear regression curve at each loading step. The derived values are shown as a function of applied strain. The Young's modulus starts with its highest value of 81.7 ± 3.3 GPa at 7 x 10^{-4} macroscopic strain and increasing strain (to 10 x 10^{-4}, 13 x 10^{-4}, 15 x 10^{-4}, 18 x 10^{-4}, 22 x 10^{-4} and 26 x 10^{-4}) until fracture with its lowest value of 68.2 ± 1.9 GPa at 30 x 10^{-4} strain. The phase sensitive micro strains during tensile testing of AlCu7 are shown in Fig. 3 in Al(311), Al(222), Al_2Cu(332) and Al_2Cu(422) lattice planes under load at defined load steps until fracture. The externally applied load is transferred into lattice planes of the different phases weighted by their Young's moduli and phase volume fractions. In the α-Al, load increases accordingly to linear Hooks law in an initially elastic region up to ~ 70 MPa. Above, the strain values split in between different Al lattice planes with higher value of 0.0023 in Al(311) and 0.0055 in Al(222) before fracture. The strains in Al_2Cu(332) and Al_2Cu(422) increase linearly in the beginning before 70 MPa and increases stronger (increasing slope) above. A saturation of micro strains can be observed after 150 MPa, staying constant (or even decrease) until fracture at ~ 200 MPa external loading. High resolution synchrotron tomography image slice of AlCu7 in T1 condition is shown in Fig. 3. A volume ~ 0.2 mm below the macroscopic fracture surface, in the tensile test specimen is shown. External loading was applied until fracture vertically to the image slice. Al_2Cu appears bright in the grey α-Al matrix by absorption contrast imaging. The primary θ-Al_2Cu interconnected network appearing with a dark micro crack formed within a coral shaped eutectic region, perpendicular to external load direction.

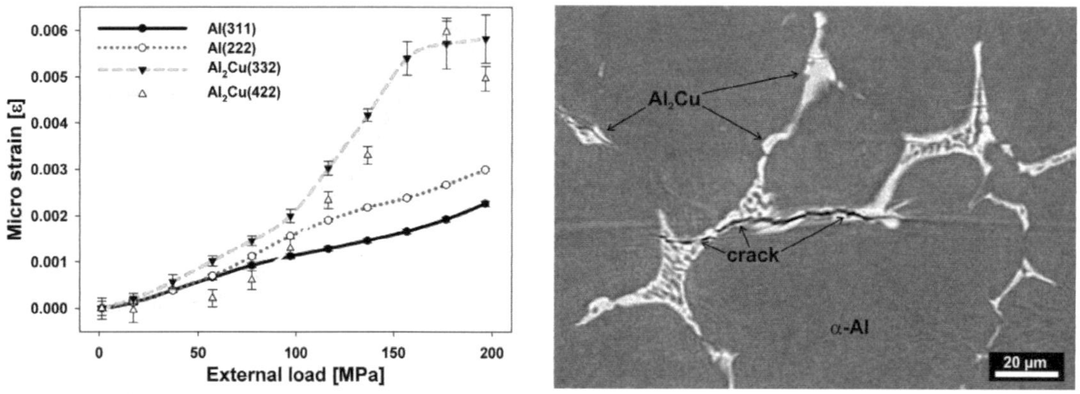

Figure 3.: Phase sensitive micro strains in AlCu7 in T1 condition (left). Micro crack initiation perpendicular to external load direction within the heterogeneous micro structure (right).

4. Discussion

AlCu7 alloys consist of a heterogeneous micro structure of α-Al dendrites and eutectic θ-Al_2Cu in the inter-dendritic areas which are formed during casting (Fig. 1). This composite-like structure forms a brittle and stiff network of Al_2Cu eutectic interconnected reinforcement which is embedded in the ductile α matrix alloy. The mismatch in yield stresses ($\sigma_{\alpha\text{-Al}} \approx 60$ MPa vs. $\sigma_{Al2Cu} \approx 150$ MPa) and Young's moduli ($E_{\alpha\text{-Al}} \approx 70$ GPa, $E_{Al2Cu} \approx 100$ GPa) are responsible for load partitioning of external stresses (macro to micro stresses) and plastic deformation within the heterogeneous micro structure. External stresses are distributed in between α and θ phase in an initially elastic region accordingly to the Young's modulus and rule of mixture until 60 MPa is reached (Fig. 3). Above 60 MPa elasto-plastic matrix deformation can be observed at Al(311) and Al(222) which provokes loading of the θ phase, as observed in Al_2Cu(332) and Al_2Cu(422). As a result, high micro stresses at the α-θ interfaces induce localized plastic deformation in ductile α and crack initiation in brittle θ phase. Accumulated damage of the intermetallic reinforcement result in a permanent reduction of the Young's modulus of AlCu7 alloys if cycling mechanical loads are applied (Fig. 2).

5. Conclusions

AlCu7 contains a composite-like micro structure which is responsible for its improved thermo-mechanical properties. The mismatch of elastic moduli between α and θ phase ($\Delta E \sim 30$ GPa) generates elasto-plastic deformation in ductile α-Al and elastically deformed stiff, brittle θ-Al_2Cu until cracking. The resulting high micro stress gradients and interface shear stresses lead to crack initiation and damage accumulation in the brittle Al_2Cu. If mechanical loads exceeds α-Al yield strength cracking starts, which is partially bridged by plastic α matrix deformation, but influence the macroscopic mechanical properties i.e. degrade the Young's modulus irreversibly. Therefore, the damage initiation and propagation mechanisms of composite-like cast alloys at small strains should be further investigated to be properly implemented in future calculation models and engine component construction to ensure operation under modified safe conditions.

Acknowledgements
The authors gratefully thank the support from Anette Danninger for mechanical tests at the Institute for Materials Science and Technology, TU Vienna, Austria. Last but not least, very much acknowledged is the financial support from Bayerische Forschungsstiftung, Germany (No.: AZ 1134-14).

References
[1] Shalev M, Zvirin Y and Stotter A 1983 *Int. J. Mech. Sci.* **25** 471-83
[2] Sigworth GK, DeHart F 2003 *AFS Trans.* **1-14** 3-315
[3] Molina R, Amalberto P, Rosso M 2011 *Metall. Sci. Tech.* **5-15** 29-1&2
[4] Schöbel M, Baumgartner G, Gerth S, Bernardi J and Hofmann M 2014 *Acta Mat.* **81** 401-8
[5] Schöbel M, Fernández R, Koos R and Bernardi J 2019 *Jour. of All. and Comp.* **775** 617-27
[6] Noreyan A, Qi Y and Stoilov V 2008 *Acta Mat.* **56** 3461-69
[7] Fernández R, Sket F, Maire E, Wilde F, Boller E and Requena G 2016 *Jour. of All. and Comp.* **697** 341-52
[8] Requena G, Garcés G. Danko S, Pirling T and Boller E 2019 *Acta Mat.* **57** 3199-210
[9] Jiang L, Li KJ Liu G, Wang RH, Chen BA, Zhang JY, Sun J, Yang MX, Yang G, Yang J and Cao XZ 2015 *Mater. Sci. Engin. A* **637** 139-54
[10] Zeren M 2005 *Jour. Mater. Process. Tech.* **169** 292-8
[11] Gosch R, Stika P 2005 *Giesserei-Rundschau* **52-7/8** 170-173
[12] http://wwwt.tuwien.ac.at 2019
[13] Hofmann M, Schneider R, Seidl GA, Rebelo-Kornmeier J, Wimpory RC, Garbe U and Brokmeier HG 2006 *Phys. B* **385-6** 1035-1037
[14] http://www.frm2.tum.de 2019
[15] Hoelzel M, Gan WM, Hofmann M, Randau C, Seidl G, Jüttner P and Schmahl WW 2013 *Nuc. Inst. A* **711** 101-105
[16] http://www.esrf.eu 2019
[17] Desjardins K, Bordessoule M, Petrache C, Menneglier C, Dalle D, Mercere P and Medjoubi K 2014 Journ. of Instr. **9**
[18] http:// rsbweb.nih.gov/ij/ 2019

IOP Conf. Series: Materials Science and Engineering **529** (2019) 012075 doi:10.1088/1757-899X/529/1/012075

Effect of Pores on Tensile Fracture of Die-cast AlSiMgMn alloys with 3D X-ray μ-CT and FE Simulation

H D Zhao [1*], F Liu [1], R S Yang [1] and F Z Sun [2]

[1] National Engineering Research Center of Near-net Shape Forming for Metallic Materials, South China University of Technology, Guangzhou 510641, China
[2] Division of Mechanical Engineering, Imperial College London, London SW7 2AZ, UK

*Corresponding author: hdzhao@scut.edu.cn

Abstract. In this study, the pores in die-cast AlSiMgMn alloys were inspected and reconstructed with high resolution three-dimensional (3D) X-ray micro computed tomograpgy (μ-CT) technique. Finite element (FE) meshes were built with consideration of the pore actural morphorloges from the CT inspection. Based on dutile damage model, the FE simulation of tensile fracture of the alloys was carried out. The simulation results were compared and verified with the tensile of in-situ scanning electron microscopy (SEM). The two results are agreement in the main crack path and pores on the fracture. With the pore-scale simulation, the effects of pore characteristics on the stress distributions and crack initiation and growth during the tensile were analyzed. It was found that the pores of lower sphericity and larger project area in tensile axis direction are prone to form microcracks and promote main crack deflection. The results also show that aggregation of brittle alpha-Fe intermetallics of the alloys also has important influence on the main crack propagation.

1. Introduction

Vehicle lightweighting is a key pursuit of automotive industries and drives carmakers to use more aluminum castings for replacing heavy iron based parts [1]. Die casting process is particularly preferred by engineers to produce aluminum components for automotive applications since it has the advantages of good surface quality, excellent dimensional precision and high producing efficiency. However, gases are inevitably entrapped in the molten alloys during high speed mold filling process, resulting in undesired porosity defects [2] which reduce the effective load-bearing areas and cause local stress concentration [3-5].

These days, 3D X-ray CT has been increasingly used to investigate the spatial characteristics of pores (e.g. pore volume, surface area, sphericity and locations) and their roles in material failure [6-8]. Weck et al. [9] applied X-ray to assess the accuracy of existing models in capturing the void growth and coalescence in metals. Nicoletto et al. [10] numerically simulated the stress concentration around an X-ray reconstructed pores in AlSi7Mg casting under push-pull testing. Our previous work on quasi-static stress concentrations around pores under fatigue load also shows that pores near casting surfaces could cause more stress concentration [11]. However, as our known simulation on dynamic fracture of die-cast Al alloys with consideration of micro-scale pore characteristics has not been reported.

In this study, the 3D pores in die-cast AlSiMnMg alloys, which with high strength and ductility [12], was reconstructed by X-ray CT. FE analysis of tensile test was then performed on the reconstructed 3D structure. The predicted crack path was compared with the experimental result identified by scanning electron microscopy (SEM). The effects of pores on stress distribution and crack propagation were discussed.

2. Experiment and simulation

2.1 Materials and experiment

The material used in this study was die-cast AlSiMnMg alloys. The chemical compositions of this alloy were measured by ICP ARL4460 including: 10.10%Si, 0.075%Fe, 0.62%Mn, 0.12%Mg, 0.012%Sr. The castings are of 2300×200×25 mm, and were fabricated by 2000 tons cold chamber die casting machine. In the die casting process, the pouring and die temperatures are 680 and 200 ℃, respectively. To investigate the effect of pores on crack propagation, the tensile samples (figure 1) were directly cut from practical pedal die-castings. The gauge length, width and thickness of the sample are 16, 2.5 and 1.3 mm respectively. Tensile tests were carried out at room temperature with the velocity at 1 μm/s. After the tensile, the fracture morphology and crack path in the sample surface were observed by SEM (Nova Nano 430).

In order to characterize the internal pores, the region near the notch as indicated by the dashed rectangle in figure 1 (about 4 mm along the tensile direction) was inspected by the X-ray CT instrument nanoVoxel-2000 (*Sanying Precision Instruments Co.,Ltd*). During the inspection, the voltage and current of the X-ray tube were 120 kV and 80 μA, respectively. Though the instrument has maximum resolution of 0.5 μm, to obtain the pores in the whole region the inspection resolution in the experiment was 3.88 μm. Then the CT data were imported into *Volume Graphic (VG) Studio* MAX 2.2 software to carry out pore 3D reconstruction, with which the volume, sphericity and spatial distribution of pores were determined.

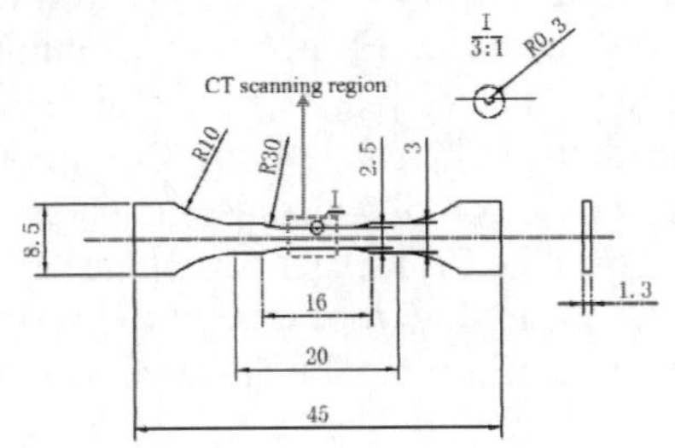

Figure 1. Shape and dimension of the in-situ tensile samples.

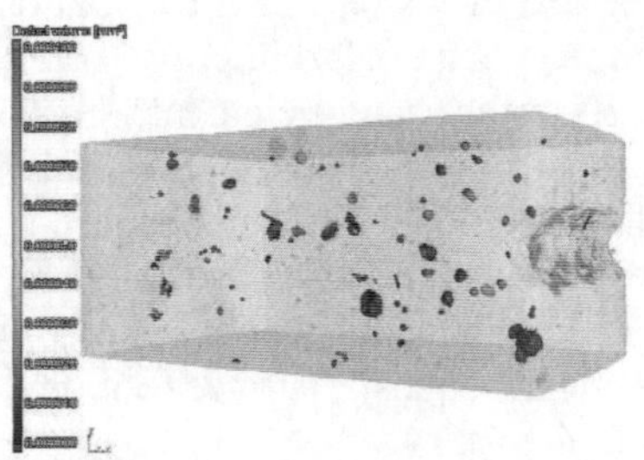

Figure 2. Reconstructed of micropores in the die-cast AlSiMgMn alloys.

2.2 FE simulation

To simulate the tensile fracture of the die-cast AlSiMgMn alloys, the ductile damage model was used since the alloy is of good ductility. It is usually used to predict the onset of damage due to nucleation, growth, and coalescence of voids in ductile metals [13]. In the model equivalent plastic strain at the onset of damage, $\bar{\varepsilon}_D^{pl}$, is a function of stress triaxiality (η) and strain rate ($\dot{\bar{\varepsilon}}^{pl}$). In the simulation, the stress triaxiality is defined as

$$\eta = \frac{-p}{q} \tag{1}$$

where, p is the pressure stress and q is the Mises equivalent.

The criteria for material damage is given by

$$\omega_d = \int \frac{d\bar{\varepsilon}^{pl}}{\bar{\varepsilon}_D^{pl}(\eta, \dot{\bar{\varepsilon}}^{pl})} = 1 \tag{2}$$

where, ω_d is a state variable, and increases with plastic deformation, the increment of $\Delta\omega_d$ during the calculation process was defined by

IOP Conf. Series: Materials Science and Engineering **529** (2019) 012075 doi:10.1088/1757-899X/529/1/012075

$$\Delta\omega_{\mathrm{d}} = \int \frac{\Delta\bar{\varepsilon}_{\mathrm{d}}^{\mathrm{pl}}}{\bar{\varepsilon}_{\mathrm{D}}^{\mathrm{pl}}(\eta,\dot{\bar{\varepsilon}}^{pl})} \geq 0 \quad (3)$$

The *Abaqus* software was used as the simulation platform, and material properties was set as isotropic hardening, which based on the von Mises type elastoplastic constitutive law. According to the tensile test conditions, the fracture strain, stress triaxial and strain rate are set to 0.07, 0.33, and 6.25x10-5 respectively. And a stress-strain curve of experiment tensile process was input into the *Abaqus* software as parameters of materials properties. The AlSiMgMn alloy density was set to 2.67 g/cm3, and young's modulus and poisson's ratio were 76 GPa and 0.33 respectively.

3. Results and discussion

3.1 FE meshing with pore 3D characteristics

Figure 2 exhibits the reconstructed pores in the inspection domain. According to the resolution (3.88 μm) of inspection, pores with the volume lager than 7×10^{-6} mm^3 could be precisely detected. The total pores number is 73 and average pore volume is 6.5×10^{-5} mm^3. Eleven of them have the volume more than 1×10^{-4} mm^3, among which the biggest one is of 1.118×10^{-3} mm^3.

To simulate effect of the pores on tensile fracture, the FE model with the pores must be established. After the CT inspection, the data of serial XY sections were imported into the *Mimics* software to generate the triangular surface meshes of specimen. As well known, large pores in the body and pores adjacent to the surfaces of the alloy were more detrimental compared to other pores. Therefore, in our simulation the large volume pores (bigger than 1×10^{-4} mm^3) and subsurface pores were kept in the computational model and relatively smaller pores were removed to improve calculation efficiency and prevent numerical diffusion in FE calculation. The ratio of height to base (H/B) in the triangle was controlled no more than 0.4 to improve grid quality, and for improving FE iteration the minimum and maximum sizes of the triangle element were set as 3 and 100 μm respectively. The surface meshes were exported as input file of *Abaqus*. Then, volumetric tetrahedral meshes were built form the triangular meshes in the *Abaqus* software as shown in figure 3a. Due to the complex details, the pores and notch part were set with smaller and denser meshes. Figure 3b shows the FE meshes for a pore. To calculate the fracture under monotonic loading of the room temperature, one side (top) was set to fixed boundary condition, while the opposite surface (bottom) was considered uniform motion at the velocity of 1 μm/s as shown in figure 3a.

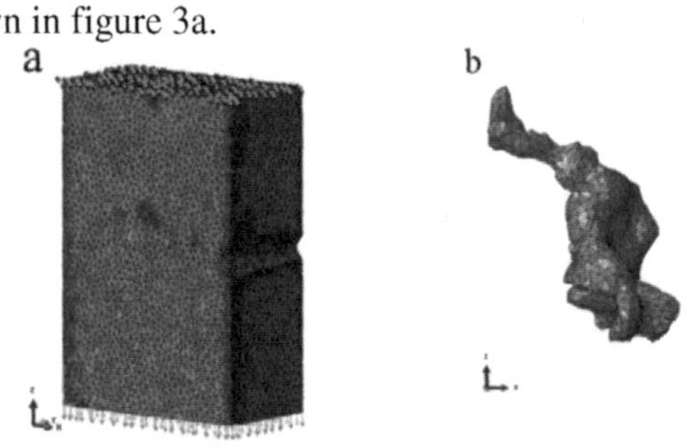

(a) 3D FE meshing and boundary conditions (b) FE meshing of a gas-shrinkage pore

Figure 3. 3D FE enmeshment of pores and boundary condition set-up.

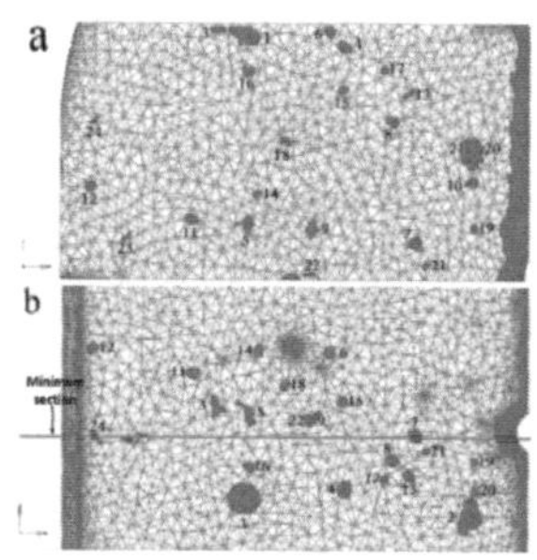

Figure 4. Distribution of pores on FE model: (a) view from the Z axis, (b) view from the Y axis.

Figure 4a and 4b show the pore distributions from Z and Y axis directions. For convenience of description later, the pores were numbered according to their volumes. Table 1 lists the details of pore characteristics in the calculation domain. The pore P1-11 are of bigger volume, among which the P1,P3, P4, and P6 are near the boundary. The pore P19, P21, and P22 are also close to the surface but they have small volumes. In table 1, PZ means the projected area of a pore in the tensile direction and Z is the distance of the pore to the minimum cross section. It is found from table 1 that the pores with larger volume usually have higher PZ values. According to the Z values, the pore P1, P2, P4, P6, P10, P11, P12, P14 and P18 are far away from the section, and occasionally the pore P7 and P23 are in the section.

Table 1. Characteristics of pores in the FE model

Pores	Volume ($10^{-6}\times$ mm^3)	sphericity	PZ (10^{-4} $\times$mm^2)	Z (mm)	Pores	Volume ($10^{-6}\times$ mm^3)	sphericity	PZ (10^{-4} $\times$mm^2)	Z (mm)
P1	1118	0.47	133	0.33	**P13**	61	0.45	20	0.18
P2	987	0.46	158	0.34	**P14**	61	0.54	16	0.44
P3	179	0.32	45	0.19	**P15**	57	0.47	23	0.17
P4	170	0.48	38	0.25	**P16**	56	0.46	28	0.16
P5	144	0.36	46	0.14	**P17**	53	0.65	18	0.21
P6	129	0.64	31	0.41	**P18**	50	0.42	22	0.24
P7	126	0.46	40	0	**P19**	41	0.6	17	0.11
P8	119	0.6	30	0.11	**P20**	33	0.42	13	0.25
P9	114	0.38	50	0.10	**P21**	32	0.58	16	0.06
P10	112	0.51	33	0.34	**P22**	31	0.5	13	0.08
P11	104	0.58	32	0.31	**P23**	30	0.49	13	0
P12	93	0.62	26	0.43	**P24**	18	0.48	10	0.003

3.2 Characteristics of fracture

Figure 5 displays the fracture observed from the front surface of the tensile specimen by SEM, and figure 5b gives the simulated fracture using the meshes generated from the CT image data. It can be seen that the simulated fracture path resembles the experimental result, indicating a good agreement between these two results. The morphology of the fracture surface is shown in figure 5c. Many pores were clearly observed on the fracture surface, confirming effect of pores on cracks. According to figure 4a, the pores P1, P4 and P21 are near the sample surface. They caused stress concentration according to our former calculation [11] hence inducing the fracture. According to the Z values in table 1, the pores P5, P7, P8, P9, P13, P17, P19 and P21 are near the minimum section. Due to the notch, the decreased area of the minimum section increased the stress. In result, even though some of them are with smaller volume, they still affected the crack propagation. For example, the pore P21 has quite small volume and high sphericity (table 1). But it is found on the fracture due to near of the surface (figure 4a) and the minimum section (0.06 Z value) that pores near the boundary and minimum section are easy to nucleation cracks. Compared to the experiment result, the simulation crack not only go through all pores in the fracture but also go through the pores P13 and P19. Other pores that with small volumes or far away from notch (the value of Z larger than 0.25 mm) did not appeared in the fracture surface.

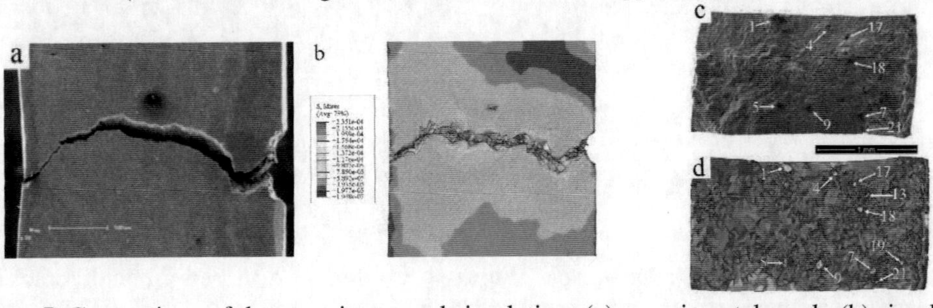

Figure 5. Comparison of the experiment and simulation: (a) experimental crack, (b) simulated crack, (c) pores on the fracture, (d) pores on the simulated fracture.

The simulation results show that the crack went through the pore P13 and P19. However, these two pores were not experimentally observed on the fracture surface, as shown in figure 5c. To further investigate the disagreement, the regions of fracture surface with similar projected location of two pores along the tensile direction were analyzed with SEM, which is shown in figure 6. It is obvious in the figure that the continuously distributed α-Fe intermetallics i.e Al15(Fe,Mn)3Si2 exist on the fracture surface as indicated by the arrows. The study on the similar alloys by Dominik [14] showed that the α-sludge particles and α-Fe intermetallic tended to initialize cracks due to their low plasticity. Therefore in the regions the crack path was controlled by these clustered intermetallics, leading to the simulation disagreement in these regions.

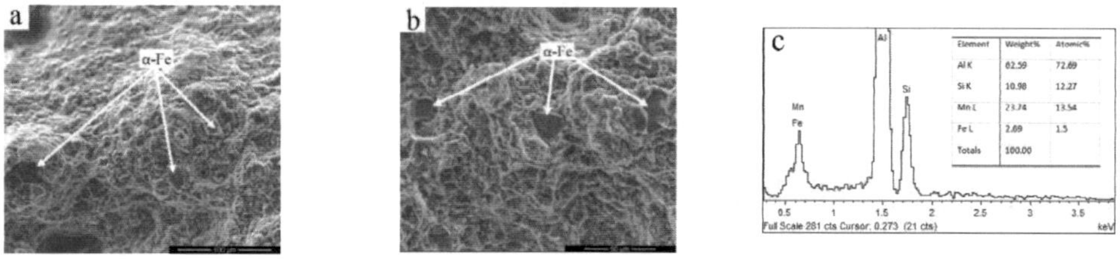

Figure 6. The difference between simulated and experimental results: (a) region of pore P13, (b) region of pore P19, (c) EDS analysis results of α-Fe phase.

3.3 Effect of pore on crack nucleation

Many researches have shown that the morphology of pore has significant effects on crack initiation. In the simulation, micro-cracks originated from the pores were found before the main crack passing through. Figure 7 plotted the times of microcrack initialization against the PZ values of pores, i.e. the projected area of a pore on the horizontal plane. A larger PZ value represents a larger reduction of the loading-bearing area in the specimen during the tensile test. Generally the initiation time of micro-crack decreases as the PZ value increases for all pores except the pore P4 and P9. Figure 8 shows the stress distribution and micro-crack formation around the pore after 71 s and 110 s. According to table 1, the sphericity of pore P4 is 0.48 (a shape close to gas pore) and the Z value (the distance to the plane of the minimum cross section) is 0.25 mm. As a result, the stress level of pore was lower at 71 s as shown in figure 8a and the crack nucleation on it is later (figure 8b). On the contrary, the pore P9 has a smaller sphericity of 0.38 and a Z value of 0.1, i.e. relatively closer to the minimum cross section. Therefore, larger stresses were found around the pore and the micro-crack nucleated at early time of 71 s as displayed in figure 8c. At 110 s, three micro-cracks formed on the pore, see figure 8d. The simulation results indicate that pores with lower sphericity and closer to the minimum sections cause more serious local stress concentration and accelerate the formation of micro-cracks nearby.

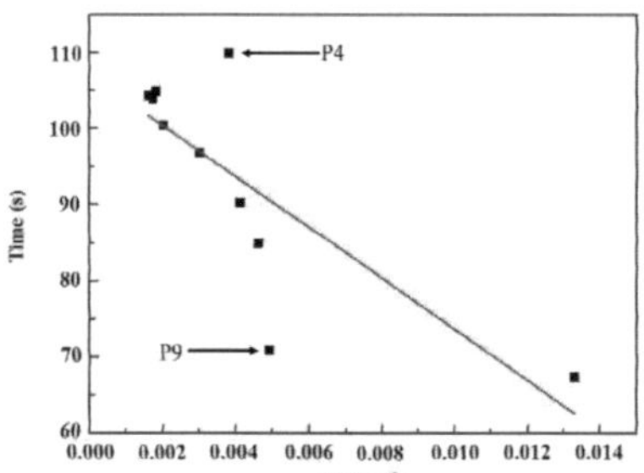

Figure 7. Times of micro-crack initialization on pores with different projection areas

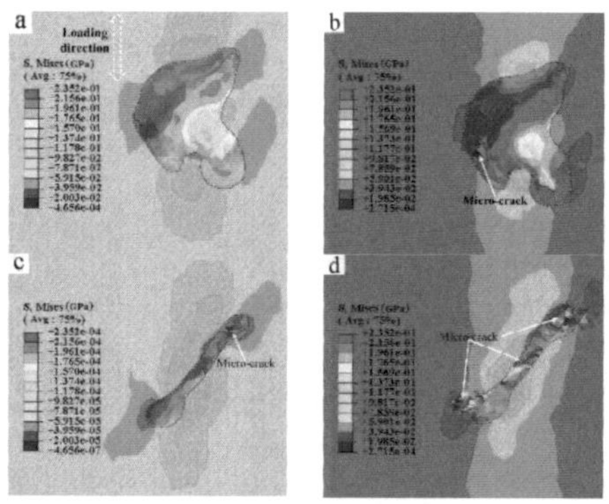

Figure 8. Stress distribution and crack initiation around:(a) P4 at 71s, (b) P4 at 110s,(c) P9 at 71s, (d) P9 at 110s

3.4 Stress distribution during tensile crack propagation

Figure 9 shows the stress distribution and crack formation on whole simulation domain at macroscopic level. In figure 9a, stress concentration develops at the root of the notch at the beginning of tensile test. With the increase of tension, the stress concentration near the notch proceeds to develop and leads to the local failure by crack. As shown in figure 5b, the crack forms at the root of the notch and propagates down to the region below the minimum cross section where pores of P2, P19 and P20 are found according to the CT data in figure 4b. However, with more increase in tension, the crack turns up and moves towards the minimum cross-section as shown in figure 9c. This is because more pores located at the center of the minimum cross-section as shown in figure 4b, where tends to develop more serious stress concentration. At the late stage of the tensile testing, the crack propagates downward again and leads to the complete failure to the specimen, as shown in figure 9d. This downward propagation was induced by the pores of P23 and P24 which are located at the minimum cross-section in figure 4b. The above FE analysis incorporating the 3D pores gives a clear picture of the significant effect of pore distribution on the crack propagation. The crack tends to propagate towards to the region where pores are located since the more serious stress concentration would develop nearby.

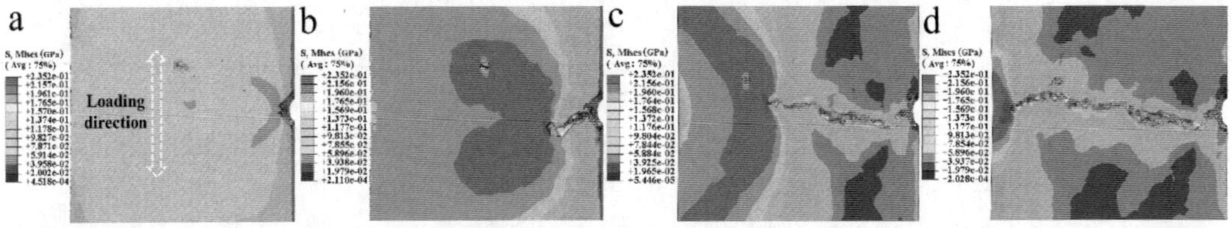

Figure 9. Mises stress distribution during tensile: (a) tensile beginning, (b) crack propagation,(c) crack expansion, (d) crack propagation in the final fracture.

4. Conclusion

The X-ray CT was used to reconstruct the pores in die-cast AlSiMgMn alloy. Using the FE meshes from the X-ray data, the tensile failure process of the alloy was simulated. The simulation was compared with the experimental results. It can be summarized:

(1) the simulated crack path agrees well with the experiment result. Most of the pores on the simulated fracture surface were experimentally observed. The disagreement between the simulation and experiment was caused by the brittle clustered α-Fe intermetallics.

(2) the simulation reveals that the porosity defects significantly affect the crack path during casting failure process. Crack tends to propagate towards the region where pores were located.

(3) the pores with lower sphericity and closer to the minimum sections cause more serious local stress concentration and therefore accelerate the formation of micro-cracks nearby.

Acknowledgments

The authors thank the support of National Industry Base Enhanced Project (TC160A310-10-01) and Guangzhou Sci. & Tech. Project (201802030012, 2014Y2-00214).

Reference

[1] Cole G S, Sherman A M 1995 *Mater. Charact.* **35** 3.
[2] Zhao H D, Wang F and Li Y Y 2009 *J Mater. Process. Tech.* **209** 4537.
[3] Tiryakioğlu M, Campbell J, Staley J T 2003 *Scripta. Mater.* **49** 873.
[4] Linder J, Axelsson M and Nilsson H 2006 *Int. J. Fatigue,* **28** 1752.
[5] Mayer H, Pakyriacou M P and Zettl B 2003 *Int. J. Fatigue*, **25** 245.
[6] Kuwazuru O, Murata Y and Hangai Y 2008 *J of Solid Mechanics & Mater. Eng.*, **2** 1220.
[7] Kastner J, Harrer B and Degischer H P 2011 *Mater. Charact.* **62** 99.
[8] Bodla K K and Murthy J Y 2010 *Mater. Sci.* **50** 622.
[9] Weck A, Wilkinson D S and Maire E 2008 *Acta Mater.* **56** 2919.
[10] Nicoletto G, Konečná R, Fintova S 2012 I*nt. J. Fatigue* **41** 39.
[11] Wan Q, Zhao H and Zou C 2014 *ISIJ Inter.* **54** 511.
[12] Niklas A, Bakedano A and Orden S 2015 *Materials Today Proceedings*, **2** 4931.
[13] Abaqus/CAE User's Manual.
[14] Dominik B, Stefan P and Marc H 2015 *Metall. Mater. Trans. A*, **46** 1035.

IOP Conf. Series: Materials Science and Engineering **529** (2019) 012076 doi:10.1088/1757-899X/529/1/012076

On the use of heterogeneous thermomechanical and thermophysical material properties in finite element analyses of cast components

Johan Jansson*, Jakob Olofsson and Kent Salomonsson

Department of Materials and Manufacturing, School of Engineering, Jönköping University, Box 1026, SE-551 11 Jönköping, Sweden

* Corresponding author: johan.jansson@ju.se

Abstract. Cast components generally show a heterogeneous distribution of material properties, caused by variations in the microstructure that forms during solidification. Variations caused by the casting process are not commonly considered in structural analyses, which might result in manufacturing of sub-optimised components with unexpected in-use behaviour. In this paper, we present a methodology which can be used to consider both thermomechanical and thermophysical variations using finite element analyses in cast components. The methodology is based on process simulations including microstructure modelling and correlations between microstructural features and material properties. Local material data are generated from the process simulation results, which are integrated into subsequent structural analyses. In order to demonstrate the methodology, it is applied to a cast iron cylinder head. The heterogeneous distribution of material properties in this component is investigated using experimental methods, demonstrating local variations in both mechanical and physical behaviour. In addition, the strength-differential effect on tensile and compressive behaviour of cast iron is considered in the modelling. The integrated simulation methodology presented in this work is relevant to both design engineers, production engineers as well as material scientists, in order to study and better understand how local variations in microstructure might influence the performance and behaviour of cast components under in-use conditions.

1. Introduction

Numerical methods such as Finite Element Analyses (FEA) and topology optimization are commonly used by engineers in order to design and optimize cast components towards high load bearing capacity while simultaneously minimizing weight. However, if the analyst does not also consider the influence of the manufacturing process, the resulting component may become sub-optimised, or show unexpected in-use behaviour.

One important and often overlooked aspect is the spatially varying material behaviour in cast components caused by microstructural variations throughout the geometry. During the solidification of pearlitic hypo-eutectic flake graphite cast iron, a dendritic austenite structure grows as the primary phase, with features governed by the chemical composition and the inoculation conditions [1]. Graphite is precipitated together with eutectic austenite in a eutectic reaction, where the local composition, cooling- and inoculation conditions determine the morphology of the graphite as well as the size of the eutectic cells [2, 3]. These features have been studied and correlated to both mechanical and physical material properties in several works. The elastic and plastic behaviour is influenced by graphite morphology [4], but also the metal matrix where a smaller eutectic cell size [5] and a denser primary dendritic network [6] has been associated with increasing ultimate tensile strength (UTS). In addition, a higher cooling rate in the subsequent solid-state transformation of austenite to pearlite refines the lamellar spacing of the pearlite, which also contributes to increased material strength with respect to plastic deformation [7]. In addition to microstructure-based variations in mechanical properties, the temperature dependency of both elastic and plastic tensile and compressive properties has been investigated and shown to vary significantly e.g. by Maraveas *et al.* [8]. Holmgren, Dioszegi and Svensson investigated the connection between thermal conductivity, carbon content and solidification conditions, and found that there exist linear relationships between thermal conductivity and the carbon content, carbon equivalent as well as the fraction of former primary austenite which transforms into pearlite [9]. More specifically, a slower cooling rate leads to a higher tendency for straight, long graphite flakes and a high thermal conductivity, while a faster cooling rate leads to an increased fraction of primary austenite and a decreased thermal conductivity [10]. The difference in thermal conductivity has also been found to be higher at room temperature and decreasing with increasing temperature [9].

A simulation strategy has previously been presented [11] which aims to predict and consider local variations of elasto-plastic material behaviour in FEA, based on microstructural features predicted by casting process simulations. The strategy, denoted *the closed chain of simulations for cast components* has been evaluated for both ductile iron and cast aluminium components, revealing that the local variations may significantly affect the distribution of stresses and strains when the casting is subjected to load [12]. The approach has been limited to temperature independent mechanical properties, which has prevented it from predicting the performance of components such as engine blocks and cylinder heads, where temperature dependence and physical properties are important aspects to consider. Furthermore, cast irons exhibit a strength-differential effect, i.e. the plastic material behaviour differs significantly between tensile and compressive loading [13], which has not been considered previously. The current work aims to extend and improve the simulation strategy by enabling consideration of temperature dependent mechanical and physical properties and the strength-differential effect of the material. The methodology is demonstrated on a cast iron cylinder head, which has been investigated experimentally for mechanical and physical property variations. In the current work, local material properties are directly linked to the solidification time of the cast material, an approach which has previously been successfully applied for mechanical properties by Ferro *et al.* [14]. The integrated simulation methodology presented in this work is relevant to design engineers, production engineers as well as material scientists, in order to study and better understand the influence local microstructural variations has on the performance and behaviour of cast components under in-use conditions.

2. Methods

2.1. Experimental procedure

Physical and mechanical testing was performed in order to evaluate the temperature dependent material properties at various locations throughout the studied component. Tensile and compression testing was performed, as well as Differential Scanning Calorimetry (DSC), Laser Flash Analysis (LFA), thermal expansion using a dilatometer and density using Archimedes principle. The sample extraction points are illustrated in Figure 1 and were selected based on solidification time as predicted by a casting process simulation, with the aim of finding regions with different microstructural characteristics. Two characteristic solidification times were selected which will be referred to as "fast" and "slow" solidification in the following analysis and correspond to thinner and thicker sections of the casting. Further details regarding the experimental investigation will be presented in an accompanying work.

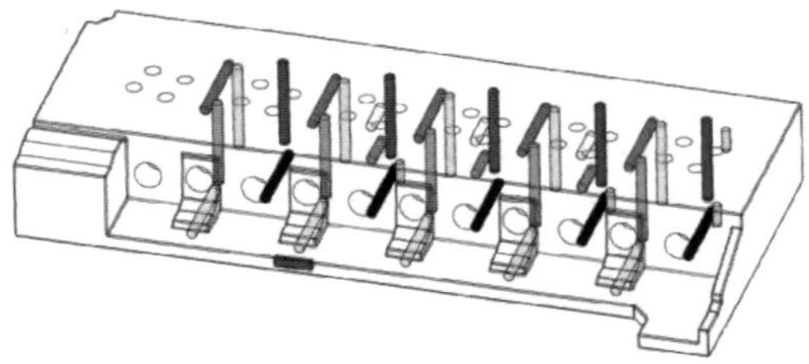

Figure 1. Extraction points for tensile testing (dark grey, orange, green, teal, purple and black), compressive testing (light grey) and physical properties (yellow, red and blue).

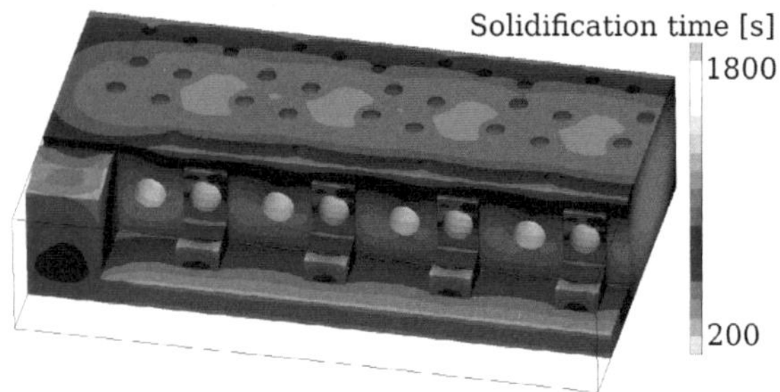

Figure 2. Solidification time distribution and variation in the simplified cylinder head model.

2.2. Material data modelling

The non-linear mechanical material behaviour was modelled using Hollomon's equation, $\sigma = K^*\varepsilon_{pl}^n$, where the flow stress σ (MPa) is given in terms of plastic strain ε_{pl}, and the strain hardening parameters n (-) and K (MPa). Curve-fitting was performed using a Hooke and Jeeves search pattern-based optimization algorithm, where the difference between experimental stress-strain curves and Hollomon's equation is minimized until n, K and Young's modulus are found [15]. This methodology allows the modelling and expression of the elasto-plastic material behaviour in terms of microstructure or solidification time. Linear material data models on the form $f(t) = A(Bt + C)$ were created based on the optimization results, where t (s) is the solidification time from casting process simulations, A is a simple temperature scaling factor based on the data presented by Maraveas *et al.* [8], B is the change in property per second of solidification time and C is the base property value at 200 seconds solidification time. Models for temperature dependent thermal expansion, density and specific heat were extracted from the experimental data in the same way.

2.3. Simulations

The MAGMAiron™ module of MAGMASOFT™ version 5.3.1 was used in order to study the heat extraction and microstructure evolution of the cylinder head during the casting process. Only solidification was simulated. The chemical composition of the cast alloy and the sand mould was provided by the cylinder head manufacturer. A temperature dependent Heat Transfer Coefficient was used between sand mould, cooling plates and the cast component. The MAGMAlink™ module of MAGMASOFT™ was then used in order to map the solidification time to each element of a structural mesh consisting of approximately 300 000 first order tetrahedral elements. The overall methodology is illustrated principally in Figure 3.

The static thermomechanical and transient thermophysical behaviour of the cylinder head was studied using Abaqus™ 6.14.3. An in-house Python code was applied in order to generate element specific material properties, based on the results of the casting process simulation. The code is an extension of what has previously been presented by Olofsson and Svensson in [11]. The main difference is that we now also consider local variations and temperature dependency in density, elasticity, plasticity and thermal conductivity, as well as the strength-differential effect, i.e. that both elastic and plastic material behaviour differs depending on if the local stress state is tensile or compressive. In order to reduce the complexity of the simulation model, a reduction of the number of material definitions was performed. This reduction is based on a small tolerance, where elements with similar material properties are grouped together using the same material definition. By letting the Young's modulus, Hollomon parameters and thermal conductivity vary using 100 discrete values each, the initial element specific material definitions were reduced down to a total of 1750. The influence of this simplification has previously been investigated and demonstrated to be low [12].

A transient thermal analysis was performed using a temperature load of 400°C applied to all cylinder surfaces during 5000 seconds in order to reach thermal equilibrium. The temperature field of the complete geometry was compared to an identical model using only one homogeneous material definition. This homogeneous material definition was taken as the average of the initial 300 000 element-specific properties, i.e. based on the variation established experimentally, and therefore a very good approximation. The same comparison was made with and without material property temperature dependency. A series of static analyses were performed using the same methodology, where a pressure linearly increasing from 0 to 150 MPa was applied to all cylinder surfaces. In these models, the variables are material homogeneity, temperature dependency of material properties and inclusion of compressive material data, which resulted in 2^3 combinations and models. The stress-field difference was then analysed by successively subtracting simulation results from the each other.

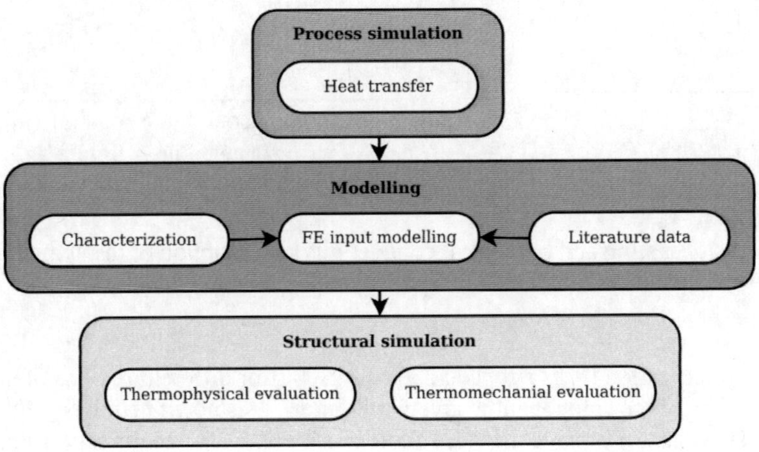

Figure 3. Methodology flowchart.

3. Results and Discussion

3.1. Material properties

The results from the physical property testing are presented in Table 1. Figure 4a shows the compressive strength of the material as measured from one sample, suggesting that the ultimate compressive stress (UCS) increases with the refinement of microstructural features caused by a lower solidification time, in agreement with Sjögren [4]. The apparent UCS of 700 - 950 MPa and maximum compressive strain of approximately 10% may be compared to the Ultimate Tensile Strength (UTS) of 180 - 220 MPa and 0.7% tensile failure strain from the same alloy and component. These values are also in agreement with findings from Maraveas *et al.* [8]. The Coefficient of Thermal Expansion (CTE), the density ρ and the specific heat C_p were not found to be significantly influenced by solidification time and are therefore presented as one value per temperature. Thermal conductivity however was found to increase with increasing solidification time at all measured temperatures, which is illustrated in Figure 4b. The difference is in agreement with previous findings such as Holmgren [10]. Previous studies [9] have however found the difference in thermal conductivity between solidification conditions to be larger at room temperature than at elevated temperatures, which was not seen in the current study. The coefficient of thermal expansion and specific heat were found to increase with temperature, while density decreased.

Table 1. Temperature dependency of the coefficient of thermal expansion CTE, density ρ, specific heat C_p, thermal conductivity λ for slow (1800 s) and fast (200 s) solidification. Values are given with an error of one standard deviation.

Temperature (°C)	CTE (μm/(m°K))	ρ (kg/m³)	C_p (J/(kg*K))	λ_{slow} (W/(m*K))	λ_{fast} (W/(m*K))
25	10±0.79	7300±19	470±3.1	49±0.23	48±0.20
100	12±0.29	7200±19	510±4.0	47±0.35	46±0.14
300	15±0.14	7200±19	600±1.1	43±0.49	42±0.19
400	15±0.57	7100±19	640±3.2	41±0.56	40±0.08
500	16±0.72	7100±19	700±6.7	40±0.46	38±0.17

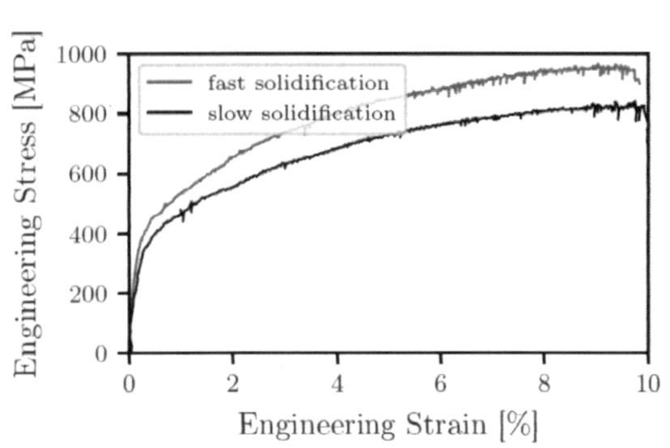

a)

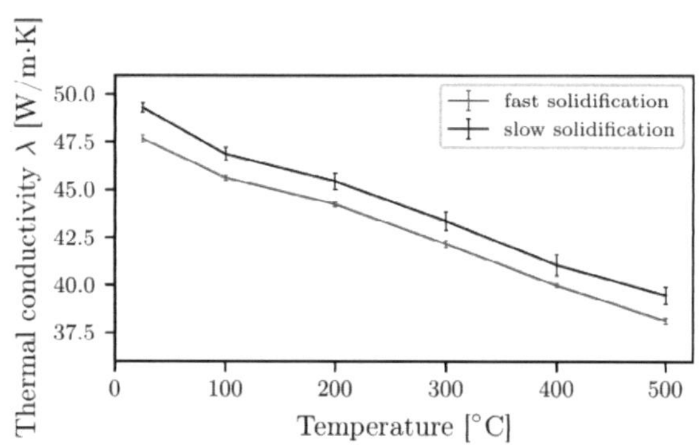

b)

Figure 4. Results for a) compression testing and b) thermal conductivity as a function of test temperature, for samples with slow and fast solidification.

3.2. *Simulation results*

The process simulation predicts solidification times between 200 and 1800 s for different regions of the cylinder head. A cross-sectional view of the component is given in Figure 2. It is noted that most of the volume, which is under both thermal and mechanical load, solidifies slowly, i.e. a solidification time above 1000 s. It should also be pointed out that the geometry presented in this work is a simplification of the actual component due to confidentiality.

To evaluate the influence of the different types of material data on the FEA simulation results, the model containing all the available data was set as the reference. Comparisons were then made where the results of the other models with reduced material data are subtracted from the reference model. The error in von Mises stress, taken as the difference between the reference and the reduced model is summarized in Table 2. Also, the difference in the predicted temperature field was evaluated in the same way. The comparison between models containing homogeneous (one material definition) and heterogeneous microstructure-based material properties shows only a small difference in stress of about 5 MPa. This is explained by the very small difference in microstructure in the volume under stress. The second comparison was made between models using temperature dependent data versus room temperature data using an elevated ambient temperature of 500°C. By neglecting temperature-dependency, one under-estimates the local stress by 23 MPa (30%), and over-estimates it by 27 MPa (42%) in other regions. If the strength-differential effect is not considered, the stress is under-estimated by 82 MPa (69%) in some regions, while it is over-estimated in other regions by 30 MPa (50%). For all of these comparisons, the error distribution was found to be non-intuitive, i.e. it is not distributed with any intuitive pattern with respect to the load-case.

Table 2. Comparison of the difference in von Mises stress and temperature obtained in the FEA simulations when different types of material data is removed.

	$\Delta\sigma_{vM,under}$		$\Delta\sigma_{vM,over}$		ΔT_{under}		ΔT_{over}	
Removed data	(MPa)	(%)	(MPa)	(%)	(°C)	(%)	(°C)	(%)
Heterogeneous distribution	4.7	9	3.6	7	0.29	0.08	0.04	0.02
Temperature dependency	23	30	27	42			51	17
Strength-differential effect	82	69	30	50				

Regarding temperature distribution, the difference between homogeneous and heterogeneous simulation models was found to be very small, 0.3°C (0.08%). Again, this is explained by the high level of microstructural homogeneity in the evaluated part of the current casting. If the temperature dependency of the material data is ignored, an overestimation in temperature of 51°C (17%) at the outermost surface from the applied temperature load was seen. In this case the error in temperature distribution is much more intuitive, increasing with distance from the applied load.

4. Conclusions

By evaluating and comparing simulation models with varying degrees of material heterogeneity, it was shown that small errors are introduced by neglecting the microstructural differences due to the casting process in the studied component. Much larger errors are introduced by neglecting the strength-differential effect or the temperature-dependency of the material properties. The error distribution is also non-intuitive in most cases, highlighting the relevancy of the method. While these conclusions are specific to the studied component, load-cases and casting conditions, it should be noted that the methodology been is general to any load-case, geometry or even manufacturing process which induces variations in material behaviour, i.e. not necessarily limited only to cast components.

Acknowledgement

The Swedish Knowledge foundation is greatly acknowledged for financing the CompCAST+ and ODISSEE2 projects at the School of Engineering, Jönköping University. Aron Brehmer, Fredrik Beckius and Robin Gustafsson are also acknowledged for their contributions to the experimental part of this work.

References

[1] Dioszegi A, Fourlakidis V and Lora R 2015 Austenite dendrite morphology in lamellar graphite iron *Int. J. Cast Metal R.* **28** 310–17

[2] Elmquist L, Salera S and Dioszegi A 2010 Inoculation and its effect on primary solidification structure of hypoeutectic grey cast iron *Int. J Cast Metal R.* **23** 124–9

[3] Collini L, Nicoletto G and Konecna R 2008 Microstructure and mechanical properties of pearlitic gray cast iron *Mater. Sci. Eng. A* **488** 529–39

[4] Sjögren T and Svensson I 2004 Modelling the effect of graphite morphology on the modulus of elasticity in cast irons *Int. J. Cast Metal R.* **17** 271–9

[5] Fourlakidis V, Diaconu L and Dioszegi A 2010 Effects of Carbon Content on the Ultimate Tensile Strength in Gray Cast Iron *Mat. Sci. Forum* **649** 511–6

[6] Fourlakidis V and Dioszegi A 2014 A generic model to predict the ultimate tensile strength in pearlitic lamellar graphite iron *Mater. Sci. Eng. A* **618** 161–7

[7] Sjögren T and Svensson H 2011 Study of the eutectoid transformation in grey cast irons and its effect on mechanical properties *Key Eng. Mater.* **457** 157–62

IOP Conf. Series: Materials Science and Engineering **529** (2019) 012076 doi:10.1088/1757-899X/529/1/012076

[8] Maraveas C, Wang Y C, Swailes T and Sotiriadis G 2015 An experimental investigation of mechanical properties of structural cast iron at elevated temperatures and after cooling down *Fire Safety J.* **71** 340–52
[9] Holmgren D, Dioszegi A and Svensson I L 2007 Effects of carbon content and solidification rate on thermal conductivity of grey cast iron *China Foundry* **4** 210–4
[10] Holmgren D and Svensson I L 2013 Thermal conductivity–structure relationships in grey cast iron *Int. J. Cast Metal R.* **18** 321–30
[11] Olofsson J and Svensson I L 2012 Incorporating predicted local mechanical behaviour of cast components into finite element simulations *Mater. Des.* **34** 494–500
[12] Olofsson J and Svensson I L 2013 The effects of local variations in mechanical behaviour – Numerical investigation of a ductile iron component *Mater. Des.* **43** 264–71
[13] Altenbach H, Stoychev G B and Tushtev K N 2001 On elastoplastic deformation of grey cast iron *Int. J. Plast.* **17** 719–36
[14] Ferro P, Borsato T, Bonollo F and Padovan S 2018 A rapid approach to estimate the mechanical properties of grey cast iron castings Acta Metall. Slovaca **24** 213 213–22
[15] Svensson I L and Salomonsson K 2018 Mathematical Characterization of the Tensile Deformation Curve of Cast Iron Materials *Mat. Sci. Forum* **925** 444–450

Determination of pattern allowances for a steel casting using an inverse elastoplastic deformation analysis

D Galles[1], J Lu[2], and C Beckermann[2]

[1]Oak Ridge Associated Universities, Oak Ridge, TN 37830, USA

[2]Department of Mechanical Engineering, University of Iowa, Iowa City, IA 52242, USA

Corresponding author: daniel.j.galles.ctr@mail.mil

Abstract. The determination of pattern dimensions using simulation is an inefficient trial-and-error process that requires several design iterations. In this study, the finite element inverse elastoplastic analysis is utilized to calculate the pattern geometry in a single iteration for a plastically deformed body. A simplified casting system is simulated to demonstrate the feasibility of the inverse method. An inverse simulation is performed first to calculate the pattern shape. This configuration is then used as the input geometry for a forward simulation, which is shown to successfully recover the original as-cast shape used for the inverse analysis. Through this sequence, the inverse deformation method is shown to be a viable technique for the determination of pattern allowances in production castings.

1. Introduction

Casting distortions are unintended deformations that occur during casting. Their presence creates dimensional inaccuracies in the as-cast part, which may lead to feature dimensions that do not adhere to customer specifications. For this reason, pattern design is inherently an iterative process for which the patternmaker alters the pattern through a trial-and-error process until the desired dimensions are achieved. For each iteration, a casting is produced and then evaluated through a dimensional analysis. Such a design strategy is not only expensive but also time-consuming. Furthermore, the efficiency of the process heavily relies on the patternmaker's expertise, which can vary considerably.

In recent years, the remarkable increase in computer speeds has coincided with advancements in computational codes suitable for predicting stresses and distortions during casting. Such developments have paved the way for a paradigm shift from physically to digitally-based design within industry. Today's casting simulations can be performed with good accuracy and in a fraction of the time needed to produce and analyze a physical casting. As a result, considerable time and cost savings are realized.

Despite the benefits, digitally-based pattern design is also an iterative procedure. This is due to the fact that the kinematics of FEM codes are implemented in a forward framework. In other words, the simulations input the pattern shape and calculate the as-cast shape. Hence, digitally-based pattern design remains a trial-and-error approach whose efficiency relies on the expertise of the design engineer.

Pattern design belongs to a class of engineering applications in which the unknown reference geometry must be determined. To this end, Yamada [1] and Govindjee et al. [2] pioneered an inverse method that directly solves the equilibrium boundary condition for an elastic material. These studies spurred the development of inverse finite element implementations capable of determining the reference configuration firstly for elastic deformations and later for elastoplastic deformations. A review of studies concerning the elastic formulation is given by Lu and Li [3]. Unfortunately, deformations for the elastoplastic problem are history dependent and thus, the inverse deformation problem is generally ill-posed. A unique solution is only achieved when the loading history or the plastic strain in the deformed state is known, as demonstrated by Germain et al. [4]. However, Lu and Li [5] demonstrated that an inverse boundary value problem gives reasonably accurate results for both displacement and plastic variables for the case of moderately large deformations and nearly monotonic loading.

In this study, the direct approach of inverse analysis developed by Lu and Li [5] is utilized for plastically-deformed material bodies to determine pattern dimensions for the desired as-cast geometry shown Figure 1. The mold is excluded from the stress analyses to prevent numerical difficulties encountered with contact interactions. Consequently, deformations are induced exclusively by thermal stresses. A sequential thermal-mechanical analysis is performed in which the calculated transient temperature fields are inputted into the stress analyses. Loop tests are conducted, in which the inverse problem is solved first to determine the pattern shape, which is subsequently used as the input geometry for the forward analysis. The inverse technique is evaluated by the forward problem's ability to recover the original geometry used for the inverse analysis. Finally, the accuracy of the inverse elastoplastic solution is evaluated through an error analysis.

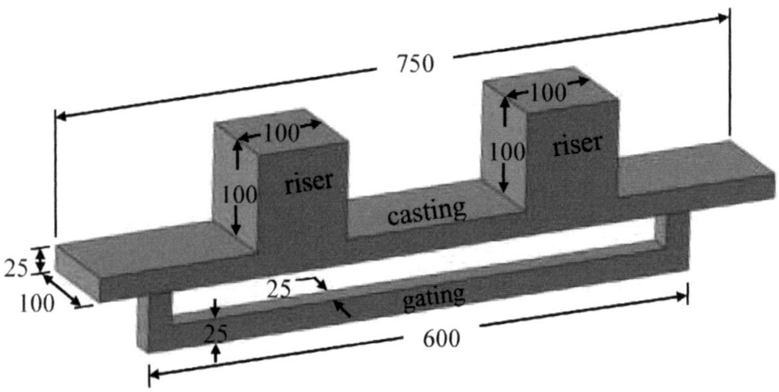

Figure 1. Desired as-cast plate casting geometry. Dimensions in mm.

2. Inverse elastoplastic problem

The inverse elastoplastic problem presented here calculates the pattern geometry from the desired as-cast shape. The model reviewed here is the general quasi-static formulation presented by Lu and Li [5] with the following simplifications. Due to the relatively small casting sizes, body forces will have a negligible impact on distortions and therefore, are not considered. Also, the exclusion of the mold from the mechanical problem precludes the need to include surface tractions in the formulation. These simplifications reduce the boundary value problem to

$$\begin{aligned} \sigma_{ij,j} &= 0 \text{ in } \Omega \\ \boldsymbol{\Phi} &= \overline{\boldsymbol{\Phi}} \quad \text{on } \partial\Omega_u \end{aligned} \tag{1}$$

where σ_{ij} is the Cauchy stress and $\overline{\boldsymbol{\Phi}}$ is a prescribed displacement on boundary $\partial\Omega_u$. Equation 1 determines an inverse motion $\boldsymbol{\Phi}: \Omega \rightarrow B \in R^3$ for the sought reference configuration, *B*, when starting from the given current configuration, Ω. Such a relation is the kinematic inverse of the usual forward deformation $\boldsymbol{\varphi}(\mathbf{X}, t)$ of a material point that progresses from the reference coordinate to the current coordinate. Additional details of the formulation can be found elsewhere [5].

During the inverse solution procedure, the material response obeys a Hencky constitutive relation, which is very close to a linear elastic response at small strains, provided the stress levels do not exceed the yield strength of the casting. For the elastic properties, a constant value of 0.3 was specified for Poisson's ratio and a temperature-dependent Young's modulus was taken from Koric and Thomas [6].

For the case of yielding, the following elasto-plastic constitutive relation is invoked:

$$\sigma = \sigma_0 \left(1 + \varepsilon_p\right)^n \tag{2}$$

where σ is the von Mises stress, σ_0 is the initial yield stress, ε_p is the equivalent plastic strain, and n is the hardening exponent. Temperature-dependent model parameters (σ_0 and n) were determined using a Levenberg-Marquardt non-linear least squares algorithm. Thermal strains were calculated from [7].

3. Numerical example

3.1. *Introduction*

During casting, cooling induces thermal strains that may lead to considerable stresses and associated distortions. For simple casting geometries without internal features (e.g., holes), minimal mechanically-induced distortions can be expected after solidification, as the casting quickly gains the necessary strength to prevent any distortions created by mold-metal interactions. This rationale motivated the design shown in Figure 1. Dimensions are shown in mm. The inclusion of simplified risers and a gating system produced the varying section thicknesses that naturally led to uneven cooling.

The numerical examples presented in this section were conducted utilizing a sequential thermal-mechanical coupling. Heat transfer simulations were performed first to calculate the transient temperature fields, which were then used as inputs for the forward and inverse stress analyses. A loop test served to evaluate the inverse method. For this test, the inverse analysis was performed first to determine the pattern shape, which was subsequently used as the input geometry for the forward analysis. Validation of the inverse technique is based on the loop test's ability to recover the initial geometry used for the inverse analysis.

3.2. *Heat transfer simulations*

For this study, spatial temperature gradients drive the thermal loading that generates distortions. Temperatures were calculated using the casting simulation software MAGMASOFT® [8]. In order to generate realistic casting temperatures, the bonded sand mold was included in the simulation. MAGMASOFT® uses the finite volume method to calculate casting and mold temperatures during filling and cooling. Simulation inputs include temperature-dependent thermophysical properties for the casting and mold. In addition, the temperature-dependent solid volume fraction during solidification and latent heat of solidification are specified in order to calculate heat transfer due to latent heat effects. Finally, an interfacial heat transfer coefficient is inputted to simulate the formation of an air gap at the mold-metal interface. This parameter allows for the partial decoupling of the thermal-mechanical problem. The entire calibration procedure for the heat transfer model can be found elsewhere [9].

Temperature fields at various times are shown at the casting mid-plane of the plate casting in Figure 3. Spatial temperature gradients, caused by variations in section thicknesses throughout the geometry, can be seen throughout the casting, most notably at $t = 5$ min and $t = 21$ min. After these times, the plate temperature fields are nearly isothermal. The calculated temperature fields were written at a sufficient number of time steps to ensure a smooth profile and copied onto the finite element mesh.

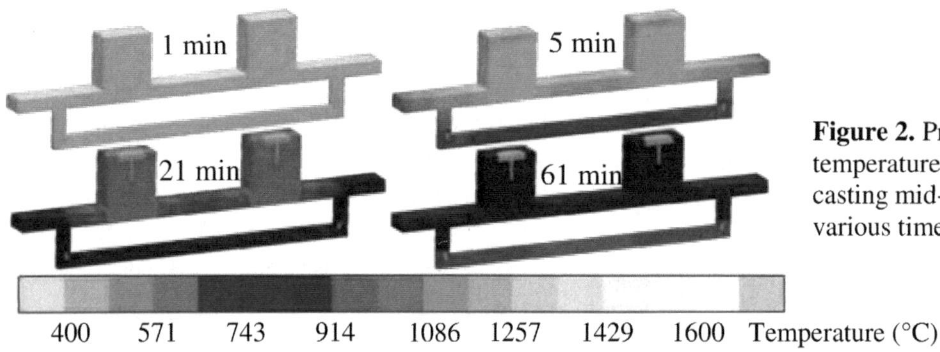

Figure 2. Predicted temperatures at the casting mid-plane at various times.

3.3. Stress simulations

3D Stress simulations were performed using the general purpose finite element code FEAP. In order to prevent rigid body translations and rotations, minimal boundary conditions were specified. Due to symmetry, only ¼ of the geometry was modelled.

Displacement differences between the forward and inverse analyses were quantified through a configuration error, $err(conf)$, which was computed by normalizing differences in the current position **x** by the forward displacement ***u***:

$$err(conf)[\%] = \frac{\int_{\Omega} \left\| \mathbf{x}^{Inverse} - \mathbf{x}^{Forward} \right\| dv}{\int_{\Omega} \left\| \boldsymbol{u} \right\| dv} \times 100 \tag{3}$$

In Equation 3, $\mathbf{x}^{Inverse}$ refers to the input geometry for the inverse simulation, whereas $\mathbf{x}^{Forward}$ relates to the calculated coordinates of the forward simulation.

In addition to the computed errors, the final vertical profile of the casting's top edge is plotted to help the reader visualize the deformations. Also, the temporal evolution of vertical displacement for a single point located at the upper left corner of each casting is plotted.

For the loop test, the inverse analysis was performed first. The uneven cooling calculated by the heat transfer analysis resulted in the calculated configuration shown in Figure 3(a), which reveals 1) the ends of the plate are distorted upward and 2) the casting length increases. Keep in mind that this deformation was calculated within the inverse framework and is therefore reverted. In reality, the plate ends will distort downward. Similarly, the usual thermal contractions that occur during casting are manifested as expansions for the inverse problem. The shape calculated by the forward simulation (see Figure 3(b))

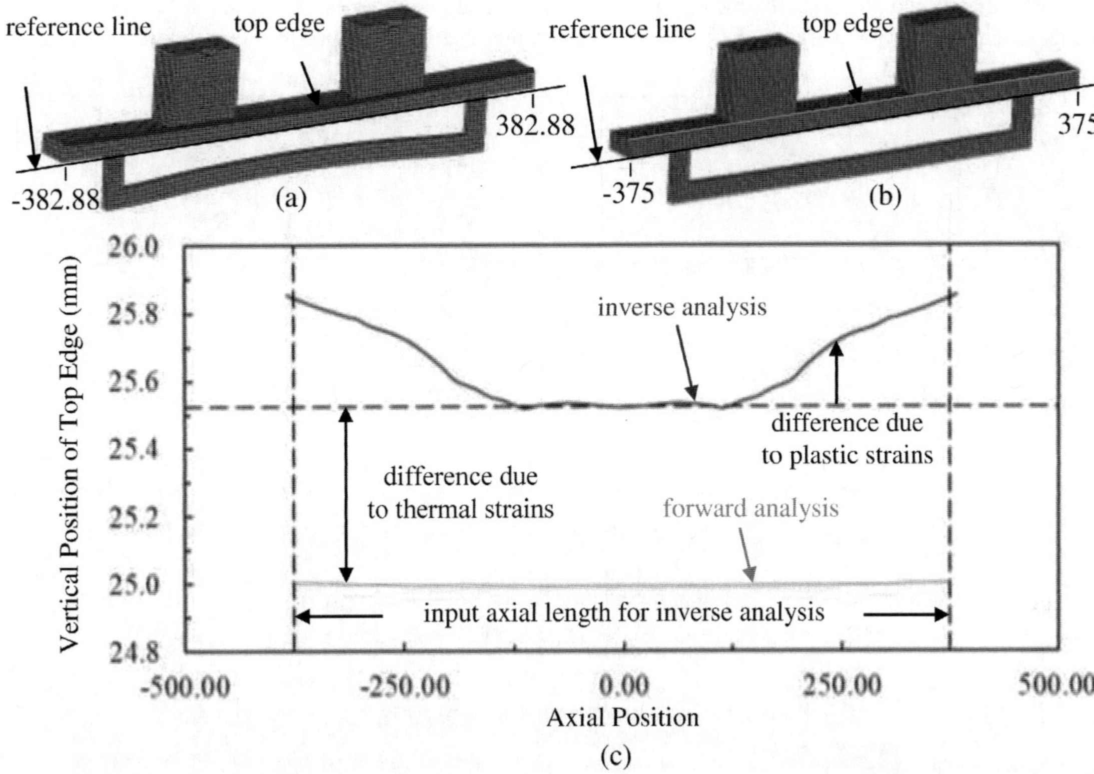

Figure 3. Plate configurations calculated by the inverse (a) and forward (b) simulations. The vertical position along the top edge of the plates is plotted as a function of axial position in (c). A 10x deformation factor is used in (a) and (b).

recovers the input geometry used for the inverse analysis. The final vertical displacement at the top edge for both configurations is plotted as a function of axial position in Figure 3(c). Thermal and plastic strains cause the vertical position of the top edge to evolve from a constant vertical height of 25 mm to the profile shown by the blue curve for the inverse simulation. The forward simulation subsequently recovers the constant vertical profile of 25 mm, as demonstrated by the red curve in Figure 3(c).

The temporal evolution of vertical displacement for a point at the top left corner of the plate is plotted on complete (80,000 s) and 5000 s time scales in Figure 4. At 100 s, the forward curve begins to decrease downward before reaching a minimum value at 600 s. This is caused by rapid cooling and in the gating, which triggers the onset of thermal contraction. Meanwhile, the plate is still mostly liquid and therefore, relatively weak. As a result, thermal contractions in the gating create a bending moment that distorts the plate ends downward. Eventually, the plate solidifies and begins to thermally contract, which counteracts the bending moment that caused the initial downward distortion. Consequently, the plate ends deflect upward until approximately 4500 s. Subsequent negative changes in forward displacement are due to additional thermal strains. The final casting shape is characterized by a slight downward distortion at its ends, which is manifested as a negative vertical displacement at 80,000 s for the forward curve. The forward curve increase from 7000-10,000 s is attributed to a volumetric expansion that accompanies the decomposition of austenite to ferrite and cementite. The final casting shape is characterized by a slight downward distortion at its ends, which is manifested as a negative vertical displacement at 80,000 s for the forward curve. This final shape was anticipated due to the generation of plastic strains at early times that increased the length at the top of the bar and prevented the bar ends from returning to their original vertical position. In general, the configuration error shown in Figure 4(c) is reasonably small (< 2%). The average configuration error for the entire plate casting system is 1.78%.

A comment is warranted regarding the vertical displacement plots in Figure 4, for which the displacement evolutions were equal but opposite. This observation may prompt the reader to suggest

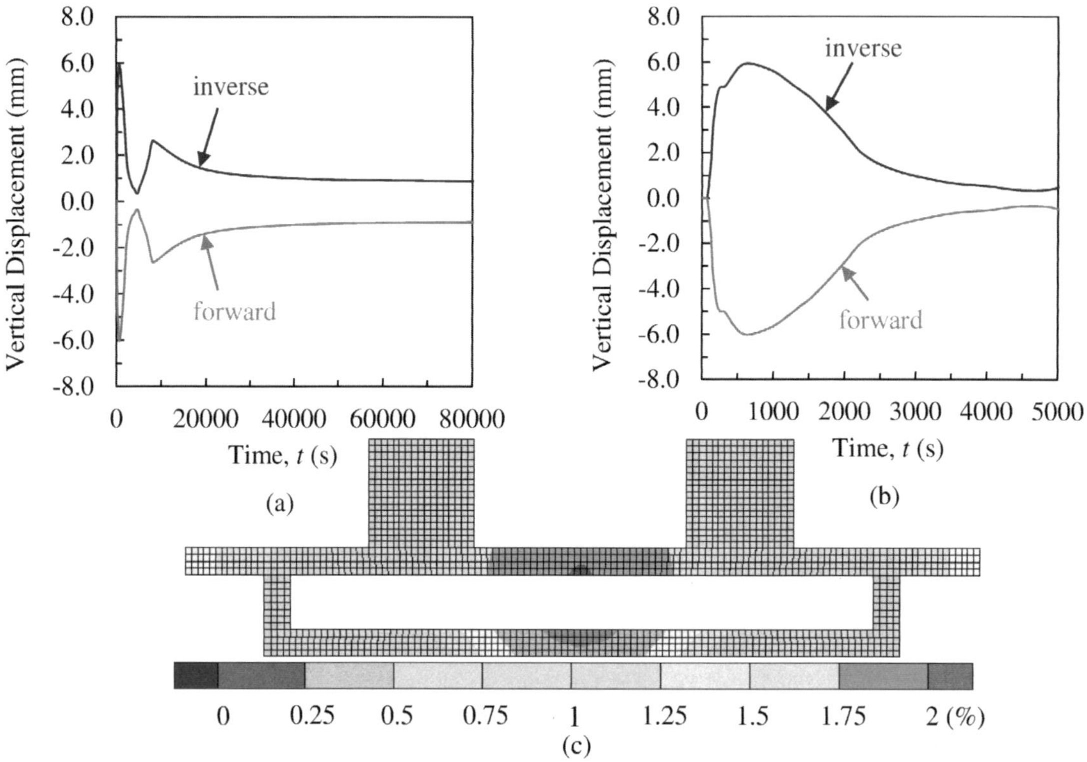

Figure 4. Vertical deflection at the upper left corner of the plate plotted as a function of time on complete (a) and 5000 s (b) time scales. The configuration error is shown in (c).

that one could simply subtract the forward displacement from the original configuration in Figure 1 to determine the pattern shape. Such a strategy will predict the pattern shape with good accuracy, provided that the displacements are small. However, for moderately large displacements, the initial configurations for the forward and inverse cases may differ considerably, for which case the calculated displacements will not mirror each other. Thus, the inverse deformation analysis is preferable, as it provides a robust and general method for accurate calculation of pattern dimensions over a wide range of deformations.

Finally, spatial limitations prevented the authors from adding additional stress and strain contour plots. It is notable, however, that the average equivalent plastic strain error ($< 2.5\%$) was similar to the displacement error. Additionally, the simulations calculated monotonic plastic strain evolution, a necessary condition specified by Lu and Li [5].

4. Conclusions

This study explored the viability of the finite element inverse deformation analysis to calculate pattern allowances for steel castings. The basic concept involves finding, by solving a boundary value problem, a deformation that is the kinematic inverse of that for the usual forward analysis As a result, the inverse analysis essentially works backward, starting from the desired as-cast configuration, to determine pattern dimensions in a single design iteration. For this study, the accuracy of the inverse analysis was demonstrated through a loop test, for which the average difference between inverse and forward configurations was found to be less than 2%.

The finite element inverse deformation analysis provides an efficient means to calculate pattern dimensions. Although this study represents an important step towards achieving this goal, further work

IOP Conf. Series: Materials Science and Engineering **529** (2019) 012077 doi:10.1088/1757-899X/529/1/012077

is needed before this technique can be applied to production castings. Primarily, mechanical interactions between the casting and mold must be considered. Currently, the authors only considered distortions caused by thermal stresses. Overcoming this issue is paramount to the success of the inverse deformation analysis. It should also be noted that accuracy of the inverse analysis was evaluated based on comparisons to the forward analysis. Thus, the importance of the forward simulation's predictive capability cannot be overstated. The present material model is based one that was previously calibrated for a low alloy steel using in situ data. If the finite element inverse deformation analysis is used for other casting materials, care must be taken to properly calibrate the associated material model parameters for the forward problem. Despite these concerns, the present results lend confidence to the inverse deformation analysis for its ability to determine pattern allowances accurately and efficiently.

Acknowledgements
Research was sponsored by the Army Research Laboratory and was accomplished under Cooperative Agreement Number W911NF-18-2-0161. The views and conclusions contained in this document are those of the authors and should not be interpreted as representing the official policies, either expressed or implied, of the Army Research Laboratory or the U.S. Government. The U.S. Government is authorized to reproduce and distribute reprints for Government purposes notwithstanding any copyright notation herein.

References
[1] Yamada T 1995 *Technical Report No. 20*
[2] Govindjee S and Mihalic PA 1996 *Comput. Methods Appl. Mech. Eng.* **136** pp 47-57
[3] Lu J and Li L 2016 *Comput. Methods Appl. Mech. Eng.* **310** pp 189–207
[4] Germain S, Steinmann P, and Menary G 2011 *AIP Conf. Proc.-American Institute of Physics*.**1353** pp 1169-74
[5] Lu J and Li L 2017 *Finite Elements in Analysis and Design* **130** pp 1-11
[6] Koric S and Thomas BG 2006 *Int. J. Numer. Methods Eng.* **66** pp 1955-89
[7] Galles D and Beckermann C 2016 *Metall. Mater. Trans. A.* **47** pp 811-29
[8] Magma GmbH, MAGMAsoft, Kackerstrasse 11, 52072. Aachen, Germany
[9] Galles D and Beckermann C 2011 *Proc. of the 65th SFSA Tech. and Operating Conf.* Chicago

IOP Conf. Series: Materials Science and Engineering **529** (2019) 012078 doi:10.1088/1757-899X/529/1/012078

Three-dimensional study of nodule clustering and heterogeneous strain localization for tailored material properties in ductile iron

Jakob Olofsson[1,*], Kent Salomonsson[1], Arne K. Dahle[1] and Ragnvald H. Mathiesen[2]

[1] Department of Materials and Manufacturing, School of Engineering, Jönköping University, Box 1026, SE-551 11 Jönköping, Sweden

[2] Department of Physics, Faculty of Natural Sciences, Norwegian University of Science and Technology, NTNU, NO-7491 Trondheim, Norway

* Corresponding author: jakob.olofsson@ju.se

Abstract. Tailored heterogeneous distributions of microstructural features enable extraordinary material performance in biological and physiological structures such as trees, the aortic arch, human teeth and dinosaur skulls. In ductile iron, a heterogeneous distribution in size and morphology of graphite nodules and variations of the fractions of ferrite and pearlite are created during solidification, and varies as a function of parameters such as local cooling rate, segregation and flow. In the current work, the size distribution as well as the orientation and relation between graphite nodules is obtained by a three-dimensional reconstruction of a ductile iron microstructure from X-ray tomography. The effect of the nodule morphology and clustering on the localization of plastic strains is studied numerically using finite element analysis of the reconstructed microstructure. Real castings have a variation in geometry, solidification conditions and are subjected to variations in loads. A framework for optimized geometry and solidification conditions in order to design and deliver castings with tailored local material performance is proposed.

1. Introduction

Nature is filled with examples of structures where heterogeneous distributions of sub-scale features have been tailored to enable extraordinary performance when the structure is subjected to load. In trees, local microstructure variations are visibly seen as annual rings, but are also present as large-scale variations throughout the tree. This enable large deformations and absorption of energy when the tree is subjected to strong winds, and also makes the tree insensitive to damage initiation and crack propagation [1]. In the skulls of dinosaurs as well as humans, tailored local variations in the structure of bone enable efficient energy absorption when chewing, and makes the skull highly tolerant to damage [2,3].

Heterogeneous distributions of sub-scale features are also found in synthetic structures made in materials as concrete [4], injection moulded thermoplastics [5], and, not least, metal castings. Most metallic alloys solidify in a non-isothermal transformation through the nucleation and growth of phases, where factors such as chemical composition and local cooling rate affects the final morphology, refinement and amount of different phases. In shape casting, this inevitably leads to heterogeneous microstructures.

In ductile iron castings, the formation and growth of graphite nodules is highly related to the local solidification conditions, leading to spatial variations in nodule characteristics (size, morphology nodule count etc.) as well as in the iron matrix (e.g. volume fractions and refinements of ferrite and pearlite) [6]. In recent research, the formation and growth of graphite nodules have received much attention. Ghassemali *et al.* [7] used Focused Ion Beam (FIB) in a Scanning Electron Microscopy (SEM) to create serial sectioning images to reveal new insights behind the formation mechanism and growth of nodules. Azeem *et al.* [8] used synchrotron X-rays to investigate the graphite nucleation and growth in-situ during solidification. The size and morphology of the graphite nodules has been shown to be strongly related to the initiation and propagation of cracks when the material is subjected to load. Kasvayee *et al.* [9] used Digital Image Correlation (DIC) and in-situ tensile testing with optical microscopy to show that there is a strain localization between nodules, and that clustered nodules interact to form bands of high strain between them. As also pointed out by the authors, the study was performed as a 2D analysis, while the nodules actually interact in 3D. The results are in line with other studies showing that the crack propagation path is highly affected by the graphite nodules [10], and that the fracture process of ductile iron cannot be explained by theoretical models that consider the material as homogeneous [11].

In the current work, 3D reconstruction based on X-ray micro-tomography has been used to investigate the size distribution and clustering of nodules in ductile iron. Numerical simulations have been performed to demonstrate the strain localization and interaction between graphite nodules. If the effects of these variations can be better understood and controlled, there is a large potential to optimise these effects to enable tailored material performance also in shape castings. This paper aims to highlight current and future research in the area and demonstrate how increased knowledge about the formation and effect of graphite nodules can be utilized in an integrated simulation methodology to enable the use of tailored heterogeneous material behaviour in the design process for ductile iron castings.

2. Methods

The preparation of the material, the 3D reconstruction procedure as well as the creation of FE models is further described in accompanying work [12]. In brief, cylindrical samples with a diameter of 1 mm were machined from an industrial casting containing a pearlitic-ferritic matrix. X-ray tomography was performed using a Nikon XT H 225 CT scanner, with the samples positioned so that a voxel size of about 1 um was obtained. A reconstructed volume was generated where the nodules, ferrite and pearlite were segmented into separate volumes. From the reconstructed sample volume, a sub-volume was selected as a Representative Volume Element (RVE) in the shape of a cubic volume with a physical side length of about 0.7 mm. This RVE forms the basis for a Finite Element (FE) model, consisting of about 5 million linear tetrahedral elements. Material data for the different phases are assigned, assuming each phase to be homogeneous and isotropic. The boundary conditions for the load simulation of the micro-scale RVE is taken from a macro-scale load simulation of the ductile iron casting, a heavy truck engine support, performed using the so-called *closed chain of simulations for cast components* [6]. The load is applied in the form of a prescribed displacement vector of $\boldsymbol{u} \approx (5.2,\ -3.5,\ 6.9)$ um on the top surface of the RVE, while the bottom surface is fixed. See [12] for additional details.

3. Results and Discussion

3.1. Nodule characteristics

The tomography data provides a lot of information about the morphology and 3D distribution of the nodules. In total, the volume is found to contain 135 nodules, corresponding to a volume fraction of 26.9%. An analysis of the sizes of the nodules, here measured by the nodules surface area, was first performed. Figure 1 shows the frequency plot for the nodule sizes, grouped into ranges of 0.002 mm^2 width. A representation of the nodule distribution is also shown in the figure, where the nodules are represented by idealised spheres of the same surface area as the real nodules. Most of the nodules are found to have a surface area in the range 0.002-0.004 mm^2; 43 of the 135 nodules, or 32%, are found in this range. There is also a slight peak of nodules with a surface area of about 0.015 mm^2, where 13 nodules are found (9.6%). Only 7 nodules have a surface area larger than this, thus in total about 15% of the nodules are found in the range with a surface area of 0.015 mm^2 or larger.

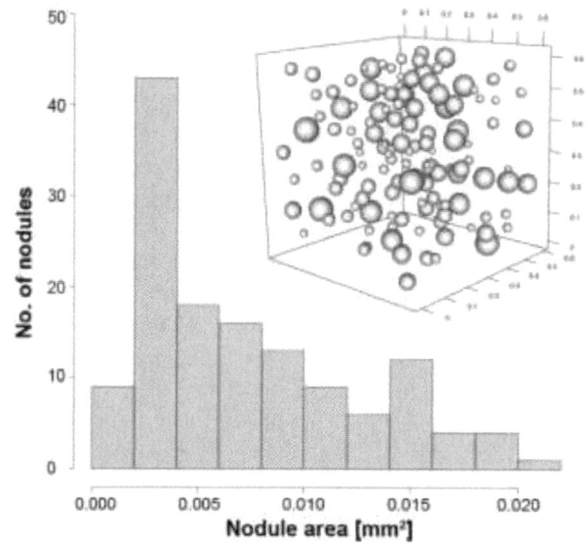

Figure 1. Frequency plot of the nodule sizes found in the RVE. The inserted graphic shows the distribution of the idealised spherical nodule representations in the volume.

Looking at the volumetric nodule size distribution, the frequency plot looks very different. Figure 2a shows the nodule volume fraction for the groups with different nodular diameters. The figure shows that there is one group that stands out; the group with a diameter of about 35 μm contains 22% of the graphite volume. The largest group of nodules in terms of numbers, with an average diameter of 15μm, only contain about 7% of the total graphite volume.

The results highlight the difficulty in describing the 3D characteristics and size distribution of the nodules. The average nodule diameter in this dataset is 23 μm, but that is a value which only describes about 12% of the number of nodules. The largest group in number (diameter around 15 μm) only describes 7% of the graphite volume, and the largest graphite volume is found in only 9% of the nodules.

By plotting the cumulative fractions of both the number of nodules and their volumes, Figure 2b is obtained. It can be seen that though 80% of the number of nodules are found in the groups with an average diameter up to 30 μm, these nodules only contain 45% of the graphite volume. Nodules larger than or equal to 35 μm are found to contain the same volume fraction graphite (45%), though only being 15% of the number of nodules. The cumulative distribution also shows that there is a clear step in volume fractions at a nodule diameter of 35 μm. Comparing the cumulative distributions of the number of nodules and the nodular volume fraction in the same figure, Figure 3a is obtained. It can be seen that both curves have a clear step at a nodule diameter of 35 μm. In Figure 3b, the data is visualised as a bubble diagram, where the size of each bubble represents the number of nodules in each group.

These figures indicate that the nodules can be seen as two different populations; one with diameters larger than 35 μm, and one with smaller diameter, which is interesting from a solidification and nodule growth perspective. There is clearly a small group of large nodules (larger than or equal to 35 μm) that contain 45% of the graphite volume. It is suggested that this population may have nucleated before the second population with smaller nodules. It is possible that the use of nucleation agents and local cooling conditions have led to preferential nucleation sites in the melt where nodules have nucleated freely, maybe already before the nucleation of austenite. These nodules would then grow without the influence of surrounding austenite, which would be the case if the nodules formed between dendrite arms. This was not observed in the before-mentioned synchrotron study at a very low cooling rate [8], but is likely to occur at higher cooling rates, depending also on the use and effect of nucleation agents. It is not in the scope of the current work to further elaborate on the mechanisms behind this, but we here highlight that there are still aspects of the nucleation and growth of nodules that needs to be addressed in future work. Additional studies of several different volumes are required, as well as further studies of the nucleation mechanisms, solidification paths and nodule growth behaviour.

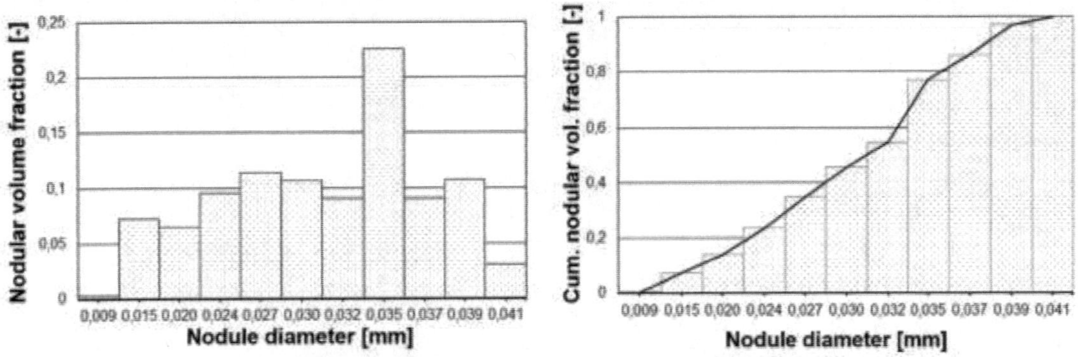

Figure 2. a) The nodular volume fraction for the different nodule sizes. c) The cumulative nodular volume fractions for the different nodule sizes.

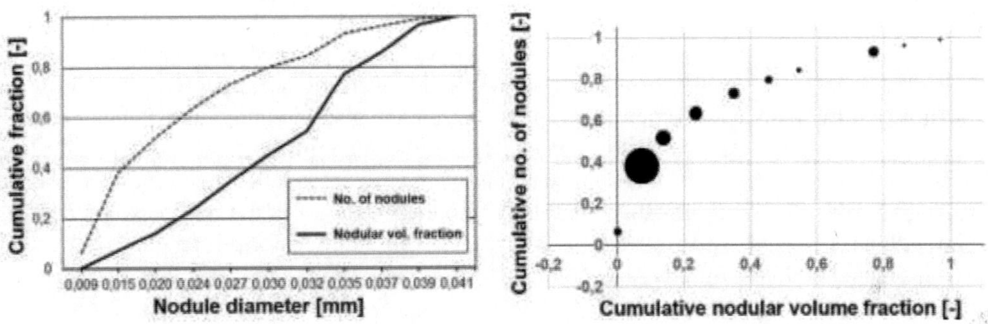

Figure 3. a) Comparison of cumulative fractions of the number of nodules and the nodular volume fraction. b) Bubble diagram of the cumulative fractions, where the size of the bubble represents the number of nodules in the different groups.

Since there is a clear nodule size variation, it is also important to know if nodules of different sizes are evenly distributed or if they are clustered. In the current study, the large nodules, with a diameter of 35um or more, are selected for further investigation of the three-dimensional arrangement of the nodules. By investigating how closely they are positioned to each other, a dendrogram can be constructed. The obtained dendrogram is shown in Figure 4a, and the location of the large nodules in the volume is shown in Figure 4b. The dendrogram shows that these large nodules are indeed not evenly distributed in the volume, but instead positioned in three distinct main clusters, with a distance of about 100 μm between the nodules in the clusters. There can be different reasons for this clustering, and indeed the investigated volume is not sufficient to draw any real conclusions. However, we suggest that if these nodules are indeed nucleated freely in the melt, they may have been pushed to these areas by the growing austenite, and/or affected by other forces as convection or gravity. Additional studies are required to understand and control the nucleation and growth mechanisms properly.

The results also highlight the limited capacity of 2D sectioning to characterise the nodule characteristics; randomly selected 2D sections through this 3D volume would yield very different views of the nodule characteristics depending on their position and would not necessarily reveal the clustering of large nodules or the complex size/volume distribution of graphite.

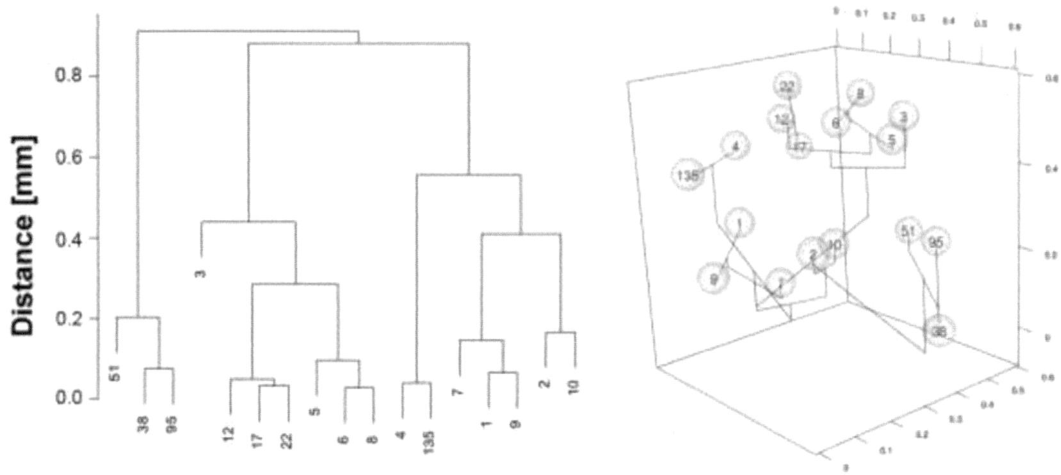

Figure 4. Dendrogram showing the clustering of nodules, as well as their distribution and connections throughout the volume. Only nodules with diameter 35 μm and larger are considered.

3.2. *FE results*

Turning to the FE simulation results, it is observed that the nodules of different sizes interact to form bands of localised strains between them. The results are further described in [12]. In the current simulations, no initial stresses are prescribed in the microstructure. As recently shown by Andriollo *et al.* [13], local residual stresses are formed around the nodules during solidification due to the larger thermal contraction by the matrix compared to the nodules during solid-state cooling. This is expected to contribute to the strain localization, but here the effect and magnitude of this contribution has not been investigated or considered. In addition, the phases are here modelled as homogeneous, which particularly for graphite is a simplification that not completely resembles the real behaviour [14]. The present simulation is thus simplified, but useful to reveal the three-dimensional interaction between non-uniformly distributed nodules. Future work could focus on combining these aspects to investigate how the internal structure of the graphite and the residual stresses around the nodules act in combination with the non-uniform spatial distribution to affect the mechanical performance.

3.3. *Simulation methodology*

In previous work, a closed chain of simulations for cast components has been developed, where casting process simulation is applied to model the variations in microstructure and mechanical behaviour throughout a casting, which is incorporated into a load analysis of the casting in use [6]. This enables an integrated simulation approach which spans casting process, microstructure formation and solidification modelling as well as in-use behaviour of castings. By adopting a knowledge-based perspective, this approach serves as the basis for a framework for an integrated optimization loop [15]. Here the geometry as well as casting process parameters act as variables to tailor the heterogeneous material behaviour in the casting in a multi-objective optimization perspective. This enables the prediction and optimization of heterogeneous material behaviour in the design and development. However, to fully utilize the potential of tailored material properties in industrial castings, there are still research challenges related to understanding and optimizing the nucleation and growth of microstructural features, e.g. graphite nodules, as highlighted in the current work.

4. Conclusions

The three-dimensional characteristics of nodules in a ductile iron is investigated, where it is found that:

- Though there is a large population of small nodules, they only represent a small part of the volume fraction nodules.

- There is a small population of large nodules, 15% of the number of nodules, that contain about 45% of the total volume fraction nodules.
- The large nodules are found to be non-uniformly distributed, clustered into three main groups.

The findings are discussed and elaborated, and topics to be addressed in future research are suggested. An integrated simulation-based design approach to tailor and optimize the local variations in microstructure of industrial castings have been highlighted.

Acknowledgement
The authors would like to acknowledge the Swedish Knowledge Foundation for financial support of the research project ODISSEE2 and the research profile CompCAST+, both part of the research environment SPARK at Jönköping University.

References

[1] Barnett J R and Bonham V A 2004 Cellulose microfibril angle in the cell wall of wood fibres *Biol. Rev. Camb. Philos. Soc.* **79** 461–72
[2] Rayfield E J, Norman D B, Horner C C, Horner J R, Smith P M, Thomason J J and Upchurch P 2001 Cranial design and function in a large theropod dinosaur *Nature* **409** 1033–7
[3] Wroe S, Moreno K, Clausen P, McHenry C and Curnoe D 2007 High-resolution three-dimensional computer simulation of hominid cranial mechanics *Anat. Rec.* **290** 1248–55
[4] Bazant Z P 1976 Instability, ductility, and size effect in strain-softening concrete *ASCE J. Eng. Mech. Div.* **102** 331–44
[5] Michaeli W and Baranowski T 2010 Simulation of the microstructure formation in injection molded semicrystalline thermoplastic parts *J. Polym. Eng.* **30** 29–41
[6] Olofsson J and Svensson I L 2012 Incorporating predicted local mechanical behaviour of cast components into finite element simulations *Mater. Des.* **34** 494–500
[7] Ghassemali E, Hernando J C, Stefanescu D M, Dioszegi A, Jarfors A E W, Dluhoš J and Petrenec M 2019 Revisiting the graphite nodule in ductile iron *Scr. Mater.* **161** 66–9
[8] Azeem M A, Bjerre M K, Atwood R C, Tiedje N and Lee P D 2018 Synchrotron quantification of graphite nodule evolution during the solidification of cast iron *Acta Mater.* **155** 393–401
[9] Kasvayee K A, Salomonsson K, Ghassemali E and Jarfors A E W 2016 Microstructural strain distribution in ductile iron; comparison between finite element simulation and digital image correlation measurements *Mater. Sci. Eng. A* **655** 27–35
[10] Iacoviello F, Di Bartolomeo O, Di Cocco V and Piacente V 2008 Damaging micromechanisms in ferritic–pearlitic ductile cast irons *Mater. Sci. Eng. A* **478** 181–6
[11] Berdin C, Dong M J and Prioul C 2001 Local approach of damage and fracture toughness for nodular cast iron *Eng. Fract. Mech.* **68** 1107–17
[12] Salomonsson K and Olofsson J 2017 Analysis of localized plastic strain in heterogeneous cast iron microstructures using 3D finite element simulations *Proc. of the 4th World Congress on Integrated Computational Materials Engineering (ICME 2017, Ypsilanti)* ed P Mason *et al.* (Springer International Publishing) pp 217–25
[13] Andriollo T, Hellström K, Sonne M R, Thorborg J, Tiedje N and Hattel J 2018 Uncovering the local inelastic interactions during manufacture of ductile cast iron: How the substructure of the graphite particles can induce residual stress concentrations in the matrix *J. Mech. Phys. Solids* **111** 333–57
[14] Andriollo T, Thorborg J and Hattel J 2016 Modeling the elastic behavior of ductile cast iron including anisotropy in the graphite nodules *Int. J. Solids Struct.* **100**–101 523–35
[15] Olofsson J, Salomonsson K, Johansson J and Amouzgar K 2017 A methodology for microstructure-based structural optimization of cast and injection moulded parts using knowledge-based design automation *Adv. Eng. Softw.* **109** 44–52

IOP Conf. Series: Materials Science and Engineering **529** (2019) 012079 doi:10.1088/1757-899X/529/1/012079

Shape of the in-situ (Al,Zn)-Ti reinforcing particles and their influence on structure and structural stability of selected Zn-Al and Al-Zn cast alloys

P K Krajewski[1], A L Greer[2], W K Krajewski[1a]

[1]AGH University of Science and Technology, Faculty of Foundry Engineering,23 Reymonta Street, 30-059 Krakow, Poland
[2]University of Cambridge, Department of Materials Science & Metallurgy, 27 Charles Babbage Road, Cambridge CB3 0FS, UK

Email: [a]krajwit@agh.edu.pl

Abstract. The foundry engineering still needs new cast materials of improved properties. These can be achieved by elaborating completely new alloys and metal matrix composites or by elaborating the alloys/composites basing on the already very well-known matrixes. The good example of the latter solution are high-aluminium zinc (H-Al Zn) and high-zinc aluminium cast alloys (H-Zn Al). Both of these groups show good damping and strength properties but rather low ductility and insufficient structural (and dimensional) stability, caused by long term transformation of the Cu-containing bearing phases. The performed works were devoted, among others, to building composites of improved structural stability reinforced with ternary (Al,Zn)-Ti aluminides. The in-situ reinforcement was built by needle-shape ternary aluminides based on the DO_{22} $TiAl_3$ binary phase introduced with AlTi-based master alloys or by compacted semi-globular ones based on the $L1_2$ Zn_3Ti particles introduced with a ZnTi-based master alloy. The mentioned particles substitute partly or totally for Cu-based bearing phase and the influence of this substitution on the structural stability and tribological properties is also discussed in the paper.

1. Introduction

Cast alloys and composites with Al and Zn based matrices show an increasing range of uses in the global economy. In the years 2007–2016, the world production of Al castings nearly increased by 50%, i.e. from 12.7 to 17.8 million metric tons Fig. 1 [1]. This trend is expected to continue in the prognosis of increased demand for Al alloys, including the Al-based cast composites. As mentioned above, among the cast alloys based on the Al-Zn system, there is a unique group, which has not been thoroughly researched, namely the high-zinc aluminium alloys containing 20% – 40% Zn, by mass. Its uniqueness lies in its wide range of potential applications, such as manufacturing alloys for castings with strength of around 300 MPa and simultaneously good tribological and damping properties. Although the high-zinc aluminium cast alloys were the subject of some early studies [2, 3], the recent literature provides little evidence of current research: what literature deals with mainly are the tribological properties of these alloys [4, 5]. Efforts to optimize the composition, structure and technology of these alloys are under way, aiming at improving their plasticity, e.g. through grain refinement [6-11]. Examinations on grain-refinement of the cast Al-Zn alloys showed that Al-Ti based and Zn-Ti based master alloys are very effective grain-refiners of these alloys [12, 13]. Additionally, a surplus addition of these master alloys can form in-situ reinforcing particles of the matrix.

The changes in the fineness of the refined structure can cause also changes in the damping properties, which in many applications are the key properties of an alloy [14, 15]. Thus, optimization should be aimed at forming a set of characteristics, suitable for high-technology applications, i.e. simultaneously having good strength and plasticity, as well as good tribological and damping properties.

It should be noted that recent efforts focus on energy saving and environmental protection at the same time. The readily contribution to these efforts is development of the new group of comparatively low-melting Al-Zn based composites of high damping, wear and strength properties, reinforced with in-situ particles built with Al-Ti-Zn based intermetallics. It is well known that the matrix of the Al-Zn based alloys (solid solution of Zn in Al) has fcc A1 crystal structure. The in-situ $Ti(Al,Zn)_3$ reinforcing phase formed directly in molten high-zinc aluminium matrix has a similar crystal structure, i.e. $L1_2$ superstructure and very low disregistry of lattice parameter with the matrix alloy [6-8, 11]. The α(Al) and $L1_2$ $Ti(Al,Zn)_3$ nearest-neighbour atomic spacing in their {111} planes are, accordingly, 0.2859 and 0.2798 nm . The percentage difference is only about 2.1%. Thus, the α(Al) is nucleated with its {111} plane on $\{111\}_{Ti(Al,Zn)3}$. This feature should be beneficial when avoiding detrimental interior stresses is considered.

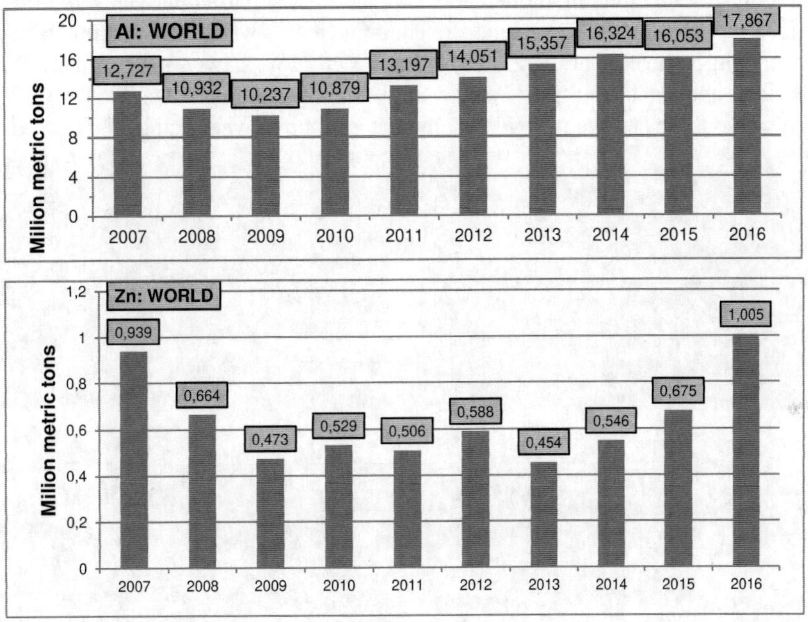

Figure 1. World production of Al and Zn based castings, 2007 to 2016 [1].

The present paper summarizes work on the development of high-aluminium zinc cast alloys and high-zinc aluminium cast alloys on the basis of joint investigations performed over the last ten years at AGH University of Science and Technology – Faculty of Foundry Engineering and at University of Cambridge, UK – Department of Materials Science and Metallurgy [14-17].

2. Experimental

The alloys Zn–(25-26)%Al-(1-2.5)%Cu-(0-1.5)%Ti, Al-27%Zn-2%Cu-0.8%Mn and master alloys Al-33%Cu, Zn-4%Ti and Al-12%Ti were melted from elements of minimum purity 99.9 % (all compositions in wt.%). Melting was performed in an electrical furnace with an argon protective atmosphere, in a clay-graphite crucible of 2-litre capacity. The melted charges were flushed with pure argon for 10 min. Then, 10 min after finishing the flushing, the dross was removed from the melt surface and the (Al,Zn)-Ti master alloy was added to the melted alloys. Five minutes after the master alloy addition, the bath was stirred for 2 minutes with an alumina rod, and the alloys were cast into a green sand mold (90% silica matrix, 7% bentonite, 3% water) with vertical cylindrical cavity ∅30/32 x 80 mm. The mean cooling rate of the samples was about 1K/s, which is typical for green sand casting. From these castings samples ∅30 x 25 mm were cut for LM and SEM examinations. Optical light metallography (LM) was performed using a Leica DM IRM microscope. Scanning electron microscopy (SEM) was performed on unetched samples with a Philips XL30 microscope equipped with an energy dispersive X-ray EDX spectrometer (Link-Isis). Wear-resistance pin-on-disc (T01M device, Poland) investigations were performed using sample rods ∅8 x 24 mm. Dry sliding wear tests were performed against a rotating steel disc of 50 HRC, at a load giving 0.8 MPa pressure and at a sliding speed of about 0.7 m/s, for a total sliding distance of 10 km. Dilatometry was performed using samples ∅5 x 35 mm, which were homogenized in air in an annealing furnace at 370°C for 48 h, and then quenched into water at room temperature. During the first 48 hours after quenching, measurements were made using a DI-105 dilatometer; thereafter the length changes of the samples were manually measured using a screw-micrometer of accuracy of 0.001 mm.

3. Results and Discussion

Grain refinement of the high-aluminium molten alloys by the Zn-4Ti master alloy is very effective, as can be clearly seen in Fig. 2 (a) to (f). At the same time elongation increases while tensile strength remains basically preserved [16]. It should be noted that the Zn-Ti based master alloys have a mass density very close to the inoculated Zn-Al alloys and that they dissolve very quickly already at temperatures of 500 °C. This allows zinc alloys to be treated without the detrimental overheating that is required when using Al-Ti based refiners.

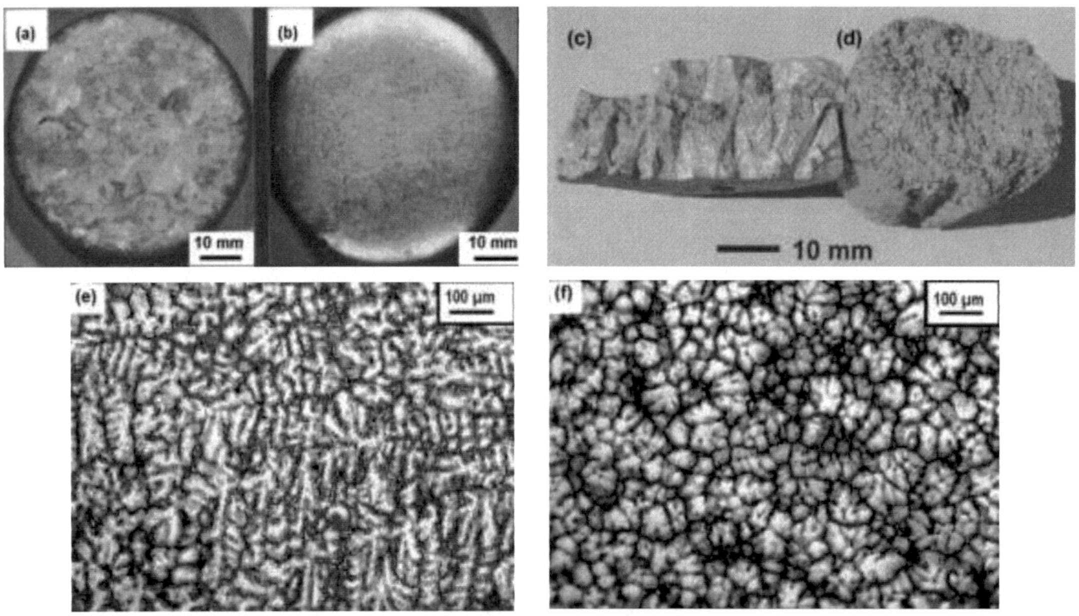

Figure 2. LM pictures of macrostructures and microstructures of the Zn-25Al alloy. (a), (c) and (e) initial alloy; (b), (d) and (f) alloy inoculated with Zn-Ti MA – 0.04 wt.% Ti; (a) and (b) surfaces of the samples; (c) and (d) crystalline fractures; (e) and (f) ground, polished and etched surfaces of a section (Leica DM IRM LM) [16].

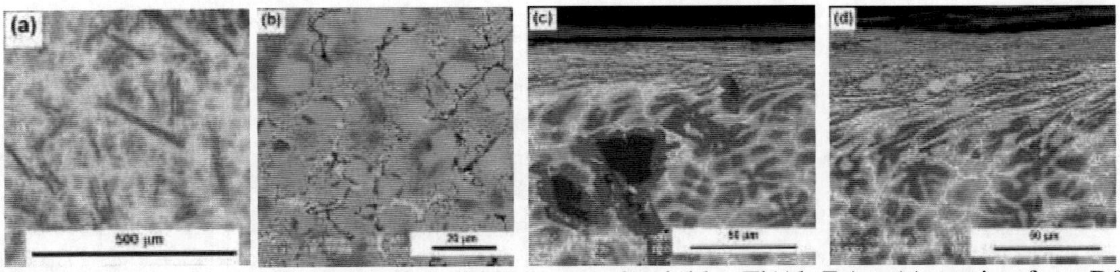

Figure 3. Examples of the master alloys with ternary aluminides Ti(Al, $Zn)_3$: (a) coming from $D0_{22}$ Al_3Ti phase; (b) coming from $L1_2$ $TiZn_3$ phase. (c) and (d): Pin-on-disc examined samples of Zn-25Al-(1.5-2) Ti composites with visible Ti(Al, $Zn)_3$ reinforcing particles.

Fig. 3 shows the in-situ composites with matrix of high-aluminium zinc based alloy. Recently we demonstrated that the binary phases $D0_{22}$ $TiAl_3$ and $L1_2$ phase Zn_3Ti undergo transformation into the $L1_2$ ternary Ti(Al,$Zn)_3$ phase in the binary Al-Zn alloys [6-8, 11, 15]. The mentioned particles play role of the bearing phases which positively influence tribological properties of the examined alloys increasing their wear resistance while preserving coefficient of friction [18]. Fig. 4 shows significant structure refinement of the high-zinc aluminium alloy doped with small addition of the traditional Al-Ti based master alloy. The observed refinement positively influences plastic properties of these alloys [18].

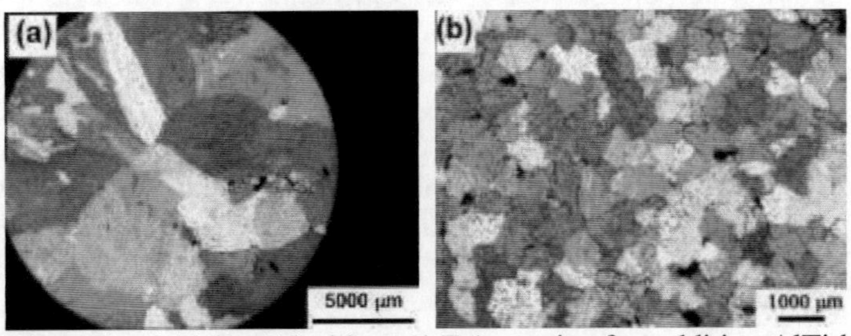

Figure 4. Example of the refined Al-Zn (Al - 20 wt.% Zn) matrix after addition AlTi-based master alloy. (a) initial alloy; (b) alloy doped with the master alloy.

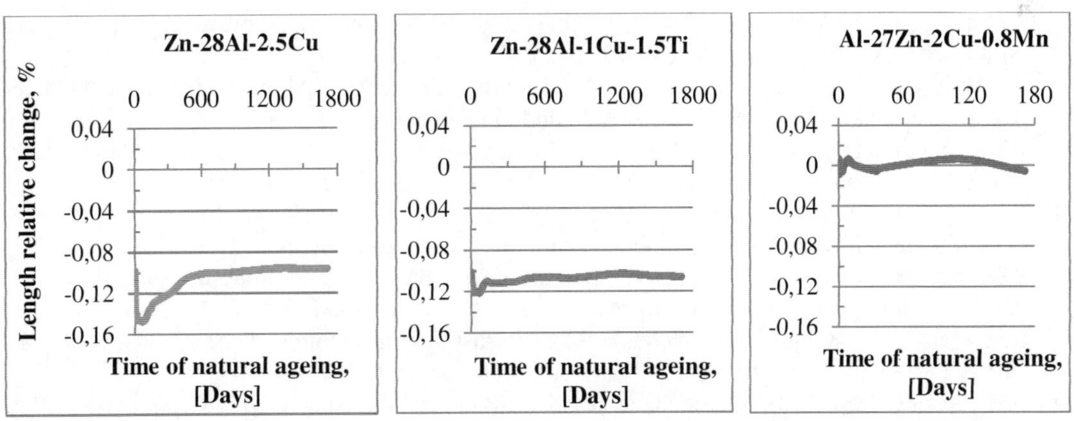

Figure 5. Reducing dimensional changes after partial replacing of Cu with Ti or Mn [15, 17].

As mentioned above, adding copper to the high-aluminium alloys leads to good tribological properties on one hand, but also causes detrimental structural instability on the other. Fig. 5 shows that a partial replacement of Cu with Ti and/or Mn significantly decreases the structural instability while preserving the good tribological properties [18].

4. Conclusion

Based on the studies described above, the following conclusions can be drawn:

Grain refinement of the examined high-aluminium zinc alloys is a promising process leading to improvement of their properties. At the same time, using the low-melting-point ZnTi-based master alloys avoids the excessive melt overheating needed for the TiCAl or TiBAl refiners and reduces the possibility of gas pick-up and material loss. In the ternary alloys Zn-26Al-Cu, partially replacing Cu with Ti allows the dimensional changes to be reduced while preserving good tribological properties.

Grain refinement of the high-zinc aluminium binary alloys with a traditional Al-Ti based master alloy causes significant refinement of the primary α(Al) dendrites, which should be beneficial considering increase of ductility measured by elongation changes.

Partial replacement of Cu with Ti and/or Mn in the examined alloys leads to obtaining practically stable structure during long term natural ageing after homogenization and quenching.

Taking the mentioned conclusions into account, it is also concluded that further studies of other property changes, e.g. creep, damping, tribological and casting properties, would be desirable.

Acknowledgements

The authors are grateful to the Polish Science National Centre for financial support under grant No. UMO-2017/25/B/ST8/00150 and to the University of Cambridge, Department of Materials Science and Metallurgy for provision of laboratory facilities.

References

[1] 38-51st Censuses of World Casting Production, Modern Casting: Issues December: 2004, pp. 25-27; 2005, pp. 27-29; 2006, pp. 28-31; 2007, pp. 22-27; 2008, pp. 24-27; 2009, pp. 17-21; 2010, pp. 24-27; 2011, pp. 16-19; 2012, pp. 26-29; 2013, pp. 18-23; 2014, pp. 17-21; 2015, pp. 26-31; 2016, pp. 24-28

[2] E. Peltzel, Aluminium-Zinc-Legierungen, Metall 3 (1954) 83-88 (in German)

[3] A .E.Wol, Structure and Properties of Metallic Systems, Fitmaszgiz, Moscow, 1959 (in Russian)

[4] Y. Alemdag, T. Savaskan, Mechanical and tribological properties of Al–40Zn–Cu alloys, Tribology International 42 (2009) 176-182

[5] T. Savaskan, Y. Alemdag, Effect of nickel additions on the mechanical and sliding wear properties of Al–40Zn–3Cu alloy, Wear 268 (2010) 565-570

[6] W. Krajewski, Phases of Heterogeneous Nucleation in the ZnAl25 Alloy Modified by Zn-Ti and Al-Ti Master Alloys. Zeitschrift für Metallkunde 87 (1996) 645-655

[7] W. Krajewski, Transformation of binary $TiZn_3$ into ternary $Ti(Al,Zn)_3$ during inoculation in the ZnAl25 alloy modified by the ZnTi4 master alloy, Archives of Metallurgy 41 (1996) 385-395

[8] W. Krajewski, Transformation of binary DO_{22} Ti-Al into ternary $L1_2$ Ti-Al-Zn during inoculation in the ZnAl25 alloy modified by Al-Ti master alloy, Archives of Metallurgy 42 (1997) 273-285

[9] W.K. Krajewski, J. Buraś, M. Zurakowski, A.L. Greer, Structure and properties of grain-refined Al-20wt% Zn sand cast alloy, Archives of Metallurgy and Materials 54 (2009) 329-335

[10] K. Haberl, W.K. Krajewski, P. Schumacher, Microstructural features of the grain-refined sand cast AlZn20 alloy, Archives of Metallurgy and Materials 55 (2010) 837-841

[11] W.K. Krajewski, A.L. Greer, EBSD study of ZnAl25 alloy inoculated with ZnTi4 master alloy, Materials Science Forum 508 (2006) 281-286

[12] K.K. Alaneme, B.). Fatile, J.O. Borode, Mechanical and corrosion behaviour of Zn-27Al based composites reinforced with groundnut shell ash and silicon carbide, Tribology in Industry 36 (2014) 195-203

[13] T. Jaglinski, R.S. Lakes, Zn–Al-based metal–matrix composites with high stiffness and high viscoelastic damping, Journal of Composite Materials 46 (2012) 755–763

[14] W.K. Krajewski, J. Buraś, P.K. Krajewski, A.L. Greer, K. Faerber, P. Schumacher, New developments of Al-Zn cast alloys, Materials Today: Proceedings 2 (2015) 4978-4983
[15] W.K. Krajewski, A.L. Greer, P.K. Krajewski, Trends in developments of high-aluminum zinc alloys of stable structure and properties, Archives of Metallurgy and Materials 58 (2013) 845-847
[16] W.K. Krajewski, A.L. Greer, G. Piwowarski, P.K. Krajewski, Property enhancement by grain-refinement of zinc-aluminium foundry alloys, Proceedings of 4th International Conference in Advances of Solidification Processes (ICASP-4), IOP Conference Series: Materials Science and Engineering 117 (2016), paper No. 012004. DOI:10.1088/1757-899X/117/1/012004
[17] W.K. Krajewski, A.L. Greer, J. Buras, G. Piwowarski, P.K. Krajewski, New developments of high-zinc Al-Zn-Cu-Mn cast alloys, Aluminium Two Thousand World Congress and International Conference on Extrusion and Benchmark ICEB, Verona, June 2017, Materials Today: Proceedings (2018) (in printing)
[18] W.K. Krajewski, A.L. Greer, P.K. Krajewski, K. Faerber, Main directions of recent works on AlZn based alloys for foundry engineering, Proceedings of the 73rd World Foundry Congress, Krakow, Septrmber 2018, ISBN: 978-83-904306-3-8, pp. 261–262

IOP Conf. Series: Materials Science and Engineering **529** (2019) 012080 doi:10.1088/1757-899X/529/1/012080

An investigation of the chill-casting interface dynamics in production of sand-cast A319 engine blocks

Farzaneh Farhang Mehr[1], Steve Cockcroft[1] and Daan Maijer[1]

[1] Materials Engineering Department, The University of British Columbia, 309, 6350 Stores Road, Vancouver, BC, Canada, V6T 1Z4

farzaneh.fmehr@ubc.ca

Abstract. In recent years, the automotive industry has been increasing the production of small, high-power gas engines as part of several strategies to achieve the new "Corporate Average Fuel Economy" (CAFE) standards. This trend requires an improvement in the thermal and mechanical fatigue durability of the aluminium alloys used in the production of the cylinder blocks in these engines. Conventionally, solid chills are employed in areas of these castings subject to high cyclic loading to enhance the mechanical performance of the cast material – i.e. in the main bearing bulkhead. A potential means of improving the efficacy of these chills is to incorporate water cooling. To assess the effectiveness of this method, a water-cooled chill was designed and installed in a bonded-sand engine block mould package (1/4 section). The moulds were instrumented with thermocouples, to measure the evolution of temperature at key locations in the casting, and "Linear Variable Displacement Transducers" (LVDTs), to measure the gap formation at the interface between the chill and the casting. This paper summarizes, at a high level, some of the findings of this work.

1. Introduction

The automotive industry is working to meet the stricter efficiency requirements laid out within the CAFE (Corporate Average Fuel Economy) regulations [1], while at the same time not sacrificing performance. One approach is to increase the use of lightweight materials, which leads to a decrease in the overall vehicle weight and fuel consumption [2,3]. Other approaches include smaller displacement engines and gas-electric hybrid drivetrains. However, one trend that is becoming clear, is a greater emphasis on force induction gasoline engines. This allows for weight reduction to be achieved with smaller displacement engines without sacrificing performance. One consequence of this approach is that the engine block materials are subject to higher in-service thermal and mechanical cyclic loads [3,4].

The fatigue properties of an alloy are strongly affected by a number of the as-cast microstructure features including, Secondary Dendrite Arm Spacing (SDAS), porosity and eutectic lamella spacing (including intermetallic β-Al5FeSi platelets) [5-7]. It is generally accepted these features can act as fatigue initiation sites and that a reduction in their size will result in an improvement in fatigue performance. For a given alloy chemistry, one of the key parameters affecting the length scale of all three is cooling rate [8].

Engine blocks and cylinder heads include thick and thin sections, at different distances from the gating system. In the absence of chills, the thick sections will undergo slower cooling rates during

IOP Conf. Series: Materials Science and Engineering **529** (2019) 012080 doi:10.1088/1757-899X/529/1/012080

solidification, have coarser microstructures, and will exhibit relatively poorer fatigue properties than thinner, more rapidly solidified sections. By design, the thicker sections also correspond with highly stressed areas, such as for example the engine block main bearing bulkheads. Hence, particular attention needs to be applied to these areas to achieve suitable in-service performance. The application of a well-designed metallic chill, which is directly in contact with the solidifying metal, can increase the heat flow from the thick sections resulting in a finer microstructure and better fatigue performance. Typically, solid H13 tool steel chills are used for this purpose in the precision sand casting industry. Often they are preheated, to avoid formation of a columnar structure directly adjacent to the chill. However, this approach limits the capacity of the chill to refine the structure deep into the bulkhead, which is needed for the high-power density engines.

This paper presents some of the preliminary results obtained with a water-cooled H13 chill using a casting geometry based on a section of an engine block including the main bearing bulk head and a cylinder. The cooling strategy explored is to initially achieve a low cooling rate directly adjacent to the chill, to avoid a columnar structure and cold shuts, and then to achieve a high, sustained, cooling rate to reduce the microstructural length scales deep into the casting, thereby improving component fatigue performance.

2. Experimental

The sand moulds used in these experiments were one-quarter sections of a 2.0 liter turbocharged 4-cylinder engine block mould package, designed at General Motors Co. to approximate the cooling conditions in a production engine block for experimental and research purposes. Figure 1 shows a side view of the assembled mould package with the refractory board removed from the face oriented toward the camera (the light blue refractory board can be seen fixed to the back of the of the other side). The solid version of the H13 chill can be seen sitting on top of the mould cavity, directly above the main bearing bulk head. Figure 2 shows the casting including the feeding system following cooling and extraction from the sand mould.

Figure 1. Mould package for a section of a 2.0 liter turbocharged 4-cylinder engine block

Figure 2. Cast engine block section – the lines show the approximate locations where metallography samples were taken

Figure 3 shows the geometries of the solid chill and the water-cooled chill fabricated at UBC for this investigation. The water-cooled chill was designed based on preliminary experimental and modelling results as a part of this research at UBC. Both chills were made of H13 tool steel.

Prior to casting, eleven Type-K thermocouples, 0.5 mm diameter wire, sheathed in a nickel–chrome-based alloy (10mm Super OMEGACLAD XL), were installed in the mould package at various locations including in the mould cavity – see Figure 4, which shows the post-cast location of the thermocouples.

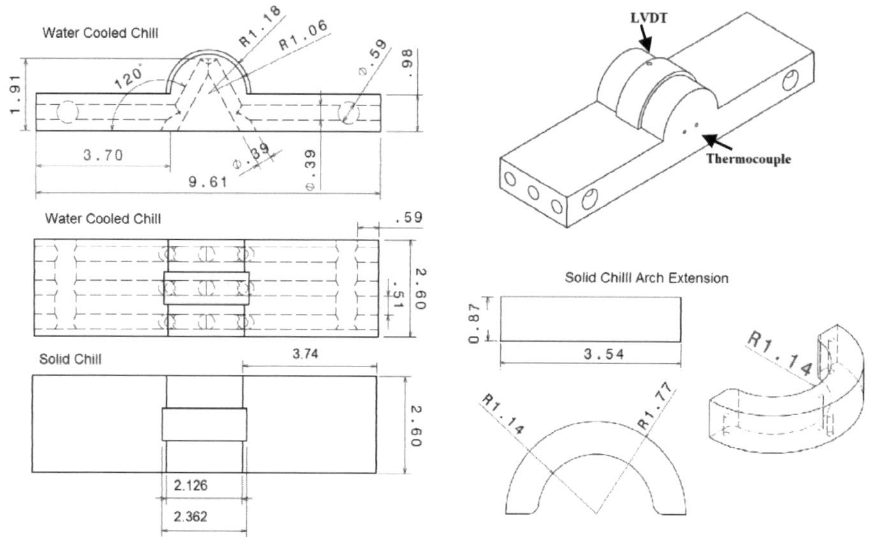

Figure 3. Configuration of the water-cooled chill, the solid chill, and the solid chill arch extension

The thermocouple wires were exposed at the tip and twisted to form the thermocouple junction. A total of two thermocouples were inserted in the bulkhead area at 10 and 30 mm away from the chill, underneath the flat section of the chill. Also, three thermocouples were inserted in the cylinder area at 10, 60, and 110 mm from the bulkhead, and four thermocouples were inserted in the cylinder water jacket area (Casting TC6-TC9). All these thermocouples were placed at the centre of the mould cavity. These measurements enabled the investigation of the variation in cooling rate with distance from the chill. Further, in order to monitor the evolution of temperature in the chill, two thermocouples (also Type-K, sheathed in alumina tubes) were placed in the middle of the chill, in the arch section – see Figures 3 and 4.

The evolution in the displacement of the casting and chill at the casting/chill interface was measured by using two LVDTs (Schaevitz –HR 050), one placed in contact with the chill and the other passing through the chill, and protruding about 2mm into the mould cavity (Figure 3). Both LVDTs were also equipped with a thermocouple (Type K, sheathed in alumina tubes) to record the temperature at both sides of the interface. The temperature-time data for the in-mould thermocouples, in-chill thermocouples and the displacement data from the two LVDTs were recorded at a frequency of 5Hz, with the use of "Lab View" software.

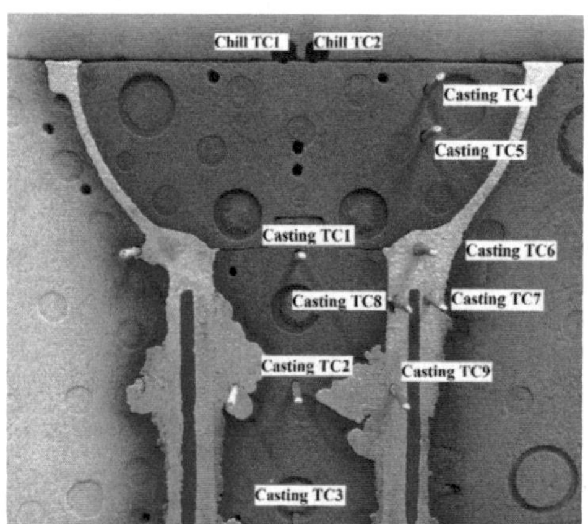

Figure 4. Thermocouple Locations in the Casting and the Chill

In this research, A319 aluminium alloy was used, which is a widely used alloy for the commercial production of engine blocks and cylinder heads in North America. The alloy composition is presented in Table 1. The pouring process started when the temperature of molten metal was 750±5°C. For the water-cooled chill, the water was switched on 10 seconds after the completion of pouring process and the water flow rate was set to 35 L·min^{-1} (measured with an inline water flow meter). The temperature of the main water supply was fairly constant at ~10°C.

Table 1. Chemical Composition of A319 Alloy

Alloying Element	Si	Fe	Cu	Mn	Mg	Zn	Ti	Sr	Al
Balance (wt%)	8.10-8.75	0.27-0.59	2.60-2.95	0.19-0.44	0.31-0.4	0.4-0.8	0.12-0.16	0.004 max	Bal.

In order to investigate the effect of variation in cooling rate on the SDAS, thirty 10×10×5 mm metallographic specimens were taken at 5, 10, 15, 20, and 25 mm from the chill at three angular locations on the arch: 1) along a line normal to the tangent at 45° to the top surface of the chill; 2) along a line normal to the tangent at 90°; and 3) along a line normal to the tangent at 135° - see Figure 2. The metallographic samples were polished down to 0.1μm and etched using 0.5% HF solution. Metallographic images were taken with a Nikon Epiphot 300 optical microscope, equipped with a QImaging digital camera. The SDAS was estimated in each sample (100 measurements per sample) using the line intercept procedure and was taken as a representative measurement of the microstructure length scale.

3. Results and Discussion

Figure 5 shows selected thermocouple data for both the water-cooled and solid chill castings from 4 different regions of the engine block - the casting/chill interface, the bulkhead area (Casting TC5), the cylinder barrel area (Casting TC1) and cylinder water-jacket area (Casting TC6) – see also Figure 4. The solidus and liquidus lines have also been added to the plots to aid with identifying the solidification temperature range. It can be observed that the water-cooled chill does not significantly alter the solidification time relative to the solid chill in the cylinder barrel (TC1) and the water-jacket areas (TC6), as would be expected. These areas are far away from the chill and will only be affected at longer times, as shown. However, at the casting/chill interface and 30mm away from the chill in the bulk head (TC5) there is a significant effect. The cooling curves for the water-cooled chill reach a lower measured peak and show a substantially higher cooling rate, compared to the data from the solid chill. Three cooling curves show that the peak measured temperature falls below the solidus temperature. The reason for this behaviour is unclear and may be related to a combination of the loss of superheat in the metal by the time it reaches the top of the bulk and slow response time of the thermocouple, albeit the TC5 temperature data for solid chill does not show this problem.

A comparison between the thermocouple data extracted from the chills - water-cooled and solid - is shown in Figure 6. The solid chill temperature at 2mm from the interface rapidly increases to approximately 250°C in the first 50 seconds, then levels off followed by a slight increase to ~300°C up to 800 seconds, when it becomes thermally saturated. The other two thermocouples farther from the interface in the solid chill (TC1 and 2) show a less rapid initial increase in temperature to ~180°C during the first 200 seconds, followed by a marginal increase to ~260°C at 800 seconds. In contrast, in the water-cooled chill, there is an initial increase in temperature to a maximum of ~290°C at the chill/casting interface (2mm from the interface) in the period when the water cooling is off. At the onset of water cooling, the temperature drops to ~75°C and then rebounds to 150 °C prior to slowly declining. It is unclear what is responsible for the rebound in temperature.

These results indicate that water-cooling has the potential to greatly increase the thermal extraction capacity of the chill. Hence, water-cooling may be an effective method for maintaining a large temperature difference between the chill and the casting– i.e. sustaining a large driving force for heat transfer at the casting/chill interface and enhancing heat transport.

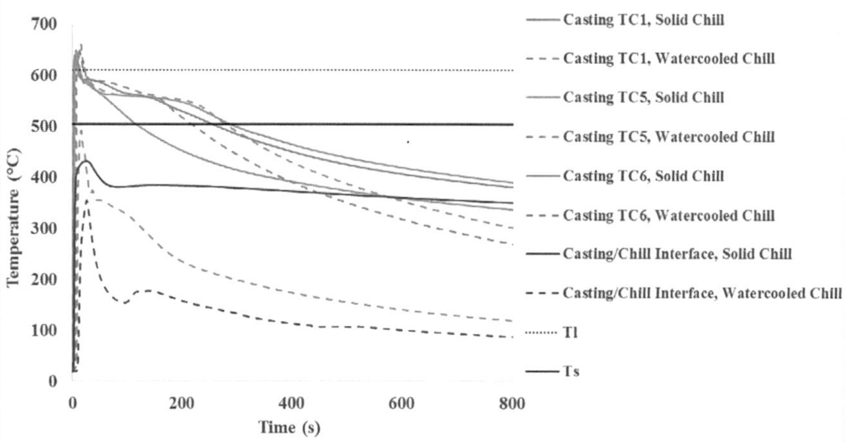

Figure 5. Comparison between thermocouple data from the casting chill interface, bulkhead (Casting TC5), cylinder barrel (Casting TC1) and water jacket (Casting TC6) for the water-cooled and solid chill castings

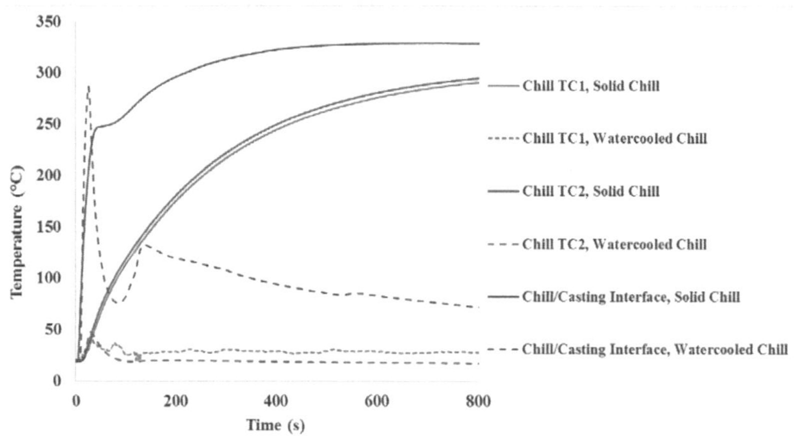

Figure 6. Comparison between thermocouple data for the Chill/Casting Interface, Chill TC1 and Chill TC2 for the water-cooled chill and the solid chill.

Figure 7 shows a comparison between the evolution in the casting/chill interface gap in the two castings - water-cooled and solid chill. The results suggest that the interface gap is about 10 times larger in the solid chill casting. Consequently, the resistance to heat transfer at the casting/chill would be larger in the case of solid chill casting further reducing the efficacy of the chill. The reason for the difference in gap formation is still under investigation and may be related to the expansion of the arch forcing the chill upward.

Figure 8 shows a comparison between SDAS in the two castings. In each casting, the SDAS increases with increasing distance from the chill – i.e. decrease in the cooling rate – as would be expected. Also, the measurements suggest that SDAS in water-cooled chill samples were approximately 20% smaller than SDAS in the solid chill samples. This result supports the fact that employing the water-cooling can increase the solidification rate and as a result, refine the microstructure of the casting, which should lead to an improvement in mechanical fatigue properties of the cast component.

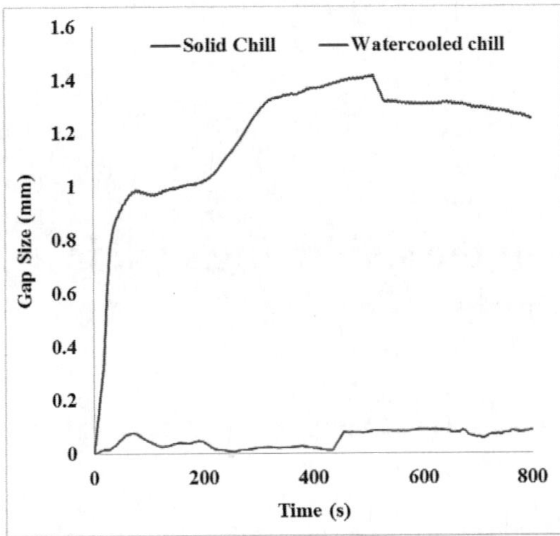

Figure 7. Comparison between the casting/chill interface gap in the water-cooled chill and the solid chill castings.

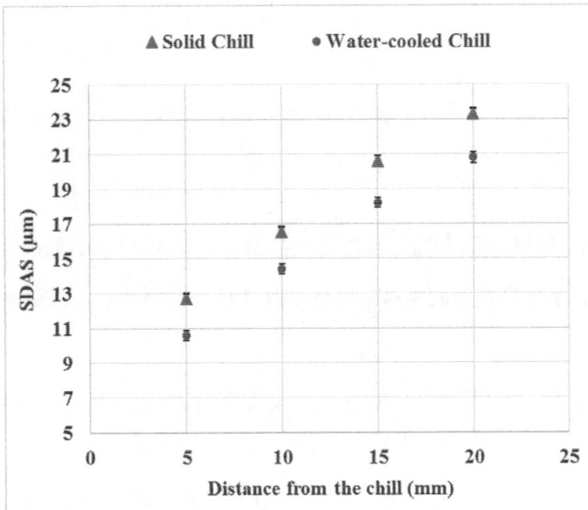

Figure 8. Comparison between the SDAS Size in the water-cooled chill and solid chill castings (the error bars represent the standard error)

4. Summary and Conclusion

The present work involved conducting a series of experiments to determine the efficacy of a water-cooled chill in refining the microstructure of a 2.0 liter turbocharged, 4-cylinder engine block, in the bulkhead section. For this purpose, a section of an engine block was cast employing both a solid H13 chill and a water-cooled H13 chill. The results show that using the water-cooled chill was effective in increasing the cooling rate in the bulkhead and reducing the SDAS and therefore has the potential to improve the mechanical properties in the highly stressed bulkhead region of the cylinder block. The main conclusions from this work are:

1. Using the water-cooled chill was found to be effective in decreasing the cooling rate in the bulkhead area of the engine block section.
2. A 20% reduction in the SDAS size in the bulkhead area of the engine block was achieved by using the water-cooled chill.
3. The solid chill was found to generate a larger interfacial gap than the water-cooled chill.

Further experimental and modeling work is needed in order to characterize the casting/chill interface heat transfer behavior in both the solid chill casting and the water-cooled chill casting.

Acknowledgement
The authors would like to acknowledge General Motors Co., and in particular, Mike Walker for providing the sand moulds.

References

[1] Bastani, P., Heywood, J., Hope, C. (2012). *MIT Energy Initiative Report*.
[2] Mackay, R., Byczynski, G. (2011). *Shape Castings: The 4th International Symposium in Honor of Prof. John T Berry*,191-198.
[3] Miller, W. Zhuang, L. Bottema, J. Wittebrood, A. De Smet, P., Haszler, A., & Vieregge, A. (2000). *Materials Science and Engineering A*, 280, 37-49.
[4] Heysler, L., Feikus, F., & Otte, M. (2001). *AFS Transactions*, 50, 1-9.
[5] Mackay, R., & Szablewski, D. (2010). *International Journal of Metalcasting*, 4, 31-45.
[6] Mackay, R., & Sokolowski, J. (2010). *International Journal of Metalcasting*, 4, 33-50.
[7] Viet-Duc, L., Morel, F., Bellett, D., Pessard, E., Saintier, N., & Osmond, P. (2015). *Procedia Engineering*, 133, 562-575.
[8] Mackay, R., Djurdjevic, M., & Sokolowski, J. (2000). *AFS Transactions*, 25, 521-530.

IOP Conf. Series: Materials Science and Engineering **529** (2019) 012081 doi:10.1088/1757-899X/529/1/012081

On modelling viscoplastic behavior of the solidifying shell in the funnel-type continuous casting mold

A Vakhrushev[1], A Kharicha[1], M Wu[2], A Ludwig[2], G Nitzl[3], Y Tang[3], G Hackl[3], J Watziger[4], C M G Rodrigues[2]

[1]Christian-Doppler Lab for Metallurgical Applications of Magnetohydrodynamics, University of Leoben, Franz-Josef-Str. 18, A-8700 Leoben, Austria
[2]Chair of Simulation and Modeling of Metallurgical Processes, University of Leoben, Franz-Josef-Str. 18, A-8700 Leoben, Austria
[3]RHI Magnesita, Austria
[4]Primetals Technologies, Austria

E-mail: alexander.vakhrushev@unileoben.ac.at

Abstract. As it is known from literature, the metals tend to follow a viscoplastic law at high temperatures. Thereby, based on the authors' previous developments to simulate the behavior of the equiaxed crystals packed bed, a viscoplastic stress model is applied to the thin slab casting process. The model is reduced to the single phase mixture formulation for faster and robust simulations. In this idea, the solidifying shell represents a 'creeping solid' and the Norton-Hoff type stress model is formulated with the model parameters obtained experimentally. A coupling procedure is established to converge the non-linear terms. The influence of the stress model parameters is investigated. Next the model is applied for the simulation of the thin slab casting to improve a previously developed technique: a viscoplastic rheology is applied to calculate the motion of the solidifying melt instead of imposing the velocities of the mush. The simulation results show that the most deformations happen at the funnel part of the mold, causing highest strain rates and the significant drop of the viscoplastic 'apparent viscosity' according to the Norton-Hoff law. The solid shell velocities are mostly uniform at the straight parts of the strand but a slight acceleration of the shell is observed along the funnel surface. Strong compression / expansion zones are detected at the funnel part, which could lead to defects formation. The solid shell thickness was successfully predicted as well and compared to the previous work by the authors.

1. Introduction

Following the pioneering work of Merton C. Flemings [1] it is well known that the properties of the alloys can vary drastically at their semi-solid state, thereby allowing the deformations to form and accumulate under mechanical strains during the casting process. The aim of the current work is to make an attempt to track the mechanical behavior of the solidifying shell during continuous casting in a funnel-type thin slab mold by employing a viscoplastic stress model.

Very recently the co-authors simulated the twin-roll casting process using viscoplastic approach [2, 3] within frames of the two-phase Eulerian–Eulerian volume-averaging model [4]. An inoculated Al–4wt%Cu alloy (typically showing globular solidification) was studied in regard of the partially solidified shell compression and the corresponding effects on the macrosegregation distribution.

Compression-induced expulsion of the melt against casting direction was observed, which led to remelting of the mixture and strong contribution to the centerline macrosegregation. A 2D simplification was done due to large width / thickness ratio of the twin-roll strip, which allowed keeping the mesh size bellow 10^4 cells.

In current work the viscoplastic Norton-Hoff type stress is applied to model a complex process of the thin slab continuous casting and the first results are discussed. The advantage of the proposed approach is using finite-volume formulation for both solid shell mechanical behavior and for the strong melt flow. The latter is typically excluded in most studies on the mechanical stress calculations by simplifying to a plug-type. In the presented study the flow phenomenon is fully considered. To make the numerical procedure stable, computationally effective and robust the following simplifications are done: (i) constant parameters of the viscoplastic model are assumed; (ii) mixture formulation is employed resulting in a single phase formulation.

The presented work considers a full 3D model of a submerged entry nozzle (SEN), a funnel-type thin slab mold (880 mm in height) and strand part (3000 mm along casting direction). The boundary layers are introduced to correctly predict near-wall flow and heat flux. Thus the typical numerical grid in the present work can reach up to 10 million finite volume (FV) elements. Additionally a strong flow in the SEN region drastically limited the efficiency of the time marching algorithm. Taking this into account, the multiphase approach was found inappropriate for transient full-scale industrial simulation.

The mixture volume averaging, based on the previous report in Vakhrushev et al. [5], was extended by combining a liquid viscous stress with a viscoplastic one for the solid phase. Next the details of the numerical model and algorithm as well as the simulation results are presented and discussed.

2. Numerical model

2.1. Viscoplastic model for the solidified shell

As it was previously described, a numerical Eulerian-Eulerian model was suggested to simulate the behavior of the packed bed of equiaxed crystals [2, 3]. The metals at the temperatures higher than 2/3 of their melting absolute (according to the book of Rappaz, Bellet and Deville, p. 309 [6]) are known to follow a viscoplastic law. Thereby based on the authors' developments and experience the viscoplastic stress model was applied to the thin slab casting process. However, to make it more robust it was simplified to a single phase 'mixture' approach. The compressibility of the solid skeleton suggested by Fachinotti et al. [7] and used in Bellet [8] was excluded at this stage from the model as well. Thus the solidifying shell was represented as a 'creeping solid' in incompressible formulation. The Norton-Hoff viscoplastic stress model [6] is formulated as follows:

$$\boldsymbol{\sigma}^{\mathrm{vp}} = 2K\left(\sqrt{3}\,\dot{\varepsilon}_{\mathrm{eq}}\right)^{m-1}\mathrm{dev}(\dot{\varepsilon}), \quad (1)$$

where the viscoplastic consistency K and strain rate sensitivity m are the model parameters obtained experimentally; $\dot{\varepsilon}$ is the strain rate tensor (the symmetric part of the velocity gradient). The equivalent strain rate is defined from the deviatoric part of $\dot{\varepsilon}$ and reflects the dynamic load rate at the local point:

$$\dot{\varepsilon}_{\mathrm{eq}} = \sqrt{\frac{2}{3}\mathrm{dev}(\dot{\varepsilon}):\mathrm{dev}(\dot{\varepsilon})}. \quad (2)$$

For a more convenient formulation the 'effective' or 'apparent' viscoplastic viscosity is introduced:

$$\mu^{\mathrm{vp}} = K\left(\sqrt{3}\,\dot{\varepsilon}_{\mathrm{eq}}\right)^{m-1}. \quad (3)$$

The main difference between the liquid and the viscoplastic viscosity is that the latter depends on the current flow situation (also called rate of load), expressed in the current model by the equivalent strain rate. Moreover since the strains and stresses are integrated into the 'apparent' viscosity calculation, the coupling procedure is required to converge the non-linear terms. This involves the following steps: (i) estimate the strain rate; (ii) update the equivalent strain and the viscoplastic viscosity; (iii) come to a new solution of the flow / heat transfer equation system based on the latest estimation of the stress distribution; (iv) repeat procedure applying under-relaxation on pressure and velocity fields as well as on the effective viscosity. An additional step is to limit the maximum viscoplastic viscosity value due to stability issues. Enhanced treatment of the stresses model will be applied in the future to remove any restrictions.

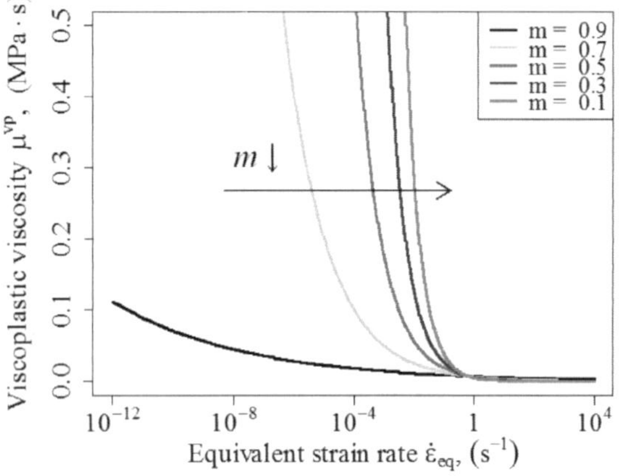

Figure 1. Viscoplastic viscosity μ^{vp} vs. equivalent strain rate $\dot{\varepsilon}_{\mathrm{eq}}$ as in (3) for different strain rate sensitivities *m*; consistency *K*=65.16 (Table 1).

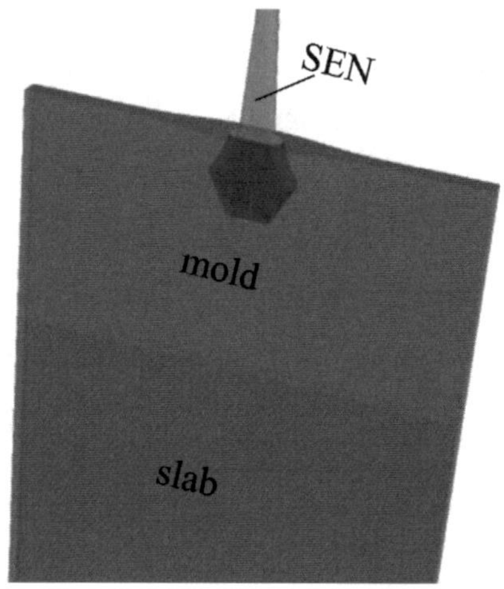

Figure 2. Simulation domain: the boundary conditions are given in Vakhrushev et. al [5]. Inlet / casting parameters are in Table 1.

The basic parameters of the viscoplastic deformation model were selected according to the model from Kozlowski et. al [10] (see Table 1). As it is indicated in Figure 1, for equivalent strain rates below 1 s^{-1}, the viscoplastic viscosity increases dramatically as the parameter *m* decreases.

2.2. General equations of motion

In the incompressible mixture volume average approach the momentum equation is formulated as

$$\rho\left[\frac{\partial\vec{u}}{\partial t}+\nabla\bullet(\vec{u}\otimes\vec{u})\right]=\nabla\bullet[\mathrm{dev}(\boldsymbol{\sigma}_{\mathrm{tot}})-p\mathbf{I}], \tag{4}$$

where the total stress $\boldsymbol{\sigma}_{\mathrm{tot}}$ is decomposed into deviatoric and hydrostatic parts; the pressure equilibrium ($p_\ell = p_s = p$) is assumed between the phases. The deviatoric part of the total stress tensor is calculated by combining the viscoplastic formulation (1) and one for the Newtonian liquid, yielding:

$$\mathrm{dev}(\boldsymbol{\sigma}_{\mathrm{tot}})=f_s\cdot\boldsymbol{\sigma}^{\mathrm{vp}}+f_\ell\cdot 2\mu_\ell\cdot\mathrm{dev}(\dot{\varepsilon}). \tag{5}$$

It is worth mentioning that for the sake of simplification, turbulence modelling using RANS, LES and further approaches was excluded in this work. Instead the 'coarse DNS' approach was utilized on a fine numerical grid (~10 million cells) by assuming the liquid viscosity μ_ℓ to be constant.

2.3. Latent heat advection

The latent heat advection is crucial for the solid shell formation [5, 9]; its formulation is re-derived here since the mixture approach gives no solid phase velocities distribution. For the total enthalpy *H*

$$H=\int C_{\mathrm{p}}\cdot\mathrm{d}T+f_\ell\cdot L \tag{6}$$

the energy equation is taken in its general advection-diffusion form

$$\rho\left[\frac{\partial H}{\partial t}+\nabla\bullet(\vec{u}H)\right]=\nabla\bullet\lambda\nabla T. \tag{7}$$

By utilizing (6) the latent heat advection term can be derived from (7) in a following form:

$$S_e = \rho L \left[\frac{\partial f_s}{\partial t} + \nabla \bullet (f_s \cdot \vec{u})\right]. \tag{8}$$

The comparison is done with the results obtained using the previous formulation [5, 9]:

$$S_e = \rho L \left[\frac{\partial f_s}{\partial t} + \nabla \bullet (f_s \cdot \vec{u}_s)\right] \tag{9}$$

The difference between (8) and (9) is in using averaged velocity $\vec{u}$ instead of the solid ones $\vec{u}_s$.

3. Simulation results

Next the model was applied for the simulation of the thin slab caster to enhance a previously developed model [5]. Instead of the imposed solid shell velocities a viscoplastic rheology of the mush was applied to calculate the motion of the solidifying melt. The material parameters of the alloy 0.06C 0.1Ni 0.13Mn 0.1Si 0.08Cu 0.035Al 0.015P 0.012S are summarized in Table 1: the strain rate sensitivity coefficient m is taken from Ludwig et al [3] and corresponds to the temperature mid-range of the experimental measurements for the Model IA in Kozlowski et. al [10]. The same model is selected to estimate the viscoplastic consistency K. Since a strong non-linear term can 'destroy' the convergence of the solution, the viscoplastic kinematic viscosity $\eta^{vp} = \mu^{vp}/\rho$ was limited to 10^{10} m^2/s allowing the numerical procedure to converge without any special technique. Additionally, the increased casting speed of 5.5 m/min was used in this study.

Table 1. Material properties used in the study. For the domain geometry and casting conditions details please refer to Vakhrushev et al. [5].

Properties	Symbols	Units	Quantities
Density	$\rho = \rho_\ell = \rho_s$	$kg \cdot m^{-3}$	6998.49
Specific heat	$C_p = C_p^\ell = C_p^s$	$J \cdot kg^{-1} \cdot K^{-1}$	838.2
Thermal conductivity	$\lambda = \lambda_\ell = \lambda_s$	$W \cdot m^{-1} \cdot K^{-1}$	35
Liquid dynamic viscosity	μ_ℓ	Pa· s	0.0054
Latent heat of fusion	L	$J \cdot kg^{-1}$	243000
Viscoplastic consistency	K	$Pa \cdot s^m$	65.16
Strain rate sensitivity	m	–	0.138
Casting velocity	$\vec{u}_{pull}$	m/min	5.5
Casting (inlet) temperature	T_{cast}	K (°C)	1825 (1552)
Liquidus temperature	$T_{liqudus}$	K (°C)	1798 (1525)
Solidus temperature	$T_{solidus}$	K (°C)	1755 (1482)

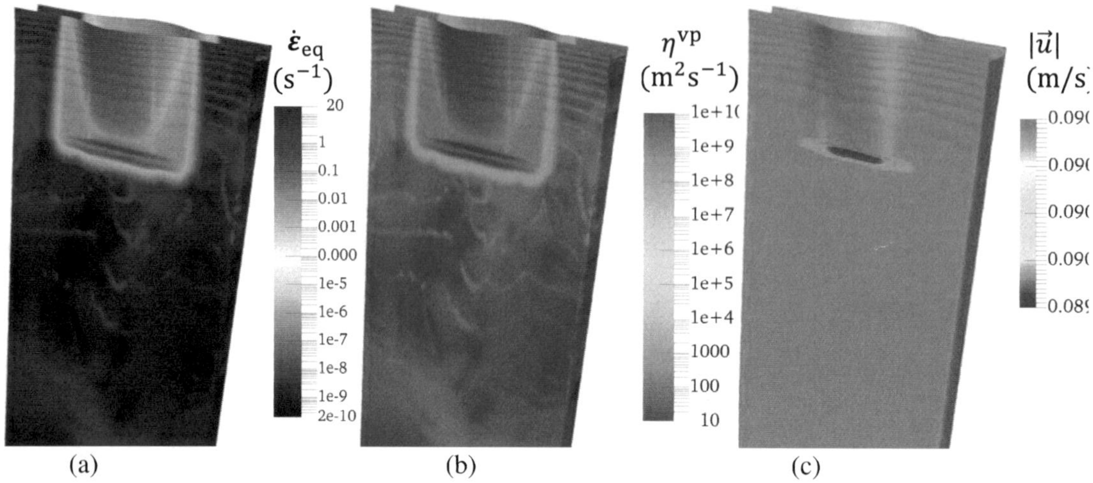

Figure 3. The simulation results of the thin slab casting using viscoplastic deformation model: (a) equivalent strain rate; (b) kinematic viscoplastic viscosity η^{vp}; (c) solid shell velocities.

The results of the thin slab simulation are shown in Figure 3: the distribution of the model parameters is displayed at the iso-surface of 90% solid fraction. The solid shell iso-surface is scaled up in thickness direction by a factor of 4 for better results representation. It was found that the most deformation happens at the funnel part of the mold, causing highest strain rate as it is observed in Figure 3-a. Viscoplastic viscosity drops at the region of the highest strain according to the Norton-Hoff law. Solid shell velocities are uniform at the straight parts of the mold and strand (see Figure 3-c); slight acceleration of the shell is observed along the funnel surface. Solid shell thickness was successfully predicted as well and is analyzed next in Figure 4.

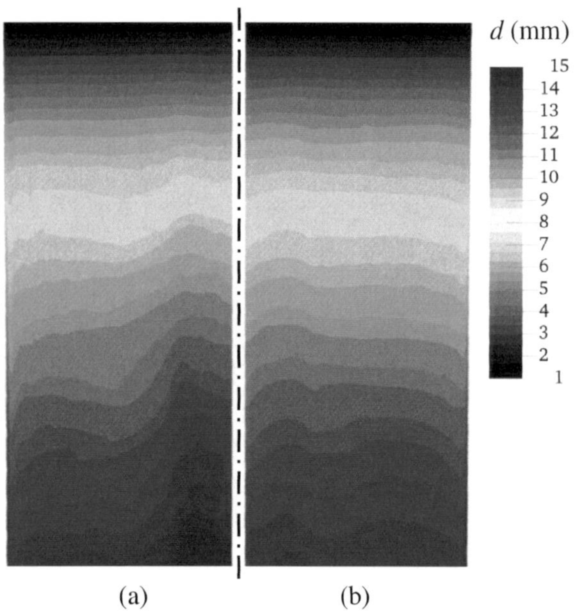

Figure 4. Comparison of the solid shell thickness d distribution along the wide face: (a) corresponds to the previous model [5]; (b) shows the results for the new viscoplastic stress model been applied.

Figure 5. Distribution of the transversal component of the viscoplastic stress tensor $\boldsymbol{\sigma}^{vp}$: compression is in red colors and expansion is in blue.

In Figure 4 the solid shell thickness is compared for both models. The comparison is shown at the wide face of the slab. It is observed that along the casting (vertical) direction the shell thickness is generally lower when the viscoplastic model is applied, but only a slight difference occurs. However a more distinct pattern between both cases is detected in the transversal distribution: in the case of the pre-estimated solid velocities [5] the shell is thicker in the central part of the wide face, whereas in the new approach it is slightly thicker in the lower part of the domain at the 1/4 of the width from the narrow side. It should be noted that at the mold region of the domain no differences between the two models could be detected. The deviation can be described by recalling the previously derived formulation (9) of the latent heat release term [5, 9], where the advection term is based on the prescribed solid velocities $\vec{u}_s$ and the new one, which according to (8) depends on the mixture velocity. Despite the model differences the shell thickness deviation is not dramatic.

The transversal component of the stress tensor acting along the wide direction of the shell surface (90% solid) is investigated as shown in Figure 5. Strong compression / tension areas are detected especially along the curved funnel part of the mold cavity. It was found that the shell is under tension horizontally at the central and outer parts of the funnel (check for $\boldsymbol{\sigma}_{\text{trans}}^{\text{vp}} < 0$ in Figure 5).

The critical tensile values in the simulation lie in the range of 250 kPa. The ultimate tensile stress (UTS) for the steel alloys is typically 500-1000 MPa at room temperature conditions. However the UTS is strongly temperature and alloy dependent. As reported by Outinen and Mäkeläinen[11] and recently reviewed by Wang and Lui[12], the strength properties of the steel drop to 5-10% level at the temperature 1000°C. Extrapolating experimental data[11, 12] up-to 1500°C we obtain the lower range for the UTS of 0,5-1 MPa which makes a simulation value (250 kPa) close to the critical one. Thereby at the upper part of the mold the hot jet impingement and local remelting can lead to the weakening of the solidifying shell strength and end-up with its rapture.

However, as mentioned previously, in the developed model the viscoplastic viscosity μ^{vp} is limited for the simulation stability purposes, thus the maximum tensile stress values currently cannot be judged realistically. Thereby it is a key point for the next studies.

4. Conclusions

In the presented study a viscoplastic stress model was newly introduced to model the thin slab casting process based on the volume-averaged mixture approach. The Norton-Hoff type stress tensor was incorporated for the two-phase region. The influence of the strain rate sensitivity parameter on the viscoplastic effective viscosity was studied.

Full 3D simulation was performed for the numerical domain including the submerged entry nozzle, the funnel-type thin slab mold and 3 m strand part along casting direction below the molds exit. The liquid flow was combined with the heat transfer and solidification based on the viscoplastic stress calculations. The modelling results were verified with previous simulations obtained using static solid velocities distribution; the comparison showed slight deviations due to different latent heat advection term formulations. However the advantage of the newly suggested approach is that it 'naturally' tracks the evolution of the solid shell motion based on the local load rate status.

The algorithm's stability was investigated and an enhanced numerical treatment of the strongly non-linear terms (reported in the literature) has to be applied in the future to realistically estimate the tensile stress values and to compare it against the temperature dependent ultimate tensile stress.

The developed numerical model can be directly applied for industrial process simulations.

5. Acknowledgments
The financial support by the Austrian Federal Ministry of Economy, Family and Youth and the National Foundation for Research, Technology and Development is gratefully acknowledged.

References

[1] Flemings M C 1991 *Metall. Trans. A* **22A** 957
[2] Ludwig A, Vakhrushev A, Holzmann T, Wu M and Kharicha A 2015 *IOP Conf. Ser.: Mater. Sci.*

Eng. **84** 012102
[3] Ludwig A, Vakhrushev A, Wu M, Holzmann T and Kharicha A 2015 *Trans. Indian Inst. Met.* **68** 1087
[4] Rodrigues C M G, Ludwig A, Kharicha A and Wu M. 2018 *Trans. Indian Inst. Met.* **71** 2645
[5] Vakhrushev A, Wu M, Ludwig A, Tang Y, Hackl G and Nitzl G 2014 *Metall. Mater. Trans. B* **45B** 1024
[6] Rappaz M, Bellet M and Deville M. *Numerical modeling in materials science and engineering*. 2003 New York: Springer
[7] Fachinotti V D, Le Corre S, Triolet N, Bobadilla M and Bellet M 2006 *Int. J. Numer. Methods Eng.* **67** 1341
[8] Bellet M 2007 AIP Conf. Proc. **908:1** 1369
[9] Wu M, Vakhrushev A, Ludwig A and Kharicha A 2016 *IOP Conf. Ser.: Mater. Sci. Eng.* **117** 012045
[10] Kozlowski P F, Thomas B G, Azzi J A, Wnag H 1992 *Metall. Mater. Trans B* **23A** 903
[11] Outinen J, Mäkeläinen P 2004 *Fire Mater.*, **28** 237
[12] Wang F, Lui E M 2016 *J. Steel Struct. Constr.* **2:2** 1000123

IOP Conf. Series: Materials Science and Engineering **529** (2019) 012082 doi:10.1088/1757-899X/529/1/012082

A partitioned solution algorithm for fluid flow and stress-strain computations applied to continuous casting

Shaojie Zhang, Gildas Guillemot, Charles-André Gandin and Michel Bellet

MINES Paristech, PSL Research University, CEMEF – Centre de Mise en Forme des Matériaux, CNRS UMR 7635, CS 10207, 1 Rue Claude Daunesse, 06904 Sophia Antipolis cedex, France

michel.bellet@mines-paristech.fr

Abstract. Continuous casting is currently the main industrial process for steel production. Since long time, industries search for efficient simulation methods, by which macrosegregation and deformation induced cracks can be predicted. As a first step this requires achieving concurrent simulation of fluid flow and stress-strain. Therefore, a partitioned solution algorithm is developed for such simulation with application to continuous casting. Liquid flow induced by natural convection or filling step, solidification shrinkage as well as thermally induced deformation of solid phase are accounted for.

1. Introduction

When modelling the continuous casting process, one of the critical issues is to concurrently achieve simulation of fluid flow in the liquid region, stress-strain analysis in the solidified region and solute transportation. For industry, it is of crucial importance to predict the macrosegregation phenomena and deformation related defects in order to optimize cast products. Therefore, the first and most important step of this solidification modelling issue is to achieve a concomitant computation of fluid flow and stress-strain, i.e. the full mechanical problem.

In the present work, a partitioned two-step solution algorithm is proposed. In one time increment, the momentum and mass conservation equations are solved twice. The first one is solid-oriented, performed with an arbitrary high liquid viscosity. From this first solution are deduced, the velocity and pressure field in the solidified regions and its associated stress-strain fields. The second solution is fluid-oriented, resuming to the Navier-Stokes equations, giving access to the liquid velocity and pressure field in the liquid regions. In addition, an arbitrary Lagrangian-Eulerian (ALE) formulation is necessary for a non-stationary simulation of the continuous casting process, in which the Lagrangian treatment applies to solid regions while the Eulerian treatment applies to liquid regions. The volume averaging methodology and the Darcy's law are used to model the interactions between solid and liquid phases in the mushy zone. All conservation equations are formulated in the framework of the level set method in order to track the metal/gas interface. The current algorithm is coupled with an existing non-linear energy solver to calculate the temperature field [1]. Finally, application to a quasi 2D continuous casting case test is presented and discussed.

2. Numerical model

2.1. Level set method

The level set method was initially proposed by Osher and Sethian [2] to capture moving interfaces. In the present work, the metal/gas interface Γ is represented by the zero-isovalue of the signed distance function $\varphi(\boldsymbol{x}, t)$, with $\boldsymbol{x}$ a point of domain Ω at time t. Noting Ω^M and Ω^G, two subdomains of Ω representing respectively the metal and gas subdomains, the smoothed Heaviside function $\mathcal{H}^M$ is then introduced, indicating the belonging of the metal subdomain with respect to the interface:

$$\mathcal{H}^{\mathrm{M}}(\varphi) = \begin{cases} 0 & \text{if } \varphi < -\varepsilon \\ 1 & \text{if } \varphi > \varepsilon \\ \frac{1}{2}\left(1 + \frac{\varphi}{\varepsilon} + \frac{1}{\pi} sin\left(\frac{\pi\varphi}{\varepsilon}\right)\right) & \text{if } -\varepsilon \leq \varphi \leq \varepsilon \end{cases} \tag{1}$$

where ε is the half thickness of a transition zone artificially distinguishing the subdomains. Respectively denoting ψ^M and ψ^G the physical property relative to the metal and gas subdomains, the mixed property $\hat{\psi}$ is given by:

$$\hat{\psi} = \mathcal{H}^M \psi^M + (1 - \mathcal{H}^M)\psi^G \tag{2}$$

Thus, over the thickness [-ε, ε] around the interface Γ, a smooth and continuous transition of properties is defined.

2.2. Solid-oriented solution

In the solid-oriented solution, the approach proposed elsewhere is applied [3]. In addition with the level set formulation, the following momentum and mass conservation equations are solved using a finite element method:

$$\begin{cases} \nabla \cdot \hat{\boldsymbol{s}} - \nabla p + \hat{\rho} \boldsymbol{g} = 0 \\ \nabla \cdot \boldsymbol{v} = \mathcal{H}^M\big(H(T_C - T)\mathrm{tr}(\dot{\boldsymbol{\varepsilon}}_{el}) + \mathrm{tr}(\dot{\boldsymbol{\varepsilon}}_{th})\big) \end{cases} \tag{3}$$

where $\hat{\boldsymbol{s}}$ and $\hat{\rho}$ are respectively the mixed stress tensor and the density in domain Ω, defined in Table 1. $\boldsymbol{s}^M$, $\langle\rho\rangle^M$ are respectively the stress tensor and average density in the metal subdomain. $\boldsymbol{s}^G$, ρ^G are the stress tensor and density in the gas subdomain. p is the pressure and $\boldsymbol{g}$ is the constant gravity vector. In this solid oriented resolution, the average density in metal is assumed to be the intrinsic density of solid $\langle\rho\rangle^s$ in the entire metal subdomain, to approximate the thermally induced solid deformation without solidification shrinkage effect.

Notation	Definition
$\hat{\boldsymbol{s}}$	$\mathcal{H}^M \boldsymbol{s}^M + (1 - \mathcal{H}^M)\boldsymbol{s}^G$
$\hat{\rho}$	$\mathcal{H}^M \langle\rho\rangle^M + (1 - \mathcal{H}^M)\rho^G$

Table 1. Mixing properties in the solid-oriented solution.

$\dot{\boldsymbol{\varepsilon}}_{el}$ and $\dot{\boldsymbol{\varepsilon}}_{th}$ are respectively the elastic part and thermal part of the strain rate tensor $\dot{\boldsymbol{\varepsilon}}$. $H(T_C - T)$ is the standard Heaviside function, taken for the temperature difference $(T_C - T)$. It is introduced as an indicator relative to the use of the thermo-elastic-viscoplastic (TEVP) constitutive model for elements with average temperature lower than the critical transition temperature T_C. Otherwise the thermo-viscoplastic (TVP) model is used for temperature higher than T_C. The corresponding relationships between the the von Mises equivalent stress, $\bar{\sigma}$, the generalized plastic strain, $\bar{\varepsilon}$, and the equivalent strain rate, $\dot{\bar{\varepsilon}}$, is given by:

$$\begin{cases} \bar{\sigma} = \sigma_Y + k\big(\sqrt{3}\big)^{m+1} \dot{\bar{\varepsilon}}^m \bar{\varepsilon}^n & \text{(TEVP)} \\ \bar{\sigma} = k\big(\sqrt{3}\big)^{m+1} \dot{\bar{\varepsilon}}^m & \text{(TVP)} \end{cases} \tag{4}$$

where n is the strain hardening coefficient, m, the strain rate sensitivity, k, the solid consistency and σ_Y, the yield stress.

In the present work T_C is taken below the solidus temperature to model the TEVP behaviour of material in the solid state; the TVP model is used for metal at high temperature, either in the solid state, mushy state or liquid state. At the critical transition temperature, the continuity of the flow stress is obtained by taking $\sigma_Y(T_C) = 0$ and $n(T_C) = 0$. Note that the mushy zone is considered as a homogenized non-Newtonian fluid, in which the strain rate sensitivity is continuously increasing with the liquid fraction and eventually with a pure Newtonian behaviour in the fully liquid state. In addition, in order to prevent numerical instabilities due to the huge difference between solid consistency and liquid viscosity, the liquid viscosity is artificially augmented for the solid-oriented solution (typically 1 to 100 $\mathrm{Pa \cdot s}$ instead of nominal value 5×10^{-3} $\mathrm{Pa \cdot s}$). In addition, the gas is taken as an incompressible Newtonian fluid in the gas subdomain.

Finally, a velocity-pressure solution is performed, the nodal velocity and pressure fields obtained from this first step being denoted $(\boldsymbol{v}_I, p_I)$ as solution of Eq. (3).

2.3. *Fluid-oriented solution*

The fluid-oriented solution takes the solid-oriented solution as an entry. While liquid is incompressible with a Newtonian behaviour, the formulation is extended to deal with a compressible formulation in the mushy zone. An effective two phase approach is used with a volume averaged method [4]. Solidification is assumed to take place with a purely columnar structure. Interactions between solid and liquid phases in the mushy zone are modelled by the Darcy's law with a permeability coefficient K approximated by the Carman-Kozeny relationship [5]. The momentum and mass conservation equations relative to the liquid phase in the level set formulation are given by:

$$\begin{cases} \widehat{\rho}_0^F \left(\frac{\partial_{msh}}{\partial t} (\boldsymbol{v} - \boldsymbol{v}_{msh}) + \frac{1}{\widehat{g}^F} (\boldsymbol{v} - \boldsymbol{v}_{msh}) \nabla \boldsymbol{v} \right) - \nabla \cdot \widehat{g\boldsymbol{s}}^F + \widehat{g}^F \nabla p - \widehat{g\rho}^F \boldsymbol{g} + \widehat{g}^F \widehat{\mu}^F \left(\widehat{K}^F\right)^{-1} (\boldsymbol{v} - \boldsymbol{v}_I) = 0 \\ \nabla \cdot \boldsymbol{v} = -\frac{\mathcal{H}^M}{\langle \rho \rangle^l} \left(\frac{\partial}{\partial t} \left(g^l \langle \rho \rangle^l + g^s \langle \rho \rangle^s \right) + (\boldsymbol{v} - \boldsymbol{v}_{msh}) \cdot \nabla \langle \rho \rangle^l + \nabla \cdot \left(g^s \langle \rho \rangle^s \boldsymbol{v}_I \right) \right) \end{cases}$$

(5)

with $\widehat{\rho}_0^F$, $\widehat{g}^F$, $\widehat{g\boldsymbol{s}}^F$, $\widehat{g\rho}^F$, $\widehat{\mu}^F$, $\widehat{K}^F$ corresponding respectively the mixing properties of fluid density, fluid fraction, averaged fluid stress tensor, averaged fluid density with Boussinesq approximation [1], averaged fluid viscosity and averaged fluid permeability coefficient between liquid and gas phases, defined in Table 2.

Notation	Definition
$\widehat{\rho}_0^F$	$\mathcal{H}^M \langle \rho \rangle_0^l + (1 - \mathcal{H}^M) \rho^G$
$\widehat{g}^F$	$\mathcal{H}^M g^l + (1 - \mathcal{H}^M)$
$\widehat{g\boldsymbol{s}}^F$	$\mathcal{H}^M g^l \langle \boldsymbol{s} \rangle^l + (1 - \mathcal{H}^M) \boldsymbol{s}^G$
$\widehat{g\rho}^F$	$\mathcal{H}^M g^l \langle \rho \rangle^l + (1 - \mathcal{H}^M) \rho^G$
$\widehat{\mu}^F$	$\mathcal{H}^M \mu^l + (1 - \mathcal{H}^M) \mu^G$
$\widehat{K}^F$	${\lambda_2}^2 {\widehat{g}^F}^3 / (180(1 - \widehat{g}^F)^2)$

Table 2. Mixing properties for the fluid-oriented solution.

$\langle \rho \rangle_0^l$, $\langle \rho \rangle^l$, g^l, $\langle \boldsymbol{s} \rangle^l$, μ^l, g^s, $\langle \rho \rangle^s$, λ_2 are respectively the intrinsic reference liquid density, intrinsic liquid density with Boussinesq approximation, liquid fraction, intrinsic liquid stress tensor, liquid viscosity, solid fraction, intrinsic solid density and the secondary dendrite spacing. μ^G is the viscosity of gas, assumed to be Newtonian and incompressible. $\boldsymbol{v}_{msh}$ is the nodal mesh velocity field, computed with the following simplified algorithm:

$$\boldsymbol{v}_{msh} = \begin{cases} 0 & \text{in liquid} \\ g^s \boldsymbol{v}_{II} & \text{in mushy zone and at boundaries} \\ \boldsymbol{v}_{II} & \text{in solid zone} \end{cases} \tag{6}$$

Let us note that the fluid-oriented solution does not need to be operated in the fully solid regions, in which the relevant information is already calculated in the first solid-oriented solution. A Dirichlet condition is thus applied for the fully solid regions, with values taken from the solid-oriented solution.

Finally, Eq. (5) is solved with a stabilized SUPG-PSPG finite element method [6]. The solution fields for nodal velocity and pressure deduced from this second step are denoted $(\boldsymbol{v}_{II}, p_{II})$.

2.4. Solution algorithm

The partitioned solution algorithm for continuous casting is presented hereunder, decomposed in 6 successive stages in one time increment Δt.

- 1st stage: **Thermal solution**. Energy conservation equation is solved, providing the temperature distribution in the metal and in the gas.
- 2st stage: **STEP I** solid-oriented solution. The first folder of the momentum solution focuses on the stress-strain analysis in the solid region, with an augmented liquid viscosity and a continuity of solid density in the mushy and liquid regions.
- 3st stage: **STEP II** fluid-oriented solution. The second folder of the momentum solution consists of the fluid flow computation in the liquid and mushy regions, with the correct value of the fluid viscosity including for solidification shrinkage.
- 4st stage: **Mesh updating**. The position of each mesh node is updated with the mesh velocity, following updating scheme: $\boldsymbol{x}^{new} = \boldsymbol{x}^{old} + \Delta t \cdot \boldsymbol{v}_{msh}$.
- 5st stage: **Metal/gas interface tracking**. The updating of the level set function permits interface tracking. It is achieved by the convection-reinitialization scheme [7].
- 6st stage: **Adaptive remeshing**. The remeshing is guided by an error estimator of Hessian type, based on the interpolation error, which allows the adaptation of mesh to different fields [8].

3. Computational configuration

3.1. Model description

The partitioned algorithm is applied to a quasi-2D continuous casting application. The initial configuration is given in Figure 1a. A simple parallelepipedic geometry is considered. Heat is extracted from the metal through the lateral vertical surface. A flux is imposed at contact with the cold mould and a convection type expression at contact with water cooling. An insulation zone is artificially created in the present model, in order to make sure that the metal/gas surface at the meniscus with the vertical boundary is always in the liquid state. The metal is supposed to slide along the surrounding vertical surfaces and has a free surface boundary condition at the upper surface. Velocity is imposed for both the liquid inlet surface (horizontal velocity $\boldsymbol{v}_{inlet}$) and the bottom surface (vertical casting velocity $\boldsymbol{v}_{cc}$). The same mechanical boundary conditions are applied at both STEP I and STEP II. A symmetry plane is defined at the vertical right-hand surface, as well as the front and back surfaces, which makes the problem planar. The initial mesh is defined in Figure 1a. The material used in the simulation is a steel grade Fe – 0.4 wt.% C. Material and simulation parameters are given in Table 3. Thermal properties of gas (except the thermal conductivity) are approximated to those of the liquid phase in the metal subdomain. All other properties are given in the Appendix.

3.2. Results

Figure 2a illustrates the results corresponding to time $t = 10$ s. The zero-isovalue of the level set function is shown with the thick white line, representing the metal/gas interface. The blue and cyan lines are respectively the solidus and liquidus isotherms. A solid shell has already formed due to primary mould cooling. The von Mises equivalent stress is shown, revealing a maximum value of 10 MPa. The fluid flow is concurrently computed, represented by coloured arrows: besides the inlet liquid flow, two vortices are created, one in upper liquid region, departing the metal/gas interface from an horizontal flat surface and another at the bottom position, slightly preventing the formation of the mushy zone due to heat transport. Figure 2b shows the concurrent results of the von Mises equivalent stress (left symmetry part) and the solid fraction (right symmetry part) at $t = 100$ s. The dotted white line represents the left-hand symmetry plane. The green line represents the exit of the mould cooling. Water sprays continuing to cool the metal, a solid region reaching a thickness of several centimetres has been formed with a maximum value of the equivalent stress of 11 MPa. Figure 2c shows the vertical stress profile 300 mm below the mould exit, at the middle of the slice thickness, as a function of the distance to the symmetry plane. Traction is formed at the cooling surface. As a consequence, a compression at inner place is observed.

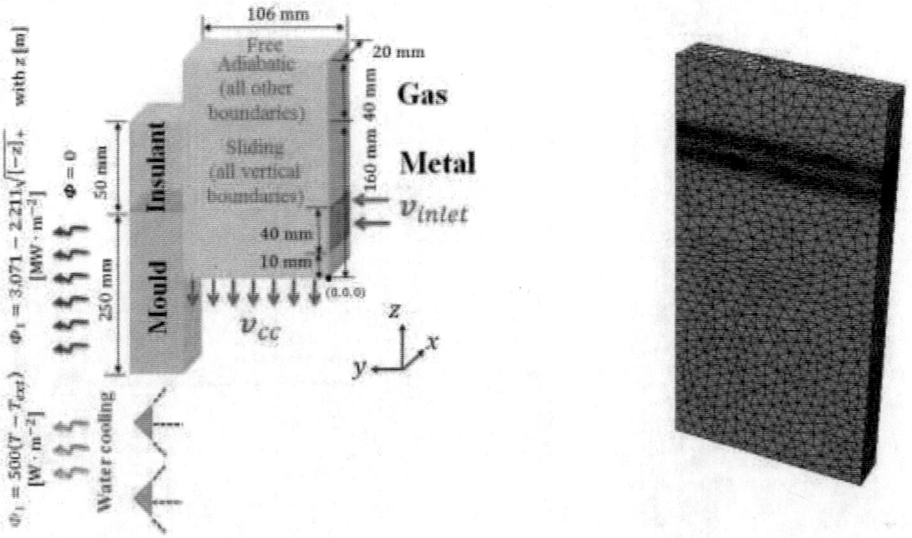

(a) (b)

Figure 1. Representation of (a) a schematic for the quasi-2D continuous casting model; and (b) the initial mesh configuration.

Parameters	Symbol	Value	Unit
Solidus temperature (nominal)		1452	°C
Liquidus temperature(nominal)		1505.5	°C
Reference temperature		1505.5	°C
Reference liquid density	$\langle\rho\rangle_0^l$	7036.5	$kg \cdot m^{-3}$
Solid density	ρ^S	*Tabulation*	$kg \cdot m^{-3}$
Gas density	ρ^G	6900	$kg \cdot m^{-3}$
Liquid thermal dilatation coefficient		8.96×10^{-5}	K^{-1}
Secondary dendrite arm spacing	λ_2	500	μm
Gas viscosity	STEP I|STEP II	μ^G	$100 \mid 1 \times 10^{-1}$ $Pa \cdot s$
Liquid viscosity	STEP I|STEP II	μ^l	$100 \mid 5 \times 10^{-3}$ $Pa \cdot s$
Solid consistency		*Tabulation*	$Pa \cdot s$
Thermal conductivity in gas		0.1	$W \cdot m^{-1} \cdot K^{-1}$

Thermal conductivity in metal		30	$W \cdot m^{-1} \cdot K^{-1}$
Initial temperature		1535.5	°C
Critical temperature	T_C	1431	°C
External temperature		20	°C
Time step	Δt	0.1	s
Casting velocity	$\boldsymbol{v}_{CC}$	0.02667	$m \cdot s^{-1}$
Liquid inlet velocity	$\boldsymbol{v}_{inlet}$	0.07068	$m \cdot s^{-1}$

Table 3. Simulation and material parameters in the quasi-2D continuous casting test case.

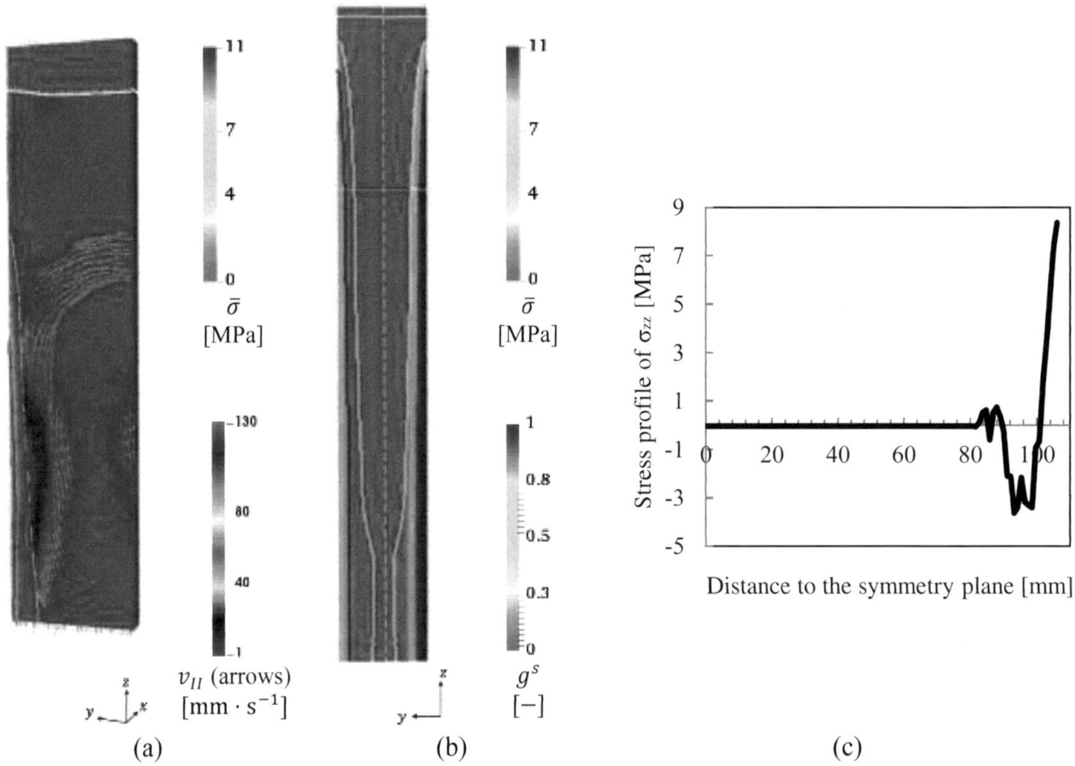

Figure 2. Snapshot of results (a) at time $t = 10$ s, showing (arrows, $\boldsymbol{v}_{II}$) the fluid flow and ($\bar{\sigma}$) the von Mises equivalent stress (represented domain width 106 mm); (b) at time $t = 100$ s, the left symmetry part presents ($\bar{\sigma}$) the von-Mises equivalent stress and the right symmetry part presents (g^s) the solid fraction (represented domain width: 212 mm); (c) at time $t = 100$ s, vertical stress profile over the width from right-hand symmetry plan to the left-hand cooling surface 300 mm below the mould exit (green line) .

4. Conclusions

An algorithm of concurrent simulation of fluid flow and stress-strain analysis for continuous casting is developed. The algorithm consists of a partitioned solution strategy, solving at one increment a solid-oriented solution and a fluid oriented solution. All solutions are formulated with the finite element method, using a level set tracking of the moving interface between the solidifying metal subdomain and the gas subdomain. This algorithm is tested in a quasi-2D continuous casting process. Simulation gives promising results, in which the fluid flow in the liquid and mushy regions and the equivalent stress formed in the solid regions are concomitantly computed.

The main perspectives of this work are twofold. Firstly, coupling of this algorithm with macrosegregation modelling for application to a 3D industrial casting process, requiring a more elaborated mesh velocity computation. Secondly, the algorithm could be coupled with equiaxed grain motion to give access to enhanced predictions of macrosegregation due to thermosolutal convection, shrinkage flow, transport of equiaxed grains [9] and thermomechanical deformation of solid and mushy regions.

5. Acknowledgements

This work is funded by the European Space Agency under project CCEMLCC (AO-2004-017 contract 4200020277) as well as by industrial partners ArcelorMittal Maizières Research, INDUSTEEL France and APERAM ALLOYS IMPHY.

References

[1] A. Saad, Ch.-A. Gandin, M. Bellet, Temperature-based energy solver coupled with tabulated thermodynamic properties - Application to the prediction of macrosegregation in multicomponent alloys, Computational Materials Science 99 (2015) 221-231

[2] S. Osher, J.A. Sethian, Fronts Propagating with Curvature Dependent Speed: Algorithms Based on Hamilton-Jacobi Formulations, Journal of Computational Physics 79 (1988) 12-49

[3] M. Bellet, V. D. Fachinotti, ALE method for solidification modelling, Computer Methods in Applied Mechanics and Engineering 193 39-41 (2004) 4355-4381

[4] J. Ni, C. Beckermann, A Volume-Averaged Two-Phase Model for Transport phenomena during Solidification, Metallurgical Transactions B 22 (1991) 349-361

[5] P. C. Carman, Fluid flow through granular beds, Chemical Engineering Research and Design: Transactions of the Institution of Chemical Engineers Part A 37 (1937) 2069-2079.

[6] E. Hachem, B. Rivaux, T. Kloczko, H. Digonnet, T. Coupez, Stabilized finite element method for incompressible flows with high Reynolds number, J. Comp. Phys. 229 (2010) 8643-8665

[7] M. Shakoor, B. Scholtes, P.-O. Bouchard, M. Bernacki, An efficient and parallel level set reinitialization method – Application to micromechanics and microstructural evolutions, Applied Mathematical Modelling 39 (2015) 7291-7302

[8] T. Coupez, Metric construction by length distribution tensor and edge based error for anisotropic adaptive meshing, Journal of Computational Physics 230 (2011) 2391-2405

[9] T.-T.-M. Nguyen, Ch.-A. Gandin, H. Combeau, M. Založnik, M. Bellet, Finite element multi-scale modelling of chemical segregation in steel solidification taking into account the transport of equiaxed grains, Metallurgical and material transactions A 49 (2018) 1725-1748

Appendix: Thermomechanical properties of steel Fe – 0.4 wt.% C

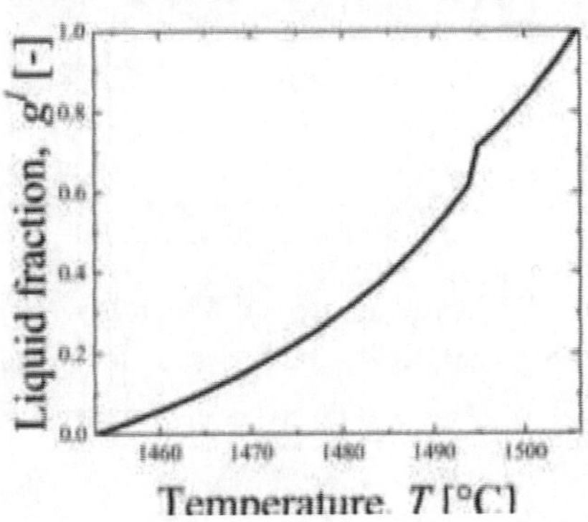

Figure A1. The liquid fraction in the solidification interval.

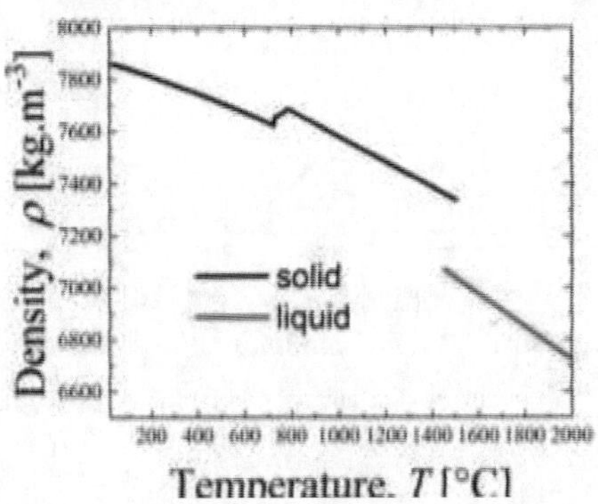

Figure A2. The solid density (until liquidus temperature) and the liquid density (until solidus temperature) as a function of temperature.

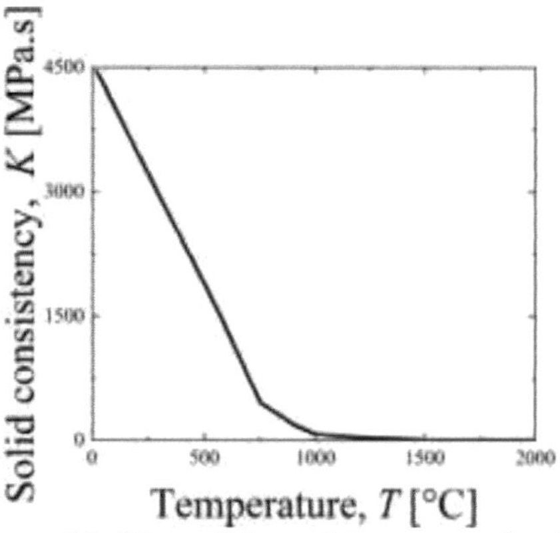

Figure A3. The solid consistency as a function of temperature.

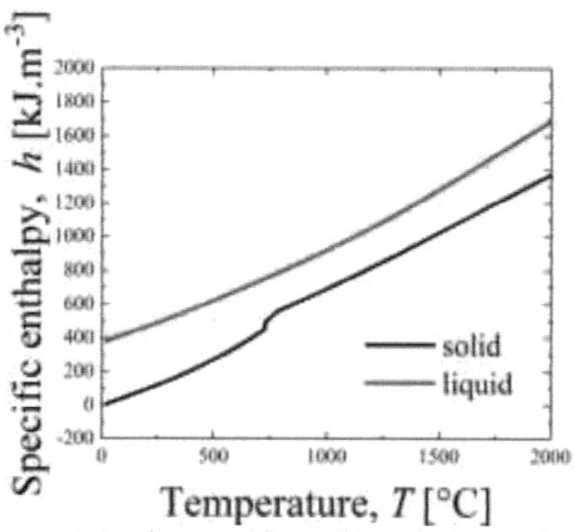

Figure A4. The specific solid enthalpy and specific liquid enthalpy as a function of temperature.

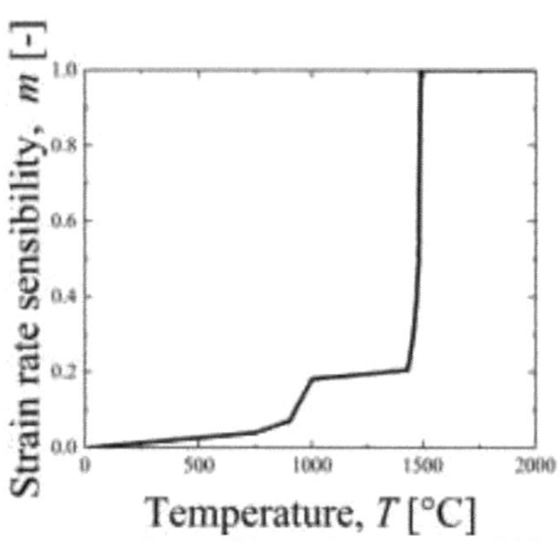

Figure A5. The solid strain rate sensibility *as* a function of temperature.

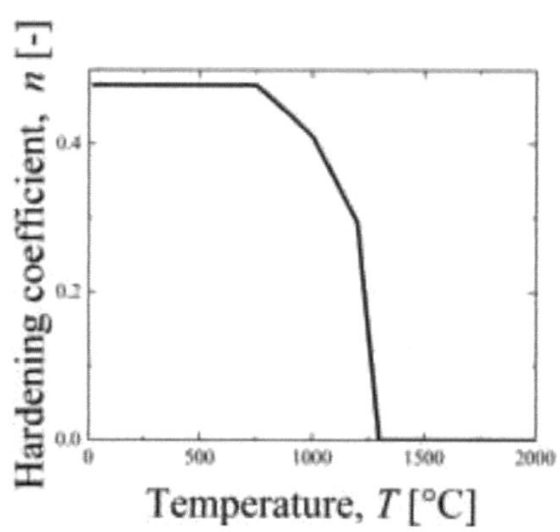

Figure A6. The solid hardening coefficient as a function of temperature.

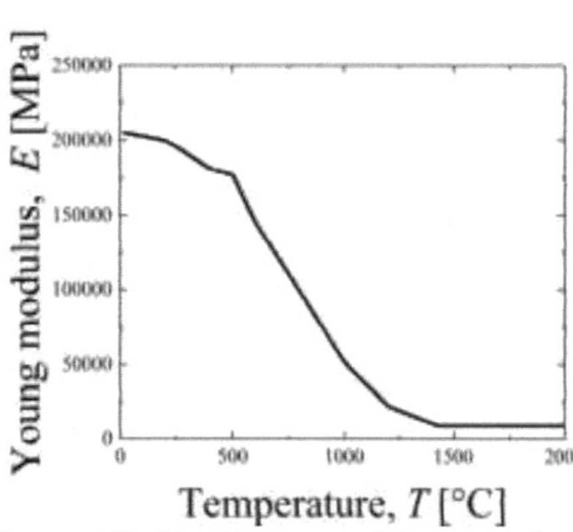

Figure A7. The Young modulus as a function of temperature.

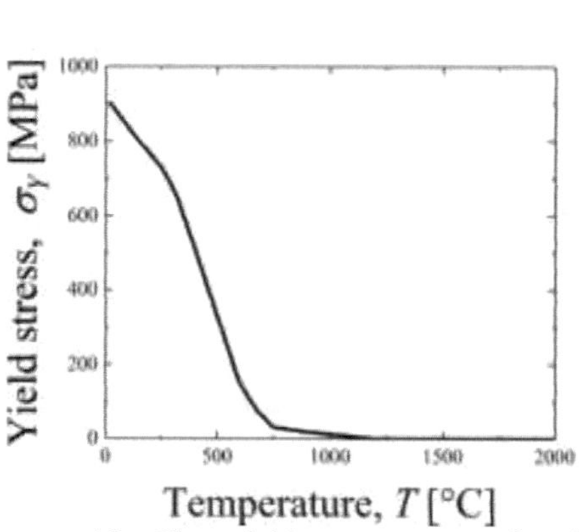

Figure A8. The yield stress as a function of temperature.

Thermo-elasto-viscoplastic constitutive laws for metallic alloys during their solidification

G M Laschet and H Behnken

ACCESS e.V., RWTH Aachen University, Intzestrasse 5, D-52072 Aachen, Germany

{g.laschet, h.behnken}@access.rwth-aachen.de

Abstract. To describe the thermomechanical behaviour of the semi-solid state during alloy solidification, specific viscoplastic constitutive laws are derived for the coherent and non-coherent regime of the mush. For the coherent semi-solid regime an original viscoplastic flow potential is derived, which includes micro-structural parameters. This potential takes also isotropic hardening, pressure dependence of yielding and the strength difference in tension and compression into account. For the non-coherent mushy state of suspensions under shear loading, a simplified viscoplastic potential is adopted by neglecting pressure effects on the solid dendrites. Continuity at mechanical coherency is imposed and allows the reduction of the model parameters of the non-coherent regime. The developed constitutive laws are implemented in Abaqus via a user-defined CREEP routine. Simulations of uniaxial tension, compression or pure shear experiments allow identifying the model parameters of an A356 aluminium alloy.

1. Introduction

During casting, most technical alloys solidify with the formation of a two-phase region known as the mushy zone. They undergo a continuous transition from the liquid to the solid state. The solidification starts below the liquidus temperature by local undercooling and the nucleation of solid germs takes place. At low solid fractions, dendritic germs growth independently and are free to move in the melt. As solidification progresses, interactions between dendrites occur and the mush develops the first shear strength due to dendrites entanglement. For higher solid fractions, dendrites interlock, solid skeletons are built and the mush reaches mechanical coherence. At that state, the solid skeleton densifies under external pressure or dilates under tensile loading. Toward the end of solidification, solid grains start to coalesce and liquid feeding drops drastically. The complexity of the solid-liquid interactions at the micro-scale makes their characterization and rheological modelling difficult and delicate.

Deformations in the coherent mushy zone are influenced by the viscoplastic solid skeleton and the flow of remaining viscous melt in the pore space. In the literature some models were proposed to describe the coherent mushy zone behaviour. Martin et al [1, 2] developed a two-phase isothermal model, which is able to characterize the compressibility and hardening/softening behaviour of the fluid saturated solid and to address the hydrostatic pressure dependence of the yield surface. Ludwig et al. [3] generalized Martin's rheological models by deriving it from a viscoplastic potential and by introducing the coalescence solid fraction g_{coal} in its formulation. To address hot tears initiation, Mihanyar et al [4] extended Ludwig's model by introducing an additional internal variable measuring the decohesion of the solid skeleton under tensile deformations. All these rheological models of the coherent mushy zone [1-4] were derived from models describing the hot deformation of porous materials [5]. These models have been validated for porosities up to 10%, which corresponds to high solid fractions

($0.9 \leq g_s \leq 1$), where hot tears can appear at the end of the solidification. For lower solid fractions between traction coherency ($g_{C,tr} \approx 0.51$ for $AlSi_7Mg$ alloy) and $g_s \approx 0.80$, their validity has not yet been shown. These models are also not linked to a viscoplastic constitutive law for the non-coherent mushy regime.

Therefore, the aim of this paper is to develop for each semi-solid regime (coherent and non-coherent) a specific constitutive law depending on microstructural features like fraction solid and grain size, which are extracted from a coupled multi-phase field simulation of the microstructure evolution [6]. Effects observed experimentally for Al-Cu alloys [2,3] and AlSi7Mg [7], like isotropic hardening, pressure dependence of yielding and the strength difference effect in tension and compression, have to be addressed in the formulation of the constitutive law for the coherent mushy regime. However, the observed softening behaviour under tensile loading is not yet addressed here.

Based on the theory of porous media, the specific thermo-elasto-viscoplastic constitutive laws for both mushy regimes are derived in section 2 and linked together at traction coherency. In section 3, the parameters of both constitutive laws are identified for an A356 aluminium alloy via FE simulations of uniaxial tension, compression and pure shear tests.

2. Thermomechanical constitutive laws

In casting and welding processes, the temperature varies strongly from above the melting point to room temperature. Therefore, the material response must reproduce either elastoplastic behaviour at room temperature or pure viscous behaviour of the melt above the liquidus temperature T_{liq}. The transition can be simulated by developing appropriate thermo-elasto-viscoplastic constitutive laws for each material state (liquid, mushy and solid) along the cooling path. Moreover, each constitutive law for a specific regime has to vary *continuously* to the next one at specific temperatures like T_{sol} (solidus temperature), T_{liq} or at specific solid fractions like $g_{C,sh}$ (shear coherency) and $g_{C,tr}$ (traction coherency).

Since the deformation in casting and welding is mainly driven by thermal deformation we adopt here the geometrical linear approach and use the porous media formalism, detailed elsewhere [8]. In this framework, the constitutive law is formulated on the solid skeleton and written using the effective solid stress tensor $\boldsymbol{\sigma}_E^s$, which takes the effect of the liquid pressure p_l on the solid phase into account via expression: $\boldsymbol{\sigma}_E^s = \boldsymbol{\sigma} + p_l \mathbf{I}$, where $\boldsymbol{\sigma}$ is the total (applied) stress in a point and $\mathbf{I}$ is the second order identity tensor. The total strain tensor is computed in terms of the displacement field of the solid phase $\mathbf{u}^s$ by: $\boldsymbol{\epsilon}(\mathbf{u}^s) = \frac{1}{2}(\nabla \mathbf{u}^s + \nabla^T \mathbf{u}^s)$ and its volumetric and deviatoric parts are given respectively by:

$$e^{vol} = \mathrm{tr}(\boldsymbol{\epsilon}) = \nabla . \mathbf{u}^s \quad ; \quad \mathbf{E} = \mathrm{dev}(\boldsymbol{\epsilon}) = \boldsymbol{\epsilon} - \frac{1}{3} e^{vol} \mathbf{I} . \tag{1}$$

Moreover, the total volumetric strain e^{vol} can be additively decomposed into elastic, viscoplastic and thermal contributions; whereas the total deviatoric strain $\mathbf{E}$ is split only in an elastic $\mathbf{E}^e$ and viscoplastic part $\mathbf{E}^{vp}$. Similarly, the effective stress tensor $\boldsymbol{\sigma}_E^s$ on the solid can be split into its hydrostatic and deviatoric parts p_s and $\mathbf{S}_s$, where p_s is the effective pressure on the skeleton (taken positive in compression): $p_s = -\frac{1}{3}\mathrm{tr}(\boldsymbol{\sigma}_E^s)$ and $\mathbf{S}_s = \boldsymbol{\sigma}_E^s - \frac{1}{3}\mathrm{tr}(\boldsymbol{\sigma}_E^s)\mathbf{I}$.

As suggested in ref. [9], the effective stress of the solid phase is governed by following split of the isotropic Hooke law:

$$p_s = \frac{B(T, \boldsymbol{V}_k)}{g_s} (e^{vol} - e^{th}) , \tag{2}$$

$$\mathbf{S}_s = \frac{2G(T, \boldsymbol{V}_k)}{g_s} \mathbf{E}^e = \frac{2G(T, \boldsymbol{V}_k)}{g_s} (\mathbf{E} - \mathbf{E}^{vp}) , \tag{3}$$

where $G(T, \boldsymbol{V}_k)$ is the shear modulus, $B(T, \boldsymbol{V}_k)$ the bulk modulus and e^{th} denote the volumetric thermal strain, which is composed of the thermal expansion and shrinkage terms. The shrinkage term expresses the liquid to solid phase transformation. Both modules depend on the temperature T and on the internal variable vector $\boldsymbol{V}_k = \{V_1, V_2, \dots, V_k\}$, which involves fraction solid g_s and additional microstructural features like grain size, dendrite arm spacing, and so on. Their effective values for a given fraction solid are determined by homogenization runs for equiaxed and directional microstructures respectively. These microstructures have been predicted by the multi-phase field model directly linked to the thermomechanical casting simulation, as outlined in ref. [6].

N.B.: During the liquid to solid phase change, the solid fraction starts from $g_s = 0$, while the stresses remain bounded. As a consequence, both the volumetric and deviatoric elastic strains vanish at T_{liq} and Eq. (2) reduces to $e^{vol} = e^{th}$; whereas Eq. (3) becomes $\mathbf{E} = \mathbf{E}^{vp}$ so that only pure viscoplastic shear deformations occur above T_{liq}.

2.1 Constitutive model for the coherent mushy zone

a) To simulate deformations during solidification, the underlying viscoplastic model has to address the compressibility of the mush at the macro-scale due to the change of pore structure as well as the strain rate and hydrostatic pressure dependency. Moreover, it has to describe continuously the coherent mushy behaviour from traction coherency up to the solid state. In general, for porous materials, their viscoplastic potential Ω^{vp} can be written in terms of the effective stress tensor $\boldsymbol{\sigma}_E^s$. Relying on ideas developed in [8, 10], we define the following viscoplastic potential for the coherent mush:

$$\Omega^{vp}(e^{vp}, T, \boldsymbol{V}_k) = \frac{1}{(n+1)[C(Tr,\boldsymbol{V}_k).K]^n} \langle \sigma_{vM}^2 - a_0^2(e^{vp}, \boldsymbol{V}_k) + a_2(Tr, \boldsymbol{V}_k)\, p_s^2 \rangle^{\frac{n+1}{2}} , \quad (4)$$

where σ_{vM} is the equivalent von Mises stress of the effective stress tensor $\boldsymbol{\sigma}_E^s$; e^{vp} is the equivalent viscoplastic strain; K the viscoplastic consistency of the mush; n the strain rate sensitivity parameter; $Tr = \frac{p_s}{\sigma_{vM}}$ specifies the stress triaxiality factor and C an internal variable describing the state of cohesion of the mush introduced firstly by Martin et al [2]. The scalar term a_0 of the generalized yield function (in Macauly brackets) defines the isotropic strain hardening behaviour by:

$$a_0(e^{vp}, T, \boldsymbol{V}_k) = Y_0 + [q(1 - \exp(-b\, e^{vp})) + h\, e^{vp}].g_s , \quad (5)$$

where Y_0 is the initial yield stress; b and q are the Voce exponent and coefficient respectively and h specifies a linear hardening coefficient.

N.B: All material parameters of the alloy depend on the temperature, fraction solid g_s and on relevant microstructural features $\boldsymbol{V}_k$. To facilitate the paper reading, these dependencies will be not mentioned explicitly in each expression hereafter.

b) The hydrostatic pressure coefficient a_2 is defined as a function of g_s by following expression:

$$a_2(Tr) = AA(Tr).\frac{(g_s - g_{C,tr})}{(1 - g_{C,tr})}.(1 - g_s)^d . \quad (6)$$

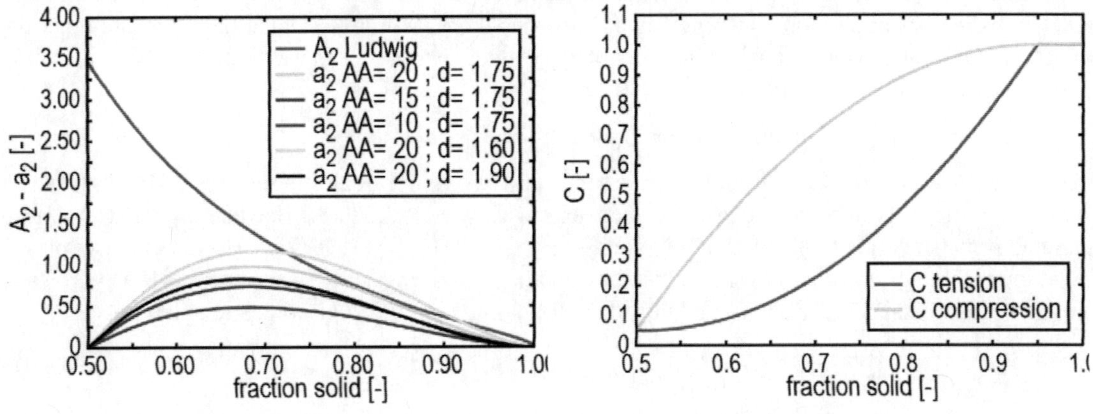

Figure 1: Variation of the hydrostatic coefficient a_2 with the fraction solid g_s for different combinations of coefficient AA and exponent d and for Ludwig's coefficient A_2 (left). Variation of internal cohesion variable C with the fraction solid (right).

Expression (6) guarantees the continuity of a_2 to both extremal cases: the solid state on one side and the non-coherent regime on the other side. Indeed, $a_2(g_s = 1) = 0$ and $a_2(g_s = g_{C,tr}) = 0$. This expression differs from Ludwig's pressure coefficient A_2[3]: $A_2 = \frac{9}{4}\left[n\left(\frac{1}{(1-g_s)^n} - 1\right)\right]^{\frac{-2n}{n+1}}$. Indeed, as shown in Figure 1.left for the Al-Cu4% alloy with $n = 3.8$, Ludwig's A_2 coefficient leads to unrealistic large pressure contributions nearly the traction coherency; whereas expression (6) reduces to zero near $g_{C,tr}$. Impact of varying exponent d and coefficient AA in their respective definition range on the pressure coefficient a_2 is outlined also on Figure 1.left. The results show that the combination $AA = 20; d = 1.75$ provides a similar pressure term as Ludwig's A_2 at high solid fractions: $g_s \in [0.75, 1]$.

c) An internal variable C has been introduced in the viscoplastic potential (4) in order to take partial cohesion of the solid skeleton into account. $C = 1$ corresponds to a fully cohesive mush, when coalescence happens at the end of the solidification. Indeed, at coalescence ($g_{coal} \approx 0.95$), liquid is still present in isolated pockets at grain boundaries. Following Ludwig [3], we assume that $C = 1$ when $g_s \geq g_{coal}$. The much weaker response in tension than in compression of the partially solidified alloy is rendered in the present model via a different evolution of C with g_s in tension and compression. Contrary to Ludwig's et al. [3] approach, in which C is expressed by a complex evolution law of the deformation rate $\mathbf{D}^{vp}$, we introduce following simple expressions for $C(g_s, Tr)$ in the solid fraction range $g_{C,tr} \leq g_s \leq g_{coal}$ (see Figure 1.right):

- if $Tr > 0$: $$C = 1 - (1 - g_{C,tr}).\left(\frac{g_{coal} - g_s}{g_{coal} - g_{C,tr}}\right)^2 \quad , \tag{7.a}$$

- if $Tr \leq 0$: $$C = \left(\frac{g_s - g_{C,tr}}{g_{coal} - g_{C,tr}}\right)^2 + C_{coh}.\frac{(g_{coal} - g_s)}{(g_{coal} - g_{C,tr})} \quad , \tag{7.b}$$

where C_{coh} is an initial cohesion value at traction coherency: we assume $C_{coh} \approx 0.05$.
Figure 1.right shows that, above $g_s = 0.75$, the cohesion term C decreases slowly in compression due to interlocking dendrites; whereas in tension, a stronger decrease happens as liquid regions expand at grain boundaries.

d) As classical for J2 viscoplastic behaviour of metals, the inelastic strain rate $\mathbf{D}^{vp}$ is given here also by *an associated flow rule*:

$$\mathbf{D}^{vp} = -\frac{1}{3}\frac{\partial\Omega^{vp}}{\partial p_s}\mathbf{I} + \frac{\partial\Omega^{vp}}{\partial\sigma_{vM}}.\left(\frac{3}{2}\frac{\mathbf{S}_s}{\sigma_{vM}}\right) = \frac{1}{(C.K)^n}\langle\sigma_{vM}^2 - a_0^2 + a_2 p_s^2\rangle^{\frac{n-1}{2}}.\left[-\frac{a_2}{3}p_s\mathbf{I} + \frac{3}{2}\frac{\mathbf{S}_s}{\sigma_{vM}}\right] \quad . \tag{8}$$

As equation (8) depends linearly from the hydrostatic pressure, it addresses well densification in compression and swelling in tension of the fluid saturated skeleton. Note that the pressure dependence of $\mathbf{D}^{vp}$ disappears at traction coherency as a_2 vanishes there. Moreover, equation (8) insures that the strain rate $\mathbf{D}^{vp}$ varies continuously if the local tensile state becomes compressive during solidification or if the inverse situation happens.

2.2 Constitutive model for the non-coherent mushy zone

Below the traction coherency, the solidifying dendrites move individually in the mush. However, shear ramp and jump tests in a rheometer on suspensions with $g_s \in [0.25, 0.35]$ have shown that the mush presents some shear resistance with a low yield stress (in kPa) (see §3.2) To describe the viscoplastic behaviour of suspensions, we consider only shear loading, implying $Tr = 0$, assume that the hydrostatic pressure effect disappears ($a_2 = 0$) and introduce following simplified viscoplastic potential:

$$\Omega^{vp}(e^{vp}) = \frac{A}{(n+1)K^n}\langle\sigma_{vM} - a_0(e^{vp})\rangle^{(n+1)} , \tag{9}$$

where $$A = \frac{\exp[w.f(g_s)]}{U} \quad \text{with} \quad f(g_s) = \frac{g_{C,tr} - g_s}{g_{C,tr} - g_{C,sh}} \tag{10}$$

and $g_{C,sh}$ specifies the shear coherency fraction solid at which first small shear stresses are measured experimentally: shear jump tests on A356 suspensions reveal $g_{C,sh} \approx 0.22$ [11].

At traction coherency, $f(g_{C,tr}) = 0$ and A reduces to $\frac{1}{U}$. By multiplying the Macauly brackets, it becomes: $\frac{1}{(n+1)U.K^n}$. On the other side, the same term of the viscoplastic potential (4) for the coherent mushy zone writes at $g_{C,tr}$: $\frac{1}{(n+1)[C_{coh}K]^n}$. The *continuity* at traction coherency of both viscoplastic constitutive laws implies that both factors are identical, which implies: $U = C_{coh}^n$. Thus, only *one additional material parameter* w is introduced to specify the viscoplastic potential (9) and the associated viscoplastic flow rule becomes:

$$\mathbf{D}^{vp} = \frac{\exp[w(\boldsymbol{V}_k).f(g_s)]}{(C_{coh}.K)^n}\langle\sigma_{vM} - a_0\rangle^n.\left[\frac{3}{2}\frac{S_s}{\sigma_{vM}}\right]. \tag{11}$$

2.3 Liquid-like constitutive law

At shear coherency $g_{C,sh}$, we have $f(g_s) = 0$ and assume that the hardening coefficients q, h become zero and disappear from coefficient a_0 (see Eq. (5)), so that a_0 equals the low initial yield stress $Y_{0,sh}$. Below this fraction solid, Y_0 reduces strongly with temperature increase and vanishes near T_{liq}, so that the viscoplastic flow rule (11) reduces to:

$$\mathbf{D}^{vp} = \frac{3}{2(C_{coh}.K)^n}\sigma_{vM}{}^{n-1}.\mathbf{S} = \eta^m \sigma_{vM}{}^{n-1}.\mathbf{S} \tag{12}$$

where η is the viscosity of the suspensions and m = 1/n.

As the rate sensitivity coefficient n differs usually from one in presence of growing dendritic germs, law (12) expresses a non-Newtonian Norton-Hoff viscoplastic flow. Above T_{liq}, coefficient n is set to one, so that the melt becomes a Newtonian fluid and law (12) reduces to a pure *viscous law* given by:

$$\mathbf{S} = \eta\,.\mathbf{D}^{vp} \tag{13}$$

2.4 Constitutive law for the solid phase

Below the solidus temperature $T_{sol}(g_s = 1)$, the hydrostatic pressure dependence disappears ($a_2 = 0$) and the viscoplastic potential (Eq. 4) reduces to a classical Lemaitre & Chaboche potential [12]:

$$\Omega^{vp}(e^{vp}, T) = \frac{1}{(n+1)K^n}\langle\sigma_{vM} - a_0(T, e^{vp})\rangle^{(n+1)} \tag{14}$$

This potential describes a coupled isotropic *strain hardening – viscosity* material behaviour, whose associated flow rule is given by:

$$\mathbf{D}^{vp} = \langle\frac{\sigma_{vM} - a_0}{K}\rangle^n.\left[\frac{3}{2}\frac{S_s}{\sigma_{vM}}\right] \tag{15}$$

3. Model parameter identification for uniaxial tests on A356 alloy

The constitutive laws for each semi-solid regime have been implemented as user defined CREEP routine in the FE program Abaqus. For each experimental flow curve (σ_{vM}, e^{vp}) at a given temperature (see ref. [7], [11]), the uniaxial strain rate D^{vp} and the fraction solid g_s are known. Note that g_s differs strongly in remelting and solidification tests, indicating a pronounced hysteresis behaviour. In the coherent mushy regime, the rate sensitivity exponent n is quasi constant for Al alloys: n = 3.8 (see ref. [3]). Under quasi-static conditions and by neglecting the hydrostatic pressure term, we get following expression:

$$\sigma_{vM} = D^{vp}(C.K)^n + Y_0 + [q(1 - \exp(-be^{vp})) + h\,e^{vp}].g_s\,, \tag{16}$$

where the first two terms correspond to the experimental yield stress Y_{exp}. Knowing internal cohesion $C(g_s)$ via expressions (7) and taking a 1st guess for K(T) from measurements [13], we calculate the yield stress Y_0 .
At next, a least-square fit is performed in order to identify the unknown isotropic hardening parameters: q, b and h. Then, the uniaxial tension, compression and shear experiments are simulated in Abaqus by using mixed hybrid elements with the implemented constitutive laws. This way, the 1st guess of viscoplastic consistency K will be improved and the parameters AA (coherent mush model), w and n (non-coherent mush model) will be identified.

3.1 Remelting tests in compression and tension

Round as cast A356 bars (ϕ = 20 mm) are heated up at a rate of 0.5 °Cs^{-1} to testing temperature. After a holding time of 120 s, constant tensile or compressive strain rates of 0.005 s^{-1} are applied. In Figure 2 are drawn the measured flow curves in uniaxial tension and compression tests at 565°C and 570°C respectively. These curves illustrate the strong different behaviour in tension and compression of remelted specimens. In table 1, the fitted quasi-static material parameters for each uniaxial test are outlined. In tension, the yield stress is larger than in compression, but the maximum tensile stress occurs at low strains. Moreover, strong softening is observed due to decohesion at grain boundaries. In compression, the initial yield stresses are small, but significant strain hardening happens, which is characterized by larger Voce coefficients q, mainly at higher solid fractions. Additionally, the creep FE simulations permit to quantify the impact of the pressure term a_2 (Eq. (6)) by varying coefficient AA within AA $\in [10, 20]$ and taking exponent d fixed (d = 1.75). At 570°C and g_s= 0.791, the impact of coefficient AA is significant (see Figure 2): coefficient AA = 10 induces too large hardening increase mainly in compression; whereas AA=15 and AA=20 lead to accurate flow curve predictions in tension and compression. Impact of coefficient AA is strongly reduced at higher fractions solid, as flow curve predictions at 565°C illustrate on Figure 2.right. Moreover, these creep FE simulations allow us, at each temperature, to derive a tuned viscoplastic consistency K, outlined in table 1.

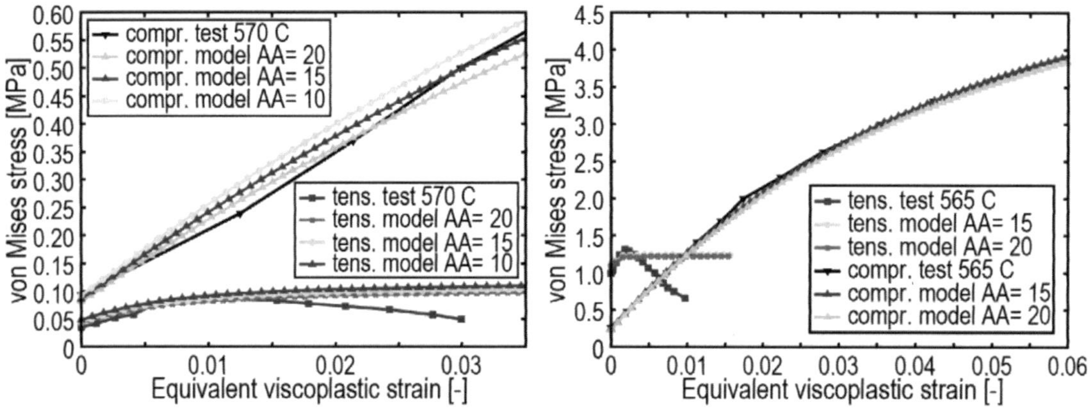

Figure 2: Identification of uniaxial compression and tension tests at strain rate $D^{vp} = 0.005\ s^{-1}$ on remelted A356 semi-solid alloys at 570°C (left) and 565°C (right) respectively.

Table 1*: Fitted material parameters of the coherent mush model for compression and tension tests. Superscripts C and T indicate the compression and tensile test respectively.*

T [°C]	K [kPa.s^{-n}]	g_S [-]	Y_0^T [kPa]	q^T [kPa]	b^T [-]	h^T [kPa]	Y_0^C [kPa]	q^C [kPa]	b^C [-]	h^C [kPa]
565	1050	.907	583	585	3000	920	0.	**5450**	24.8	1560
570	500	.7907	4.	75.	159	400	0.	**2006**	12.3	20

3.2 Shear tests on A356 suspensions

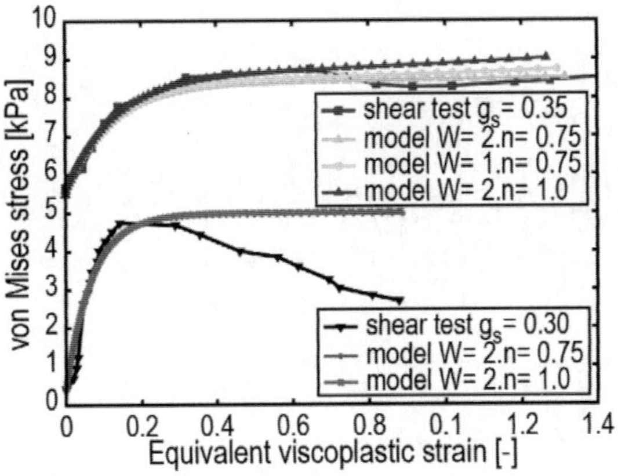

Figure 3: Model identification of shear jump tests on suspensions (g_S = 0.3 and g_S=0.35) .

A Searle rheometer was used to perform shear rate jump experiments on semi-solid suspensions. Results at $D^{vp} = 0.005\,s^{-1}$ for $g_s = 0.35$ and 0.35 are displayed in Figure 3. Linear stress increase occurs until yielding happens at few kPa, then stress decreases due breaking of dendrite entanglements. Model predictions have been realized for different combinations of unknown exponents w and n. They show clearly that combination $w = 2$ and $n = 1$ leads to the most accurate flow curve for both simulated suspensions. Moreover, the fact, that $n = 1$ is more accurate, underlies that both identified A356 suspensions behave like a Newton fluid.

4. Conclusions

Specific viscoplastic constitutive laws have been derived for the coherent and non-coherent regimes of a semi-solid alloy during solidification and remelting. Continuity between the laws has been expressed at traction and shear coherency and at T_{liq} and T_{sol}, such that a continuous transition is guaranteed from pure viscous melt flow to an elastoplastic behaviour at room temperature. The model parameters have been identified successfully for isothermal uniaxial tension, compression and shear tests of an A356 aluminium alloy. In near future, this model will be used in casting simulations and extended to handle also kinematic hardening (e.g. backstress) and decohesion in tension. However, the real challenge will be to identify for each material parameter of the presented constitutive laws, the relevant microstructural features by adoption a multiscale simulation strategy for casting and welding processes.

Acknowledgments
The authors acknowledge the financial support from the DFG in the framework of the collaborative research centre SFB 1120 “Precision Melting Engineering” and B. Zhou and M. Budnitzki (TU Bergakademie Freiberg) for the interesting discussions.

References

[1] Martin C L, Favier D and Suéry M 1997 *Int. J of Plasticity* **13** 215-235.-
[2] Martin C L, Braccini M and Suéry M 2002 *Mater. Scien. & Engng. A* **325** 292-301.
[3] Ludwig O, Drezet J-M, Martin C and Suéry M 2005 *Metall. & Mater. Trans.* A **36** 1525-35.
[4] Mihanyar S, Mo A, M’Hamdi M, Ellingsten K 2011 *Metall. & Mater. Trans.* A **42** 1887-95.
[5] Zavaliangos L and Anand L 1993 J. Mech. & Phys. Of Solids 41 1087-1118.
[6] Laschet G, Vossel T, Wolff N, Apel M and Bührig-Polaczek A 11 2017 *Proc. 6th Int. Conf. on Solidification Processing (SP17)* 576-580
[7] Dziallach S, Benke S, Prahl U and Bleck W 2009 *Int. J of Cast Metals Res.* **22** 248-51.
[8] Benke S, Dziallach S, Laschet G, Prahl U and Bleck W 2009 *Comput. Mat. Sci.* **45** 633-37.
[9] Chiumenti M, Cervera M, Dialani N, Wu B and Jinwei J 2016 FE in Anal. & Des. **121** 118-133.
[10] Benke S, Laschet G 2008 *Solid State Phenomena* **141-143** 653-658.
[11] Moll A, Modigell M., Caylak I, Mahnken R 2010, final report of DFG project about “experimental characterization and modelling of suspension in casting processes”
[12] Lemaitre J, Chaboche J.-L. 1990 *Mechanics of Solid Materials* (Cambridge, Cambridge University Press).
[13] Decultieux F, Ph D thesis at ParisTech, 1996.

IOP Conf. Series: Materials Science and Engineering **529** (2019) 012084 doi:10.1088/1757-899X/529/1/012084

Solid-liquid interfacial energy of Al-Zn solid-solutions in equilibrium with Al-Zn liquid

E Yılmaz[1*], H Erol[2], E Acer Erol[3] and M Gündüz[4]

[1] Department of Automotive, Vocational College, Kayseri University, Kayseri/Turkey
[2] Department of Physics, Faculty of Science, Ç. Karatekin University, Çankırı/Turkey
[3]Department of Physics, Institute of Science, Erciyes University, Kayseri/Turkey
[4]Department of Physics, Faculty of Science, Erciyes University, Kayseri/Turkey

E-mail:[*]ecengiz@erciyes.edu.tr

Abstract. The grain boundary groove method has been successfully used to measure solid-liquid interfacial energies, σ_{SL}, experimentally for binary eutectic and peritectic systems, multi-component systems as well as pure materials and for opaque materials as well as transparent materials. It was shown that the grain boundary groove method can be use to obtain σ_{SL} for any alloy system provided that the prepared alloy sample can be held at the evaluated temperature for a long enough time with a very stable temperature gradient. In order to show the applicability of the groove method to any system, a part of the Al-Zn phase diagram was chosen. Equilibrated grain boundary groove shapes for solid Alα solution (Al-30wt%Zn) in equilibrium with AlZn liquid (Al-60wt%Zn) have been directly observed with a radial heat flow apparatus. The Gibbs-Thomson coefficient, Γ, was determined with a numerical method using observed groove shapes. The measured thermal conductivities of the solid Alα solution and AlZn liquid phases and the temperature gradient in the solid phase at the solid-liquid interface were used for the calculation of Γ and then σ_{SL} was determined using the Gibbs-Thomson equation. The grain boundary energy for the same system was also obtained from the observed groove shapes. The results of the work were compared with the results of the related experimental works.

1. Introduction

The solubility of Zn in Al is the largest among all elements, showing a maximum of 67at% at 654K [1]. In lower Zn content (Co=40at%Zn), Al and Zn do not form intermetallic phases since the interaction between Al and Zn is fairly week [2]. Al has a FCC structure and is slightly anisotropic whereas Zn has a HCP structure and is strongly anisotropic. This difference in anisotropy involves a variation of solid-liquid interfacial energy, σ_{SL}, modifying morphologies and directions of dendritic growth. σ_{SL} plays a key role in a wide range of materials and metallurgical phenomena, such as nucleation, solidification processing, welding and sintering, through to phase transformation, wetting and coarsening etc. [3-11]. Thus, a quantitative knowledge of σ_{SL} values is necessary. The present most useful and powerful technique for the experimental measurement of σ_{SL} in multi-component systems as well as pure materials and opaque materials as well as transparent materials, has been found to be the "grain boundary groove, GBG," method. The method has been successfully used to measure σ_{SL} in eutectic, peritectic and monotectic systems [12-21]. In this study, it has been shown that the GBG method is not limited to these systems only. This method can be used to obtain σ_{SL} for any alloy system. Over the last 50 years, determining the values of the σ_{SL} in a variety of materials has been used through various attempts. There are various techniques for estimating σ_{SL}. The GBG consist of intersection of grain boundaries with the equilibrated planar solid-liquid interfaces (Figure 1).

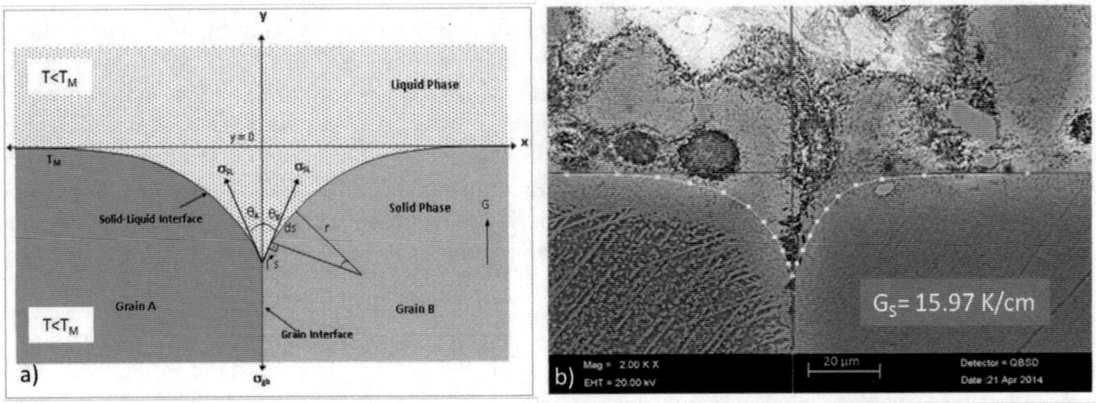

Figure 1. Equilibrated GBG shapes: **(a)** Schematic illustration of an equilibrated GBG shape showing the x, y coordinates and angle θ. **(b)** Alα solid solution equilibrated with the liquid AlZn

σ_{SL} might be calculated from the equilibrated GBG shapes formed at the solid-liquid interface under a gradient G. When σ_{SL} is isotropic the interface at the groove must everywhere satisfy;

$$\Delta T_r = \frac{\sigma_{SL}}{\Delta S_f}\left(\frac{1}{r_1} + \frac{1}{r_2}\right) \tag{1}$$

where ΔT_r is the curvature undercooling, ΔS_f is the entropy of fusion per unit volume, r_1 and r_2 are the principal radii of the curvature. For the case of a GBG shape intersecting a planar solid–liquid interface, $r_2 = \infty$ and Gibbs-Thomson equation, (Equation 1), becomes;

$$\Gamma = r\Delta T_r \tag{2}$$

where Γ is the Gibbs–Thomson coefficient. The GBG shape for an applied gradient in two dimensions has been calculated theoretically using Equation 2 by Bolling and Tiller for the case of equal thermal conductivities in the solid-liquid phases [22]. The theoretical description of Bolling and Tiller model was modified by Nash and Glicksman [13] to account for different thermal conductivities in the solid and liquid phases. Gündüz and Hunt [6] solved the heat flow problem numerically through the experimentally observed GB shape using different thermal conductivities for the solid and liquid phases and the G in the solid phase at the solid-liquid interface, G_S value with appropriate boundary and temperature distribution conditions. A finite difference method was used to calculate the difference in temperature between the flat surface and points on the GBG shape (Figure 1). The numerical method allows Γ to be obtained for any equilibrated GBG shape provided that G_S, K_L and K_S values and the size of the groove are known. Then, σ_{SL} is obtained from Equation 1. Over the last 30 years, the equilibrated GBG shapes in variety of eutectic, peritectic and monotectic systems have been observed and measurements of σ_{SL} have been made from the observed GBG shapes [12-21].

The aims of the present work were to observe the GBG shapes for the solid Alα solution, in equilibrium with the AlZn liquid, and to determine the Γ, σ_{SL} and grain boundary energy, σ_{gb}.

IOP Conf. Series: Materials Science and Engineering **529** (2019) 012084 doi:10.1088/1757-899X/529/1/012084

2. Experimental Procedure

2.1. Sample Preparation

The Al-30wt%Zn samples were prepared in a vacuum furnace from 4-9s Al and Zn. The molten alloy was mixed several times and then poured into a graphite crucible held in a specially manufactured casting furnace which is heated to above 100K of the liquidus temperature of the alloy. Subsequently, the samples were directionally solidified from the bottom to top. The cast sample was placed in the radial heat flow apparatus and the thermocouples were placed in the alumina tubes and then the sample was placed in the radial heat flow apparatus. The apparatus can give a very stable temperature, (± 0.1K), for a long period of time and allows a wide range of G. The detail of the experimental procedure, sample preparation and obtaining of the GBG shapes are given in references [6, 7, 14-21].

2.2. Temperature Gradient Measurements

In the steady-state condition, the temperature gradient in the solid, G_S, at the radius, r, of a long cylindrical sample is given by;

$$G_S=\left(\frac{dT}{dr}\right)_S=-\frac{Q/l}{2\pi r K_S} \tag{3}$$

where Q is the input power, r is the distance of the planar solid-liquid interface from the center, l is the length of central heating element, and K_S is the thermal conductivity in the solid phase. The average G_S was calculated for each GBG shape. The Q was determined by measuring the voltage drop across the heating element and the current flowing through the wire and the r value measured on the polished sample surface where the GBG shape was observed.

2.3 Thermal Conductivities of the Phases

In order to calculate the Γ from the GBG shapes, the solid and liquid phases, thermal conductivities of K_S and K_L as well as G_S and coordinates of the GBG shapes at the solid-liquid interface must be known as accurately as possible. The K_S and K_L values of the phases are generally unknown. The radial heat flow apparatus can be used to measure the K_S value. Equation 3 may be integrated to give;

$$K_S=\frac{1}{2\pi}\ln\left(\frac{r_2}{r_1}\right)\frac{Q/l}{T_2-T_1} \tag{4}$$

where T_1 and T_2 are the temperatures at two fixed distances, that is, at r_1 and r_2 from the centre. The specimen is heated with a central heating wire to 10K below the solidus temperature in 50K steps. The specimen was kept at least two hours at each measurement temperature. In the steady-state conditions, Q, T_1, T_2 and vertical temperature change were measured. The K_S value was obtained by using the measured T_1, T_2, r_1, r_2, l and Q values in Equation 4 and shown in Figure 2.

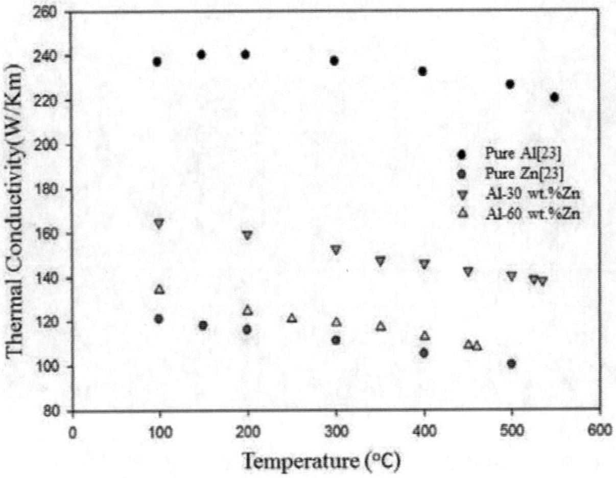

Figure 2. Plots of thermal conductivity of the solid phases with temperature

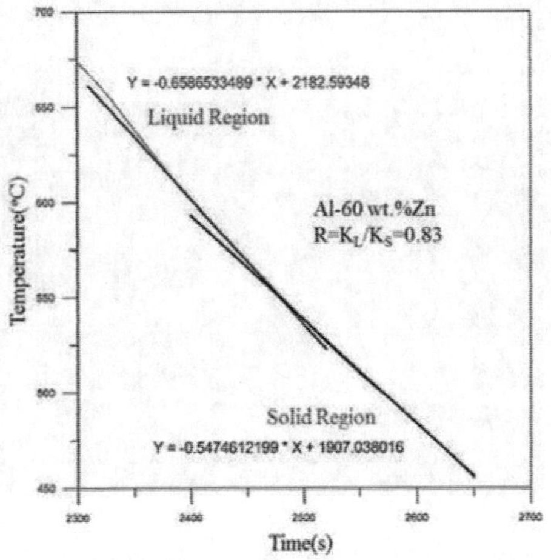

Figure 3. The time-temperature trace and R value for Al-60wt% Zn alloy

K values used in the calculations were obtained by extrapolation to the equilibrium temperature. The K_S of the Alα solution was found to be as 138.61 W/Km and K_S value of the Al-60wt%Zn alloy was found to be as 108.80 W/Km at the equilibrium temperature. The temperature-time trace enables $R=K_L/K_S$ ratio of the phases to be calculated. The $R=K_L/K_S$ ratio of the Al-60wt%Zn alloy was obtained to be as 0.83 by using the trace obtained with a Bridgman type directional growth apparatus (Figure 3). The K_L value was calculated to be as 89.64 W/Km by using the measured K_S and R values for the Al-60wt% Zn alloy. Thus, the ratio of the equilibrated Alα solid phase to AlZn liquid phase was obtained as 0.65.

3. Result and Discussion

3.1. Determination of the Gibbs-Thomson Coefficient

The Gibbs-Thomson constant, Γ, is one of the basic parameters required in the solidification theories. To obtain the Γ using the numerical method, G_S, K_S, K_L values and the coordinates of the equilibrated GBG shape must be known. The numerical method is described in detail in Ref [6]. The Γ values for the equilibrium solid Alα (Al-30wt%Zn) and liquid AlZn (Al-60wt%Zn) were calculated using at least 10 equilibrated symmetrical GBG shapes (e.g. Figure 1b and Figure 4). The determined average values of Γ for the solid Alα solution in equilibrium with the AlZn liquid was found (Table 1). The experimental error in the Γ obtained by considering the fractional error in each of the measured quantities was estimated to be in the order of 7-9%.

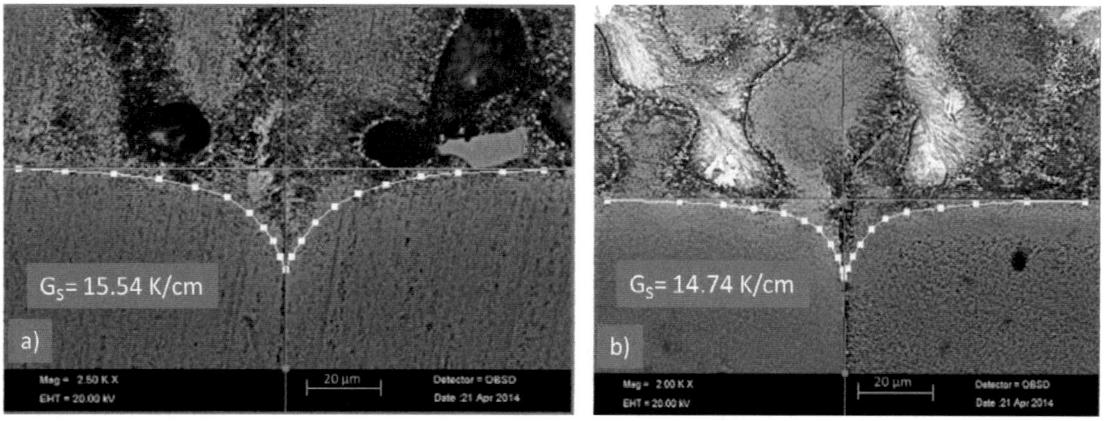

Figure 4. Typical GBG shapes for the solid Alα solution in equilibrium with the liquid AlZn alloy

3.2. Determination of Entropy of Fusion

In order to determine $\sigma_{SL,}$ it is also necessary to know the entropy of fusion per unit volume, ΔS_f. The entropy of fusion for a binary alloy, is written as [6,7];

$$\Delta S_f = \frac{RT_M}{m_L V_S} \frac{C_S - C_L}{(1-C_L)C_L} \tag{5}$$

where R is the gas constant, T_M is the melting temperature, C_S is the composition of the solid phase, C_L is the composition of the liquid phase, m_L is the slope of liquidus and V_S is the molar volume of the solid phase. The values of the constants used in the determination of the ΔS_f were obtained from the Al-Zn phase diagram [1]. The ΔS_f value was obtained by using the R, T_M, m_L, V_S, C_S, and C_L values (Table 1). The error in the determination of ΔS_f is estimated to be about 3-4% [6,7].

Table 1. A comparison of Γ, ΔS_f and σ_{SL} measured in the present work (pw) with previous work

Alloy	Solid Phase	Liquid Phase	Γx10^{-6} (K.cm)	ΔS_f x10^7 (erg/K.cm^3)	σ_{SL} (erg/cm^2)	σ_{gb} (erg/cm^2)	$\frac{\sigma_{gb}}{\sigma_{SL}}$	Ref.
Al-Zn	Al-30wt.%Zn	Al-60wt.%Zn	5.25	3.01	158.03	304.7	1.93	pw
Al-Zn	Al-40wt.%Zn	Al-70wt.%Zn	4.18	3.52	161.58	316.14	1.96	[24]
Al-Zn	Al-66.5at.%Zn	Al-88.7at.%Zn	3.41	3.14	106.94	204.72	1.91	[15]
Zn-Al	Zn-1wt.%1Al	Zn-5wt.%5Al	5.80	1.61	93.496	182.3	1.95	[25]

3.3. Solid-Liquid Interfacial Energy

The σ_{SL} can be obtained from the thermodynamic definition of the Gibbs-Thomson equation for isotropic σ_{SL} and it is expressed as;

$$\Gamma = \frac{\sigma_{SL}}{\Delta S_f} \tag{6}$$

If the values of Γ and ΔS_f are measured or known, then the σ_{SL} can be obtained from Equation 6. The σ_{SL} of the solid Alα solution in equilibrium with the AlZn liquid was determined by using the values of

Γ and ΔS_f (Table 1). Thus, the total experimental error of the σ_{SL} evaluation in the present study is approximately 12%. A comparison of the σ_{SL} with previous work is shown in Table 1.

3.4. Grain Boundary Energy

The GBG shapes allow the grain boundary energies, σ_{gb} to be determined once σ_{SL} have been obtained. The grain boundary energy can be expressed as;

$$\sigma_{gb} \leq 2\sigma_{SL}\cos\theta \tag{7}$$

where $\theta = (\theta_A + \theta_B)/2$ is the angle that the GBG interfaces make with the y axis (Figure 1). The angle, θ was obtained from the cusp coordinates, x, y, using a Taylor expansion for parts at the base of the grooves. According to Equation 7, the value of σ_{gb} should be smaller or equal to twice the σ_{SL} i.e. $\sigma_{gb} \leq 2\sigma_{SL}$. The value of σ_{gb} for the solid Alα solution was found to be as 304.7±43.8 erg/cm^2 using the values of the σ_{SL} and θ in Equation 7. σ_{gb} / σ_{SL} ratio suggest that small θ, so that Cos$\theta \approx 1$ (Table 1). The estimated error in determination of angles θ was found to be 1%. Thus, the total experimental error in the resulting σ_{gb} is approximately 13%. A comparison of the values of ΔS_f, Γ, σ_{SL} and σ_{gb} for the solid Alα solution in equilibrium with the AlZn liquid determined in the present work with the values of previous ΔS_f, Γ, σ_{SL}, and σ_{gb} for different Al-Zn alloy systems has been given in Table 1 and refs [6,7,14-21,25]. The value of ΔS_f is in reasonably good agreement with the value of Keslioglu and Marasli [15], but Γ, σ_{SL} and σ_{gb} were found to be higher than their values. A reasonably good agreement was obtained with the other Al based alloys for σ_{SL} and σ_{gb} values. It should be noted that all previous works considered eutectic and peritectic systems having different compositions of solid and liquid phases and different equilibrium temperatures. The alloying elements of the phases and the equilibrium temperatures in the systems probably affect the Γ and thus σ_{SL} values.

4. Conclusion

1. The equilibrated GBG shapes for the solid Al solution (Al-30wt%Zn) in equilibrium with the Al-Zn liquid (Al-60wt%Zn) were observed.
2. From the observed GBG shapes, the Γ, σ_{SL}, and σ_{gb} have been determined for the solid Al solution in equilibrium with the Al-Zn liquid.
3. The K_S values of the solid phases for Al-30wt.%Zn and the Al-60wt%Zn alloys have been measured. The K_L /K_S ratio of the liquid phase to the solid phase has been obtained by using a Bridgman type directional apparatus for the Al-60wt%Zn alloy, and K_L value of Al-60wt%Zn was measured.
4. It has been shown that the radial heat flow apparatus can be used to obtain equilibrated solid-liquid interfaces and GBG shapes in not only eutectic, peritectic and monotectic systems but also any other alloy system. Thus, the GBG method may be use to determine σ_{SL} as a function of temperature and composition for any alloy system. σ_{SL} energies might be used to measure at different compositions and temperatures in the Al-Zn system by using GBG method as shown in this work. In this way it might be possible to extrapolate the σ_{SL} obtained for alloys to that for pure Al. So, the experimentally obtained pure Al, σ_{SL}, could be compared with the σ_{SL} obtained by using the homogeneous nucleation method, theoretical methods and other methods.

Acknowledgments
The authors thank Erciyes University Scientific Research Project Unit for their financial support of the project (Contract No: FDK-2013-4756).

References

[1] Massalaski T B, Okamoto H, Subramanian P R and Kacprzak L 1990 *Binary Alloy Phase Diagrams 2nd Edition* (ASM International) p 240

[2] Skoko Z, Popović S and Štefanić G 1999 *Croatica Chemica Acta* **72** 621
[3] Turnbull D and Cech R E 1950 *Journal of Applied Physics* **21** 804
[4] Jones D R H 1974 *Journal of Materials Science* **9** 1
[5] Eustathopoulos N 1983 *International Metals Reviews* **28** 189
[6] Gündüz M and Hunt J D 1985 *Acta Metallurgica* **33** 1651
[7] Maraşlı N and Hunt J D 1996 *Acta Materialia* **44** 1085
[8] Jiang Q and Lu H M 2008 *Surface Science Reports* **63** 427
[9] Jones H 2007 *Metallurgical and Materials Transactions A* **38A** 1563
[10] Kaptay G 2018 *Journal of Mater.* **53** 3767
[11] Glicksman M E and Ankit K 2018 *J. Mater. Sci.* **53** 10955
[12] Schaefer R J, Glicksman M E and Ayers J D 1975 *Philosophical Magazine* **32** 725
[13] Nash G E and Glicksman M E 1971 *Philosophical Magazine* **24** 577
[14] Meydaneri F, Saatci B, Gündüz M and Özdemir M 2007 *Surface Science* **601** 2171
[15] Keslioglu K and Marasli N 2004 *Mater.Sci. Eng. A* **369** 294
[16] Keslioglu K, Gündüz M, Kaya H and Çadırlı E 2004 *Mater. Letters* **58** 3067
[17] Bulla A, et. al. 2007 *Metall. and Materials Trans. A* **38A** 1956
[18] Keslioglu K, et al. 2010 *Met. Mater. Int.* **16** 51
[19] Aksöz S, Ocak Y, Maraşlı N and Keşlioğlu K 2010 *Fluid Phase Equilibria* **293** 32
[20] Akbulut S, et al. *J. Phys. Condenced Matter* **18** 8403
[21] Boyuk U,et al. 2009 *Surface Review and Letters* **16(2)** 203
[22] Bolling G F and Tiller WA 1960 *J. Appl. Phys.* **31** 1345
[23] Touloukian YS, Powel RW, Ho CY, Klemens PG, 1970, Thermal Conductivity Metallic Elements and Alloys, New York, Washington
[24] Gündüz M, Aydemir S G, Acer E and Erol H 2011 *Solidification Science and Technology (London)*, ed Fan Z and Stone I C (Brunel /Bcast) p395
[25] Keslioglu K and Marasli N 2004 *Metall. and Materials Trans. A* **35A** 3665